GWAS and QTL Mapping in Horticultural Crops

Shyam Sundar Dey • Mir Asif Iquebal
Sarika • Reeta Bhatia Dey
Editors

GWAS and QTL Mapping in Horticultural Crops

Vegetable Crops

Springer

Editors
Shyam Sundar Dey
Division of Vegetable Science
ICAR-Indian Agricultural Statistics
Research Institute
New Delhi, Delhi, India

Mir Asif Iquebal
Division of Agricultural Bioinformatics
ICAR-Indian Agricultural Statistics
Research Institute
New Delhi, Delhi, India

Sarika
Division of Agricultural Bioinformatics
ICAR-Indian Agricultural Statistics
Research Institute
New Delhi, Delhi, India

Reeta Bhatia Dey
Division of Floriculture and Landscaping
ICAR-Indian Agricultural Research Institute
New Delhi, India

ISBN 978-981-95-2305-4 ISBN 978-981-95-2306-1 (eBook)
https://doi.org/10.1007/978-981-95-2306-1

This Springer imprint is published by the registered company Springer Nature Singapore Pte Ltd.
The registered company address is: 152 Beach Road, #21-01/04 Gateway East, Singapore 189721, Singapore

Preface

Let food be thy medicine and medicine be thy food. (Hippocrates)

This enduring principle by "Father of Medicine" highlights the role of diet in maintaining health and combating diseases. The plant-based, nutrient-dense foods, like vegetables, foster healing and promoting health. Consequently, a comprehensive understanding of advanced breeding technologies is critical for the development of vegetables that are not only more nutritious but also more resilient and sustainable, thereby contributing to long-term food security for future generations.

Vegetables become ever more important to provide a consistent supply of critical nutrients and support global food security. Being a foundation for diversification policies in agriculture, vegetable production is not only responsible for human well-being but also presents great economic potential, especially in alleviating food shortages, malnutrition, and rural poverty in developing nations. In spite of their unequivocal value as the primary suppliers of vitamins and minerals, vegetables' full nutritional and economic potential is generally left untapped, particularly against the backdrop of underexploited genetic and genomic potential.

The search for plant genetics and genomics has for years played a central role in driving our comprehension of vegetable biology as well as developing sustainable farming. Over the last several decades, bringing advanced genomic technology together with classical breeding has radically changed how complex traits in vegetables can be researched and engineered. Among the most influential methodologies driving this change are Genome-Wide Association Studies (GWAS) and Quantitative Trait Locus (QTL) mapping, which offer valuable tools for illuminating the genetics of major agronomic traits. These developments allow breeders and researchers to improve yield, nutritional quality, disease resistance, and environmental tolerance in vegetable crops.

This book aims to provide an in-depth exploration of these two essential tools, with a specific focus on their application in vegetables—an agriculturally important group of plants that contribute to human nutrition and food security worldwide. The book attempts to close the gap between basic genetic research and practical application in vegetable improvement. By drawing attention to the most recent advances in genomic technologies and their ability to unravel complicated traits, we aim to create a useful reference for researchers, plant breeders, and policymakers striving for a more resilient and nutritious food system worldwide. Emphasis is also placed on the challenges and opportunities of translating genetic findings into breeding programs that address the needs of farmers, consumers, and the broader agricultural sector.

In compiling this book, we have drawn on the expertise of leading scientists in plant genetics, genomics, and breeding who are associated with vegetable crop improvement programs. Their contributions provide a comprehensive, accessible resource for researchers, breeders, and students alike, with the goal of advancing the scientific understanding and practical application of genetic mapping tools in vegetable crops.

We hope that this book will serve as both a reference and a source of inspiration for the next generation of plant geneticists and breeders working to unlock the genetic potential of vegetables, ultimately contributing to the sustainable development of global food systems.

New Delhi, Delhi, India

Shyam Sundar Dey
Mir Asif Iquebal
Sarika
Reeta Bhatia Dey

Contents

1 **Development of Mapping Population in Vegetable Crops for GWAS and QTL Mapping** 1
N. Pradeepkumara, Saheb Pal, B. G. Supreetha, Reeta Bhatia Dey, A. D. Munshi, T. K. Behera, and Shyam Sundar Dey

2 **Tools and Techniques Used for Quantitative Trait Loci (QTL) and Association Mapping Studies** 61
Babita Joshi, Sanjay Singh, Mir Asif Iquebal, Sarika, Dinesh Kumar, Samir V. Sawant, and Satya Narayan Jena

3 **GWAS and QTL Mapping in Potato (*Solanum tuberosum* L.)** 115
Jelli Venkatesh, Abhinandan Surgonda Patil, Jagesh Kumar Tiwari, Dalamu, Tanuja Buckseth, Satish Kumar Luthra, and Ajay Kumar Thakur

4 **Advances in GWAS and QTL Mapping in Tomato (*Solanum lycopersicum* L.)** 157
Jagesh Kumar Tiwari, Gaurav Mishra, Nagendra Rai, Y. S. Reddy, Priyanka Sharma, Rajat Singh, Manish Singh, P. K. Singh, and T. K. Behera

5 **QTL Mapping and GWAS Triumphs in Tackling Economic and Environmental Challenges for *Capsicum annuum* L.** 185
Arpita Srivastava, Nikita Baliyan, Priti Upadhyay, Sikha Manoharan, and Manisha Mangal

6 **Advances in QTL Mapping and GWAS: Implications for Cucumber Breeding** 231
Gayatri B. K. and Shyam Sundar Dey

7 **GWAS and QTL Mapping in Muskmelon and Watermelon** 261
Raman Selvakumar

8 **GWAS and QTL Mapping in Cauliflower (*Brassica oleracea* var. *botrytis* L.)** 295
Srija Priyadarsini, Saurabh Singh, Alok Nandi, Shyam Sundar Dey, and Reeta Bhatia Dey

9 **GWAS and QTL Mapping in Cabbage (*Brassica oleracea* var. *capitata* L.)** 307
Srija Priyadarsini, Saurabh Singh, Alok Nandi, Reeta Bhatia Dey, and Shyam Sundar Dey

10 **Advances in GWAS and QTL Mapping in Gourd Vegetables** 321
Shreyas Aradhya C. S., K. K. Gangopadhyay, and Shyam Sundar Dey

11 **GWAS and QTL Mapping in Carrot (*Daucus carota* L.)** 343
Raman Selvakumar

12 **QTL Mapping in Onion and Garlic** 373
Priyanka Anand, Karanveer Kaur, Neha Verma, and Jiffinvir Khosa

13 **Advances in GWAS and QTL Mapping in Garden Pea** 399
Brij Bihari Sharma, Akanksha Singh, and Vinod Kumar Sharma

14 **GWAS and QTL Mapping in Radish, Carrot and Beetroot** 421
N. Pradeepkumara, Bichhinna Maitri Rout, Saheb Pal, L. C. Sushmitha, Reeta Bhatia Dey, A. D. Munshi, T. K. Behera, and Shyam Sundar Dey

15 **Genome-wide Association Study (GWAS) and QTL-Based Improvement of Lettuce (*Lactuca* spp.)** 457
Khaled F. M. Salem, Tilak Chandra, Essam Fathy El-Hashash, and Amira A. Ibrahim

16 **Genome-wide Association Study (GWAS) and QTL-Assisted Improvement of Squash (*Cucurbita pepo* L.)** 475
Khaled F. M. Salem, Tilak Chandra, Essam Fathy El-Hashash, and Amira A. Ibrahim

17 **A Compendium of Advances on GWAS and QTL Atlas for Accelerating Trait Discovery in Eggplant** 489
Tilak Chandra, Sarika, Shyam Sundar Dey, Mir Asif Iquebal, and Dinesh Kumar

18 **GWAS and QTL Mapping in Broccoli** 505
Shrawan Singh, Raman Selvakumar, Monica Sharma, and Anamika Chandel

About the Editors and Contributors

About the Editors

Shyam Sundar Dey is presently working as Principal Scientist at the Division of Vegetable Science, ICAR-Indian Agricultural Research Institute, New Delhi. He obtained his MSc and PhD degrees at ICAR-Indian Agricultural Research Institute, New Delhi. He has made significant contributions to the development of a large number of doubled haploid (DH) lines in cucumber, cabbage, and cauliflower and the development of CMS lines in cauliflowers. He is involved in the development of 18 hybrids in varieties of different vegetable crops. He has registered eight germplasm with NBPGR, New Delhi. Presently, he is working on development of molecular markers for economically important traits, allele mining, marker-assisted back-cross breeding, and development of genomic resources using the indigenous cucumber germplasm. He has received a number of awards like Fellowship of National Academy of Agricultural Sciences (NAAS), Associateship of National Academy of Agricultural Sciences (NAAS), Early Career Science and Technology Visiting Scholar by Indian National Science Academy (INSA), selected for post-Doctoral research fellowship at University of Western Australia (UWA), Perth, besides several other national level awards by professional societies and agencies.

Reeta Bhatia Dey is currently currently serving as a Principal Scientist at the Division of Floriculture and Landscaping, ICAR-Indian Agricultural Research Institute, New Delhi. With over 17 years of experience in research and teaching, she has gained extensive expertise in the field of floriculture. She is particularly skilled in applying modern biotechnological tools to improve horticultural crops and has pioneered the development of doubled haploids in various horticultural species.She completed her master's degree at Punjab Agricultural University, Ludhiana, and her doctoral studies at the Indian Agricultural Research Institute, New Delhi. She was awarded the prestigious Endeavor Research Fellowship by the Government of Australia, which enabled her to pursue postdoctoral research at the University of Western Australia.Throughout her academic and research career, she has received numerous accolades, including the D. P. Ghosh Young Scientist Award, Fellowship of the Indian Academy of Horticultural Sciences, IARI Best PhD Student Award, Santosh Shiksha Puraskar, the Lal Bahadur Shastri Memorial Gold Medal, and the Dr. Y.S. Parmar University of Horticulture and Forestry Gold Medal, recognizing her exceptional academic performance. She has published over 85 research papers in high-impact scientific journals, edited 5 books, and authored more than 50 review, popular, and scientific articles. Additionally, she serves as a review editor for *Frontiers in Horticulture* and as an editor for the *Journal of Ornamental Horticulture*.

Mir Asif Iquebal is Principal Scientist at the Division of Agricultural Bioinformatics, ICAR-Indian Agricultural Statistics Research Institute, New Delhi. With over 17 years of research and teaching experience, his expertise spans computational biology, bioinformatics, and agricultural statistics. A graduate in Agriculture from Uttar Banga Krishi Viswavidyalaya, West Bengal, he earned his MSc and PhD from ICAR-IARI, New Delhi. He has led multidisciplinary bioinformatics research across crop, animal, fish, and microbial systems, handling projects funded by organizations such as BMGF, DBT, and ICAR and various state governments. Currently, he has over 140 publications with more than 1,600 citations. His specialization includes genomics, genome assembly, transcriptomics, pangenome construction, genome editing, and AI/ML applications in bioinformatics. Notable contributions include methodologies for the world's first crop variety signature server and domestic animal breed signature server, AI-based tools for AMP and miRNA discovery, and the development of over 36 genomic databases and tools. He holds several copyrights for his innovations. He has been honored as a Fellow by leading scientific academies, including the West Bengal Academy of Science and Technology, National Environmental Science Academy, and National Academy of Biological Sciences. He is a member of NASI, NAAS Associate (Social Sciences), owns the prestigious Jawahar Lal Nehru Young Scientist Award, and serves on editorial boards of several reputed journals.

Sarika, Professor (Bioinformatics) and Principal Scientist at the Division of Agricultural Bioinformatics, ICAR-Indian Agricultural Statistics Research Institute, New Delhi, has research and teaching experience of over 17 years. Her experience in statistics, bioinformatics, and computational genomics has led to significant contributions to multidisciplinary bioinformatics research in crop, animal, fish, and microbial domains. She obtained her BSc (Agriculture) degree from Banaras Hindu University, while MSc and PhD degrees from ICAR-Indian Agricultural Research Institute, New Delhi. She has expertise in data handling in the field of genome and transcriptome assembly, genome editing, Copy Number Variations, and AI/ML applications in bioinformatics. She has led multiple research projects supported by both national and international funding agencies, including the Bill & Melinda Gates Foundation (BMGF), the Department of Biotechnology (DBT), the Indian Council of Agricultural Research (ICAR), and the National Agricultural Science Fund (NASF), among others. She has developed over 50 genomic databases, webservers, and AI/ML-based tools for DNA marker discovery, GWAS, and trait prediction and holds 17 copyrights for the same, in her name. Internationally trained at Iowa State University, she actively contributes to human resource development through teaching, guiding PhD/MSc students, and leading training programs regularly. She has received prestigious awards, including NABS-Best Woman Scientist Award (2020), NAAS Associate (2019), NASI Membership (2020), NABS Fellow (2023), and NESA Fellow (2023), reflecting her excellence and leadership in computational genomics and agricultural bioinformatics.

Contributors

Priyanka Anand Department of Vegetable Science, Punjab Agricultural University, Ludhiana, Punjab, India

Shreyas Aradhya C. S. Division of Vegetable Science, ICAR-Indian Agricultural Research Institute, New Delhi, India

Mir Asif Iquebal Division of Agricultural Bioinformatics, ICAR-Indian Agricultural Statistics Research Institute, New Delhi, India

Nikita Baliyan Division of Vegetable Science, ICAR-IARI, New Delhi, India

T. K. Behera ICAR-Indian Institute of Horticulture Research, Bengaluru, Karnataka, India

ICAR-Indian Institute of Vegetable Research, Varanasi, Uttar Pradesh, India

Reeta Bhatia Dey Division of Floriculture and Landscaping, ICAR-Indian Agricultural Research Institute, New Delhi, India

Division of Floriculture and Landscaping, ICAR-Indian Agricultural Research Institute, New Delhi, India

Tanuja Buckseth ICAR-Central Potato Research Institute, Shimla, Himachal Pradesh, India

Anamika Chandel Division of Vegetable Science, ICAR-Indian Agricultural Research Institute, New Delhi, India

Tilak Chandra Division of Agricultural Bioinformatics, ICAR-Indian Agricultural Statistics Research Institute, New Delhi, India

Dalamu ICAR-Central Potato Research Institute, Shimla, Himachal Pradesh, India

Shyam Sundar Dey Division of Vegetable Science, ICAR-Indian Agricultural Research Institute, New Delhi, India

ICAR-Central Institute of Temperate Horticulture, Regional Station, Dirang, Arunachal Pradesh, India

Essam Fathy El-Hashash Department of Agronomy, Al-Azhar University, Cairo, Egypt

K. K. Gangopadhyay Division of Vegetable Science, ICAR-Indian Agricultural Research Institute, New Delhi, India

B. K. Gayatri Division of Vegetable Science, ICAR-Indian Agricultural Research Institute, New Delhi, India

Amira A. Ibrahim Arid land Cultivation Research Institute, Cairo, Egypt

Sarika Division of Agricultural Bioinformatics, ICAR-Indian Agricultural Statistics Research Institute, New Delhi, India

Satya Narayan Jena Plant Genetic Resources and Improvement, CSIR-National Botanical Research Institute, Lucknow, Uttar Pradesh, India

Babita Joshi Plant Genetic Resources and Improvement, CSIR-National Botanical Research Institute, Lucknow, Uttar Pradesh, India

Academy of Scientific and Innovative Research (AcSIR), Ghaziabad, Uttar Pradesh, India

Karanveer Kaur Department of Vegetable Science, Punjab Agricultural University, Ludhiana, Punjab, India

Jiffinvir Khosa Department of Vegetable Science, Punjab Agricultural University, Ludhiana, Punjab, India

Dinesh Kumar Division of Agricultural Bioinformatics, ICAR-Indian Agricultural Statistics Research Institute, New Delhi, India

Satish Kumar Luthra ICAR-Central Potato Research Institute Research Station, Modipuram, Uttar Pradesh, India

Manisha Mangal Division of Vegetable Science, ICAR-IARI, New Delhi, India

Sikha Manoharan Division of Vegetable Science, ICAR-IARI, New Delhi, India

Gaurav Mishra ICAR-Indian Institute of Vegetable Research, Varanasi, Uttar Pradesh, India

A. D. Munshi ICAR-Division of Vegetable Science, Indian Agricultural Research Institute, New Delhi, India

Alok Nandi Institute of Agricultural Sciences, Siksha 'O' Anusandhan (Deemed to be University), Bhubaneswar, Odisha, India

Saheb Pal ICAR-Indian Agricultural Research Institute, Hazaribagh, Jharkhand, India

Abhinandan Surgonda Patil Agarkar Research Institute, Pune, Maharashtra, India

N. Pradeepkumara ICAR-Central Institute of Temperate Horticulture, Regional Station, Dirang, Arunachal Pradesh, India

Srija Priyadarsini Department of Vegetable Science, Odisha University of Agriculture and Technology (OUAT), Bhubaneswar, Odisha, India

Nagendra Rai ICAR-Indian Institute of Vegetable Research, Varanasi, Uttar Pradesh, India

Y. S. Reddy ICAR-Indian Institute of Vegetable Research, Varanasi, Uttar Pradesh, India

Bichhinna Maitri Rout ICAR-Division of Vegetable Science, Indian Agricultural Research Institute, New Delhi, India

Khaled F. M. Salem Department of Plant Biotechnology, Genetic Engineering and Biotechnology Research Institute, Shaqra University, Shaqraa, Saudi Arabia

Samir V. Sawant Plant Genetic Resources and Improvement, CSIR-National Botanical Research Institute, Lucknow, Uttar Pradesh, India

Raman Selvakumar Centre for Protected Cultivation Technology, ICAR-Indian Agricultural Research Institute, New Delhi, India

Brij Bihari Sharma Department of Vegetable Science, College of Horticulture and Forestry, Rani Lakshmi Bai Central Agricultural University, Jhansi, Uttar Pradesh, India

Monica Sharma Department of Plant Pathology, College of Horticulture & Forestry, Dr YS Parmar University of Horticulture & Forestry (Solan), Hamirpur, Himachal Pradesh, India

Priyanka Sharma ICAR-Indian Institute of Vegetable Research, Varanasi, Uttar Pradesh, India

Vinod Kumar Sharma ICAR- National Bureau of Plant Genetic Resources, New Delhi, India

Akanksha Singh Department of Vegetable Science, College of Horticulture and Forestry, Rani Lakshmi Bai Central Agricultural University, Jhansi, Uttar Pradesh, India

Manish Singh ICAR-Indian Institute of Vegetable Research, Varanasi, Uttar Pradesh, India

P. K. Singh Department of Vegetable Science, Chandra Shekhar Azad University of Agriculture & Technology, Kanpur, Uttar Pradesh, India

Rajat Singh ICAR-Indian Institute of Vegetable Research, Varanasi, Uttar Pradesh, India

Sanjay Singh Division of Agricultural Bioinformatics, ICAR-Indian Agricultural Statistics Research Institute, New Delhi, India

Saurabh Singh Department of Vegetable Science, Rani Lakshmi Bai Central Agricultural University, Jhansi, Uttar Pradesh, India

Shrawan Singh Division of Vegetable Science, ICAR-Indian Agricultural Research Institute, New Delhi, India

Arpita Srivastava Division of Vegetable Science, ICAR-IARI, New Delhi, India

B. G. Supreetha ICAR-Central Institute of Temperate Horticulture, Dirang, Arunachal Pradesh, India

L. C. Sushmitha ICAR-Division of Vegetable Science, Indian Agricultural Research Institute, New Delhi, India

Ajay Kumar Thakur ICAR-Central Potato Research Institute, Shimla, Himachal Pradesh, India

Jagesh Kumar Tiwari ICAR-Central Potato Research Institute, Shimla, Himachal Pradesh, India

ICAR-Indian Institute of Vegetable Research, Varanasi, Uttar Pradesh, India

Jagesh Kumar Tiwari ICAR-Indian Institute of Vegetable Research, Varanasi, Uttar Pradesh, India

Priti Upadhyay Division of Vegetable Science, ICAR-IARI, New Delhi, India

Jelli Venkatesh Department of Agriculture, Forestry and Bioresources, Research Institute of Agriculture and Life Sciences, Plant Genomics Breeding Institute, College of Agriculture and Life Sciences, Seoul National University, Seoul, Republic of Korea

Neha Verma Department of Vegetable Science, Punjab Agricultural University, Ludhiana, Punjab, India

Development of Mapping Population in Vegetable Crops for GWAS and QTL Mapping

1

N. Pradeepkumara, Saheb Pal, B. G. Supreetha, Reeta Bhatia Dey, A. D. Munshi, T. K. Behera, and Shyam Sundar Dey

Abstract

Over the last 15 years, there has been a notable increase in the application of molecular genetics to improve global vegetable crop production. This has led to the creation of two comprehensive molecular linkage maps for vegetable crops, facilitating the identification of numerous genes and quantitative trait loci (QTLs). Through primary studies, QTLs associated with disease resistance, tolerance to abiotic stresses, and yield potential across different environments have been successfully pinpointed. The ability to precisely locate, manipulate, and potentially clone individual genes responsible for inherited traits, along with the observed conservation of numerous linkage blocks among various plant families like Solanaceous, Cucurbitaceous, Alliaceae, and Leguminous families, highlights the significant impact of map-based genetic analyses on both applied and fundamental crop research. Conventional phenotyping methods aimed at achieving consistent high yields entail evaluating yield performance across various environments and seasons, a process characterized by its labor intensity, high

N. Pradeepkumara · B. G. Supreetha
ICAR-Central Institute of Temperate Horticulture, Regional Station, Dirang, Arunachal Pradesh, India

S. Pal
ICAR-Indian Agricultural Research Institute, Hazaribagh, Jharkhand, India

R. Bhatia
Division of Floriculture and Landscaping, ICAR-Indian Agricultural Research Institute, New Delhi, India

A. D. Munshi · S. S. Dey (✉)
ICAR-Division of Vegetable Science, Indian Agricultural Research Institute, New Delhi, India

T. K. Behera
ICAR-Indian Institute of Horticulture Research, Bengaluru, Karnataka, India

S. S. Dey et al. (eds.), *GWAS and QTL Mapping in Horticultural Crops*,
https://doi.org/10.1007/978-981-95-2306-1_1

costs, and time consumption. Mapping populations plays a fundamental role in identifying markers associated with traits of interest. Marker-assisted selection (MAS) emerges as a pivotal tool in expediting the breeding cycle and serves as a potent molecular method for selecting fruit traits, ultimately boosting yield. High-resolution mapping of QTLs contributes to understanding the molecular basis of phenotypic variation in economically and horticulturally significant traits, thereby facilitating efforts in marker-assisted breeding.

Keywords

Linkage map · Mapping population · Molecular markers · Marker-assisted selection · Quantitative trait loci

The agricultural sector plays a crucial role in addressing food security concerns amid a growing global population (Marcinkowski and Piniewski 2024; Mendelsohn 2008; Fischer et al. 2005). However, the latest Global Hunger Index (GHI) reveals limited progress in curbing hunger worldwide since 2015, with the 2023 score standing at 18.3, only slightly lower than the 2015 score of 19.1. Despite notable strides in some nations, the prevalence of undernourishment has surged, with the number of undernourished individuals climbing from 572 million to approximately 735 million since 2017 (FAO 2023). Projections based on various socio-economic scenarios indicate a significant rise in global food demand by 2050, ranging from 35% to 56%, while the population at risk of hunger is expected to increase by 91% to 98%. This underscores the pressing need to address global hunger, a critical sustainable development goal, and ensure food security for future generations. To effectively confront these challenges, there is a crucial requirement to gain a better understanding of potential future outcomes and the key driving forces behind them. This understanding can aid in identifying and developing superior breeding materials and impactful policies (Wheeler and Von Braun 2013; Calzadilla et al. 2013). Climate-related factors play a significant role in crop production, and forecasts suggest that they will become more pronounced in the coming years, characterized by significant shifts in temperature, rainfall patterns, and increased occurrence of extreme weather events. It is essential to comprehend how future crop yields might respond to these changing climatic conditions to ensure the stability and security of food production (Marcinkowski and Piniewski 2024).

Plant breeding plays a crucial role in transforming wild and weedy plant species into the crop plants essential for human sustenance. Its ongoing objective is to adapt the genetic makeup of these plants to better serve evolving human needs and changing environmental conditions. Throughout history, plant breeders have achieved notable success in meeting the growing global demand for food while also catering to consumer preferences (Amanullah et al. 2022; Casañas et al. 2017; Guo et al. 2017; Ceccarelli and Grando 2024; Lazaridi et al. 2024). Innovations such as hybrid varieties and semi-dwarf fruit and vegetable crops have significantly increased crop productivity. The advent of the "green revolution," propelled by these advancements, notably enabled countries like India to achieve nearly self-sufficient production in essential nutrient-rich and high-yielding vegetable hybrids and varieties within a remarkably short span. Landraces and indigenous varieties serve as crucial repositories of crop species diversity (Lee et al. 2016; Lin et al. 2023; Muleke et al. 2017; Pal et al. 2023; Pavan et al. 2017; Truco et al. 2007; Shakya and Khwaounjoo 2013). Integrating them into plant breeding endeavors holds the potential to enhance yields, improve quality traits, and fortify resilience against a broad spectrum of environmental stresses, both abiotic and biotic.

The Expert Committee (EC) of the Indian Council of Medical Research (ICMR), considering nutritional requirements, advises a daily vegetable intake of at least 300 g per individual, inclusive of green leafy vegetables (50 g), other vegetables (200 g), and roots and tubers (50 g). Pregnant women, requiring adequate iron and folic acid, are recommended to consume a minimum of 100 g of leafy vegetables daily (Kumar et al. 2019; Gupta et al. 2018). Vegetable production is estimated at 212.91 million tons, cultivated across 11.35 million hectares (NHB 2022–2023).

The global contribution to vegetable production is substantial, yet the industry faces imminent threats from climate change. Factors such as rising temperatures, erratic rainfall patterns, and extreme weather events present formidable challenges to vegetable cultivation worldwide, posing unprecedented risks to farmers (Maurya et al. 2019). While vegetable farming offers advantages over traditional cereal crops, including higher yields, shorter growth cycles, and increased job opportunities, it also exposes crops to climate change-related hazards such as heatwaves, droughts, and floods. To secure the future of vegetable production, it is crucial to prioritize research into the impacts of climate change and devise solutions to mitigate its adverse effects on vegetable growth, yield, and quality.

Micronutrient malnutrition presents a significant global public health challenge, impacting individuals across diverse geopolitical regions and socio-economic strata. To address this pervasive issue of "hidden hunger," interventions focus on advocating dietary diversity, implementing food supplementation initiatives, fortifying food products, and biofortifying commonly consumed vegetables, fruits, and crops. Vegetables are classified into distinct groups such as fruit vegetables, leafy vegetables, and root and bulb vegetables, each offering unique characteristics and nutritional benefits. Despite their bulkiness, vegetables serve as vital sources of essential dietary elements, including minerals, vitamins, and antioxidants (Verhoeven et al. 2006). Their consumption contributes to overall well-being by supplying crucial nutrients and antioxidants that combat free radicals, thereby helping to safeguard against immune-related illnesses. Additionally, certain vegetables exhibit favorable shelf-life properties, rendering them suitable for storage and transportation. Hence, it is imperative to develop high-yielding varieties with enhanced quality traits to meet future agricultural demands. Molecular tools and genome sequencing techniques have proven instrumental in expediting traditional breeding methods to augment intricate pathways associated with these traits. Furthermore, genome editing techniques such as QTL mapping, GWAS, RNA sequencing, BSA, omics, transgenics, RNA interference, and CRISPR/Cas-9 hold significant promise for unraveling complex traits in vegetables, enhancing health-promoting constituents, and mitigating antinutritional factors (Verhoeven et al. 2006).

Quantitative Trait Loci (QTL) mapping plays a pivotal role in bridging genomics with trait enhancement (Saraiva 2018). This method involves identifying genomic regions associated with specific traits, offering insights into the genetic factors influencing variations in those traits. By discerning the genetic markers correlated with traits such as yield, disease resistance, and nutritional content, breeders can strategically select plants harboring the most favorable gene combinations. QTL mapping thus serves as a guiding tool, aiding breeders in their efforts to cultivate crops with targeted enhancements (Razzaq et al. 2021).

Recent advancements in molecular marker analysis have revolutionized the study of both simple inherited and quantitative traits, enabling the identification of individual genes responsible for these traits (Zafar et al. 2023; Hamid and Maria 2023). Molecular markers are employed to tag QTLs, aiding in the assessment of their impact on phenotype by selecting favorable alleles at these loci through MAS schemes, with the aim of expediting genetic progress. Previous QTL mapping studies have addressed critical questions concerning the accuracy of QTLs, particularly focusing on the identification and isolation of genes governing essential traits, which is of significant interest to molecular breeders. Two main types of QTLs have been identified: main-effect QTLs and epistatic QTLs, each exerting varying effects on specific phenotypes. It is anticipated that the development of high-resolution maps will facilitate the direct isolation of genes rather than merely markers. The cloning of genes of interest can be achieved through three primary approaches: positional cloning, insertional mutagenesis, and the candidate gene approach.

Quantitative traits display continuous variation due to a combination of polygenic inheritance and environmental influences. Polygenes individually contribute minor effects to the trait phenotype, but their collective impact is cumulative. The concept of genes with cumulative action was proposed by Yule in 1906, and experimental validation was provided by Nilsson-Ehle in 1908. Extensive data collected between 1910 and 1916 by East and Emerson in maize and tobacco further supported the concept of polygenic inheritance. These studies clarified that the continuous variation in quantitative traits results from the combined influence of numerous polygenes and environmental factors on phenotypic expression (Singh 2009). Initially, polygenes were believed to exert only additive effects, but it is now recognized that they can also exhibit dominance and epistatic effects. Classical Mendelian methods are insufficient for studying polygenic inheritance; hence, various statistical tools have been developed, giving rise to the field of quantitative genetics.

QTL mapping provides a direct approach to investigate the number of genes influencing a trait, identify their locations, and evaluate the dosage effect on trait variation. It serves as a foundational step toward map-based cloning and finds applications in DNA-based marker-assisted selection (MAS) and the study of gene linkage (Tierney and Lamour 2005; White et al. 2007; Pekosz et al. 1999). Detecting, localizing, and estimating the effective size of genes with minor effects can be challenging through QTL mapping, which is more effective in identifying single genes with significant effects (Li 2000; Zeng 1994). The reliability of QTL maps relies on the repetition of phenotyping experiments and demands high heritability in the screening system. The primary objectives of QTL mapping include identifying genomic regions affecting the target trait, analyzing QTL effects, determining the proportion of trait variation attributable to specific regions, identifying gene action types associated with QTLs (additive/dominant), and identifying favorable alleles linked to the trait. Emphasis should be placed on traits easily observable in robust screening systems, derived from crosses between highly resistant and susceptible lines. Utilizing highly reliable screening methods and analyzing means of repeated screenings rather than single trials is crucial. Basic prerequisites for QTL mapping include phenotypically contrasting parental lines, a saturated linkage map based on molecular markers, dependable phenotypic screening, and suitable statistical packages for QTL detection. The fundamental principle underlying QTL mapping is the co-segregation of marker loci and QTLs across generations, indicating tight linkage between them. Instances where genes do not segregate independently exhibit "linkage disequilibrium," upon which QTL analysis relies to determine if a QTL is linked to a marker (Wang et al. 1999).

Conventional methods of phenotypic selection are limited due to the substantial impact of environmental factors on phenotype, making the selection of suitable breeding materials challenging. However, since the 1980s, technological advancements such as high-throughput sequencing and molecular mapping have transformed vegetable breeding. These advancements allow for the exploration of genetic variation, identification and tagging of genetic loci, development of stress-resistant cultivars, enhancement of yield, improvement of health-promoting compounds, and the utilization of marker-trait information for MAS. This technological revolution provides breeders with the tools to manipulate plant genomes and effectively use molecular markers in breeding programs.

Advanced genetic methodologies, including biparental mapping, agro-biotechnology/omics applications, and innovative breeding approaches such as traditional QTL mapping, association mapping, RNA sequencing, bulked segregant analysis, and genomics, are anticipated to tackle the challenges encountered by future breeders. In vegetable crops, both yield and quality are economically crucial, with most quantitative traits primarily controlled by multiallelic or polygenic genes. These traits are significantly impacted by environmental factors.

A fundamental objective in quantitative genetics and molecular genomics is to uncover the genetic basis of traits that hold evolutionary significance. However, this endeavor is complex because many of these traits likely have a polygenic nature. They are governed by numerous major and minor effective genes, each exerting a small individual effect. Consequently, characterizing them genomically using traditional QTL mapping methods.

Two common mapping techniques, namely linkage mapping/QTL mapping and association mapping, are widely utilized to identify genetic variations associated with economically important traits in vegetable crops. These approaches rely on functional polymorphisms and molecular markers to identify QTLs linked to yield and quality.

Steps Followed in the QTL Mapping

1. Creating a mapping population involves selecting two parental strains with distinct phenotypic traits, such as high disease resistance and susceptibility.
2. Afterward, choosing an appropriate molecular marker and constructing a comprehensive linkage map is essential.
3. Subsequently, phenotyping and genotyping the mapping population is carried out.
4. Finally, utilizing a suitable statistical package, linkage analysis is performed by integrating genotypic and phenotypic data.

Mapping populations are generated by crossing genetically diverse lines in a controlled manner to create a suitable population for linkage mapping of genetic markers (Fig. 1.1). Typically, these lines are from the same species, but in cases of limited intraspecific variation, related species may be utilized. These populations aid in determining genetic distances between loci/genes and mapping them to specific genomic locations. Additionally, they facilitate the identification of molecular markers linked to genes/loci of interest, crucial for MAS. Thus, mapping populations are fundamental for identifying genomic regions containing genes/QTLs and estimating their effects. The choice of parents, mating design, and marker system used for mapping depends on study objectives, cost, marker accessibility, and the availability of a molecular map. It is essential that the selected parents exhibit maximum divergence at both the DNA sequence and phenotypic levels to trace recombination events accurately. Greater DNA sequence variation enhances the likelihood of finding polymorphic molecular markers. When identifying and mapping genes

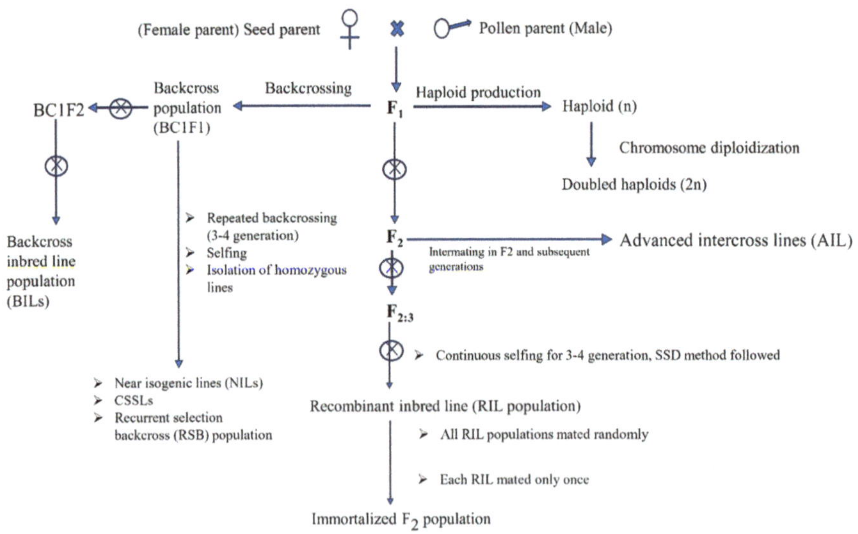

Fig. 1.1 Mapping the population's development in Vegetable crops for mapping studies

governing specific traits, parents must display genetic variation for those traits. However, environmental effects can sometimes inflate phenotypic variation without a genetic basis, while the absence of phenotypic variation between parents does not necessarily imply a lack of genetic variation for the trait, as different gene sets can produce similar phenotypes.

There are two main types of mapping populations: primary and secondary. Primary mapping populations are formed by crossing two homozygous lines with differing traits of interest. Conversely, secondary mapping populations are created by crossing selected individuals from a primary mapping population and are primarily for fine mapping genomic regions of interest. Primary mapping populations encompass various types, including F_2, F_2-derived F_3 ($F_{2:3}$), backcross (BC), backcross inbred lines (BILs), doubled haploids (DHs), recombinant inbred lines (RILs), near-isogenic lines (NILs), chromosomal segment substitution lines (CSSLs), immortalized F2, advanced intercross lines, recurrent selection backcross (RSB) populations, and interconnected populations. The choice of mapping population depends on factors such as the ability of the plant species to undergo self-fertilization without significant inbreeding depression, the time available for population development, and the trait(s) to be mapped (Schneider 2005).

Types of Mapping Population Utility in Vegetable Crop Improvement

1.1 Mapping Populations Used in Mapping Studies

1.1.1 F_2 Population

Crossing two distinct parents, the first filial generation (F_1) is produced. When the resulting population is self-pollinated or sib-mated, the progeny is called the second filial generation (F_2). This generation is used as the most basic and predominant mapping population across vegetable crops for the development of linkage maps and also mapping of oligogenes, as it shows maximum genetic variation, and each plant/individual carries a unique combination of chromosomal segments coming from each parent. Such combinations are considered for estimating the linkage between a pair of loci. This population takes a minimum time and effort for its development (two seasons) and is also cost-effective when compared to the other populations. It can also be produced in crops that show very high inbreeding depression. All the plants are grown and individually phenotyped for the target trait in this population. It is also preferred as all three components of genetic variance, viz., additive, dominance, and epistatic components, can be estimated. Moreover, this population represents the recombination events from both male and female parents, enabling easy identification of heterotic QTLs (Table 1.1).

However, it results in low mapping resolution (markers are generally located at a greater distance from the loci) as the population is the result of a single meiotic or recombination cycle. As all the plants carry a different combination of the chromosomes from the parents, replication over seasons and locations is not possible in this

Table 1.1 Summary and characteristic features of some of the mapping populations utilized in the GWAS and QTL studies in vegetable crops

	Mapping population							
Features/ characteristics	F_2 segregating population	Backcross population	Recombinant inbred lines	Near-isogenic lines	Chromosomal substitution lines	Immortalized F_2 population	NAM population	MAGIC population
Mortal or immortal population	Mortal	Mortal	Immortal	Immortal	Immortal	Immortal	Immortal	Immortal
Biparental or multiparental	Biparental	Biparental	Biparental	Biparental	–	Biparental	Multiparent	Multiparent
Genetic constitution	Homozygotes and heterozygotes	Homozygotes and heterozygotes	Homozygotes	Homozygotes	Homozygotes	Homozygotes and heterozygotes	Homozygotes and heterozygotes	Homozygotes and heterozygotes
Genotype representative	Single parent	Single parent	Single parent	Single parent	Single parent	Single parent	Multiparent	Multiparent
Generations/ breeding cycles need to be developed	Two generations (one parental cross followed by F_1 selfing)	Two generations (one parental cross followed by an F_1 cross with either of the parents)	7–8 generations (one parental cross followed by F_1 selfing till $F_{7:8}$ to get ~99.99% homozygosity)	8–10 generations	7–10 generations	Single generation after the development of the RIL population	–	–
Crosses to be made	One cycle of crossing (F_1)	Two cycles of crossing (F_1 and BC)	One cycle of crossing (F_1)	6–7 cycles of crossing (F_1 and BC)	6–7 cycles of crossing (F_1 and BC)	As per the need, many crosses can be made (Hundreds/ population)	–	–

Selection can be made during population development (No/Yes)	No	No	No	Yes (Foreground and Background)	Yes (Foreground and Background)	No	–	–
The number of recombination happened during population development	Single cycle	Single cycle	Two cycles	One parental cross and 6–7 backcrosses	One parental cross and 6–7 backcrosses	Two cycles	–	–
Oligogene mapping	Possible	Possible	Possible	Possible	Possible	Possible	Possible	Possible
QTL mapping	course map construction possible, but not use full much	No	Yes	Yes	Yes	Yes	Yes	Yes
Fine mapping/ high-resolution mapping	No	No	Yes (after attaining 7–8 generations of selfing and getting homozygosity)	Yes	Yes	No	–	–
Heterosis loci mapping	Possible	No	No	No	No	Yes	–	–
Positional cloning	No	No	No	Yes	yes	No	–	–

(continued)

Table 1.1 (continued)

	Mapping population							
Features/ characteristics	F_2 segregating population	Backcross population	Recombinant inbred lines	Near-isogenic lines	Chromosomal substitution lines	Immortalized F_2 population	NAM population	MAGIC population
Assessment of QTL x Genotype interaction	No	No	Yes	Yes	No	No	–	–
Mapped gene/ loci belongs to	Any of the parents	Any of the parents	Any of the parents	Donor/male parent	Donor/male parent	Either of the parents	–	–
Mapping coverage	Whole genome	Whole genome	Whole genome	A genomic segment	A genomic segment	Whole genome	–	–

population, limiting the precise estimation of genotype × environment interactions required for fine mapping. Further, this population is a mortal population that can be maintained further only through asexual reproduction, like tissue culture, cutting, budding, etc. However, the response to regeneration in such asexual methods of propagation varies across vegetable crops. For tissue culture, a standardized protocol becomes necessary with the involvement of additional cost and effort. To date, the tissue culture protocol for all the vegetables has not been standardized. Though this population has such limitations, different mapping experiments have been conducted using the F_2 population across vegetable crops.

1.1.2 F_2-Derived F_3 ($F_{2:3}$) Population

To minimize the major demerits of the F_2 population, the $F_{2:3}$ population is generated by individually self-pollinating the F_2 plants, followed by harvesting the seeds separately. As a result, each F_2 plant is represented by the individual plant progeny. For mapping experiments, the phenotyping is done in the $F_{2:3}$ progenies, and the mean of the population (multiple $F_{2:3}$ plants) is designated as the phenotype of the parental F_2 plant from which the progeny was derived. The genotyping of the population is done using the DNA extracted from the individual F_2 plants or bulked DNA extracted from at least 20 $F_{2:3}$ progenies to represent the parental genotype. As phenotyping data is recorded from multiple plants of the family, sampling error is minimized. It enables the breeders to map QTLs for mapping recessive gene-governed oligogenic traits. The deployment of this population is generally limited by its mortal nature, i.e., like F_2, this population cannot be perpetuated to the next generation through sexual reproduction. Additionally, its production needs extra effort and time. Replication of this population is also not possible, as the majority of the F_3 families are heterogeneous. As a result, the means of the individuals of the F_3 families do not precisely correspond to the performance of the particular F_2 plant due to segregation. In crops with high or very high inbreeding depression, like onion, carrot, early-season cauliflower, etc., the phenotyping of the $F_{2:3}$ population generally underestimates dominance, overdominance, and some part of epistatic genetic variation due to inbreeding. Various mapping experiments using the $F_{2:3}$ population for different horticultural and yield traits of different vegetables have been summarized in the current chapter.

1.1.3 Backcross Inbred Line (BILs) Population

The crossing of the F_1 to either of the homozygous parents is called backcrossing, which produces the BC_1F_1 progeny. BILs are developed by forwarding BC_1F_1 progeny up to BC_1F_6 or BC_1F_7 until appreciable homozygosity is achieved in the population. This population is advantageous over the F_2 population, as it increases the allele frequency of the parent's traits to which the F_1 is backcrossed. Thus, it is generally followed to backcross the F_1 with the donor parent for resistance traits or with the parent having better yield and quality for mapping traits contributing to yield and quality. A BIL possessing resistance to Watermelon Bud Necrosis Virus in watermelon was developed at the ICAR-Indian Institute of Horticultural Research, Hesaraghatta, Bengaluru, India. Using this line to develop an $F_{2:3}$ population, Maragal et al. (2021) mapped three QTLs each on chromosomes 2, 3, and 7, conferring resistance to WBNV. Çolak et al. (2020) developed a BIL population from the LA 1589 interspecific inbred of tomato and carried out mapping for antioxidant compounds like carotenoids, vitamin C, and syringic acid. They identified a QTL, *lyc8.1* on chromosome 8, involved in the production of carotenoids; *vite9.2* on chromosome 9 for vitamin C content; and *sya6.1* on chromosome 6 for the production of syringic acid.

1.2 Doubled Haploid (DH) Population

The doubling of chromosomes of haploid F_1 plants generated from anther/pollen grain culture gives rise to the DH plants. Different chemicals like colchicine, colchamine, oryzaline, etc., to double the chromosomes by blocking spindle development. As a result, the DH lines possess complete homozygosity for all the loci of the genome. The absolute homozygosity of all loci makes it advantageous over the RILs, in which some residual heterozygosity remains. Although from the meiosis point of view, DH lines are comparable to F_2, as both are the product of a single meiotic cycle while developing the F_1 plants, the frequency of occurrence of recombinants in the DH population always remains greater than the corresponding F_2 population. The DH population can be prepared within the same time required for the development of the F_2 population. However, a standardized tissue culture protocol and technical skills are necessary. This population cannot be developed in crops where a haploid plant production protocol has not been standardized. Even though haploid plant production is standardized for a particular crop, different genotypes of the same crop often show differential responses toward tissue culture. After development, the seeds should be harvested from individual plants and can be maintained through sexual reproduction indefinitely. Thus, being an immortal population, this helps researchers map both quantitative and qualitative traits through precise phenotyping in replicated trials over locations and seasons and for the sharing of germplasm accessions. This population serves as one of the best populations to map recessive and quantitative traits in crops like carrots, onions, and early-season cauliflowers, which suffer from severe inbreeding depression, making it extremely difficult to achieve complete homozygosity through self-pollination or sib-mating. However, it is worth noting that this population is unsuitable for mapping heterotic QTLs. In such a homozygous population, estimation of only additive and additive × additive components of genetic variation can be estimated. Treatment with chemicals to double the chromosomes of haploid plants may sometimes result in the development of undesirable genetic variation.

A DH population is mainly developed and utilized in crops that show high inbreeding depression, which limits the development of a homozygous population for conducting mapping experiments. This population is primarily preferred in cole crops to map different quantitative and qualitative traits. Zhao et al. (2020a, b) utilized a DH population derived from IL4305 × ZN198 to map two QTL contributing to stalk length (*qSL.C6-1* and *qSL.C6*-2) and two QTL for curd solidity (*qCS.C6-1* and *qCS.C6-2*), all on chromosome 6. In another study, Li et al. (2021a, b, c, d, e) utilized a DH population derived from lines 86101 × 90196 and mapped four QTL on chromosome 3 involved in sulforaphane metabolism.

1.3 Recombinant Inbred Lines (RILs)

Recombinant Inbred Lines (also referred to as F_2-derived inbred lines) are homozygous lines with random chromosomal blocks produced through inbreeding (generally self-pollination, rarely by sib-mating) of individual F_2 plants and forwarded up to the F_7 or F_8 generation through the single seed descent or double seed descent method unless complete homozygosity is obtained. Additionally, for the advancement of the F_2 population, both pedigree and bulk methods without selection can be used to advance the generations. It is important to note that the bulk or pedigree method is to be followed only in regions suitable for growing the particular vegetable to avoid survival differences across the population due to the effect of the environment (i.e., natural selection). Forwarding the F_2 population following single seed descent methods is advantageous as it enables the breeders to maintain the maximum genetic variation within the limited resources; however, it is limited by the loss of valuable lines due to problems during seed germination, survival due to variable environmental parameters, or extreme weather events. This problem is prominent in vegetables as well as other crops which are grown in close spacing. For a particular locus, the RIL population consists mainly of two homozygotes (e.g., AA or aa) and a very small proportion of heterozygotes (Aa). Thus, this population results in a better mapping resolution, i.e., identification of molecular markers in the close vicinity of the target locus, when compared to that of the F_2, DH, and backcross populations, as these latter populations are the result of a single meiotic cycle, and a single recombination event between the target locus and marker is possible as compared to multiple crossover events in the RIL population. Additionally, the RIL populations can be maintained indefinitely through controlled self-pollination, can be multiplied, and are amenable to phenotyping over seasons and locations. However, the use of this population is limited by the time and resources required to develop. This population is challenging to develop and maintain in cross-pollinated crops, which show moderate to severe inbreeding depression. Further, like the DH population, the RIL population can only detect additive and additive × additive components of genetic variation.

1.4 Near-Isogenic Lines

Homozygous lines, otherwise identical except for a single locus/gene, are known as NILs. However, the individuals of the NIL may also differ for other random genomic regions randomly across the genome. The lines are generally produced by back-crossing of the F_1 with the recurrent parent, followed by the backcross of the resulting progeny with the recurrent parent for seven to eight generations. After this, the backcross progenies are self-pollinated, and the homozygous plants with the desirable allele of the donor parent coupled with most of the traits identical to the recurrent parent are identified to constitute the NIL. The NIL population is homozygous and an immortal population like RILs and DHs. However, over these two populations, NILs, harboring the QTLs or loci in the recurrent parent background, are advantageous as they enable assessing the effects of QTL with more resolution, accuracy, and reliability, and subsequently for map-based cloning of such QTLs. However, the deployment of this population in different vegetable crops is often limited by the poor diversity of molecular markers between the recurrent and donor parents. It should also be noted that if any wild relative or unadapted germplasm accession is used as a donor parent, the breeder may incur problems of linkage drag due to preferential inheritance of the genome of the unadapted/wild germplasm. Using two F_8 NIL populations, Tikunov et al. (2020) mapped the *FLORAL4* gene contributing to fruit flavor in tomato.

1.5 Advanced Intercross Lines (AILs)

The F_2 population maintains maximum genetic variability. Thus, intermating among individuals from the F_2 population helps maintain better heterozygosity in the population and increases the chances of recombination between the QTLs and the closely linked markers. The advanced intercross lines help in the precise mapping of QTLs, reducing the confidence interval by up to fivefold in AILs compared to the F_2 population (Darvasi and Soller 1995). However, this population is not commonly used in mapping, as the required statistical methods for this population are yet to be developed.

1.6 Chromosome Segment Substitution Lines (CSSLs)

A series of homozygous lines, each with a particular and distinct segment of chromosome from the donor parent in the background of the recurrent parent, is known as CSSL. Further, the haploid chromosome complement of the donor parent is represented by the total of all the segments of the donor parent's genome. These are also known as intervarietal substitution lines or introgression lines (ILs). Lines with chromosome segments of a variety of the same crop are termed as intervarietal substitution lines, while lines with chromosomal segments from other crossable/wild species are termed as ILs (Schneider 2005). These lines are also called segment introgression/segment substitution, chromosome segment substitution, or chromosome segment ILs. These lines are produced by backcrossing the F_1 with the recurrent parent for a minimum of six generations, and the subsequent two or more generations isolate the desirable lines that are homozygous for the introgressed chromosomal segments. In order to confirm the content of the distinct but slightly overlapping genome of the donor parent, selection based on molecular markers evenly distributed throughout the genome is used. Similar to DHs, RILs, and NILs, these lines are also immortal. Thus, these can be deployed to map both oligogenes as well as preliminary and fine mapping of the QTLs, even if the phenotypic variation explained is very minute. Such QTLs with minor contributions to the phenotypic variation often get masked by the QTLs with more significant phenotypic effects and remain undetected in F_2 and RIL populations. Further, this population avoids the common interference during the QTL mapping, i.e., QTL × QTL interactions or epistatic effects of the genome of the donor parent on the causal QTLs. Identification of QTL using CSSL does not require the development of a linkage map, and the QTLs can be mapped into smaller confidence intervals. These lines provide better insights into the inheritance of the trait and distribution of the genes over the genome and can be directly used for mapping and cloning of candidate genes or QTLs. However, the main disadvantage of these lines is the linkage between the desirable and undesirable traits in the close vicinity of the introgressed chromosomes, mainly when any wild relative or unadapted germplasm lines are used as donors. Unlike linkage drag incurred in other populations, it may create much more difficulty in alien chromosomal substitution lines, as there are reduced chances of pairing and recombination between the chromosomes of donor and recurrent parents. A set of alien CSSLs was developed by introgression of the overlapping chromosomal segments with an average size of ~33 cM from *Solanum pennellii* (LA716: green fruited) into *Solanum lycopersicum* (M82: red fruited) (Eshed and Zamir 1994). The resultant population, representing the entire genome of *Solanum pennellii,* was further used for QTL mapping, and 23 loci for total soluble solids and 18 loci for fruit weight were identified, which was around double that reported earlier using other mapping populations.

1.7 Immortalized F_2 Population

The term immortalized F_2 population was first used by Gardiner et al. (1993) in maize, in which the F_3 progenies obtained from the F_2 population were intermated into two groups, and seeds from at least 20 plants were harvested in bulk to represent each of the parental F_2 plants from which the F_3 progenies were produced. This population is advantageous over the others, as a true representation of all possible genotypes, including the heterozygotes of this population, is possible. Although the name of the population is "immortalized," it is not a perpetual or immortal population, and it is like the F_2 population. However, the same immortalized population can be reconstructed from the immortal and parental RILs. Thus, it is possible to get data from the replicated trials over seasons and locations and enable the breeders to map the QTL with higher precision, including the detection of the epistatic QTLs. However, the deployment of this population in vegetable improvement programs is limited by the time and resources required for making a large number of crosses, particularly in strictly self-pollinated crops like legumes, tomatoes, and lettuce. Gur and Zamir (2015) developed an immortalized F_2 population in tomato derived from the inbred variety M82 (*S. lycopersicum*) and the inbred accession of *S. pennellii* (LA716) to study the QTL-QTL interaction for brix yield per unit area.

1.8 Multiparent Advance Generation Intercross (MAGIC) Population

The collections of RILs generated by involving several parental accessions are commonly known as multiparent advanced generation intercross (MAGIC) populations. The MAGIC name was proposed by Mackay and Powell (2007). This population is also regarded as an extended version of Advanced Intercross Lines (AILs) proposed by Darvasi and Soller (1995). However, it differs from the AILs, as it involves only two parents for its development, while the MAGIC population involves multiple parents. For the development of the RILs, inbreds, clones, or individual plants are selected based on the origin of these lines. Typically, eight parental lines are crossed in pairs simultaneously to develop four single crosses. In the next step, these single crosses are again crossed in pairs to develop two double crosses, which are subsequently crossed together to produce the complex cross resulting from eight parents. Then, each of the plants resulting from this last step is forwarded following the single seed descent method for developing the desired number of RILs to constitute the MAGIC population. This population is advantageous over most other populations, as it is immortal and can be utilized for both linkage analysis and association analysis for mapping multiple traits for which the parental accessions have contrasting phenotypes, as well as the effects of their cytoplasm. This population is also suitable for the construction of high-density maps and genomic selection. Further, mapping of multiple QTLs for each trait and more than two alleles of each QTL is also possible using this population. As an added advantage, the RILs developed from the elite and diverse parents generally possess a combination of many desirable traits, making them suitable to be used as a variety. With an objective to combine the most favorable QTL alleles for white mold resistance, Agir (2023) selected eight snap bean lines, viz., Banga, Flo, Stayton, Top Crop, Unidor, US Refugee 5, A195, and USPT-WM-12, and developed a MAGIC population in snap bean. These parental genotypes possessed different QTL alleles, which contributed to resistance.

1.9 Nested Association Mapping (NAM) Population

Linkage mapping is done in the populations derived from the crossing of two distinct parents. Across the populations, the recombination events between the marker and the loci remain limited and pose difficulty in achieving high resolution during

QTL mapping. In association mapping, after assessing their population structure and estimation of relatedness (kinship values), the germplasm panel is directly used as a mapping population. The historical recombination happening in the genotypes over the generations gives rise to a better mapping resolution when compared to biparental QTL mapping. The NAM is conceptualized to take advantage of both association and linkage mapping. In this, a set of diverse parents is crossed to one or two common parents (known as nested parents), and the population is forwarded following a single seed descent method to develop 250 RILs from each cross (Yu et al. 2008). Thus, this population enables efficient utilization of both genetic and genomic resources for the dissection of complex traits. Agir (2023) developed four NAM populations of the snap beans by crossing them to a common parent to deploy the population for mapping of QTLs conferring resistance to white mold. Further, a total of 22 QTLs for white mold resistance on nine different chromosomes of snap bean were detected using the NAM populations.

1.10 Recurrent Selection Backcross Population

The idea of RSB population was put forth by Wright (1952) for the identification of QTL with more significant phenotypic effects. For the development of this population, one parent having the higher value of the quantitative trait of interest is crossed as a donor parent possessing the low value for the same trait. The resultant F_1 is backcrossed to the recurrent parent. From the resulting backcross progeny, the plants with a higher magnitude of the phenotypic traits are selected to backcross with the recurrent parent again. As a result, a rapid fixation of homozygosity of the alleles contributed by the recurrent parent happens except for the loci contributed by the donor parent. Such selection for a higher magnitude of the phenotypic trait will result in the maintenance of the QTLs from the donor parent, which will have more effects on the phenotype in the population. In contrast, the QTLs with low to moderate phenotypic effects will be lost. However, it should be kept in mind that a sufficiently large number of backcross generations should be generated and evaluated to ensure precise QTL mapping. Thus, it requires considerable time and resources. It is also important to note that this population is particularly suitable for detecting QTLs with significant phenotypic effects. In cases where quantitative traits like yield are governed by many QTLs with low to moderate additive contribution to yield, deployment of this population may not give the best results.

1.11 Markers Utility in QTL Mapping

A diverse range of random and sequence-based markers, such as simple sequence repeats (SSRs)/microsatellites and single-nucleotide polymorphisms (SNPs), have been extensively utilized in various genomics-assisted breeding applications of crop plants despite the availability of various marker types, SSRs remain the most popular and preferred choice for plant genetic analysis due to their advantageous genetic attributes. SSR markers are favored for their co-dominant inheritance, high reproducibility, robust amplification, multi-allelic nature, and wide distribution across different genomic regions, including both coding and non-coding sequences.

Microsatellites, also known as SSRs, are particularly suitable for genetic mapping and marker-assisted breeding due to their abundance, widespread presence in diverse genomes, and co-dominant inheritance. SSRs consist of repeat motifs ranging from 1 to 6 base pairs in length, making them highly prone to mutations, primarily caused by DNA polymerase errors, leading to the addition or subtraction of repeat units. Consequently, SSR sequences exhibit high levels of polymorphism, enabling the detection of allelic variation within populations.

These repeats are found in both genic and nongenic regions and are sometimes transcribed, making them identifiable in expressed sequence tags (ESTs) as well as in non-genic DNA sequences. Additionally, the ability of SSR markers to be multiplexed for high-throughput genotyping renders them invaluable for various plant structural, functional, and comparative genomics applications.

It has been determined that for various experimental designs, moderately wide marker spacings, approximately around 50 cm, are optimal or near-optimal for initial investigations into marker-QTL linkage. For constructing linkage maps and conducting linkage mapping of traits in specific crops, a minimum of 2000–3000 markers, with at least 100 polymorphic markers, is considered most suitable. All markers, including SSRs and other types, chosen for mapping experiments should be evenly distributed across all linkage groups of the crop in question. Each linkage group should ideally have a minimum of 50 markers, densely placed with minimal distances between adjacent markers. The proximity and minimal distance (measured in cM/bp) between markers hold significant importance for MAS of the mapped traits.

1.12 QTL Mapping of Horticultural Traits in Vegetable Crops

1.12.1 Tomato (*Solanum lycopersicum* L.)

In a study focusing on horticultural traits, genetic loci influencing various attributes were mapped using a Recombinant Inbred Line (RIL) population and 353 molecular markers. The *FrW7.1* QTL, associated with fruit size, was found to overlap with QTLs for soluble solids, vitamin C, glucose, and dry weight/fresh weight, indicating a shared genetic control mechanism potentially mediated by the Sucrose Phosphate Synthase gene. Additionally, candidate genes for beta-carotene and vitamin C levels were identified as 1-deoxy-D-xylulose 5-phosphate synthase and tocopherol cyclase genes, respectively (Capel et al. 2015). Another investigation on salad tomatoes revealed yield-related QTLs on chromosome 11, where specific alleles increase fruit number and yield, while adjacent regions decrease fruit size. Additionally, on chromosome 9, certain alleles were found to elevate fruit sweetness (Brix) by 25% (Brekke et al. 2019) (Table 1.2).

Table 1.2 Status of QTL mapping in vegetable crops for horticulture traits

Trait	Mapping progeny	Technique	Molecular markers	QTLs/genes	Chromosome no.	References
Tomato (*Solanum lycopersicum* L.)						
Fruit traits						
Fruit weight	F_2 progeny	QTL-Seq Analysis	SNP	*fw6.3*	Ch6	Ning et al. (2023)
Fruit diameter	F_2 and F_3 progeny (Caimanta × LA0722)	QTL mapping using SNP studies		*FD1, FD2, FD7*	Ch1, Ch2, Ch7	Cabodevila et al. (2021)
Fruit weight				*FW2*	Ch2	
Fruit shape index				*FSI2, FSI6, FSI 9*	Ch2, Ch6, Ch9	
Fruit flavor	Two F_8 NILs	QTL analysis	KASP and SNP	*FLORAL4*	Ch4	Tikunov et al. (2020)
Early ripening	F2 progeny Bone MM × 071-440)	QTL-seq	SNP	*Solyc11g071510.1.1*	Ch11	Ruangrak et al. (2019)
Fruit shape	F2 progenies (Yellow Stuffer × Heinz 1439 [F2YSxH] and Voyage × Old Brooks [F2VxOB])	QTL-seq analysis	SNPs and InDels	*ld6, ld11, and ld8*	Ch6, Ch11, and Ch8	Vazquez et al. (2022)
Antioxidant compounds (carotenoids, vitamin C, and syringic acid)	LA 1589 interspecific inbred backcross line (IBL)	QTL mapping	SNP	*lyc8.1, vite9.2, and sya6.1*	Ch8, Ch9, and Ch6	Çolak et al. (2020)
Agronomic traits						
Plant height	F_2 (NC 11212) (NC 84173 × CLN1466EA)	QTL mapping	SNP	*qph-08*	Ch8	Siddique et al. (2024)
Early flowering	F_2 progeny (Bone MM' × 071-440)	QTL-seq analysis	SNP	*EF1 (Solyc01g017060)*	Ch1	Ruangrak et al. (2018)

(continued)

Table 1.2 (continued)

Trait	Mapping progeny	Technique	Molecular markers	QTLs/genes	Chromosome no.	References
Trichomes **IV** Density	F2 progeny (*S. lycopersicum* var. *cerasiforme* × *S. lycopersicum* var. *lycopersicum*)	QTL mapping and QTL-seq	SNP	*TIVd9*	Ch9	Mata-Nicolás et al. (2021)
Brinjal (*Solanum melongena* L.)						
Fruit traits						
Fruit length	121 plants of F_2 progeny (1836 × 1809)	**QTL analysis**	SNP	*fl12.1*	Ch12	Wei et al. (2020a, b)
Fruit diameter				*fd3.1*	Ch3	
Fruit shape				*fs3.1*, *fs1.3*	Ch3, Ch1	
Fruit weight	F_2 progeny (Pusa Safed Baingan 1 × Pusa Uttam)	QTL mapping	SSR	*qsfwpu1*, *qsfwpu3* and *qsfwpu2*	Ch1, Ch2 and Ch3	Bhanushree et al. (2020)
Plant anthocyanin pigmentation						
Adaxial leaf Lamina Anthocyanin	RIL progeny (305E40 × 67/3)	QTL mapping	SNP	*adlan10.1*	Ch10	Toppino et al. (2020)
Stem Anthocyanins				*stean10.1*	Ch10	
Anthocyanin Tonality				*toan5.1*	Ch5	
Hypocotyl Anthocyanins				*hyan5.1*	Ch5	
Corolla Color				*corcol5.1*	Ch5	
Leaf prickle color	121 plants of F_2 progeny (1836 × 1809)	QTL analysis	SNP	*lpc10.1*	Ch10	Wei et al. (2020a, b)
Vein color				*vc10.1*	Ch10	
Leaf traits						

Leaf lobing	121 plants of F_2 progeny (1836 × 1809)	QTL analysis	SNP	*llob6.1* and *llob7.1*	Ch6 and Ch7	Wei et al. (2020a, b)
Leaf prickle number				*lpn7.1*	Ch7	
Plant characteristics						
Main stem height	121 plants of F_2 progeny (1836 × 1809)	QTL analysis	SNP	*msh5.1*		Wei et al. (2020a, b)
Other traits						
Seed Vigor	RIL progeny (305E40 × 67/3)	QTL mapping	SNP	*sei2.1*	Ch2	Toppino et al. (2020)
Chilli (*Capsicum annuum* L.)						
Fruit traits						
Fruit weight	F_2 and $F_{2:3}$ progenies	QTL mapping	SSR and InDel markers	*ftw2.1*	Ch2	Ma et al. (2022)
Fruit shape				*fts1.1*	Ch1	
Fruit length				*ftl2.1*	Ch2	
Fruit diameter				*ftd2.1*	Ch2	
Locule number				*lcn1.1*	Ch1	
Fruit length	F2 progenies (*C. annuum* L. cv. 'FL 201' × *C. galapagoense* cv. 'TC 07245')	QTL mapping	SSR	*paufl2.1 and paufl2.2*	Ch2	Arjun et al. (2018)
Fruit weight	F2 progenies (*Capsicum annuum* '007EA' × *Capsicum frutescens* 'P1512')	QTL mapping	SNP	*qFW11*	Ch11	Wei et al. (2020a, b)
Fruit length				*qFL3*	Ch3	
Fruit orientation	F2 progenies (*C. annuum* 'BA3') × (*C. frutescens* 'YNXML')	QTL mapping	SNP	*Up12.1*	Ch12	Cheng et al. (2016)

(continued)

Table 1.2 (continued)

Trait	Mapping progeny	Technique	Molecular markers	QTLs/genes	Chromosome no.	References
Fruit length	RILs (Perennial × Dempsey)	QTL mapping	SNP	*FL-3.2, FL-3.4*	Ch3	Han et al. (2016)
Fruit shape				*FS-3.1, FS-3.2*	Ch3	
Fruit weight				*FW-6.1, FW-1*	Ch6, Ch1	
Fruit position				*FP-12.1, FP-12.2, FP-12.3*	Ch12	
Calyx shape				*CS-1.1, CS-1.2*	Ch1	
Immature fruit color				*IFC-10.2, IFC-10.3*	Ch10	
Unripe fruit pigmentation	$F_{2:3}$ population (Puya × Chiltepin)	QTL mapping	SNP	*UFP10.1*	Ch10	Lopez-Moreno et al. (2023)
Fruit orientation				*F012.1*	Ch12	
Seedless fruit				*SF1.1 and SF6.1*	Ch1, Ch6	
Fruit weight				*FWE2.1*	Ch2	
Fruit length				*FL4.1*	Ch4	
Fruit area				*FA2.1*	Ch2	
Fruit shape				*FS2.1*	Ch2	
Plant characteristics						
Plant height	$F_{2:3}$ progenies (Puya × Chiltepin)	QTL mapping	SNP	*PH8.1*	Ch8	Lopez-Moreno et al. (2023)
Growth habit				*GH1.1*	Ch1	
Stem pigmentation				*SP10.1*	Ch10	
Leaf width	F_2 progenies (*Capsicum annuum* '007EA' × *Capsicum frutescens* 'P1512')	QTL mapping	SNP	*qLW4, qLW9*	Ch4, Ch9	Wei et al. (2020a, b)
Leaf length				*qLL4*	Ch4	
Main stem length				*qMSL10*	Ch10	
Internode length				*qIL11*	Ch11	
Offshoot number				*qON4 and qON12.1*	Ch4, Ch12	

Main stem length	RILs (*C. annuum* 'Perennial' and *C. annuum* 'Dempsey'	QTL mapping	SNP	*MSL-8.1*	Ch8	Han et al. (2016)
Plant height				*PH-2*	Ch2	
Lateral branch number				*LBN-2.1, LBN-2.2*	Ch2	
Other traits						
Initial blooming time	F2 populations (*Capsicum annuum* ' 007EA' × *Capsicum frutescens* 'P1512')	QTL mapping	SNP	*qIBT7*	Ch7	Wei et al. (2020a, b)
First flower node	F_{10} RIL (PM702 × FS871)	QTL mapping	SLAF	*Ffn2.1 and Ffn2.2*	Ch2	Zhang et al. (2019a, b)
Flowering time	F_2 populations (*C. chinense* '740' × (*C. annuum* 'CA1')	QTL analysis using the high-density genetic map	SLAF	*Ft2.1*, *Ft6.1* and *Ft6.2*	Ch2, Ch6	Zhu et al. (2019)
Flower number per node				*Mf2.1*, *Mf7.1* and *Mf10.1*,	Ch2, Ch7, Ch10	
Microspore embryogenesis	RIL (FS871 × PM702)	QTL mapping	SLAF	*Me2.2*	Ch2	Bai et al. (2023)
Cucumber (*Cucumis sativus* L.)						
Fruit traits:						
Fruit shape	$F_{2:5}$ (Q16 × Q24)	QTL mapping	SSR and InDel marker	*FS1.1, FS1.2, FS2.1*	Ch1, Ch2	Zhang et al. (2019a, b)
Ovary length	F_2 and BC_1F_1 (CNS21 × RNS7)	QTL mapping	InDel markers	*ol2.1,*	Ch2	Gao et al. (2020)
Ovary diameter				*od1.1, od1.2, od2.1*	Ch1, Ch2	
Ovary index				*osi1.2, osi2.1*	Ch1, Ch2	
Fruit length				*fl1.2, fl2.1*	Ch1, Ch2	
Fruit diameter				*fd1.2, fd2.1*	Ch1, Ch2	
Fruit shape index				*fsi1.2, fsi2.1*	Ch1, Ch2	

(continued)

Table 1.2 (continued)

Trait	Mapping progeny	Technique	Molecular markers	QTLs/genes	Chromosome no.	References
Green flesh color	F2 and BC1	QTL mapping	SNP	*qgf5.1*	Ch5	Bo et al. (2019)
Fruit length	RILs (*Cucumis sativus* L. *hardwickii* × *Northern Chinese inbred line XinTaiMiCi*)	QTL mapping	SNP	*Fl4.1*	Ch4	Xing et al. (2023)
Shape of fruit apex	$F_{2:3}$ and $F_{2:6}$ (EC1 × 8419s-1)	QTL mapping	SSR and InDels	*Bfal4.1*, *Bfai4.1*, *Bfad6.1* and *Bfai6.1*	Ch4, Ch6	Zhu et al. (2022a, b)
Other traits:						
Pre-harvest sprouting	F2 (Jinyan No.4' × Sipingcigua)	QTL-seq method	SNP and InDel	qPHS4.1	Ch4	Cao et al. (2021)
Low temperature germination ability	F_2 (7088D × Coolgreen)	QTL mapping	SSRs, SNPs, and indels	*qLTG1.2*, *qLTG2.1 and qLTG4.1*	Ch1, Ch2, Ch4	Yagcioglu et al. (2019)
Musk melon (*Cucumis melo* L.)						
Fruit traits						
Type of exocarp	F2 (M4-5 × M1-15)	QTL mapping	SNP-derived CAPS markers	*TOP9.1*	Ch9	Zhang et al. (2020)
Fruit length				*FL5.1*	Ch5	
Fruit shape index				*FSI7.1*	Ch7	
Bitter gourd (*Momordica charantia* L.)						
Fruit length	$F_{2:3}$ (DBGy-201 × Pusa Do Mausami)	Genetic analysis and QTL mapping	SNP	*qFL5*, *qFL14*	Ch5, Ch14	Rao et al. (2021)
Fruit diameter				*qFD16*	Ch16	
Fruit weight				*qFW1*	Ch1	
Number of fruits per plant				*qFN20*	Ch20	
Yield per plant				*qYD1*	Ch1	
Pumpkin (*Cucurbita moschata* L.)						
Fruit traits						

Fruit length	F2 (N137 × N29)	QTL mapping	SNP	*fl2.1*	Ch2	Han et al. (2022)
Fruit diameter				*fd1.1, fd5.1*	Ch1, Ch5	
Fruit shape index				*fsi2.1*	Ch2	
Fruit weight				*fw1.1, fw16.1*	Ch1, Ch16	
TSS				*tss4.1, tss5.1, tss7.1*	Ch4, Ch5, Ch7	
Other traits						
First female flowering node	F2 (CMO-E × CMO-X)	QTL mapping	SNP	*qfffn9-a*	Ch9	Abbas et al. (2022)
First male flowering node				*qfmfn4-a, qfmfn9-a*	Ch4, Ch9	
Trichome density	F2 (inbred line 16 x 63)	QTL-seq and genetic map-based QTL analyses	SNPs and InDels	*CpTD3.1 and CpTD15.1*	Ch3, Ch15	
Melon (*Cucumis melo* L.)						
Fruit traits						
Bitterness	F_2 and BC_1 (C68 × C69)	QTL mapping	SNP	*BT-1, BT-2, and BT-3*	Ch2 and Ch5	Shang et al. (2020)
Soluble solid content	F_2 (M4-5 × M1-15)	QTL mapping	SNP-CAPS	SSC6.1 and SSC11.1	Ch6, Ch11	Zhao et al. (2023)
Exocarp firmness				*EF12.1*	Ch12	
Edible pericarp firmness				EPF3.1, EPF3.2, and EPF7.1	Ch3, Ch7	
Rind Firmness	Recombinant Inbred Lines (Dulce × Tam Dew)	QTL mapping	SNPs and short InDels	*RF2.1*	Ch2	Oren et al. (2022)
Fruit firmness				*FF8.3*	Ch8	

(continued)

Table 1.2 (continued)

Trait	Mapping progeny	Technique	Molecular markers	QTLs/genes	Chromosome no.	References
Other traits						
Stigma color	F2 and F2:3 (MR-1 × M4-7)	Genetic analysis and mapping	CAPS	*SC2.1* and *SC8.1*	Ch2, Ch8	Lv et al. (2022)
Days to harvest	Recombinant Inbred Lines (Dulce × Tam Dew)	QTL mapping	SNPs and short InDels	*DtH3.3* and DtH8.2	Ch3, Ch8	Oren et al. (2022)
Flowering to harvest				*FtH3.3* and FtH8.2	Ch3, Ch8	
Ethylene emission				*EtE3.3* and EtE8.2	Ch3, Ch8	
Cauliflower (*Brassica oleracea* var. *botrytis*)						
Curd setting height	F_2 (ZAASC4101 × ZAASR06) and (ZAASC4101 × ZAASJ1401)	QTL mapping	SNP	*BocDWARF1*	Ch1	Sheng et al. 2024
Stalk length	DH lines (IL4305 × ZN198)	QTL mapping	SSR and SRAP	*qSL.C6–1, qSL.C6–2*	Ch6	Zhao et al. (2020a, b)
Curd solidity				*qCS.C6–1 and qCS.C6–2*	Ch6	
Oil seed *(Brassica napus. L)*						
Soluble solids content	RIL (APL01 × Holly)	QTL mapping	SNP	qSSC/21GY.A04-1 and qSSC/21NJ.A08-1	Ch4, Ch8	Wu et al. (2021a, b, c)
Turnip (*Brassica rapa. L)*						
Swollen root	F_2 and BC_1F_2 *Brassica rapa* ssp. *rapifera* (10601) × Brassica rapa ssp. pekinensis (10603)	QTL mapping	SNP	*FR1.1 and* FR7.1	Ch1, Ch7	Wu et al. (2021a, b, c)
Chinese cabbage (*Brassica rapa L. ssp. pekinensis*)						
Main flower stalk length	RIL (06–247 × He102)	QTL mapping	SNP	*MFSL4, MFSL10*	Ch4, Ch10	Liu et al. (2019a, b, c)

Broccoli (*Brassica oleracea L.* var. *italica*)						
Plant height	F1 (inbred lines 86101 and 90196)	QTL mapping	SSR	*phc1, phc2, phc4-1, phc4-2*	Ch1, Ch2, Ch4	Huang et al. (2020)
Petiole length				*plc6*	Ch6	
Outer leaf width				*lwc1, lwc3*	Ch1, Ch3	
Sulforaphane metabolism	DH (86, 101 × 90,196)	QTL mapping	SSR	*qSF-C3-0, qSF-C3-3, qSF-C3-6, and qSF-C3-5*	Ch3	Li et al. (2021a, b)
Anthocyanin biosynthesis	F2 (BT 126 × SN 60)	QTL-seq	SNP	*BoPur7.1 and BoPur9.1*	Ch7, Ch9	Liu et al. (2021a, b, c, d, e)
Spinach (*Spinacia oleracea* L.)						
Leaf length	**BC_1**	QTL analysis	SLAF and KASP	*qLL5.3 and qLL5.2*	Ch5	Liu et al. (2021a, b, c, d, e)
Leaf width				*qLW.3.3*	Ch3	
Petiole length				*qPL5.2 and qPL5.1*	Ch5	
Ratio of leaf length to width				*qLR5.2 and qLR5.1*	Ch5	
Bolting	BC_1 and F_2 (12S3 (late-bolting) × and 12S4 (early bolting))	QTL mapping	KASP	*qBT1.1*	Ch1	Meng et al. (2022)
Onion (*Allium cepa* L.)						
Anthocyanin synthesis	F_2 (SP3B (yellow bulb) × H6 (red bulb))	Genotyping-by-sequencing (GBS) analysis	SNP and HRM	*qAS7.1*	Ch7	Choi et al. (2020)
Anthocyanin content				qAC4.1 and qAC4.2	Ch4	

(continued)

Table 1.2 (continued)

Trait	Mapping progeny	Technique	Molecular markers	QTLs/genes	Chromosome no.	References
Sugar content						
Sucrose	F_2 (16P118 × Sweet Green)	Genotyping-by-Sequencing Analysis	SNP	*qSC4.1*	Ch4	Lee et al. (2021a, b)
Glucose				*qGC5.1*	Ch5	
Fructose				*qFC5.1*	Ch5	
Total sugar content				*qTSC5.1*	Ch5	
Adzuki bean (*Vigna angularis*)						
Epicotyl length	F2 (Tokei1121 × Erimo167)	QTL mapping	SSR and SNP	*qECL10*	Ch10	Mori et al. (2021)
Yardlong bean (*Vigna unguiculcta* L.)						
Pod shattering	$F_{2:3}$ (JP81610 × TVnu-457)	QTL mapping	SSR and InDels	*Shat7.1.1*	Ch7	Watcharatpong et al. (2020)
Lignin content				*Lig7.1.1*	Ch7	
Hemicellulose				*Hem7.1.2*	Ch7	

These tomato QTL analyses were based on populations with introgressions from *Solanum pennellii* and *Solanum habrochaites* integrated into the cultivated tomato *S. lycopersicum* background. Across four populations, a total of 95 QTLs in 34 genomic regions were identified, with a notable proportion being specific to particular populations, especially for fruit weight, pericarp thickness, and fruit shape (Chaim et al. 2006). More recent research on tomato genetics unveiled the molecular basis of the classical rosette (ro) mutation, characterized by severely reduced internodes and complete sterility. This mutation was attributed to a tomato gene orthologous to *BIG/TRANSPORT INHIBITOR RESPONSE 3 (TIR3), affecting the activity of PIN-FORMED (PIN)* auxin transport proteins and resulting in reduced auxin transport rates and absence of adventitious roots in stems. Additionally, the extreme adventitious root phenotype of the tomato aer mutant was investigated, elucidating the involvement of genes related to auxin biosynthesis, transport, and signaling.

A study focused on the development of ILs and mapping of major agronomic QTLs in tomato utilized 447 Cleaved Amplified Polymorphic Sequences (CAPS) and 525 SSR markers. It identified a total of 11 QTLs across 6 chromosomes, including 3 QTLs for plant height, 1 for leaf size, 1 for fruit color, 4 for fruit weight, and 2 for soluble solids content (SSC) in ripening fruits (Chen et al. 2023). Additionally, previous reports on QTLs controlling SSC in tomato fruits revealed 95 QTLs associated with SSC, distributed across 14 IL populations constructed from eight wild tomato cultivars (Labate et al. 2007). It is noteworthy that SSC in fruits typically shows a negative correlation with fruit size. Moreover, fine mapping of *fw6.3*, a major-effect QTL controlling fruit weight, revealed two QTLs, *qSSC7a* and *qSSC9a,* related to SSC. qSSC7a significantly increased SSC by 16–32% and reduced fruit weight by 6–13%, providing potential for enhancing fruit quality (Ning et al. 2023). *qSSC9a* corresponded to the previously cloned gene *brix9-2-5*, influencing sucrose transport and regulating SSC in fruits (Labate et al. 2007; Fridman et al. 2000). Furthermore, QTL analysis for bacterial wilt (BW) resistance detected three QTLs on chromosomes 1, 2, and 3, and for plant height (PH), three QTLs on chromosomes 1, 8, and 11 were identified (Siddique et al. 2024). Another study mapped QTLs associated with early blight resistance on chromosomes 2, 8, and 11 (Adhikari et al. 2023). Liu et al. (2023a, b) investigated genetic regions related to nine major agronomic traits, identifying significant signals on chromosomes 1, 5, 9, and 11, particularly associated with soluble sugar content, fruit firmness, and gray leaf spot resistance. Additionally, fine mapping studies identified thermotolerant trait genomic regions, including genes involved in heat shock proteins (HSPs), flowering, and pollen activity, mapped to chromosome 11.

1.12.2 Brinjal (*Solanum melongena* L)

In a study focusing on eggplant ILs under two nitrogen fertilization levels, 36 QTLs associated with various traits were identified. Among these, 12 QTLs were specific to high nitrogen (HN) fertilization, while 17 were specific to low nitrogen (LN) fertilization. *Solanum incanum* generally exerted a negative effect on most traits, except for specific instances such as QTLs for fruit dry matter and fruit length on chromosome 10 under HN and for fruit pedicel length on chromosome 9 under LN (Rosa-Martínez et al. 2023). Furthermore, the genes controlling anthocyanin biosynthesis in eggplant peel color, named D and P, were found to be epistatic and were mapped and validated by You et al. (2023). Lv et al. (2023) conducted candidate gene and fine mapping studies of the Gv1 locus controlling green-peel color in brinjal, speculating that *SmAPPR2*-like is the candidate gene for *Gv1*. Moreover, molecular mapping and identification of SSR markers linked to phomopsis blight resistance in eggplant revealed two markers (*emk03O04* and *emf11A03*) capable of differentiating resistant and susceptible bulk and co-segregating with the resistance gene. The genetic distance between these markers was determined to be 18.12 cM. Additionally, Narayanswami et al. (2023) identified two new QTLs on chromosome 6, which are significant for marker-assisted breeding programs. Lastly, evaluation of eggplant lines with introgressions of *S. incanum* under water stress conditions showed that wild alleles for two QTLs related to stem and root dry weight conferred improved tolerance to water stress (Flores-Saavedra et al. 2023).

1.12.3 Chilli (*Capsicum annuum* L)

Bacterial wilt resistance-associated QTL mapping, utilizing biparental mapping and GWAS association analysis, revealed significant findings. Notably, a prominent QTL (*QTL.Bw5*) demonstrating resistance properties was identified on the telomeric region of chromosome 5. Additionally, GWAS analysis identified four loci (*GWAS.Bw.4, GWAS.Bw.5, GWAS.Bw.8.1,* and *GWAS.Bw.8.2*) associated with BW resistance on chromosomes 4, 5, and 8 (Lee et al. 2024). In constructing the linkage map, a comprehensive set of 1023 polymorphic, high-quality SNP markers was employed, resulting in the clustering of all markers into 12 linkage groups, consistent with the Capsicum chromosome haploid number (Mimura et al. 2012). Several QTLs related to fruit size and shape on chromosome 2 have been previously reported in various genetic backgrounds in chili peppers (Han et al. 2016). The presence of the ovate gene, associated directly with fruit shape and size in pepper, co-localizes with a QTL on chromosome 2. Conversely, individual QTLs for fruit-related traits on chromosome 4 have been documented previously, although no cluster of QTLs has been indicated in this region as observed in this study. Similarly, a major QTL (FO12.1) for the fruit orientation trait on chromosome 12 was identified, consistent with previously reported associations (Han et al. 2016). Furthermore, the study noted discrepancies in QTLs between different studies, potentially attributed to variations in parental lines, including the inclusion of a wild relative in our study (Han et al. 2016). The adaptation of domesticated plants to agricultural environments often results in the loss of valuable alleles crucial for survival in harsher, wilder conditions. Specifically, organ pigmentation found in wild plants serves as a protective mechanism against abiotic factors like radiation. A detailed tabulation on mapped QTL has been provided as Table 1.3 and 1.4. In a recent study, fruit and

Table 1.3 Status of QTL mapping in vegetable crops for abiotic stress

Abiotic stress	Mapping population	Technique	Molecular markers	QTLs/Genes	Chromosome no.	References
Tomato (*Solanum lycopersicum* L.)						
High temperature	F_2 seedlings	QTL mapping, QTL-seq, and RNA-seq	SNP and small InDel markers	*qHII-1-1*, *qHII-1-2*, *qHII-1-3*, *qHII-2-1*,*qCC-1-5*	Ch1, Ch 2	Wen et al. (2019)
Physiological disorders						
Blossom end rot	F_2, $F_{3:4}$ and BC_1	linkage-based QTL mapping	KASP	*BER3.2* *BER11.1*	Ch3 Ch11	Topcu et al. (2021)
Fruit Cracking	F_1 and F_2 ('S189' × 'R91')	QTL analysis	–	*qCR2*, *qCR5*	Ch2, Ch5	Chen et al. (2021)

(continued)

Table 1.3 (continued)

Abiotic stress	Mapping population	Technique	Molecular markers	QTLs/Genes	Chromosome no.	References
Chilli (*Capsicum annuum* L.)						
Heat tolerance						
Plant height	F_2 population (DLS-161-1 × DChBL-240)	QTL mapping	SSR	*qPH3.1*	Ch3	Aruna et al. (2023)
Fruit length				*qFL1.1, qFL7.1*	Ch1, ch7	
Average fruit weight				*qAFW4.1, qAFW6.1*	Ch4, ch6	
Fruit yield per plant				*qFYP4.2, qFYP4.3*	Ch4	
Number of healthy seeds per fruit				*qNS3.1*	Ch3	
Leaf length				*qLL1.1, qLL6.1*	Ch1, Ch6	
Leaf area				*qLA3.1*	Ch3	
Canopy temperature depression				*qCTD11.1, qCTD11.2,*	Ch11	
Stomatal density				*qSD9.1*	Ch9	
Relative chlorophyll content				*qCC11.1*	Ch11	
SOD activity				*qSOD 10.2*	Ch10	

Musk melon (*Cucumis melo* L.)						
Tolerance to sulfur phytotoxicity	170 RIL plants (MR-1 × Ananas Yok'neum)	QTL mapping	KASP	*qSulf-1*	Ch1	
Cucumber (*Cucumis sativus* L.)						
Heat tolerance	RILs (99281 × 931)	QTL mapping	SSR	*qHT1.1*	Ch1	
Low temperature tolerance in seedlings	F_2 (CG104 × CG37)	QTL mapping	SNP	*qLTT6.2*	Ch6	Dong et al. (2019)
Salt tolerance in seedlings	RIL (CG104 × CG37)	QTL mapping	SSR	*qST6.2*	Ch6	
Cabbage (*Brassica oleracea var capitata* L.)						
Head splitting	F_2 (G274 × G279)	QTL-seq	SNP and InDel	*Bol016058*	Ch3	

Table 1.4 Status of QTL mapping in vegetable crops for biotic stress

Trait	Mapping population	Technique	Molecular markers	QTLs/genes	Chromosome no.	References
Tomato (*Solanum lycopersicum* L.)						
Fungal diseases						
Late blight (*Phytophthora infestans*)	F2 progenies	QTL mapping	SNP	*Ph-3* *Ph-2*	Ch9 Ch10	Brekke et al. (2019)
	F2 and F2-derived progenies (NC 1CELBR × Fla. 7775)	QTL mapping	SNP	*Ph-3* *Ph-2*	Ch9 Ch10	Panthee et al. (2017)
Leaf mold (*Cladosporium fulvum*)	F_1, F_2, and BC_1F_1	SNP-index, InDel-index analyses, and local QTL mapping	KASP, InDel, and SNP	*Cf-10*	Ch1	Liu et al. (2019a, b, c)
Bacterial diseases						
Bacterial wilt (*Ralstonia solanacearum*)	RIL parents, NIL parents, and BC_3F_3 NIL lines	QTL mapping	SNP and CAPS	*Bwr*-6 and *Bwr*-12	Ch 6 Ch 12	Shin et al. (2020)
	F_2 population (NC 11212) (NC 84173 × CLN1466EA)	QTL mapping	SNP	*qbw-03*	Ch3	Siddique et al. (2024)
Bacterial canker (*Clavibacter michiganensis*)	F_2 population (E6203 × Hawaii 7998)	QTL-seq analysis	InDel Markers	*Rcm6*	Ch6	Abebe et al. (2022)
Brinjal (*Solanum melongena* L.)						
Fungal diseases						
Fusarium oxysporum f. sp. melongenae	RILs population (305E40 × 67/3)	QTL mapping	SNP	*FomCH02 and FomCH11*	Ch2 and Ch11	Tassone et al. (2022)

Chilli (*Capsicum annuum* L.)						
Fungal diseases						
Fruit and root rot (*Phytophthora capsici*)	RIL or an F_2 biparental progenies	Meta-QTL analysis	AFLP, SSR, and SNP	*MQTL5.1 and MQTL5.2*	Ch5	Lozada et al. (2021)
Phytophthora capsici) Isolate KPC-7	$F_{7:8}$ progenies (*C. annuum* line CM334 × ECW30R)	Genotyping-by-sequencing-based QTL mapping with GWAS	SNP	*E1Kpc-5.2, E2Kpc-5.2*	Ch5	Siddique et al. (2019)
Phytophthora capsici) Isolate JHAI1–7				*E1Jha-5.2, E2Jha-5.1, E2Jha-5.2*	Ch5	
Phytophthora capsici) Isolate MY-1				*E1My-5.1, E1My-5.2*	Ch5	
Anthracnose (*Colletotrichum scovillei*)	F_2 progenies (*Capsicum baccatum* (PBC80 × CA1316))	QTL mapping	SNP	*RA80f6_r1*	Ch4	Kethom and Mongkolporn (2021)
Anthracnose (*Colletotrichum species*)	F2 progenies (PBC80 × CA1316)	QTL mapping	SNP	*RA80rP2, RA80rP3.1 and RA80rHP1*		Mahasuk et al. (2016)
Bacterial diseases						
Bacterial wilt (*Ralstonia solanacearum*)	F_2 progenies (KC352 × 14F6002-14)	QTL mapping using Genotyping-by-Sequencing	SNP	*pBWR-1*	Ch1	Chae et al. (2022)
	F2 progenies (BVRC 25 × BVRC 1)	QTL mapping	SNP	*qRRs-10.1*	Ch10	Du et al. (2019)
	F_2 progenies (Konesian Hot × Geonchowang)	QTL mapping	SNP	*Bwr6w-5.1*	Ch5	Lee et al. (2024)

(continued)

Table 1.4 (continued)

Trait	Mapping population	Technique	Molecular markers	QTLs/genes	Chromosome no.	References
Viral diseases						
Pepper yellow leaf curl virus	F_2 population (*C. annuum* line LP97 × *C. annuum* line ECW30R)	QTL analysis	SNP	*peplcv-1*, *peplcv-7*, and *peplcv-12*	Ch1, Ch7, Ch12	Siddique et al. (2022)
Pepper yellow leaf curl virus	F1 and F2 populations (Habanero × GR1)	QTL sequencing	SNP	*Pepy-chr3 and Pepy-chr11*	Ch3, Ch11	Mori et al. (2022)
Pepper yellow leaf curl virus	132 $F_{2:3}$ populations	QTL mapping	AFLP	*QPlcv.teri.8.1, QPlcv.teri.8.2, and QPlcv.teri.8.3*	Ch8	Dwivedi et al. (2023)
Cucumber mosaic virus	F2 ('BJ0747' × 'XJ0630')	SLAF-seq approach and QTL mapping	SLAF	*qcmv11.1, qcmv11.2*	Ch11	Li et al. (2018)
	F_2, BC_1, and $F_{2:3}$ (PBC688 × G29)	QTL mapping	SLAF	*qCmr2.1*	Ch2	
Pests						
Thrips (*Thrips palmi*)	F2 progenies (*C. annuum* AC 1979 × *C. annuum* Berceo)	QTL analysis	SNP	*QTL1 and QTL2*	Ch3 and Ch12	Saeko et al. (2020)
Bitter gourd (*Momordica charantia* L.)						
Yellow mosaic disease	$F_{4:5}$ (Punjab-14 × PAUBG-6)	QTL analysis using the genetic map and phenotyping data	SNP	*qYMD.pau_5.1*	Ch5	
Citrus melon (*Citrullus amarus*)						
Fungal diseases						
Fusarium wilt race 1 (*Fr. f. sp. Niveum*)	$F_{2:3}$ and recombinant inbred lines (USVL246-FR2 × USVL114)	QTL mapping	SNP	*qFon1-9*	Ch1	Branham et al. (2019a, b)
Bacterial disease						

Bacterial fruit blotch	F_2 (USVL246-FR2 × USVL114)	QTL mapping	SNP	*Ac-1.1*, *qAc-2.1*, *qAc-3.1*, *qAc-3.2*, and *qAc-8.1*	Ch1, Ch2, Ch3, Ch8	Branham et al. (2019a, b)
Watermelon (*Citrullus lanatus* L.)						
Bacterial disease						
Gummy stem blight	F_2 (920533 × PI 189225)	QTL mapping	SNP	*qLL8.1 and qSB8.1*	Ch8	Lee et al. (2021a, b)
Bacterial fruit blotch	F_2 (PI 189225 (*Citrullus amarus*) × W 26 (*C. lanatus*))	GBS-Based QTL Mapping	SNP	*ClBFB10.1 and ClBFB10.2*	Ch10	Yeo et al. (2022)
Melon (*Cucumis melo* L.)						
Fungal disease						
Downy mildew	F_2 (PI 442177 × Huangtu) and (PI 442177 × Huangdanzi)	GBS-Based QTL Mapping	KASP	*DM9.1*	Ch9	Zhang et al. (2023a, b)
Squash (*Cucurbita moschata L.)*						
Fungal disease						
Phytophthora crown rot	F_2 progenies (#394-1-27-12 × Butterbush)	QTL-seq	SNP	*QtlPC-C04, QtlPC-C11, and QtlPC-C14*	Ch 4, 11, and 14	Ramos et al. (2020)
Radish (*Raphanus sativus* L.)						
Fungal diseases						
Clubroot (*Plasmodiophora brassicae*)	F_2 (BJJ × XNQ)	RAD-Seq	SNP	*RsCr1, RsCr2, RsCr4, and RsCr5*	Ch 8 and 9	Gan et al. (2019)

(continued)

Table 1.4 (continued)

Trait	Mapping population	Technique	Molecular markers	QTLs/genes	Chromosome no.	References
Bacterial diseases						
Black rot (*Xanthomonas campestris pv. Campestris*)	F2 (KB12-1 × PP12-1)	RAD-seq approach and **QTL mapping**	SNPs and indels	*qBRR2*	Ch2	Wu et al. (2020)
Cabbage (*Brassica oleracea* var. *capitata* L.)						
Pest						
Onion thrips (*Thrips tabaci*)	RILs	**QTL mapping**	SNP	*Ttq1 and Ttq2 Thrips*	Ch2 and Ch7	Roth et al. (2021)

stem pigmentation were analyzed, revealing a colocalized QTL on chromosome 10. While previous studies have focused solely on unripe fruit pigmentation (Han et al. 2016), they have identified the A gene as responsible for this trait. Additionally, molecular mapping efforts have successfully identified a major genomic region on chromosome 9 associated with resistance to Chilli veinal mottle virus (ChiVMV) in hot pepper. Within the GWAS regions, a cluster of defense-related genes and elongation translation factors (*eIFLE*) implicated in disease resistance were predicted (Ponnam et al. 2023). Genetic analysis and QTL identification for resistance to Pepper leaf curl virus (PepLCV), a Begomovirus, revealed the presence of eight stable QTLs across five linkage groups (LG1, LG3, LG5, LG7, and LG8) out of 132 evaluated F2:3 families (Hu et al. 2023).

Moreover, refining the genetic map highlighted potential genes associated with fruit color in peppers. Notably, inactive shikimate kinase-like 2 (*BC332_00870*) emerged as a strong candidate gene regulating immature fruit color, while a premature termination mutation in the capsanthin–capsorubin synthase (*BC332_17237*) gene potentially causes yellow coloration in *HNUCC16* pepper fruits (Shu et al. 2023). Additionally, a major QTL for flavonoid content was identified on chromosome 5, with a *CaMYB12*-like transcription factor gene pinpointed as a candidate (Wu et al. 2023). Similarly, a QTL for ripe fruit resistance from 'PBC80'-derived RILs was located on chromosome 4 (123 Mb), containing over 80 defense-related genes (Kethom et al. 2023). In a separate study on peppers, eight QTLs linked to PepLCV disease were identified across five linkage groups (LG1, LG3, LG5, LG7, and LG8) (Dwivedi et al. 2023).

1.12.4 Okra (*Abelmoschus esculentus* L.)

A comprehensive genome-wide assessment of genetic diversity and association mapping for salt tolerance traits in okra was conducted using GBS. The study identified 77 SNP loci associated with four salt tolerance traits. Among these loci, *c3341.graph_c0* was observed in multiple SNP loci, suggesting its potential significance as a key unigene regulating okra seedlings' salt tolerance (Sun et al. 2023). Another investigation utilized a genome-wide association study to identify single-nucleotide polymorphic loci and candidate genes associated with seed germination percentage (GP) in okra under both salt and no-salt stresses. *Ae59G004900* and *Ae44G005470* were identified as regulators of GP in okra seeds subjected to salt and no-salt stresses (Xu et al. 2024). In a separate study, an SSR marker (SSR54) linked to resistance against Yellow Vein Mosaic Virus was identified in RILs developed from a cross between a YVMV-resistant breeding line (NS25581) and a susceptible line (NS17556) (Shwetha et al. 2023).

Additionally, association mapping was employed to assess abiotic stress tolerance in okra using heat- and drought-related syntenic markers. Relative water content was found to be associated with four markers (*Unigene10673, Unigene99547, Unigene152901*, and *Unigene129684*) under drought conditions. In contrast, electrolyte leakage was associated with three markers (*Unigene109922*, *Unigene28667*, and *Unigene146907*) under heat stress conditions (Sun et al. 2023; Ahmad et al. 2022).

1.12.5 Cucumber (*Cucumis sativus* L)

Pradeepkumara et al. (2022) conducted fruit transcriptional profiling of contrasting cucumber genotypes for shelf life, identifying 1364 DEGs involved in cell wall degradation, chlorophyll, and ethylene metabolism. Downregulation of Polygalacturonase (PG), expansin (EXP), and xyloglucan determined fruit firmness, while low expression of chlorophyll catalytic enzymes (CCEs) retained fresh green color. The study revealed hub genes and cross-talk among biological processes and identified numerous SSRs, SNPs, and InDels for cucumber improvement. Nandi et al. (2024) detected a QTL associated with retention of green color on chromosome 4 through QTL-seq. Identified major QTLs (*Parth6.1* and *Parth6.2*) linked to parthenocarpic fruit development in slicing cucumber. Similarly, Wu et al. (2016) found a major QTL (*Parth 2.1*) for parthenocarpy in a European greenhouse-slicing cucumber genotype. Lietzow et al. (2016) identified seven QTLs for parthenocarpy in a North American pickling cucumber, while Niu et al. (2020) detected four novel QTLs in a South China ecotype cucumber. Sun et al. (2024) identified *CsPM5.2* and *Cspm5.1* genes for powdery mildew resistance, with transcriptome analysis implicating the salicylic acid (SA) pathway. Zhang et al. (2013) found five QTLs for resistance to downy mildew and identified linked SSR markers. Li et al. (2023a, b) fine-mapped *qPFR5.1* on chromosome 5 associated with resistance to Phytophthora fruit rot. Wang et al. (2023a, b) detected 27 QTLs for ovary size, fruit size, and flowering time, including major-effect QTL *qOHN4.1* on chromosome 4. Lastly, Li et al. (2018) identified five QTLs for downy mildew resistance, with *dm4.1* on chromosome 4 explaining 27% of phenotypic variance.

1.12.6 Musk Melon (*Cucumis melo* L.)

Adedze et al. (2024) utilized bulked segregant analysis (BSA-seq) in melon to identify the genomic region associated with powdery mildew on chromosome 12, designated *LVpm12.1* (QTL). Perchepied et al. (2005) identified a major QTL, *pcXII.1*, explaining 12–38% of the phenotypic variation for Pseudoperonospora cubensis resistance, along with two independent major genes, *PmV.1* and *PmXII.1*, involved in simple resistance to powdery mildew. Cui et al. (2022a, b) highlighted the significance of *C. melo* var. *cantalupensis* accessions PMR 45, PMR 5, PMR 6, and WMR 29, as well as *C. melo* var. *momordica* accessions PI 124111, PI 124112, and PI 414723, in melon resistance breeding due to the presence of multiple resistance genes. Yuste-Lisbona et al. (2011) mapped a major QTL (Pm-R) on LG V for resistance to powdery mildew races 1, 2, and 5, closely linked to the PM4-CAPS marker at a genetic distance of 1.9 cM. Branham et al. (2021)

identified two major (*qPx1-5* and *qPx1-12*) and two minor (*qPx1-4* and *qPx1-10*) QTLs for resistance to *P. xanthii race 1* using an RIL population in melon. Toporek et al. (2021) associated two major QTLs, *qPcub-10.1* and *qPcub-8.2*, with resistance to CDM (*Pseudoperonospora cubensis clade 1*, mating type A2), which harbor MLO-like protein-coding genes known to confer resistance to powdery mildew and Phytophthora in melon plants.

1.12.7 Bottle Gourd and Bitter Gourd (*Lagenaria siceraria* L.)

Seven heat-tolerant QTL (*qHT1.1, qHT2.1, qHT2.2, qHT5.1, qHT6.1, qHT7.1,* and *qHT8.1*) were identified, with *qHT2.1* being a particularly noteworthy major-effect QTL, mapped using heat-tolerant parent L1 (P17) and heat stress-sensitive parent L6 (P23) in an F2 segregating population (Song et al. 2020). Li et al. (2021a, b) mapped genomic regions (*HG_GLEAN_10001030* and *HG_GLEAN_10001042*) associated with Fusarium wilt resistance in bottle gourd using GWAS. Zhang et al. (2023a, b) investigated the defense response of resistant and susceptible traits in bottle gourd against powdery mildew disease through transcriptome sequence analysis. Kaur et al. (2022) conducted mapping for earliness, fruit, and seed related traits in bitter gourd using a high-density genotyping-by-sequencing-based genetic map. In another study, Song et al. (2019) explored seven heat-tolerant loci (*qHT1.1, qHT2.1, qHT2.2, qHT5.1, qHT6.1, qHT7.1*, and *qHT8.1*) exhibiting high Δ(SNP-index) values (ranging from 0.19 to 0.38) in bottle gourd. Gangadhara et al. (2018) identified the gynoecious (*gy-1*) locus and major QTLs related to earliness traits in bitter gourd, facilitating marker development and MAS for rapid development of gynoecious lines. Vinay et al. (2023) identified several genes, including Ethylene-responsive transcription factor 12, Auxin response factor 6, and others, potentially involved in gynoecious sex expression in bitter gourd. Mishra et al. (2014) associated the RAPD primer OPZ 13 with the gynoecious trait ('*gy-1*' gene) in bitter gourd. Zhong et al. (2023) identified the candidate gene *Mcgy1* responsible for the gynoecious trait in bitter gourd. Matsumura et al. (2014) constructed a linkage map using RAD-seq analysis of F_2 individuals, with GTFL-1 being the closest SNP locus to the putative gynoecious locus. Behera et al. (2009) inferred from the segregation pattern in F_2 that the gynoecious sex form in bitter gourd is governed by a single recessive gene. Finally, Cui et al. (2022a, b) conducted fine mapping and candidate gene analysis of the *Mcgy1* locus responsible for gynoecy in karela.

1.12.8 Watermelon (*Citrullus lanatus* L.)

Lee et al. (2021a, b) identified three QTLs associated with Gummy Stem Blight (GSB) resistance on chromosomes 8 and 6. Among them, *qLL8.1* and *qSB8.1* on chromosome 8 explained 10.5% and 10.0% of the phenotypic variations, respectively, while a minor QTL, *qSB6.1,* accounting for 9.7% of phenotypic variation, was detected on chromosome 6. Yang et al. (2021) identified five putative genetic regions associated with watermelon rind hardness and related traits using BSA-seq, with two colocalized focal QTLs, "RH10 and RTO10," mapped for horticultural traits. Osae et al. (2021) found QTLs for seed length, seed width, thickness of seed coat, and 100 seed weight (*qSL-6-1, qSW-6-1, qTSC-6-1, q100SWT-6-1*) all located on chromosome 6 within a genetic distance of less than 5 cM. Additionally, Cho et al. (2021a, b) conducted mapping of a candidate gene (*sdw-1*) on chromosome 9 controlling semi-dwarfism.

1.12.9 Carrot (*Daucus carota.* L.)

Turner et al. (2018), employing an image analysis pipeline to investigate root morphology, observed the colocalization of *QTLs* on chromosomes 1, 2, and 7 associated with interrelated carrot root traits. This implies the potential genetic linkage and quantitative inheritance control over these traits. Linking the OFP-TRM and IQD regulon to root shape control is in radish (*Raphanus sativus*) (Wang et al. 2020). The plant-specific OFP-TRM and IQD regulatory pathway plays a crucial role in shape patterning and has been extensively researched across various plant organs such as fruit, leaves, stems, and tubers (van der Knaap et al. 2014; Li et al. 2023a, b). In their study, Vega et al. (2024) investigated how members of the OFP-TRM and IQD regulon overlap with root shape *QTLs*. Understanding the genetic architecture of root traits has significant implications for breeding decisions and can unveil new opportunities in crop improvement. Identified a novel set of QTL for four of the most relevant morphological components of root market class in carrot. A highly significant SNP on chromosome 2 associated with root fill was identified (Brainard et al. 2022). Regarding length, the gene *DCv3_Chr5.21023*, situated at the peak on chromosome 5, is identified as a piezo-type mechanosensitive ion channel according to predictions. This gene has been associated with the ability of roots to vertically penetrate the soil profile (Mousavi et al. 2023). A gene located at the peak on chromosome 9, known as *DCv3_Chr9.36166*, is identified as a homolog of protein terminal ear1. This protein has been linked to the regulation of root growth mediated by abscisic acid (Wang et al. 2018).

1.12.10 Radish (*Raphanus sativus* L.)

A high-density genetic map was built for Yunnan red radish using specific-locus amplified fragment (SLAF) sequencing, revealing loci controlling anthocyanin pigmentation. The F2 population was utilized alongside SLAF-seq, resulting in a comprehensive genetic map spanning 1230.16 cM. Further analysis of gene structure and expression validated the findings. Specifically, *RsMYB1.3* within *QAC2* emerged as a probable candidate gene influencing the pigmented/non-pigmented trait in Yunnan red radish (Tao et al. 2022). Transcriptional analyses indicate higher expression of *RsDFR, RsANS, RsUFGT, RsF3H, RsCHS3*, and *RsF3′H* genes in red fleshy radishes compared to other varieties. Notably, RsUFGT is crucial for anthocyanin accumulation, supported by studies by Gao et al. (2019). Structural genes and the *AtPAP1* homolog control anthocyanin production with varied expression across red radish varieties, as observed in studies by Gao et al. (2019).

InDel markers were created based on gene differences in YAAS-WR1 and YAAS-RR1, as reported by Tao et al. (2022). A highly detailed genetic map has offered valuable insights into genome organization, recombination patterns, and the genetic basis of taproot skin color in radish (*Raphanus sativus* L.). Analysis through map-based cloning has identified the *R2R3-MYB* transcription factor *RsMYB90* as a key candidate gene responsible for determining taproot skin color in radish (Luo et al. 2020). Researchers employed QTL-seq and traditional linkage analysis to map genes related to purple skin in radish taproots. They identified a single dominant gene, *Rsps*, through segregation analysis. Whole-genome sequencing and linkage mapping narrowed the *Rsps* locus to a 238.51-kb region containing 18 genes. Interestingly, *RsMYB1.1* and *RsMYB1.4*, known for red skin, were located on different chromosomes. These findings illuminate the complex regulation of anthocyanin biosynthesis in radish, offering insights for molecular breeding to improve taproot appearance and anthocyanin content (Liu et al. 2019a, b, c). Gene structure analysis revealed that *RsF3′H* has no introns, while most anthocyanin biosynthesis genes (ABGs) display asymmetrical exon patterns, except *RsSAM*. Anthocyanin accumulation in red radish leaves and colored cultivar taproots increases gradually during development, with a significant rise at 30 days after sowing (DAS) and peak levels reached at maturity, as reported by Muleke et al. in 2017. A pioneering GWAS uncovered the vital role of the *RsPAP2* gene in red-skinned *Raphanus sativus* L. It identified a key SNP on chromosome 2 linked to the *R2R3-MYB* transcription factor responsible for anthocyanin pigment 2 (PAP2) production. This study showcases GWAS's efficacy in identifying radish candidate genes. It also suggests *RsPAP2*, potentially working with *RsTT8*, can regulate gene transcripts associated with anthocyanin biosynthesis, enhancing its production (Fan et al. 2020). An SNP mutation in the *RsPP* gene alters petal color from purple to white in radish (Liu et al. 2021a, b, c, d, e).

1.12.11 Beet Root (*Beta vulgaris* subsp. *vulgaris* L.)

In a wide-ranging GWAS involving 977 sugar beet accessions, 170,750 high-quality SNPs were identified using GBS. Within the 25 kb linkage disequilibrium range, several genes were associated with taproot traits like RGD, RS, CS, and FC, including *NRT1/PTR FAMILY 6.3, nudix hydrolase 15, TRANSPORT INHIBITOR RESPONSE 1, MYB77,* and *ERF014*. For aboveground traits such as HC, PT, PL, C, and FLT, genes like UDP-glycosyltransferase 79B6, NAC domain-containing protein 7, F-box protein *At1g10780, PHO1, CPR1, MYB77*, and alcohol acyltransferase 9 were implicated. Additionally, genes like *AP-2* complex subunit mu, trihelix transcription factor *ASIL2*, and late embryogenesis abundant protein 18 were linked to PQ variation. These genes play roles in hormone response, nitrogen and phosphorus transportation, secondary metabolism, fertilization, and embryo maturation (Liu et al. 2023a, b).

1.12.12 Lettuce (*Lactuca sativa* L.)

Genetic mapping studies identified four significant QTLs associated with powdery mildew resistance, located on linkage groups LG 1 (*pm-1.1*), LG 2 (*pm-2.1* and *pm-2.2*), and LG 7 (*pm-7.1*), explaining between 35% and 42% of the phenotypic variation (Simko et al. 2014). Hartman et al. (2013) mapped two main regions with clusters of QTLs: one on linkage group 3, where the crop allele induced lower seed output, and another on linkage group 7, where the crop allele caused a delay in flowering time. An integrated map for lettuce, comprising 2744 markers, was developed from seven intra- and interspecific mapping populations. Among the wild species in the subsection *Lactuca, including L. serriola, L. saligna,* and *L. virosa,* which are $2n = 2x = 18$ and self-fertilizing, *L. serriola*, particularly, has been a valuable source of disease resistance genes (Kesseli et al. 1991; Crute 1988). Additionally, Rosental et al. (2021) identified three QTLs associated with genetic loci, with one each on chromosome 8 (*qBLT8.1*), chromosome 2 (*qBFr2.1*), and chromosome 6 (*qBFr6.1*), where the allele contributing to the delayed flowering phenotype at *qBFr2.1* originated from the early flowering parent PI 251246.

1.12.13 Onion (*Allium cepa* L.)

Late-bolting is a crucial agronomic trait in onions, particularly valuable for breeding late-bolting genotypes. Wako et al. (2016) conducted studies and successfully mapped the genetic loci associated with late-bolting, *qBlt1a* and *qBlt2a*, on chromosome 1 and chromosome 2, respectively, in bunching onion (*Allium fistulosum* L.). In bulb onion, research has focused on investigating *FLOWERING LOCUS T (FT)*-like genes for functional characterization. The up-regulation of *AcFT2* gene expression appears to play a role in the vernalization-responsive initiation of flowering, while *AcFT1* and *AcFT4* expression are correlated with bulb formation (Lee et al. 2013). Additionally, Baldwin et al. (2014) mapped a QTL named *AcBlt1* on chromosome 1 that governs bolting, although its connection to flowering candidate genes remains unclear. Despite limited genetic studies on flowering in bunching onion, efforts have been made to initiate molecular studies by identifying approximately 2000 genomic simple sequence repeats (gSSRs) and constructing genetic linkage maps (Song et al. 2004; Tsukazaki et al. 2007; Ohara et al. 2005; Tsukazaki et al. 2008, 2011, 2012). Tsukazaki et al. (2012) revealed a major QTL for pungency using the SaT03 map, emphasizing the importance of investigating QTLs for bolting time to understand the genetics of flower initiation in bunching onion. In a study by the impact of day length and bulbs on the bulbing trait was examined. A genetic map of 112 SNPs identified two main QTLs affecting bulbing in response to longer daylengths, with significant additive effects and no dominance. One major QTL significantly influenced bulb shape, explaining 30% of the variance, while three others, slightly above the significance threshold, collectively accounted for over 50% of the variance in bulb shape. However, no significant QTLs were found for bulb volume. Previous research on the physiological bases of bulbing in onion has primarily focused on evaluating long-day (LD) and short-day (SD) cultivars planted under inductive and non-inductive daylengths (Cheng et al. 2021). In onion, six FT-like transcripts (*AcFT1* through *AcFT6*) have been identified, two of which are involved in bulbing. Under non-inductive daylengths, *AcFT4* is expressed and acts as a repressor of bulbing. Conversely, when plants are exposed to daylengths inductive for bulbing, the expression of *AcFT4* is reduced, while the expression of *AcFT1* is increased to promote bulbing (Cheng et al. 2021). Cho et al. (2021a, b) discovered four major loci (*C, I, R*, and *L*) determining onion bulb colors, each located on separate chromosomes. Investigation of the F1 cultivar 'Santero', resistant to downy mildew, revealed a chromosome fragment from Allium roylei, containing the resistance gene, estimated at 27.6 cM at the end of chromosome 3. Wako et al. (2016) repeatedly detected two QTLs associated with bolting time on linkage groups 1a (Chr. 1a) and 2a (Chr. 2a) on chromosome 1 and chromosome 2, respectively. Through GBS analysis, Choi et al. (2020) and Lee et al. (2021a, b) identified QTLs (*qAS7.1, qAC4.1, qAC4.2*, and *qAS7.1*) governing anthocyanin synthesis and sugar content, including sucrose (*qSC4.1*), glucose (*qGC5.1*), fructose (*qFC5.1*), and total sugar content (*qTSC5.1*). Additionally, Tsukazaki et al. (2012) located a major pungency QTL (*AFAT04B03*) within a 24.2-cM interval on Chr. 2a.

1.12.14 Brassica sp.

Sheng et al. (2024) conducted major QTL identification and fine mapping of the genetic loci associated with curd setting height (*BocDWARF1*) in cauliflower, pinpointing the locus to chromosome 1 and narrowing it down to a specific interval defined by SNP markers *1_40298271* and *1_40400283*. Mangal and Singh (2024) analyzed candidate genes, including *FRUITFULL-c (BoFUL-c), FLOWERING LOCUS T (BoFT), SUPPRESSOR OF OVEREXPRESSION OF CONSTANS1 (SOC1)*, and *LEAFY (BoFLY)*, among others, distinguishing regulatory mechanisms for phenological transitions in Indian and Snowball cauliflower. *BolC.Pb9.1*, located on chromosome 9 and controlling clubroot resistance in a wild Brassica relative, was investigated using map-based cloning and functional analysis by Zhang et al. (2024). Wu et al. (2024) demonstrated that overexpression of *BoMAM1 (BolI0108790)* in broccoli led to high accumulation and a high ratio of *C4-GSLs*, highlighting *BoMAM1* as a key enzyme in *C4-GSL* biosynthesis. mapped curd architecture-related QTLs, such as stalk length (*qSL.C6–1, qSL.C6–2*) and curd solidity (*qCS.C6–1* and *qCS.C6–2*) in cauliflower, providing insights into the molecular mechanisms of curd development and enabling MAS in cauliflower breeding. Tan et al. (2020) detected two QTLs for the green curding trait on chromosomes 5 (*Gr5.1*) and 7 (*Gr7.1*) using F2 bulk population DNA sequencing, with *Gr5.1* identified as a major QTL for green curding. Zhao et al. (2020a, b) identified candidate genes (*SOC1, Bo4g024850*) involved in curd riceyness, suggesting an upstream SNP likely influences the riceyness phenotype by affecting the expression levels of *Bo4g024850*. Studies on flowering-time regulators and QTL co-localizations suggested that allelic variation of *FLOWERING LOCUS C (FLC)* might play a role in flowering-time regulation in B. oleracea, although some studies provided conflicting evidence regarding the functional relevance of FLC homologs in cauliflower (Koornneef et al. 1994; Razi et al. 2008; Ridge et al. 2015; Matschegewski et al. 2015; Brown et al. 2007; Hasan et al. 2016).

1.12.15 Garden Pea (*Pisum sativum* L.)

Leonforte et al. (2013) conducted a study on QTLs associated with enhanced salinity tolerance in field pea, discovering a total of 705 SNPs (91.7%). Ubayasena et al. (2011) investigated visual quality traits in two recombinant inbred line populations (RILs) of yellow pea and green pea, constructing a genetic linkage map based on 224 AFLP and SSR markers spanning 890 cM of the pea genome. Major QTLs were detected on LG IV and LG V, with additional QTLs on LG II and LG III associated with green cotyledon bleaching resistance. Irzykowska and Wolko (2004) utilized 204 markers in an F2 population derived from a cross between Wt10245 and Wt11238. Wu et al. (2021a, b, c) studied an F8 RILs population derived from the cross between 'Reward' and '00-2067', identifying a genomic region on chromosome 4 responsible for partial resistance to A. euteiches, along with regions on chromosomes 5, 6, and 7. Identified 14 QTLs, including two for super-earliness, one for PH, six for number of pods per plant, one for number of seeds per pod, one for pod length, and three for harvest index in an elite line derived from *Pisum sativum* L. and *P. fulvum*. Zheng et al. (2018) mapped two QTLs for leaf shape traits on linkage group 7 (LG7) in *Pisum sativum* using SLAF sequencing. Lavaud et al. (2024) conducted fine mapping of *Ae-Ps4.5*, a major locus for resistance to pathotype III of *Aphanomyces euteiches* in pea.

1.13 Conclusion

Concluding QTL mapping studies in vegetable crops reveals significant advancements in understanding the genetic architecture underlying various agronomically important traits. Through the integration of molecular markers and phenotypic data, researchers have successfully identified and characterized QTLs associated with traits such as yield, quality, disease resistance, and stress tolerance.

One key finding from this chapter is the compilation of all identified genomic regions harboring major-effect QTLs, which can serve as targets for MAS in breeding programs. By utilizing molecular markers linked to these QTLs, breeders can expedite the development of improved cultivars with desirable traits, leading to enhanced crop productivity and resilience in the face of changing environmental conditions. Moreover, QTL mapping studies have shed light on the genetic basis of complex traits, uncovering the polygenic nature of many important agricultural traits. This understanding allows for more precise breeding strategies, including genomic selection approaches, which leverage genome-wide marker information to predict phenotypic performance and facilitate the selection of superior individuals across breeding populations. Furthermore, the availability of high-throughput genotyping technologies and increasingly sophisticated statistical methodologies has accelerated QTL discovery and fine mapping, enabling researchers to dissect the genetic basis of traits with higher resolution and accuracy. QTL mapping in vegetable crops has provided invaluable insights into the genetic control of important agronomic traits, paving the way for more efficient and targeted breeding efforts aimed at developing improved varieties with enhanced yield, quality, and resilience to biotic and abiotic stresses. Continued advancements in genomic technologies and bioinformatics tools will further enhance our ability to unravel the complex genetic networks governing vegetable crop traits, ultimately contributing to sustainable agriculture and food security.

References

Abbas HMK, Zhou YY, Huang HX, Qamar MTU, Wang MM, Lu S, Zhong YJ (2022) QTL mapping, whole genome resequencing, and marker-assisted selection provide basics of early flowering in pumpkin. Plant Breed 141(2):266–276. https://doi.org/10.1111/pbr.12996

Abebe AM, Oh CS, Kim HT, Choi G, Seo E, Yeam I, Lee JM (2022) QTL-seq analysis for identification of resistance loci to bacterial canker in tomato. Front Plant Sci 12:809959. https://doi.org/10.3389/fpls.2021.809959

Adedze YMN, Lu X, Fan W, Zhang W, Yang X, Deng Z, Li W (2024) Development of PCR-based markers associated with powdery mildew resistance using bulked segregant analysis (BSA-seq) in melon. Czech J Genet Plant Breed 60:25–33

Adhikari TB, Siddique MI, Louws FJ, Sim SC, Panthee DR (2023) Molecular mapping of quantitative trait loci for resistance to early blight in tomatoes. Front Plant Sci 14:1135884

Agri Ahmet R (2023) Development of a multi-parent advanced generation inter-cross (MAGIC) population and analysis of a nested association mapping (NAM) population for improvement of genetic resistance to white mold in snap bean. M.Sc. Thesis submitted to the Oregon State University, USA. Available at: https://ir.library.oregonstate.edu/concern/graduate_thesis_or_dissertations/0v838825v. Assessed 23 Nov 2024

Ahmad I, Rana RM, Hassan MU, Khan MA, Sajjad M (2022) Association mapping for abiotic stress tolerance using heat-and drought-related syntenic markers in okra. Mol Biol Rep 49:11409–11419

Amanullah S, Osae BA, Yang T, Li S, Abbas F, Liu S, Liu S, Song Z, Wang X, Gao P, Luan F (2022) Development of whole genome SNP-CAPS markers and preliminary QTL mapping of fruit pedicel traits in watermelon. Front Plant Sci 13:879919. https://doi.org/10.3389/fpls.2022.879919

Arjun K, Dhaliwal MS, Jindal SK, Fakrudin B (2018) Mapping of fruit length related QTLs in interspecific cross (*Capsicum annuum* L.× *Capsicum galapagoense* Hunz.) of chilli. Breed Sci 68(2):219–226. https://doi.org/10.1270/jsbbs.17073

Aruna TS, Srivastava A, Tomar BS, Behera TK, Krishna H, Jain PK, Pandey R, Singh B, Gupta R, Mangal M (2023) Genetic analysis of heat tolerance in hot pepper: insights from comprehensive phenotyping and QTL mapping. Front Plant Sci 14:1232800. https://doi.org/10.3389/fpls.2023.1232800

Bai J, Chen B, Qi M, Du H, Wang Q, Geng S, Zhang X (2023) QTL mapping and candidate gene analysis of microspore embryogenesis in Capsicum. Euphytica 219(1):3. https://doi.org/10.1007/s10681-022-03100-1

Baldwin S, Revanna R, Pither-Joyce M, Shaw M, Wright K, Thomson S, Moya L, Lee R, Macknight R, McCallum J (2014) Genetic analyses of bolting in bulb onion (*Allium cepa* L.). Theor Appl Genet 127:535–547

Behera TK, Dey SS, Munshi AD, Gaikwad AB, Pal A, Singh I (2009) Sex inheritance and development of gynoecious hybrids in bitter gourd (*Momordica charantia* L.). Sci Hortic 120:130–133

Bhanushree N, Saha P, Tomar BS, Gopala Krishnan S, Gurung B, Ghoshal C, Das Munshi A (2020) Single marker analysis and mapping of QTLs governing fruit weight in eggplant (*Solanum melongena* L.). Proc Natl Acad Sci India B - Biol Sci 90:769–775. https://doi.org/10.1007/s40011-019-01149-y

Bo K, Wei S, Wang W, Miao H, Dong S, Zhang S, Gu X (2019) QTL mapping and genome-wide association study reveal two novel loci associated with green flesh color in cucumber. BMC Plant Biol 19:1–13. https://doi.org/10.1186/s12870-019-1835-6

Brainard SH, Ellison SL, Simon PW, Dawson JC, Goldman IL (2022) Genetic characterization of carrot root shape and size using genome-wide association analysis and genomic-estimated breeding values. Theor Appl Genet:1–18. https://doi.org/10.1007/s00122-021-03988-8

Branham SE, Levi A, Wechter WP (2019a) QTL mapping identifies novel source of resistance to Fusarium wilt race 1 in *Citrullus amarus*. Plant Dis 103(5):984–989. https://doi.org/10.1094/PDIS-09-18-1677-RE

Branham SE, Levi A, Katawczik ML, Wechter WP (2019b) QTL mapping of resistance to bacterial fruit blotch in *Citrullus amarus*. Theor Appl Genet 132:1463–1471. https://doi.org/10.1007/s00122-019-03292-6

Branham SE, Kousik C, Mandal MK, Wechter WP (2021) Quantitative trait loci mapping of resistance to powdery mildew race 1 in a recombinant inbred line population of melon. Plant Dis 105:3809–3815

Brekke TD, Stroud JA, Shaw DS, Crawford S, Steele KA (2019) QTL mapping in salad tomatoes. Euphytica 215(7):115. https://doi.org/10.1007/s10681-019-2440-3

Brown AF, Jeffery EH, Juvik JA (2007) A polymerase chain reaction-based link-age map of broccoli and identification of quantitative trait loci associated with harvest date and head weight. J Hortic Sci 132:507–513

Cabodevila VG, Cambiaso V, Rodríguez GR, Picardi LA, Pratta GR, Capel C, Lozano R, Capel J (2021) A segregating population from a tomato second cycle hybrid allows the identification of novel QTL for fruit quality traits. Euphytica 217(1):6. https://doi.org/10.1007/s10681-020-02731-6

Calzadilla A, Rehdanz K, Betts R, Falloon P, Wiltshire A, Tol RS (2013) Climate change impacts on global agriculture. Clim Chang 120:357–374. https://doi.org/10.1007/s10584-013-0822-4

Cao M, Li S, Deng Q, Wang H, Yang R (2021) Identification of a major-effect QTL associated with pre-harvest sprouting in cucumber (*Cucumis sativus* L.) using the QTL-seq method. BMC Genomics 22:1–11. https://doi.org/10.1186/s12864-021-07548-8

Capel C, Fernández del Carmen A, Alba JM, Lima-Silva V, Hernández-Gras F, Salinas M, Boronat A, Angosto T, Botella MA, Fernández-Muñoz R, Granell A (2015) Wide-genome QTL mapping of fruit quality traits in a tomato RIL population derived from the wild-relative species *Solanum pimpinellifolium* L. Theor Appl Genet 128:2019–2035. https://doi.org/10.1007/s00122-015-2563-4

Casañas F, Simó J, Casals J, Prohens J (2017) Toward an evolved concept of landrace. Front Plant Sci 8:245305

Ceccarelli S, Grando S (2024) Diversity as a plant breeding objective. Agronomy 14(3):550

Chae SY, Lee K, Do JW, Hong SC, Lee KH, Cho MC, Yang EY, Yoon JB (2022) QTL mapping of resistance to bacterial wilt in pepper plants (*Capsicum annuum*) using genotyping-by-sequencing (GBS). Horticulturae 8(2):115. https://doi.org/10.3390/horticulturae8020115

Chaim AB, Borovsky Y, Rao GU, Gur A, Zamir D, Paran I (2006) Comparative QTL mapping of fruit size and shape in tomato and pepper. Israel J Plant Sci 54:191–203

Chen B, Wu Z, Wen JQ, Lin HW, Yu L, Xue LZ, Zhou R, Jiang FL (2021) QTL mapping and candidate genes analysis of irregular fruit cracking in tomato. Acta Hortic Sin 48(7):1329–1339. https://doi.org/10.16420/j.issn.0513-353x.2020-1032

Chen Y, Qiu S, Zhou H, Gao W, Cui L, Qiu Z, Wang X (2023) Construction of a tomato (*Solanum lycopersicum* L.) introgression line population and mapping of major agronomic quantitative trait loci. Horticulturae 9:823

Cheng J, Qin C, Tang X, Zhou H, Hu Y, Zhao Z, Cui J, Li B, Wu Z, Yu J, Hu K (2016) Development of a SNP array and its application to genetic mapping and diversity assessment in pepper (*Capsicum* spp.). Sci Rep 6(1):33293. https://doi.org/10.1038/srep33293

Cheng W, Rashid H, Stark R, Thomas B (2021) The role of organ- and daylength-specific gene expression in bulb development and resource management in onion (*Allium cepa* L.). Sci Hortic 286:110223. https://doi.org/10.1016/j.scienta.2021.110223

Cho Y, Kim B, Lee J, Kim S (2021a) Construction of a high-resolution linkage map and chromosomal localization of the loci determining major qualitative traits in onion (Allium cepa L.). Euphytica 217:1–12

Cho Y, Lee S, Park J, Kwon S, Park G, Kim H, Park Y (2021b) Identification of a candidate gene controlling semi-dwarfism in watermelon, Citrullus lanatus, using a combination of genetic linkage mapping and QTL-seq. Hortic Environ Biotechnol 62:447–459

Choi Y, Kim S, Lee J (2020) Construction of an onion (*Allium cepa* L.) genetic linkage map using genotyping-by-sequencing analysis with a reference gene set and identification of QTLs controlling anthocyanin synthesis and content. Plants 9(5):616. https://doi.org/10.3390/plants9050616

Çolak NG, Eken NT, Ülger M, Frary A, Doğanlar S (2020) Mapping of quantitative trait loci for antioxidant molecules in tomato fruit: carotenoids, vitamins C and E, glutathione and phenolic acids. Plant Sci 292:110393. https://doi.org/10.1016/j.plantsci.2019.110393

Crute IR (1988) The impact of breeding on pest and disease control in lettuce. Asp Appl Biol 17:305–312

Cui J, Liu J, Zhong C, Hu F, Dong J, Cheng J, Hu K (2022a) Fine-mapping and candidate gene analysis of the Mcgy1 locus responsible for gynoecy in bitter gourd (*Momordica* spp.)

Cui L, Siskos L, Wang C, Schouten HJ, Visser RG, Bai Y (2022b) Breeding melon (*Cucumis melo*) with resistance to powdery mildew and downy mildew. Hortic Plant J 8:545–561

Darvasi A, Soller M (1995) Advanced intercross lines, an experimental population for fine gene mapping. Genetics 141:1199–1207

Dong S, Wang W, Bo K, Miao H, Song Z, Wei S, Zhang S, Gu X (2019) Quantitative trait loci mapping and candidate gene analysis of low temperature tolerance in cucumber seedlings. Front Plant Sci 10:467691. https://doi.org/10.3389/fpls.2019.01620

Du H, Wen C, Zhang X, Xu X, Yang J, Chen B, Geng S (2019) Identification of a major QTL (*qRRs-10.1*) that confers resistance to *Ralstonia solanacearum* in pepper (*Capsicum annuum*) using SLAF-BSA and QTL mapping. Int J Mol Sci 20(23):5887. https://doi.org/10.3390/ijms20235887

Dwivedi N, Mishra M, Sharma SS, Singh RK (2023) Genetic analysis and QTLs identification for resistance to the Begomovirus causing pepper leaf curl virus (PepLCV) disease. J Plant Biochem Biotechnol 1-11. https://doi.org/10.1007/s13562-023-00855-z

Eshed Y, Zamir D (1994) A genomic library of *Lycopersicon pennellii* in *L. esculentum*: a tool for fine mapping of genes. Euphytica 79:175–179

Fan L, Wang YXL, Tang M, Zhang X, Ying J, Liu L (2020) A genome-wide association study uncovers a critical role of the *RsPAP2* gene in red-skinned *Raphanus sativus* L. Hortic Res 7. https://doi.org/10.1038/s41438-020-00385-y

FAO. (2023) FAOSTAT emissions database. Retrieved from Food and Agriculture Organization of the United Nations (FAO). https://www.globalhungerindex.org/pdf/en/2023.pdf

Fischer G, Shah MN, Tubiello F, Van Velhuizen H (2005) Socio-economic and climate change impacts on agriculture: an integrated assessment. Philos Trans R Soc B Biol Sci 360:2067–2083. https://doi.org/10.1098/rstb.2005.1744

Flores-Saavedra M, Gramazio P, Vilanova S, Mircea DM, Ruiz-González MX, Vicente Ó, Plazas M (2023) Evaluation of a set of eggplant (*Solanum melongena*) lines with introgressions of *S. incanum* under water stress conditions. bioRxiv:2023–05

Fridman E, Pleban T, Zamir D (2000) A recombination hotspot delimits a wild-species quantitative trait locus for tomato sugar content to 484 bp within an invertase gene. Proc Natl Acad Sci USA 97:4718–4723

Gan C, Deng X, Cui L, Yu X, Yuan W, Dai Z, Yao M, Pang W, Ma Y, Yu X, Choi SR (2019) Construction of a high-density genetic linkage map and identification of quantitative trait loci associated with clubroot resistance in radish (*Raphanus sativus* L.). Mol Breed 39:1–12. https://doi.org/10.1007/s11032-019-1020-5

Gangadhara RP, Behera TK, Gaikwad AB, Munshi AD, Jat GS, Boopalakrishnan G (2018) Mapping and QTL analysis of gynoecy and earliness in bitter gourd (*Momordica charantia* L.) using genotyping-by-sequencing (GBS) technology. Front Plant Sci 9:396931

Gao J, Li WB, Liu HF, Chen FB (2019) De novo transcriptome sequencing of radish (Raphanus sativus L.) fleshy roots: analysis of major genes involved in the anthocyanin synthesis pathway. BMC Mol Cell Biol 20:1–12. https://doi.org/10.1186/s12860-019-0228-x

Gao Z, Zhang H, Cao C, Han J, Li H, Ren Z (2020) QTL mapping for cucumber fruit size and shape with populations from long and round fruited inbred lines. Hortic Plant J 6(3):132–144. https://doi.org/10.1016/j.hpj.2020.04.004

Gardiner JM, Coe EH, Melia-Hancock S et al (1993) Development of a core RFLP map in maize using an immortalized F_2 population. Genetics 154:917–930

Guo G, Wang S, Liu J, Pan B, Diao W, Ge W, Gao C, Snyder JC (2017) Rapid identification of QTLs underlying resistance to Cucumber mosaic virus in pepper (*Capsicum frutescens*). Theor Appl Gene 130:41–52. https://doi.org/10.1007/s00122-016-2790-3

Gupta N, Yadav KK, Kumar V, Kumar S, Chadd RP, Kumar A (2018) Trace elements in soil-vegetables interface: translocation, bioaccumulation, toxicity and amelioration-a review. Sci Total Environ 65:2927–2942. https://doi.org/10.1016/j.scitotenv.2018.10.047

Gur A, Zamir D (2015) Mendelizing all components of a pyramid of three yield QTL in tomato. Front Plant Sci 6:1096. https://doi.org/10.3389/fpls.2015.01096

Hamid MR, Maria SK (2023) Harnessing nature's tiny warriors: plant viruses as innovative bio-herbicides. Trends Hortic 6:3146

Han K, Jeong HJ, Yang HB, Kang SM, Kwon JK, Kim S (2016) An ultra-high-density bin map facilitates high-throughput QTL mapping of horticultural traits in pepper (*Capsicum annuum*). DNA Res 23:81–91. https://doi.org/10.1093/dnares/dsv038

Han X, Min Z, Wei M, Li Y, Wang D, Zhang Z, Hu X, Kong Q (2022) QTL mapping for pumpkin fruit traits using a GBS-based high-density genetic map. Euphytica 218(8):106. 10.21203/rs.3.rs-1142200/v2

Hartman Y, Hooftman DA, Eric Schranz M, van Tienderen PH (2013) QTL analysis reveals the genetic architecture of domestication traits in Crisphead lettuce. Genet Resour Crop Evol 60:1487–1500

Hasan Y, Briggs W, Matschegewski C, Ordon F, Stützel H, Zetsche H (2016) Quantitative trait loci controlling leaf appearance and curd initiation of cauliflower in relation to temperature. Theor Appl Genet 129:1273–1288. https://doi.org/10.1007/s00122-016-2702-6

Hu F, Jichi D, Shaobin Z, Zhao S, Wendong G, Fanchong Y, Jian Z, Jia L, Kailin H, Jiaowen C (2023) Fine mapping and gene silencing pinpoint *Capana10g002229* as a strong candidate gene regulating the deciduous character of ripe pepper fruit (*Capsicum spp*.). Theor Appl Genet 5:107

Huang J, Sun J, Liu E, Liu Y, Fang Z, Yang L, Zhuang M, Zhang Y, Lv H, Wang Y, Han F (2020) Mapping of QTLs detected in a broccoli double diploid population for plant height and leaf characteristics. Res Sq. https://doi.org/10.21203/rs.3.rs-25053/v1

Irzykowska L, Wolko B (2004) Interval mapping of QTLs controlling yield-related traits and seed protein content in Pisum sativum. J Appl Genet 45:297–306

Kaur G, Pathak M, Singla D, Chhabra G, Chhuneja P, Kaur Sarao N (2022) Quantitative trait loci mapping for earliness, fruit, and seed related traits using high density genotyping-by-Sequencing-Based genetic map in bitter gourd (*Momordica charantia* l.). Front Plant Sci 12:799932

Kesseli RV, Ochoa O, Michelmore RW (1991) Variation at RFLP loci in *Lactuca* spp. and origin of cultivated lettuce. Genome 34:430–436

Kethom W, Mongkolporn O (2021) New QTLs for anthracnose resistance identified in *Capsicum baccatum* 'PBC80'-derived recombinant inbred lines. Euphytica 217(6):128. https://doi.org/10.1007/s10681-021-02866-0

Kethom W, Taylor PW, Mongkolporn O (2023) Expression of genes involved in anthracnose resistance in chili (*Capsicum baccatum*)'PBC80'-derived recombinant inbred lines. Pathogens 12:1306

Koornneef M, Blankestijn-de Vries H, Hanhart C, Soppe W, Peeters T (1994) The phenotype of some late-flowering mutants is enhanced by a locus on chromosome 5 that is not effective in the Landsberg erecta wild-type. Plant J 6:911–919. https://doi.org/10.1046/j.1365-313X.1994.6060911.x

Kumar S, Prasad S, Yadav KK, Shrivastava M, Gupta N, Nagar S, Bach QV, Kamyab H, Khan SA, Yadav S, Malav LC (2019) Hazardous heavy metals contamination of vegetables and food chain: Role of sustainable remediation approaches-a review. Environ Res:108792. https://doi.org/10.1016/j.envres.2019.108792

Labate JA, Grandillo S, Fulton T, Muños S, Caicedo AL, Peralta I, Ji Y, Chetelat RT, Scott JW, Gonzalo MJ et al (2007) Tomato. In: Kole C (ed) Vegetables; Genome mapping and molecular breeding in plants, vol 5. Springer, Berlin/Heidelberg

Lavaud C, Lesne A, Leprevost T, Pilet-Nayel ML (2024) Fine mapping of *Ae-Ps4. 5*, a major locus for resistance to pathotype III of *Aphanomyces euteiches* in pea. Theor Appl Genet 137:47

Lazaridi E, Kapazoglou A, Gerakari M, Kleftogianni K, Passa K, Sarri E, Papasotiropoulos V, Tani E, Bebeli PJ (2024) Crop landraces and indigenous varieties: a valuable source of genes for plant breeding. Plants 13(6):758. https://doi.org/10.3390/plants13060758

Lee R, Baldwin S, Kenel F, McCallum J, Macknight R (2013) FLOWERING LOCUS T genes control onion bulb formation and flowering. Nat Commun 4:2884

Lee J, Park SJ, Hong SC, Han JH, Choi D, Yoon JB (2016) QTL mapping for capsaicin and dihydrocapsaicin content in a population of *Capsicum annuum* 'NB 1'× *Capsicum chinense* 'Bhut Jolokia'. Plant Breed 135(3):376–383. https://doi.org/10.1111/pbr.12355

Lee ES, Kim DS, Kim SG, Huh YC, Back CG, Lee YR, Lee J (2021a) QTL mapping for gummy stem blight resistance in watermelon (*Citrullus spp.*). Plants 10:500

Lee YR, Kim CW, Han J, Choi HJ, Han K, Lee ES, Kim DS, Lee J, Siddique MI, Lee HE (2021b) Genotyping-by-Sequencing derived genetic linkage map and quantitative trait loci for sugar content in onion (*Allium cepa* l.). Plants 10(11):2267. https://doi.org/10.3390/plants10112267

Lee JH, Siddique MI, Jang S, Kim GW, Choi GJ, Kwon JK, Kang BC (2024) Identification of QTLs associated with resistance to bacterial wilt in pepper (*Capsicum annuum* L.) through biparental QTL mapping and genome-wide association analysis. Sci Hortic 329:112987. https://doi.org/10.3390/plants11121551

Leonforte A, Sudheesh S, Cogan NO, Salisbury PA, Nicolas ME, Materne M, Kaur S (2013) SNP marker discovery, linkage map construction and identification of QTLs for enhanced salinity tolerance in field pea (*Pisum sativum* L.). BMC Plant Biol 13:1–14

Li ZK (2000) QTL mapping in rice: a few critical considerations. Rice Genetics IV, Proceedings of the fourth international rice genetics symposium, 22–27 October, Los Baños, Philippines. Science Publishers, Inc./International Rice Research Institute, New Delhi/Los Baños, 488 p

Li N, Yin Y, Wang F, Yao M (2018) Construction of a high-density genetic map and identification of QTLs for cucumber mosaic virus resistance in pepper (*Capsicum annuum* L.) using specific length amplified fragment sequencing (SLAF-seq). Breed Sci 68(2):233–241. https://doi.org/10.1270/jsbbs.17063

Li YF, Zhang SC, Yang XM, Wang CP, Huang QZ, Huang RZ (2021a) Generation of a high-density genetic map of pepper (*Capsicum annuum* L.) by SLAF-seq and QTL analysis of Phytophthora capsici resistance. Horticulturae 7(5):92. https://doi.org/10.3390/horticulturae7050092

Li Z, Liu Y, Yuan S, Han F, Fang Z, Yang L, Zhuang M, Zhang Y, Lv H, Wang Y, Ji J (2021b) Fine mapping of the major QTLs for biochemical variation of sulforaphane in broccoli florets using a DH population. Sci Rep 11(1):9004. https://doi.org/10.1038/s41598-021-88652-3

Li J, Liu X, Yao Q, Xu L, Li W, Tan W, Liu D (2023a) Tolerance and adaptation characteristics of sugar beet (*Beta vulgaris* L.) to low nitrogen supply. Plant Signal Behav 18:2159155. https://doi.org/10.1080/15592324.2022.2159155

Li N, Kong S, Zhou D, Li N, Shang J, Wang J, Ma S (2023b) Mapping and validation of a new quantitative trait locus (QTL) for fruit size in watermelon (*Citrullus lanatus*). Sci Hortic 318:112054. https://doi.org/10.1016/j.scienta.2023.112054

Lietzow CD, Zhu H, Pandey S, Havey MJ, Weng Y (2016) QTL mapping of parthenocarpic fruit set in North American processing cucumber. Theor Appl Genet 129:2387–2401. https://doi.org/10.1007/s00122-016-2778-z

Lin YC, Mansfeld BN, Tang X, Colle M, Chen F, Weng Y, Grumet R (2023) Identification of QTL associated with resistance to Phytophthora fruit rot in cucumber (*Cucumis sativus* L.). Front Plant Sci 14:1281755

Liu G, Zhao T, You X, Jiang J, Li J, Xu X (2019a) Molecular mapping of the Cf-10 gene by combining SNP/InDel-index and linkage analysis in tomato (*Solanum lycopersicum*). BMC Plant Biol 19:1–11. https://doi.org/10.1186/s12870-018-1616-7

Liu S, Jia M, Chen J, Wan H, Dong R, Nie S, Yu Q (2019b) Removal of bound polyphenols and its effect on antioxidant and prebiotics properties of carrot dietary fiber. Food Hydrocoll 93:284–292. https://doi.org/10.1016/j.foodhyd.2019.02.047

Liu S, Wang R, Zhang Z, Li Q, Wang L, Wang Y, Zhao Z (2019c) High-resolution mapping of quantitative trait loci controlling main floral stalk length in Chinese cabbage (*Brassica rapa* L. ssp. *pekinensis*). BMC Genomics 20:1–10. https://doi.org/10.1186/s12864-019-5810-2

Liu C, Yao X, Li G, Huang L, Wu X, Xie Z (2021a) Identification of major loci and candidate genes for anthocyanin biosynthesis in broccoli using QTL-Seq. Horticulturae 7(8):246. https://doi.org/10.3390/horticulturae7080246

Liu D, Dong S, Bo K, Miao H, Li C, Zhang Y, Zhang S, Gu X (2021b) Identification of QTLs controlling salt tolerance in cucumber (*Cucumis sativus* L.) seedlings. Plants 10(1):85. https://doi.org/10.3390/plants10010085

Liu Y, Dong S, Wei S, Wang W, Miao H, Bo K, Gu X, Zhang S (2021c) QTL mapping of heat tolerance in cucumber (*Cucumis sativus* L.) at adult stage. Plants 10(2):324. https://doi.org/10.3390/plants10020324

Liu Z, She H, Xu Z, Zhang H, Li G, Zhang S, Qian W (2021d) Quantitative trait loci (QTL) analysis of leaf related traits in spinach (*Spinacia oleracea* L.). BMC Plant Biol 21(1):290. https://doi.org/10.1186/s12870-021-03092-5

Liu D, Wei X, Sun D, Yang S, Su H, Wang Z, Yuan Y (2021e) An SNP mutation of gene RsPP converts petal color from purple to white in radish (*Raphanus sativus* L.). Front Plant Sci 12:643579

Liu P, Yuan W, Fu J, Jiang Z, Hayashi H, Neubig G (2023a) Pre-train, prompt, and predict: A systematic survey of prompting methods in natural language processing. ACM Comput Surv 55:1–35. https://doi.org/10.1145/3560815

Liu Z, Jiang J, Li J (2023b) GWAS characterization of the genetic regions associated with nine important agronomic traits in tomato (*Solanum lycopersicum* L.). Agronomy 13:1191

Lopez-Moreno H, Basurto-Garduño AC, Torres-Meraz MA, Diaz-Valenzuela E, Arellano-Arciniega S, Zalapa J, Sawers RJ, Cibrián-Jaramillo A, Diaz-Garcia L (2023) Genetic analysis and QTL mapping of domestication-related traits in chili pepper (*Capsicum annuum* L.). Front Genet 14:1101401. https://doi.org/10.3389/fgene.2023.1101401

Lozada DN, Whelpley M, Acuña-Galindo A (2021) Genetic architecture of Chile pepper (*Capsicum* spp.) QTLome revealed using Meta-QTL analysis. Horticulturae 7(8):227. https://doi.org/10.3390/horticulturae7080227

Luo X, Xu L, Wang Y, Dong J, Chen Y, Tang M, Liu L (2020) An ultra-high-density genetic map provides insights into genome synteny, recombination landscape and taproot skin colour in radish (*Raphanus sativus* L.). Plant Biotechnol J 18:274–286. https://doi.org/10.1111/pbi.13195

Lv Y, Gao P, Liu S, Fang X, Zhang T, Liu T, Amanullah S, Wang X, Luan F (2022) Genetic mapping and QTL analysis of stigma color in melon (*Cucumis melo* L.). Front Plant Sci 13:865082. https://doi.org/10.3389/fpls.2022.865082

Lv Z, Jin Q, Li Z, Li T, Wang Y, You Q, Sun B (2023) Fine mapping and candidate gene analysis of the Gv1 locus controlling green-peel color in eggplant (*Solanum melongena* L.). Horticulturae 9:888

Ma X, Qiao YM, Li Y, Yu YN, Gong ZH (2022) Identification of fruit traits related QTLs and a candidate gene, CaBRX, controlling locule number in pepper (*Capsicum annuum* L.). Horticulturae 8(2):146. https://doi.org/10.3390/horticulturae8020146

Mackay L, Powell W (2007) Methods for linkage disequilibrium mapping in crops. Trends Plant Sci 12:57–63

Mahasuk P, Struss D, Mongkolporn O (2016) QTLs for resistance to anthracnose identified in two *Capsicum* sources. Mol Breed 36:1–10. https://doi.org/10.1007/s11032-016-0435-5

Mangal M, Singh S (2024) Candidate gene analysis distinguishes regulatory mechanisms for phenological transitions in Indian and Snowball cauliflower. J Hortic Sci Biotechnol 99:20–32

Maragal S, Rao ES, Reddy DCL, Samuel D (2021) Mapping resistance to WBNV through CIM, JICIM and QTL-meta-analysis approaches in watermelon. Sci Hortic 289:110475. https://doi.org/10.1016/j.scienta.2021.110475

Marcinkowski P, Piniewski M (2024) Future changes in crop yield over Poland driven by climate change, increasing atmospheric CO_2 and nitrogen stress. Agric Syst 213:103813. https://doi.org/10.1016/j.agsy.2023.103813

Mata-Nicolás E, Montero-Pau J, Gimeno-Paez E, García-Pérez A, Ziarsolo P, Blanca J, Van der Knaap E, Díez MJ, Cañizares J (2021) Discovery of a major QTL controlling trichome IV density in tomato using K-seq genotyping. Gene 12(2):243. https://doi.org/10.3390/genes12020243

Matschegewski C, Zetsche H, Hasan Y, Leibeguth L, Briggs W, Ordon F (2015) Genetic variation of temperature-regulated curd induction in cauliflower: elucidation of floral transition by genome-wide association mapping and gene expression analysis. Front Plant Sci 6:720. https://doi.org/10.3389/fpls.2015.00720

Matsumura H, Miyagi N, Taniai N, Fukushima M, Tarora K, Shudo A, Urasaki N (2014) Mapping of the gynoecy in bitter gourd (*Momordica charantia*) using RAD-seq analysis. PLoS One 9:e87138

Maurya PK, Malik DS, Yadav KK, Kumar A, Kumar S, Kamyab H (2019) Bioaccumulation and potential sources of heavy metal contamination in fish species in River Ganga basin: possible human health risks evaluation. Toxicol Rep. https://doi.org/10.1016/j.toxrep.2019.05.012

Mendelsohn R (2008) The impact of climate change on agriculture in developing countries. J Natl Resour Policy Res 1:5–19. https://doi.org/10.1080/19390450802495882

Meng Q, Liu Z, Feng C, Zhang H, Xu Z, Wang X, Wu J, She H, Qian W (2022) Quantitative trait locus mapping and identification of candidate genes controlling bolting in spinach (*Spinacia oleracea* L.). Front Plant Sci 13:850810. https://doi.org/10.3389/fpls.2022.850810

Mimura Y, Inoue T, Minamiyama Y, Kubo N (2012) An SSR-based genetic map of pepper (*Capsicum annuum* L.) serves as an anchor for the alignment of major pepper maps. Breed Sci 62:93–98. https://doi.org/10.1270/jsbbs.62.93

Mishra S, Behera TK, Munshi AD, Gaikwad K, Mohapatra T (2014) Identification of RAPD marker associated with gynoecious trait (*'gy-1'gene*) in bitter gourd (*'Momordica charantia'*L.). Aust J Crop Sci 8

Mori M, Maki K, Kawahata T, Kawahara D, Kato Y, Yoshida T, Nagasawa H, Sato H, Nagano AJ, Bethke PC, Kato K (2021) Mapping of QTLs controlling epicotyl length in adzuki bean (Vigna angularis). Breed Sci 71(2):208–216. https://doi.org/10.1270/jsbbs.20093

Mori N, Hasegawa S, Takimoto R, Horiuchi R, Watanabe C, Onizaki D, Shiragane H, Nagano AJ, Kesumawati E, Koeda S (2022) Identification of QTLs conferring resistance to begomovirus isolate of PepYLCIV in *Capsicum chinense*. Euphytica 218(2):20. https://doi.org/10.1007/s10681-022-02970-9

Mousavi SN, Parchami M, Ramamoorthy SK, Soufiani AM, Hakkarainen M, Zamani A (2023) Bioconversion of carrot pomace to value-added products: rhizopus delemar fungal biomass and cellulose. Fermentation 9:374. https://doi.org/10.3390/fermentation9040374

Muleke E, Fan L, Xu L, Zhu X, Zhang W, Liu L (2017) Coordinated regulation of anthocyanin biosynthesis genes confers varied phenotypic and spatial-temporal anthocyanin accumulation in radish (*Raphanus sativus* L.). Front Plant Sci 8:268743

Nandi L, Suresh P, Pradeepkumara N, Munshi AD, Sharma PK, Boopalakrishnan G, Dey SS (2024) Elucidating the genetics of post-harvest shelf-life of cucumber fruits and identification of associated QTLs and candidate genes. Sci Hortic 327:112800

Narayanswami B, Tomar BS, Akhtar J, Behera TK, Ellur RK, Sevanthi RMV, Saha P (2023) Genetic analysis and identification of SSR marker linked to Phomopsis blight resistance in eggplant (*Solanum melongena* L.). Plant Breed 142:700–710

Ning Y, Wei K, Li S, Zhang L, Chen Z, Lu F, Yang P, Yang M, Liu X, Liu X, Wang X (2023) Fine mapping of fw6.3, a major-effect quantitative trait locus that controls fruit weight in tomato. Plants 12(11):2065. https://doi.org/10.3390/plants12112065

Niu ZH, Song XF, Li XL, Guo XY, He SQ, He LJZ (2020) Inheritance and QTL mapping for parthenocarpy in cucumber. Sci Agric Sin 53:160–171. https://doi.org/10.3864/j.issn.0578-1752.2020.01.015

Ohara T, Song YS, Tsukazaki H, Wako T, Nunome T, Kojima A (2005) Construction of a genetic linkage map of Japanese bunching onion (*Allium fistulosum*) based on AFLP and SSR markers. Euphytica 144:255–263

Oren E, Tzuri G, Dafna A, Rees ER, Song B, Freilich S, Elkind Y, Isaacson T, Schaffer AA, Tadmor Y, Burger J (2022) QTL mapping and genomic analyses of earliness and fruit ripening traits in a melon recombinant inbred lines population supported by de novo assembly of their parental genomes. Hortic Res 9:uhab081

Osae BA, Liu S, Amanullah S, Gao P, Fan C, Luan F (2021) QTL mapping for significant seed traits of watermelon (*Citrullus lanatus* Schrad.). Pak J Bot 53(3):923–931. https://doi.org/10.30848/PJB2021-3(24)

Pal S, Rao ES, Reddy DCL, Dayanandhi E (2023) QTL-seq identifies and QTL mapping refines genomic regions conferring resistance to *Fusarium oxysporum* f. sp. *niveum* race 2 in cultivated watermelon [*Citrullus lanatus* (Thunb.) Matsum & Nakai]. Sci Hortic 319:112180. https://doi.org/10.1016/j.scienta.2023.112180

Panthee DR, Piotrowski A, Ibrahem R (2017) Mapping quantitative trait loci (QTL) for resistance to late blight in tomato. Int J Mol Sci 18(7):1589. https://doi.org/10.3390/ijms18071589

Pavan S, Marcotrigiano AR, Ciani E, Mazzeo R, Zonno V, Ruggieri V, Lotti C, Ricciardi L (2017) Genotyping-by-sequencing of a melon (*Cucumis melo* L.) germplasm collection from a secondary center of diversity highlights patterns of genetic variation and genomic features of different gene pools. BMC Genomics 18:59. https://doi.org/10.1186/s12864-016-3429-0

Pekosz A, He B, Lamb RA (1999) Reverse genetics of negative-strand RNA viruses: closing the circle. PNAS 96(16):8804–8806

Perchepied L, Bardin M, Dogimont C, Pitrat M (2005) Relationship between loci conferring downy mildew and powdery mildew resistance in melon assessed by quantitative trait loci mapping. Phytopathology 95:556–565

Ponnam N, Reddy M, Reddy K, Reddy AC (2023) Molecular mapping of *Chilli veinal mottle virus* (*ChiVMV*) resistance in hot pepper (*Capsicum annuum* L.). Plant Gene 33:100396

Pradeepkumara N, Sharma PK, Munshi AD, Behera TK, Bhatia R, Kumari K, Dey SS (2022) Fruit transcriptional profiling of the contrasting genotypes for shelf life reveals the key candidate genes and molecular pathways regulating post-harvest biology in cucumber. Genomics 114:110273

Ramos A, Fu Y, Michael V, Meru G (2020) QTL-seq for identification of loci associated with resistance to Phytophthora crown rot in squash. Sci Rep 10(1):5326. https://doi.org/10.1038/s41598-020-62228-z

Rao PG, Behera TK, Gaikwad AB, Munshi AD, Srivastava A, Boopalakrishnan G, Vinod (2021) Genetic analysis and QTL mapping of yield and fruit traits in bitter gourd (*Momordica charantia* L.). Sci Rep 11(1):4109. https://doi.org/10.1038/s41598-021-83548-8

Razi H, Howell EC, Newbury HJ, Kearsey MJ (2008) Does sequence polymorphism of FLC paralogues underlie flowering time QTL in *Brassica oleracea*? Theor Appl Genet 116:179–192. https://doi.org/10.1007/s00122-007-0657-3

Razzaq A, Zafar MM, Li P, Qun G, Deng X, Ali A, Yuan Y (2021) Transformation and overexpression of primary cell wall synthesis-related zinc finger gene *Gh_A07G1537* to improve fiber length in cotton. Front Plant Sci 12:777794

Ridge S, Brown PH, Hecht V, Driessen RG, Weller JL (2015) The role of *BoFLC2* in cauliflower (*Brassica oleracea* var. *botrytis* L.) reproductive development. J Exp Bot 66:125–135. https://doi.org/10.1093/jxb/eru408

Rosa-Martínez E, Villanueva G, Şahin A, Gramazio P, García-Martínez MD, Raign MD, Plazas M (2023) Characterization and QTL identification in eggplant introgression lines under two N fertilization levels. Hortic Plant J 9:971–985

Rosental L, Still DW, You Y, Hayes RJ, Simko I (2021) Mapping and identification of genetic loci affecting earliness of bolting and flowering in lettuce. Theor Appl Genet 134:3319–3337

Roth F, Briggs WH, van der Ploeg S, Galli Z (2021) Genetic mapping of QTLs associated with resistance against onion thrips and related morphological characteristics in white cabbage. Int J Agric Biol 26:935–943

Ruangrak E, Su X, Huang Z, Wang X, Guo Y, Du Y, Gao J (2018) Fine mapping of a major QTL controlling early flowering in tomato using QTL-seq. Can J Plant Sci 98(3):672–682. https://doi.org/10.1139/cjps-2016-0398

Ruangrak E, Du Y, Htwe NM, Pimorat P, Gao J (2019) Identification of early tomato fruit ripening loci by QTL-seq. J Agric Sci 11(2):51. https://doi.org/10.5539/jas.v11n2p51

Saeko A, Lithanatudom SK, Struss D, Pongjaroenkit S (2020) QTL mapping specific to *Thrips palmi* resistance in *Capsicum annuum*. Asia-Pac J Sci Technol 25(3):1–8

Saraiva T (2018) Fascist pigs: technoscientific organisms and the history of fascism. MIT Press

Schneider K (2005) Mapping populations and principles of genetic mapping. In: Meksem K, Kahl G (eds) The handbook of plant genome mapping, genetic and physical mapping. WILEY-VCH Verlag GmbH & Co KGaA, Weinheim, pp 3–21

Shakya PR, Khwaounjoo NM (2013) Heavy metal contamination in green leafy vegetables collected from different market sites of Kathmandu and their associated health risks. Sci World 11:37–42

Shang J, Kong S, Li N, Wang J, Zhou D, Li N, Ma S (2020) Genetic mapping and localization of major QTL for bitterness in melon (*Cucumis melo* L.). Sci Hortic 266:109286. https://doi.org/10.1016/j.scienta.2020.109286

Sheng X, Cai S, Shen Y, Yu H, Wang J, Qiao S, Lin F, Gu H (2024) QTL analysis and fine mapping of a major QTL and identification of candidate genes controlling curd setting height in cauliflower. Veg Res 4(1):10.48130/vegres-0024-0002

Shin IS, Hsu JC, Huang SM, Chen JR, Wang JF, Hanson P, Schafleitner R (2020) Construction of a single nucleotide polymorphism marker-based QTL map and validation of resistance loci to bacterial wilt caused by *Ralstonia solanacearum* species complex in tomato. Euphytica 216:1–20. https://doi.org/10.1007/s10681-020-2576-1

Shu H, He C, Mumtaz MA, Hao Y, Zhou Y, Jin W, Wang Z (2023) Fine mapping and identification of candidate genes for fruit color in pepper (*Capsicum chinense*). Sci Hortic 310:111724

Shwetha A, Gunnaiah R, Basavaraja N, Krishna KA, UdayKumar HR, Sameer S (2023) Discovery of an SSR marker linked to Yellow Vein Mosaic Virus resistance in Okra (*Abelmoschus esculentus* L. Moench). Genet Resour Crop Evol:1–12

Siddique MI, Lee HY, Ro NY, Han K, Venkatesh J, Solomon AM, Patil AS, Changkwian A, Kwon JK, Kang BC (2019) Identifying candidate genes for *Phytophthora capsici* resistance in pepper (*Capsicum annuum*) via genotyping-by-sequencing-based QTL mapping and genome-wide association study. Sci Rep 9(1):9962. https://doi.org/10.1038/s41598-019-46342-1

Siddique MI, Lee JH, Ahn JH, Kusumawardhani MK, Safitri R et al (2022) Genotyping-by-sequencing-based QTL mapping reveals novel loci for Pepper yellow leaf curl virus (PepYLCV) resistance in *Capsicum annuum*. PLoS ONE 17(2):e0264026. https://doi.org/10.1371/journal.pone.0264026

Siddique MI, Silverman E, Louws F, Panthee DR (2024) Quantitative trait loci mapping for bacterial wilt resistance and plant height in tomatoes. Plants 13(6):876. https://doi.org/10.3390/plants13060876

Simko I, Rauscher G, Sideman RG, McCreight JD, Hayes RJ (2014) Evaluation and QTL mapping of resistance to powdery mildew in lettuce. Plant Pathol 63:344–353

Singh BD (2009) Genetics, 2nd edn. Kalyani Publishers, New Delhi

Song YS, Suwabe K, Wako T, Ohara T, Nunome T, Kojima A (2004) Development of microsatellite markers in bunching onion (*Allium fistulosum* L.). Breed Sci 54:361–365

Song H, Huang Y, Gu B (2019) QTL-seq identifies QTL for relative electrical conductivity associated with heat tolerance in bottle gourd (*Lagenaria Siceraria* L.). bioRxiv:2019–2012

Song H, Huang Y, Gu B (2020) QTL-Seq identifies quantitative trait loci of relative electrical conductivity associated with heat tolerance in bottle gourd (*Lagenaria siceraria*). PLoS One 15:e0227663

Sun J, Gaowen X, Yudie H, Yujing C, Xiaoqiu W, Jing Y, Meng Y, Dongwei X, Zhigang D (2023) Genome-wide assessment of genetic diversity and association mapping for salt tolerance traits in okra (*Abelmoschus esculentus* L. Moench) using genotyping-by-sequencing. Sci Hortic 313:111922

Sun J, Nie J, Xiao T, Guo C, Lv D, Zhang K, Wang G (2024) *CsPM5. 2* a phosphate transporter protein-like gene, promotes powdery mildew resistance in cucumber. Plant J 117:1487–1502

Tan H, Wang X, Fei Z, Li H, Tadmor Y, Mazourek M, Li L (2020) Genetic mapping of green curd gene Gr in cauliflower. Theor Appl Genet 133:353–364

Tao J, Li S, Wang Q, Yuan Y, Ma J, Xu M, Sun Y (2022) Construction of a high-density genetic map based on specific-locus amplified fragment sequencing and identification of loci controlling anthocyanin pigmentation in Yunnan red radish. Hortic Res 9:uhab031. https://doi.org/10.1093/hr/uhab031

Tassone MR, Bagnaresi P, Desiderio F, Bassolino L, Barchi L, Florio FE, Sunseri F, Sirangelo TM, Rotino GL, Toppino L (2022) A genomic BSAseq approach for the characterization of QTLs underlying resistance to *Fusarium oxysporum* in eggplant. Cells 11(16):2548. https://doi.org/10.3390/cells11162548

Tierney MB, Lamour KH (2005) An introduction to reverse genetic tools for investigating gene function. Plant Health Instruct. https://doi.org/10.1094/PHI-A-2005-1025-01

Tikunov YM, Roohanitaziani R, Meijer-Dekens F, Molthoff J, Paulo J, Finkers R, Capel I, Carvajal Moreno F, Maliepaard C, Nijenhuis-de Vries M, Labrie CW (2020) The genetic and functional analysis of flavor in commercial tomato: the *FLORAL4* gene underlies a QTL for floral aroma volatiles in tomato fruit. Plant J 103(3):1189–1204. https://doi.org/10.1111/tpj.14795

Topcu Y, Sapkota M, Illa-Berenguer E, Nambeesan SU, van der Knaap E (2021) Identification of blossom-end rot loci using joint QTL-seq and linkage-based QTL mapping in tomato. Theor Appl Genet 134(9):2931–2945. https://doi.org/10.1007/s00122-021-03869-0

Toporek SM, Branham SE, Katawczik ML, Keinath AP, Patrick Wechter W (2021) QTL mapping of resistance to *Pseudoperonospora cubensis clade* 1, mating type A2, in *Cucumis melo*. Theor Appl Genet 134:577–2586

Toppino L, Barchi L, Mercati F, Acciarri N, Perrone D, Martina M, Gattolin S, Sala T, Fadda S, Mauceri A, Ciriaci T (2020) A new intra-specific and high-resolution genetic map of eggplant based on a RIL population, and location of QTLs related to plant anthocyanin pigmentation and seed vigour. Gene 11(7):745. https://doi.org/10.3390/genes11070745

Truco MJ, Antonise R, Lavelle D, Ochoa O, Kozik A, Witsenboer H, Peleman J (2007) A high-density, integrated genetic linkage map of lettuce (*Lactuca spp.*). Theor Appl Genet 115:735–746

Tsukazaki H, Nunome T, Fukuoka H, Kanamori H, Kono I, Yamashita K, Wako T, Kojima A (2007) Isolation of 1,796 SSR clones from SSR-enriched DNA libraries of bunching onion (*Allium fistulosum*). Euphytica 157:83–94

Tsukazaki H, Yamashita K, Yaguchi S, Masuzaki S, Fukuoka H, Yonemaru J, Kanamori H, Kono I, Tran Thi Minh H, Shigyo M, Kojima A, Wako T (2008) Construction of SSRbased chromosome map in bunching onion (*Allium fistulosum*). Theor Appl Genet 117:1213–1223

Tsukazaki H, Yamashita K, Yaguchi S, Yamashita K, Hagihara T, Shigyo M, Kojima A, Wako T (2011) Direct determination of the chromosomal location of bunching onion and bulb onion markers using bunching onion-shallot monosomic additions and allotriploid-bunching onion single alien deletions. Theor Appl Genet 122:501–510

Tsukazaki H, Yaguchi S, Yamashita KI, Shigyo M, Kojima A, Wako T (2012) QTL analysis for pseudostem pungency in bunching onion (*Allium fistulosum*). Mol Breed 30:1689–1698

Turner SD, Ellison SL, Senalik DA, Simon PW, Spalding EP, Miller ND (2018) An automated image analysis pipeline enables genetic studies of shoot and root morphology in carrot (*Daucus carota* L.). Front Plant Sci 9:1703. https://doi.org/10.3389/fpls.2018.01703

Ubayasena L, Bett K, Taran B, Warkentin T (2011) Genetic control and identification of QTLs associated with visual quality traits of field pea (*Pisum sativum* L.). Genome 54:261–272

Van der Knaap E, Chakrabarti M, Chu YH, Clevenger JP, Illa-Berenguer E, Huang Z, Keyhaninejad N et al (2014) What lies beyond the eye: the molecular mechanisms regulating tomato fruit weight and shape. Front Plant Sci 5:89256. https://doi.org/10.3389/fpls.2014.00227

Vazquez DV, da Costa JH, Godoy FN, Cambiaso V, Rodriguez GR (2022) Genetic basis of the lobedness degree in tomato fruit morphology. Plant Sci 319:111258. https://doi.org/10.1016/j.plantsci.2022.111258

Vega A, Brainard SH, Goldman IL (2024) Linkage mapping of root shape traits in two carrot populations. G3 Genes Genomes Genetics:jkae041. https://doi.org/10.1093/g3journal/jkae041

Verhoeven KJF, Jannink JL, McIntyre LM (2006) Using mating designs to uncover QTL and the genetic architecture of complex traits. Heredity 96:139–149. https://doi.org/10.1038/sj.hdy.6800763

Vinay ND, Matsumura H, Munshi AD, Ellur RK, Chinnusamy V, Singh A, Behera TK (2023) Molecular mapping of genomic regions and identification of possible candidate genes associated with gynoecious sex expression in bitter gourd. Front Plant Sci 14:1071648

Wako T, Tsukazaki H, Yaguchi S, Yamashita KI, Ito SI, Shigyo M (2016) Mapping of quantitative trait loci for bolting time in bunching onion (*Allium fistulosum* L.). Euphytica 209:537–546

Wang DL, Zhu J, Li ZK, Paterson AH (1999) Mapping QTLs with epistatic effects and genotype x environment interactions by mixed linear model approaches. Theor Appl Genet 99:1255–1264

Wang Q, Zhang Y, Zhang L (2018) A naturally occurring insertion in the *RsFLC2* gene associated with late-bolting trait in radish (*Raphanus sativus* L.). Mol Breed 38:1–12. https://doi.org/10.1007/s11032-018-0897-8

Wang Y, Jiali Y, Yang Z, Liang X, Wanting Z, Meng N, Yuelin Z, Liwang L (2020) Genome-wide identification and functional characterization of the cation proton antiporter (CPA) family related to salt stress response in radish (Raphanus sativus L.). Int J Mol Sci 21:8262. https://doi.org/10.3390/ijms21218262

Wang Y, Fang Y, Ning S, Xia L, Zhan J, Yang Z, Chen J (2023a) QTL mapping for ovary-and fruit-related traits in *Cucumis sativus-c. hystrix* introgression line IL52. Gene 14:1133

Wang Y, Wang G, Lin D, Luo Q, Xu W, Qu S (2023b) QTL mapping and stability analysis of trichome density in zucchini (*Cucurbita pepo* L.). Front Plant Sci 14:1232154. https://doi.org/10.3389/fpls.2023.1232154

Watcharatpong P, Kaga A, Chen X, Somta P (2020) Narrowing down a major QTL region conferring pod fiber contents in yardlong bean (*Vigna unguiculata*), a vegetable cowpea. Gene 11(4):363. https://doi.org/10.3390/genes11040363

Wei Q, Wang W, Hu T, Hu H, Wang J, Bao C (2020a) Construction of a SNP-based genetic map using SLAF-Seq and QTL analysis of morphological traits in eggplant. Front Genet 11:178. https://doi.org/10.3389/fgene.2020.00178

Wei J, Li J, Yu J, Cheng Y, Ruan M, Ye Q, Yao Z, Wang R, Zhou G, Deng M, Wan H (2020b) Construction of high-density bin map and QTL mapping of horticultural traits from an interspecific cross between *Capsicum annuum* and Chinese wild *Capsicum frutescens*. Biotechnol Biotechnol Equip 34(1):549–561. https://doi.org/10.1080/13102818.2020.1787863

Wen J, Jiang F, Weng Y, Sun M, Shi X, Zhou Y, Yu L, Wu Z (2019) Identification of heat-tolerance QTLs and high-temperature stress-responsive genes through conventional QTL mapping, QTL-seq and RNA-seq in tomato. BMC Plant Biol 19:1–17. https://doi.org/10.1186/s12870-019-2008-3

Wheeler T, Von Braun J (2013) Climate change impacts on global food security. Science 341:508–513. https://doi.org/10.1126/science.1239402

White TL, Adams WT, Neale DB (2007) Forest genetics. CABI, 682p

Wright S (1952) Quantitative genetics. Her Majesty's Stationery Office, London

Wu Z, Zhang T, Li L, Xu J, Qin X, Zhang T (2016) Identification of a stable major-effect QTL (*Parth 2.1*) controlling parthenocarpy in cucumber and associated candidate gene analysis via whole genome re-sequencing. BMC Plant Biol:16–182. https://doi.org/10.1186/s12870-016-0873-6

Wu C, Qiu Y, Duan Y, Guo Y, Wang H, Zhang X, Song J, Li X (2020) Construction of high-density linkage map and identification of QTLs associated with resistance to black rot in radish (*Raphanus sativus*) by RAD sequencing. Plant Breed 139(3):660–671. https://doi.org/10.1111/pbr.12794

Wu L, Fredua-Agyeman R, Hwang SF, Chang KF, Conner RL, McLaren DL, Strelkov SE (2021a) Mapping QTL associated with partial resistance to Aphanomyces root rot in pea (Pisum sativum L.) using a 13.2 K SNP array and SSR markers. Theor Appl Genet 134:2965–2990

Wu X, Chen F, Zhao X, Pang C, Shi R, Liu C, Sun C, Zhang W, Wang X, Zhang J (2021b) QTL mapping and GWAS reveal the genetic mechanism controlling soluble solids content in *Brassica napus* shoots. Foods 10(10):2400. https://doi.org/10.3390/foods10102400

Wu Y, Zhang S, Zhang H, Li F, Li G, Fan C, Sun R, Zhang S (2021c) QTL mapping and candidate gene identification of swollen root formation in turnip. Int J Mol Sci 22(2):653. https://doi.org/10.3390/ijms22020653

Wu Y, Popovsky-Sarid S, Tikunov Y, Borovsky Y, Baruch K, Visser RG, Bovy A (2023) CaMYB12-like underlies a major QTL for flavonoid content in pepper (*Capsicum annuum*) fruit. New Phytol 237:2255–2267

Wu Q, Mao S, Huang H, Liu J, Chen X, Hou L, Huang K (2024) Chromosome-scale reference genome of broccoli (*Brassica oleracea var. italica* Plenck) provides insights into glucosinolate biosynthesis. Hortic Res:uhae063

Xing Y, Cao Y, Ma Y, Wang F, Xin S, Zhu W (2023) QTL mapping and transcriptomic analysis of fruit length in cucumber. Front Plant Sci 14:1208675. https://doi.org/10.3389/fpls.2023.1208675

Xu G, Cheng Y, Wang X, Dai Z, Kang Z, Ye Z, Sun J (2024) Identification of single nucleotide polymorphic loci and candidate genes for seed germination percentage in okra under salt and no-salt stresses by genome-wide association study. Plants 13:588

Yagcioglu M, Jiang B, Wang P, Wang Y, Ellialtioglu SS, Weng Y (2019) QTL mapping of low temperature germination ability in cucumber. Euphytica 215:1–11. https://doi.org/10.1007/s10681-019-2408-3

Yang T, Amanullah S, Pan J, Chen G, Liu S, Ma S, Wang X (2021) Identification of putative genetic regions for watermelon rind hardness and related traits by BSA-seq and QTL mapping. Euphytica 217:1–18

Yeo SM, Hong J, Hossain MR, Jung HJ, Choe P, Nou IS (2022) Genotyping by sequencing (GBS)-based QTL mapping for bacterial fruit blotch (BFB) in watermelon. Gene 13(12):2250. https://doi.org/10.3390/genes13122250

You Q, Li H, Wu J, Li T, Wang Y, Sun G, Sun B (2023) Mapping and validation of the epistatic D and P genes controlling anthocyanin biosynthesis in the peel of eggplant (*Solanum melongena* L.) fruit. Hortic Res 10:uhac268

Yu J, Holland JB, McMullen MD et al (2008) Genetic design and statistical power of nested association. Genetics 178:539–551

Yuste-Lisbona FJ, Capel C, Sarria E, Torreblanca R, Gómez-Guillamón ML, Capel J, López-Sesé AI (2011) Genetic linkage map of melon (*Cucumis melo* L.) and localization of a major QTL for powdery mildew resistance. Mol Breed 27:181–192

Zafar MM, Chattha WS, Khan AI, Zafar S, Anwar Z, Seleiman MF, Razzaq A (2023) Drought and heat stress on cotton genotypes suggested agro-physiological and biochemical features for climate resilience. Front Plant Sci 14:1265700

Zeng ZB (1994) The precision mapping of quantitative trait loci. Genetics 136:1457–1468

Zhang SP, Liu MM, Miao H, Zhang SQ, Yang YH, Xie BY, Gu XF (2013) Chromosomal mapping and QTL analysis of resistance to downy mildew in *Cucumis sativus*. Plant Dis 97:245–251

Zhang T, Li X, Yang Y, Guo X, Feng Q, Dong X, Chen S (2019a) Genetic analysis and QTL mapping of fruit length and diameter in a cucumber (*Cucumber sativus* L.) recombinant inbred line (RIL) population. Sci Hortic 250:214–222. https://doi.org/10.1016/j.scienta.2019.01.062

Zhang XF, Wang GY, Dong TT, Chen B, Du HS, Li CB, Zhang FL, Zhang HY, Xu Y, Wang Q, Geng SS (2019b) High-density genetic map construction and QTL mapping of first flower node in pepper (*Capsicum annuum* L.). BMC Plant Biol 19:1–13. https://doi.org/10.1186/s12870-019-1753-7

Zhang T, Ding Z, Liu J, Qiu B, Gao P (2020) QTL mapping of pericarp and fruit-related traits in melon (*Cucumis melo* L.) using SNP-derived CAPS markers. Sci Hortic 265:109243. https://doi.org/10.1016/j.scienta.2020.109243

Zhang X, Ling Y, Yang W, Wei M, Wang Z, Li M, Yang Y, Liu B, Yi H, Guo YD, Kong Q (2023a) Fine mapping of a novel QTL *DM9.1* conferring downy mildew resistance in melon. Front Plant Sci 14:1202775. https://doi.org/10.3389/fpls.2023.1202775

Zhang Z, Wang K, Chen C, Tian S, Wu J, Li J, Xiao D (2023b) Transcriptome sequence analysis of defense response of resistant and susceptible bottle gourd to powdery mildew. Agronomy 13:1406

Zhang X, Han F, Li Z, Wen Z, Cheng W, Shan X, Liu Y (2024) Map-based cloning and functional analysis of a major quantitative trait locus, BolC. Pb9. 1, controlling clubroot resistance in a wild Brassica relative (*Brassica macrocarpa*). Theor Appl Genet 137:41

Zhao Z, Sheng X, Yu H, Wang J, Shen Y, Gu H (2020a) Identification of candidate genes involved in curd riceyness in cauliflower. Int J Mol Sci 21:1999

Zhao ZQ, Sheng XG, Yu HF, Wang JS, Shen YS, Gu HH (2020b) Identification of QTLs associated with curd architecture in cauliflower. BMC Plant Biol 20:1–8. https://doi.org/10.1186/s12870-020-02377-5

Zhao H, Zhang T, Meng X, Song J, Zhang C, Gao P (2023) Genetic mapping and QTL analysis of fruit traits in melon (*Cucumis melo* L.). Curr Issues Mol Biol 45(4):3419–3433. https://doi.org/10.3390/cimb45040224

Zheng Y, Xu F, Li Q, Wang G, Liu N, Gong Y, Xu S (2018) QTL mapping combined with bulked segregant analysis identify SNP markers linked to leaf shape traits in Pisum sativum using SLAF sequencing. Front Genet 9:426389

Zhong J, Cui J, Liu J, Zhong C, Hu F, Dong J, Hu K (2023) Fine-mapping and candidate gene analysis of the Mcgy1 locus responsible for gynoecy in bitter gourd *(Momordica spp.)*. Theor Appl Genet 136:81

Zhu Z, Sun B, Wei J, Cai W, Huang Z, Chen C, Cao B, Chen G, Lei J (2019) Construction of a high-density genetic map of an interspecific cross of *Capsicum chinense* and *Capsicum annuum* and QTL analysis of floral traits. Sci Rep 9(1):1054. https://doi.org/10.1038/s41598-018-38370-0

Zhu X, Tai X, Ren Y, Wang M, Chen J, Bo T (2022a) QTL-seq and marker development for resistance to head splitting in cabbage. Euphytica 218(4):41. https://doi.org/10.1007/s10681-022-02982-5

Zhu P, Meng Y, Zhang K, Wang X, Liang K, Wang T, Xu J, Qin X, Wu Z, Cheng C, Yu X (2022b) Mapping of fruit apex shape related QTLs across multi-genetic backgrounds in cucumber (*Cucumis sativus* L.). Hortic Plant J 8(3):328–340. https://doi.org/10.1016/j.hpj.2021.12.001

Tools and Techniques Used for Quantitative Trait Loci (QTL) and Association Mapping Studies

2

Babita Joshi, Sanjay Singh, Mir Asif Iquebal, Sarika, Dinesh Kumar, Samir V. Sawant, and Satya Narayan Jena

Abstract

Quantitative traits are measurable phenotypic traits influenced by the combined effects of multiple genes and environments, leading to phenotypic variation among individuals. Linking phenotypes and genotypes using statistical techniques to underline the genetic factors that regulate the quantitative variation of complex traits with agricultural and evolutionary significance is crucial to crop improvement for plant breeders. Most of the agronomically important traits show quantitative inheritance, which is regulated by quantitative trait loci (QTL), environment, and the interaction between environmental factors and QTL. At present, QTL mapping and association mapping are the two most frequently utilized tools for interpreting the relationship between genes and complex traits. Both techniques emerge as an indispensable tool in the breeder's toolkit, a beacon of hope in the quest to developing resilient, high-yielding, and quality crop varieties capable of sustaining the growing global population. Successful QTL and association mapping rely on the precision of marker-trait association, which largely

B. Joshi · S. N. Jena (✉)
Plant Genetic Resources and Improvement, CSIR-National Botanical Research Institute, Lucknow, Uttar Pradesh, India

Academy of Scientific and Innovative Research, Ghaziabad, Uttar Pradesh, India
e-mail: satyanarayan.nbri@csir.res.in

S. Singh · M. Asif Iquebal · Sarika · D. Kumar
Division of Agricultural Bioinformatics, ICAR-Indian Agricultural Statistics Research Institute, New Delhi, India

S. V. Sawant
Molecular Biology and Biotechnology, CSIR-National Botanical Research Institute, Lucknow, Uttar Pradesh, India

Academy of Scientific and Innovative Research, Ghaziabad, Uttar Pradesh, India

S. S. Dey et al. (eds.), *GWAS and QTL Mapping in Horticultural Crops*,
https://doi.org/10.1007/978-981-95-2306-1_2

depends on the marker density, its quality, and accuracy in the phenotypic data. Recent advancements in genomic technology and the development of robust statistical analysis methods have made QTL and association mapping an attractive and cost-effective approach for plant research programs. Increased marker density supplemented with accurate and precise phenotyping using high-throughput tools and techniques improves the power of QTL and significant marker identification. This chapter gives an overview of the various tools and techniques used in QTL and association mapping studies. We have discussed comprehensive details on the development of various mapping populations, recent tools and techniques for high-throughput phenotyping and genotyping, and different statistical methods and software employed in molecular plant breeding research.

Keywords

QTL mapping · Mapping population · GWAS · High throughput phenotyping · Genotyping · Linkage disequilibrium · Population structure

2.1 Introduction

The structure and functions of all living organisms are controlled by genes composed of nucleotide sequences in the form of DNA or RNA, which pass down genetic information from one generation to the next. Phenotypic variation in many quantitative traits of agricultural importance crops is governed by networks of multiple genes and interactions between genetic and environmental factors. Considering the present condition, many challenges, such as biotic and abiotic stress, and climate change, has adversely affected the quality and production of many important crop plants. With the growing population and complex climate change, plant breeders should increase the yield potential and maintain the quality of the crop plants to sustain the food requirements in the future. Thus, improving the productivity while maintaining the quality and stability in crop plants is quite a challenging task for plant breeders. Conventional breeding has played a major role in crop improvement: e.g., through conventional breeding many rice varieties (KDML105, Minghui 63, IR5, IR8, IR20) were produced against blast disease (Korinsaka et al. 2011; Peng and Khush 2003) and tomato varieties (Pusa Ruby, Arka Vikas, Roma, Sioux, Rakshita, NDT 5), seedless watermelon, apriums, pluots, and tangelos are the products of conventional breeding techniques (Das et al. 2011). However, in conventional breeding, selection depends on the phenotypic data which are less heritable and easily influenced by the environment. Also, improvement of complex or quantitative traits is difficult due to genetic correlation and environmental influence and conventional breeding requires a long period of about 10 years to develop varieties (Falconer 1996). Thus, the genetic improvement of crops using traditional breeding techniques for such quantitative traits is a difficult task. However, the advancement of high-throughput sequencing and genotyping technology have significantly lowered the cost of identifying molecular markers (especially single-nucleotide polymorphisms [SNPs]); plant geneticists have shown a growing interest in utilizing high-resolution marker-based tools for investigating complex quantitative traits (Zhu et al. 2008). The two most frequently utilized tools for dissecting complex traits are quantitative trait loci (QTL) mapping, which is based on linkage analysis, and association mapping, which is based on linkage disequilibrium (Fig. 2.1). The key difference between association and QTL mapping is that, QTL mapping depends on recombination occurring infrequently within families and pedigrees with known ancestry, leading to relatively low mapping resolution. In contrast, association mapping leverages historical recombination and naturally diverse genetic populations, allowing for high-resolution mapping. Both QTL and association mapping studies depend on the inheritance of functional polymorphisms and DNA variants.

The main aim of this chapter is to highlight comprehensive details on the type of mapping population, high-throughput phenotyping (HTP), genotyping techniques, various statistical methods, and software tools used in QTL mapping and marker-trait association studies. The content of this chapter will guide the new aspiring breeders in understanding how to analyze and utilize natural variations for crop improvement.

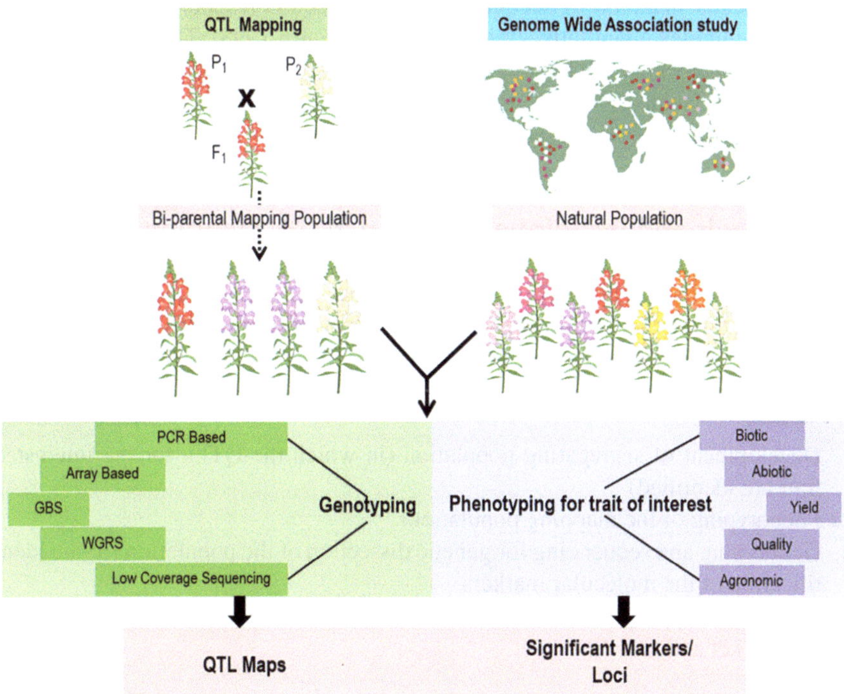

Fig. 2.1 Schematic representation of QTL and GWAS for dissecting complex traits

2.2 QTL Mapping

Quantitative traits are measurable phenotypic traits that are influenced by the combined effects of multiple genes and environments, leading to phenotypic variation among individuals. Quantitative traits are complex traits that differ from monogenic traits because, in this case, many genes are involved together to produce traits and usually it does not behave according to the simple Mendelian concept of inheritance. Chromosomal segments which contain genetic factors/genes that are responsible for variation in these complex quantitative traits are known as quantitative trait loci (QTL), a termed coined by Geldermann (1975). In short, QTL may be a single gene or a network of linked genes in chromosomal segments that affect a particular trait. According to Semagn et al. (2010), for quantitative traits, the genotype of particular individuals is difficult to infer from its phenotype due to the following factors: (1) the behavior of plants having same QTLs may vary under different environmental conditions and locations; (2) different alleles at many QTLs may be present for plants showing similar phenotypes; and (3) the effects of one QTL may influence the allelic combination of another QTLs of the plants. The basic concept of QTL detection was initially developed from the study conducted by Sax (1923) on the relationship between seed coat pattern and pigmentation in beans (*Phaseolus vulgaris*). QTL mapping, also known as linkage mapping, is a statistical analysis that links the phenotypic data (measurement for the trait of interest) to the genotypic data (molecular markers) to decipher the genetic factors responsible for the variation in the complex quantitative traits (Lynch and Walsh 1998). The primary goal of this analysis is to answer whether the phenotypic variation in complex traits among the population is due to few loci or many loci, how is the interaction between these loci, their action, and the precise location of these loci.

2.3 Tools and Techniques Involved in QTL Mapping

The primary emphasis in the domain of quantitative genetics has been on employing advanced statistical tools alongside biometric mathematical methods to explore the connections between genotypic and phenotypic diversity among individuals. To begin with the QTL mapping, information on the following factors is required:

1. Development of segregating population (in which the QTLs for the interested trait are identified)
2. Phenotyping of the mapping population
3. Genotyping and sequencing for genetic dissection of the population for the identification of the molecular markers
4. Software or appropriate statistical packages for the detection of linkage between the marker and trait

2.3.1 Mapping Population

The key step in QTL mapping is establishing a mapping population derived from parents with distinct phenotypes for the traits of interest. Selecting genetically diverse parents with desirable economic and agronomic traits is essential, as it increases the possibility of identifying a wide range of polymorphic markers across the genome. However, it is important to strike a balance, as excessive genetic divergence can lead to progeny sterility and segregation distortion during linkage analysis. Breeders typically follow a standard process for creating a mapping population: selecting two genetically distinct parent lines with noticeable phenotypic differences in the target traits, crossing them to generate heterozygous F_1 individuals, and subsequently selfing the F_1 generation to obtain the mapping population. Several types of mapping populations such as F_2, restricted inbred lines (RILs), backcross population (BC), near-isogenic lines (NILs), and double haploids (DHs) could be utilized for QTL mapping analysis depending on the objectives of the study undertaken. Basically, mapping populations are categorized into two types: (a) mortal mapping population and (b) immortal mapping population (Fig. 2.2).

Mortal mapping population: This is a type of mapping population where the individuals cannot be maintained for a long period of time. They are typically used for short-term experiments (mostly for single experiments), and data is collected from the population over a specific period, where the genetic constitutions are not fixed. Some of the mortal mapping populations are as follows:

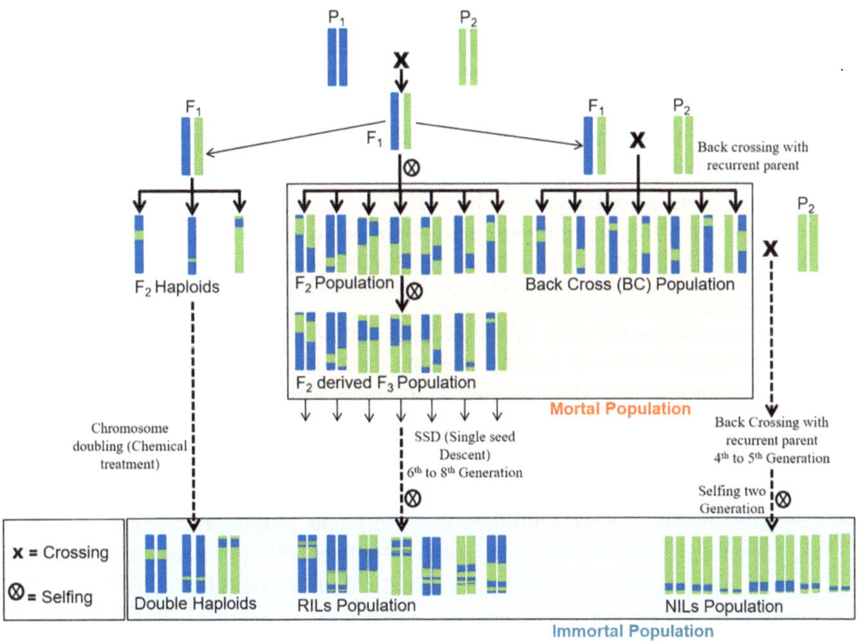

Fig. 2.2 Schematic representation of the development of different types of mapping populations

1. *F_2 population*: The simplest mapping population for genetic and QTL studies is the F_2 generation, produced by crossing two parental lines. Each F_2 individual carries a unique combination of parental linkage blocks, forming the basis for linkage detection. As the F_2 results from a single meiotic cycle (in the F_1 plants), recombination only occurs once between two loci which serves as a reference point for estimating recombination frequency. Dominant and co-dominant markers in F_2 exhibit expected ratios of 3:1 and 1:2:1, respectively. F_2 facilitates gene effect, epistasis, gene–environment interaction, identification of heterotic QTLs, and pleiotropy estimation, shedding light on a trait's genetic architecture and underlying control mechanisms that control the trait. However, the main constraint of this population is that it cannot be sustained beyond one generation except through asexual reproduction. Also, it is not possible to replicate an F_2 population precisely or to boost seed quantities for F_2 plant genotypes (Singh et al. 2015). In some cases, the F_2 population may show non-Mendelian segregation with distorted ratios. QTL mapping using F_2 mapping populations are reported in many plants such as cotton (Li et al. 2021), maize (Wang et al. 2018a, b, c), rice (Stephen et al. 2023), sugarcane (Gutierrez et al. 2018), and *Brassica napus* (Liu et al. 2022).
2. *F_2-derived F_3 population ($F_{2:3}$ population)*: Obtaining an F_2-derived F_3 or $F_{2:3}$ population involves single-generation selfing of F_2 individuals and separate seed collection from each F_2 plant, representing them individually. This approach is suitable for mapping recessive genes underlying the quantitative trait. $F_{2:3}$ populations have several limitations: (1) construction takes an extra season compared to F_2 populations; (2) most F_3 families are heterogeneous due to gene segregation, making it impossible to use genotypically identical $F_{2:3}$ plants as replicates; (3) the F_3 genotype and phenotype do not precisely match the F_2 generation due to additional segregation, recombination, and inbreeding; (4) the F_3 average phenotype is related but not strictly comparable to the parent F_2 plant; and (5) F_3 data tend to underestimate dominance, overdominance, and some epistatic gene components due to increased inbreeding (Hua et al. 2002). Some examples of QTL mapping using $F_{2:3}$ mapping population are maize (Guo et al. 2022), pigeon-pea (Kumawat et al. 2012), melon (Lv et al. 2022), tobacco (Hoshyardel et al. 2021), and soybean (Kumar et al. 2023a, b).
3. *Backcross (BC) population*: Backcross is a type of experimental population in QTL mapping that is created by crossing the hybrid F_1 individual with either of its recurrent parents. The resulting individuals from the cross will have half of their genetic material from the hybrid individual and half from the recurrent parent. However, for genetic analysis using a backcross population there should be detectable phenotypic segregation among the individuals; thus, by rule, the F_1 hybrid is mostly crossed with the recessive plants (having the recessive form of the target trait). Such a type of crossing (between F_1 and a recessive recurrent parent) is also called test cross and is usually denoted by B2 which exhibits a 1:1 ratio for trait phenotype. The backcross populations provide a particular benefit by enabling their subsequent utilization in marker-assisted backcrossing to introduce the desired traits, as outlined in the advanced backcross QTL method

(Tanksley and Nelson 1996). The two key characteristics of BC progenies are the presence of unlinked donor fragments (due to segregation) and the absence of linked donor fragments (through recombination) with the recurrent parent. To effectively diminish both the number and size of donor fragments, repeated backcrossing is performed. With each successive backcross, the proportion of the donor genome is reduced by half, achieving a 50% decrease in each iteration (Boopathi and Boopathi 2020). Backcrossing is a valuable method for dissecting the genetic architecture of quantitative traits by isolating genes or chromosomal regions in the recurrent parent's genetic background and validating the additive effects of identified QTLs or candidate genes. In addition, employing backcrossing before detection enhances the precision of QTL mapping and allows for the exploration of wild genetic resources. Introgressing a gene into a different genetic background eliminates or alters epistatic interactions, enabling the study of its additive effects. Combining introgression lines with different genes and backgrounds enables the investigation of epistatic interactions (Hospital 2005). Despite its numerous advantages, backcross mapping populations share similar drawbacks to those of the F_2 population. They are temporary, and in the case of backcrosses, recombination information relies on only one parent, making it challenging to distinguish between additive and dominance effects, and certain types of epistatic effects may become confounded (Zhu et al. 2012). QTL mapping using backcross mapping population was studied in many crops such as rice (Saxena et al. 2017), cotton (Yu et al. 2013; Ma et al. 2019), tomato (Fulop et al. 2016), and wheat (Buerstmayr et al. 2012; Kumar et al. 2023a, b).

Immortal mapping population: An immortal mapping population, on the other hand, consists of individuals or lines that can be maintained indefinitely over generations. These mapping populations can reproduce themselves without changes in the genetic constituents of an individual. Some of the immortal mapping populations are as follows:

1. Recombinant *inbred lines (RILs)*: RILs are homozygous lines resulting from continuous selfing/inbreeding of F_2 plants via single seed descent process (Burr et al. 1988), also commonly known as single seed descent (SSD)/F_2-derived inbred lines. The technique of linkage mapping using RILs was first pioneered in mice, involving approximately 20 generations of sib-mating to attain significant homozygosity levels (Schneider 2005). The use of RILs in QTL mapping offers several advantages, such as increased power and accuracy in detecting QTLs. It also aids in studying gene–gene interactions, the replication and validation of QTLs across environments, and the availability of a renewable resource for further genetic studies. RILs enable the creation of genetic maps with higher resolution compared to F_2 populations, allowing for the precise determination of map positions, even for closely linked markers. This is achievable due to the increased degree of recombination in RILs compared to F_2 populations. Although RILs have several advantages, they also have drawbacks such as reduced genetic diversity, resource-intensive establishment and maintenance, constraints in

studying gene–environment interactions, and difficulties in identifying minor-effect QTLs. QTL mapping using RIL population has been previously reported in several plant species (Leon et al. 2016; Hu et al. 2020; Shukla et al. 2021, Boopathi et al. 2022).

2. *Double haploid (DH) population*: DH lines are the plants achieved through the chromosome doubling of haploid plants containing two identical sets of chromosomes. They are typically derived from the culture of anthers or pollen grains produced by F_1 plants. Haploid lines can either spontaneously occur (e.g., in rapeseed and maize) or can be artificially induced. Artificially, double-haploid plants are primarily achieved by treating haploid plants with colchicine. Alternatively, they can also be induced by culturing immature anthers on specialized media, with the subsequent regeneration of haploid plants from the cells of the gametophyte. Seeds from the DH plants are collected separately and maintained in a similar way as that of RILs. DH plants exhibit homozygosity across all the loci of the genome. Doubled haploids have various advantages, including the acceleration of the breeding cycle through the immediate establishment of homozygosity and providing a high level of selection efficiency. Although doubled haploids offer distinct advantages, they also pose several limitations, including the high cost of haploid production, limited capacity for long-term maintenance through repeated passages, restricted applicability across populations, and complications arising from segregation distortion (Niemirowicz-Szczytt 1997). Despite this, DH is widely used as a resource for genetic mapping since they do not contain any residual heterozygosity. Some examples of DH utilized in QTL mapping are *Brassica* species (Aakanksha et al. 2021), wheat (Tehseen et al. 2022; Guo et al. 2023), and rice (Lee et al. 2020).
3. *Near-isogenic lines (NILs)*: NILs are pairs of homozygous lines that share an identical genotype, differing only at a single gene or locus. The development of NILs requires multiple generations of backcrossing. Molecular markers are mostly used during backcrossing to maximize the recovery of the recurrent genome. In addition, two extra rounds of self-fertilization are necessary at the end of the backcrossing process to stabilize the donor segments and observe traits influenced by recessive genes. Some advantages of using near-isogenic mapping populations are reduced genetic background variation, validation of a putative QTL, isolation of targeted genes, and identification of epistatic interactions. Although NILs offer significant advantages, they also have limitations, such as the potential for unintended genetic variations and the labor-intensive process of developing and maintaining these populations. Some of the QTL mapping studies using NILs in plants was reported by (Morales et al. 2020; Yaobin et al. 2018; Guan et al. 2019; Jiang et al. 2019).

The above-discussed mapping populations are the most widely used mortal and immortal mapping populations for QTL mapping. Besides these, other types of mapping populations are also used in the field of genetics and breeding such as backcross inbred lines, chromosomal segment substitution lines, recurrent selection backcross populations, and advanced inter-cross lines; however, reports on their uses for QTL mapping studies are scant.

Besides the types of mapping populations, the size of the mapping populations should also be considered. The population size plays a crucial role in quantitative trait mapping studies. It has been observed that using a relatively large population provides sufficient genetic variation, which is essential for identifying significant QTLs (Huang et al. 2017). However, the limited population sizes employed in numerous QTL detection experiments might have resulted in an underestimation of the number of QTLs, an overestimation of QTL effects, an increase in the false-positive rate, and the inability to quantify interactions among QTLs (Beavis 1998; Utz et al. 2000; Vales et al. 2005). In QTL mapping, the Beavis effect arises from population size, indicating that as the population size decreases, there is a reduction in the number of detected QTLs associated with a specific trait. According to Bernardo (2008), in order to minimize the Beavis effect, the population size used in the mapping study should be ≥500 individuals. In the majority of QTL experiments, the size of the population is constrained by the expenses associated with marker genotyping and/or trait phenotyping. The effect of the number of populations used for QTL identification was demonstrated by Bradshaw et al. (1995), where they examined the floral morphology of monkeyflower (*Mimulus* spp.) comprising 93 and 465 F_2 mapping populations. Studies using 93 individuals (smaller mapping population) identified 12 QTLs having large effects, while the studies conducted using 465 individuals (larger mapping population) identified 11 common QTLs (as in the case of 93 individuals) along with an addition of 16 more QTLs with smaller effects. Also, the common QTLs in both populations had smaller effects in the case of a larger mapping population, indicating that the magnitude of QTL effects tends to be overestimated in small populations. Population size ranging from 94 to 1500 (Atkinson et al. 2015; Karam et al. 2019; Sun et al. 2012; Wang et al. 2018a, b, c) has been previously reported for QTL mapping in many crop plants.

2.4 Phenotyping

The plant phenotype comprises all the biochemical, physiological, and morphological traits that reflect the structure, growth, and composition of a plant (Fiorani and Schurr 2013). It arises due to the interaction between the genotype and the environment which leads to plants productivity, performance, and temporal expression in a specific geographical region (Zhao 2019). Plant phenotyping is a set of methodologies or protocols that is used for the determination of quantitative or qualitative values for different morphological, biochemical, or physiological traits of cells, tissues, organs, plants, or a population. Phenotypic expression in plants represents a critical focus for plant breeders, as it emerges from the complex interactions between the plant's genome and the diverse micro- and macro-environmental factors encountered throughout its lifespan. Traditionally, plant breeders perform phenotyping artificially based on the weight, color, taste, appearance, size, number, and touch of the crop. However, this conventional phenotyping is often performed manually, which is undoubtedly time-consuming, costly, labor-intensive, inefficient, requires huge human resources for phenotyping large populations, prone to errors, and is mostly destructive (Lebedev et al. 2020). This limitation on phenotyping efficiency, particularly concerning the time required to acquire traits in diverse environmental conditions, has been the key constraint for the advancement in applied genetics (Guzman et al. 2015). Recently, with the advancement in genomic high-throughput technologies, traditional phenotyping methods have emerged as a significant bottleneck, impeding the ability to leverage these high-throughput molecular advancements fully. Thus, addressing this bottleneck has given rise to a growing and increasingly significant field within the biological sciences known as "phenomics" (Furbank and Tester 2011).

In recent years, significant progress has been achieved in the field of genome sequencing, greatly expediting advancements in functional genomics (Huang et al. 2010; Werner 2010). Thus, this progress has facilitated research on integrating genotyping and phenotyping to identify the genetic architecture governing economically important traits through various quantitative trait mapping methods contributing to crop improvement. Various high-throughput phenotyping platforms have been developed since 2000, and these are now widely used tools in both commercial and research settings (Granier and Vile 2014). As imaging sensors are becoming more sophisticated, capable, and compact, the imaging-based approach is rapidly emerging as the primary strategy for numerous phenotyping applications. The capacity for high-throughput phenotyping through image analysis has sparked interest in automated methods to quantify intricate traits, aiming to screen germplasm, enhance cultivation practices, and identify and quantify both biotic and abiotic stresses.

High-throughput phenotyping (HTP) systems are often defined as a platform capable of automating the collection of extensive phenotypic data from hundreds of plants on a daily basis and play a crucial role in advancing the understanding of phenomics and their genetic foundations (Fahlgren et al. 2015). HTP is a novel and potent tool enabling the non-destructive, rapid monitoring and quantification of plant growth and production-related traits providing valuable spatial and temporal information with accuracy and precision thereby exceeding manual phenotyping. With HTP, genomics-assisted breeding (GAB) by employing methods such as quantitative trait loci (QTL) mapping, genomic selection (GS), marker-assisted selection (MAS), genome-wide association studies (GWAS), and candidate gene association study has become fast, highly repeatable, and accurate, thereby assisting crop growers in adapting to evolving climate conditions and meeting market demands for increased yield.

The phenotyping platforms incorporate diverse imaging methodologies, automation, and new sensors to acquire high-throughput, nondestructive phenotype data. Depending on the sensor and imaging design, the phenotyping system can be categorized as sensor-to-plant, where the plants remain in a fixed position with the imaging setup moving (Skirycz et al. 2011), and plant-to-sensor in which plants are transported to an imaging station (Golzarian et al. 2011). This data is utilized for quantitative investigations of intricate traits, encompassing aspects such as growth, resistance, physiology, tolerance, yield, architecture, and the fundamental measurement of individual quantitative parameters that constitute the foundation of complex traits. Various imaging techniques, including visible light imaging/red–green–blue (RGB) imaging, fluorescence imaging, infrared imaging, hyperspectral imaging, and others, are employed to gather multilevel phenotype data, ranging from macroscopic to molecular scales, over durations spanning seconds to weeks (Sozzani et al. 2014). An overview of high-throughput phenotyping integrated with QTL and GWAS in plants using various techniques or platforms is given in Table 2.1.

Visible light/RGB imaging has found widespread use in plant science owing to its cost-effectiveness and simplicity. This imaging system generates growth profiles by capturing time-series data using cameras sensitive to the visible range of the electromagnetic spectrum (400–700 nm) (Tackenberg 2007; Golzarian et al. 2011). It enables the analysis of various phenotypic characteristics and the recording of changes in a plant's biomass through the capture of two-dimensional (2D) images. It represents the raw phenotype image data as spatial matrices, based on intensity values corresponding to photon fluxes within the visible light spectrum, including red (~600 nm), green (~550 nm), and blue (~450 nm). The configuration of imaging setups varies significantly based on the cultivation format. In the case of cereals, multiple view angles achieved by rotating the plants are commonly employed to minimize image occlusions (Golzarian et al. 2011). RGB imaging offers a rapid method for quantifying parameters related to greenness. In practical applications such as wheat breeding programs in the field, RGB digital cameras serve as a cost-effective alternative to more expensive sensors for estimating green biomass, utilizing specially developed indices. Campbell et al. (2020) employed an RGB imaging system for phenotypic analysis of individual rice plants under drought stress.

Like RGB imaging, fluorescence is also widely used for phenotyping plants from the field to the laboratory. This imaging technique gives information about the metabolic status of a plant, which is acquired through the artificial excitation of the plant's photosystems. This system utilizes charge-coupled device (CCD) cameras capable of detecting sensitive fluorescence signals produced by illuminating samples with visible or ultraviolet light. This approach allows simultaneous capture of fluorescence emissions and provides a rapid method for investigating the in vivo status of photosystem II (Maxwell and Johnson 2000). Fluorescence cameras are being widely used to screen the chloroplast protein mutants with different photosynthetic pigment compositions in Arabidopsis (Lu et al. 2011). However, the application of these fluorescence cameras is limited to the seedling level of the model crop such as Arabidopsis (Jansen et al. 2009).

Infrared imaging technologies are used to screen objects exhibiting internal molecular movements that emit infrared radiation. Two widely used infrared imaging devices, near-infrared (NIR) and far-infrared (far-IR, also known as thermal IR), are commonly employed for capturing radiation images. It provides benefits like high spatial resolution and more accurate measurements under diverse environmental conditions. In field trials, these technologies enable the simultaneous imaging of a large number of plots at the same time. RGB along with NIR imaging techniques was used to screen nitrogen content and multiple traits under salinity tolerance

Table 2.1 Overview of high-throughput phenotyping integrated with QTL and GWAS in plants

Sl. no.	Plant species	Techniques	Sample size	Traits	Methods	References
1	*Arabidopsis thaliana*	Red–green–blue (RGB) imaging (Phenoscope)	358 (RILs)	Relative expansion rate, evaporation, projected rosette size	QTL mapping	Tinse et al. (2013)
2	*Arabidopsis thaliana*	Near-infrared reflectance spectroscopic	164 (RILs)	Oil, protein, carbon, and nitrogen content in seed	QTL mapping	Jasinski et al. (2016)
3	*Arabidopsis thaliana*	11 charge-coupled device cameras each equipped with a macro zoom lens	162 (RILs)/92 (NILs)	Plant root gravitropism, root tip angle	QTL mapping	Moore et al. (2013)
4	*Arabidopsis thaliana*	Phenopsis	324 (AP)	Growth model parameters, projected leaf area, fresh weight, and rosette traits	GWAS	Bac-Molenaar et al. (2015)
5	*Arabidopsis thaliana*	Fluorescence imaging, visible light imaging	382 (AP)	Projected leaf area, relative growth rates, seed area, seed length, and seed width	GWAS	Meyer et al. (2021)
6	*Zea mays*	High-throughput Rice Phenotyping Facility (HRPF), incorporated with color imaging, X-ray	167 (RILs)	Biomass-related traits, color traits, morphological traits, growth traits, and leaf architecture traits (a total of 106 traits)	QTL mapping	Zhang et al. (2017)
7	*Zea mays*	Visible light/ RGB imaging	942 inbred lines	15 tassel morphological traits	GWAS	Gage et al. (2018a, b)
8	*Zea mays*	Bayer color, RGB imaging	228 (AP)	Plant height, growth rate, biomass	GWAS	Wang et al. (2023)

(continued)

Table 2.1 (continued)

Sl. no.	Plant species	Techniques	Sample size	Traits	Methods	References
9	*Zea mays*	Unmanned aerial vehicle (aerial visible light)/ RGB imaging	252 inbred lines	Plant height and growth rate at four different stages	GWAS	Wang et al. (2019)
10	*Zea mays*	visible light/RGB imaging	942 inbred lines	Tassel length, spike length, branch number, and tassel weight	GWAS	Gage et al. (2018a, b)
11	*Zea mays*	RGB and fluorescence imaging near-infrared, visible light	252 inbred lines	Plant fresh weight, growth traits, biomass estimation at 11 different developmental time	GWAS	Muraya et al. (2017)
12	*Zea mays*	RGB imaging	255 hybrids lines	Radiation interception efficiency and radiation use efficiency	GWAS	Chen et al. (2019)
13	*Zea mays*	PhenoArch	254 (AP)	Transpiration rate, plant leaf area, fresh biomass	GWAS	Prado et al. (2018)
14	*Zea mays*	RGB imaging	468 inbred lines	Bulliform cell patterning traits	GWAS	Qiao et al. (2019)
15	*Brassica napus*	Root images using a flatbed scanner	190 (DH)	Biomass traits under high and low phosphate condition, primary lateral root density	QTL mapping	Shi et al. (2013)
16	*Brassica napus*	Gas chromatography, near-infrared reflectance spectroscope, nuclear magnetic resonance	238 cultivars	Three seed-quality traits	GWAS	Wang et al. (2018a, b, c)
17	*Brassica napus*	Visible light/ RGB imaging	248 (AP)	Nine seed vigor and germination traits	GWAS	Hatzig et al. (2015)

(continued)

Table 2.1 (continued)

Sl. no.	Plant species	Techniques	Sample size	Traits	Methods	References
18	*Brassica napus*	Fluorescence imaging, visible light imaging	477 (AP)	Estimated bio-volume, projected leaf area, early plant height, color uniformity	GWAS	Knoch et al. (2020)
19	*Oryza sativa*	3D root imaging, GiA roots 2D and 3D image analysis system	177 (RILs)	Root system architecture (RSA)	QTL mapping	Topp et al. (2013)
20	*Oryza sativa*	Tractor-based high-throughput phenotyping	986 (RILs)	Canopy height, canopy temperature depression, and three reflectance ratios	QTL mapping	Tanger et al. (2017)
21	*Oryza sativa*	Hydroponic system (rhizoscope)	167 (AP)	Root system architecture and biomass traits (total 15 traits)	GWAS	Courtois et al. (2013)
22	*Oryza sativa*	Hyperspectral imaging system	529 (AP)	1540 hyperspectral indices, chlorophyll content	GWAS	Feng et al. (2017)
23	*Oryza sativa*	Visible light imaging (PANorama)	242 (AP)/168 (RILs)	49 panicle phenotypes	QTL and GWAS	Crowell et al. (2016)
24	*Oryza sativa*	High-throughput rice phenotyping facility (RGB imaging)	507 (AP/RILs)	51 image-based traits	QTL and GWAS	Guo et al. (2018)
25	*Oryza sativa*	Visible light/RGB imaging	378 (AP)	Projected shoot area	GWAS	Campbell et al. (2020)
26	*Oryza sativa*	High-throughput rice phenotyping facility (RGB imaging)	529 (AP)	Morphological traits, biomass, and yield related traits	GWAS	Yang et al. (2014)
27	*Triticum aestivum*	PATON electronic belt driven planimeter (Australian Plant Phenomics Facility, The Plant Accelerator)	250 (RILs)	Water-use efficiency, growth rate, leaf area Biomass, and transpiration	QTL mapping	Parent et al. (2015)

(continued)

Table 2.1 (continued)

Sl. no.	Plant species	Techniques	Sample size	Traits	Methods	References
28	*Triticum aestivum*	A germination paper-based "pouch and wick" phenotyping system, coupled with digital image analysis	226 (RILs)	Yield-related traits, seedling root traits, and phenology in field trials	QTL mapping	Xie et al. (2017)
29	*Triticum aestivum*	LiDAR (light detection and ranging)	335 (AP)	Stem elongation rate, canopy height	GWAS	Kronenberg et al. (2021)
30	*Gossypium hirsutum*	RGB imaging	200 (AP)	63 texture-related traits and 56 morphological traits	GWAS	Li et al. (2020a, b)
31	*Sorghum bicolor*	Image acquisition using a Microsoft Kinect, RGB imaging	98 (RILs)	Leaf and shoot morphology-related traits	QTL mapping	McCormick et al. (2016)
32	*Sorghum bicolor*	Leaf surfaces were imaged using Nanfocus µsurf explorer optical topometers (optical tomography)	869 (AP)	Stomatal density, specific leaf area, water use efficiency (WUE), plant height, biomass, gas exchange	GWAS and TWAS	Ferguson et al. (2021)
33	*Hordeum vulgare*	Plant images using a LemnaTec 3D Scanalyzer/RGB imaging	47 (ILs)	Traits related to drought tolerance; shoot area, height, growth, biomass, water use efficiency, etc.	QTL mapping	Honsdorf et al. (2014)
34	*Hordeum vulgare*	Chlorophyll fluorescence	109 (AP)	23 traits related to chlorophyll fluorescence induction parameters	GWAS	Rapacz et al. (2019)

(continued)

Table 2.1 (continued)

Sl. no.	Plant species	Techniques	Sample size	Traits	Methods	References
35	*Capsicum annuum*	SPICY: a sensor-to-plant phenotyping device equipped with RGB cameras and time-of-flight depth sensor	148 (RILs)	Plant height, leaf area, leaf size, and leaf angle	QTL mapping	van der Heijden et al. (2012)

AP association panel, *DH* double haploid population, *GWAS* genome-wide association study, *ILs* introgression lines, *NILs* near isogenic lines, *QTL mapping* quantitative trait loci mapping, *RILs* recombinant inbred lines

(Rajendran et al. 2009). The application of infrared imaging has enhanced research on drought resistance and/or salinity resistance by quantifying osmotic tolerance (Munns et al. 2010). RGB fluorescence along with NIR cameras was used for studying the spatiotemporal accumulation of biomass under drought conditions in barley (Neumann et al. 2015).

Hyperspectral imaging/spectroscopy imaging system measures the interaction between plants and solar radiation and the concept originated from remote sensing of vegetative crops. In addition to visible and infrared imaging techniques, hyperspectral imaging divides images into multiple bands, capturing a significant portion of the electromagnetic spectrum. Ge et al. (2016) used RGB and hyperspectral cameras to study water use, leaf water content, and maize growth. Furthermore, hyperspectral imaging at the organ level of maize and sorghum was used to identify the genetic association (Miao et al. 2020).

Recently, many research institutes have emerged to efficiently integrate the current HTP system, data accessibility, reducing duplicates, phenotypic methods, and resources, to expedite the development of high-quality phenotypic accomplishments and accelerate crop breeding. From the database of the International Plant Phenotyping Network (IPPN), 17 research institutes were established in November 2015, with an increase in number to 44 in 2019 (https://www.plant-phenotyping.org/). Some of the notable research institutes contributing to plant phenotyping include (1) Australia Plant Phenomics Facility (APPF, https://www.plantphenomics.org.au/), (2) High-throughput Rice Phenotyping Facility (HRPF, developed by Huazhong University of Science and Technology and Huazhong Agricultural University, Wuhan, China), (3) CropDesign (www.cropdesign.com, Belgium), (4) French Plant Phenomic Network (FPPN/PHENOME, https://www6.dijon.inrae.fr/umragroecologie_eng/ResearchPrograms/Investissement-Avenir/PHE), (5) Italian Plant Phenotyping Network (PHEN-ITALY, http://www.phen-italy.it/index.php), (6) The North American Plant Phenotyping Network (NAPPN, http://nappn.plant-phenotyping.org/), (7) European Plant Phenotyping Infrastructure (EMPHASIS, https://emphasis.plant-phenotyping.eu/), and (8) The Cereal Systems Initiative for South Asia (CSISA, https://csisa.org/). Although a diversified range of commercial phenotyping platforms has been developed, this system generates extensive multi-dimensional datasets, which require effective data management and analytical frameworks for their interpretation. These platforms are primarily developed by commercial companies, with proprietary software and hardware protected by patents, limiting their customization for specific research needs. The key challenges faced by the new generation of phenotyping include data management, image processing, and trait analysis (Hickey et al. 2019).

Developing an effective phenotyping protocol requires consideration of several factors: (1) selecting representative environments and identifying their optimal locations, (2) determining the appropriate number of replications per individual in each environment, (3) designing experiments to minimize extraneous variation in the experimental field, (4) applying suitable statistical methods for efficient data analysis, and (5) accounting for QTL × environment interactions. When traits are measured in multiple environments with replication, then it decreases the likelihood of detecting QTLs that are specific to particular, nonrepeatable environmental conditions. Before using the phenotypic data for an association study with genotypic data, it has to be evaluated to determine the significance of the trait value. Different types software, like R or SAS, are commonly available to analyze the basic quality control steps like checking outliers, normality, significance level, heritability, and correlation estimation to check the reliability of the phenotypic data.

2.4.1 Genotyping

Genetic markers are the sequences of DNA that can be used to identify the genetic variations or differences among individuals/populations. Typically, they do not directly represent the specific genes but serve as an indicator or "flag" or "tag" and are located within coding or non-coding regions of the genome. Those genetic markers that are closely located to genes are called "gene tags"; however, these markers do not directly impact the phenotype of a trait of interest. Genotyping is the process of determining variation in the genetic makeup within a population to obtain diversity patterns resulting from the differences in the specific markers (Collard et al. 2005). Genotype data can be generated through three distinct genotyping methods:

1. *Genotyping the entire mapping population*: This is a reliable and most commonly used genotyping method where the whole mapping population is genotyped to identify the markers across the whole genome although it is a time-consuming, costly, and tedious process.
2. *Selective genotyping*: It is a simple, economical, and effective approach wherein a subset of specific individuals is genotyped based on their extreme phenotypes (those exhibiting exceptionally high and/or low phenotypic values). These methods reduce the number of individuals required by focusing on genotyping only those at the extreme ends of the phenotypic distribution (Fig. 2.3a) specific for trait of interest (Lebowitz et al. 1987; Vision et al. 2000). Selective genotyping

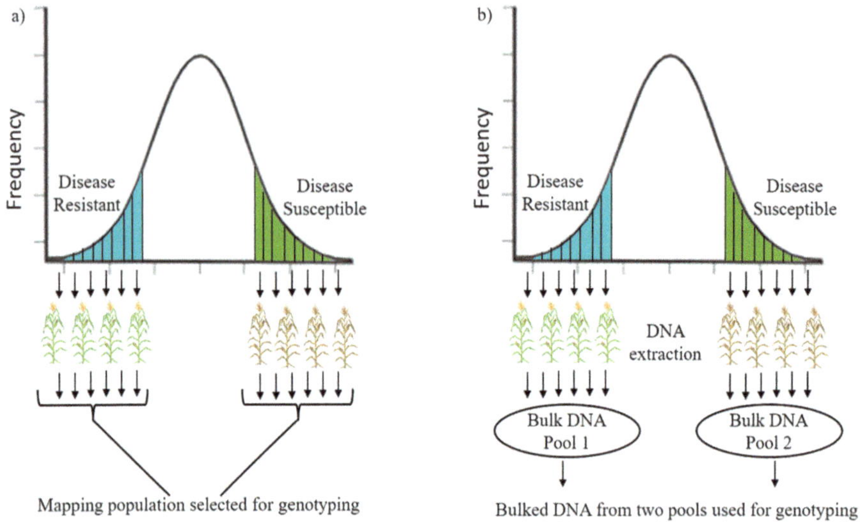

Fig. 2.3 Methods of genotyping: (**a**) selective genotyping and (**b**) bulk segregant analysis

proves to be a promising genotyping method when complete population genotyping is either too expensive or impractical and screening of superior donors focuses on one side of the tail with large effects. Yang et al. (2020) genotyped only 60 RILs (high and low grain weight) of wheat from the entire 266 population which was sufficient to identify major QTLs. Nevertheless, the limited acceptance of selective genotyping may be due to the distorted segregation observed in generating linkage maps, their efficiency majorly depends on the normal distribution of the phenotypic traits, possible to study only one trait at a time (difficult to select multiple linked traits), and increase the possibilities of identifying false positive QTLs (Bovenhuis and Spelman 2000).

3. *Bulk segregant analysis*: This technique provides a simple approach for identifying markers that are tightly linked to the causal QTLs underlying specific phenotype (Giovannoni et al. 1991). In this method, either an equal amount of leaf samples or an equal concentration of bulked DNA samples from 10 to 20 individuals of two groups (exhibiting extreme phenotypes, e.g., resistance versus susceptibility to a particular disease) is pooled (Quarrie et al. 1999). This enables randomization of all loci, except for the region harboring the gene of interest (Fig. 2.3b). BSA technologies have been used in many organisms to map important genes (Yi et al. 2006; Zhao et al. 2023). However, BSA is generally considered ineffective for identifying QTLs influenced by multiple genes with minor effects. The effectiveness of BSA in QTL mapping may be compromised by factors such as (1) inadequate marker density, (2) inaccurate estimation of allele frequencies within pools, (3) small population sizes, and (4) high likelihood of false positives.

Typically, genetic markers are divided into morphological; biochemical, and molecular/DNA markers. During the onset of the breeding technologies, plant breeders used to select and cross the individuals based on identifiable morphological or visible markers (such as flower color, quality, quantity, and shape), because these markers are easily distinguishable and correlate with the expression of agriculturally and economically significant traits. The concept of using morphological markers for constructing genetic maps dates back to the nineteenth century, when A. H. Sturtevant (1913) used six morphological traits to generate the first genetic map of *Drosophila melanogaster*. Biochemical markers or isoenzymes are the allelic variations of enzymes/proteins (identified through staining techniques or electrophoresis) that are able to detect the variation at the functional gene level. Morphological markers are few in number and strongly influenced by environmental factors or the developmental stage of the plant. Conversely, biochemical markers, despite their greater abundance, can be influenced by the environment, posing challenges in measurement and consequently resulting in inaccuracies in gene-mapping studies with both false positives and negatives. Molecular/DNA markers refer to the variation in the DNA nucleotide sequence between individuals or species, located near or closely linked to a target gene responsible for expressing a specific trait and are inherited together (Collard et al. 2005).

Genetic marker technology has rapidly evolved from traditional morphological to biochemical and presently to molecular markers. Early 1980s are reported as the era of molecular marker development and application marking a major milestone in plant genomic research. Since then, numerous molecular markers have been reported for use in various areas of plant molecular breeding and genomics. There are several types of molecular markers such as RFLP (restriction fragment length polymorphism), AFLP (amplified fragment length polymorphism), RAPD (randomly amplified polymorphic DNAs), SCAR (sequence-characterized amplified regions), SSR (microsatellites or simple sequence repeat), inter-SSR (ISSR), CAPSs (cleaved amplified polymorphic sequences), and SNP (single-nucleotide polymorphism). Details of molecular markers along with their merits and demerits have been described in numerous reviews (Semagn et al. 2006; Amiteye 2021). In the past few decades, significant efforts have been made using molecular markers to investigate the genetic factors through QTL mapping and GWAS in several crops for many important economic traits. QTL mapping using RFLP (Jordan et al. 2003), AFLP (Fang et al. 2013), SCAR (Shu et al. 2013), RAPD (Milad et al. 2011), and SSR (Chen et al. 2010) markers have been employed for many agronomic complex traits. However, these markers have limitations in terms of genomic distribution, availability, compatibility with automated platforms, and low-resolution mapping or a limited number of markers; further, the linkage between the associated marker and the genetic loci (QTL) is often weak. The major challenges using these markers in marker-assisted breeding include a high occurrence of false-positive selection among lines, lower phenotypic effects in comparison to estimated effects, and increased unwanted genetic drag when selecting desirable loci (Podlich et al. 2004). SNP markers identified using high-throughput genotyping offer numerous benefits for high-resolution mapping of genetic loci that control traits, enabling the identification of closely linked markers and improved estimation of phenotypic variation governed by these loci. This section will briefly describe several genotyping techniques used for SNP identification and their utilization in QTL mapping and GWAS. Many genotyping techniques have been developed over the past few decades for identifying SNPs for QTL and GWAS applications. The most commonly used genotyping of SNP markers can be achieved through SNP arrays, reduced representation sequencing, amplicon sequencing, whole genome resequencing, and RNA-seq analysis.

2.4.2 SNP Microarray-Based Genotyping

SNP microarray-based high-throughput genotyping identifies SNPs through DNA hybridization to oligonucleotides placed on chips and allows for the direct scanning of allelic variation across the genome, encompassing hundreds to thousands of SNPs in a short period (Bennetzen et al. 2012). Numerous SNP assay systems have been created by various companies, including Illumina's SNP chip platforms (http://www.illumina.com), GenomeLab's SNPstream genotyping system (Beckman Coulter, https://www.beckmancoulter.com), Affymetrix's SNP array platform (http://www.affymetrix.com), and the TaqMan OpenArray genotyping system (Applied Biosystems, http://www.appliedbiosystems.com). However, the Illumina and Affymetrix SNP arrays are more frequently utilized in many crops for the construction of high-density maps, mapping quantitative trait loci (QTLs), and analyzing genetic diversity studies which are cost-effective and take minimal time. Until 2017, a total of 46 SNP array platforms for 25 crops were developed by Illumina and Affymetrix (Rasheed et al. 2017). SNP chips, such as AppleSNP20K (Bianco et al. 2014), CottonSNP63K (Hulse-Kemp et al. 2015), MaizeSNP50K (Ganal et al. 2011), and RiceSNP6K (Yu et al. 2014), have been developed through Illumina Infinium bead chip-based technology. Some examples of SNP arrays developed by Affymetrix Axiom are Maize55K Axiom (Xu et al. 2017), SoyaSNP180K Axiom (Lee et al. 2015), and Rye600K (Bauer et al. 2017). SNP array genotyping offers several advantages such as genotype calls are typically accurate, even for highly heterozygous species (Bourke et al. 2018), low proportion of missing data, and they are particularly beneficial for the genotyping of polyploid crops. However, the major disadvantage of these arrays is that they are limited to known SNPs and rely on preselected SNPs.

2.4.3 Reduced Representation Sequencing

Reduced representation sequencing (RRS) is a method in which only small fragments of the genome are sequenced thereby reducing the cost of whole genome sequencing. Methods such as GBS (genotyping by sequencing) (Elshire et al. 2011), ddRAD-seq (double digest RAD sequencing) (Truong et al. 2012), and restriction site-associated DNA sequencing (RADseq) (Davey and Blaxter 2011) are widely utilized for reducing genome complexity using restriction enzymes (REs) and are currently the most favored RRS techniques for conducting QTL, GWAS, and genome selection study in crops, since they are cost-effectiveness. It also offers the advantage of scalability, allowing for the customization of the percentage of the genome captured by using different combinations of restriction enzymes. Restriction fragment ligation methods involve digesting genomic DNA from individuals or populations with restriction enzymes, ligating adapters to the fragments, and sequencing the final fragments. The polymorphisms identified in these sequences serve as genetic markers for further analysis. These methods provide a comprehensive view of the genome by generating genetic markers across the entire genome. The higher marker density from RRS improves QTL and GWAS resolution, enabling more precise mapping and genetic association identification. RRS has facilitated rapid sequencing and genotyping in breeding populations, supporting genetic diversity and marker development in various species, even without prior genomic knowledge (Poland and Rife 2012). However, RRS is limited by the SNP distribution, which depends on the restriction enzyme combination used, and uneven sequencing depth at SNP loci, leading to undercalling of heterozygous loci and missing data. Genotype imputation can address this, though its success depends on the mapping population's genetic makeup and the presence of long homozygous segments useful for haplotype reconstruction (Glaubitz et al. 2014).

2.4.4 Amplicon Sequencing

Amplicon sequencing is one of the most suitable methods for discovering SNPs and InDels of known and novel variants in genomic studies, and marker development at specific loci, thereby identifying the variation at specific regions of the genome and studying their association with the trait of interest. This process involves designing primers to amplify DNA fragments from candidate genes, Expressed Sequence Tags (ESTs), or single-copy genomic sequences. The amplified products are sequenced and compared using bioinformatics tools, enabling SNP detection in specific genome regions, such as gene families or chromosomes of interest. Amplicon sequencing is the most frequently used approach in polyploid species, where the identification and discovery of SNPs are more complex due to the need to differentiate between SNPs present between the genomes and within the genomes. Thus, for crops having complex genome and allopolyploidy, utilizing genome-specific primers for amplicon sequencing could be the most effective option for marker discovery.

2.4.5 Whole Genome Resequencing (WGR)

Whole genome resequencing involves sequencing of various individuals and populations whose complete genome sequence is already known. The advancements in next-generation sequencing technologies have not only made the de novo sequencing of crop genomes more feasible but have also opened up numerous opportunities for plant breeding and genetic studies that were previously not possible. WGR of plant genomes enables the identification of a vast array of DNA markers, including SNPs, copy number variation, structural variant, and indels in crops, offering comprehensive insights into genome evolution (Huang et al. 2010) and is mostly performed using paired-end Illumina technology. WGR-based QTL mapping has enhanced the accuracy, output, speed, and cost-effectiveness of genome-wide genotyping. Advancement in NGS technology has enable the availability of high-quality reference genome sequences which in turn can be used for re-sequencing a panel of diverse germplasm for studying genome diversity, identifying selection signatures, and discovering sequence variations. Reducing the average sequencing depth can decrease the expense of whole-genome re-sequencing; nevertheless, this could lead to an undesirable increase in genotyping errors, particularly at heterozygous loci, which are linked to a large number of genotypic combinations (Kishikawa et al. 2019). Despite being the one of the most widely used genotyping technologies, there are also challenges such as its very expensive nature, data storage and analysis, as well as the need for specialized expertise in interpreting the vast amount of genetic data generated.

2.4.6 RNA-Seq

Genomic resources for most non-model crop species are limited, and de novo assembly from whole-genome sequencing is often ineffective. Transcriptome sequencing offers a practical alternative for generating functional genomic data in these crops by extracting genotype information from RNA-*seq* data. Since genome size varies widely among crops but gene counts and sizes remain relatively stable. Transcriptome sequencing is a cost-effective strategy for SNP calling, avoiding repetitive regions ignored in whole-genome re-sequencing. Transcriptome data reported in the database accessible in public repositories (www.ncbi.nlm.nih.gov/sra) can also be used for gene expression analysis and also can be employed for SSR and SNP identification in less-studied species. Transcript data are advantageous for marker discovery because transcripts rarely include extensive repetitive DNA, improving base-calling accuracy and allele assignment. However, RNA-seq-based genotyping has limitations. SNP density is lower in genic regions compared to intergenic regions, potentially insufficient for GWAS in crops with high linkage disequilibrium (LD). SNP distribution bias is another challenge, as many genes show low or no expression in specific tissues at given time points, making them unsuitable for genotyping. Expanding RNA sampling across tissues and time points in large populations adds significant workload (Zargar et al. 2015).

2.5 Statistical Methods for QTL Mapping

Once the phenotype and genotype data are generated, the main objective of QTL mapping studies is to identify the linkage between markers and the phenotype of desired traits while reducing the likelihood of false positives (Type I errors). A variety of data analysis approaches that are commonly used to test QTL/trait associations have been developed and performed by the following methods.

1. *Single marker analysis (SMA)*: This approach is also known as the single factor analysis of variance (SF-ANOVA) or single point analysis which is the simplest and earliest employed method for detecting QTL (Soller et al. 1976). This approach involves testing each marker locus (which is independent of details of other loci) for its association with the target trait. Statistical tests including analysis of variance, *t*-test, likelihood ratio test, linear regression, or maximum likelihood estimation can be employed to determine whether differences between genotypes for marker classes are significant or not. The probability of detecting a QTL primarily depends on two key factors: (1) the magnitude of the QTL effect size and (2) QTL and the marker loci recombination rate (r) (Zou 2009). Despite its computational simplicity, this method has several major limitations. It generally cannot determine the precise position of QTLs or distinguish whether markers are linked to one or multiple QTLs. The likelihood of QTL detection decreases as the distance between the marker and QTL increases. In addition, QTL effects are underestimated due to confounding with recombination frequencies, and their accuracy is lower compared to other methods. QTL Cartographer, QGene, and MapManager QTX are widely used computer programs for conducting single-marker analysis (Basten 1994; Nelson 1997; Manly et al. 2001).
2. *Simple interval mapping (SIM)*: These methods were initially introduced by Lander and Botstein in 1989 and are also referred to as the two-marker approach. SIM method is based on linkage maps and concurrently examines the intervals defined by two linked markers at various positions on chromosomes, instead of analyzing a single marker. It evaluates the potential association between phenotypic values and the genotype of a target QTL (hypothetical QTL) at various analysis points within the interval bounded by adjacent marker loci (target interval). These methods used the maximum likelihood ratio (log of odds ratio, LOD) to determine the position of each QTL within the interval defined by both markers. When the LOD exceeds a commonly set critical threshold (often ≥ 3), it indicates the presence of a potential QTL. This method is statistically more robust than the SMA because it employs linked markers for statistical assessment, thereby accounting for any recombination events between markers and the trait associated with the genomic region, if applicable. SIM represents a significant advancement over SMA methods as it examines the potential existence of QTLs between adjacent pairs of markers. LOD score is computed at each testing site, indicating the likelihood of a QTL being present. The presence of a QTL within a chromosomal region is indicated when the LOD score, plotted along the

chromosome map, exceeds a predetermined significance threshold. The highest LOD score peak is typically interpreted as the probable QTL position. The major advantages of SIM methods are that it has superior statistical power in QTL detection compared to single-marker analysis, they generate a LOD score curve, the QTL position is represented by a support interval, and they accommodate missing marker genotype data thereby enhancing the reliability of the findings. Nevertheless, SIM has its primary limitations such as the linkage map should be constructed, estimations of QTL position in the genome and its QTL effect can be biased when multiple QTLs influencing the trait of interest are closely linked, and specialized QTL analysis software is needed. SIM is the preferred choice for many due to its ease of execution using statistical packages like MAPMAKER/QTL (Lincoln et al. 1993).

3. *Composite interval mapping (CIM)*: CIM was developed by Jansen and Stam (1994) to address the limitations of SIM and it combines SIM with regression analysis to effectively isolate individual QTL effects. By doing so, it regulates genetic variation in other genomic regions, reducing unwanted background noise that can interfere with QTL detection. To mitigate background variation, the analysis software incorporates a set of markers known as "cofactors" into the model. CIM enhances QTL detection precision by controlling the effects of QTLs not only within the same chromosome but also across different chromosomes. Initially, it conducts single-marker analysis and subsequently constructs a multiple QTL model using backward stepwise or forward regression. This process begins by selecting the marker with the highest LOD score, followed by the addition of the marker with the second-highest LOD score. The significance of these markers is then reevaluated. If they both remain significant, the marker with the next highest LOD score is included in the model, and the significance of the three markers is reevaluated. This iterative approach identifies and incorporates all markers from the entire genome for QTL identification. The major advantages of CIM methods include the ability to map multiple QTLs in one dimension and the robustness of the test against QTLs outside the region of interest due to the use of linked markers as covariates, thereby increasing the precision of QTL mapping. Furthermore, CIM reduces genetic variance introduced by other QTLs, thus enhancing the efficiency of QTL determination by reducing residual variance. As in SIM, the construction of a linkage map is a prerequisite in CIM, requiring specialized software designed for QTL analysis. The effectiveness of CIM relies significantly on background markers, and the permutation process can be time-consuming. The commonly used software for conducting CIM are QTL Cartographer, PLABQTL, and MapManager QTX (Basten 1994; Utz and Melchinger 1996; Manly et al. 2001). Methods such as inclusive composite interval mapping (ICIM) are also used for QTL mapping which uses the modification of the CIM algorithm (Li et al. 2007). The limitation of CIM methods is that the cofactors are randomly chosen, however, in the case of ICIM, significant cofactors are preselected, followed by the estimation of their associated effects through a stepwise regression analysis before interval mapping. In general, ICIM identifies more true-positive QTLs; high LOD scores

in the genomic location where QTLs are detected, and estimates the presence of epistasis and additive effects of QTLs.

4. *Multi-trait interval mapping (MIM)*: It is the mapping method when multiple traits exhibit correlations with each other by mitigating the impact of error variance, thereby enhancing the ability to map multiple QTLs. This approach incorporates multiple potential QTLs directly into the model for QTL mapping, utilizing multiple marker intervals simultaneously. The utilization of the MIM approach enhances the accuracy and effectiveness of QTL mapping. Moreover, it enables the evaluation and analysis of interactions among QTLs, the genotypic values of individuals, and the heritabilities of quantitative traits, facilitating their estimation and examination. This method is a potential tool for identifying interactions between QTLs (Collard et al. 2005). The major limitation of the MIM method is that when more QTLs are included in the model, there is a rapid proliferation of parameters, resulting in significant computational demands. A challenge arises when the model contains a large number of QTLs, leading to a problem known as saturation, especially when the number of covariates exceeds the number of available samples. Selecting the right model from the numerous possible options poses a significant challenge, as developing a dependable and appropriate criterion for model selection is a difficult task.

Newly developed statistical software packages now provide the tools to examine interactions between marker and phenotype trait values. In addition to the QTL mapping techniques mentioned, several alternative approaches are also available such as Bayesian methods and the utilization of genetic algorithms. A list of statistical packages employed for QTL analysis is given in Table 2.2.

2.6 Association Mapping

QTL mapping offers valuable insights into the genetic control of trait inheritance and has been widely used in breeding programs for many economically important crops. Nevertheless, bi-parental mapping analysis encounters its limitations such as less allelic diversity, insufficient phenotypic variation among the mapping, a higher number of alleles can be compared, restricted genomic resolution due to the insufficient recombination events that occur during the development of the mapping population, and faster completion time (Jannink et al. 2001; Zhu et al. 2008). Thus, to address the constraints of bi-parental QTL mapping, association mapping (AM) proves to be an alternative approach for mapping and identification of QTLs of complex traits, based on linkage disequilibrium (LD). This method identifies alleles in a population that are more frequently associated with specific phenotypes than expected, providing insights into the genetic basis of complex traits (Flint-Garcia et al. 2003). It also enables high-resolution identification of genetic crossovers and trait-associated loci in large natural populations. In comparison to traditional QTL mapping, AM offers several advantages such as (1) utilizing all recombination events that occurred during the evolutionary history of a crop species leads to

Table 2.2 List of software/packages used for QTL mapping analysis

Sl. no.	Software/packages	Description	Implementation in plant species
1	Join map	Used for diploid species with Monte Carlo maximum likelihood mapping algorithm. It assesses linkage group formation using metrics such as LOD scores and recombination frequencies and organizes genetic markers linearly. https://www.kyazma.nl/index.php/JoinMap/	Carrot (Iorizzo et al. 2016)
2	QTL Cartographer	User-friendly, graphical interface and supports multiple analysis methods, including single-marker regression, interval mapping, and Bayesian interval mapping. The software calculates empirical threshold LOD scores and determines QTL position confidence intervals. http://statgen.ncsu.edu/qtlcart/WQTLCart.htm	Cotton (Liu et al. 2020)
3	MapQTL	User-friendly genetic analysis software designed for diploid species, including BC_1, F_2, RIL, DH, and outbred full-sib families. The software's notable features include the analysis of multiple populations, covariance analysis, automatic marker cofactor selection, and enhanced QTL chart presentation. http://www.kyazma.nl/index.php/mc.MapQTL/sc	Tea (An et al. 2021)
4	MapManager	User-friendly interface, robust graphic capabilities, and cross-platform compatibility. Key functionalities include regression-based SMA, SIM, CIM, and interactive QTL search, calculation of empirical threshold LOD scores, and determination of QTL confidence intervals using bootstrap or the Piepho method. http://www.mapmanager.org/mmQTX.html	Wheat (Sadeque and Turner 2010)
5	R/QTL	It is an open-source extension for the R statistical software, enhancing its capabilities in quantitative trait locus (QTL) analysis. Available at the provided URLs, it supports various platforms and offers methods like SMA, SIM, CIM, and MQM, with the MQM method notably improving the statistical power for identifying QTL effects. http://www.rqtl.org/	Coffee (Moncada et al. 2016)

(continued)

Table 2.2 (continued)

Sl. no.	Software/packages	Description	Implementation in plant species
6	QGene	It is a Java-based software with a comprehensive GUI, compatible with Java-supporting computers. It specializes in QTL mapping, offering multiple methods including unique ones like SIM, CIM, MIM, and multitrait methods. QGene manages data from diverse populations and unconventional mating designs, though it does not support outcross, half-sib, and multi-cross populations. http://www.qgene.org/	Rice (Wen et al. 2020)
7	SAS	For single-marker analysis in QTL mapping, SAS version 8.1 is utilized. This software identifies QTLs by analyzing associations between marker genotypes and phenotypic traits. It supports various statistical methods including analysis of variance, *t*-tests, general linear models, and regression analysis	Maize (Lima et al. 2006)
8	MQTL	It is optimal for analyzing data from multiple environments and homozygous biparental progeny (such as recombinant inbred lines and doubled haploid lines), especially for simple interval mapping or a streamlined version of composite interval mapping. http://gnome.agrenv.mcgill.ca/pub/genetics/software/MQTL/	Maize (Jin et al. 2013)
9	TetraploidSNPMap	It is a software tool designed for linkage analysis and QTL mapping in full-sib populations of autotetraploid species. Key features include SNP clustering, efficient ordering of SNPs through multidimensional scaling analysis and phase calling, and advanced QTL mapping with a hidden Markov model. https://github.com/BiomathematicsAndStatisticsScotland/TetraploidSNPMap	Potato (Park et al. 2019)
10	MapMaker/QTL	It incorporates interval mapping and nonparametric methods to analyze non-normal phenotype data. The software, compatible with various computer platforms, features a command-driven user interface and allows users to export output graphs as PostScript files. http://www.broad.mit.edu/ftp/distribution/software/mapmaker3/	Rice (Lanceras et al. 2004)

(continued)

Table 2.2 (continued)

Sl. no.	Software/packages	Description	Implementation in plant species
11	PLABQTL	Used for analyzing top-cross progenies and is also applicable to F_2 and subsequent generations, backcrosses, and double-haploid populations. It incorporates both additive and dominant gene effects in its genetic model and efficiently performs simple interval mapping (SIM) and composite interval mapping (CIM) using a rapid multiple regression approach. PLABQTL can handle missing data, identify outliers, compute and compare LOD curves under different model assumptions, and analyze QTL-environment interactions. https://www.unihohenheim.de/plantbreeding/software/	Sorghum (Muturi et al. 2021)
12	R/QTLBIM	It is a software extension for Bayesian interval mapping, specializing in Bayesian model selection for the identification of multiple interacting QTLs. It handles both continuous and categorical traits and is designed for use with data from experimentally inbred lines. http://qtlbim.org/	Watermelon (Li et al. 2020a, b)
13	FlexQTL	Based on Bayesian theory and implemented via Markov Chain Monte Carlo simulation, is versatile for QTL mapping in various pedigreed populations, including F_2 and backcrosses. It is capable of handling missing phenotype and/or genotype data and can analyze multiple traits simultaneously, revealing pleiotropic behavior of QTLs and correlations between polygenic and residual components of quantitative traits. www.flexqtl.nl	Apple (Guan et al. 2015)
14	MCQTL	Designed for UNIX environments, specializes in QTL mapping for multi-allele QTLs in multi-cross designs and biparental populations. It offers flexibility in treating within-family QTL effects as fixed or allowing diallel modeling. Based on linear regression and using an additive genetic model, it conducts genome-wide scans via composite interval mapping (CIM). http://www.genoplante.com	Oil palm (Billotte et al. 2010)
15	QTL IciMapping	User-friendly, public software tailored for creating marker linkage maps and QTL mapping in biparental populations. The software handles input files in plain text or MS Excel formats and is proficient in mapping additive, dominant, epistatic QTLs, and QTL-environment interactions. http://www.isbreeding.net/	Maize (Almeida et al. 2013)

(continued)

Table 2.2 (continued)

Sl. no.	Software/packages	Description	Implementation in plant species
16	QTL Network	It is capable of simultaneously mapping quantitative trait loci (QTL) with individual effects, epistasis, and QTL-environment interactions. The software accommodates data from F_2, backcross, recombinant inbred lines, double-haploid populations, and specific mating designs like immortalized F_2. http://ibi.zju.edu.cn/software/qtlnetwork	Pigeon Pea (Kumawat et al. 2012)
17	fullsibQTL	Designed for composite interval QTL mapping in bi-parental populations of outcrossing populations. Estimates QTL position, effects, segregation patterns, and linkage phase with flanking markers, having several statistical and graphical tools. https://github.com/augusto-garcia/fullsibQTL	*Miscanthus sinensis* (Raverdy et al. 2022)

significantly enhanced mapping resolution, (2) cost-effective as because the genotypic data of the same association panel can be utilized for mapping various complex traits, (3) no extra time for developing specialized mapping population, as it works well with naturally occurring variety of germplasm, and (4) consideration population structure analysis to avoid false positive detection of marker trait association. Over the past two decades, numerous linkage studies have been conducted in various plant species, but only a limited number of identified QTLs have been cloned or tagged at the gene level (Price et al. 2006).

With the advancement of tools and techniques in genome sequencing and high-density SNP genotyping, the development of new approaches for finding the association between complex quantitative traits and genetic factors and subsequently identifying the genes responsible for the traits has become easier. The presence of reference genome sequencing technologies and new computational pipelines makes it feasible to undergo association mapping studies as it helps in the quick examination of genetic variations in natural crop variants. Reference genomes of several major crops have been reported (Feng et al. 2002; Paterson et al. 2009; Velasco et al. 2010; Mayer et al. 2012; Zhang et al. 2015). Reference sequencing can be done through clone-to-clone (as in rice) or whole-genome shotgun approaches (as in millets). The de novo assembly from a whole-genome shotgun often results in sequence gaps. Thus, various other approaches are complemented to construct a long super scaffold. The quality of the reference genome greatly impacts the association mapping study because it will help in capturing the full coverage of sequence polymorphisms and variants from the genotypic data of the crops.

Association mapping, also known as linkage disequilibrium (LD) mapping, involves the analysis of statistical relationships between the genotypes (usually molecular markers) of individuals, and the corresponding traits (phenotypes) within a group of individuals. It relies on the presence of LD, where the physical size of LD blocks (chromosomal regions where all pairs of adjacent loci are in LD) determines the marker density essential for association mapping study (Stich et al. 2006). The concept of association mapping was initially employed to find the genes responsible for human diseases (Stein and Elston 2009); however, in the last few decades their application is gaining popularity in the field of crop improvement and crop genetics. Recently, utilizing the LD to identify marker-trait association and key genes associated with complex traits has been widely investigated in many crop plants such as rice (Huang et al. 2012; Barik et al. 2022), chickpea (Varshney et al. 2019; Roorkiwal et al. 2022), sugarcane (Gouy et al. 2015), wheat (Arruda et al. 2016; Li et al. 2018), cotton (Fang et al. 2017; Joshi et al. 2023), maize (Wang et al. 2020), barley (Pasam et al. 2012), and soybean (Ravelombola et al. 2021). Depending on the scale and scope of a particular study, AM generally falls into two primary types:

1. *Candidate-gene association mapping*: It is a hypothetical based approach which examines the polymorphisms in selected candidate genes (based on prior knowledge) that play roles in controlling phenotypic variation for specific traits primarily focuses on identifying major genes/QTLs that govern simple biochemical pathways. This method increases the existing information about the candidate gene and its association with the trait with better understanding of the genetic architecture characterizing a trait. The marker-trait association is statistically validated, and a detailed map is created using polymorphic markers and individuals exhibiting variation in the trait of interest. Since this approach does not involve the search for new allelic variations outside the proposed candidate gene, it does not allow for the identification of new genes/QTLs. Conversely, resequencing the previously identified candidate gene across a large population increases the likelihood of identifying any rare alleles that may have been overlooked in earlier mapping attempts. These approaches of association study are previously reported in several crops (Ehrenreich et al. 2009; Talukder et al. 2014; Yu et al. 2015).
2. *Genome-wide association mapping (GWAS)*: Also known as a genome scan, GWAS examines genetic variation across the genome to identify associations with complex traits (Risch and Merikangas 1996). In numerous plants, GWA studies have been conducted (Wang et al. 2012; Kaler and Purcell 2019; Liu et al. 2019; Pang et al. 2020). However, GWAS often detects loci unrelated to the trait due to linkage disequilibrium among multiple factors, only some of which are directly linked. Challenges such as epistasis and population structure add complexity and cannot be resolved by increasing sample size or marker density but can be addressed using models like MTMM or ML MTMM (multi-trait mixed models) or MLMM (multi-trait mixed models) MM (Segura et al. 2012).

2.7 Concept of Linkage Disequilibrium (LD) and LD Decay

LD is defined as the nonrandom linkage of alleles situated at distinct loci on the same chromosomes (Mackay and Powell 2007). The concepts of linkage and linkage disequilibrium are frequently conflated, although both are interconnected, yet they are fundamentally different. Linkage refers to the tendency of two or more loci (from the same chromosomes) to remain together during the inheritance process. It refers to the physical state of correlated loci located closely in a chromosome and are always inherited simultaneously into gametes. LD denotes the correlation between alleles at different loci within a population. The concept of LD was initially by Jennings in 1917; however, the term was given by Lewontin and Kojima in 1960. Initially, LD was a subject of interest in population genetics, and later, gained its importance in evolutionary biology and human genetics as well. LD is denoted by the symbol D which is mathematically derived using the formula:

$$D = P_{AB} - P_A P_B$$

where P_{AB} is the gamete (AB) frequency and P_A and P_B are the frequencies of allele A and B.

In order to explain D, two important statistics are available viz., D' (Lewontin 1964) and r^2 (Hill and Robertson 1968). D', a standardized disequilibrium coefficient, assesses historical recombination in a population, while r^2, the correlation between alleles at two loci, reflects both recombination and mutation events and is useful for association studies. Both parameters range from 0 to 1.

During selection in a population, multiple loci regulated the important traits and recent introduction of genotypes may increase LD in a population. Other factors that contribute to the increase of LD include small population size, genetic isolation, inbreeding, epistasis, population admixture, low recombination rate, and genetic drift. LD generally extends over greater distances in self-pollinated crops (e.g., wheat) than in cross-pollinated species (e.g., maize). Population structure-induced LD must be accounted for to avoid spurious results. While LD is influenced by various factors (Ardlie et al. 2002), linkage-derived LD reflects the cumulative effect of recombination events since an allele's emergence, with a higher likelihood of recombination between closely linked loci compared to linkage analysis. After every random mating, the LD is expected to reduce in magnitude due to the recombination between the two loci which is termed as LD decay. The rate of LD decay is calculated by the product of r (recombination frequency) and d (disequilibrium). To visualize the extent of LD, LD decay is calculated and plotted which gives an idea of how LD declines with genetic (cM, centiMorgans) or physical (bp, basepair). To study the marker-trait association in plants, LD must be present in the plant species. Since the distance till which LD persists will help in determining the number and density of markers, it is therefore important to understand and determine the LD for the species used in the association study.

2.7.1 Population Used for Association Mapping Study

The success of marker-trait association mapping studies relies on the varietal population or material utilized for analysis. The population can include historical germplasm, family-based populations, and breeding populations which exhibit high genetic diversity. Primarily, an association mapping panel consists of three types of population: core or germplasm bank collections, elite breeding materials/cultivars, and synthetic populations. A comparative study on different types of association mapping populations is given in Table 2.3.

1. *Core collections/panel*: Core collections are the types of population that are expected to encompass the majority of genetic variability within a manageable number of accessions, making them suitable for genetic studies. It represents a group of genetically diverse populations of a species (having wide allele diversity). The selection methods for core collections often involve genotyping using unlinked markers to calculate genetic distances, thereby offering insights into population structure. The process of selecting a minimal sample (creating a core set of accessions) with maximum variation has a normalizing effect, which is expected to reduce population structure and linkage disequilibrium (LD) between unlinked loci, thereby creating an ideal condition for association analysis (Breseghello and Sorrells 2006). However, a challenge that may arise in this type of material is associated with genetic heterogeneity within samples. Core collections serve as valuable resources for association analysis of quantitative traits, such as disease resistance, or distinctive quality characteristics such as aroma and color.
2. *Elite breeding materials or cultivars*: In the case of elite materials, the sample may consist of lines and checks assessed in regional trials. Marker-assisted selection is expected to have the highest relative efficiency compared to phenotypic selection when heritability is low and the linked markers explain a substantial portion of the trait variation (Lande and Thompson 1990). Elite lines are ideal materials for association mapping of traits with low heritability, such as yield, yield components, and tolerance to abiotic stresses, due to their genetic stability and adaptation to standard growing conditions. Also, the advantage of

Table 2.3 Comparison of different type of association mapping panel

Features	Core panel	Elite germplasm	Synthetic population
Type of sample	Natural population or landraces	Breeding germplasm and cultivars	Progenies from bi-parental or multi-parental advance crosses
Types of traits that can be screened	Quantitative traits such as color and aroma	Quantitative traits like yield, biotic abiotic stress	Depends of the type of traits that has been introgressed
Expected LD	Low	High	Initially high with intermediate decaying
Population structure	Moderate	High	Less
Heritability	High	Low	Depend on the traits
Resolution of AM	High	Low	Intermediate elevating
Power of AM	Low	High	Intermediate declining
Utilization of identified significant markers	In MAS backcross	In forward genetics as well as genome selection (GC)	Incorporation in selection index

using elite lines is the availability of phenotypic data, often accumulated from replicated plant breeding field trials across various locations and years. Elite plant breeding material often exhibits a prominent population structure, as closely related lines are frequently derived from a limited number of founders in recent years and are expected to show high LD (Ching et al. 2002). If the pedigrees of the population are known, the relationships among the lines can be estimated and can be utilized for managing the polygenic effects. Significant advantages of using elite lines for AM in plants include a substantially higher level of polymorphism and the ability to detect favorable alleles directly within the target population.

3. *Synthetic population*: The potential of using synthetic population (such as RILs, F_2, NILs, backcross) in association mapping study is not clearly known. However, they could represent the plant breeding material that most closely approximates the assumption of random mating, as synthetics are typically designed and maintained to minimize inbreeding (Breseghello and Sorrells 2006). In the case of synthetic mapping materials, population structure is expected to be absent or very mild which will increase the potential of association mapping. Recently, multi-parent advanced synthetic populations such as nested association mapping (NAM) and multi-parent advanced generation intercrosses (MAGIC) populations are gaining popularity in AM studies. NAM involves the use of large recombinant inbred lines created from a diverse set of bi-parental mapping, enabling efficient genome-wide scans with greater resolution than QTL linkage mapping. The procedure of NAM development was first described by Yu et al. (2008) in maize. The steps involved are selection of diverse parents and developing of a huge set of mapping progenies (RILs with diverse phenotypic traits), and then sequencing of the complete set of progenies or selectively genotyping the founders of the progenies, phenotyping of the progenies for trait of interest and then conducting genome wide association mapping study for identifying trait specific markers. GWAS analysis using the NAM population has been reported in many crops (Tian et al. 2011; Hu et al. 2018). MAGIC populations comprise of crossing eight selected parents (could be landraces of cultivars) leading to generation of numerous crosses involving multiple parents and multiple rounds of recombination events and are generally developed from the germplasm having trait of interest. These populations also enable the detection of multiple alleles at target loci, combined with the trait of interest, providing opportunities to uncover chromosomal recombination and gene introgression in crops. At the beginning, single crosses between four parents (leading to the formation of four single crosses) are made; then, in the second generation, these four crossed lines are used for producing two double crosses and at the end these double crosses are crossed to produced Magic population. AM studies using MAGIC population were reported by Diaz et al. (2020) and Ayaad et al. (2021) and many others in crop plants. In synthetic populations, the initial level of linkage disequilibrium (LD) is expected to be high, allowing a genome scan to detect large chromosome segments associated with traits. As recombination occurs in subsequent generations, the decay of LD favors increasingly precise mapping.

2.8 Phenotyping and Genotyping of Association Panel

Precise phenotyping of the association mapping panel is the prerequisite for every genetic mapping study. The enhanced number of individuals or lines increases the strength of marker-trait associations more than increasing the number of markers for genotyping. The phenotypic data can be enhanced by increasing the number of replications (minimizes statistical errors) or trials across different locations and years (helps in estimating the genotype and genotype X environment interaction effects), which in turn increases the power of significant marker/QTL detection, (Zhu et al. 2008; Stich and Melchinger 2010). Genotyping of association panels for identifying polymorphic markers (SSR, SNPs) can be achieved through SNP arrays, reduced representation sequencing, amplicon sequencing, whole genome resequencing, and RNA-seq analysis. Advancement in high throughput phenotyping and various genotyping technologies as discussed in the previous section offers precise phenotyping of complex traits and higher marker density which in turn improves the resolution and efficiency of association mapping study.

2.9 Statistical Methods Used for Association Mapping

Once the high-quality phenotypic and genotypic data are generated, various statistical methods are employed to display marker-trait association. Conceptually, AM leads to the identification of SNPs where genotypic variation is significantly associated with phenotypic variation. This can be achieved by performing a separate ANOVA statistical test for each SNP. Each SNP is tested under the null hypothesis that there is no difference in the trait mean data among the genotype groups (AA, Aa, and aa) (Bush and Moore 2012). However, a major problem with these simple approaches is that they lead to a high rate of false positives. When more markers are tested, the possibility of getting a false positive increases. Thus, to correct these limiting false discovery rates (FDRs), multiple testing corrections such as Benjamini and Hochberg, or Bonferroni correction methods are used to determine the correct threshold level for a marker to be significant. Besides the determination of a significant threshold, another factor that increases the false positive is the relatedness among the association panel. In the diverse populations commonly employed for AM, certain individuals exhibit closer relations to each other than to others, thus creating subpopulations of related individuals within the overall population. It is challenging to minimize or avoid the presence of unequal relationships within a population (Yu et al. 2006; Zhu and Yu 2009). Thus, before starting the association mapping analysis, genotypic data should be used to address confounding factors such as population structure and FDR (Fig. 2.4).

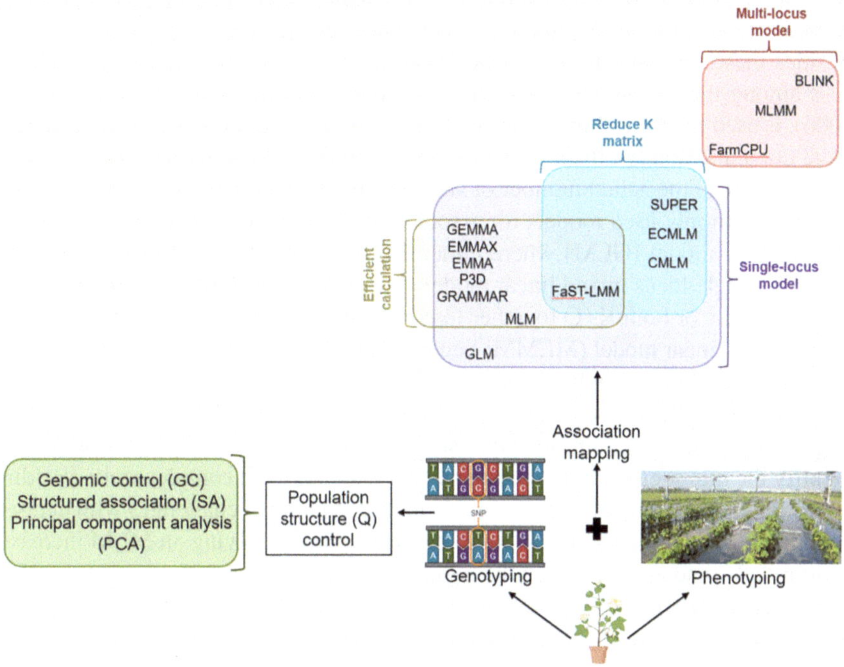

Fig. 2.4 Diagrammatic representation of association mapping with various statistical models

2.10 Population Structure (Q) Analysis

Natural and artificial selection, and geographic isolation, play a crucial role in population structure, which represents the arrangement of individuals in a population into distinct subgroups with varying gene frequencies. In most AM studies, the AM panel typically comprises a structured population consisting of several subpopulations that vary in the frequencies of alleles at numerous marker loci. Due to the existence of distinct sub-divisions in the population, the likelihood of an individual having a specific phenotype is different from one group to another. When a phenotype is correlated within the population for specific traits, it may show a statistically significant association with the loci but biologically it could be an invalid association. Thus, if the population structure study is not taken into consideration, then it can lead to confounding in the results due to genome-wide LD between unlinked marker loci. To avoid these difficulties, different models are used to reduce the effect of population structure. It can be analyzed using several available methods such as a set of unlinked markers obtained from the genotyping of the AM panel. The first method to control population structure is "genomic control" (GC) described by Devlin and Roeder (1999) and standardizes/scales the test statistics uniformly, ensuring that the test statistics are equal to the expected value. However, this method is rarely used in plants. Soon after that, the second method "structured association"

(SA) or "STRUCTURE" was developed by Pritchard and Rosenberg (1999), which is used to define the subpopulation and estimate the "Q" factor for each individual. Besides these, the kinship matrix (*K*; Yu et al. 2006), which estimates the relatedness among the individuals, and principal component analysis (PCA; Price et al. 2006), is used for controlling false positive estimation. PCA is generally regarded as more computationally efficient than STRUCTURE, as the latter has been observed to overestimate the actual number of subpopulations in certain situation.

The commonly used models for association mapping are (1) single locus generalized linear model (GLM) where either GC, Q, or PCA cofactor (only one) are used, (2) single locus mixed linear model (MLM; Yu et al. 2006), that uses either K or Q cofactors or both K+Q together to correct population structure, and (3) multilocus mixed linear model (MLMM: Segura et al. 2012). GLM model was one of the previously used models which due to its drawbacks has been replaced by MLM and MLMM. Regardless of the statistical method employed, GWAS would require a large sample size to attain adequate statistical power, especially when aiming to identify the minor effect polymorphisms that underlie most complex traits (Balding 2006). The MLM approach faces significant computational challenges with datasets containing large samples, as the computational time for solving an MLM increases substantially, and an increase in the number of individuals leads to a random effect (Zhang et al. 2010). Keeping in view of these, the MLM approach was implemented with Efficient Mixed Model Association (EMMA) which improves the computational speed by reducing the redundant matrix at each iteration of the likelihood function (Kang et al. 2008). In addition, it can also be used for calculating the kinship matrix. This method has been used in many AM analyses (Li et al. 2010), yet it finds its drawbacks when there is an increased number of data points. To overcome the drawbacks of EMMA, several other approaches have been implemented with software like TASSEL, SAS, GAPIT, and R such as Genome-wide Rapid Association using Mixed Model and Regression (GRAMMAR; Aulchenko et al. 2007), Population Parameters Previously Determined (P3D), Compressed MLM (CMLM, Zhang et al. 2010), EMMA eXpedited (EMMAX; Kang et al. 2010), Genome-wide Efficient Mixed Model Analysis (GEMMA; Zhou and Stephens 2012), FActored Spectrally Transformed Linear Mixed Models (FaST-LMM; Lippert et al. 2011), and Settlement of MLM Under Progressively Exclusive Relationship (SUPER; Wang et al. 2014).

GRAMMAR estimates the residuals adjusted for family effects, and then it treats these residuals as phenotypes and genetic polymorphisms for genome-wide quantitative trait nucleotide association analysis using rapid least-squares methods, thereby reducing the per-SNP computation time (Aulchenko et al. 2007). P3D eliminates the need to estimate population parameters, such as variance components, for each genetic marker test, while maintaining statistical power regardless of the genetic architecture of the phenotypes. The CMLM approach simplifies analysis by grouping individuals based on kinship, replacing pairwise kinship calculations between individuals with kinship calculations and between groups for the random effects in an MLM. Combining P3D and CMLM methods hugely reduces the computational demands and increases the statistical power (Zhang et al. 2010). EMMAX can significantly decrease computational time by leveraging the fact that each locus (SNP) only explains a small portion of the complex trait. This allows it to avoid repetitive variance component estimation and conduct association mapping using extensive data in a very short time. The variance component approach of EMMAX is advantageous as it considers a broad range of sample structures, including hidden relatedness and population stratification, and surpasses PCA and GC in correcting for sample structure (Kang et al. 2010). Both GEMMA and Fast-LMM improve efficiency by rewriting the likelihood function so that it becomes easy to evaluate. In the SUPER method, after sorting SNPs based on their association with the trait of interest, the genome is partitioned into bins. Within each bin, the SNP with the lowest *p*-value is designated as the pseudo-quantitative trait nucleotide. These SNPs are then used to construct a reduced kinship matrix, which is used to account for the genetic relationships between individuals in the study (Fig. 2.4).

The MLM model efficiently tests the association between a single marker locus and a single trait. However, it is widely recognized that complex traits are influenced by a large network of genes exhibiting pleiotropy. Thus, estimating complex traits with a single-locus MLM model will give biased results. To overcome these problems, a multilocus mixed linear model (MMLM) is used to determine markers having pleiotropic effects. MLMM model upgraded with Fixed and random model circulating probability unification (FarmCPU; Liu et al. 2016), and Bayesian information and LD Iteratively Nested Keyway (BLINK; Huang et al. 2019) are presently prevailing statistical models for estimating marker trait-association studies. FarmCPU (C language) reduces the kinship matrix and iterates the fixed effect mode and random effects using the restricted maximum likelihood (REML) criterion (Liu et al. 2016). FarmCPU was further modified to produce the BLINK method, which improves power by easing the constraint of random effects that markers must be uniformly distributed in bins across the genome (Fig. 2.4). This modification also enhances speed, as it eliminates the need to optimize bin size and number. In addition, BLINK enhances computational efficiency by replacing the resource-intensive random-effect model and its associated REML optimization with a fixed-effect model optimized using the Bayesian information criterion (Huang et al. 2019).

2.11 Determination of Significant Threshold

In order to correct FDR, many correction methods are used to determine the statistical threshold (*p*-value) level is set to identify significant SNPs. The various testing corrections to address the false-positive issue are as follows:

1. *Bonferroni correction*: It is calculated by multiplying the estimated *p*-value with the total number of SNPs (n) used in the AM (i.e., Corrected *p*-value = estimated *p*-value × n). For example, if the estimated *p*-value (α) is 0.05 (5% cut off) or 0.01 (1% cut off) for one million markers ($n = 10^6$), then the corrected *p*-value will be $p = 5 \times 10^6$ or 1×10^6. Thus, while plotting the marker-trait association

diagram (Manhattan plot), $-\log_{10}$ of calculated *p*-value is plotted and all markers above the threshold level are reported to be significant.

2. *Westfall and Young permutation*: It is a resampling-based approach used for controlling the familywise error rate (FWER) in multiple testing situations. This method is particularly valuable when the assumptions of traditional parametric tests are not met or when the data are non-normally distributed. The approach involves permuting the observed data to create a large number of possible permutations under the null hypothesis. For each permutation, the test statistic of interest is recalculated, and the maximum test statistic across all the permutations is recorded. This process generates a null distribution for the maximum test statistic, allowing for the calculation of adjusted *p*-values or critical values to control the FWER (Westfall et al. 1993).
3. *Benjamini and Hochberg FDR method*: Benjamini and Hochberg (1995) introduced the false discovery rate (FDR) control method, which has since become a fundamental tool in multiple testing scenarios. The FDR is defined as the expected proportion of false positives among the declared significant results. Unlike controlling the traditional Type I error rate, which is the probability of falsely rejecting a true null hypothesis, the FDR method offers a more appropriate approach for large-scale studies by controlling the rate of false discoveries. In their method, *p*-values are ranked, and adjusted *p*-values are calculated to control the FDR, thereby identifying significant results while minimizing the number of false positives.
4. *q-Value*: The q-value method, introduced by Storey in 2002, is indeed complementary to the FDR method of Benjamini and Hochberg and is a measure of significance in terms of the false discovery rate (FDR) rather than the false positive rate. It represents the minimum FDR at which a test may be called significant. The q-value method has widely gained its importance for offering a sensible balance between the number of true and false positives. It has become a valuable tool for researchers in controlling the FDR while identifying significant results in large-scale studies.

2.12 Imputation of Missing Data

Sequencing of genotypes with NGS and GBS techniques may lead to the generation of millions of SNP markers; however, the possibility of obtaining missing data is very high in some genotypes. These missing data can cause significant challenges in conducting AM studies. One effective approach to addressing this issue is data imputation, which entails replacing missing genotypes with predicted values derived from observed genotypes at adjacent markers. Imputation of missing genotypic data enhances the power of fine-map association analysis (Balding 2006; Marchini and Howie 2010).

2.13 Software and Packages Used for Association Mapping

Due to the advancement in plant genomic study along with high throughput phenotyping and genotyping methods and the generation of vast data in a cost-effective manner, interest in crop improvement programs has been increasing. Vast phenotypic and genotypic data not only improve the power and resolution of the AM study but also make the development of various new and robust software/packages to efficiently address the challenges related to AM. A list of several software and packages used for association mapping is given in Table 2.4.

2.14 Conclusion

Natural germplasm serves as a vital resource for genetic studies in crops. While the genetic diversity of cultivated crops has diminished significantly compared to their wild ancestors, a relatively high level of variation persists due to genetic drift and introgressions between domesticated crops and their closely related wild species. Such a diverse crop population serves as a potent resource that facilitates crop improvement through modern molecular breeding techniques. Both QTL and association mapping platforms are useful tools for leveraging the genetic diversity of crops cultivated by farmers to develop more comprehensive panels. It is now feasible to explore the narrow genetic diversity of these crops and identify the novel genes that govern the complex traits. Identification of markers/genes that are tightly linked with the agronomically important traits can be used as a molecular tool in marker-assisted breeding, ultimately leading to crop improvement and development of new varieties. Future challenges in enhancing plant breeding strategies through QTL and association mapping include addressing missing heritability, refining experimental designs, improving statistical analyses, and developing advanced algorithms for efficient epistasis detection. Other key areas of focus are accounting for population structure, ensuring accurate phenotyping, discovering rare alleles of genes or QTLs linked to uncommon marker alleles, conducting association studies, advancing marker-assisted selection, mapping QTLs using linkage and LD, and identifying quantitative trait nucleotides (QTNs) in crop plant species. Thus, this chapter serves as a guide for new breeders, offering a clear understanding of the complexities involved in QTL and association mapping, and how it can be effectively utilized for crop improvement. The potential of these molecular genetic techniques in identifying and leveraging genetic variations to enhance crop yield and quality cannot be overstated. As we move forward, breeders must embrace these advanced techniques, integrating them with traditional methods to meet the increasing demands of global food production and to overcome the challenges posed by a rapidly changing environment.

Table 2.4 List of software/packages used for association mapping study

S. no.	Software/ packages	Description	Examples
1	TASSEL	Calculation of LD statistics, graphic interphase, association mapping employing logistic regression, mixed model, linear model, and others; structures and kinship analysis, sequence analysis (http://www.maizegenetics.net)	Soybean (Contreras-Soto et al. 2017)
2	GAPIT	Implemented with a series of models (GLM, MLM, CMLM, SUPER, MLMM, FarmCPU, and BLINK for association mapping, R-based free software (http://www.maizegenetics.net/gapit)	Wheat (Mathew et al. 2019)
3	EMMAX	Used particularly for large-scale association mapping, corrects confounding factors due to population structure and kinship, offering high computational speed (http://genetics.cs.ucla.edu/emmax/)	Rice (Zhang et al. 2019)
4	STRUCTURE	Population structure analysis and genetic relatedness, generate Q matrix (http://pritch.bsd.uchicago.edu/structure.html)	Cotton (Huang et al. 2017)
5	ASReml	Commercial software, used for handling large data sets, calculates population structure and pedigree-based kinship using residual maximum likelihood (http://www.vsni.co.uk/products/asreml)	Maize (Kump et al. 2011)
6	GenABEL	Association mapping study for both qualitative and quantitative data (http://www.genabel.org/packages/GenABEL)	–
7	PLINK	Handle large data sets, data management, association study, summary statistics, population structure, LD analysis (http://pngu.mgh.harvard.edu/purcell/plink/)	Rice (Biscarini et al. 2016)
8	GenAMap	Visual and statistical analysis of structure association mapping, graphic interphase (http://sailing.cs.cmu.edu/genamap/)	–
9	FaST-LMM	Association mapping study for a large set of datasets (up to one million samples), Python-based software (http://fastlmm.codeplex.com/)	Rice (Yuan et al. 2020)
10	GEMMA	Estimation of proportion of variance in phenotypes explained using Bayesian sparse linear mixed model, association mapping using the univariate and multivariate linear mixed model (https://github.com/genetics-statistics/GEMMA/releases)	Maize (Xu et al. 2018)
11	EINGENSTRAT	PCA analysis to generate P matrix (instead of Q matrix); used in association mapping analysis (http://genepath.med/harvard.edu/~reich/software.html	Maize (Dwiningsih et al. 2022)
12	GenStat	Implements linear and mixed model, population structure analysis, and phenotypic data analysis (http://www.vsni.co.uk/software/genstat)	Sugarcane (Barreto et al. 2019)

References

Aakanksha, Yadava SK, Yadav BG, Gupta V, Mukhopadhyay A, Pental D, Pradhan AK (2021) Genetic analysis of heterosis for yield influencing traits in *Brassica juncea* using a doubled haploid population and its backcross progenies. Front Plant Sci 12:721631

Almeida GD, Makumbi D, Magorokosho C, Nair S, Borém A, Ribaut JM, Bänziger M, Prasanna BM, Crossa J, Babu R (2013) QTL mapping in three tropical maize populations reveals a set of constitutive and adaptive genomic regions for drought tolerance. Theor Appl Genet 126:583–600

Amiteye S (2021) Basic concepts and methodologies of DNA marker systems in plant molecular breeding. Heliyon 7(10)

An Y, Chen L, Tao L, Liu S, Wei C (2021) QTL mapping for leaf area of tea plants (*Camellia sinensis*) based on a high-quality genetic map constructed by whole genome resequencing. Front Plant Sci 12:705285

Ardlie KG, Kruglyak L, Seielstad M (2002) Patterns of linkage disequilibrium in the human genome. Nat Rev Genet 3(4):299–309

Arruda MP, Brown P, Brown-Guedira G, Krill AM, Thurber C, Merrill KR, Foresman BJ, Kolb FL (2016) Genome-wide association mapping of Fusarium head blight resistance in wheat using genotyping-by-sequencing. Plant Genome 9(1):plantgenome2015-04

Atkinson JA, Wingen LU, Griffiths M, Pound MP, Gaju O, Foulkes MJ, Le Gouis J, Griffiths S, Bennett MJ, King J, Wells DM (2015) Phenotyping pipeline reveals major seedling root growth QTL in hexaploid wheat. J Exp Bot 66(8):2283–2292

Aulchenko YS, De Koning DJ, Haley C (2007) Genomewide rapid association using mixed model and regression: a fast and simple method for genomewide pedigree-based quantitative trait loci association analysis. Genetics 177(1):577–585

Ayaad M, Han Z, Zheng K, Hu G, Abo-Yousef M, Sobeih El SS, Xing Y (2021) Bin-based genome-wide association studies reveal superior alleles for improvement of appearance quality using a 4-way MAGIC population in rice. J Adv Res 28:183–194

Bac-Molenaar JA, Vreugdenhil D, Granier C, Keurentjes JJ (2015) Genome-wide association mapping of growth dynamics detects time-specific and general quantitative trait loci. J Exp Bot 66(18):5567–5580

Balding DJ (2006) A tutorial on statistical methods for population association studies. Nat Rev Genet 7(10):781–791

Barik SR, Pandit E, Sanghamitra P, Mohanty SP, Behera A, Mishra J, Nayak DK, Bastia R, Moharana A, Sahoo A, Pradhan SK (2022) Unraveling the genomic regions controlling the seed vigour index, root growth parameters and germination per cent in rice. PLoS One 17(7):e0267303

Barreto FZ, Rosa JRBF, Balsalobre TWA, Pastina MM, Silva RR, Hoffmann HP, de Souza AP, Garcia AAF, Carneiro MS (2019) A genome-wide association study identified loci for yield component traits in sugarcane (*Saccharum* spp.). PLoS One 14(7):e0219843

Basten CJBS (1994) Zmap: a QTL cartographer. In: 5th world congress on genetics applied to livestock production: computing strategies and software. Guelph

Bauer E, Schmutzer T, Barilar I, Mascher M, Gundlach H, Martis MM, Twardziok SO, Hackauf B, Gordillo A, Wilde P, Schmidt M (2017) Towards a whole-genome sequence for rye (*Secale cereale* L.). Plant J 89(5):853–869

Beavis WD (1998) QTL analyses: power, precision, and accuracy. In: Paterson AH (ed) Molecular dissection of complex traits. CRC Press, New York, pp 145–162

Benjamini Y, Hochberg Y (1995) Controlling the false discovery rate: a practical and powerful approach to multiple testing. J R Stat Soc Ser B Methodol 57(1):289–300

Bennetzen JL, Schmutz J, Wang H, Percifield R, Hawkins J, Pontaroli AC, Estep M, Feng L, Vaughn JN, Grimwood J, Jenkins J (2012) Reference genome sequence of the model plant *Setaria*. Nat Biotechnol 30(6):555–561

Bernardo R (2008) Molecular markers and selection for complex traits in plants: learning from the last 20 years. Crop Sci 48(5):1649–1664

Bianco L, Cestaro A, Sargent DJ, Banchi E, Derdak S, Di Guardo M, Salvi S, Jansen J, Viola R, Gut I, Laurens F (2014) Development and validation of a 20K single nucleotide polymorphism (SNP) whole genome genotyping array for apple (Malus× domestica Borkh). PLoS One 9(10):e110377

Billotte N, Jourjon MF, Marseillac N, Berger A, Flori A, Asmady H, Adon B, Singh R, Nouy B, Potier F, Cheah SC (2010) QTL detection by multi-parent linkage mapping in oil palm (*Elaeis guineensis* Jacq.). Theor Appl Genet 120:1673–1687

Biscarini F, Cozzi P, Casella L, Riccardi P, Vattari A, Orasen G, Perrini R, Tacconi G, Tondelli A, Biselli C, Cattivelli L (2016) Genome-wide association study for traits related to plant and grain morphology, and root architecture in temperate rice accessions. PLoS One 11(5):e0155425

Boopathi NM, Boopathi NM (2020) Mapping population development. In: Genetic mapping and marker assisted selection: basics, practice and benefits, pp 69–106

Boopathi NM, Tiwari GJ, Jena SN, Nandhini K, Sri Subalakhshmi VKI, Shyamala P, Joshi B, Premalatha N, Rajeswari S (2022) Identification of stable and multiple environment interaction QTLs and candidate genes for fiber productive traits under irrigated and water stress conditions using intraspecific RILs of *Gossypium hirsutum* var. MCU5 X TCH1218. Front Plant Sci 13:851504

Bourke PM, Voorrips RE, Visser RG, Maliepaard C (2018) Tools for genetic studies in experimental populations of polyploids. Front Plant Sci 9:513

Bovenhuis H, Spelman RJ (2000) Selective genotyping to detect quantitative trait loci for multiple traits in outbred populations. J Dairy Sci 83(1):173–180

Bradshaw HD Jr, Wilbert SM, Otto KG, Schemske DW (1995) Genetic mapping of floral traits associated with reproductive isolation in monkeyflowers (*Mimulus*). Nature 376(6543):762–765

Breseghello F, Sorrells ME (2006) Association analysis as a strategy for improvement of quantitative traits in plants. Crop Sci 46(3):1323–1330

Buerstmayr M, Huber K, Heckmann J, Steiner B, Nelson JC, Buerstmayr H (2012) Mapping of QTL for Fusarium head blight resistance and morphological and developmental traits in three backcross populations derived from *Triticum dicoccum*× *Triticum durum*. Theor Appl Genet 125:1751–1765

Burr B, Burr FA, Thompson KH, Albertson MC, Stuber CW (1988) Gene mapping with recombinant inbreds in maize. Genetics 118(3):519–526

Bush WS, Moore JH (2012) Chapter 11: genome-wide association studies. PLoS Comput Biol 8(12):e1002822

Campbell MT, Grondin A, Walia H, Morota G (2020) Leveraging genome-enabled growth models to study shoot growth responses to water deficit in rice. J Exp Bot 71(18):5669–5679

Chen G, Geng J, Rahman M, Liu X, Tu J, Fu T, Li G, McVetty PB, Tahir M (2010) Identification of QTL for oil content, seed yield, and flowering time in oilseed rape (*Brassica napus*). Euphytica 175:161–174

Chen TW, Cabrera-Bosquet L, Alvarez Prado S, Perez R, Artzet S, Pradal C, Coupel-Ledru A, Fournier C, Tardieu F (2019) Genetic and environmental dissection of biomass accumulation in multi-genotype maize canopies. J Exp Bot 70(9):2523–2534

Ching ADA, Caldwell KS, Jung M, Dolan M, Smith O, Tingey S, Morgante M, Rafalski AJ (2002) SNP frequency, haplotype structure and linkage disequilibrium in elite maize inbred lines. BMC Genet 3:1–14

Collard BC, Jahufer MZZ, Brouwer JB, Pang ECK (2005) An introduction to markers, quantitative trait loci (QTL) mapping and marker-assisted selection for crop improvement: the basic concepts. Euphytica 142:169–196

Contreras-Soto RI, Mora F, de Oliveira MAR, Higashi W, Scapim CA, Schuster I (2017) A genome-wide association study for agronomic traits in soybean using SNP markers and SNP-based haplotype analysis. PLoS One 12(2):e0171105

Courtois B, Audebert A, Dardou A, Roques S, Ghneim-Herrera T, Droc G, Frouin J, Rouan L, Gozé E, Kilian A, Ahmadi N (2013) Genome-wide association mapping of root traits in a japonica rice panel. PLoS One 8(11):e78037

Crowell S, Korniliev P, Falcao A, Ismail A, Gregorio G, Mezey J, McCouch S (2016) Genome-wide association and high-resolution phenotyping link Oryza sativa panicle traits to numerous trait-specific QTL clusters. Nat Commun 7(1):10527

Das B, Ahmed N, Singh P (2011) Prunus diversity-early and present development: a review. Int J Biodivers Conserv 3(14):721–734

Davey J, Blaxter ML (2011) RADSeq: next-generation population genetics. Brief Funct Genomics 10:108

De Leon TB, Linscombe S, Subudhi PK (2016) Molecular dissection of seedling salinity tolerance in rice (*Oryza sativa* L.) using a high-density GBS-based SNP linkage map. Rice 9:1–22

Devlin B, Roeder K (1999) Genomic control for association studies. Biometrics 55(4):997–1004

Diaz S, Ariza-Suarez D, Izquierdo P, Lobaton JD, de la Hoz JF, Acevedo F, Duitama J, Guerrero AF, Cajiao C, Mayor V, Beebe SE (2020) Genetic mapping for agronomic traits in a MAGIC population of common bean (*Phaseolus vulgaris* L.) under drought conditions. BMC Genomics 21(1):1–20

Dwiningsih Y, Al-Kahtani J (2022) Genome-wide association study of complex traits in maize detects genomic regions and genes for increasing grain yield and grain quality. Adv Sustainable Sci EngTech 4(2):p.0220209

Ehrenreich IM, Hanzawa Y, Chou L, Roe JL, Kover PX, Purugganan MD (2009) Candidate gene association mapping of *Arabidopsis* flowering time. Genetics 183(1):325–335

Elshire RJ, Glaubitz JC, Sun Q, Poland JA, Kawamoto K, Buckler ES, Mitchell SE (2011) A robust, simple genotyping-by-sequencing (GBS) approach for high diversity species. PLoS One 6(5):e19379

Falconer DS (1996) Introduction to quantitative genetics. Pearson Education India

Fahlgren N, Feldman M, Gehan MA, Wilson MS, Shyu C, Bryant DW, Hill ST, McEntee CJ, Warnasooriya SN, Kumar I, Ficor T (2015) A versatile phenotyping system and analytics platform reveals diverse temporal responses to water availability in *Setaria*. Mol Plant 8(10):1520–1535

Fang H, Zhou H, Sanogo S, Flynn R, Percy RG, Hughs SE, Ulloa M, Jones DC, Zhang J (2013) Quantitative trait locus mapping for Verticillium wilt resistance in a backcross inbred line population of cotton (*Gossypium hirsutum*× *Gossypium barbadense*) based on RGA-AFLP analysis. Euphytica 194:79–91

Fang L, Wang Q, Hu Y, Jia Y, Chen J, Liu B, Zhang Z, Guan X, Chen S, Zhou B, Mei G (2017) Genomic analyses in cotton identify signatures of selection and loci associated with fiber quality and yield traits. Nat Genet 49(7):1089–1098

Feng Q, Zhang Y, Hao P, Wang S, Fu G, Huang Y, Li Y, Zhu J, Liu Y, Hu X, Jia P (2002) Sequence and analysis of rice chromosome 4. Nature 420(6913):316–320

Feng H, Guo Z, Yang W, Huang C, Chen G, Fang W, Xiong X, Zhang H, Wang G, Xiong L, Liu Q (2017) An integrated hyperspectral imaging and genome-wide association analysis platform provides spectral and genetic insights into the natural variation in rice. Sci Rep 7(1):4401

Ferguson JN, Fernandes SB, Monier B, Miller ND, Allen D, Dmitrieva A, Schmuker P, Lozano R, Valluru R, Buckler ES, Gore MA (2021) Machine learning-enabled phenotyping for GWAS and TWAS of WUE traits in 869 field-grown sorghum accessions. Plant Physiol 187(3):1481–1500

Fiorani F, Schurr U (2013) Future scenarios for plant phenotyping. Annu Rev Plant Biol 64:267–291

Flint-Garcia SA, Thornsberry JM, Buckler ES IV (2003) Structure of linkage disequilibrium in plants. Annu Rev Plant Biol 54(1):357–374

Fulop D, Ranjan A, Ofner I, Covington MF, Chitwood DH, West D, Ichihashi Y, Headland L, Zamir D, Maloof JN, Sinha NR (2016) A new advanced backcross tomato population enables high resolution leaf QTL mapping and gene identification. G3 Genes Genomes Genetics 6(10):3169–3184

Furbank RT, Tester M (2011) Phenomics–technologies to relieve the phenotyping bottleneck. Trends Plant Sci 16(12):635–644

Gage JL, de Leon N, Clayton MK (2018a) Comparing genome-wide association study results from different measurements of an underlying phenotype. G3 Genes Genomes Genetics 8(11):3715–3722
Gage JL, White MR, Edwards JW, Kaeppler S, de Leon N (2018b) Selection signatures underlying dramatic male inflorescence transformation during modern hybrid maize breeding. Genetics 210(3):1125–1138
Ganal MW, Durstewitz G, Polley A, Bérard A, Buckler ES, Charcosset A, Clarke JD, Graner EM, Hansen M, Joets J, Le Paslier MC (2011) A large maize (*Zea mays* L.) SNP genotyping array: development and germplasm genotyping, and genetic mapping to compare with the B73 reference genome. PLoS One 6(12):e28334
Ge Y, Bai G, Stoerger V, Schnable JC (2016) Temporal dynamics of maize plant growth, water use, and leaf water content using automated high throughput RGB and hyperspectral imaging. Comput Electron Agric 127:625–632
Geldermann H (1975) Investigations on inheritance of quantitative characters in animals by gene markers I. Methods. Theor Appl Genet 46(7):319–330
Giovannoni JJ, Wing RA, Ganal MW, Tanksley SD (1991) Isolation of molecular markers from specific chromosomal intervals using DNA pools from existing mapping populations. Nucleic Acids Res 19(23):6553–6568
Glaubitz JC, Casstevens TM, Lu F, Harriman J, Elshire RJ, Sun Q, Buckler ES (2014) TASSEL-GBS: a high capacity genotyping by sequencing analysis pipeline. PLoS One 9(2):e90346
Golzarian MR, Frick RA, Rajendran K, Berger B, Roy S, Tester M, Lun DS (2011) Accurate inference of shoot biomass from high-throughput images of cereal plants. Plant Methods 7:1–11
Gouy M, Rousselle Y, Thong Chane A, Anglade A, Royaert S, Nibouche S, Costet L (2015) Genome wide association mapping of agro-morphological and disease resistance traits in sugarcane. Euphytica 202(2):269–284
Granier C, Vile D (2014) Phenotyping and beyond: modelling the relationships between traits. Curr Opin Plant Biol 18:96–102
Guan Y, Peace C, Rudell D, Verma S, Evans K (2015) QTLs detected for individual sugars and soluble solids content in apple. Mol Breed 35:1–13
Guan P, Di N, Mu Q, Shen X, Wang Y, Wang X, Yu K, Song W, Chen Y, Xin M, Hu Z (2019) Use of near-isogenic lines to precisely map and validate a major QTL for grain weight on chromosome 4AL in bread wheat (*Triticum aestivum* L.). Theor Appl Genet 132:2367–2379
Guo Z, Yang W, Chang Y, Ma X, Tu H, Xiong F, Jiang N, Feng H, Huang C, Yang P, Zhao H (2018) Genome-wide association studies of image traits reveal genetic architecture of drought resistance in rice. Mol Plant 11(6):789–805
Guo Z, Wang S, Li WX, Liu J, Guo W, Xu M, Xu Y (2022) QTL mapping and genomic selection for Fusarium ear rot resistance using two $F_{2:3}$ populations in maize. Euphytica 218(9):131
Guo G, Xu S, Chen H, Hao Y, Mao H (2023) QTL mapping for wheat seed dormancy in a Yangmai16/Zhongmai895 double haploid population. Plants 12(4):759
Gutierrez AF, Hoy JW, Kimbeng CA, Baisakh N (2018) Identification of genomic regions controlling leaf scald resistance in sugarcane using a bi-parental mapping population and selective genotyping by sequencing. Front Plant Sci 9:877
Guzman HAR, González-Navarro FF, Selvaraj-Gomez M, Valencia M, Delgado A (2015) Field phenomics: a web based image analysis platform using open source tools. In: 2015 international conference on computational science and computational intelligence (CSCI). IEEE, pp 851–852
Hatzig SV, Frisch M, Breuer F, Nesi N, Ducournau S, Wagner MH, Leckband G, Abbadi A, Snowdon RJ (2015) Genome-wide association mapping unravels the genetic control of seed germination and vigor in *Brassica napus*. Front Plant Sci 6:221
Hickey LT, Hafeez N, Robinson H, Jackson SA, Leal-Bertioli SC, Tester M, Gao C, Godwin ID, Hayes BJ, Wulff BB (2019) Breeding crops to feed 10 billion. Nat Biotechnol 37(7):744–754
Hill WG, Robertson A (1968) Linkage disequilibrium in finite populations. Theor Appl Genet 38:226–231

Honsdorf N, March TJ, Berger B, Tester M, Pillen K (2014) High-throughput phenotyping to detect drought tolerance QTL in wild barley introgression lines. PLoS One 9(5):e97047

Hoshyardel F, Darvishzadeh R, Hatami Maleki H, Jannatdoust M (2021) Mapping QTLs associated with chloride accumulation in leaves of oriental tobacco (*Nicotiana tabacum* L.) using $F_{2:3}$ population of Basma Seres 31× SPT 406 cross. *Journal of Plant.* Mol Breed 7(2):12–18

Hospital F (2005) Selection in backcross programmes. Philos Trans R Soc B Biol Sci 360(1459):1503–1511

Hu J, Guo C, Wang B, Ye J, Liu M, Wu Z, Xiao Y, Zhang Q, Li H, King GJ, Liu K (2018) Genetic properties of a nested association mapping population constructed with semi-winter and spring oilseed rapes. Front Plant Sci 9:1740

Hu J, Wang X, Zhang G, Jiang P, Chen W, Hao Y, Ma X, Xu S, Jia J, Kong L, Wang H (2020) QTL mapping for yield-related traits in wheat based on four RIL populations. Theor Appl Genet 133:917–933

Hua JP, Xing YZ, Xu CG, Sun XL, Yu SB, Zhang Q (2002) Genetic dissection of an elite rice hybrid revealed that heterozygotes are not always advantageous for performance. Genetics 162(4):1885–1895

Huang X, Wei X, Sang T, Zhao Q, Feng Q, Zhao Y, Li C, Zhu C, Lu T, Zhang Z, Li M (2010) Genome-wide association studies of 14 agronomic traits in rice landraces. Nat Genet 42(11):961–967

Huang X, Zhao Y, Wei X, Li C, Wang A, Zhao Q, Li W, Guo Y, Deng L, Zhu C, Fan D (2012) Genome-wide association study of flowering time and grain yield traits in a worldwide collection of rice germplasm. Nat Genet 44(1):32–39

Huang C, Nie X, Shen C, You C, Li W, Zhao W, Zhang X, Lin Z (2017) Population structure and genetic basis of the agronomic traits of upland cotton in China revealed by a genome-wide association study using high-density SNPs. Plant Biotechnol J 15(11):1374–1386

Huang M, Liu X, Zhou Y, Summers RM, Zhang Z (2019) BLINK: a package for the next level of genome-wide association studies with both individuals and markers in the millions. Gigascience 8(2):giy154

Hulse-Kemp AM, Lemm J, Plieske J, Ashrafi H, Buyyarapu R, Fang DD, Frelichowski J, Giband M, Hague S, Hinze LL, Kochan KJ (2015) Development of a 63K SNP array for cotton and high-density mapping of intraspecific and interspecific populations of *Gossypium* spp. G3 Genes Genomes Genetics 5(6):1187–1209

Iorizzo M, Ellison S, Senalik D, Zeng P, Satapoomin P, Huang J, Bowman M, Iovene M, Sanseverino W, Cavagnaro P, Yildiz M (2016) A high-quality carrot genome assembly provides new insights into carotenoid accumulation and asterid genome evolution. Nat Genet 48(6):657–666

Jannink JL, Bink MC, Jansen RC (2001) Using complex plant pedigrees to map valuable genes. Trends Plant Sci 6(8):337–342

Jansen RC, Stam P (1994) High resolution of quantitative traits into multiple loci via interval mapping. Genetics 136(4):1447–1455

Jansen M, Gilmer F, Biskup B, Nagel KA, Rascher U, Fischbach A, Briem S, Dreissen G, Tittmann S, Braun S, De Jaeger I (2009) Simultaneous phenotyping of leaf growth and chlorophyll fluorescence via GROWSCREEN FLUORO allows detection of stress tolerance in Arabidopsis thaliana and other rosette plants. Funct Plant Biol 36(11):902–914

Jasinski S, Lécureuil A, Durandet M, Bernard-Moulin P, Guerche P (2016) *Arabidopsis* seed content QTL mapping using high-throughput phenotyping: the assets of near infrared spectroscopy. Front Plant Sci 7:1682

Jennings HS (1917) The numerical results of diverse systems of breeding, with respect to two pairs of characters, linked or independent, with special relation to the effects of linkage. Genetics 2(2):97

Jiang Y, Habib A, Zheng Z, Zhou M, Wei Y, Zheng YL, Liu C (2019) Development of tightly linked markers and identification of candidate genes for *Fusarium* crown rot resistance in barley by exploiting a near-isogenic line-derived population. Theor Appl Genet 132:217–225

Jin T, Zhou J, Chen J, Zhu L, Zhao Y, Huang Y (2013) The genetic architecture of zinc and iron content in maize grains as revealed by QTL mapping and meta-analysis. Breed Sci 63(3):317–324

Jordan D, Tao Y, Godwin I, Henzell R, Cooper M, McIntyre C (2003) Prediction of hybrid performance in grain sorghum using RFLP markers. Theor Appl Genet 106:559–567

Joshi B, Singh S, Tiwari GJ, Kumar H, Boopathi NM, Jaiswal S, Adhikari D, Kumar D, Sawant SV, Iquebal MA, Jena SN (2023) Genome-wide association study of fiber yield-related traits uncovers the novel genomic regions and candidate genes in Indian upland cotton (*Gossypium hirsutum* L.). Front Plant Sci 14

Kaler AS, Purcell LC (2019) Estimation of a significance threshold for genome-wide association studies. BMC Genomics 20:1–8

Kang HM, Ye C, Eskin E (2008) Accurate discovery of expression quantitative trait loci under confounding from spurious and genuine regulatory hotspots. Genetics 180(4):1909–1925

Kang HM, Sul JH, Service SK, Zaitlen NA, Kong SY, Freimer NB, Sabatti C, Eskin E (2010) Variance component model to account for sample structure in genome-wide association studies. Nat Genet 42(4):348–354

Karam MJ, Souleman D, Schvartzman MS, Gallina S, Spielmann J, Poncet C, Bouchez O, Pauwels M, Hanikenne M, Frérot H (2019) Genetic architecture of a plant adaptive trait: QTL mapping of intraspecific variation for tolerance to metal pollution in *Arabidopsis halleri*. Heredity 122(6):877–892

Kishikawa T, Momozawa Y, Ozeki T, Mushiroda T, Inohara H, Kamatani Y, Kubo M, Okada Y (2019) Empirical evaluation of variant calling accuracy using ultra-deep whole-genome sequencing data. Sci Rep 9(1):1784

Knoch D, Abbadi A, Grandke F, Meyer RC, Samans B, Werner CR, Snowdon RJ, Altmann T (2020) Strong temporal dynamics of QTL action on plant growth progression revealed through high-throughput phenotyping in canola. Plant Biotechnol J 18(1):68–82

Korinsak S, Sirithunya P, Meakwatanakarn P, Sarkarung S, Vanavichit A, Toojinda T (2011) Changing allele frequencies associated with specific resistance genes to leaf blast in backcross introgression lines of Khao Dawk Mali 105 developed from a conventional selection program. Field Crop Res 122(1):32–39

Kronenberg L, Yates S, Boer MP, Kirchgessner N, Walter A, Hund A (2021) Temperature response of wheat affects final height and the timing of stem elongation under field conditions. J Exp Bot 72(2):700–717

Kumar PK, Bellundagi A, Krishna H, Mallikarjuna MG, Thimmappa RK, Rai N, Shashikumara P, Sinha N, Jain N, Singh PK, Singh GP (2023a) Development of bread wheat (*Triticum aestivum* L) variety HD3411 following marker-assisted backcross breeding for drought tolerance. Front Genet 14:1046624

Kumar R, Saini M, Taku M, Debbarma P, Mahto RK, Ramlal A, Sharma D, Rajendran A, Pandey R, Gaikwad K, Lal SK (2023b) Identification of quantitative trait loci (QTLs) and candidate genes for seed shape and 100-seed weight in soybean [*Glycine max* (L.) Merr.]. Front Plant Sci 13:1074245

Kumawat G, Raje RS, Bhutani S, Pal JK, Mithra AS, Gaikwad K, Sharma TR, Singh NK (2012) Molecular mapping of QTLs for plant type and earliness traits in pigeonpea (*Cajanus cajan* L. Millsp.). BMC Genet 13(1):1–11

Kump KL, Bradbury PJ, Wisser RJ, Buckler ES, Belcher AR, Oropeza-Rosas MA, Zwonitzer JC, Kresovich S, McMullen MD, Ware D, Balint-Kurti PJ (2011) Genome-wide association study of quantitative resistance to southern leaf blight in the maize nested association mapping population. Nat Genet 43(2):163–168

Lanceras JC, Pantuwan G, Jongdee B, Toojinda T (2004) Quantitative trait loci associated with drought tolerance at reproductive stage in rice. Plant Physiol 135(1):384–399

Lande R, Thompson R (1990) Efficiency of marker-assisted selection in the improvement of quantitative traits. Genetics 124(3):743–756

Lander ES, Botstein D (1989) Mapping Mendelian factors underlying quantitative traits using RFLP linkage maps. Genetics 121(1):185–199

Lebedev VG, Lebedeva TN, Chernodubov AI, Shestibratov KA (2020) Genomic selection for forest tree improvement: methods, achievements and perspectives. Forests 11(11):1190

Lebowitz RJ, Soller M, Beckmann JS (1987) Trait-based analyses for the detection of linkage between marker loci and quantitative trait loci in crosses between inbred lines. Theor Appl Genet 73:556–562

Lee YG, Jeong N, Kim JH, Lee K, Kim KH, Pirani A, Ha BK, Kang ST, Park BS, Moon JK, Kim N (2015) Development, validation and genetic analysis of a large soybean SNP genotyping array. Plant J 81(4):625–636

Lee SM, Kang JW, Lee JY, Seo J, Shin D, Cho JH, Jo S, Song YC, Park DS, Ko JM, Koh HJ (2020) QTL analysis for Fe and Zn concentrations in rice grains using a doubled haploid population derived from a cross between rice (*Oryza sativa*) cultivar 93-11 and milyang 352. Plant Breed Biotechnol 8(1):69–76

Lewontin RC (1964) The interaction of selection and linkage. I. General considerations; heterotic models. Genetics 49(1):49

Li H, Ye G, Wang J (2007) A modified algorithm for the improvement of composite interval mapping. Genetics 175(1):361–374

Li Y, Huang Y, Bergelson J, Nordborg M, Borevitz JO (2010) Association mapping of local climate-sensitive quantitative trait loci in *Arabidopsis thaliana*. Proc Natl Acad Sci 107(49):21199–21204

Li F, Wen W, He Z, Liu J, Jin H, Cao S, Geng H, Yan J, Zhang P, Wan Y, Xia X (2018) Genome-wide linkage mapping of yield-related traits in three Chinese bread wheat populations using high-density SNP markers. Theor Appl Genet 131:1903–1924

Li B, Chen L, Sun W, Wu D, Wang M, Yu Y, Chen G, Yang W, Lin Z, Zhang X, Duan L (2020a) Phenomics-based GWAS analysis reveals the genetic architecture for drought resistance in cotton. Plant Biotechnol J 18(12):2533–2544

Li N, Shang J, Wang J, Zhou D, Li N, Ma S (2020b) Discovery of the genomic region and candidate genes of the scarlet red flesh color (Yscr) locus in watermelon (*Citrullus lanatus* L.). Front Plant Sci 11:116

Li H, Pan Z, He S, Jia Y, Geng X, Chen B, Wang L, Pang B, Du X (2021) QTL mapping of agronomic and economic traits for four F_2 populations of upland cotton. J Cotton Res 4:1–12

Lima MDLA, de Souza CL, Bento DAV, De Souza AP, Carlini-Garcia LA (2006) Mapping QTL for grain yield and plant traits in a tropical maize population. Mol Breed 17:227–239

Lincoln SE, Daly MJ, Lander ES (1993) Mapping genes controlling quantitative traits using MAPMAKER/QTL version 1.1: a tutorial and reference manual. Whitehead Institute for Biomedical Research Technical Report

Lippert C, Listgarten J, Liu Y, Kadie CM, Davidson RI, Heckerman D (2011) FaST linear mixed models for genome-wide association studies. Nat Methods 8(10):833–835

Liu X, Huang M, Fan B, Buckler ES, Zhang Z (2016) Iterative usage of fixed and random effect models for powerful and efficient genome-wide association studies. PLoS Genet 12(2):e1005767

Liu C, Chen K, Zhao X, Wang X, Shen C, Zhu Y, Dai M, Qiu X, Yang R, Xing D, Pang Y (2019) Identification of genes for salt tolerance and yield-related traits in rice plants grown hydroponically and under saline field conditions by genome-wide association study. Rice 12:1–13

Liu R, Xiao X, Gong J, Li J, Zhang Z, Liu A, Lu Q, Shang H, Shi Y, Ge Q, Iqbal MS (2020) QTL mapping for plant height and fruit branch number based on RIL population of upland cotton. J Cotton Res 3(1):1–9

Liu Z, Dong X, Zheng G, Xu C, Wei J, Cui J, Cao X, Li H, Fang X, Wang Y, Tian H (2022) Integrate QTL mapping and transcription profiles reveal candidate genes regulating flowering time in *Brassica napus*. Front Plant Sci 13:904198

Lu Y, Savage LJ, Last RL (2011) Chloroplast phenomics: systematic phenotypic screening of chloroplast protein mutants in *Arabidopsis*. In: Chloroplast research in arabidopsis: methods and protocols, vol II, pp 161–185

Lv Y, Gao P, Liu S, Fang X, Zhang T, Liu T, Amanullah S, Wang X, Luan F (2022) Genetic mapping and QTL analysis of stigma color in melon (*Cucumis melo* L.). Front Plant Sci 13:865082

Lynch M, Walsh B (1998) Genetics and analysis of quantitative traits, vol 1. Sinauer, Sunderland, pp 535–557

Ma J, Pei W, Ma Q, Geng Y, Liu G, Liu J, Cui Y, Zhang X, Wu M, Li X, Li D (2019) QTL analysis and candidate gene identification for plant height in cotton based on an interspecific backcross inbred line population of *Gossypium hirsutum*× *Gossypium barbadense*. Theor Appl Genet 132:2663–2676

Mackay I, Powell W (2007) Methods for linkage disequilibrium mapping in crops. Trends Plant Sci 12(2):57–63

Manly KF, Cudmore RH Jr, Meer JM (2001) Map Manager QTX, cross-platform software for genetic mapping. Mamm Genome 12:930–932

Marchini J, Howie B (2010) Genotype imputation for genome-wide association studies. Nat Rev Genet 11(7):499–511

Mathew I, Shimelis H, Shayanowako AIT, Laing M, Chaplot V (2019) Genome-wide association study of drought tolerance and biomass allocation in wheat. PLoS One 14(12):e0225383

Maxwell K, Johnson GN (2000) Chlorophyll fluorescence—a practical guide. J Exp Bot 51(345):659–668

Mayer KF, Waugh R, Langridge P, Close TJ, Wise RP, Graner A, Matsumoto T, Sato K, Schulman A, Muehlbauer GJ, Stein N (2012) A physical, genetic and functional sequence assembly of the barley genome. Nature 491:711–716

McCormick RF, Truong SK, Mullet JE (2016) 3D sorghum reconstructions from depth images identify QTL regulating shoot architecture. Plant Physiol 172(2):823–834

Meyer RC, Weigelt-Fischer K, Knoch D, Heuermann M, Zhao Y, Altmann T (2021) Temporal dynamics of QTL effects on vegetative growth in *Arabidopsis thaliana*. J Exp Bot 72(2):476–490

Miao C, Pages A, Xu Z, Rodene E, Yang J, Schnable JC (2020) Semantic segmentation of sorghum using hyperspectral data identifies genetic associations. Plant Phenomics

Milad SI, Wahba LE, Barakat MN (2011) Identification of RAPD and ISSR markers associated with flag leaf senescence under water-stressed conditions in wheat (*Triticum aestivum* L.). Aust J Crop Sci 5(3):337–343

Moncada MDP, Tovar E, Montoya JC, González A, Spindel J, McCouch S (2016) A genetic linkage map of coffee (*Coffea arabica* L.) and QTL for yield, plant height, and bean size. Tree Genet Genomes 12:1–17

Moore CR, Johnson LS, Kwak IY, Livny M, Broman KW, Spalding EP (2013) High-throughput computer vision introduces the time axis to a quantitative trait map of a plant growth response. Genetics 195(3):1077–1086

Morales L, Repka AC, Swarts KL, Stafstrom WC, He Y, Sermons SM, Yang Q, Lopez-Zuniga LO, Rucker E, Thomason WE, Nelson RJ (2020) Genotypic and phenotypic characterization of a large, diverse population of maize near-isogenic lines. Plant J 103(3):1246–1255

Munns R, James RA, Sirault XR, Furbank RT, Jones HG (2010) New phenotyping methods for screening wheat and barley for beneficial responses to water deficit. J Exp Bot 61(13):3499–3507

Muraya MM, Chu J, Zhao Y, Junker A, Klukas C, Reif JC, Altmann T (2017) Genetic variation of growth dynamics in maize (*Zea mays* L.) revealed through automated non-invasive phenotyping. Plant J 89(2):366–380

Muturi PW, Mgonja M, Rubaihayo P, Mwololo JK (2021) QTL mapping of traits associated with dual resistance to the African stem borer (*Busseola fusca*) and spotted stem borer (*Chilo partellus*) in sorghum (*Sorghum bicolor*). Int J Genomics

Nelson JC (1997) QGENE: software for marker-based genomic analysis and breeding. Mol Breed 3:239–245

Neumann K, Klukas C, Friedel S, Rischbeck P, Chen D, Entzian A, Stein N, Graner A, Kilian B (2015) Dissecting spatiotemporal biomass accumulation in barley under different water regimes using high-throughput image analysis. Plant Cell Environ 38(10):1980–1996

Niemirowicz-Szczytt K (1997) Excessive homozygosity in doubled haploids—advantages and disadvantages for plant breeding and fundamental research. Acta Physiol Plant 19:155–167

Pang Y, Liu C, Wang D, Amand PS, Bernardo A, Li W, He F, Li L, Wang L, Yuan X, Dong L (2020) High-resolution genome-wide association study identifies genomic regions and candidate genes for important agronomic traits in wheat. Mol Plant 13(9):1311–1327

Parent B, Shahinnia F, Maphosa L, Berger B, Rabie H, Chalmers K, Kovalchuk A, Langridge P, Fleury D (2015) Combining field performance with controlled environment plant imaging to identify the genetic control of growth and transpiration underlying yield response to water-deficit stress in wheat. J Exp Bot 66(18):5481–5492

Park J, Hackett CA, Dandurand LM, Wang X, De Jong WS (2019) QTL for resistance to *Globodera rostochiensis* pathotype Ro2 and *G. pallida* pathotype Pa2/3 in autotetraploid potato. Am J Potato Res 96:552–563

Pasam RK, Sharma R, Malosetti M, van Eeuwijk FA, Haseneyer G, Kilian B, Graner A (2012) Genome-wide association studies for agronomical traits in a world wide spring barley collection. BMC Plant Biol 12(1):1–22

Paterson AH, Bowers JE, Bruggmann R, Dubchak I, Grimwood J, Gundlach H, Haberer G, Hellsten U, Mitros T, Poliakov A, Schmutz J (2009) The *Sorghum bicolor* genome and the diversification of grasses. Nature 457(7229):551–556

Peng S, Khush GS (2003) Four decades of breeding for varietal improvement of irrigated lowland rice in the International Rice Research Institute. Plant Product Sci 6(3):157–164

Podlich DW, Winkler CR, Cooper M (2004) Mapping as you go: an effective approach for marker-assisted selection of complex traits. Crop Sci 44(5):1560–1571

Poland JA, Rife TW (2012) Genotyping-by-sequencing for plant breeding and genetics. Plant Genome 5(3)

Prado SA, Cabrera-Bosquet L, Grau A, Coupel-Ledru A, Millet EJ, Welcker C, Tardieu F (2018) Phenomics allows identification of genomic regions affecting maize stomatal conductance with conditional effects of water deficit and evaporative demand. Plant Cell Environ 41(2):314–326

Price AL, Patterson NJ, Plenge RM, Weinblatt ME, Shadick NA, Reich D (2006) Principal components analysis corrects for stratification in genome-wide association studies. Nat Genet 38(8):904–909

Pritchard JK, Rosenberg NA (1999) Use of unlinked genetic markers to detect population stratification in association studies. Am J Hum Genet 65(1):220–228

Qiao P, Lin M, Vasquez M, Matschi S, Chamness J, Baseggio M, Smith LG, Sabuncu MR, Gore MA, Scanlon MJ (2019) Machine learning enables high-throughput phenotyping for analyses of the genetic architecture of bulliform cell patterning in maize. G3 Genes Genomes Genetics 9(12):4235–4243

Quarrie SA, Lazić-Jančić V, Kovačević D, Steed A, Pekić S (1999) Bulk segregant analysis with molecular markers and its use for improving drought resistance in maize. J Exp Bot 50(337):1299–1306

Rajendran K, Tester M, Roy SJ (2009) Quantifying the three main components of salinity tolerance in cereals. Plant Cell Environ 32(3):237–249

Rapacz M, Wójcik-Jagła M, Fiust A, Kalaji HM, Kościelniak J (2019) Genome-wide associations of chlorophyll fluorescence OJIP transient parameters connected with soil drought response in barley. Front Plant Sci 10:78

Rasheed A, Hao Y, Xia X, Khan A, Xu Y, Varshney RK, He Z (2017) Crop breeding chips and genotyping platforms: progress, challenges, and perspectives. Mol Plant 10(8):1047–1064

Ravelombola W, Qin J, Shi A, Song Q, Yuan J, Wang F, Chen P, Yan L, Feng Y, Zhao T, Meng Y (2021) Genome-wide association study and genomic selection for yield and related traits in soybean. PLoS One 16(8):e0255761

Raverdy R, Lourgant K, Mignot E, Arnoult S, Bodineau G, Griveau Y, Taniguti CH, Brancourt-Hulmel M (2022) Linkage mapping of biomass production and composition traits in a *Miscanthus sinensis* population. BioEnergy Res 15(2):755–776

Risch N, Merikangas K (1996) The future of genetic studies of complex human diseases. Science 273(5281):1516–1517

Roorkiwal M, Bhandari A, Barmukh R, Bajaj P, Valluri VK, Chitikineni A, Pandey S, Chellapilla B, Siddique KH, Varshney RK (2022) Genome-wide association mapping of nutritional traits for designing superior chickpea varieties. Front Plant Sci 13:843911

Sadeque A, Turner MA (2010) QTL analysis of plant height in hexaploid wheat doubled haploid population. Thai J Agric Sci 43(2):91–96

Sax K (1923) The association of size differences with seed-coat pattern and pigmentation in *Phaseolus vulgaris*. Genetics 8(6):552

Saxena RK, Singh VK, Kale SM, Tathineni R, Parupalli S, Kumar V, Garg V, Das RR, Sharma M, Yamini KN, Muniswamy S (2017) Construction of genotyping-by-sequencing based high-density genetic maps and QTL mapping for fusarium wilt resistance in pigeonpea. Sci Rep 7(1):1911

Schneider K (2005) Mapping populations and principles of genetic mapping. In: The handbook of plant genome mapping: genetic and physical mapping, pp 1–21

Segura V, Vilhjálmsson BJ, Platt A, Korte A, Seren Ü, Long Q, Nordborg M (2012) An efficient multi-locus mixed-model approach for genome-wide association studies in structured populations. Nat Genet 44(7):825–830

Semagn K, Bjørnstad Å, Ndjiondjop MN (2006) An overview of molecular marker methods for plants. Afr J Biotechnol 5(25)

Semagn K, Bjørnstad Å, Xu Y (2010) The genetic dissection of quantitative traits in crops. Electron J Biotechnol 13(5):16–17

Shi L, Shi T, Broadley MR, White PJ, Long Y, Meng J, Xu F, Hammond JP (2013) High-throughput root phenotyping screens identify genetic loci associated with root architectural traits in *Brassica napus* under contrasting phosphate availabilities. Ann Bot 112(2):381–389

Shu YJ, Yu DS, Wang D, Bai X, Zhu YM, Guo CH (2013) Genomic selection of seed weight based on low-density SCAR markers in soybean. Genet Mol Res 12(3):2178–2188

Shukla RP, Tiwari GJ, Joshi B, Song-Beng K, Tamta S, Boopathi NM, Jena SN (2021) GBS-SNP and SSR based genetic mapping and QTL analysis for drought tolerance in upland cotton. Physiol Mol Biol Plants 27(8):1731–1745

Singh BD, Singh AK, Singh BD, Singh AK (2015) Mapping of quantitative trait loci. In: Marker-assisted plant breeding: principles and practices, pp 185–216

Skirycz A, Vandenbroucke K, Clauw P, Maleux K, De Meyer B, Dhondt S, Pucci A, Gonzalez N, Hoeberichts F, Tognetti VB, Galbiati M (2011) Survival and growth of *Arabidopsis* plants given limited water are not equal. Nat Biotechnol 29(3):212–214

Soller M, Brody T, Genizi A (1976) On the power of experimental designs for the detection of linkage between marker loci and quantitative loci in crosses between inbred lines. Theor Appl Genet 47:35–39

Sozzani R, Busch W, Spalding EP, Benfey PN (2014) Advanced imaging techniques for the study of plant growth and development. Trends Plant Sci 19(5):304–310

Stein CM, Elston RC (2009) Finding genes underlying human disease. Clin Genet 75(2):101–106

Stephen K, Aparna K, Beena R, Sah RP, Jha UC, Behera S (2023) Identification of simple sequence repeat markers linked to heat tolerance in rice using bulked segregant analysis in F_2 population of NERICA-L 44× Uma. Front Plant Sci 14:1113838

Stich B, Melchinger AE (2010) An introduction to association mapping in plants. CABI Rev 2010:1–9

Stich B, Melchinger AE, Piepho HP, Heckenberger M, Maurer HP, Reif JC (2006) A new test for family-based association mapping with inbred lines from plant breeding programs. Theor Appl Genet 113:1121–1130

Storey JD (2002) A direct approach to false discovery rates. J R Stat Soc Ser B Stat Methodol 64(3):479–498

Sturtevant, A.H., 1913. The linear arrangement of six sex-linked factors in Drosophila, as shown by their mode of association. J Exp Zool, 14(1), pp.43-59.

Sun FD, Zhang JH, Wang SF, Gong WK, Shi YZ, Liu AY, Li JW, Gong JW, Shang HH, Yuan YL (2012) QTL mapping for fiber quality traits across multiple generations and environments in upland cotton. Mol Breed 30:569–582

Tackenberg O (2007) A new method for non-destructive measurement of biomass, growth rates, vertical biomass distribution and dry matter content based on digital image analysis. Ann Bot 99(4):777–783

Talukder ZI, Hulke BS, Qi L, Scheffler BE, Pegadaraju V, McPhee K, Gulya TJ (2014) Candidate gene association mapping of Sclerotinia stalk rot resistance in sunflower (*Helianthus annuus* L.) uncovers the importance of COI1 homologs. Theor Appl Genet 127:193–209
Tanger P, Klassen S, Mojica JP, Lovell JT, Moyers BT, Baraoidan M, Naredo MEB, McNally KL, Poland J, Bush DR, Leung H (2017) Field-based high throughput phenotyping rapidly identifies genomic regions controlling yield components in rice. Sci Rep 7(1):42839
Tanksley SD, Nelson JC (1996) Advanced backcross QTL analysis: a method for the simultaneous discovery and transfer of valuable QTLs from unadapted germplasm into elite breeding lines. Theor Appl Genet 92:191–203
Tehseen MM, Tonk FA, Tosun M, Randhawa HS, Kurtulus E, Ozseven I, Akin B, Nur Zulfuagaoglu O, Nazari K (2022) QTL mapping of adult plant resistance to stripe rust in a doubled haploid wheat population. Front Genet 13:900558
Tian F, Bradbury PJ, Brown PJ, Hung H, Sun Q, Flint-Garcia S, Rocheford TR, McMullen MD, Holland JB, Buckler ES (2011) Genome-wide association study of leaf architecture in the maize nested association mapping population. Nat Genet 43(2):159–162
Tisné S, Serrand Y, Bach L, Gilbault E, Ben Ameur R, Balasse H, Voisin R, Bouchez D, Durand-Tardif M, Guerche P, Chareyron G (2013) Phenoscope: an automated large-scale phenotyping platform offering high spatial homogeneity. Plant J 74(3):534–544
Topp CN, Iyer-Pascuzzi AS, Anderson JT, Lee CR, Zurek PR, Symonova O, Zheng Y, Bucksch A, Mileyko Y, Galkovskyi T, Moore BT (2013) 3D phenotyping and quantitative trait locus mapping identify core regions of the rice genome controlling root architecture. Proc Natl Acad Sci 110(18):E1695–E1704
Truong HT, Ramos AM, Yalcin F, de Ruiter M, van der Poel HJ, Huvenaars KH, Hogers RC, van Enckevort LJ, Janssen A, van Orsouw NJ, van Eijk MJ (2012) Sequence-based genotyping for marker discovery and co-dominant scoring in germplasm and populations. PLoS One 7(5):e37565
Utz HF, Melchinger AE (1996) PLABQTL: a program for composite interval mapping of QTL. J Quant Trait Loci 2(1):1–5
Utz HF, Melchinger AE, Schön CC (2000) Bias and sampling error of the estimated proportion of genotypic variance explained by quantitative trait loci determined from experimental data in maize using cross validation and validation with independent samples. Genetics 154(4):1839–1849
Vales MI, Schön CC, Capettini F, Chen XM, Corey AE, Mather DE, Mundt CC, Richardson KL, Sandoval-Islas JS, Utz HF, Hayes PM (2005) Effect of population size on the estimation of QTL: a test using resistance to barley stripe rust. Theor Appl Genet 111:1260–1270
van der Heijden G, Song Y, Horgan G, Polder G, Dieleman A, Bink M, Palloix A, van Eeuwijk F, Glasbey C (2012) SPICY: towards automated phenotyping of large pepper plants in the greenhouse. Funct Plant Biol 39(11):870–877
Varshney RK, Thudi M, Roorkiwal M, He W, Upadhyaya HD, Yang W, Bajaj P, Cubry P, Rathore A, Jian J, Doddamani D (2019) Resequencing of 429 chickpea accessions from 45 countries provides insights into genome diversity, domestication and agronomic traits. Nat Genet 51(5):857–864
Velasco R, Zharkikh A, Affourtit J, Dhingra A, Cestaro A, Kalyanaraman A, Fontana P, Bhatnagar SK, Troggio M, Pruss D, Salvi S (2010) The genome of the domesticated apple (Malus× domestica Borkh.). Nat Genet 42(10):833–839
Vision TJ, Brown DG, Shmoys DB, Durrett RT, Tanksley SD (2000) Selective mapping: a strategy for optimizing the construction of high-density linkage maps. Genetics 155(1):407–420
Wang M, Jiang N, Jia T, Leach L, Cockram J, Waugh R, Ramsay L, Thomas B, Luo Z (2012) Genome-wide association mapping of agronomic and morphologic traits in highly structured populations of barley cultivars. Theor Appl Genet 124:233–246
Wang Q, Tian F, Pan Y, Buckler ES, Zhang Z (2014) A SUPER powerful method for genome wide association study. PLoS One 9(9):e107684
Wang B, Liu H, Liu Z, Dong X, Guo J, Li W, Chen J, Gao C, Zhu Y, Zheng X, Chen Z (2018a) Identification of minor effect QTLs for plant architecture-related traits using super high den-

sity genotyping and large recombinant inbred population in maize (*Zea mays*). BMC Plant Biol 18:1–12

Wang B, Wu Z, Li Z, Zhang Q, Hu J, Xiao Y, Cai D, Wu J, King GJ, Li H, Liu K (2018b) Dissection of the genetic architecture of three seed-quality traits and consequences for breeding in *Brassica napus*. Plant Biotechnol J 16(7):1336–1348

Wang J, Zhang X, Lin Z (2018c) QTL mapping in a maize F_2 population using Genotyping-by-Sequencing and a modified fine-mapping strategy. Plant Sci 276:171–180

Wang X, Zhang R, Song W, Han L, Liu X, Sun X, Luo M, Chen K, Zhang Y, Yang H, Yang G (2019) Dynamic plant height QTL revealed in maize through remote sensing phenotyping using a high-throughput unmanned aerial vehicle (UAV). Sci Rep 9(1):3458

Wang B, Lin Z, Li X, Zhao Y, Zhao B, Wu G, Ma X, Wang H, Xie Y, Li Q, Song G (2020) Genome-wide selection and genetic improvement during modern maize breeding. Nat Genet 52(6):565–571

Wang W, Guo W, Le L, Yu J, Wu Y, Li D, Wang Y, Wang H, Lu X, Qiao H, Gu X (2023) Integration of high-throughput phenotyping, GWAS, and predictive models reveals the genetic architecture of plant height in maize. Mol Plant 16(2):354–373

Wen Y, Fang Y, Hu P, Tan Y, Wang Y, Hou L, Deng X, Wu H, Zhu L, Zhu L, Chen G (2020) Construction of a high-density genetic map based on SLAF markers and QTL analysis of leaf size in rice. Front Plant Sci 11:1143

Werner T (2010) Next generation sequencing in functional genomics. Brief Bioinform 11(5):499–511

Westfall PH, Young SS, Wright SP (1993) On adjusting P-values for multiplicity. Biometrics 49(3):941–945

Xie Q, Fernando KM, Mayes S, Sparkes DL (2017) Identifying seedling root architectural traits associated with yield and yield components in wheat. Ann Bot 119(7):1115–1129

Xu C, Ren Y, Jian Y, Guo Z, Zhang Y, Xie C, Fu J, Wang H, Wang G, Xu Y, Li P (2017) Development of a maize 55 K SNP array with improved genome coverage for molecular breeding. Mol Breed 37:1–12

Xu Y, Yang T, Zhou Y, Yin S, Li P, Liu J, Xu S, Yang Z, Xu C (2018) Genome-wide association mapping of starch pasting properties in maize using single-locus and multi-locus models. Front Plant Sci 9:1311

Yang W, Guo Z, Huang C, Duan L, Chen G, Jiang N, Fang W, Feng H, Xie W, Lian X, Wang G (2014) Combining high-throughput phenotyping and genome-wide association studies to reveal natural genetic variation in rice. Nat Commun 5(1):5087

Yang L, Zhao D, Meng Z, Xu K, Yan J, Xia X, Cao S, Tian Y, He Z, Zhang Y (2020) QTL mapping for grain yield-related traits in bread wheat via SNP-based selective genotyping. Theor Appl Genet 133:857–872

Yaobin Q, Peng C, Yichen C, Yue F, Derun H, Tingxu H, Xianjun S, Jiezheng Y (2018) QTL-Seq identified a major QTL for grain length and weight in rice using near isogenic F_2 population. Rice Sci 25(3):121–131

Yi B, Chen Y, Lei S, Tu J, Fu T (2006) Fine mapping of the recessive genic male-sterile gene (Bnms1) in *Brassica napus* L. Theor Appl Genet 113:643–650

Yu J, Pressoir G, Briggs WH, Vroh Bi I, Yamasaki M, Doebley JF, McMullen MD, Gaut BS, Nielsen DM, Holland JB, Kresovich S (2006) A unified mixed-model method for association mapping that accounts for multiple levels of relatedness. Nat Genet 38(2):203–208

Yu J, Holland JB, McMullen MD, Buckler ES (2008) Genetic design and statistical power of nested association mapping in maize. Genetics 178(1):539–551

Yu J, Zhang K, Li S, Yu S, Zhai H, Wu M, Li X, Fan S, Song M, Yang D, Li Y (2013) Mapping quantitative trait loci for lint yield and fiber quality across environments in a *Gossypium hirsutum*× *Gossypium barbadense* backcross inbred line population. Theor Appl Genet 126:275–287

Yu H, Xie W, Li J, Zhou F, Zhang Q (2014) A whole-genome SNP array (RICE 6 K) for genomic breeding in rice. Plant Biotechnol J 12(1):28–37

Yu X, Pijut PM, Byrne S, Asp T, Bai G, Jiang Y (2015) Candidate gene association mapping for winter survival and spring regrowth in perennial ryegrass. Plant Sci 235:37–45

Yuan J, Wang X, Zhao Y, Khan NU, Zhao Z, Zhang Y, Wen X, Tang F, Wang F, Li Z (2020) Genetic basis and identification of candidate genes for salt tolerance in rice by GWAS. Sci Rep 10(1):9958

Zargar SM, Raatz B, Sonah H, Nazir M, Bhat JA, Dar ZA, Agrawal GK, Rakwal R (2015) Recent advances in molecular marker techniques: insight into QTL mapping, GWAS and genomic selection in plants. J Crop Sci Biotechnol 18:293–308

Zhang Z, Ersoz E, Lai CQ, Todhunter RJ, Tiwari HK, Gore MA, Bradbury PJ, Yu J, Arnett DK, Ordovas JM, Buckler ES (2010) Mixed linear model approach adapted for genome-wide association studies. Nat Genet 42(4):355–360

Zhang T, Hu Y, Jiang W, Fang L, Guan X, Chen J, Zhang J, Saski CA, Scheffler BE, Stelly DM, Hulse-Kemp AM (2015) Sequencing of allotetraploid cotton (*Gossypium hirsutum* L. acc. TM-1) provides a resource for fiber improvement. Nat Biotechnol 33(5):531–537

Zhang X, Huang C, Wu D, Qiao F, Li W, Duan L, Wang K, Xiao Y, Chen G, Liu Q, Xiong L (2017) High-throughput phenotyping and QTL mapping reveals the genetic architecture of maize plant growth. Plant Physiol 173(3):1554–1564

Zhang P, Zhong K, Zhong Z, Tong H (2019) Genome-wide association study of important agronomic traits within a core collection of rice (*Oryza sativa* L.). BMC Plant Biol 19:1–12

Zhao C (2019) Big data of plant phenomics and its research progress. J Agric Big Data 1(2):5–18

Zhao R, Liu B, Wan W, Jiang Z, Chen T, Wang L, Bie T (2023) Mapping and characterization of a novel adult-plant leaf rust resistance gene LrYang16G216 via bulked segregant analysis and conventional linkage method. Theor Appl Genet 136(1):1

Zhou X, Stephens M (2012) Genome-wide efficient mixed-model analysis for association studies. Nat Genet 44(7):821–824

Zhu C, Yu J (2009) Nonmetric multidimensional scaling corrects for population structure in association mapping with different sample types. Genetics 182(3):875–888

Zhu C, Gore M, Buckler ES, Yu J (2008) Status and prospects of association mapping in plants. Plant Genome 1(1)

Zhu Z, Hayart Y, Yang J, Cao L, Lou X, Xu H (2012) Statistical method for mapping QTLs for complex traits based on two backcross populations. Chin Sci Bull 57:2645–2654

Zou F (2009) QTL mapping in intercross and backcross populations. In: Cardiovascular genomics: methods and protocols, pp 157–173

GWAS and QTL Mapping in Potato (*Solanum tuberosum* L.)

3

Jelli Venkatesh, Abhinandan Surgonda Patil, Jagesh Kumar Tiwari, Dalamu, Tanuja Buckseth, Satish Kumar Luthra, and Ajay Kumar Thakur

Abstract

Potato (*Solanum tuberosum*) stands as a globally significant crop, serving as a vital source of sustenance for millions. In this book chapter, we uncover the essential methodologies, tools, and resources that empower researchers to map Quantitative Trait Loci (QTLs) and conduct Genome-Wide Association Studies (GWAS) in potatoes. We also examine the roles of QTLs and GWAS, highlighting how these parallel approaches independently and collectively enhance potato genetics and genomics. Through important case studies, we showcase the impact of QTL and GWAS discoveries in shaping the future of potato cultivation. Furthermore, we critically analyze the advantages and limitations inherent in these tools for potato genetic improvement. This thorough elucidation lays the base for the genetic enhancement of potatoes, with the ultimate goal of fortifying global food security through sustainable food production and strengthening nutritional resources.

J. Venkatesh
Department of Agriculture, Forestry and Bioresources, Research Institute of Agriculture and Life Sciences, Plant Genomics Breeding Institute, College of Agriculture and Life Sciences, Seoul National University, Seoul, Republic of Korea

A. S. Patil
Agarkar Research Institute, Pune, Maharashtra, India

J. K. Tiwari
ICAR-Central Potato Research Institute, Shimla, Himachal Pradesh, India

ICAR-Indian Institute of Vegetable Research, Varanasi, Uttar Pradesh, India

Dalamu · T. Buckseth · A. K. Thakur (✉)
ICAR-Central Potato Research Institute, Shimla, Himachal Pradesh, India

S. K. Luthra
ICAR-Central Potato Research Institute Research Station, Modipuram, Uttar Pradesh, India

S. S. Dey et al. (eds.), *GWAS and QTL Mapping in Horticultural Crops*,
https://doi.org/10.1007/978-981-95-2306-1_3

Keywords

QTL · GWAS · Potato · Tetraploid · Molecular markers · Genomics · Biotic and abiotic stresses

3.1 Introduction

The cultivated potato (*Solanum tuberosum* L. subsp. *tuberosum*) is a tetraploid crop with 48 chromosomes (n = 12), exhibiting high heterozygosity. It belongs to the *Solanaceae* family within the genus Solanum. In contemporary classification, the potato has been recognized as eight cultivated species (Huamán and Spooner 2002; van den Berg and Jacobs 2007). The probable center of origin for the potato can be traced back to the Andes region in Southern Peru and Northern Bolivia (Hawkes 1999). Cultivated potatoes, including *S. tuberosum* L. subsp. *tuberosum* and *andigena*, as well as other closely related species, exhibit varying levels of ploidy, ranging from diploid to hexaploid. Potatoes are cultivated extensively across diverse geographic regions, encompassing temperate zones, subtropical climates, and tropical environments. Potatoes thrive under various agroecological conditions, including lowlands and highlands. Sandy loam soil is most suitable for potato cultivation. Moreover, potatoes are cultivated in vastly different socio-economic settings, ranging from large-scale farming operations in high-input agricultural systems to smallholder farming prevalent in many developing nations.

The potato stands as one of the primary staple food crops globally in 2021, as documented by FAOSTAT (FAOSTAT 2023). Potato cultivation covered an extensive area of 18.13 million hectares, yielding an impressive production of approximately 376.12 million metric tons. China was the leading potato-producing country, contributing 94.30 million metric tons from 5.78 million hectares, followed by India with 54.23 million metric tons from 2.25 million hectares, and Ukraine produced 21.36 million metric tons from 1.28 million hectares. Other notable contributors included the United States of America with 18.58 million metric tons, the Russian Federation with 18.30 million metric tons, and Germany with 11.31 million metric tons, among others (FAOSTAT 2023). These top-ranking nations exemplify the paramount importance of potatoes as a vital staple food crop on a global scale, each making substantial contributions to meet worldwide food demands. Potatoes are a good source of iron, vitamin C, protein, and carbohydrates. The content of these nutrients differs across different cultivars. Carbohydrates make up around 75% of the total dry matter content (DMC) and serve as an important source of energy (Beals 2019). Potatoes are rich in various nutritionally valuable compounds that provide numerous health benefits (Beals 2019; Mishra et al. 2020). They provide a significant energy source and contain secondary metabolites beneficial for human well-being (Raigond et al. 2017). Potatoes possess a diverse array of phytonutrients, such as carotenoids, phenolics, anthocyanins, and flavonoids; and minerals, such as sodium, potassium, magnesium, iron, zinc, and copper; and vitamins, such as vitamins C, E, B1, B6, and B9. These elements collectively contribute to their health-promoting properties. These phytonutrients found in potatoes are particularly intriguing for human consumption due to their noteworthy antioxidant properties (Raigond et al. 2017).

Potatoes can be prepared in various ways, including boiling, baking, frying, and roasting. The primary focus of potato breeding efforts revolves around cultivars tailored for both processing and fresh table markets (Schreiber et al. 2014; Pandey et al. 2022). The choice of tuber shape is of paramount significance and is influenced by regional preferences and culinary traditions (Stark et al. 2020). Round-shaped tubers are preferred for chip production, while oblong to long tubers are the favored choice for making French fries. It is generally considered undesirable for potatoes to exhibit irregular shapes and possess deep-set eyes (Śliwka et al. 2008). Additionally, characteristics like tuber size, skin, and flesh color hold importance for consumers and the industry, with preferences varying depending on the intended use. Consequently, the breeding efforts toward enhancing tuber morphological traits in potatoes assume a pivotal role, as these attributes significantly influence potato consumption patterns. Potato breeding efforts have focused on identifying specific genomic loci linked to the tuber quality traits (Schreiber et al. 2014; Li et al. 2018; Pandey et al. 2022). These breeding efforts have also placed significant emphasis on traits of agronomic importance (Li et al. 2018). Furthermore, research has explored nitrogen use efficiency, drought tolerance, and floral, root, and stolon characteristics (Zia et al. 2020; Ospina Nieto et al. 2021; Alvarez-Morezuelas et al. 2023). Additionally, molecular markers have been developed for marker-assisted selection (MAS) to enhance resistance against certain diseases, such as common scab and *Phytophthora infestans* in breeding programs (Mosquera et al. 2016; Wang et al. 2020; Yuan et al. 2020; Koizumi et al. 2021).

3.2 Genetic Resources

Considerable resources and steadfast commitment have been invested worldwide in gathering, conserving, disseminating, and assessing potato genetic resources. Consequently, potato genetic diversity is currently safeguarded in gene banks worldwide, providing accessibility to potato breeders and researchers. It is important to highlight that a staggering number of more than 10,000 distinct potato varieties exist globally. Over the past 45 years, the International Potato Center (CIP) has had 17,326 unique variants. Currently, there are 4727 distinct potato varieties available for cultivation. This rich collection encompasses 4354 traditional varieties with deep-rooted cultivation histories spanning 17 countries, particularly within the Andean region, and improved varieties (Dolničar 2021). These diverse potato strains have evolved over centuries through meticulous selection from natural genetic reservoirs. The European Cultivated Potato Database (https://www.europotato.org/) encompasses 4174 varieties and 1513 breeding lines sourced from gene banks and collections across 13 participant nations. Furthermore, the World Potato Atlas (Hills and Pieterse 2009) catalogs over 4000 potato varieties hailing from more than a hundred countries across the globe. The remarkably diverse Colombian Central

Collection of potatoes, containing 1255 variants, is a crucial source of genetic diversity for breeding and agricultural advancement (Manrique-Carpintero et al. 2023). Nevertheless, despite this extensive diversity, a demand for novel potato varieties continues. Especially in regions such as Asia and Africa, where potato cultivation is increasing, new potato cultivars need to deliver heightened and consistent yields with minimal resource inputs. These lines should exhibit resistance to diseases and pests, resilience to environmental stresses, such as heat, cold, drought, and salt, and offer improved nutritional profiles.

3.3 Overview of QTL Studies in Potato

QTLs in potatoes have been mapped using populations derived from either diploid or tetraploid varieties. When initiating QTL mapping in potatoes, the typical procedure commences with choosing two potato parent varieties that display allelic variations influencing the trait of interest to create a biparental population. Mapping genetic traits using biparental populations depends on the occurrence of genetic recombination and segregation, as they significantly impact the diversity of alleles and the precision of genetic mapping. The potato biparental population may involve the use of various populations; F1 populations (full-sib families) of biparental crosses involving heterozygous parents are more frequently used in potato genetic studies (Bradshaw et al. 2008; da Silva Pereira et al. 2021b). The size of individuals within a mapping population commonly spans from as few as 25 to as many as 250 individuals, and these populations often exhibit limited recombination events (Bradshaw 2021). This is because the primary source of diversity in segregated alleles occurs within biparental populations. Larger populations are preferred to achieve higher resolution and more precise fine mapping. Initially, breeding efforts relied on a smaller number of markers for genotyping of the biparental mapping population, which was carried out using amplified-fragment length polymorphisms (AFLP), simple sequence repeat (SSR), and cleavage amplified polymorphism sequence (CAPS) markers; nevertheless, the effective linkage mapping requires the essential use of high-density markers (Hackett et al. 2014). Modern approaches now incorporate a significantly greater number of markers, sometimes numbering in the thousands or even millions of SNP markers.

3.4 Software Tools for QTL Analyses

The cultivated potatoes are highly heterozygous and autotetraploid, presenting complexities in line development, molecular breeding, and genetic mapping investigations. The size of the population is the primary constraining factor in full-sib families (Hackett et al. 2014). Furthermore, high-precision QTL detection in potatoes demands high-density marker coverage of all four homologues of a chromosome (Bourke et al. 2019). In previous studies, with a population size typically ranging from 200 to 400 and an array size of 8000 to 20,000 SNPs, researchers have predominantly identified QTL alleles of substantial effect, explaining approximately 10–50% of the phenotypic variation (PV) (Bradshaw 2021). However, detecting alleles of small effect is very challenging, mainly due to a lack of statistical power (Bradshaw 2021, 2022).

Nevertheless, for markers associated with QTL alleles of substantial effect, breeders can manage them similarly to markers for major genes. However, this may entail consideration of an even greater number of alleles. The recent introduction of high-throughput genotyping techniques, such as SNP arrays and genotyping-by-sequencing (GBS), along with specialized software tools, has significantly advanced the potato's genetic research and breeding initiatives. These advancements enable efficient and accurate genotyping of large populations, opening up new avenues for linkage and association mapping in potatoes for pinpointing genomic regions that contribute to valuable traits. Moreover, software tools are currently available to facilitate genotyping at the tetraploid level within a full-sib family. These software options accommodate various considerations, including scenarios where no double reduction is assumed (Hackett et al. 2017) or when accounting for double reduction and quadrivalents (Chen et al. 2021). With these remarkable developments, researchers can now more effectively identify and study genomic regions contributing to high-value traits in cultivated potatoes.

3.5 QTL Studies of Potato Tuber Qualities

Breeding for high-starch potatoes primarily focuses on increasing starch yield per unit area rather than simply raising the starch content of individual tubers. This is because achieving high tuber starch content (TSC) often comes at the expense of lower overall yield. Through QTL analysis based on field assessments conducted in diverse environments, it has been revealed that the genetic regulation of TSC is intricate, involving genes located on all 12 potato chromosomes (Bradshaw et al. 2008). Genetic factors influencing TSC may also have broader effects on tuber weight, impacting overall tuber yield. Some of these genetic factors may be commonly linked to the QTLs associated with TSC and tuber yield, found in the same genomic regions. The genetic dissection of QTLs that specifically affect TSC and those influencing both TSC and yield, as detailed in Table 3.1, forms the foundation for developing marker-assisted strategies aimed at optimizing starch yield. For instance, the time required for plant maturity and tuberization also demonstrates correlations with DMC or TSC (Bradshaw et al. 2008). Previous studies suggest that there may be gene(s) with pleiotropic effects affecting tuberization, plant maturity, TSC, and tuber yield, and these genes are located on potato Chr 5 (Table 3.1).

Over the past few decades, numerous QTLs associated with quality traits in potatoes have been identified (Table 3.1). DMC, often called underwater weight, is a key factor influencing the quality of processed tubers. Underwater weight/DMC also serves as an indicator of TSC. Higher DMC values are preferred for better processing quality. Starch accounts for approximately 60–80% of total DMC, and sugars and starch metabolites are interconnected within a common metabolic network. The

Table 3.1 Summary of QTLs for potato tuber yield and quality-related traits

Trait	Markers	Mapping population	Chr/linkage group	PV (R^2) or $-\log_{10}(P)$	References
Tuber starch content (TSC)	DArT	Interspecific F1 diploid potato clones (DG 00-683 × DG 08-28/13)	Detected eight QTLs on Chr 1, 2, 3, 8, 10, 11, 12	PV of 9.4–19.2 %	Śliwka et al. (2016)
	RFLP	Two F1 mapping populations: K31 (cross of two di-haploids) and LH (cross of a diploid and di-haploid)	Detected 18 QTLs [ts(a) to ts(r)] for TSC spread on all 12 linkage groups (LGs)	PV of 2.9–43.9%	Schäfer-Pregl et al. (1998)
	AFLP, SSR, CAPS, SCAR	A diploid mapping population (US-W5337.3 × 77.2102.37)	Two QTLs were detected on each LG of 5, 8, 9, 10, and 12	PV of 5–17.3%	Werij et al. (2012)
	AFLP, SSR	A tetraploid population derived from a cross between Longshu 8 × Zaodabai	11 QTLs on seven Chr (1, 2, 4, 6, 8, 10, 12)	PV of 9.1–27.34%	Li et al. (2019)
Dormancy	AFLP	A diploid F1 population (98-21) derived from DG 83-1520 and DG 84-195	Four QTLs located on Chr 3, 5 and 10	PV of 4.5–14.7%	Śliwka et al. (2008)
Tuber shape	AFLP	A diploid F1 population (98-21) derived from *S. verrucosum* and *S. microdontum*	Detected on Chr 2 and 11	PV of 5.6–8.0%	Śliwka et al. (2008)
	SolCAP 8K SNP array	F1 progeny of 'Rio Grande Russet' × 'Premier Russet'	A QTL on Chr 5	PV of 7.5%	Massa et al. (2018)
Shape regularity	AFLP	A diploid F1 population (98-21) derived from DG 83-1520 and DG 84-195	Three QTLs on Chr 3, 5, 11	PV of 3.8–10.4%	Śliwka et al. (2008)
	AFLP, SSR	A tetraploid population derived from 12601ab1 × Stirling	Two QTLs detected on LG V, XIa.	PV of 9.5–15.5 %	Bradshaw et al. (2008)

Eye depth	AFLP	A diploid F1 population (98-21) derived from DG 83-1520 and DG 84-195	Four QTLs on Chr 3, 5, 10	PV of 4.5–14.7%	Śliwka et al. (2008)
Flesh color (whiter)	AFLP	A diploid F1 population (98-21) derived from DG 83-1520 and DG 84-195	Two QTLs on Chr 4 and 12	PV of 5.1–5.8%	Śliwka et al. (2008)
Flesh color (purple)	AFLP, CAPS, SSRs	F1 mapping population of derived from 10618-01 (purple-diploid) × 320-02 (white flesh)	QTLs for tuber flesh pigmentation detected on Chr 5, 8, and 9	PV of 6.5–81%	Zhang et al. (2009)
Dry matter content (DMC)	AFLP, SSR	A tetraploid population derived from 12601ab1 × Stirling	Two QTLs detected on LG 5	PV of 9.1–9.7 %	Bradshaw et al. (2008)
	AFLP, SSR	B2721 population (F1) derived from Atlantic (chip processing cultivar) × B1829-5 (advanced breeding selection)	QTLs were detected on Chr 2, 3, and 5 and LG 7	PV of 3.2–17.9%	McCord et al. (2011)
Tubers/plant	582 SNPs Infinium 8303 Potato Array	A diploid mapping population of *S. tuberosum* Group Phureja DM 1–3 516 R44 × Group Tuberosum RH89-039-16	Two QTLs on Chr 5	PV of 11.9–31.4%	Manrique-Carpintero et al. (2015)
Tuber weight	582 SNPs Infinium 8303 Potato Array	A diploid mapping population of *S. tuberosum* Group Phureja DM 1–3 516 R44 × Group Tuberosum RH89-039-16	Six QTL on Chr 1, 4, 5, 6	PV of 10.3–31.6%	Manrique-Carpintero et al. (2015)
	2095 SNPs	Tetraploid family W9817 derived from Liberator × Wisconsin breeding clone W4013-1/110	A QTL on Chr 5.	PV of 20%	Rak et al. (2017)

(continued)

Table 3.1 (continued)

Trait	Markers	Mapping population	Chr/linkage group	PV (R^2) or $-\log_{10}(P)$	References
Vigor	582 SNPs Infinium 8303 Potato Array	A diploid mapping population of *S. tuberosum* Group Phureja DM 1–3 516 R44 × Group Tuberosum RH89-039-16	Three QTL on Chr 3, 5, 10	PV of 13.8–53.9%	Manrique-Carpintero et al. (2015)
Plant maturity	AFLP, SSR	a tetraploid population of 192 clones from Longshu 8 × Zaodabai	Six QTL on Chr 2, 4, 5, 7, and 11	PV of 9.52–38.41%	Li et al. (2019)
	AFLP, SSR	A tetraploid population derived from 12601ab1 × Stirling	Two QTLs detected on LG V.	PV of 54.4–55.5 %	Bradshaw et al. (2008)
	582 SNPs Infinium 8303 Potato Array	A diploid mapping population of *S. tuberosum* Group Phureja DM 1–3 516 R44 × Group Tuberosum RH89-039-16	A QTL on Chr 5	PV of 53.2–75.7%	Manrique-Carpintero et al. (2015)
	Infinium 8303 Potato Array	F1 population (MSL603) of 'Jacqueline Lee' × 'MSG227-2'	A QTL on Chr 5	PV of 14.7–16.1%	Massa et al. (2015)
	SolCAP 8K SNP array	F1 progeny of 'Rio Grande Russet' × 'Premier Russet'	A QTL on Chr 5	PV of 38.5 and 54.5%	Massa et al. (2018)
Specific gravity	582 SNPs Infinium 8303 Potato Array	A diploid mapping population of *S. tuberosum* Group Phureja DM 1–3 516 R44 × Group Tuberosum RH89-039-16	Two QTLs on Chr 5, 9	PV of 20.8–49.4%	Manrique-Carpintero et al. (2015)
	AFLP, SSR	B2721 mapping population (F1) derived from Atlantic (chip processing cultivar) × B1829-5	QTLs detected on Chr 2, 3, 5, 8, 9, and LG 7	PV of 2.4–12.5%	McCord et al. (2011)
	Illumina Infinium® 8303 Potato Array	F1 tetraploid population ('Atlantic' × B1829-5)	QTLs detected on LG 2,5,8,10	PV of 13.1–30.4%	da Silva Pereira et al. (2021b)

Chip color (Hunter scale)	2095 SNPs	Tetraploid family W9817 derived from Liberator × Wisconsin breeding clone W4013-1/110	QTL on Chr 2, 3, and 9.	PV of 9.4–12.7%	Rak et al. (2017)
Chipping color 4C (Stored at 4 °C)	AFLP, SSR, CAPS, SCAR	A diploid mapping population (US-W5337.3 × 77.2102.37)	LG 3, 5, 8, and 10	PV of 4.9–16.4%	Werij et al. (2012)
	AFLP, SSR	A tetraploid population derived from 12601ab1 × Stirling	Four QTLs detected on LGs, Xia, A, VI	PV of 6.5–8.8%	Bradshaw et al. (2008)
Chipping color 10C (Stored at 10 °C)	AFLP, SSR	A tetraploid population derived from 12601ab1 × Stirling	Four QTLs on LG VI, Xia, Ia, XIa	PV of 5.9–8.1 %	Bradshaw et al. (2008)
	Isozymes, RFLP, RAPD	A diploid population of *S. tuberosum* × *S. chaeoense*	Six QTLs on Chr 2, 4, 5, 10	PV of 43.5%	Douches and Freyre (1994)
Fry color (Stored at 5.5 °C)	SolCAP 8K SNP array	F1 progeny of Rio Grande Russet' × 'Premier Russet'	QTLs detected on Chr 4, 6,10,11	PV of 6.8–16.1%	Massa et al. (2018)
Amylose content	AFLP, SSR, CAPS, SCAR	A diploid mapping population (US-W5337.3 × 77.2102.37)	7 QTLs on LGs 2, 3, 5, 7, and 10	PV of 4.0–28.5%	Werij et al. (2012)
Glucose	SolCAP 8K SNP array	F1 progeny of Rio Grande Russet' × 'Premier Russet'	QTLs detected on Chr 4, 5, 6, 10	PV of 6.2–11.3%	Massa et al. (2018)
Sugars (Cold stored potato)	RFLP, AFLP	Two diploid F1 populations (H94A and H94C)	QTL reducing sugars were located on LG I, III, VII, VIII, IX, and XI	PV >10.0%	Menéndez et al. (2002)
Cold-induced Sweetening	AFLP, SSR, CAPS, SCAR	A diploid mapping population (US-W5337.3 × 77.2102.37)	Detected a QTLs on LG 8	PV of 6.8%	Werij et al. (2012)

(continued)

Table 3.1 (continued)

Trait	Markers	Mapping population	Chr/linkage group	PV (R^2) or $-\log_{10}(P)$	References
Skin texture	AFLP and SSR	B2721 mapping population (F1) derived from Atlantic (chip processing cultivar) × B1829-5	QTL on Chr 2, 3, 5, 6, 9, and LG 12	2.7–13.6% (Atlantic) 9.2–20.1% (B1829-5)	McCord et al. (2011)
Tuber yield	AFLP, SSR	B2721 mapping population (F1) derived from Atlantic (chip processing cultivar) × B1829-5	QTLs were detected on LG 10, 12, Chr 2, 3, 5, 6	PV of 4.2–9.3%	McCord et al. (2011)
	AFLP, SSR	A tetraploid population derived from 12601ab1 × Stirling	Two QTLs on LGs VI, Ia	PV of 5.4–13.3%	Bradshaw et al. (2008)
	582 SNPs Infinium 8303 Potato Array	A diploid mapping population of *S. tuberosum* Group Phureja DM 1–3 516 R44 × Group Tuberosum RH89-039-16	Three QTLs on Chr 2, 5, 12	PV of 12.6–27.4%	Manrique-Carpintero et al. (2015)
Tuber carotenoids	AFLP, SSR, DArT	Two diploid hybrid populations (01H15, 03TR2)	QTLs identified on Chr 3 and 9	PV of 20–71%	Campbell et al. (2014)
Sprouting	AFLP, SSR	A tetraploid population derived from 12601ab1 × Stirling	Four QTLs detected on LGs V, VIII, C	PV of 4.8–9.6 %	Bradshaw et al. (2008)
Nitrogen use efficiency (low/high)	SSR, AFLP, CAPS	Diploid mapping population derived from C (USW5337.3) × E (77.2102.37)	A total of 52 putative QTLs were identified for ten traits (28 QTLs under low N and 24 QTLs under high N conditions)	PV of 11.9–38.4%	Getahun et al. (2020)

TSC (interchangeably used as DMC) comprises a significant proportion of the tuber, typically ranging from 70% to 85%. Previous investigations have revealed important QTLs related to TSC located on Chr 1, 2, 5, 7, 9, 10, and 11 (Chen et al. 2001; Gebhardt et al. 2005; Li et al. 2005; Werij et al. 2012; Śliwka et al. 2016; Li et al. 2019). QTLs associated with DMC were identified on Chr 2, 3, 5, 8, 9, and 11 (Bradshaw et al. 2008; McCord et al. 2011; Manrique-Carpintero et al. 2015). Genes associated with carbohydrate synthesis and transport, including *ADP-glucose pyrophosphorylase* (*AGPase*), *invertases*, *sucrose phosphate synthase*, *sucrose synthase*, and *starch phosphorylases*, were co-located with QTLs associated with starch and sugar contents. Specifically, *AGPase*, identified on Chr 1 and 8, was correlated with both TSC and chip quality (Gebhardt et al. 2005; Li et al. 2008). QTLs harboring genes that function as a sucrose sensor and a sucrose transporter were found on Chr 5 and 11, respectively (Barker et al. 2000).

Variations in chip color are primarily influenced by genetic factors, with reducing sugars playing a significant role in contributing to this variation. Allelic variants associated with these traits can be found on nearly all Chr (Table 3.1), which comprises genes that play roles in carbohydrate metabolism (Menéndez et al. 2002). QTLs related to potato chip color were detected on the short arms of Chr 2 and 9 (Rak et al. 2017). A QTL with a moderate effect on chip color (at 3 months) was mapped on Chr 9 and harbors candidate genes, such as *invertase inhibitor* (*INH2*), *vacuolar invertase* (*VInv*), and *apoplastic invertase* (*Inv ap-b*) (Rak et al. 2017). A significant QTL for short-duration chip color was also observed on Chr 3. A QTL associated with vine maturity was identified on Chr 5, accounting for a significant portion of the PV of approximately 56% (Bradshaw et al. 2008). In contrast, other QTLs for traits, such as yield, DMC, tuber size, shape, and fry color (stored at 4°C and 10°C), PV varied from 5.4% to 16.5% individually (Bradshaw et al. 2008). By employing a multiple-QTL model approach, an important cluster of QTLs was identified on Chr 5 (da Silva Pereira et al. 2021b) for traits such as maturity, specific gravity (SG), tuber yield, and resistance to internal heat necrosis (Table 3.1). Notably, among the genes located near the peaks of these QTLs on Chr 5 were *StCDF1* and *POTH1*, associated with traits related to maturity and tuber formation (da Silva Pereira et al. 2021b). Massa et al. (2015) also identified a QTL related to vine maturity located on Chr 5 (Duplex-simplex, c2_22986) with a PV of 14.7–16.1%. A potential SNP marker associated with vine maturity was located approximately 0.25 Mb away from the *StCDF1* gene. QTLs associated with tuber length, width, and weight were found to co-localize with a QTL on Chr 5 linked to plant maturity (Rak et al. 2017).

Genetic analyses revealed that the presence of QTLs on Chr 3, 5, 8, and 9 influences tuber flesh pigmentation (Bonierbale et al. 1988; Campbell et al. 2014; Zhang et al. 2009). Among these identified QTLs, candidate genes pinpointed include *chalcone synthase* and *chalcone isomerase*, both playing key roles in the anthocyanin biosynthesis. Notably, a bHLH transcription factor associated with the anthocyanin synthesis pathway was found to be colocalized on Chr 9 (Zhang et al. 2009). Considering consumer preferences for tubers with red and purple flesh, primarily due to their high antioxidant content, genetic investigations have demonstrated that the *R* locus, which co-segregates with *dihydroflavonol 4-reductase*, is crucial for the synthesis of red colored anthocyanins (De Jong et al. 2003). On the other hand, the purple flesh trait is controlled by the *P* locus which harbors a candidate gene encoding flavonoid 3′,5′-hydroxylase (Jung et al. 2005). Additionally, previous studies have mapped QTLs for flesh color pigmentation on Chr 10, linked to the *Pf* locus (De Jong et al. 2003). The yellow flesh color of potatoes is determined by the presence of the dominant allele of the *Y* locus. B-carotene, mainly responsible for the yellow flesh, is located on Chr 3 (Campbell et al. 2014). The possible candidate gene for the yellow flesh color is *β-carotene hydroxylase 2* (*Chy2*), located on Chr 3 (Campbell et al. 2014). It is worth noting that among the 11 *Chy2* alleles, only one is dominant, leading to the conversion of white flesh to yellow. Whereas the orange colored tuber flesh is attributed to the increased accumulation of zeaxanthin, mainly due to the presence of the recessive *zeaxanthin epoxidase* (*zep*) allele (Wolters et al. 2010).

3.6 Potato QTL Studies for Disease Resistance

In potato breeding programs, disease resistance is achieved through the utilization of single *R* genes as well as polygenic quantitative resistance. The molecular mechanisms underlying interactions between plants and pathogens are extensively studied using experimental systems that focus on single *R* genes. The quantitative resistance is generally regarded as more durable over time. However, genes responsible for this type of resistance remain largely unidentified in potatoes. The quantitative resistance of potato to *P. infestans* is weak in late-maturity potato cultivars, which is an undesirable agronomic trait. Recently, a major QTL for late blight resistance on Chr 10 (at 26.5 cM) and a minor QTL on Chr 5 were detected using a diploid potato population derived from an interspecific cross (MS × 902) between an *S. tuberosum* × *Solanum chacoense* hybrid (84SD22) and a late blight-resistant *Solanum berthaultii* hybrid (Ber83) (Manrique-Carpintero et al. 2020). The statistically significant correlation between tuber yield and susceptibility to *P. infestans* indicates the presence of common QTL loci on Chr 5 (Visker et al. 2003). Massa et al. (2015) identified a late blight-resistant QTL on Chr 9 (Simplex, h4, c1_10590) with 25.7–39.3 PV. QTL mapping revealed 'Jacqueline Lee' as a major source of late blight resistance. The significant SNP marker was located approximately 0.54 Mb away from the resistance locus on the long arm of Chr 9. Previously, Collins et al. (1999) also showed the existence of a QTL associated with resistance to late blight located on Chr 9 (linkage group (LG) IX).

Common scab of potato is a globally significant disease with economic implications, caused by soil-borne bacteria from various species of *Streptomyces* (Wanner 2009; da Silva Pereira et al. 2021a). While certain management practices have been suggested to alleviate the impacts of common scab (Dees and Wanner 2012), presently, no chemical or cultural management strategies offer effective disease management. Despite its detrimental impact on the potato industry, our understanding of the genetic basis of resistance to common scab in potatoes remains limited. There have been limited studies conducted to date regarding the identification of QTLs associated with resistance to common scab in potatoes. Bradshaw et al. (2008), using a mapping population derived from the tetraploid cross '12601ab1' × 'Stirling,' consisting of 227 F_1 clones, common scab resistance QTLs were identified on homology groups II and IV. Subsequently, a single QTL on Chr 11 was identified using a diploid 'US-W4' × '524-8' mapping population of 49–91 F_2 clones (Braun et al. 2017). In another study, QTLs located on Chr 1, 2, 3, 5, 6, 10, and 11 were identified for scab resistance using a tetraploid mapping population of 151 F_1 clones from the 'Atlantic' × 'Superior' cross (Zorrilla et al. 2021). QTL mapping using a 'Tundra' × 'Kalkaska' full-sib population of 198 F_1 clones identified QTL on Chr 1 and 2 (Zorrilla et al. 2021). Recently, da Silva Pereira et al. (2021a) identified a single QTL associated with common scab resistance, located within LG 3, using a mapping population comprising around 150 full-sibling individuals obtained from a cross between two tetraploid potato varieties, Atlantic and B1829-5. This QTL contributed to approximately 22–30% of the PV (da Silva Pereira et al. 2021a).

3.7 QTLs for Abiotic Stress Tolerance

Drought constitutes a significant abiotic stress factor that adversely affects potato tubers' growth, yield, and quality. Schumacher et al. (2021) conducted QTL analyses in cultivated potato plants to assess their drought tolerance index, referred to as DRYM (deviation of relative starch yield from the experimental median). During QTL mapping, 19 DRYM QTLs, eight located on Chr of the drought-tolerant parent (Albatros) and 11 on chromosomes of the drought-sensitive parent (Ramses), were associated with drought tolerance and susceptibility. Within these DRYM QTLs from the resistant parent, candidate genes were identified, including genes related to ethylene, abscisic acid, and brassinosteroid signaling pathways, as well as genes encoding enzymes involved in cell-wall remodeling.

3.8 Potato Genome-Wide Association Studies (GWAS)

Recently, potato research has seen a significant expansion in association mapping (AM) studies, emphasizing their increasing importance in unraveling the genetic basis of agricultural traits and understanding adaptive significance, as the AM offers notable benefits compared with conventional biparental mapping. These include increased mapping precision, the ability to analyze a greater variety of traits within a single study, and an increased likelihood of discovering MTAs that remain robust across a broader range of germplasm. Initial efforts in AM for potatoes primarily consisted of candidate gene-based approaches (Li et al. 2005; Malosetti et al. 2007; Achenbach et al. 2009; Pajerowska-Mukhtar et al. 2009; Urbany et al. 2011). While the candidate gene approach has proven successful in pinpointing MTAs, it comes with important drawbacks, such as biased selection of candidate genes based on prior genetic, biochemical, or molecular foundations and the potential for overlooking essential yet undiscovered genes related to the target traits (Hall et al. 2010; Stich et al. 2013). Nevertheless, despite these limitations, AM utilizing candidate genes has demonstrated its effectiveness as a robust approach for analyzing complex traits in potatoes.

Moreover, there have been AM studies in potatoes employing genome-wide markers (D'hoop et al. 2008; Fischer et al. 2013; D'hoop et al. 2014; Lindqvist-Kreuze et al. 2014; Rosyara et al. 2016; Vos et al. 2017). GWAS offers high mapping resolution and is valuable for identifying candidate alleles/genes, making it a powerful tool for associating phenotypes with genotypes. It provides an alternative to biparental crosses, especially for tetraploid crops like potatoes. Indeed, these comprehensive GWAS have yielded noteworthy outcomes by identifying marker alleles linked to crucial potato characteristics (Table 3.2). These include traits like tuber shape, flesh color, post-cooking discoloration, and browning (D'hoop et al. 2014); late blight resistance (Lindqvist-Kreuze et al. 2014); and starch accumulation (Schönhals et al. 2016). A noteworthy achievement was the pivotal role of GWAS in uncovering the *StCDF1* gene, a component of the circadian clock linked to a major QTL related to plant maturity (Kloosterman et al. 2013).

GWAS takes advantage of natural variation and quantitative trait mapping, but several factors, including PV, allele frequency, number of markers, sample size, population structure (PS), kinship, and linkage disequilibrium (LD), should be considered to avoid spurious associations. A few important considerations have been discussed here.

3.9 Molecular Markers Used in GWAS

In the early stages of potato AM, the evaluation of PS and GWAS relied on traditional marker types, such as AFLPs and SSRs, and a limited number of markers (D'hoop et al. 2008; Achenbach et al. 2009; Pajerowska-Mukhtar et al. 2009; D'hoop et al. 2014). However, for more comprehensive studies, the preference has shifted toward using high-throughput genotypic data with high-density markers. The first such platform in potato, which included a set of 8303 SNPs (Hamilton et al. 2011; Felcher et al. 2012), has been employed to analyze diploid and tetraploid germplasm (Hirsch et al. 2013; Stich et al. 2013; Hardigan et al. 2015; Kolech et al. 2016). One of the challenges when conducting genetic mapping studies in polyploid species has been the lack of appropriate methods for accurately determining marker allele dosage. Recently, methods have been developed to address this issue, particularly in codominant markers like SNPs (Voorrips et al. 2011; Hackett et al. 2013; Schmitz Carley et al. 2017). These advancements, combined with high-throughput assays, have established SNPs as the ideal markers for conducting GWAS. As a result, the utilization of high-throughput SNP markers in AM studies for potatoes is on the rise (Kloosterman et al. 2013; Lindqvist-Kreuze et al. 2014; Rosyara et al. 2016; Berdugo-Cely et al. 2017; Schmitz Carley et al. 2017; van Eck et al. 2017; Vos et al. 2017).

Table 3.2 Summary of QTLs for potato biotic and abiotic stress-related traits

Trait	Markers	Mapping population	Chr/Linkage group	PV(R^2) or $-\log_{10}(P)$	References
Potato cyst nematodes (PCN), *Globodera rostochiensis* (Ro5-22), *G. pallida* (Pa2 and Pa3)	AFLP, CAPS, RFLP	A F1 mapping population of 122 diploid potato lines from 3778-16 (AM, resistant) × RH89-039-16 (RH, susceptible)	*Grp1* located on Chr 5	PV of 66% (against Pa2), 45% (against Pa3), and 77% (against Ro5-22)	van der Voort et al. (1998)
G. pallida (Ro and Pa)	AFLP, CAPS	A diploid mapping population of from a cross between 3704–76 × RH89–039–16	*Gpa5* on Chr 5, *Gpa6* on Chr 9	Gpa5, PV of 61% *Gpa6*, PV of 24%	Rouppe van der Voort et al. (2000)
G. pallida (Pa2/3)	AFLP	Hybrid clones developed from susceptible *S. tuberosum* Caspar H3 × resistant *S. sparsipilum* spl329	$GpaV^{s}_{spl}$ and $GpaXI^{s}_{spl}$ were detected on Chr 5 and 11	PV of 76.6% for GpaV^{s}_{spl} PV of 12.7% for GpaXI^{s}_{spl} [R2]	Caromel et al. (2005)
	AFLP, RFLP	A F1 diploid population from *S. tuberosum* 2x × wild *S. spegazzinii*	*GpaM1*, *GpaM2*, and *GpaM3* on Chr 5, 6, and 12, respectively.	PV of 15–56%	Caromel et al. (2003)
	SSRs, AFLP	Two diploid populations derived from *S. vernei* (Stirling × 12288af23 and PDH538 × IVP48)	QTL on linkage groups V and IX	PV of 15–20%	Bryan et al. (2002)

(continued)

Table 3.2 (continued)

Trait	Markers	Mapping population	Chr/Linkage group	PV(R^2) or $-\log_{10}(P)$	References
G. rostochiensis (Ro1)	RFLP	A backcross population, derived from the cross (*S. tuberosum* × *S. spegazzinii*) × *S. tuberosum*	*Gro1.2* and *Gro1.4*, located on Chr 10 and 3, respectively	PV of 22.5–51.0%	Kreike et al. (1996)
	RFLP	F1 population derived from a *S. tuberosum* SH 78-88-1320 × *S. spegazzinii* BGRC	QTL on Chr 10 and 11	PV of 14–22%	Kreike et al. (1993)
G. pallida (Pa2 and Pa3)	RFLPs	F1 population derived from a *S. tuberosum* SH 78-88-1320 × *S. spegazzinii* BGRC	*Gpa* on Chr 5 and two minor QTLs on Chr 4 and 7	PV of 45–51% and 5-9% for major and minor QTLs, respectively	Kreike et al. (1994)
G. pallida (Pa3)	AFLP	Diploid BC1 population RHAM-061 (RH90-011-4 × *S. tuberosum* RH89-039-16)	$GpaXI^{l}_{tar}$ on Chr 11, and a minor QTL on Chr *9*	PV of 81.3% and 5.3%, respectively	Adillah Tan et al. (2009)
G. rostochiensis (Ro2) and *G. pallida* (Pa2/3)	8303 SNP markers	Autotetraploid mapping population Y36 [NY121 (resistant to Ro1, Ro2 and Pa2/3) × NY115 (resistant to Ro1, susceptible to Ro2)]	Two Ro2 resistance QTLs on Chr 5	PV of 23.8–24.4% (for Ro2) and 9.2% (for Pa2/3)	Park et al. (2019)
G. pallida Pa2/3 (Lindley)	AFLP, SSR	SCRI clone 12601ab1 × cv Stirling	A QTL on Chr 4	PV of ~27.8%	Bradshaw et al. (1998)
Ralstonia solanacearum (Bacterial wilt)	4139 SNPs	Diploid population [clone 10-03-30 × susceptible clone F_1-1]	Five QTLs (*qBWR-1* to *-5*) detected on Chr 1, 3, 7, 10, and 11	PV of 9.3–18.4%	Habe et al. (2019)

Early Blight	SolCAP 8K SNP array	F1 progeny of Rio Grande Russet' × 'Premier Russet'	QTLs detected on Chr 5	PV of 41.3%	Massa et al. (2018)
Verticillium wilt	SolCAP 8K SNP array	F1 progeny of Rio Grande Russet' × 'Premier Russet'	QTLs detected on Chr 5	PV of 46.5%	Massa et al. (2018)
Common scab (*Streptomyces* spp.)	AFLP, SSR	A tetraploid population derived from 12601ab1 × Stirling	Two QTLs detected on LGs 2, 6	PV of 6.9–8.1%	Bradshaw et al. (2008)
Late blight (*P. infestans*)	AFLP	Diploid backcross population [*Solanum tuberosum* PDH247 × *S. phureja* clone DB226(70) × DB226(70)]	A QTL on Chr 4	PV of 78% and 51% for greenhouse and field experiments, respectively	Bradshaw et al. (2006)
	AFLP	Tetraploid Population [3681ad1 × Maris Piper]	QTL, *R10* was detected on Chr 11	PV of 56.9%	Bradshaw et al. (2006)
	3,013 SNPs	Diploid population (MSX902) [(*S. tuberosum* × *S. chacoense* hybrid (84SD22) × *S. berthaultii* hybrid (Ber83)]	Major QTL on Chr 10 and a minor QTL on Chr 5	PV of 26.8–71.7% (Chr 10) and 18% (Chr 5)	Manrique-Carpintero et al. (2020)
	Infinium 8303 Potato Array	F1 population (MSL603) of 'Jacqueline Lee' × 'MSG227-2'	A QTL on Chr 9	PV of 25.7–39.3%	Massa et al. (2015)
Tuber end rot (physiological disorder)	582 SNPs Infinium 8303 Potato Array	A diploid population of *S. tuberosum* Group Phureja DM 1–3 516 R44 × Group Tuberosum RH89-039-16	Four QTLs detected on Chr 1, 3, 5, 6	PV of 11.3–34.1%	Manrique-Carpintero et al. (2015)
Internal heat necrosis	Illumina Infinium® 8303 Potato Array	F1 tetraploid population ('Atlantic' × B1829-5)	Two QTLs detected on LG 1, 5	PV of ~13.0%	da Silva Pereira et al. (2021b)

3.10 SNP Marker Quality Control

Many data preprocessing and quality control steps are done before the GWAS analysis to ensure the dataset is well-prepared for association analysis. Various methods of marker reduction, such as filtering SNPs based on criteria such as minor allele frequency (MAF), retaining only one SNP per haplotype block, and employing LD-based marker pruning. Typically, an MAF threshold of 1.0–1.5%, equivalent to 4–6% in diploid organisms, is utilized for SNP filtering, as demonstrated in a number of potato genomic studies (Sverrisdóttir et al. 2017; Klaassen et al. 2019; Prodhomme et al. 2020; Díaz et al. 2021; Selga et al. 2021). Additionally, markers are commonly removed when the proportion of missing values surpasses 10–20%.

The level of LD in a population impacts the choice of markers and population size for GWAS. High LD allows for fewer markers and smaller populations, while low LD may require more markers and larger populations to achieve meaningful results (Angelin-Bonnet et al. 2023). Researchers assess LD patterns in their specific study population to make informed decisions about study design. Vos et al. (2017) predict that 40,000 SNPs are required for successful potato QTL detection. However, obtaining the necessary high-throughput genotyping data for a large breeding population could pose significant financial challenges. Selga et al. (2021) demonstrated that genotyping a carefully chosen subset of the breeding population, followed by SNP pruning for LD, is an effective strategy to decrease the quantity of genetic markers without sacrificing the intricacy of the genetic information. Furthermore, utilizing LD pruning for marker selection is a reliable method for reducing the number of markers needed for potato genotyping, thus reducing associated costs. The LD pruning process ensures an even distribution of SNP markers throughout the genome and eliminates data redundancy, diminishing the pronounced impact of SNP clusters (Laurie et al. 2010).

3.11 Genome-Wide LD Decay

LD represents the non-random linkage or association between alleles at distinct loci within a breeding population. LD can be measured by assessing the correlation among linkage values assigned to SNP markers at those loci. The degree of LD between these loci bears significance in genetic investigations like GWAS because it dictates the requisite count of SNP markers and the precision of mapping outcomes (Flint-Garcia et al. 2003). Well-known elements of population genetics, including selection, migration, admixture, non-random mating, genetic drift, mutation, and population size, play a role in shaping the LD and its subsequent decay (Flint-Garcia et al. 2003). In the case of heterozygous outbreeding species like potatoes, SNP alleles located on the same haplotype can exhibit substantial LD values, and the decline or decay of LD depends on the frequency of recombination and the number of generations undergone.

In the case of potato genetics, LD plays a pivotal role in the success of GWAS. This is due to the fact that the majority of cultivated tetraploid potato genotypes exhibit limited genetic diversity. This constrained genetic base primarily comprises a small set of haplotypes due to pronounced LD. To bolster the effectiveness of MTAs, it is imperative to conduct LD assessments and simultaneously correct for PS, meticulously considering subpopulations, which ensures the reliability and robustness of GWAS findings, which are vital for advancing potato breeding and crop improvement efforts (Simko et al. 2004).

Self-fertilizing plants, like potatoes, often exhibit slower LD decay, characterized by persistently high LD values. This phenomenon arises primarily due to their extended breeding cycles with a limited number of historical recombination events and homozygous genetic background, where recombination events have a limited impact on LD decay (reference). The presence of large LD blocks on various chromosomes in potato populations, notably on Chr 8, suggests that unique selection pressures have influenced the genetic makeup of this chromosome over the last 150 years of potato breeding. Additionally, a previous study underscores that differences in LD estimates can stem from various factors, including population characteristics, marker spacing, genotyping methods, distance units, and significance thresholds (Vos et al. 2017; Sharma et al. 2018). The extent of LD decay in potato genetics varies widely among studies. Hence, direct comparisons between LD patterns across studies may be challenging (Vos et al. 2017; Sharma et al. 2018). Nevertheless, central trends in LD decay can still be compared to draw broad insights into genetic structure and evolution. A study by Vos et al. (2017) revealed a pattern of moderate LD decay across the 12 Chr of the potato. Notably, interchromosomal disparities in LD decay are apparent, with each Chr exhibiting a unique range of physical distances beyond which LD drops below the commonly utilized significance threshold of 0.1 (Simko et al. 2006; D'hoop et al. 2010; Stich et al. 2013). The minimal LD decay observed in the centromeric heterochromatin across all chromosomes is attributed to recombination suppression in this genomic region, making it impractical to model LD decay for this portion (Vos et al. 2017).

3.12 Analysis of PS and Kinship

In general, PS is a significant obstacle in GWAS. Many spurious associations might be identified without an appropriate correction for the relationship between individuals (Vos et al. 2022). Addressing PS and kinship in GWAS studies is crucial to minimizing false associations. Different methods, including STRUCTURE, principal component analysis (PCA), and non-metric multidimensional scaling, have been employed to estimate PS and mitigate its effects. The STRUCTURE tool is often used for PS analysis, which estimates cluster proportions (Q matrix) based on genotypic data (Pritchard et al. 2000). The kinship matrix (K) incorporates both pairwise relatedness and genetic covariance among individuals. Combining Q and K matrices in a mixed-model approach helps control spurious associations. The number of subpopulations or population clusters is determined using delta K values, which are vital for GWAS in potatoes. Overall, accounting for PS and kinship is essential in GWAS studies to distinguish true MTAs from spurious ones and to ensure the accuracy of genetic analyses. It found kinship correction was more effective than PS adjustment for cultivated potato GWAS, given the absence of significant PS in cultivated potatoes (Sharma et al. 2018). It is important to emphasize that correcting for PS and relatedness is particularly critical when these factors influence the trait under investigation in the germplasm. Nevertheless, overcompensation for PS and relatedness can also lead to false negatives. Hence, it is crucial to carefully choose appropriate covariates for each trait to prevent spurious associations and overfitting in GWAS models.

3.13 Statistical Correction and Quality Assessment in Post-GWAS Analysis

Various post-GWAS tests and methodologies are typically employed to ensure SNPs' reliability while minimizing the risk of spurious results. Researchers often employ statistical corrections to identify significant SNPs. The Bonferroni correction and False Discovery Rate (FDR) control are two widely used correction methods. The Bonferroni correction is particularly valuable when controlling the overall Type I error rate. It adjusts the significance threshold for individual tests by dividing it by the total number of tests, typically equivalent to the number of markers. The significance threshold is commonly set at 0.05–0.1 (5–10%) for a single association test (Angelin-Bonnet et al. 2023). While the Bonferroni correction effectively controls Type I errors, it can become excessively conservative when dealing with a large number of tests, as is often the case in GWAS with thousands or millions of SNPs. In such cases, alternative methods like FDR control or permutation testing may be preferred, as they offer a more balanced approach between controlling errors and avoiding excessive conservatism. FDR calculations may involve assessing the random percentage, with FDR values typically being less than a specified threshold (5–10%) (Angelin-Bonnet et al. 2023; Díaz et al. 2021; Khlestkin et al. 2019). These post-GWAS strategies enhance the credibility and reliability of significant SNPs, reducing the likelihood of reporting spurious genetic associations. Additionally, Quantile-Quantile (QQ) plots are commonly employed to visually assess the distribution of observed p-values compared to the expected null distribution (Díaz et al. 2021; Yuan et al. 2020). These plots assist in detecting deviations from the predicted distribution, mitigating false positives stemming from factors like population stratification, and facilitating an evaluation of the overall quality of the GWAS.

3.14 Software Tools Used in Potato GWAS Studies

The application of statistical methods in GWAS has evolved significantly in recent decades. To achieve this, tools like STRUCTURE and EIGENSTRAT were utilized, resulting in the generation of Q and P matrices (Pritchard et al. 2000; Price et al. 2006). Additionally, Yu et al. (2006) calculated the kinship (K) matrix, which incorporates random polygenic effects. These techniques have played a crucial role in enhancing the speed and effectiveness of GWAS models, and they have also been seamlessly integrated into widely adopted software tools, such as Trait Analysis by Association, Evolution, and Linkage (Bradbury et al. 2007), GAPIT (Lipka et al. 2012; Tang et al. 2016), and GEMMA (Zhou and Stephens 2012).

TASSEL is a widely employed software package for LD mapping, PS analysis, graphical representation of LD, and determining genetic distances in plants. TASSEL supports various types of markers, including SSR and SNPs. However, TASSEL faces a limitation in handling diploid allele dosage, making it unable to differentiate heterozygotes in a codominant marker system. This limitation is particularly relevant when conducting association analysis in autotetraploid species like potatoes, where five allele dosage states exist (nulliplex, simplex, duplex, triplex, and quadriplex). In response to this limitation, Rosyara et al. (2016) developed the GWASpoly R package, which accommodates tetraploid allele dosage and various forms of polyploid gene action. GWASpoly (Rosyara et al. 2016) and SHEsis (Shen et al. 2016) are tools specifically designed to handle polyploid genomic data. These tools encompass GAPIT, which accepts polyploid genomic data. They include GAPIT (Lipka et al. 2012; Tang et al. 2016) and TASSEL (Bradbury et al. 2007), originally designed for diploid GWAS but requiring modifications for tetraploid data.

Recently, Garreta et al. (2021) introduced MultiGWAS, a tool capable of conducting GWAS in both diploids and tetraploids concurrently by running four software packages in parallel. Two of these (GWASpoly and SHEsis) are tailored for polyploid data, while the others are tailored for diploid data (GAPIT and TASSEL). Garreta et al. (2021) showed that MultiGWAS is more adept at detecting true associations in tetraploid potatoes. For polyploid species, tools that integrate and compare different gene action models among software are crucial for gaining insights into how redundancy or allele dosage influences phenotype expression and the emergence of new traits. Furthermore, ongoing advancements in Bayesian and mixed-model GWAS approaches aim to enhance computational efficiency further, addressing this constraint.

3.15 Genome-Wide Association Mapping Studies Potatoes for Quality Traits

Potatoes have been a focal point of molecular research due to their nutritional importance and significant role in the processing industry. In potato research, various successful studies have employed the GWAS approach to identify significant marker loci associated with tuber traits. Several noteworthy studies are summarized in Table 3.3, as discussed here.

Table 3.3 Genome-wide AM studies for yield and quality-related traits in potato

Trait	Genotyping	Population	Chr (position)	PV(R^2) or -$\log_{10}(P)$	References
Underwater weight	AFLP	221 tetraploid cultivars	6 loci on Chr 1; 4; 5; 9	With *P*-value ≤ 0.01	D'hoop et al. (2008)
Specific gravity	33 candidate genes + 29 SSR loci	205 tetraploids	6 associations on 2, 3, 5, 8	$P < 0.001$	Urbany et al. (2011)
Enzymatic browning	AFLP	221 tetraploid cultivars	6 loci on Chr 1, 4, 7, 8, 11	With *P*-value ≤ 0.01	D'hoop et al. (2008)
Tuber bruising	33 candidate genes + 29 SSR loci	205 tetraploids	9 associations on 2, 3, 8	$P < 0.001$	Urbany et al. (2011)
	Capture-seq genotyping	158 progeny panel [half-sibling populations of Crop20 × its progeny Crop52]	Five significant markers on Chr 1 (12.8 Mb), 7 (5 Mb), and 8 (45.9 Mb, 46 Mb, 49.1 Mb)	$-\log_{10}(p)$ values of 6.85 (Chr 1), 3.97 (Chr 7), 3.86–4.12 (Chr 8)	Angelin-Bonnet et al. (2023)
Potato starch phosphorus content	Illumina 22 K SNP potato array	90 *S. tuberosum* L. varieties	Out of 17 significant SNPs identified, 14 were assigned to Chr 1, 4, 5, 7, 8, 10, 11	$-\log_{10}(p)$ values of 4.18–7.19	Khlestkin et al. (2019)
Tuber starch preparative yield and granule morphology	Illumina 22 K SNP potato array	90 *S. tuberosum* L. varieties	A total of 53 novel SNPs were identified on Chr 1, 2, 4, 5, 6, 7, 9, 11, and 12	$-\log_{10}(p)$ of 4.32–5.64	Khlestkin et al. (2020)

Tuber starch content	RADseq and SolCAP 8.3K array	264 varieties and breeding clones	127 loci spread on 12 Chr	$p < 0.0001$	Schönhals et al. (2017)
	SNP	40 cultivars	Nine SNPs on Chr 3	PV of 11.2–7.5% (with $p < 0.001$)	Fischer et al. (2013)
	SNP	243 tetraploid cultivars	Significant associations were detected on Chr 1, 3	PV of 1.8–11.4% (with $p < 0.05$)	Li et al. (2013)
	SolCAP 20 K SNP array	277 genotypes	Four SNPs on Chr 3 (61.49, 61.55, 61.55, and 61.89 Mbp)	PV of 13–15% ($-\log_{10}(P)$ of 3.91–4.48)	Klaassen et al. (2019)
	SNPs (GBS)	762 clones (MASPOT population)	612 SNPs located mainly on Chr 10	PV of 6.5–24.8%	Sverrisdóttir et al. (2017)

(continued)

Table 3.3 (continued)

Trait	Genotyping	Population	Chr (position)	PV(R^2) or -$\log_{10}(P)$	References
Tuber shape	SolCAP 8,303 SNPs	809 accessions of the Central Colombian Collection	Chr 7	4.47 (−$\log_{10}(P)$)	Berdugo-Cely et al. (2017)
	AFLP +SSR	205 academic panel (a), 299 industrial panel (b)	With panel (a), 4 SNPs on Chr 2, 10, 11, 12 With panel (b) SNP on Chr 6	$p \leq 0.001$	D'hoop et al. (2014)
	Infinium SNP array of 8303 markers	SolCAP potato diversity panel	Chr 10 (c1_8019)	−log10p of 9.5	Rosyara et al. (2016)
	22 K SNP potato array	Diversity panel of 214 advanced clones	Chr 10	−$\log_{10}(p)$ of 6.2–8.7	Pandey et al. (2022)
	AFLP	221 tetraploid cultivars.	5 loci on Chr 2, 4, 11	With P-value ≤ 0.01	D'hoop et al. (2008)
	8303 Illumina Infinium SNP array	351 diverse tetraploid genotypes	Chr 10 (solcap_snp_c1_8019)	14.7(K model), 13.85 (QK model)	Sharma et al. (2018)
	Set of 27,494,422 SNPs	108 autotetraploid potato accession	Chr 6 (49.12 Mb), chr07 (55.66 Mb), chr09 (21.82 Mb), Chr 10 (8.23 Mb, 50.37 Mb), Chr 12 (57.23 Mb)	−log10p of 7.63–13.41	Zhang et al. (2022) (Tuber aspect ratio)

Tuber Flesh color	SolCAP 8,303 SNPs	809 accessions of Central Colombian Collection	Chr 2, 4, 7	4.6–5.33 ($-\log_{10}(P)$	Berdugo-Cely et al. (2017)
	8303 Illumina Infinium SNP array	351 diverse tetraploid genotypes	Chr3 (solcap_snp_c2_20285)	8.67 (K model), 7.84 (QK model)	Sharma et al. (2018)
	GBS	83 tetraploid cultivars	Chr 3, 4	PV of 61% ($-\log_{10}(p) \geq 7$), PV of 9% ($-\log_{10}(p) = 3.7$)	Uitdewilligen et al. (2013)
	AFLP	221 tetraploid cultivars.	6 loci on Chr 1, 2, 3, 4, 6, 7	With P-value ≤ 0.01	D'hoop et al. (2008)
	AFLP +SSR	205 academic panel (a), 299 industrial panel (b)	With (a) panel, 5 SNPs on Chr 1, 5, 6, 9, 12 With (b) panel, 3 SNPs on Chr 3, 6	$p \leq 0.001$	D'hoop et al. (2014) (Flesh Color (Y))
	22 K SNP potato array	Diversity panel of 214 advanced clones	Chr 1, Chr 3	$-\log_{10}(p)$ of 5.8 (Chr 1), 10.5–28 (Chr 3)	Pandey et al. (2022) (chroma value)

(continued)

Table 3.3 (continued)

Trait	Genotyping	Population	Chr (position)	PV(R^2) or -$\log_{10}(P)$	References
Maturity	GBS	83 tetraploid cultivars	Chr 5	$-\log10(p) = 6.0$) and PV of 44%	Uitdewilligen et al. (2013)
	Infinium 22 K V3 Potato Array	384 chipping potato clones	A significant marker on Chr 5	Accounted for 21% of the breeding value ($-\log_{10}(p) > 5$)	Pandey et al. (2023)
	AFLP	221 tetraploid cultivars.	11 loci on Chr 1, 4, 5, 6, 9, 10	$P \leq 0.01$	D'hoop et al. (2008)
	AFLP +SSR	205 academic panel (a), 299 industrial panel (b)	With (a) panel, 2 SNPs on Chr 1, 3 With panel (b) 1 SNPs on Chr 2	$p \leq 0.001$	D'hoop et al. (2014)
	Infinium SNP array of 8303 markers	SolCAP potato diversity panel	Chr 1 (c2_34548), Chr 9 (c2_13133), Chr 11 (c1_9183)	–log10p values 4.5, 5.2, 4.7	Rosyara et al. (2016) (at 95 days)
	Infinium SNP array of 8303 markers	SolCAP potato diversity panel	Chr 7 (c2_25219)	–log10p of 5.0	Rosyara et al. (2016) (at 120 days)
Frying color (Apr 4C)	AFLP +SSR	205 academic panel (a), 299 industrial panel (b)	With (a) 1 SNP on Chr 9 With panel (b) 5 SNPs on Chr 1, 6, 7, 9, 12	$p \leq 0.001$	D'hoop et al. (2014)
Frying color (Apr 8C)	AFLP +SSR	205 academic panel (a), 299 industrial panel (b)	With (a) 7 SNPs on Chr 1, 2, 4, 5, 6, 9 With panel (b) 2 SNPs on Chr 6, 9	$p \leq 0.001$	D'hoop et al. (2014)

Fry Color	SNPs (GBS)	219 tetraploid genotypes	Significant SNPS on Chr 4, Chr 10	−log10(p) values of 5.91–6.23, 9.69, respectively	Byrne et al. (2020)
Chipping color	Infinium 22 K V3 Potato Array	384 chipping potato clones	Identified QTL loci for chip color on Chr 1, 3, and 7 ($-\log_{10}(p) > 5$)	Accounted for 28% of the breeding value	Pandey et al. (2023)
	AFLP	221 tetraploid cultivars.	6 loci on Chr 1, 2, 4, 7, 10	With P-value ≤ 0.01	D'hoop et al. (2008)
Tuber eye depth	Set of 27,494,422 SNPs	108 autotetraploid potato accession	Chr 10 (at 49.59 Mb)	−log10p of 9.44	Zhang et al. (2022)
	8303 Illumina Infinium SNP array	351 diverse tetraploid genotypes	Chr 10 (solcap_snp_c1_8019)	–log10p 13.15 (K model), 13.54 (QK model)	Sharma et al. (2018)
	Infinium SNP array of 8303 markers	SolCAP potato diversity panel	Chr 5 (solcap_snp_c2_11685), Chr10 (solcap_snp_c1_8019)	−log10p of 4.7, 7.2	Rosyara et al. (2016)
	22 K SNP potato array	Diversity panel of 214 advanced clones	Chr 3, Chr 5, Chr 10	$-\log_{10}(p)$ of 5.9, 5.6, 6.2	Pandey et al. (2022)
Plant height	8303 Illumina Infinium SNP array	351 diverse tetraploid genotypes	chr5 (solcap_snp_c2_50302)	5.3 (K model)	Sharma et al. (2018)
Reducing sugar	Set of 27,494,422 SNPs	108 autotetraploid potato accessions	Chr 4 (5.12 Mb)	8.06	Zhang et al. (2022)
Sucrose	Array of 47K SNPs	104 diploid potato accessions	Chr 1, 5, 12	−log10p of 3.30–3.56 (R^2 = 13.61–14.57)	Díaz et al. (2021) (Drought conditions)
Glucose	Array of 47K SNPs	104 diploid potato accessions	Chr 9, 12	3.60–3.74 (R^2 = 14.73–15.25)	Díaz et al. (2021) (Drought conditions)

(continued)

Table 3.3 (continued)

Trait	Genotyping	Population	Chr (position)	PV(R^2) or -$\log_{10}(P)$	References
Fructose	Array of 47K SNPs	104 diploid potato accessions	Chr 10	3.17 (R^2 = 13.0)	Díaz et al. (2021) (Drought conditions)
	Set of 27,494,422 SNPs	108 autotetraploid potato accessions	Chr 1 (1.67), Chr 4 (68.5), Chr 6 (51.9)	−log10p of 7.4–11.2	Zhang et al. (2022)
Amylopectin	Set of 27,494,422 SNPs	108 autotetraploid potato accessions	Chr 1 (76.5) and Chr 9 (39.6)	7.4–8.2	Zhang et al. (2022)
Amylose	Set of 27,494,422 SNPs	108 autotetraploid potato accessions	Chr 1 (76.5) and Chr 9 (39.6)	7.4–8.2	Zhang et al. (2022)
Protein	SolCAP 20 K SNP array	277 genotypes	12 SNPs on Chr 3, 5, 7	PV of 10–12% (−$\log_{10}(P)$ of 4.07–5.84)	Klaassen et al. (2019)
Anthocyanidins	7,520 SNPs (GBS)	109 diploid landraces (Group *Phureja*)	Three QTL on Chr 10, one on Chr I, II, XI, and XII	PV of 31.3–44.4%	Parra-Galindo et al. (2019)
Tuber size	AFLP +SSR	205 academic panel (a), 299 industrial panel (b)	With (a) panel, 1 SNP on Chr 8 With panel (b) 1 SNP on Chr 4	$p \leq 0.001$	D'hoop et al. (2014)
Tuber length	Infinium SNP array of 8303 markers	SolCAP potato diversity panel	chr10 (c1_8019)	−log10p of 6.5	Rosyara et al. (2016)
	Set of 27,494,422 SNPs	108 autotetraploid potato accession	Chr 9 (25.59 Mb, 45.76 Mb), Chr 10 (50.06 Mb)	−log10p of 7.47–7.77	Zhang et al. (2022)
Tuber weight	Set of 27,494,422 SNPs	108 autotetraploid potato accessions	Chr 6 (39.5 Mb)	−log10p of 7.6	Zhang et al. (2022) Commercial potato weight)
Tuber number	Set of 27,494,422 SNPs	108 autotetraploid potato accession	Chr 8 (8.40)	−log10p of 7.80	Zhang et al. (2022)

3.16 Tuber Quality Traits

Tuber flesh color is a trait of significant importance, as the natural variation in this attribute, ranging from white to yellow or even dark yellow, is indirectly linked to the presence of carotenoids in the tuber tissue (D'hoop et al. 2014). In GWAS investigations, a genomic region from 8 to 38.5 cM on Chr 1, 3, 5, 6, 9, and 12 was identified as associated with tuber flesh color (D'hoop et al. 2014). The *Y* locus, which determines the yellow flesh color, has been mapped to Chr 3 and co-locates with the *beta-carotene hydroxylase* (*BCH*) gene, also known as *Chy2* (Bonierbale et al. 1988). Wolters et al. (2010) have demonstrated that the *Chy2* allele 3 is the primary determinant of flesh color in tetraploid potatoes. In contrast, Śliwka et al. (2008) associated Chr 12 with tuber flesh color in their study. Other QTLs linked to flesh color were also detected on Chr 1, 2, 3, 4, 6, and 7 (D'hoop et al. 2008). These GWAS investigations suggested that the gene governing the yellow coloration in potatoes is almost all potato chromosomes.

Tuber skin color, an important feature, greatly influences consumers' choices when purchasing potatoes. Carotenoid and anthocyanin pigments, responsible for a spectrum of tuber colors, also play a role in human nutrition. By pinpointing the genetic loci associated with these traits, desirable tuber skin colors can be efficiently selected for breeding programs, thus reducing the need for extensive field-based selection. To uncover tuber skin color MTAs, Zia et al. (2020) performed a comprehensive GWAS and identified 16 SNPs associated with tuber skin color. These SNPs were found within candidate genes that play crucial roles in carotenoid pathways, including *lycopene epsilon-cyclase*. The genomic regions linked to skin color were distributed across Chr 1, 4, 5, 7, 8, 10, 11, and 12 (Zia et al. 2020). In a similar study performed by Berdugo-Cely et al. (2017), genomic regions on Chr 2 (at 25.05 Mbp), 7 (at 53.65 Mbp), and 10 (at 58.43 Mbp) were identified in association with tuber skin color. Additionally, the study detected the presence of four SNPs associated with flesh color on Chr 2, 4, and 7.

Fry color is another crucial quality trait in potato processing, indicating the degree of discoloration that potato chips or French fries undergo due to the Maillard reaction. Previous studies (D'hoop et al. 2008; Menéndez et al. 2002) successfully identified MTAs for fry color on Chr 1, 2, and 7. Additionally, chipping color was associated with Chr 1, 2, 4, 7, and 10 (D'hoop et al. 2008; Douches and Freyre 1994). Both of these traits serve as indicators of cold-induced sweetening (CIS) in potatoes; notably, the detected loci on Chr 2 and 7 that were linked to both traits appeared to exhibit pleiotropic effects since they were associated with the same AFLP markers (D'hoop et al. 2008). A QTL peak on Chr 3 was found to be within 1 cM vicinity of *vacuolar acid invertase* (*VInv*), a major candidate gene for CIS (Rak et al. 2017). Byrne et al. (2020) conducted an association study and revealed a major QTL on Chr 10 associated with fry color, notably the SNP cluster at the telomeric region of Chr 10 (spanning from 49 to 59 Mbp), which is well-known in the potato genome for its notably higher gene density. In this specific genomic region, numerous genes were identified, establishing connections with sucrose biosynthesis, cleavage, metabolism, and starch storage. These include genes encoding for a cell-wall invertase, invertase inhibitors, a fructose-1,6-bisphosphatase, and a sucrose-phosphatase (Byrne et al. 2020). In addition, significant associations with fry color were also detected on Chr 4, specifically at positions 67 and 68 Mbp. Additionally, Sverrisdóttir et al. (2017) also highlighted the clustering of 372 SNPs on Chr 10, spanning the region between 50 and 60 Mbp, associated with fry color. D'hoop et al. (2014) also successfully identified an MTA for fry color, pinpointing it at 57.6 Mbp on Chr 10. These findings collectively underscore the genetic complexity underlying the fry color trait in potatoes, particularly its strong association with Chr 10 and the gene-rich region encompassing crucial metabolic genes (Sverrisdóttir et al. 2017; Byrne et al. 2020).

Multiple AM studies have been undertaken to uncover MTAs and candidate genes linked to essential potato traits. Multiple GWAS studies aimed to identify markers and loci associated with TSC, DMC, and sugars. Schönhals et al. (2017) identified a total of 127 SolCAP SNPs, linked to 117 genes related to TSC, and unevenly distributed across all chromosomes in clusters ranging from 0.5 to 4 Mbp in width, potentially involved in regulatory and signaling functions. In GWAS studies conducted by D'hoop et al. (2008, 2014), specific loci associated with underwater weight were identified on Chr 1 (at 8.4, 9.2, and 24.9 cM), 2 (at 62.8 cM), 4 (at 21.3 cM), 5 (at 16.2 cM), and 9 (at 46 cM). Additionally, these studies revealed a negative correlation between higher underwater weight and early tuber maturity. In another AM study, which accounted for genotype-environment interactions (GEI) concerning soil type and rainfall patterns, 21 QTLs were identified in a panel of 205 tetraploid cultivars. These QTLs were associated with underwater weight, after-cooking types, flesh color, tuber size, shape, and maturity (D'hoop et al. 2014). These QTLs, referred to as GEI QTLs, were linked to frying color, flesh color, and underwater weight, with positions on Chr 9 (at 16.6 cM), 10 (at 14.1 cM), 6 (at 15.2 cM), 7 (at 56.5 cM), and 12 (at 46.6 cM). These QTLs were located on Chr 3 (60.84 Mbp), 5 (4.40–5.69 Mbp), and 7 (50.15 Mbp) for protein content. Additionally, four significant MTAs for starch content were found on Chr 3. The contents of starch and proteins usually positively correlate due to interconnected carbon and nitrogen assimilation pathways in photosynthesis. Molecular studies suggested that genes affecting starch in potatoes might similarly impact protein content (Klaassen et al. 2019). Using a combination of microsatellites, SNPs, and known markers, Schönhals et al. (2016) identified MTAs on several Chr, implicating genes like *ADP-glucose pyrophosphorylase S* (*AGPase S*), transcription factor (*StCDF1*), vacuolar invertase (*Pain-1*), *starch synthase* (*SSIV*), *apoplastic invertas*e (*InvCD141*), and *leucine aminopeptidase* (*LapN*) in regulating TSC and yield.

Another study focused on diploid populations from *S. tuberosum* Group *Phureja* and discovered seven significant QTLs associated with anthocyanidins, primarily on Chr 1 and 10, governing purple pigmentation of tubers (Parra-Galindo et al. 2019). The GWAS analysis revealed the presence of seven QTLs associated with the anthocyanidins located on Chr 1, 10, 11, and 12. Notably, the most stable QTLs were identified on Chr 1 (at 50.7 Mbp) and 10 (52–57 Mbp). The SNPs within candidate genes were found to have functional significance in the synthesis of purple-colored delphinidin pigments, which are primarily located on locus *P*, known as flavonoid 3′,5′-hydroxylase (F3′5′H). Some of the identified SNPs are associated with the *PAL* gene, which plays a key role in the anthocyanin biosynthesis by mediating the conversion of phenylalanine into coumarate-CoA (Parra-Galindo et al. 2019).

Angelin-Bonnet et al. (2023) identified five significant markers that are associated with tuber bruising: on Chr 1 (at 12.8 Mb), on Chr 7 (5 Mb), and on Chr 8 (at 45.9, 46, and 49.1 Mb). Notably, two of these markers on Chr 8 around 46 Mb were closely linked to genes coding for PPOs (polyphenol oxidases), including the *POT32* gene, responsible for melanin pigment production contributing to bruising color. A significant marker on Chr 7 (at 50.7 Mb) was within 1 Mb of a *phosphate adenylyltransferase* gene, while the other two highly significant markers on Chr 8 (3.2–3.4 Mb) were near the *patatin* genes. Additionally, a significant marker on Chr 3 (at 52 Mb) was linked to a *proteinase inhibitor type-2* gene. A number of other GWAS markers with high scores were identified within a 5 Mb region of previously identified candidate genes (Angelin-Bonnet et al. 2023).

Studies on the regulation and genetics of amino acid content, particularly involving asparagine (ASN) and glutamine (GLU), are significant in understanding nitrogen transport and storage in tubers, which is limited (Chawla et al. 2012). The genes underlying the biosynthesis pathway have been identified; nevertheless, their biosynthesis is influenced by the environmental cues and is polygenic in nature (Chawla et al. 2012). An AM analysis by Fischer et al. (2013) identified an SNP_{2746} marker (located on Chr 12) within the *LapN* gene associated with TSC and chip quality during cold storage. *LapN* is essential in the conversion of starch into sugars, leading to the high accumulation of reducing sugars responsible for CIS. An AM study using markers linked to putative candidate genes revealed a significant association of *UGPase* and *apoplastic invertase* genes with CIS and glucose content (Fischer et al. 2013). Additionally, an interaction was observed between the *apoplastic invertase* and *invertase inhibitor* genes.

In another study, by employing a candidate gene AM approach, SNPs within eight candidate genes associated with TSC and the quality of chipping were identified (Schreiber et al. 2014): *AGPase S*, *plastidial phosphoglucomutase* (*PGM1-3*), *soluble starch synthase I* (*SssI*), *L-type starch phosphorylase b (PHO1b)*, invertase (*INV-8/2*), *β-amylase* (*BMY-8/2*), *phosphoglucan water dikinase* (*PWD*), and *InvCD141*. All these genes are involved in converting starch to sugar. In particular, *INV-8/2* plays a key role in plastidial starch biosynthesis and degradation. Additionally, it was observed that an increased TSC is linked to the *PHO1b* gene (Schreiber et al. 2014). Polymorphisms in candidate genes, specifically *AGPase S* (located on Chr 1) and the *Pain-1* promoter region (located on Chr 3), were associated with TSC, starch yield, and chip quality. The *Pain-1* candidate gene is responsible for encoding soluble acid invertases, which play a critical role in converting sucrose into reducing sugars, and it is located on Chr 3. These genetic associations are particularly significant for chipping quality. Moreover, the study observed the presence of epistatic interactions between candidate genes and their alleles, particularly in the context of TSC and starch yield traits (Li et al. 2013).

An AM study employing a candidate gene approach successfully pinpointed allelic variations within genes responsible for the process of browning and tuber bruising (Urbany et al. 2011). This study identified 21 MTAs associated with traits, such as SG, tuber shape, yield, tuber bruising, and plant maturity. The genes associated with bruising and SG include L-type starch phosphorylases (*PHO1A* and *PHO1B*), *PPO isoform potpoloxA*, *lipase class III* (*LIPIII-27-1h*), and *4-coumarate-CoA ligase* genes (Urbany et al. 2011).

3.17 Abiotic and Biotic Stress-Related GWAS

GWAS have emerged as pivotal tools in unraveling the genetic foundations of resistance to pathogens and responses to environmental stresses. Despite significant strides in recent years, it is noteworthy that the landscape of comprehensive studies in potato genetics remains relatively limited, as underscored by the insights presented in Table 3.4. This highlights the untapped potential of GWAS in fully exploring the genetic basis of complex traits and stress responses in potatoes.

Potato common scab, caused by the bacterial pathogen *Streptomyces scabies*, stands out as one of the most significant potato diseases globally. Yuan et al. (2020) performed a GWAS for potato common scab resistance in a tetraploid diversity panel and successfully located three QTLs responsible for resistance. These QTLs are located on potato Chr 2, 4, and 12, respectively, and account for the PV of approximately 1%, 19%, and 26%, respectively. The QTL on Chr 2 displays a single-dominant inheritance pattern, whereas the QTLs on Chr 4 and 12 reveal double-dominant inheritance in their genetic makeup.

Synchytrium endobioticum is a soil-borne obligate biotrophic fungal pathogen responsible for potato wart disease. In various genetic backgrounds, QTLs associated with resistance to multiple or specific pathotypes distributed throughout the potato genome have been identified as responsible for potato wart disease resistance (Ballvora et al. 2011; Groth et al. 2013; Obidiegwu et al. 2015). However, only a few markers associated with *Sen1* or the other single-gene loci and QTLs have undergone validation across a diverse range of potato varieties, which are considered representative of the overall genetic diversity found in commercial potato crops. Prodhomme et al. (2020) performed a GWAS using 330 commercial potato varieties and discovered a haplotype-specific SNP marker on Chr 11. This marker is significantly associated with wart disease (pathotype 1(D1)) resistance linked to the dominant monogenic *Sen1* locus. The *Sen1* locus encompasses a cluster of multiple nucleotide-binding leucine-rich repeat-encoding (NLR) genes (Hehl et al. 1999).

The growing risk of drought stress resulting from climate change represents a severe challenge to agricultural sustainability worldwide. It is crucial to dissect the genetic architecture of drought stress responses, as this is highly crucial for developing drought-tolerant potato varieties and the enhancement of the resilience of existing ones. Recently, Díaz et al. (2021) analyzed the genetic variation in response to drought stress within a group of 104 diploid potato plants and identified the QTL responsible for drought stress (expressed drought stress index, DTI) using GWAS

Table 3.4 Summary of GWAS for potato biotic and abiotic stress-related traits

Trait	Genotyping	Population	Chr (position)	PV(R^2) or $-\log_{10}(P)$	References
Potato wart disease (*Synchytrium endobioticum*)	10,968 SNP markers	A panel of 330 varieties	A significant association on Chr 11 (0.78–4.34 Mb)	$-\log_{10}(p)$ value of 8.97	Prodhomme et al. (2020)
Resistance to late blight	22,489 SNPs	A panel of 284 cultivars	964 loci associated with the trait	$-\log_{10}(p)$ values of 1.30–3.95	Wang et al. (2020, 2021)
	120,622 SNPs (GBS)	A panel of 222 diverse accessions	Seven significant SNPs detected on Chr 3, 4, 5, 8, 11 (3 SNPs)	PV of 9.9–17.1%	Sood et al. (2023)
G. rostochiensis	120,622 SNPs (GBS)	A panel of 222 diverse accessions	11 SNPs on Chr 1, 2, 3, 4, 5, 10, 12	PV of 4.5–21.7%	Sood et al. (2023)
G. pallida	120,622 SNPs (GBS)	A panel of 222 diverse accessions	Nine SNPs on Chr 1, 3, 5, 10, 11	PV of 4.0–15.7%	Sood et al. (2023)
Common scab resistance	13,002 SNPs	165 diverse genotypes	A locus on Chr 1 (0.43- 1.07 Mb)	~ 14.7% PV with $-\log_{10}(p)$ >5	Koizumi et al. (2021)
	12K SolCAP SNPs	A panel of 143 clones	Three QTLs were identified on Chr 2, 4, 12	PV of 21%, 19%, and 26%, respectively	Yuan et al. (2020)
Relative water content (RWC)	Array of 47K SNPs	104 diploid potato accessions	Chr 6	3.09 ($R^2 = 12.79$)	Díaz et al. (2021)
Tuber number (Well-watered)	Array of 47K SNPs	104 diploid potato accessions	A significant locus on Chr 2	5.3 ($R^2 = 20.9$)	Díaz et al. (2021) (Well-watered)
Tuber number (Drought conditions)	Array of 47K SNPs	104 diploid potato accessions	Two loci on Chr 1	4.41-4.86 ($R^2 = 17.5–19.36$)	Díaz et al. (2021)
Tuber number (Drought stress index)	Array of 47K SNPs	104 diploid potato accessions	A locus on Chr 12	4.18 ($R^2 = 16.89$)	Díaz et al. (2021)

and identified a total of 38 QTLs: 7 associated with well-watered conditions, 22 associated with water deficit conditions, and 9 related to the drought stress index (DTI). These QTLs account for 12.6%–44.1% of the PV. Among the nine QTLs identified for the increased drought tolerance across Chr 2, 3, 5, 8, 10, and 12, three candidate genes appeared to play a plausible role in responding to drought stress. These findings hold potential utility in MAS strategies for breeding drought-tolerant potato varieties. A total of 22 potential candidate genes were linked to the response under water deficit conditions. These genes encompass a variety of functions, including encoding transcription factors, sugar transporters, antioxidant enzymes, and heat shock proteins. Notably, the majority of these candidate genes are known to play roles in responding to abiotic stress (Díaz et al. 2021).

3.18 Future Prospects and Conclusions

While conventional breeding has significantly advanced potato varieties, recent years have witnessed a remarkable genomic transformation, marked by the shift from short-read to long-read sequencing approaches. These advancements in genomics have collectively paved the way for developing the pan-genome within the Solanum species (Bozan et al. 2023). Advances in genomics and bioinformatics tools have addressed various technical hurdles in understanding the genetic basis of traits in autopolyploid crops. Increasing marker density is pivotal to improving the accuracy of genetic loci identification for polygenic tuber traits. This can be achieved through SNP genotyping platforms such as SolCAP and the GBS platform. Strong analysis of multi-allelic haplotypes and the differentiation of five heterozygous states in potatoes enhances the power of MTAs.

Surprisingly, many genes responsible for critical agronomic and economic traits in potatoes remain undiscovered, a surprising gap in our understanding. This limitation is, in part, due to the imprecise determination of the genetic locations of QTLs within the complex genetic architectures and the use of highly heterozygous biparental populations. Despite the advancements, the genetic complexities introduced by autopolyploidy have led most potato genetic studies to focus on the diploid level, utilizing biparental populations. However, this approach comes with its own set of challenges, as markers closely linked to valuable trait alleles in biparental crosses may exhibit limited transferability to broader germplasm pools. Furthermore, recombination events in biparental populations are restricted to a single meiotic event, reducing the precision of QTL mapping. Nonetheless, recent developments have seen the emergence of sophisticated algorithms and software tools specifically tailored for linkage and QTL analysis in polyploid crop species.

The periderm plays a crucial role in safeguarding tubers against various stresses through suberization and lignification. Understanding the genetic markers associated with periderm traits through GWAS can provide valuable insights into periderm formation and facilitate the selection of genotypes with enhanced resistance to damage, pathogens, and dehydration. In the case of breeding for resistance, scab resistance in potatoes presents unique challenges. While scab symptoms exhibit a quantitative distribution, the genetic basis is believed to involve both dominant and recessive loci (Yuan et al. 2020). Achieving a quadruplex recessive state through conventional breeding is notoriously challenging due to the autotetraploidy of potatoes. Here, the use of MTAs could prove invaluable in introgressing recessive alleles. Once MTAs are identified through GWAS, they can be converted into high-throughput genotyping markers such as TaqMan(®) and Kompetitive Allele-Specific PCR assays for MAS.

Furthermore, optimizing genetic gains from GWAS involves the comprehensive integration of high-throughput phenotyping methods, extensive field testing across various environments, and advanced mixed models that account for kinship, PS, and pedigree relationships. However, it is crucial to exercise caution in the excessive application of methods to correct for PS and individual relatedness. While these corrections are vital for controlling false positives, an overreliance on them can inadvertently reduce statistical power. In summary, GWAS endeavors to discern genuine MTAs from spurious effects introduced by confounding factors such as kinship and PS. While autotetraploidy and high heterozygosity can potentially weaken population structuring, addressing weak PS necessitates careful analysis. AM offers several advantages over biparental populations, including heightened mapping precision, the ability to analyze multiple traits simultaneously, and the increased likelihood that identified MTAs hold across a broader range of genetic resources. The quest to unlock the genetic mysteries of tetraploid potatoes, spanning diverse germplasm through advanced molecular marker platforms, is a significant and worthwhile goal.

Acknowledgments The support provided by the Director, ICAR-Central Potato Research Institute, Shimla, HP, India, and the CABin Project is highly acknowledged.

References

Achenbach U, Paulo J, Ilarionova E et al (2009) Using SNP markers to dissect linkage disequilibrium at a major quantitative trait locus for resistance to the potato cyst nematode *Globodera pallida* on potato chromosome V. Theor Appl Genet 118(3):619–629. https://doi.org/10.1007/s00122-008-0925-x

Adillah Tan MY, Park T-H, Alles R et al (2009) $GpaXI_{tar}^{l}$ originating from *Solanum tarijense* is a major resistance locus to *Globodera pallida* and is localised on chromosome 11 of potato. Theor Appl Genet 119(8):1477–1487. https://doi.org/10.1007/s00122-009-1149-4

Alvarez-Morezuelas A, Barandalla L, Ritter E et al (2023) Genome-wide association study of agronomic and physiological traits related to drought tolerance in potato. Plants 12(4):734. https://doi.org/10.3390/plants12040734

Angelin-Bonnet O, Thomson S, Vignes M et al (2023) Investigating the genetic components of tuber bruising in a breeding population of tetraploid potatoes. BMC Plant Biol 23(1):238. https://doi.org/10.1186/s12870-023-04255-2

Ballvora A, Flath K, Lübeck J et al (2011) Multiple alleles for resistance and susceptibility modulate the defense response in the interaction of tetraploid potato (*Solanum tuberosum*) with *Synchytrium endobioticum* pathotypes 1, 2, 6 and 18. Theor Appl Genet 123(8):1281–1292. https://doi.org/10.1007/s00122-011-1666-9

Barker L, Kühn C, Weise A et al (2000) SUT2, a putative sucrose sensor in sieve elements. Plant Cell 12:1153–1164. https://doi.org/10.1105/tpc.12.7.1153

Beals KA (2019) Potatoes, nutrition and health. Am J Potato Res 96(2):102–110. https://doi.org/10.1007/s12230-018-09705-4

Berdugo-Cely J, Valbuena RI, Sánchez-Betancourt E et al (2017) Genetic diversity and association mapping in the Colombian Central Collection of *Solanum tuberosum* L. Andigenum group using SNPs markers. PLoS One 12(3):e0173039. https://doi.org/10.1371/journal.pone.0173039

Bonierbale MW, Plaisted RL, Tanksley S (1988) RFLP maps based on a common set of clones reveal modes of chromosomal evolution in potato and tomato. Genetics 120(4):1095–1103. https://doi.org/10.1093/genetics/120.4.1095

Bourke PM, Hackett CA, Voorrips RE et al (2019) Quantifying the power and precision of QTL analysis in autopolyploids under bivalent and multivalent genetic models. G3 (Bethesda) 9(7):2107–2122. https://doi.org/10.1534/g3.119.400269

Bozan I, Achakkagari SR, Anglin NL et al (2023) Pangenome analyses reveal impact of transposable elements and ploidy on the evolution of potato species. Proc Natl Acad Sci USA 120(31):e2211117120. https://doi.org/10.1073/pnas.2211117120

Bradbury PJ, Zhang Z, Kroon DE et al (2007) TASSEL: software for association mapping of complex traits in diverse samples. Bioinformatics 23(19):2633–2635. https://doi.org/10.1093/bioinformatics/btm308

Bradshaw JE (2021) Potato breeding: theory and practice. Springer

Bradshaw JE (2022) A brief history of the impact of potato genetics on the breeding of tetraploid potato cultivars for tuber propagation. Potato Res 65(3):461–501. https://doi.org/10.1007/s11540-021-09517-w

Bradshaw JE, Hackett CA, Meyer RC et al (1998) Identification of AFLP and SSR markers associated with quantitative resistance to *Globodera pallida* (Stone) in tetraploid potato (*Solanum tuberosum* subsp. tuberosum) with a view to marker-assisted selection. Theor Appl Genet 97(1):202–210. https://doi.org/10.1007/s001220050886

Bradshaw JE, Bryan GJ, Lees AK et al (2006) Mapping the *R10* and *R11* genes for resistance to late blight (*Phytophthora infestans*) present in the potato (*Solanum tuberosum*) R-gene differentials of black. Theor Appl Genet 112(4):744–751. https://doi.org/10.1007/s00122-005-0179-9

Bradshaw JE, Hackett CA, Pande B et al (2008) QTL mapping of yield, agronomic and quality traits in tetraploid potato (*Solanum tuberosum* subsp. tuberosum). Theor Appl Genet 116(2):193–211

Braun SR, Endelman JB, Haynes KG et al (2017) Quantitative trait loci for resistance to common scab and cold-induced sweetening in diploid potato. Plant Genome 10(3):plantgenome2016.2010.0110. https://doi.org/10.3835/plantgenome2016.10.0110

Bryan G, McLean K, Bradshaw J et al (2002) Mapping QTLs for resistance to the cyst nematode Globodera pallida derived from the wild potato species Solanum vernei. Theor Appl Genet 105(1):68–77. https://doi.org/10.1007/s00122-002-0873-9

Byrne S, Meade F, Mesiti F et al (2020) Genome-wide association and genomic prediction for fry color in potato. Agronomy 10(1):90. https://doi.org/10.3390/agronomy10010090

Campbell R, Pont SD, Morris JA et al (2014) Genome-wide QTL and bulked transcriptomic analysis reveals new candidate genes for the control of tuber carotenoid content in potato (*Solanum tuberosum* L.). Theor Appl Genet 127:1917–1933. https://doi.org/10.1007/s00122-014-2349-0

Caromel B, Mugniéry D, Lefebvre V et al (2003) Mapping QTLs for resistance against Globodera pallida (Stone) Pa2/3 in a diploid potato progeny originating from Solanum spegazzinii. Theor Appl Genet 106(8):1517–1523. https://doi.org/10.1007/s00122-003-1211-6

Caromel B, Mugniéry D, Kerlan M-C et al (2005) Resistance quantitative trait loci originating from *Solanum sparsipilum* Act independently on the sex ratio of *Globodera pallida* and together for developing a necrotic reaction. Mol Plant-Microbe Interact 18(11):1186–1194. https://doi.org/10.1094/mpmi-18-1186

Chawla R, Shakya R, Rommens CM (2012) Tuber-specific silencing of asparagine synthetase-1 reduces the acrylamide-forming potential of potatoes grown in the field without affecting tuber shape and yield. Plant Biotechnol J 10(8):913–924. https://doi.org/10.1111/j.1467-7652.2012.00720.x

Chen X, Salamini F, Gebhardt C (2001) A potato molecular-function map for carbohydrate metabolism and transport. Theor Appl Genet 102:284–295. https://doi.org/10.1007/s001220051645

Chen J, Leach L, Yang J et al (2021) A tetrasomic inheritance model and likelihood-based method for mapping quantitative trait loci in autotetraploid species. New Phytol 230(1):387–398. https://doi.org/10.1111/nph.16413

Collins A, Milbourne D, Ramsay L et al (1999) QTL for field resistance to late blight in potato are strongly correlated with maturity and vigour. Mol Breed 5:387–398. https://doi.org/10.1023/A:1009601427062

D'hoop BB, Paulo MJ, Mank RA et al (2008) Association mapping of quality traits in potato (*Solanum tuberosum* L.). Euphytica 161(1):47–60. https://doi.org/10.1007/s10681-007-9565-5

D'hoop BB, Paulo MJ, Kowitwanich K et al (2010) Population structure and linkage disequilibrium unravelled in tetraploid potato. Theor Appl Genet 121(6):1151–1170. https://doi.org/10.1007/s00122-010-1379-5

D'hoop BB, Keizer PLC, Paulo MJ et al (2014) Identification of agronomically important QTL in tetraploid potato cultivars using a marker–trait association analysis. Theor Appl Genet 127(3):731–748. https://doi.org/10.1007/s00122-013-2254-y

da Silva Pereira G, Mollinari M, Qu X et al (2021a) Quantitative trait locus mapping for common scab resistance in a tetraploid potato full-sib population. Plant Dis 105(10):3048–3054. https://doi.org/10.1094/PDIS-10-20-2270-RE

da Silva Pereira G, Mollinari M, Schumann MJ et al (2021b) The recombination landscape and multiple QTL mapping in a *Solanum tuberosum* cv.'Atlantic'-derived F1 population. Heredity 126(5):817–830. https://doi.org/10.1038/s41437-021-00416-x

De Jong W, De Jong D, De Jong H et al (2003) An allele of dihydroflavonol 4-reductase associated with the ability to produce red anthocyanin pigments in potato (*Solanum tuberosum* L.). Theor Appl Genet 107:1375–1383. https://doi.org/10.1007/s00122-003-1395-9

Dees MW, Wanner LA (2012) In search of better management of potato common scab. Potato Res 55:249–268. https://doi.org/10.1007/s11540-012-9206-9

Díaz P, Sarmiento F, Mathew B et al (2021) Genomic regions associated with physiological, biochemical and yield-related responses under water deficit in diploid potato at the tuber initiation stage revealed by GWAS. PLoS One 16(11):e0259690. https://doi.org/10.1371/journal.pone.0259690

Dolničar P (2021) Importance of potato as a crop and practical approaches to potato breeding, *Solanum tuberosum*. Methods Mol Biol 2354:3–20. https://doi.org/10.1007/978-1-0716-1609-3_1

Douches DS, Freyre R (1994) Identification of genetic factors influencing chip color in diploid potato (Solanum spp.). Am Potato J 71(9):581–590. https://doi.org/10.1007/BF02851523

Felcher KJ, Coombs JJ, Massa AN et al (2012) Integration of two diploid potato linkage maps with the potato genome sequence. PLoS One 7(4):e36347. https://doi.org/10.1371/journal.pone.0036347

Fischer M, Schreiber L, Colby T et al (2013) Novel candidate genes influencing natural variation in potato tuber cold sweetening identified by comparative proteomics and association mapping. BMC Plant Biol 13(1):113. https://doi.org/10.1186/1471-2229-13-113

Flint-Garcia SA, Thornsberry JM, Buckler ES IV (2003) Structure of linkage disequilibrium in plants. Annu Rev Plant Biol 54(1):357–374. https://doi.org/10.1146/annurev.arplant.54.031902.134907

Garreta L, Cerón-Souza I, Palacio MR et al (2021) MultiGWAS: An integrative tool for genome wide association studies in tetraploid organisms. Ecol Evol 11(12):7411–7426. https://doi.org/10.1002/ece3.7572

Gebhardt C, Menendez C, Chen X et al (2005) Genomic approaches for the improvement of tuber quality traits in potato. Acta Hortic 684:85–92. https://doi.org/10.17660/ActaHortic.2005.684.11

Getahun BB, Visser RGF, van der Linden CG (2020) Identification of QTLs associated with nitrogen use efficiency and related traits in a diploid potato population. Am J Potato Res 97(2):185–201. https://doi.org/10.1007/s12230-020-09766-4

Groth J, Song Y, Kellermann A et al (2013) Molecular characterisation of resistance against potato wart races 1, 2, 6 and 18 in a tetraploid population of potato (*Solanum tuberosum* subsp. tuberosum). J Appl Genet 54(2):169–178. https://doi.org/10.1007/s13353-013-0141-5

Habe I, Miyatake K, Nunome T et al (2019) QTL analysis of resistance to bacterial wilt caused by *Ralstonia solanacearum* in potato. Breed Sci 69(4):592–600. https://doi.org/10.1270/jsbbs.19059

Hackett CA, McLean K, Bryan GJ (2013) Linkage analysis and QTL mapping using SNP dosage data in a tetraploid potato mapping population. PLoS One 8(5):e63939

Hackett CA, Bradshaw JE, Bryan GJ (2014) QTL mapping in autotetraploids using SNP dosage information. Theor Appl Genet 127:1885–1904

Hackett CA, Boskamp B, Vogogias A et al (2017) TetraploidSNPMap: software for linkage analysis and QTL mapping in autotetraploid populations using SNP dosage data. J Hered 108(4):438–442. https://doi.org/10.1093/jhered/esx022

Hall D, Tegström C, Ingvarsson PK (2010) Using association mapping to dissect the genetic basis of complex traits in plants. Brief Funct Genomics 9(2):157–165. https://doi.org/10.1093/bfgp/elp048

Hamilton JP, Hansey CN, Whitty BR et al (2011) Single nucleotide polymorphism discovery in elite North American potato germplasm. BMC Genomics 12(1):302. https://doi.org/10.1186/1471-2164-12-302

Hardigan MA, Bamberg J, Buell CR et al (2015) Taxonomy and genetic differentiation among wild and cultivated germplasm of Solanum sect. Petota. Plant Genome 8(1):plantgenome2014.2006.0025. https://doi.org/10.3835/plantgenome2014.06.0025

Hawkes JG (1999) The evidence for the extent of NI Vavilov's new world Andean centres of cultivated plant origins. Genet Resour Crop Evol 46:163–168. https://doi.org/10.1023/A:1008642630144

Hehl R, Faurie E, Hesselbach J et al (1999) TMV resistance gene N homologues are linked to Synchytrium endobioticum resistance in potato. Theor Appl Genet 98(3):379–386. https://doi.org/10.1007/s001220051083

Hills U, Pieterse L (2009) World catalogue of potato varieties. Agrimedia GmbH, Bergen/Dumme, p 326

Hirsch CN, Hirsch CD, Felcher K et al (2013) Retrospective view of North American potato (*Solanum tuberosum* L.) breeding in the 20th and 21st centuries. G3 (Bethesda) 6:1003–1013. https://doi.org/10.1534/g3.113.005595

Huamán Z, Spooner DM (2002) Reclassification of landrace populations of cultivated potatoes (Solanum sect. Petota). Am J Bot 89(6):947–965. https://doi.org/10.3732/ajb.89.6.947

Jung CS, Griffiths HM, De Jong DM et al (2005) The potato *P* locus codes for flavonoid 3′, 5′-hydroxylase. Theor Appl Genet 110:269–275. https://doi.org/10.1007/s00122-004-1829-z

Khlestkin VK, Rozanova IV, Efimov VM et al (2019) Starch phosphorylation associated SNPs found by genome-wide association studies in the potato (*Solanum tuberosum* L.). BMC Genet 20(1):29. https://doi.org/10.1186/s12863-019-0729-9

Khlestkin VK, Erst TV, Rozanova IV et al (2020) Genetic loci determining potato starch yield and granule morphology revealed by genome-wide association study (GWAS). PeerJ 8:e10286. https://doi.org/10.7717/peerj.10286

Klaassen MT, Willemsen JH, Vos PG et al (2019) Genome-wide association analysis in tetraploid potato reveals four QTLs for protein content. Mol Breed 39(10):151. https://doi.org/10.1007/s11032-019-1070-8

Kloosterman B, Abelenda JA, Gomez MMC et al (2013) Naturally occurring allele diversity allows potato cultivation in northern latitudes. Nature 495(7440):246–250. https://doi.org/10.1038/nature11912

Koizumi E, Igarashi T, Tsuyama M et al (2021) Association of genome-wide SNP markers with resistance to common scab of potato. Am J Potato Res 98:149–156

Kolech SA, Halseth D, Perry K et al (2016) Genetic diversity and relationship of Ethiopian potato varieties to germplasm from North America, Europe and the International Potato Center. Am J Potato Res 93(6):609–619. https://doi.org/10.1007/s12230-016-9543-3

Kreike CM, de Koning JRA, Vinke JH et al (1993) Mapping of loci involved in quantitatively inherited resistance to the potato cyst-nematode Globodera rostochiensis pathotype Ro1. Theor Appl Genet 87(4):464–470. https://doi.org/10.1007/BF00215092

Kreike CM, de Koning JRA, Vinke JH et al (1994) Quantitatively-inherited resistance toGlobodera pallida is dominated by one major locus inSolanum spegazzinii. Theor Appl Genet 88(6):764–769. https://doi.org/10.1007/BF01253983

Kreike CM, Kok-Westeneng AA, Vinke JH et al (1996) Mapping of QTLs involved in nematode resistance, tuber yield and root development in Solanum sp. Theor Appl Genet 92(3):463–470. https://doi.org/10.1007/BF00223694

Laurie CC, Doheny KF, Mirel DB et al (2010) Quality control and quality assurance in genotypic data for genome-wide association studies. Genet Epidemiol 34(6):591–602. https://doi.org/10.1002/gepi.20516

Li L, Strahwald J, Hofferbert H-R et al (2005) DNA variation at the invertase locus invGE/GF is associated with tuber quality traits in populations of potato breeding clones. Genetics 170(2):813–821

Li L, Paulo M-J, Strahwald J et al (2008) Natural DNA variation at candidate loci is associated with potato chip color, tuber starch content, yield and starch yield. Theor Appl Genet 116:1167–1181

Li L, Tacke E, Hofferbert H-R et al (2013) Validation of candidate gene markers for marker-assisted selection of potato cultivars with improved tuber quality. Theor Appl Genet 126:1039–1052. https://doi.org/10.1007/s00122-012-2035-z

Li Y, Colleoni C, Zhang J et al (2018) Genomic analyses yield markers for identifying agronomically important genes in potato. Mol Plant 11(3):473–484. https://doi.org/10.1016/j.molp.2018.01.009

Li J, Wang Y, Wen G et al (2019) Mapping QTL underlying tuber starch content and plant maturity in tetraploid potato. Crop J 7(2):261–272

Lindqvist-Kreuze H, Gastelo M, Perez W et al (2014) Phenotypic stability and genome-wide association study of late blight resistance in potato genotypes adapted to the tropical highlands. Phytopathology 104(6):624–633. https://doi.org/10.1094/phyto-10-13-0270-r

Lipka AE, Tian F, Wang Q et al (2012) GAPIT: genome association and prediction integrated tool. Bioinformatics 28(18):2397–2399. https://doi.org/10.1093/bioinformatics/bts444

Malosetti M, van der Linden CG, Vosman B et al (2007) A mixed-model approach to association mapping using pedigree information with an illustration of resistance to *Phytophthora infestans* in potato. Genetics 175(2):879–889. https://doi.org/10.1534/genetics.105.054932

Manrique-Carpintero NC, Coombs JJ, Cui Y et al (2015) Genetic map and QTL analysis of agronomic traits in a diploid potato population using single nucleotide polymorphism markers. Crop Sci 55(6):2566–2579. https://doi.org/10.2135/cropsci2014.10.0745

Manrique-Carpintero N, Islam MS, Enciso-Rodríguez F et al (2020) Mapping *Solanum berthaultii*-based late blight (*Phytophthora infestans*) resistance in a diploid population. Am J Potato Res 97:596–604. https://doi.org/10.1007/s12230-020-09808-x

Manrique-Carpintero NC, Berdugo-Cely JA, Cerón-Souza I et al (2023) Defining a diverse core collection of the Colombian Central Collection of potatoes: a tool to advance research and breeding. Front Plant Sci 14:1046400. https://doi.org/10.3389/fpls.2023.1046400

Massa AN, Manrique-Carpintero NC, Coombs JJ et al (2015) Genetic linkage mapping of economically important traits in cultivated tetraploid potato (*Solanum tuberosum* L.). G3 (Bethesda) 5(11):2357–2364

Massa AN, Manrique-Carpintero NC, Coombs J et al (2018) Linkage analysis and QTL mapping in a tetraploid russet mapping population of potato. BMC Genet 19(1):1–13. https://doi.org/10.1186/s12863-018-0672-1

McCord PH, Sosinski BR, Haynes KG et al (2011) Linkage mapping and QTL analysis of agronomic traits in tetraploid potato (*Solanum tuberosum* subsp. tuberosum). Crop Sci 51(2):771–785. https://doi.org/10.2135/cropsci2010.02.0108
Menéndez CM, Ritter E, Schäfer-Pregl R et al (2002) Cold sweetening in diploid potato: mapping quantitative trait loci and candidate genes. Genetics 162(3):1423–1434. https://doi.org/10.1093/genetics/162.3.1423
Mishra T, Raigond P, Thakur N et al (2020) Recent updates on healthy phytoconstituents in potato: a nutritional depository. Potato Res 63:323–343. https://doi.org/10.1007/s11540-019-09442-z
Mosquera T, Alvarez MF, Jiménez-Gómez JM et al (2016) Targeted and untargeted approaches unravel novel candidate genes and diagnostic SNPs for quantitative resistance of the potato (*Solanum tuberosum* L.) to *Phytophthora infestans* causing the late blight disease. PLoS One 11(6):e0156254. https://doi.org/10.1371/journal.pone.0156254
Obidiegwu JE, Sanetomo R, Flath K et al (2015) Genomic architecture of potato resistance to *Synchytrium endobioticum* disentangled using SSR markers and the 8.3k SolCAP SNP genotyping array. BMC Genet 16(1):38. https://doi.org/10.1186/s12863-015-0195-y
Ospina Nieto CA, Lammerts van Bueren ET, Allefs S et al (2021) Association mapping of physiological and morphological traits related to crop development under contrasting nitrogen inputs in a diverse set of potato cultivars. Plants 10(8):1727. https://doi.org/10.3390/plants10081727
Pajerowska-Mukhtar K, Stich B, Achenbach U et al (2009) Single nucleotide polymorphisms in the *allene oxide synthase 2* gene are associated with field resistance to late blight in populations of tetraploid potato cultivars. Genetics 181(3):1115–1127. https://doi.org/10.1534/genetics.108.094268
Pandey J, Scheuring DC, Koym JW et al (2022) Genomic regions associated with tuber traits in tetraploid potatoes and identification of superior clones for breeding purposes. Front Plant Sci 13:952263. https://doi.org/10.3389/fpls.2022.952263
Pandey J, Scheuring DC, Koym JW et al (2023) Genomic selection and genome-wide association studies in tetraploid chipping potatoes. Plant Genome 16(1):e20297. https://doi.org/10.1002/tpg2.20297
Park J, Hackett CA, Dandurand L-M et al (2019) QTL for resistance to *Globodera rostochiensis* pathotype Ro2 and *G. pallida* Pathotype Pa2/3 in autotetraploid potato. Am J Potato Res 96(6):552–563. https://doi.org/10.1007/s12230-019-09745-4
Parra-Galindo M-A, Piñeros-Niño C, Soto-Sedano JC et al (2019) Chromosomes I and X harbor consistent genetic factors associated with the anthocyanin variation in potato. Agronomy 9(7):366. https://doi.org/10.3390/agronomy9070366
Price AL, Patterson NJ, Plenge RM et al (2006) Principal components analysis corrects for stratification in genome-wide association studies. Nat Genet 38(8):904–909. https://doi.org/10.1038/ng1847
Pritchard JK, Stephens M, Donnelly P (2000) Inference of population structure using multilocus genotype data. Genetics 155(2):945–959. https://doi.org/10.1093/genetics/155.2.945
Prodhomme C, Vos PG, Paulo MJ et al (2020) Distribution of P1(D1) wart disease resistance in potato germplasm and GWAS identification of haplotype-specific SNP markers. Theor Appl Genet 133(6):1859–1871. https://doi.org/10.1007/s00122-020-03559-3
Raigond P, Singh B, Dutt S et al (2017) Potential of Indian potatoes for the management of hyperglycemia. Indian J Hortic 74(01):103–108. https://doi.org/10.5958/0974-0112.2017.00023.8
Rak K, Bethke PC, Palta JP (2017) QTL mapping of potato chip color and tuber traits within an autotetraploid family. Mol Breed 37:1–15. https://doi.org/10.1007/s11032-017-0619-7
Rosyara UR, De Jong WS, Douches DS et al (2016) Software for genome-wide association studies in autopolyploids and its application to potato. Plant Genome 9(2):plantgenome2015.2008.0073. https://doi.org/10.3835/plantgenome2015.08.0073
Rouppe van der Voort J, van der Vossen E, Bakker E et al (2000) Two additive QTLs conferring broad-spectrum resistance in potato to Globodera pallida are localized on resistance gene clusters. Theor Appl Genet 101(7):1122–1130. https://doi.org/10.1007/s001220051588

Schäfer-Pregl R, Ritter E, Concilio L et al (1998) Analysis of quantitative trait loci (QTLs) and quantitative trait alleles (QTAs) for potato tuber yield and starch content. Theor Appl Genet 97(5):834–846. https://doi.org/10.1007/s001220050963

Schmitz Carley CA, Coombs JJ, Douches DS et al (2017) Automated tetraploid genotype calling by hierarchical clustering. Theor Appl Genet 130(4):717–726. https://doi.org/10.1007/s00122-016-2845-5

Schönhals EM, Ortega F, Barandalla L et al (2016) Identification and reproducibility of diagnostic DNA markers for tuber starch and yield optimization in a novel association mapping population of potato (*Solanum tuberosum* L.). Theor Appl Genet 129(4):767–785. https://doi.org/10.1007/s00122-016-2665-7

Schönhals EM, Ding J, Ritter E et al (2017) Physical mapping of QTL for tuber yield, starch content and starch yield in tetraploid potato (*Solanum tuberosum* L.) by means of genome wide genotyping by sequencing and the 8.3 K SolCAP SNP array. BMC Genomics 18(1):642. https://doi.org/10.1186/s12864-017-3979-9

Schreiber L, Nader-Nieto AC, Schönhals EM et al (2014) SNPs in genes functional in starch-sugar interconversion associate with natural variation of tuber starch and sugar content of potato (*Solanum tuberosum* L.). G3 (Bethesda) 4(10):1797–1811. https://doi.org/10.1534/g3.114.012377

Schumacher C, Thümecke S, Schilling F et al (2021) Genome-wide approach to identify quantitative trait loci for drought tolerance in tetraploid potato (*Solanum tuberosum* L.). Int J Mol Sci 22(11):6123. https://doi.org/10.3390/ijms22116123

Selga C, Koc A, Chawade A et al (2021) A bioinformatics pipeline to identify a subset of SNPs for genomics-assisted potato breeding. Plants 10(1):30. https://doi.org/10.3390/plants10010030

Sharma SK, MacKenzie K, McLean K et al (2018) Linkage disequilibrium and evaluation of genome-wide association mapping models in tetraploid potato. G3 (Bethesda) 8(10):3185–3202. https://doi.org/10.1534/g3.118.200377

Shen J, Li Z, Chen J et al (2016) SHEsisPlus, a toolset for genetic studies on polyploid species. Sci Rep 6(1):24095. https://doi.org/10.1038/srep24095

Simko I, Haynes KG, Ewing EE et al (2004) Mapping genes for resistance to *Verticillium alboatrum* in tetraploid and diploid potato populations using haplotype association tests and genetic linkage analysis. Mol Gen Genomics 271(5):522–531. https://doi.org/10.1007/s00438-004-1010-z

Simko I, Haynes KG, Jones RW (2006) Assessment of linkage disequilibrium in potato genome with single nucleotide polymorphism markers. Genetics 173(4):2237–2245. https://doi.org/10.1534/genetics.106.060905

Śliwka J, Wasilewicz-Flis I, Jakuczun H et al (2008) Tagging quantitative trait loci for dormancy, tuber shape, regularity of tuber shape, eye depth and flesh colour in diploid potato originated from six Solanum species. Plant Breed 127(1):49–55. https://doi.org/10.1111/j.1439-0523.2008.01420.x

Śliwka J, Sołtys-Kalina D, Szajko K et al (2016) Mapping of quantitative trait loci for tuber starch and leaf sucrose contents in diploid potato. Theor Appl Genet 129:131–140. https://doi.org/10.1007/s00122-015-2615-9

Sood S, Bhardwaj V, Bairwa A et al (2023) Genome-wide association mapping and genomic prediction for late blight and potato cyst nematode resistance in potato (*Solanum tuberosum* L.). Front Plant Sci 14. https://doi.org/10.3389/fpls.2023.1211472

Stark JC, Thornton M, Nolte P (2020) Potato production systems. Springer

Stich B, Urbany C, Hoffmann P et al (2013) Population structure and linkage disequilibrium in diploid and tetraploid potato revealed by genome-wide high-density genotyping using the SolCAP SNP array. Plant Breed 132(6):718–724. https://doi.org/10.1111/pbr.12102

Sverrisdóttir E, Byrne S, Sundmark EHR et al (2017) Genomic prediction of starch content and chipping quality in tetraploid potato using genotyping-by-sequencing. Theor Appl Genet 130(10):2091–2108. https://doi.org/10.1007/s00122-017-2944-y

Tang Y, Liu X, Wang J et al (2016) GAPIT Version 2: An enhanced integrated tool for genomic association and prediction. Plant Genome 9(2):plantgenome2015.2011.0120. https://doi.org/10.3835/plantgenome2015.11.0120

Uitdewilligen JGAML, Wolters A-MA, D'hoop BB et al (2013) A next-generation sequencing method for genotyping-by-sequencing of highly heterozygous autotetraploid potato. PLoS One 8(5):e62355. https://doi.org/10.1371/journal.pone.0062355
Urbany C, Stich B, Schmidt L et al (2011) Association genetics in *Solanum tuberosum* provides new insights into potato tuber bruising and enzymatic tissue discoloration. BMC Genomics 12(1):7. https://doi.org/10.1186/1471-2164-12-7
van den Berg RG, Jacobs MM (2007) Molecular taxonomy. In: Potato biology and biotechnology. Elsevier, pp 55–76
van der Voort JR, Lindeman W, Folkertsma R et al (1998) A QTL for broad-spectrum resistance to cyst nematode species (Globodera spp.) maps to a resistance gene cluster in potato. Theor Appl Genet 96(5):654–661. https://doi.org/10.1007/s001220050785
van Eck HJ, Vos PG, Valkonen JPT et al (2017) Graphical genotyping as a method to map Ny(o,n) stoand Gpa5 using a reference panel of tetraploid potato cultivars. Theor Appl Genet 130 (3):515-528. doi:https://doi.org/10.1007/s00122-016-2831-y
Visker M, Keizer L, Van Eck H et al (2003) Can the QTL for late blight resistance on potato chromosome 5 be attributed to foliage maturity type? Theor Appl Genet 106:317–325. https://doi.org/10.1007/s00122-002-1021-2
Voorrips RE, Gort G, Vosman B (2011) Genotype calling in tetraploid species from bi-allelic marker data using mixture models. BMC Bioinf 12(1):172. https://doi.org/10.1186/1471-2105-12-172
Vos PG, Paulo MJ, Voorrips RE et al (2017) Evaluation of LD decay and various LD-decay estimators in simulated and SNP-array data of tetraploid potato. Theor Appl Genet 130(1):123–135. https://doi.org/10.1007/s00122-016-2798-8
Vos PG, Paulo MJ, Bourke PM et al (2022) GWAS in tetraploid potato: identification and validation of SNP markers associated with glycoalkaloid content. Mol Breed 42(12):76. https://doi.org/10.1007/s11032-022-01344-2
Wang F, Zou M, Zhao L et al (2020) Genome-wide association analysis of late blight resistance traits in potato germplasm resources. https://doi.org/10.21203/rs.3.rs-113697/v1
Wang F, Zou M, Zhao L et al (2021) Genome-wide association mapping of late blight tolerance trait in potato (*Solanum tuberosum* L.). Front Genet 12. https://doi.org/10.3389/fgene.2021.714575
Wanner LA (2009) A patchwork of Streptomyces species isolated from potato common scab lesions in North America. Am J Potato Res 86:247–264. https://doi.org/10.1007/s12230-009-9078-y
Werij JS, Furrer H, van Eck HJ et al (2012) A limited set of starch related genes explain several interrelated traits in potato. Euphytica 186:501–516. https://doi.org/10.1007/s10681-012-0651-y
Wolters A-MA, Uitdewilligen JG, Kloosterman BA et al (2010) Identification of alleles of carotenoid pathway genes important for zeaxanthin accumulation in potato tubers. Plant Mol Biol 73:659–671. https://doi.org/10.1007/s11103-010-9647-y
Yu J, Pressoir G, Briggs WH et al (2006) A unified mixed-model method for association mapping that accounts for multiple levels of relatedness. Nat Genet 38(2):203–208. https://doi.org/10.1038/ng1702
Yuan J, Bizimungu B, De Koeyer D et al (2020) Genome-wide association study of resistance to potato common scab. Potato Res 63:253–266. https://doi.org/10.1007/s11540-019-09437-w
Zhang Y, Jung CS, De Jong WS (2009) Genetic analysis of pigmented tuber flesh in potato. Theor Appl Genet 119:143–150. https://doi.org/10.1007/s00122-009-1024-3
Zhang F, Qu L, Gu Y et al (2022) Resequencing and genome-wide association studies of autotetraploid potato. Mol Hortic 2(1):6. https://doi.org/10.1186/s43897-022-00027-y
Zhou X, Stephens M (2012) Genome-wide efficient mixed-model analysis for association studies. Nat Genet 44(7):821–824. https://doi.org/10.1038/ng.2310
Zia MAB, Demirel U, Nadeem MA et al (2020) Genome-wide association study identifies various loci underlying agronomic and morphological traits in diversified potato panel. Physiol Mol Biol Plants 26:1003–1020. https://doi.org/10.1007/s12298-020-00785-3
Zorrilla C, Navarro F, Vega-Semorile S et al (2021) QTL for pitted scab, hollow heart, and tuber calcium identified in a tetraploid population of potato derived from an Atlantic × Superior cross. Crop Sci 61(3):1630–1651. https://doi.org/10.1002/csc2.20388

Advances in GWAS and QTL Mapping in Tomato (*Solanum lycopersicum* L.)

4

Jagesh Kumar Tiwari, Gaurav Mishra, Nagendra Rai,
Y. S. Reddy, Priyanka Sharma, Rajat Singh, Manish Singh,
P. K. Singh, and T. K. Behera

Abstract

Tomato is a major vegetable crop valued by both consumers and the processing industry, with strong demand for traits such as high yield, improved quality, and resistance to pathogens, physiological disorders, and environmental stresses. Modern tomato cultivars have been developed through targeted introgression to incorporate desirable traits, resulting in high-yielding, disease-resistant varieties that also meet consumer preferences for superior quality and nutritional content. Research efforts have increasingly focused on identifying genes and quantitative trait loci (QTLs) associated with key quality attributes such as soluble sugars, total soluble solids, titratable acidity, pH, β-carotene, and lycopene. These components are critical indicators of table and processing quality, as well as potential health benefits due to their antioxidant properties. Despite this progress, the identification and functional characterization of useful QTLs for these traits from intraspecific crosses remain limited. This chapter aims to investigate the genetic mechanisms underlying fruit quality in a tomato population derived from an intraspecific cross, with the goal of supporting breeding efforts to enhance fruit quality and identify specific QTLs associated with key quality attributes. Understanding the genetics of quantitative traits is essential in plant biotechnology, as many agronomically and commercially important traits exhibit quantitative inheritance. QTLs are genomic regions associated with the variation of quantitative traits and may consist of individual genes or clusters of functionally related genes that collectively influence trait expression. QTL mapping is a

J. K. Tiwari (✉) · G. Mishra · N. Rai · Y. S. Reddy · P. Sharma · R. Singh · M. Singh · T. K. Behera
ICAR-Indian Institute of Vegetable Research, Varanasi, Uttar Pradesh, India

P. K. Singh
Department of Vegetable Science, Chandra Shekhar Azad University of Agriculture & Technology, Kanpur, Uttar Pradesh, India

S. S. Dey et al. (eds.), *GWAS and QTL Mapping in Horticultural Crops*,
https://doi.org/10.1007/978-981-95-2306-1_4

widely used approach to locate these regions and can be applied to diverse types of mapping populations, including F_2 populations, double haploids, backcross populations, and recombinant inbred lines. This chapter focuses on the application of QTL mapping and genome-wide selection in tomato, providing insights that will aid researchers in identifying current research gaps and deepening their understanding of fruit quality trait improvement in tomato breeding programs.

Keywords

Tomato · Genetic resources · Genotyping · Genome wide association studies · Quantitative trait loci · Recombinant inbred lines

4.1 Introduction

Tomato is a major vegetable crop valued by both consumers and the food processing industry, with strong demand for a wide range of traits, including yield-related characteristics and quality attributes. In addition, there is a significant focus on developing cultivars with enhanced resistance to pathogens, physiological disorders, and various environmental stresses. To address these needs, modern tomato cultivars have been developed through targeted introgression to incorporate desirable traits. These efforts aim to deliver high-yielding, disease-resistant varieties that also meet consumer expectations for superior quality and nutritional value. Consumers place particular importance on fruit quality traits, including nutrient content and flavor components, which strongly influence breeding objectives. Extensive research has been conducted to identify genes and quantitative trait loci (QTLs) associated with key quality-related parameters such as soluble sugars, total soluble solids (TSSs), titratable acidity, pH, β-carotene, and lycopene. These biochemical components are critical indicators of suitability for fresh consumption and processing, and they also contribute to the health-promoting properties of tomatoes due to their high antioxidant content. However, the current understanding of useful QTLs associated with these traits, particularly those derived from intraspecific crosses, remains limited. To address this knowledge gap, our research aims to investigate the genetic mechanisms governing fruit quality in a tomato population developed from an intraspecific cross. Through this approach, we seek to support breeding programs by identifying specific QTLs that contribute to the enhancement of overall fruit quality in tomatoes (Kimbara et al. 2018).

The study of quantitative trait genetics in various plant systems is a fundamental aspect of plant biotechnology research. Such studies, conducted across numerous crop species, have significantly advanced our understanding of the inheritance of complex traits. It is well established that most commercially important traits in crop plants, domesticated animals, and even humans exhibit a quantitative nature, meaning they are influenced by multiple genes, commonly referred to as polygenes. By employing molecular markers in combination with appropriate statistical methodologies, researchers can identify specific chromosomal regions associated with these traits. These regions are known as QTLs, which are defined as genomic regions linked to the expression of quantitative traits. A QTL may consist of a single gene or a cluster of functionally related genes that collectively influence a trait (Veeresha et al. 2015). Linkage or QTL mapping is a widely used method for identifying genomic regions that control specific traits. This family-based mapping approach relies on genetic recombination and segregation in mapping populations derived from bi-parental crosses. These factors directly influence the mapping resolution and the range of allelic diversity captured. QTL mapping has proven to be a powerful tool for detecting genomic regions that co-segregate with traits of interest within breeding populations. It can be applied to a variety of mapping populations, including F_2 populations, double haploid (DH) lines, backcross populations, and recombinant inbred line (RIL) families. A range of molecular markers—such as restriction fragment length polymorphisms (RFLPs), amplified fragment length polymorphisms, microsatellites or simple sequence repeats (SSRs), and single nucleotide polymorphisms (SNPs)—can be utilized in these analyses (Alqudah et al. 2019). QTL mapping studies across various crop species have documented the involvement of QTLs in controlling traits related to yield, quality, disease and insect resistance, abiotic stress tolerance, and environmental adaptation. In this book chapter, we aim to explore QTL mapping and genome-wide association studies (GWAS), with a particular focus on their applications and future prospects in tomato crop improvement. We are confident that the insights presented will provide a comprehensive overview of QTL and GWAS methodologies, aiding researchers in identifying existing research gaps and advancing their understanding of genetic mechanisms underlying key traits in tomato.

4.2 Tomato Genetic Resources

Tomato (*Solanum lycopersicum*) is one of the major vegetable crops belonging to the Solanaceae family, which comprises more than 3000 species (Causse et al. 2016). Within the Lycopersicon clade of the Solanum section, there are 13 closely related tomato species, including one cultivated species, *S. lycopersicum*, and 12 wild relatives: *Solanum arcanum*, *Solanum cheesmaniae*, *Solanum chilense*, *Solanum chmielewskii*, *Solanum corneliomulleri*, *Solanum galapagense*, *Solanum habrochaites*, *Solanum huaylasense*, *Solanum neorickii*, *Solanum pennellii*, *Solanum peruvianum*, and *Solanum pimpinellifolium* (Liedl et al. 2013). The tomato is believed to have originated in the Andean regions of South America, including Peru, Bolivia, Ecuador, Colombia, and Chile. However, the exact location of tomato domestication is debated, with evidence pointing to both the Andean regions and Mexico. The cultivated tomato represents only about 5% of the total genetic diversity found across all tomato species. Tomato was first introduced to Europe by the Spanish in 1523, spreading subsequently to Italy in 1544, England in 1597, and to the Philippines and Malaysia by 1650. By the eighteenth century, tomato cultivation had expanded from Europe to North America (Liedl et al. 2013).

Tomato is well represented in global genebank collections, with approximately 36,431 accessions documented on the Genesys PGR platform (https://www.genesys-pgr.org) (Table 4.1). Among these, the largest number of accessions is held by institutions in the United States of America (15,992), followed by the World Vegetable Center (WVC), Taiwan, which maintains the largest single institutional collection (8497). Other notable contributors include Ukraine (2178), the Czech Republic (1391), Bulgaria (1365), the Netherlands (1337), and Australia (1276). In terms of species representation, *S. lycopersicum* dominates the collection (28,435 accessions), followed by *Lycopersicon esculentum* (5319), *S. pimpinellifolium* (971), *S. peruvianum* (470), *S. habrochaites* (362), *S. chilense* (160), *S. pennellii* (132), and *S. cheesmaniae* (71), among others. The collection comprises various types of genetic materials, including advanced/improved cultivars (10,808), traditional cultivars/landraces (2680), wild species (1942), breeding/research materials (702), breeder lines (584), genetic stocks (569), hybrids (234), mutants (186), and inbred lines (79), along with many unspecified entries.

Germplasm storage types include seed collections (14,516), long-term seed collections (6670), medium-term collections (3678), short-term collections (99), and unspecified storage types (14,465). After pepper, tomato is the most widely

Table 4.1 Cultivated and wild species of tomato

Species	Distribution
Section: Lycopersicon 'Lycopersicon group'	
Solanum lycopersicum	Globally cultivated tomato
Solanum cheesmaniae	Galápagos Islands
Solanum galapagense	Galápagos Islands
Solanum pimpinellifolium	South-western Ecuador to Northern Chile
Section: Lycopersicon 'Eriopersicon group'	
Solanum chilense	Coastal Chile and Southern Peru
Solanum peruvianum	Central Peru to Northern Chile
Solanum corneliomulleri	Southern Peru (Lima southwards)
Solanum habrochaites	Andean Ecuador and Peru
Solanum huaylasense	Río Santa river drainage, north-central Peru
Section: Lycopersicon ' Arcanum group'	
Solanum arcanum	Northern Peru
Solanum chmielewskii	Southern Peru and northern Bolivia
Solanum neorickii	Southern Ecuador to southern Peru
Section: Lycopersicon 'Neolycopersicon group'	
Solanum pennellii	Northern Peru to northern Chile
Section: Juglandifolia	
Solanum ochranthum	Andean Colombia, Ecuador, and Peru
Solanum juglandifolium	Andean Colombia, Ecuador, and Peru
Section: Lycopersicoides	
Solanum lycopersicoides	Southern Peru and northern Chile
Solanum sitiens	Northern Chile

Source: Modified from Liedl et al. (2013) and Causse et al. (2016)

distributed crop maintained by the WVC. Wild relatives of tomato serve as an important genetic reservoir for biotic and abiotic stress resistance. In fact, many of the most effective resistance genes used in tomato breeding programs have been sourced from wild species. At the WVC, major breeding efforts focus on the introgression of resistance genes against late blight, bacterial wilt, tomato leaf curl virus (Begomovirus), and tolerance to heat stress. Tomato lines developed by the WVC have been utilized globally in the development of improved cultivars (https://www.genesys-pgr.org).

4.3 Biotic Stresses in Tomato

4.3.1 Disease Resistance

Diseases remain a primary concern for both the processing and fresh market tomato industries worldwide. The economic impact of crop losses and the cost of disease management are substantial. Tomatoes are susceptible to over 200 diseases caused by a range of pathogens, including fungi, bacteria, viruses, and nematodes, all of which pose serious threats to plant health and productivity (Lukyanenko 1991). One of the most significant achievements in modern tomato breeding has been the development of cultivars with enhanced resistance to these diseases. Resistance has been identified—and in many cases genetically characterized—for over 30 major tomato diseases. These include *fusarium* wilt, *verticillium* wilt, root-knot nematode, Alternaria stem canker, gray leaf spot, and several bacterial and viral pathogens. In addition to vertical (monogenic) resistance governed by major genes, there are also reports of horizontal (polygenic or field) resistance in tomatoes. In these cases, resistance is not attributed to a single major gene but rather to the cumulative effect of multiple genes, and it offers broader, often more durable, protection. Horizontal resistance has been observed in response to diseases such as early blight, powdery mildew, bacterial canker, and bacterial wilt. Wild tomato species have served as an invaluable reservoir of resistance genes against a wide range of tomato diseases. Species such as *S. pimpinellifolium*, *S. peruvianum*, and *S. hirsutum* have been particularly instrumental in providing resistance traits. Both vertical (monogenic) and horizontal (polygenic) resistance mechanisms have been identified in these wild species for diseases such as late blight (*Phytophthora infestans*) and powdery mildew (*Oidium lycopersicum*). Initially, resistance gene identification, characterization, and incorporation were primarily conducted through phenotypic selection and conventional breeding methods. Even today, a significant portion of disease-resistance breeding in tomato continues to rely on these traditional approaches.

However, the introgression of resistance genes from wild species into cultivated tomato varieties has been met with several challenges using conventional breeding techniques. These include issues such as linkage drag, reduced fertility, and the genetic complexity of some traits, which have limited the efficient transfer of resistance into elite cultivars (Foolad 2007). In addition to limitations in screening facilities and technical expertise, many tomato breeding programs face challenges in developing cultivars with multiple disease resistances. To address these issues, breeders have increasingly sought more efficient strategies for resistance breeding. Over the past two decades, the use of molecular markers and marker-assisted selection (MAS) has significantly advanced the identification, mapping, and introgression of numerous disease-resistance genes and QTLs in tomatoes. The development of disease-resistant varieties remains a central goal of both public and private breeding programs, particularly in light of emerging diseases and the evolution of new pathogen strains. Ultimately, the strategic incorporation of host resistance aims to reduce or eliminate dependence on chemical pesticides, thereby promoting more sustainable tomato cultivation.

4.3.2 Insect Resistance

Cultivated tomato is highly susceptible to a range of insect pests, including mites, whiteflies, aphids, tomato fruit worms, beet armyworms, cotton bollworms, leaf miners, fruit flies, thrips, stinkbugs, and cutworms, all of which can cause significant yield and quality losses. Compared to disease resistance, insect resistance in tomato has received relatively less research attention, leading to the development of only a limited number of commercial cultivars with targeted insect resistance. However, several wild tomato species—particularly *S. hirsutum* and *S. pennellii*—have demonstrated strong resistance to major insect pests (Kennedy et al. 1994). In addition to these species, insect resistance has also been documented in *S. lycopersicum var. cerasiforme, S. pimpinellifolium, S. cheesmanii, S. chmielewskii, S. peruvianum,* and *S. chilense* (Mutschler et al. 1996). Breeding insect-resistant tomato varieties has proven to be more challenging than breeding for disease resistance. A major obstacle is the phenomenon of linkage drag, where undesirable traits from wild species are co-inherited with resistance genes, complicating the breeding process and reducing agronomic performance. The identification of molecular markers and QTLs associated with insect resistance, combined with the use of MAS and QTL mapping, holds promise for overcoming these challenges and facilitating the efficient development of insect-resistant tomato cultivars.

4.4 Abiotic Stresses in Tomato

Cultivated tomato is widely recognized for its adaptability to diverse climatic conditions; however, it remains highly sensitive to various abiotic stresses, including salinity, drought, waterlogging, extreme temperatures, mineral toxicities and deficiencies, and environmental pollution. Providing optimal environmental conditions is therefore essential for its successful growth and productivity. Within the cultivated tomato gene pool, genetic variability for abiotic stress tolerance is limited, and most commercial cultivars are considered moderately to highly sensitive to such stresses. In contrast, several wild tomato species—such as *S. chilense*, *S. peruvianum*, *S. pennellii*, *S. pimpinellifolium*, *S. hirsutum*, *S. cheesmanii*, *S. chmielewskii*, and *S. parviflorum*—have been identified as valuable sources of genetic tolerance or resistance to various abiotic stresses (Foolad 2005). These wild relatives have been extensively studied for their genetic, physiological, and biochemical traits related to stress tolerance and have been incorporated into breeding programs aimed at improving abiotic stress resilience in cultivated tomato.

However, there have been relatively few successful reports of stress-tolerant tomato cultivars developed through traditional breeding approaches. This is largely due to the complex nature of abiotic stress tolerance, which involves multiple physiological, biochemical, and agronomic traits. The expression of these traits is governed by the interaction of numerous genes, and their manifestation is often strongly influenced by environmental conditions (Scott 1993). Achieving sustained tomato production under stress conditions requires tolerance across all developmental stages—ranging from seed germination and vegetative growth to flowering and fruit set. Each developmental stage may necessitate distinct screening protocols, and conducting simultaneous or sequential assessments can be both logistically challenging and technically demanding. Moreover, accurately quantifying stress tolerance is inherently difficult. Phenotypic selection under field conditions is often hampered by environmental variability, which reduces the precision and reproducibility of screening results. These challenges highlight the need for more precise, efficient tools—such as molecular markers and controlled-environment phenotyping platforms—to enhance the selection process in stress-resilient tomato breeding. Developing a reliable and consistent screening technique for the long-term selection of stress-tolerant tomato plants has been a significant challenge. Phenotypic evaluation for stress tolerance often requires trained personnel, sophisticated protocols, and substantial investment in field nurseries or controlled greenhouse facilities. Moreover, the variability of environmental conditions further complicates the accuracy and reproducibility of such assessments. To overcome these limitations, researchers have increasingly focused on genetic approaches, particularly the use of molecular markers associated with stress tolerance traits. One such approach, QTL mapping, has been extensively applied in tomato to identify specific genomic regions linked to tolerance against various environmental stresses. This strategy enables the dissection of complex traits into their genetic components and facilitates MAS for more efficient breeding of stress-resilient cultivars.

4.5 Fruit Quality

For many years, tomato breeding programs have prioritized the improvement of fruit quality traits that are essential for both fresh market consumption and processing industries. Key characteristics targeted include fruit size and shape, TSSs content, color, firmness, ripening behavior, nutritional value, and flavor profile.

4.5.1 Total Soluble Solids

Among these traits, total solid content is particularly critical for the processing industry and has received considerable attention in tomato breeding programs. In cultivated tomatoes, total solids typically range from 4% to 7.5% of the fruit's fresh weight, although certain wild species exhibit even higher levels (Stevens 1972). Total solids encompass all non-water and non-volatile components of the fruit. Approximately 75% of the total solids in cultivated tomatoes are soluble, while the remaining 25% are insoluble. The soluble solids are primarily composed of reducing sugars such as glucose and fructose, with sucrose present in lower concentrations; however, some wild tomato species contain higher levels of sucrose. Other constituents of the soluble solids include organic acids, lipids, minerals, and pigments. Insoluble solids consist of proteins, cellulose, hemicellulose, pectin, and various polysaccharides, all of which contribute to the viscosity of tomato-based products such as juice, ketchup, sauces, soups, and pastes. The proportions of both soluble and insoluble solids are closely associated with the yield and quality of processed tomato products. The sugar content of the fruit, in particular, significantly influences both processing yield and flavor quality. In commercial processing, where products are often valued based on their solids content, an increase of just 1% in total solids can result in up to a 20% increase in product yield. Additionally, elevated sugar levels enhance the flavor profile of fresh-market tomatoes, thereby improving consumer appeal (Foolad 2007). Given the critical role of fruit solids content in tomato processing quality, many breeding programs have prioritized increasing this trait. Commercial tomato cultivars generally exhibit soluble solids content between 4.6% and 6.3% of fresh weight, with processing varieties typically showing higher levels (Berry and Uddin 1991). Notably, several wild tomato species—such as *S. pimpinellifolium*, *S. chmielewskii*, and *S. cheesmanii*—demonstrate considerably higher soluble solids concentrations, ranging from 9% to 15%. Despite this valuable genetic variation, breeders have encountered significant challenges in enhancing fruit soluble solids or simultaneously achieving high soluble solids and high yield. These difficulties stem from the complex quantitative inheritance of the trait and the often observed negative correlation between yield and soluble solids content.

4.5.2 Fruit Color

Tomato fruit color is a key quality attribute that has attracted considerable attention. The primary pigments responsible are carotenoids and chlorophylls, with the final fruit color determined by the total amount and relative proportions of various carotenoids. Lycopene, the dominant red carotenoid in tomatoes, is the major contributor to the vibrant red color prized in both fresh market and processing tomatoes. In processing tomatoes, fruit color influences product grading and quality standards, while in the fresh market, it strongly affects consumer appeal and marketability. Recently, interest in fruit color has intensified due to increasing awareness of the health benefits associated with different carotenoids. Fresh tomatoes and tomato products serve as important dietary sources of lycopene, a potent natural antioxidant. Numerous studies have linked lycopene consumption to reduced risks of certain cancers—including prostate, lung, oral, and colon cancers—as well as coronary heart disease. Lycopene's antioxidant activity is estimated to be about twice that of β-carotene, justifying the focus on enhancing its content. As scientific understanding of carotenoids' health impacts has grown, many breeding programs have prioritized developing tomato cultivars with elevated lycopene levels. In addition to lycopene, ripe tomato fruit contains β-carotene and smaller quantities of other carotenoids such as phytoene, phytofluene, γ-carotene, neoxanthin, and lutein (Khachik et al. 2002).

4.5.3 Flavor

Flavor in tomatoes is a highly complex trait influenced by numerous genetic and non-genetic factors, many of which remain unidentified or poorly understood. The perception of flavor involves an interplay of taste, aroma, texture, appearance, fruit temperature, and mouthfeel. A critical determinant of tomato flavor is the balance between sugars and acids, with higher soluble solids levels positively correlated with a desirable, characteristic tomato flavor. Research has shown that breeding for increased soluble solids and acidity can enhance flavor quality. Among sugars and acids, fructose and citric acid have a stronger influence on sweetness and sourness, respectively, than glucose and malic acid. For measuring tartness, pH serves as a more precise objective indicator than titratable acidity (Stevens et al. 1977). A notable discovery is the *Fgr* gene in *Lycopersicum hirsutum*, which increases fruit sweetness by raising the fructose-to-glucose ratio (Levin et al. 2000). Although tomatoes produce over 400 aromatic volatile compounds, only about 16 significantly impact flavor (Buttery 1993). Tomato fruit quality traits, including flavour, vary with environmental conditions, and the ripeness stage at harvest greatly affects taste—fruits harvested at later ripening stages tend to be sweeter and exhibit a more pronounced tomato flavor compared to those picked at mature green or breaker stages. Additionally, environmental stresses during growth and ripening can influence flavor both positively and negatively; for example, elevated salinity at certain growth stages may enhance flavor but reduce fruit size. Furthermore, post-harvest handling and storage conditions prior to marketing play a crucial role in preserving or altering fresh tomato flavour.

4.5.4 Miscellaneous Fruit Traits

Tomato fruit quality is influenced by several factors, including pH, titratable acidity, fruit firmness, and vitamin content. Acidity plays a vital role in the storability of processed tomato products, as a lower pH inhibits pathogen growth by facilitating the heat inactivation of thermophilic organisms (Rice and Pederson 1994).

Specifically, the growth of *Bacillus coagulans*, the bacterium responsible for flat sours in tomato products, is completely suppressed when the pH falls below 4.1 (Lambeth et al. 1966). Titratable acidity has a limited effect unless the pH is low. Consequently, an ideal processing tomato should have a pH below 4.5 and a citric acid content exceeding 0.35 g per 100 g of fresh tomato weight. Breeding programs targeting processing tomatoes have made progress toward these goals. Although tomatoes contain intermediate levels of vitamins A and C compared to other vegetables, they rank highly as dietary sources of vitamin A in the United States due to their widespread consumption and significantly contribute to vitamin C intake. Plant carotenoids, especially β-carotene—the predominant carotenoid in orange-yellow tomatoes—are the main sources of vitamin A in tomatoes. The identification and use of genes associated with enhanced vitamin content in breeding programs can yield both economic and nutritional benefits.

4.6 Genome-Wide Association Studies

GWAS, also referred to as whole-genome association studies, examines a comprehensive set of genetic variations across the entire genome to identify variants associated with specific traits (Manolio 2010). GWAS has emerged as a powerful and efficient approach for uncovering genome-to-phenotype associations and pinpointing causative loci or genes. It is widely acknowledged for its effectiveness in elucidating the genetic basis of traits and identifying precise genomic regions linked to phenotypic variation. In GWAS, the core methodology involves testing the association between each genetic marker and a phenotype of interest. This analysis is typically conducted using a diverse panel of unrelated individuals, characterized by genetic heterogeneity and distant relatedness. By assessing the correlation between markers and phenotypes, GWAS provides valuable insights into the genetic architecture underlying complex traits (Huang and Han 2014). GWAS has demonstrated robustness and efficacy in dissecting the genetic basis of complex traits in crops. It is expected that the power of GWAS will continue to increase in identifying specific loci or genes responsible for quantitative traits, facilitated by large population sizes and advances in high-throughput sequencing technologies. The high-resolution mapping capability of GWAS is largely attributed to historical recombination events and the incorporation of a broad allelic diversity in the analysis (Rafalski 2010).

The application of advanced analytical methodologies has begun to enhance the utilization of diverse genetic resources for investigating inherent variations with potential for crop improvement. While extensively explored in human genetics, these approaches have also been adopted in plant research since the early twenty-first century (Flint-Garcia et al. 2003). Recently, GWAS has emerged as a prominent tool for mapping quantitative traits and examining natural genetic variation. The use of high-density genotyping platforms in GWAS enables comprehensive marker coverage, facilitating detailed dissection of the genetic architecture underlying traits of interest across various crops. By thoroughly screening diverse and large germplasm collections with sufficient marker density, GWAS can identify specific loci responsible for phenotypic variation. The primary objectives of GWAS are to pinpoint causative factors influencing specific traits and to elucidate the complex genetic architecture governing them. The number of loci contributing to phenotypic variation can vary widely; some traits are controlled by a relatively simple genetic architecture involving few loci with large effects, whereas others are regulated by numerous loci, reflecting a more complex genetic basis.

GWAS is a method used to correlate genotype and phenotype in individuals to dissect complex traits using a reference genome sequence. This approach enables the collection of diverse accessions or breeding/inter-cross lines, followed by sequencing and comparative analysis of genomic diversity. To address the challenges associated with complex traits, modern genomics tools have been integrated with GWAS to discover novel genes underlying traits such as fruit flavor and quality, metabolic attributes, and agronomic characteristics (Bauchet et al. 2017) (Tables 4.2 and 4.3). A substantial number of polymorphisms have been identified between cultivated and wild species genomes through GWAS (Zhu et al. 2018). The successful identification of genes or QTLs via GWAS depends on the extent of phenotypic variation within the population structure and the resolution of the SNP genotyping system. Rare variants and loci with small effects are often difficult to detect. In tomato, GWAS has been applied to traits such as fruit flavour, sugars, and organic acid volatiles, leading to the identification of markers linked to flavor-related genes. However, GWAS-identified markers may yield false positives and may lack reproducibility. Therefore, markers identified through GWAS must be validated further through genotyping and polymorphism analysis. To facilitate the dissection of complex traits, tomato breeders have developed large breeding populations involving both wild and cultivated species. For example, biparental populations such as RILs, introgression lines (ILs), and backcross inbred lines have been extensively used for QTL and gene identification. One notable example is the development of ILs derived from a cross between the green-fruited wild species *S. pennellii* and the cultivated variety M82 (Ofner et al. 2016). Each IL contains specific introgressions from *S. pennellii* that exhibit phenotypic variation, allowing identification of QTLs associated with plant phenotype and yield (Caruso et al. 2016).

Identifying genomic regions associated with economically important traits is essential for genomics-assisted breeding. GWAS represents a powerful approach that utilizes genome-wide markers through high-throughput SNP genotyping to detect significant associations between phenotypes and genomic variants, thereby identifying linked markers for MAS. Currently, GWAS is performed using SNP array genotyping and/or reduced-representation sequencing technologies such as genotyping-by-sequencing (GBS), restriction site-associated DNA sequencing, and double digest RAD sequencing. With the increasing availability of reference genomes and pan-genomes from many wild species, GWAS is being extensively employed to dissect the genetic basis of complex trait variation, including yield and yield-related traits, fruit quality, and tolerance or resistance to biotic and abiotic stresses in both fresh-market and processing tomatoes. Numerous candidate loci have been identified for traits such as flavor and nutrition-associated metabolites, agronomic performance, fruit quality, and root architecture (Kim et al. 2021), as

Table 4.2 Identification and mapping of QTLs associated with biotic, abiotic, and fruit quality traits in tomato

Traits	Sources	Genes/QTLs	Location of chromosome	Mapping population	References
(a) Disease and insect resistance					
ToYLCV	*L. chilence*	*Ty-1*	6	–	Zamir et al. (1994)
	L. pimpinellifolium	–	6	F_4	Chagué et al. (1997)
	L. hirsutum	*Ty-1*	11	–	Hanson et al. (2000)
	L. hirsutum	*TY-2*	11	F_6 RILs	Hanson et al. (2006)
	L. Chilence	*TY-3*	6	F_2 NILs	Ji and Scott (2006)
Fusarium wilt	*L. pennellii*	*I3*	7	–	Bournival et al. (1989)
	L. Pimpinellifolium	*I, I2*	8	–	Bournival et al. (1990)
	L. pennellii	*I1*	11	–	Sarfatti et al. (1991)
Verticillium wilt	*L. cheesmanii*	*Ve*	9	F_2	Diwan et al. (1999)
Early blight	*L. hirsutum*	11 (QTLs)	1,2,5,8,9,10,11,12	BC_1	Foolad et al. (2002)
		13 (QTLs)	1,2,3,5,8,9, 10,11,12	BC_1S_1	Foolad et al. (2002)
	L. hirsutum	7 (QTLs)	3,4,5,6,8,10,11	BC_1	Zhang et al. (2003)
	S. habrochaites & S. pimpinellifolium	6 (QTLs)	2,8,11	$F_{2:3}$	Adhikari et al. (2023)
Late blight	*S. pimpinellifolium*	*Ph-1,* *Ph-2,* *Ph-3*	– 10 9	F_2	Panthee et al. (2017)
	S. pimpinellifolium	*Ph-5*	–	F_2 and F_3	Foolad et al. (2008)
Bacterial wilt	–	*Bwr-4,* *Bwr-6,* *and Bwr-12*	4 6 12	F_2, F_3 RILs	Nguyen et al. (2021)
Tomato spotted wilt virus	*L. peruvianum*	*Sw-5*	9	F_2 NILs	Stevens et al. (1995)

Powdery mildew	*L. parviflorum*	*Ol-qtl1* *Ol-qtl2 and Ol-qtl3*	6 12	F_2	Bai et al. (2003)
Aphid (potato)	*L. peruvianum*	*Meu-1*	6	F_2 NILs	Foolad (2007)
Nematode (potato cyst)	*S. pimpinellifolium*	*Hero*	4	NILs F_2	Ganal et al. (1995)
Nematode (root knot)	*L. peruvianum*	*Mi*	6	F_2	Kaloshian et al. (1998)
(b) Abiotic stress tolerance					
Drought	*L. pimpinellifolium*	4	1,8,9,12	BC_1S_1	Foolad et al. (2003)
	L. pennellii	3	–	BC_1S_1, F_3	Martin et al. (1989)
	S. habrochaites	11	9	Sub NILs	Lounsbery et al. (2016)
Salinity	*L. pennellii*	5	1,3,7,8,12	F_2	Foolad and Jones (1993)
	L. pimpinellifolium	7	1,2,5,7,9,12	BC_1S_1	Foolad et al. (1998a, b)
	L. pennellii	6	1,2,4,5,6,12	F_2	Zamir and Tal (1987)
	L. pimpinellifolium	4	1,5,9	BC_1S_1	Foolad and Chen (1999)
Cold	*L. hirsutum*	3	6,7,12	BC_1	Vallejos and Tanksley (1983)
	L. pimpinellifolium	3	1,4	BC_1S_1	Foolad et al. (1998a, b)
Heat	*L. pimpinellifolium*	5	1,2	F_2	Wen et al. (2019)
	S. esculentum	21	1,2,3,4,7,8	F_2	Lin et al. (2010)

(continued)

Table 4.2 (continued)

Traits	Sources	Genes/QTLs	Location of chromosome	Mapping population	References
(c) *Fruit quality traits*					
Fruit color	*L. pimpinellifolium*	2	2,6	BC_1	Grandillo and Tanksley (1996)
	L. peruvianum	8	1,6,7,8,9,10,12	BC_3, BC_4	Fulton et al. (1997)
	L. chmielewskii	1	1	NILs	Frary et al. (2003)
Fruit color (β-carotene)	*L. parviflorum*	6	2,4,8,9,10,11	BC_3	Fulton et al. (2000)
Fruit color (carotene content)	*L. esculentum*	3	2,3,8	RILs	Saliba et al. (2001)
Fruit color (Yellow)	*L. parviflorum*	1	12	BC_3	Fulton et al. (2000)
Fruit cracking (Radial)	*L. pimpinellifolium*	2	2,9	BC_2F_6	Doganlar et al. (2002)
Fruit cracking (concentric)	*L. pimpinellifolium*	3	2,8,9	BC_2F_6	Doganlar et al. (2002)
Fruit firmness	*L. pennellii*	3	2,10	BC_2/ BC_2F_1	Frary et al. (2004)
	L. esculentum	3	4,9	RIL	Saliba et al. (2001)
Fruit pH	*L. chmielewskii*	5	3,6,7,8,10	BC_1, BC_2	Paterson et al. (1988)
	L. cheesmanii	9	1,3,4,6,7,8,10	F_2, F_3	Paterson et al. (1991)
	L. parviflorum	10	2,3,4,5,6,7,9,12	BC_3	Fulton et al. (2000)
TSS	*L. chmielewskii*	1	–	NIL F_2	Osborn et al. (1987)
	L. pennellii	3	1,5,7	IL_s	Eshed and Zamir (1994)
	L. pimpinellifolium	3	3,6,9	BC_1	Grandillo and Tanksley (1996)
Titrable acidity	*L. esculentum*	6	1,2,3,9,12	RIL	Saliba et al. (2001)
	L. pennellii	15	2,3,4,5,7,8,9,10,11,12	ILs	Causse et al. (2004)
Viscosity	*L. pimpinellifolium*	1	9	BC_2, BC_3	Tanksley et al. (1996)
	L. pennellii	4	2,3,9,12	BC_2/ BC_2F_1	Frary et al. (2004)

Table 4.3 Identification and mapping of QTLs associated with biotic, abiotic, and various fruit quality traits through GWAS in tomato

Traits	Mapping population	QTLs	References
(a) ***Disease and insect resistance***			
Bacteria wilt	191 core-collection	8 MTAs	Nguyen et al. (2021)
Anthracnose	RILs	20 QTLs	Lopez-Ortiz et al. (2023)
Fusarium wilt	340 accessions	6 unique SNPs	Kawicha et al. (2023)
Early blight	NILs	2 QTLs	Anderson et al. (2021)
(b) ***Fruit quality traits***			
Fruit quality traits	162 diverse genotypes	30 MTAs (marker traits association)	Kim et al. (2021)
	212 accessions	28 MLM	Liu et al. (2023)
	174 diverse genotypes	22 MTAs	Zhang et al. (2016)
	244 improved accessions	59 MLM	Tripodi et al. (2021)
	96 genotypes	19 putative genes	Ruggieri et al. (2014)
	174 accessions	125 significant association	Zhang et al. (2015)
Agronomic traits	605 diverse accessions	239 significant association	Ye et al. (2021)
(c) ***Abiotic stress tolerance***			
Water deficit and Salinity stress	250 lines	54 QTLs	Diouf et al. (2018)
Drought	119 RILs	56 QTLs	Albert et al. (2016a, b)
	141	31 QTLs	Albert et al. (2016a, b)
Heat Stress	265 (magic population) and 139 (core-collection)	69 pQTLs	Bineau et al. (2021)
	81 genotypes	35 MTAs	Ruggieri et al. (2019)
	Solanum lycopersicum var. *cerasiforme* (line E9) × *S. pimpinellifolium* (line L5)	31 QTLs	Gonzalo et al. (2022)

well as salinity tolerance (Wang et al. 2020), among others. For instance, Kim et al. (2021) identified QTLs for eight fruit traits in cultivated tomato using GWAS based on 34,550 high-confidence SNPs from the 51K Axiom® tomato array. Similarly, Yang et al. (2022) conducted a GWAS of eigenvectors, providing genetic insights into selective breeding for tomato metabolites across 331 accessions. They identified two independent gene sets associated with fruit metabolites and 57 candidate genes related to polyphenol and polyamine biosynthesis.

4.7 GWAS Analysis Tools

In tomato breeding, the focus has been on developing versatile cultivars capable of meeting diverse requirements for both fresh market consumption and processing industries. Breeding objectives are shaped by the cultivation environment—whether field or greenhouse—and the intended end use of the product, be it fresh consumption or processing (Stevens and Rick 1986; Tigchelaar 1986). A primary goal has been to enhance tomato fruit yield per unit area while addressing the distinct quality criteria specific to fresh market and processing cultivars. These cultivars are tailored to satisfy the unique demands of their respective uses. Key traits of importance across both sectors include disease resistance, environmental adaptability, early maturity, fruit set under adverse temperatures, resistance to fruit cracking and rots, vine coverage, fruit firmness, and other quality attributes. Processing cultivars, in particular, require characteristics such as a compact, determinate growth habit, synchronized flowering and fruit set conducive to mechanical harvesting, easy fruit separation (jointless pedicel), and fruit quality parameters including color, pH, acidity, soluble solids, total solids, and viscosity (Foolad 2007). From a scientific perspective, fresh market cultivars are desired to possess uniform fruit size, high firmness and shelf life, attractive color, pleasant taste and flavor, and resilience against biotic and abiotic stresses. In the following sections, we will provide a detailed review of key achievements in various breeding domains, highlighting significant breakthroughs and advancements attained through dedicated breeding efforts.

GWAS has become an invaluable tools for identifying correlations between genetic variations within a population and specific traits or diseases. These studies aim to detect statistically significant associations between single-nucleotide polymorphisms (SNPs)—variations at a single DNA base pair—and phenotypes such as height, disease susceptibility, or other traits of interest. The primary objective is to pinpoint SNPs that are significantly linked to the phenotype under investigation. The GWAS experimental workflow typically involves several key steps: collection of DNA samples and phenotypic data; genotyping individuals using available GWAS arrays or sequencing technologies; haplotype phasing and variant imputation; application of appropriate statistical tests; and conducting meta-analyses when applicable. Finally, results are interpreted through various post-GWAS analyses to validate findings and explore biological implications.

4.7.1 Phenotypic Data Collection

The first step in a GWAS is selecting a diverse set of germplasm that represents the target plant species or population. Ideally, the germplasm should capture a broad spectrum of genetic variation and phenotypic diversity. Accurate and comprehensive phenotypic data collection is essential to ensure reliable association analysis. Moreover, phenotypic measurements should be obtained across multiple environments to account for genotype-by-environment interactions.

4.7.2 Genotyping

Genotyping is the process of identifying and cataloguing genetic variations, primarily SNPs, across the genome of the plant population under study. High-throughput genotyping technologies—such as GBS, SNP arrays, and whole-genome re-sequencing—are commonly employed to efficiently genotype large populations. The resulting genotypic data form the foundation for subsequent association analyses between genetic markers and phenotypic traits.

4.7.3 Population Structure and Kinship Analysis

Plant populations often exhibit complex patterns of population structure and relatedness, which can lead to spurious associations in GWAS. To address these confounding factors, statistical approaches such as principal component analysis and kinship modeling are employed to account for population structure and genetic relatedness. These methods help reduce false-positive associations and enhance the power to detect true genotype-phenotype correlations.

4.7.4 Statistical Analysis

Statistical methods are essential in GWAS for identifying significant associations between genetic markers and traits. Commonly used approaches include single-locus and multi-locus models, mixed linear models (MLMs), and compressed mixed linear models. These methods effectively control for population structure and relatedness while quantifying the contribution of individual genetic variants to phenotypic variation.

4.7.5 Validation

To validate GWAS findings, identified marker-trait associations must undergo further investigation. This process may include independent replication studies in diverse populations, functional characterization of candidate genes, fine-mapping of associated genomic regions, and integration of validated markers into MAS breeding programs. In plant GWAS, a comprehensive workflow encompasses germplasm selection, precise phenotypic evaluation, advanced genotyping technologies, population structure analysis, robust statistical modeling, and rigorous validation studies.

Together, these methodologies facilitate the accurate identification of genetic variants underlying complex traits.

4.7.6 Tools and Software

GWAS is a vital tool in plant genetics research for identifying genetic variations associated with complex traits and phenotypes. Throughout the GWAS process, various computational tools are employed to facilitate data analysis and interpretation. Some of the commonly used tools in plant GWAS include:

- ***PLINK***: PLINK is a widely used open-source software tool for GWAS analysis. It offers a comprehensive suite of functions for managing and analyzing genetic data, including quality control, population structure assessment, and association testing.
- ***TASSEL:*** TASSEL (Trait Analysis by Association, Evolution, and Linkage) is a versatile software package that provides a wide range of statistical genetic methods for GWAS. It supports diverse population structures, accommodates various types of genetic markers, and enables comprehensive association analyses.
- ***GAPIT:*** **GAPIT** (genome association and prediction integrated tool) is an R package designed for performing GWAS using MLMs. It integrates various algorithms for association mapping and offers functions for downstream analyses, including genomic prediction and estimation of genetic relatedness.
- ***FarmCPU:*** It is an R package specifically developed for GWAS in crop plants. It employs an MLM approach that effectively controls for population structure and relatedness. FarmCPU is particularly well-suited for multi-environment studies, making it ideal for agricultural research.
- ***EMMA:*** Efficient mixed-model association (EMMA) is a widely used software tool for GWAS across various organisms, including plants. EMMA applies a mixed-model approach that effectively accounts for population structure and kinship in association analyses.
- ***GEMMA:*** Genome-wide efficient mixed-model association (GEMMA) is a widely used software for GWAS in plants. It employs a linear mixed model to control for population structure and genetic relatedness, enabling accurate and efficient association analyses.
- ***EMMAX:*** **EMMAX** (Efficient Mixed-Model Association eXpedited) is a widely used software tool for GWAS that efficiently handles large-scale datasets. It employs a linear mixed-model approach to account for population structure and relatedness. Together with other tools, EMMAX provides essential functionalities for data pre-processing, quality control, statistical analysis, and visualization, thereby facilitating the GWAS workflow in plants. Researchers often integrate multiple tools to gain comprehensive insights into the genetic architecture underlying complex traits.

4.8 Genome Sequencing/Re-sequencing of Tomato Species

Tomato, a key Solanaceous vegetable alongside potato, pepper, and eggplant, has a fully sequenced genome, making it a valuable model crop for studying fleshy fruit traits. The International Tomato Genome Sequencing Project was launched in 2004 by a multinational consortium comprising scientists from 14 countries, including the United States, Korea, China, the United Kingdom, India, the Netherlands, France, Japan, Spain, and Italy. Initially, the project employed a BAC-by-BAC sequencing approach. However, in 2009, the strategy shifted to a whole-genome shotgun method, which produced high-quality genome assemblies. The International Tomato Annotation Group (ITAG) subsequently annotated the genome sequence data. By 2012, the high-quality tomato genome assembly SL2.40, along with ITAG2.3 annotations for the common processing variety 'Heinz 1706', was published. The tomato genome was estimated to be approximately 900 Mb in size, derived through a combination of Sanger and next-generation sequencing technologies (The Tomato Genome Consortium 2012). Additionally, the genome of *Solanum pimpinellifolium* (739 Mb), the closest wild relative of cultivated tomato, was sequenced. This comparison revealed only 0.6% nucleotide diversity between the wild and cultivated species, highlighting the role of plant adaptation to various stress environments (The Tomato Genome Consortium 2012).

Within the tomato genome, numerous disease resistance (R) genes have been identified. These include genes encoding pathogenesis-related (PR) proteins containing domains such as nucleotide-binding site (*NBS*), coiled-coil (*CC*), leucine-rich repeats (*LRR*), and toll/interleukin-1 receptor (TIR) motifs. These genes are classified into different groups including *CC-NBS-LRR, CC-NBS, LRR* receptor-like kinases (*LRR-RLK*), *NBS-LRR, NBS, TIR-NBS-LRR*, and *TIR-NBS*. Furthermore, several transcription factors (TFs), lipoxygenase (LOX) family genes, and protein kinases involved in regulating responses to biotic and abiotic stresses have been characterized (The Tomato Genome Consortium 2012). The availability of the tomato genome sequence has significantly advanced research on biotic stress resistance (Foolad and Panthee 2012) and has been recently updated by Tiwari et al. (2022). In 2014, the genome sequence of the stress-tolerant wild tomato species *S. pennellii* was deciphered and compared with the reference tomato genome, revealing the presence of stress tolerance genes in the wild species (Bolger et al. 2014). The study also identified several cis-acting elements that play crucial roles in stress tolerance and the biosynthesis of flavor-associated compounds.

More recently, an improved version of the tomato reference genome assembly, SL4.0, was released. This assembly was generated de novo using long-read sequencing technology from PacBio, combined with scaffold construction aided by Hi-C contact maps. The updated reference genome reduced contig gaps by 11 Mb compared to the previous assembly, resulting in a total genome size of 782.6 Mb with 71.77% repeat content. The corresponding updated annotation, ITAG4.0, reported 34,075 protein-coding genes based on RNA-seq, resistance gene enrichment sequencing (RenSeq), and additional expression datasets (Hosmani et al. 2019). Functional annotations are available for 29,532 genes, including 4794 novel genes associated with disease resistance (Hosmani et al. 2019). The latest tomato genome sequence and associated resources are accessible at https://solgenomics.net.

Numerous vegetable crop genomes have been sequenced and utilized for the identification of genes and molecular markers essential for breeding new varieties. Extensive sequencing of crop wild relatives and species, followed by comparison with their respective reference genomes, has greatly enhanced our understanding of crop evolution, domestication, genetic relatedness, phylogeny, and the genetic regulation of multiple traits. The modern large-fruited tomato evolved from a small-fruited cherry tomato through artificial selection. To better understand the processes of evolution, domestication, and the genes involved in tomato trait improvement, pan-genomes have been constructed covering 725 cultivated and wild tomato species. This effort revealed 4873 novel genes related to disease resistance, fruit weight, and flavour compounds that were absent from the reference tomato genome, highlighting their crucial roles in domestication and breeding (Gao et al. 2019). More recently, long-read sequencing of 1000 tomato accessions uncovered 238,490 structural variations (SVs) critical for trait improvement and domestication in tomato (Alonge et al. 2020). The availability of the tomato reference genome and pan-genomes from diverse wild species paves the way for the identification of novel genes, QTLs, and trait-linked markers for marker-assisted breeding. This has significantly accelerated tomato breeding programs through genome sequencing approaches. Furthermore, a variety of tomato genomics web tools have been developed worldwide and are freely accessible to the research community. These databases serve as valuable resources for data mining and gene/marker discovery, supporting next-generation breeding applications.

4.9 SNP Genotyping and Mapping

Most genetic maps of tomato have been developed through interspecific crosses between cultivated tomato and wild relatives, utilizing polymorphic markers. For example, *Solanum pennellii* has been widely used for constructing high-density genetic maps (Pillen et al. 1996). In contrast, breeding populations are typically developed through intraspecific crosses within cultivated varieties or between cultivated tomato and closely related species such as *S. pimpinellifolium*. Therefore, the availability of polymorphic markers that segregate in interspecific crosses, the tight linkage between the gene/QTL of interest and the marker, and cost-effectiveness are critical factors in marker-assisted breeding applications (Arens et al. 2010). Recently, SNPs have gained popularity as molecular markers in tomato breeding because they detect both major and minor genetic variations with high resolution. SNP markers offer several advantages, including their abundance throughout the genome and compatibility with high-throughput genotyping platforms (Martino et al. 2010). The Illumina-based SNP array developed under the USDA SolCAP initiative (http://solcap.msu.edu) provides a valuable resource, containing approximately 9200 SNP markers derived from diverse germplasm and breeding lines for tomato research. More recently, GBS technology has emerged as a preferred marker system and is being widely deployed in tomato breeding programs (Elshire et al. 2011).

Advancements in molecular marker technologies have enabled the development of ultra-high-density genetic maps in tomato, primarily utilizing genome-based markers such as SNPs and InDels. These molecular markers facilitated the construction of high-density genetic maps in the *S. lycopersicum* × *S. pennellii* mapping population (Fulton et al. 2002). Later, Hamilton et al. (2012) generated ultra-high-density genetic maps comprising over 3000 SNP markers across three interspecific F_2 mapping populations derived from *S. lycopersicum* × *S. pennellii* and *S. lycopersicum* × *S. pimpinellifolium*. Several other mapping populations, involving both cultivated and wild tomato species, have been developed and utilized for the identification of QTLs and genes associated with important agronomic traits, including yield (Ohyama et al. 2017), fruit quality (Rambla et al. 2017), nutrient content (Capel et al. 2017), and stress tolerance. Using a genome resequencing approach, Zhang et al. (2018) identified 104 loci linked to 18 agronomic traits in an RIL population derived from *S. pimpinellifolium* × *S. lycopersicum*, highlighting the potential of high-density polymorphisms for high-resolution genome mapping. However, the phenotypic variation and QTL mapping power in this study were limited by the use of only two parental lines. To overcome these limitations, multi-parent advanced generation inter-cross (MAGIC) populations, composed of multiple parents and diverse recombination events over generations, have been developed. For example, Diouf et al. (2018) created a MAGIC population of 397 lines from eight diverse parents, which facilitated the identification of QTLs for fruit weight and stress tolerance traits. Furthermore, bulk-segregant analysis combined with genome sequencing has been successfully applied to detect QTLs governing quantitative traits. The discovery of millions of SNPs through whole-genome sequencing and transcriptomics has accelerated the construction of high-density genetic maps and marker development for tomato breeding.

Due to the limited genetic diversity in cultivated tomato, wild species serve as a rich reservoir of genetic variation for both simple and complex traits. Recent genome sequencing of 360 accessions—including wild species such as *S. pimpinellifolium*, *S. lycopersicum* var. *cerasiforme*, and cultivated *S. lycopersicum*—identified 186 domestication sweeps and 133 improvement sweeps, covering approximately 15% of the genome assembly. From these regions, 18 QTLs associated with fruit quality, fruit size, and stress tolerance were discovered, underscoring their significance in tomato domestication (Ye et al. 2017). Following the sequencing of the tomato genome and its wild relatives, tomato breeding has entered the post-genomics era, with genome-based technologies increasingly employed to discover genes using the reference and other tomato genomes. The availability of the reference tomato genome (The Tomato Genome Consortium 2012) has significantly enhanced genetic mapping of traits related to plant morphology, root and leaf development, fruit characteristics, disease and pest resistance, and abiotic stress tolerance (Xie et al. 2019). Notably, disease resistance genes and their associated markers have been identified using resistance gene enrichment sequencing (RenSeq), which targets *NBS-LRR* gene-enriched libraries in tomato (Andolfo et al. 2014). These recent genomics advances have greatly improved the resolution of fine mapping, facilitating the discovery of new genes and QTLs for tomato breeding applications.

4.10 Challenges and Future Perspective

When conducting GWAS and QTL mapping in tomato, researchers face several challenges. Firstly, the tomato genome is complex, containing numerous genes and regulatory elements, which complicates the identification of specific genetic variants responsible for trait variation. Additionally, population structure and relatedness within tomato populations can lead to false positive associations in GWAS, requiring careful statistical control to avoid spurious results. Another significant challenge is the phenotypic variation in tomatoes. Complex traits such as fruit quality and yield are influenced by multiple genes as well as environmental factors, necessitating large sample sizes and precise phenotypic measurements to effectively dissect their genetic basis. Furthermore, linkage disequilibrium in tomatoes means that associated genetic variants may be physically distant from causal loci, making it difficult to pinpoint the exact causal variants. To address this, fine-mapping approaches and functional validation are essential for identifying true causal variants linked to important tomato traits.

Looking towards the future, GWAS and QTL mapping in tomato hold exciting prospects. Advances in high-throughput sequencing technologies now allow for more comprehensive genomic coverage, enabling the discovery of rare genetic variants that contribute to trait variation. Integrating multiple omics datasets—such as genomics, transcriptomics, and metabolomics—provides a holistic understanding of the genetic architecture underlying tomato traits. This integrative approach will help identify key regulatory networks and candidate genes linked to desirable characteristics. The emergence of genome editing tools like CRISPR-Cas9 offers great promise for accelerating tomato trait improvement. With precise genome editing capabilities, researchers can functionally validate candidate genes and unravel their roles in trait expression, thereby significantly speeding up breeding programs. Moreover, the growing field of phenomics, which focuses on high-throughput phenotyping, combined with advanced data integration and machine learning, can enhance the identification of novel QTLs and improve predictions of complex trait performance. Collaborative efforts among researchers, breeders, and industry stakeholders will be critical to this progress. By sharing data, resources, and expertise, the tomato research community can accelerate advancements and develop improved tomato varieties that meet agricultural and consumer needs.

References

Adhikari TB, Siddique MI, Louws FJ, Sim SC, Panthee DR (2023) Molecular mapping of quantitative trait loci for resistance to early blight in tomatoes. Front Plant Sci 14:1135884

Albert E, Gricourt J, Bertin N, Bonnefoi J, Pateyron S, Tamby JP, Bitton F, Causse M (2016a) Genotype by watering regime interaction in cultivated tomato: lessons from linkage mapping and gene expression. Theor Appl Genet 129(2):395–418

Albert E, Segura V, Gricourt J, Bonnefoi J, Derivot L, Causse M (2016b) Association mapping reveals the genetic architecture of tomato response to water deficit: focus on major fruit quality traits. J Exp Bot 67(22):6413–6430

Alonge M, Wang X, Benoit M, Soyk S, Pereira L, Zhang L, Suresh H, Ramakrishnan S, Maumus F, Ciren D, Levy Y, Harel TH, Shalev-Schlosser G, Amsellem Z, Razifard H, Caicedo AL, Tieman DM, Klee H, Kirsche M, Aganezov S, Lippman ZB (2020) Major impacts of widespread structural variation on gene expression and crop improvement in tomato. Cell 182:145–161.e23

Alqudah AM, Sallam A, Stephen Baenziger P, Börner A (2019) GWAS: fast-forwarding gene identification and characterization in temperate Cereals: lessons from Barley—a review. J Adv Res 4(22):119–135

Anderson TA, Zitter SM, De Jong DM, Francis DM, Mutschler MA (2021) Cryptic introgressions contribute to transgressive segregation for early blight resistance in tomato. Theor Appl Genet 134(8):2561–2575

Andolfo G, Jupe F, Witek K, Etherington GJ, Ercolano MR, Jones JDG (2014) Defining the full tomato NB-LRR resistance gene repertoire using genomic and cDNA RenSeq. BMC Plant Biol 14:120

Arens P, Mansilla C, Deinum D, Cavellini L, Moretti A, Rolland S, van der Schoot H, Calvache D, Ponz F, Collonnier C, Mathis R, Smilde D, Caranta C, Vosman B (2010) Development and evaluation of robust molecular markers linked to disease resistance in tomato for distinctness, uniformity and stability testing. Theor Appl Genet 120:655–664

Bai Y, Huang CC, van der Hulst R, Meijer-Dekens F, Bonnema G, Lindhout P (2003) QTLs for tomato powdery mildew resistance (*Oidium lycopersici*) in Lycopersicon parviflorum G1.1601 co-localize with two qualitative powdery mildew resistance genes. Mol Plant-Microbe Interact 16(2):169–176

Bauchet G, Grenier S, Samson N, Bonnet J, Grivet L, Causse M (2017) Use of modern tomato breeding germplasm for deciphering the genetic control of agronomical traits by genome wide association study. Theor Appl Genet 130:875–889

Berry SZ, Uddin MR (1991) Breeding tomato for quality and processing attributes. In: Genetic Improvement of tomato. Springer, Berlin/Heidelberg, pp 197–206

Bineau E, Diouf I, Carretero Y, Duboscq R, Bitton F, Djari A, Zouine M, Causse M (2021) Genetic diversity of tomato response to heat stress at the QTL and transcriptome levels. Plant J 107(4):1213–1227

Bolger A, Scossa F, Bolger ME, Lanz C, Maumus F, Tohge T, Quesneville H, Alseekh S, Sørensen I, Lichtenstein G, Fich EA, Conte M, Keller H, Schneeberger K, Schwacke R, Ofner I, Vrebalov J, Xu Y, Osorio S, Aflitos SA, Fernie AR (2014) The genome of the stress-tolerant wild tomato species *Solanum pennellii*. Nat Genet 46:1034–1038

Bournival BL, Scott JW, Vallejos CE (1989) An isozyme marker for resistance to race 3 of Fusarium oxysporum f. sp. lycopersici in tomato. Theor Appl Genet 78(4):489–494

Bournival BL, Vallejos CE, Scott JW (1990) Genetic analysis of resistances to races 1 and 2 of Fusarium oxysporum f. sp. lycopersici from the wild tomato Lycopersicon pennellii. Theor Appl Genet 79(5):641–645

Buttery RG (1993) Quantitative and sensory aspects of flavor of tomato and other vegetables and fruits. In: Flavor science: sensible principles and techniques, pp 259–286

Capel C, Yuste-Lisbona FJ, López-Casado G, Angosto T, Heredia A, Cuartero J, Fernández-Muñoz R, Lozano R, Capel J (2017) QTL mapping of fruit mineral contents provides new chances for molecular breeding of tomato nutritional traits. Theor Appl Genet 130:903–913

Caruso G, Gomez LD, Ferriello F, Andolfi A, Borgonuovo C, Evidente A, Simister R, McQueen-Mason SJ, Carputo D, Frusciante L, Ercolano MR (2016) Exploring tomato *Solanum pennellii* introgression lines for residual biomass and enzymatic digestibility traits. BMC Genet 17:56

Causse M, Duffe P, Gomez MC, Buret M, Damidaux R, Zamir D, Gur A, Chevalier C, Lemaire-Chamley M, Rothan C (2004) A genetic map of candidate genes and QTLs involved in tomato fruit size and composition. J Exp Bot 55(403):1671–1685
Causse M, Giovannoni J, Bouzayen M, Zouine M (eds) (2016) The tomato genome. Compendium of plant genomes. Springer, Berlin, p 259
Chagué V, Mercier JC, Guenard M, De Courcel A, Vedel F (1997) Identification of RAPD markers linked to a locus involved in quantitative resistance to TYLCV in tomato by bulked segregant analysis. Theor Appl Genet 95:671–677
Diouf IA, Derivot L, Bitton F, Pascual L, Causse M (2018) Water deficit and salinity stress reveal many specific QTL for plant growth and fruit quality traits in tomato. Front Plant Sci 9:279
Diwan N, Fluhr R, Eshed Y, Zamir D, Tanksley SD (1999) Mapping of Ve in tomato: a gene conferring resistance to the broad-spectrum pathogen, Verticillium dahliae race 1. Theor Appl Genet 98:315–319
Doganlar S, Frary A, Ku HM, Tanksley SD (2002) Mapping quantitative trait loci in inbred backcross lines of Lycopersicon pimpinellifolium (LA1589). Genome 45(6):1189–1202
Elshire RJ, Glaubitz JC, Sun Q, Poland JA, Kawamoto K, Buckler ES, Mitchell SE (2011) A robust, simple genotyping-by-sequencing (GBS) approach for high diversity species. PLoS One 6:e19379
Eshed Y, Zamir D (1994) Introgressions from Lycopersicon pennellii can improve the soluble-solids yield of tomato hybrids. Theor Appl Genet 88(6–7):891–897
Flint-Garcia SA, Thornsberry JM, Buckler ES 4th (2003) Structure of linkage disequilibrium in plants. Annu Rev Plant Biol 54:357–374
Foolad MR (2005) Breeding for abiotic stress tolerance in tomato. In: Abiotic stresses: plant resistance through breeding and molecular approaches. The Haworth press, New York, pp 613–684
Foolad MR (2007) Genome mapping and molecular breeding of tomato. Int J Plant Genomics 2007:64358
Foolad MR, Chen FQ (1999) RFLP mapping of QTLs conferring salt tolerance during the vegetative stage in tomato. Theor Appl Genet 99:235–243
Foolad MR, Jones RA (1993) Mapping salt-tolerance genes in tomato (Lycopersicon esculentum) using trait-based marker analysis. Theor Appl Genet 87(1–2):184–192
Foolad MR, Panthee DR (2012) Marker-assisted selection in tomato breeding. Crit Rev Plant Sci 31:93–123
Foolad MR, Chen FQ, Lin GY (1998a) RFLP mapping of QTLs conferring salt tolerance during germination in an interspecific cross of tomato. Theor Appl Genet 97:1133–1144
Foolad MR, Chen FQ, Lin GY (1998b) RFLP mapping of QTLs conferring cold tolerance during seed germination in an interspecific cross of tomato. Mol Breed 4:519–529
Foolad R, Zhang P, Khan AA, Niño-Liu D, Lin Y (2002) Identification of QTLs for early blight (Alternaria solani) resistance in tomato using backcross populations of a Lycopersicon esculentum x L. hirsutum cross. Theor Appl Genet 104(6-7):945–958
Foolad MR, Zhang LP, Subbiah P (2003) Genetics of drought tolerance during seed germination in tomato: inheritance and QTL mapping. Genome 46(4):536–545
Foolad MR, Merk HL, Ashrafi H (2008) Genetics, genomics and breeding of late blight and early blight resistance in tomato. Crit Rev Plant Sci 27(2):75–107
Frary A, Doganlar S, Frampton A, Fulton T, Uhlig J, Yates H, Tanksley S (2003) Fine mapping of quantitative trait loci for improved fruit characteristics from Lycopersicon chmielewskii chromosome 1. Genome 46(2):235–243
Frary A, Fulton TM, Zamir D, Tanksley SD (2004) Advanced backcross QTL analysis of a Lycopersicon esculentum x L. pennellii cross and identification of possible orthologs in the Solanaceae. Theor Appl Genet 108(3):485–496
Fulton TM, Beck-Bunn T, Emmatty D, Eshed Y, Lopez J, Petiard V, Uhlig J, Zamir D, Tanksley SD (1997) QTL analysis of an advanced backcross of Lycopersicon peruvianum to the cultivated tomato and comparisons with QTLs found in other wild species. Theor Appl Genet 95:881–894

Fulton TM, Grandillo S, Beck-Bunn T, Fridman E, Frampton A, Lopez J, Petiard V, Uhlig J, Zamir D, Tanksley SD (2000) Advanced backcross QTL analysis of a Lycopersicon esculentum x Lycopersicon parviflorum cross. Theor Appl Genet 100:1025–1042

Fulton TM, Van der Hoeven R, Eannetta NT, Tanksley SD (2002) Identification, analysis, and utilization of conserved ortholog set markers for comparative genomics in higher plants. Plant Cell 14:1457–1467

Ganal MW, Simon R, Brommonschenkel S, Arndt M, Phillips MS, Tanksley SD, Kumar A (1995) Genetic mapping of a wide spectrum nematode resistance gene (Hero) against Globodera rostochiensis in tomato. Mol Plant-Microbe Interact 8(6):886–891

Gao L, Gonda I, Sun H, Ma Q, Bao K, Tieman DM, Burzynski-Chang EA, Fish TL, Stromberg KA, Sacks GL, Thannhauser TW, Foolad MR, Diez MJ, Blanca J, Canizares J, Xu Y, van der Knaap E, Huang S, Klee HJ, Giovannoni JJ, Fei Z (2019) The tomato pan-genome uncovers new genes and a rare allele regulating fruit flavor. Nat Genet 51:1044–1051

Gonzalo MJ, da Maia LC, Nájera I, Baixauli C, Giuliano G, Ferrante P, Granell A, Asins MJ, Monforte AJ (2022) Genetic control of reproductive traits under different temperature regimes in inbred line populations derived from crosses between *S. pimpinellifolium* and *S. lycopersicum* accessions. Plants (Basel) 11(8):1069

Grandillo S, Tanksley SD (1996) QTL analysis of horticultural traits differentiating the cultivated tomato from the closely related species Lycopersicon pimpinellifolium. Theor Appl Genet 92(8):935–951

Hamilton JP, Sim SC, Stoffel K, Van Deynze A, Buell CR, Francis DM (2012) Single nucleotide polymorphism discovery in cultivated tomato via sequencing by synthesis. Plant Genome 5:17–29

Hanson PM, Bernacchi D, Green S, Tanksley SD, Muniyappa V, Padmaja AS, Chen HM, Kuo G, Fang D, Chen JT (2000) Mapping a wild tomato introgression associated with tomato yellow leaf curl virus resistance in a cultivated tomato line. J Am Soc Hortic Sci 125(1):15–20

Hanson P, Green SK, Kuo G (2006) Ty-2, a gene on chromosome 11 conditioning geminivirus resistance in tomato. Tomato Genet Coop Rep 56:17–18

Hosmani PS, Flores-Gonzalez M, van de Geest H, Maumus F, Bakker LV, Schijlen E, van Haarst J, Cordewener J, Sanchez-Perez G, Peters S, Fei Z, Giovannoni JJ, Mueller LA, Saha S (2019) An improved de novo assembly and annotation of the tomato reference genome using single-molecule sequencing, Hi-C proximity ligation and optical maps. bioRxiv:767764

Huang X, Han B (2014) Natural variations and genome-wide association studies in crop plants. Annu Rev Plant Biol 65:531–551

Ji Y, Scott JW (2006) Ty-3, a begomovirus resistance locus linked to Ty-1 on chromosome 6 of tomato. Tomato Genet Coop Rep 56:22–25

Kaloshian I, Yaghoobi J, Liharska T, Hontelez J, Hanson D, Hogan P, Jesse T, Wijbrandi J, Simons G, Vos P, Zabel P, Williamson VM (1998) Genetic and physical localization of the root-knot nematode resistance locus mi in tomato. Mol Gen Genet 257(3):376–385

Kawicha P, Tongyoo P, Wongpakdee S, Rattanapolsan L, Duangjit J, Chunwongse J, Suwor P, Sangdee A, Thanyasiriwat T (2023) Genome-wide association study revealed genetic loci for resistance to fusarium wilt in tomato germplasm. Crop Breed Appl Biotechnol 23:e43532311

Kennedy GG, Farrar RR Jr, Barbour JD, Kennedy GG (1994) Field evaluation of insect resistance in a wild tomato and its effects on insect parasitoids. Entomol Exp Appl *71*(3):211–226

Khachik F, Carvalho L, Bernstein PS, Muir GJ, Zhao DY, Katz NB (2002) Chemistry, distribution, and metabolism of tomato carotenoids and their impact on human health. Exp Biol Med (Maywood) 227(10):845–851

Kim M, Nguyen T, Ahn JH, Kim GJ, Sim SC (2021) Genome-wide association study identifies QTL for eight fruit traits in cultivated tomato (*Solanum lycopersicum* L.). Hortic Res 8(1):203

Kimbara J, Ohyama A, Chikano H, Ito H, Hosoi K, Negoro S, Miyatake K, Yamaguchi H, Nunome T, Fukuoka H, Hayashi T (2018) QTL mapping of fruit nutritional and flavor components in tomato (*Solanum lycopersicum*) using genome-wide SSR markers and recombinant inbred lines (RILs) from an intra-specific cross. Euphytica 214:1–12

Lambeth VN, Straten EF, Fields ML (1966) Fruit quality attributes of 250 foreign and domestic tomato accessions
Levin I, Gilboa N, Yeselson E, Shen S, Schaffer AA (2000) Fgr, a major locus that modulates the fructose to glucose ratio in mature tomato fruits. Theor Appl Genet 100:256–262
Liedl BE, Labate JA, Stommel JR, Slade A, Kole C (2013) Genetics, genomics and breeding of tomato. CRC Press, New York, p 480
Lin KH, Yeh WL, Chen HM, Lo HF (2010) Quantitative trait loci influencing fruit-related characteristics of tomato grown in high-temperature conditions. Euphytica 174:119–135
Liu Z, Jiang J, Li J (2023) GWAS characterization of the genetic regions associated with nine important agronomic traits in tomato (Solanum lycopersicum L.). Agronomy *13*(5):1191
Lopez-Ortiz C, Reddy UK, Zhang C, Natarajan P, Nimmakayala P, Benedito VA, Fabian M, Stommel J (2023) QTL and PACE analyses identify candidate genes for anthracnose resistance in tomato. Front Plant Sci 4(14):1200999
Lounsbery JK, Arms EM, Bloom AJ, St. Clair DA (2016) Quantitative trait loci for water-stress tolerance traits localize on chromosome 9 of wild tomato. Crop Sci *56*(4):1514–1525
Lukyanenko AN (1991) Disease resistance in tomato. In: Genetic improvement of tomato. Springer, Berlin/Heidelberg, pp 99–119
Manolio TA (2010) Genomewide association studies and assessment of the risk of disease. N Engl J Med 363(2):166–176
Martin B, Nienhuis J, King G, Schaefer A (1989) Restriction fragment length polymorphisms associated with water use efficiency in tomato. Science 243(4899):1725–1728
Martino A, Mancuso T, Rossi A (2010) Application of high-resolution melting to large-scale, high-throughput SNP genotyping: a comparison with the TaqMan (R) method. J Biomol Screen 15:623–629
Mutschler MA, Doerge RW, Liu SC, Kuai JP, Liedl BE, Shapiro JA (1996) QTL analysis of pest resistance in the wild tomato Lycopersicon pennellii: QTLs controlling acylsugar level and composition. Theor Appl Genet 92(6):709–718
Nguyen TT, Le NT, Sim SC (2021) Genome-wide association study and marker development for bacterial wilt resistance in tomato (Solanum lycopersicum L.). Sci Hortic 289:110418
Ofner I, Lashbrooke J, Pleban T, Aharoni A, Zamir D (2016) *Solanum pennellii* backcross inbred lines (BILs) link small genomic bins with tomato traits. Plant J 87:151–160
Ohyama A, Shirasawa K, Matsunaga H, Negoro S, Miyatake K, Yamaguchi H, Nunome T, Iwata H, Fukuoka H, Hayashi T (2017) Bayesian QTL mapping using genome-wide SSR markers and segregating population derived from a cross of two commercial F1 hybrids of tomato. Theor Appl Genet 130:1601–1616
Osborn TC, Alexander DC, Fobes JF (1987) Identification of restriction fragment length polymorphisms linked to genes controlling soluble solids content in tomato fruit. Theor Appl Genet 73(3):350–356
Panthee DR, Piotrowski A, Ibrahem R (2017) Mapping quantitative trait loci (QTL) for resistance to late blight in tomato. Int J Mol Sci 18(7):1589
Paterson AH, Lander ES, Hewitt JD, Peterson S, Lincoln SE, Tanksley SD (1988) Resolution of quantitative traits into Mendelian factors by using a complete linkage map of restriction fragment length polymorphisms. Nature 335(6192):721–726
Paterson AH, Damon S, Hewitt JD, Zamir D, Rabinowitch HD, Lincoln SE, Lander ES, Tanksley SD (1991) Mendelian factors underlying quantitative traits in tomato: comparison across species, generations, and environments. Genetics 127(1):181–197
Pillen K, Pineda O, Lewis C, Tanksley SD (1996) Status of genome mapping tools in the taxon *Solanaceae*. In: Paterson et al (eds) Genome mapping in plants, Austin, pp 281–308
Rafalski JA (2010) Association genetics in crop improvement. Curr Opin Plant Biol 13(2):174–180
Rambla JL, Medina A, Fernández-Del-Carmen A, Barrantes W, Grandillo S, Cammareri M, López-Casado G, Rodrigo G, Alonso A, García-Martínez S, Primo J, Ruiz JJ, Fernández-Muñoz R, Monforte AJ, Granell A (2017) Identification, introgression, and validation of fruit volatile QTLs from a red-fruited wild tomato species. J Exp Bot 68:429–442
Rice AC, Pederson CS (1994) Factors influencing growth of *Bacillus coagulans* in canned tomato juice. II. Acidic constituents of tomato juice and specific organic acids. Food Res 19:124–133

Ruggieri V, Francese G, Sacco A, D'Alessandro A, Rigano MM, Parisi M, Milone M, Cardi T, Mennella G, Barone A (2014) An association mapping approach to identify favourable alleles for tomato fruit quality breeding. BMC Plant Biol 3(14):337

Ruggieri V, Calafiore R, Schettini C, Rigano MM, Olivieri F, Frusciante L, Barone A (2019) Exploiting genetic and genomic resources to enhance heat-tolerance in tomatoes. Agronomy 9(1):22

Saliba-Colombani V, Causse M, Langlois D, Philouze J, Buret M (2001) Genetic analysis of organoleptic quality in fresh market tomato. 1. Mapping QTLs for physical and chemical traits. Theor Appl Genet 102:259–272

Sarfatti M, Abu-Abied M, Katan J, Zamir D (1991) RFLP mapping of I1, a new locus in tomato conferring resistance against Fusarium oxysporum f. sp. lycopersici race 1. Theor Appl Genet 82(1):22–26

Scot JW (1993) Breeding tomatoes for resistance to high temperatures, biotic and abiotic diseases. Hortic Bras 11:167–170

Stevens MA (1972) Relationships between components contributing to quality variation among tomato lines 1. J Am Soc Hortic Sci 97(1):70–73

Stevens MA, Rick CM (1986) Genetics and breeding. In: The tomato crop: a scientific basis for improvement. Springer, Dordrecht, pp 35–109

Stevens MA, Kader AA, Albright-Holton M, Algazi M (1977) Genotypic variation for flavor and composition in fresh market tomatoes 1. J Am Soc Hortic Sci 102(5):680–689

Stevens MR, Lamb EM, Rhoads DD (1995) Mapping the Sw-5 locus for tomato spotted wilt virus resistance in tomatoes using RAPD and RFLP analyses. Theor Appl Genet 90(3–4):451–456

Tanksley SD, Grandillo S, Fulton TM, Zamir D, Eshed Y, Petiard V, Lopez J, Beck-Bunn T (1996) Advanced backcross QTL analysis in a cross between an elite processing line of tomato and its wild relative L. pimpinellifolium. Theor Appl Genet 92(2):213–224

The Tomato Genome Consortium (2012) The tomato genome sequence provides insights into fleshy fruit evolution. Nature 485:635–641

Tigchelaar TC (1986) Tomato breeding. In: Bassett MJ (ed) Breeding for vegetable crops, pp 135–171

Tiwari JK, Reddy YS, Rai N, Singh DP, Singh AK, Karkute SG, Singh PM, Behera TK (2022) Progress in marker-assisted selection to genomics-assisted breeding in tomato. Crit Rev Plant Sci 41(5):321–350

Tripodi P, Soler S, Campanelli G, Díez MJ, Esposito S, Sestili S, Figàs MR, Leteo F, Casanova C, Platani C, Soler E, Bertone A, Pereira-Dias L, Palma D, Burguet R, Pepe A, Rosa-Martínez E, Prohens J, Cardi T (2021) Genome wide association mapping for agronomic, fruit quality, and root architectural traits in tomato under organic farming conditions. BMC Plant Biol 21(1):481

Vallejos CE, Tanksley SD (1983) Segregation of isozyme markers and cold tolerance in an interspecific backcross of tomato. Theor Appl Genet 66(3–4):241–247

Veeresha BA, Rudra Naik V, Chetti MB, Desai SA, Biradar SS (2015) QTL mapping in crop plants: principles and applications. Int J Dev Res 5(1):2961–2965

Wang X, Gao L, Jiao C, Stravoravdis S, Hosmani PS, Saha S, Zhang J, Mainiero S, Strickler SR, Catala C, Martin GB, Mueller LA, Vrebalov J, Giovannoni JJ, Wu S, Fei Z (2020) Genome of *Solanum pimpinellifolium* provides insights into structural variants during tomato breeding. Nat Commun 11:5817

Wen J, Jiang F, Weng Y, Sun M, Shi X, Zhou Y, Yu L, Wu Z (2019) Identification of heat-tolerance QTLs and high-temperature stress-responsive genes through conventional QTL mapping, QTL-seq, and RNA-seq in tomato. BMC Plant Biol 19(1):398

Xie L, Klein P, Crosby K, Jifon J (2019) A genotyping- by-sequencing single nucleotide polymorphism–based map and genetic analysis of root traits in an interspecific tomato population. J Am Soc Hortic Sci 144:394–404

Yang J, Liang B, Zhang Y, Liu Y, Wang S, Yang Q, Geng X, Liu S, Wu Y, Zhu Y, Lin T (2022) Genome-wide association study of eigenvectors provides genetic insights into selective breeding for tomato metabolites. BMC Biol 20:120

Ye J, Wang X, Hu T, Zhang F, Wang B, Li C, Yang T, Li H, Lu Y, Giovannoni JJ, Zhang Y, Ye Z (2017) An InDel in the promoter of *Al-ACTIVATED MALATE TRANSPORTER9* selected during tomato domestication determines fruit malate contents and aluminium tolerance. Plant Cell 29:2249–2268

Ye J, Wang X, Wang W, Yu H, Ai G, Li C, Sun P, Wang X, Li H, Ouyang B, Zhang J, Zhang Y, Han H, Giovannoni JJ, Fei Z, Ye Z (2021) Genome-wide association study reveals the genetic architecture of 27 agronomic traits in tomato. Plant Physiol 186(4):2078–2092

Zamir D, Tal M (1987) Genetic analysis of sodium, potassium and chloride ion content in Lycopersicon. Euphytica 36:187–191

Zamir D, Ekstein-Michelson I, Zakay Y, Navot N, Zeidan M, Sarfatti M, Eshed Y, Harel E, Pleban T, van-Oss H, Kedar N, Rabinowitch HD, Czosnek H (1994) Mapping and introgression of a tomato yellow leaf curl virus tolerance gene, TY-1. Theor Appl Genet 88(2):141–146

Zhang LP, Lin GY, Nino-Liu D, Foolad MR (2003) Mapping QTLs conferring early blight (Alternaria solani) resistance in a Lycopersicon esculentum× L. hirsutum cross by selective genotyping. Mol Breed 12:3–19

Zhang J, Zhao J, Xu Y, Liang J, Chang P, Yan F, Li M, Liang Y, Zou Z (2015) Genome-wide association mapping for tomato volatiles positively contributing to tomato flavor. Front Plant Sci 27(6):1042

Zhang J, Zhao J, Liang Y, Zou Z (2016) Genome-wide association-mapping for fruit quality traits in tomato. Euphytica 207:439–451

Zhang X, Wang G, Chen B, Du H, Zhang F, Zhang H et al (2018) Candidate genes for first flower node identified in pepper using combined SLAF-seq and BSA. PLoS One 13:e0194071

Zhu G, Wang S, Huang Z, Zhang S, Liao Q, Zhang C et al (2018) Rewiring of the fruit metabolome in tomato breeding. Cell 172:e212

QTL Mapping and GWAS Triumphs in Tackling Economic and Environmental Challenges for *Capsicum annuum* L.

5

Arpita Srivastava, Nikita Baliyan, Priti Upadhyay, Sikha Manoharan, and Manisha Mangal

Abstract

The pepper, belonging to the genus *Capsicum*, is one of the oldest and most widely utilized vegetables, condiments, and spices worldwide. Originating from Central and South America, Mexico, and Southwest Asia, it has served diverse purposes, including use as a vegetable, natural colorant, and traditional medicine. As one of the earliest domesticated plants in the Mesoamerican subcontinent, archaeological evidence indicates its consumption dates back to approximately 7500 B.C. Over time, *Capsicum* has evolved into numerous cultivated varieties exhibiting a spectrum of flavors ranging from mild to hot. Currently, sweet and pungent peppers are primarily employed as vegetables and spices. Genetic resources within the *Capsicum* genus have been effectively harnessed in modern breeding programs, resulting in the development and commercialization of various sweet and hot pepper cultivars with diverse market applications. These cultivars are bred to express resistance to both biotic and abiotic stresses, often incorporating genes derived from landraces and cultivated species within the genus. This breeding strategy not only aligns with market demands but also enhances crop resilience to environmental challenges. The introgression of genetic material from landraces and cultivated species imparts valuable traits such as resistance to pests, diseases, and adverse environmental conditions. Quantitative trait loci (QTL) represent genomic regions associated with complex traits, including yield, heat tolerance, fruit quality, and disease resistance in pepper. QTL mapping involves identifying associations between measurable phenotypes and genetic markers linked to these traits. In *Capsicum* breeding, molecular markers are utilized to detect QTLs that influence trait expression. Numerous QTL mapping studies have identified significant loci

A. Srivastava (✉) · N. Baliyan · P. Upadhyay · S. Manoharan · M. Mangal
Division of Vegetable Science, ICAR-IARI, New Delhi, India

S. S. Dey et al. (eds.), *GWAS and QTL Mapping in Horticultural Crops*,
https://doi.org/10.1007/978-981-95-2306-1_5

accounting for a substantial proportion of phenotypic variance within the genus, underscoring the potential of marker-assisted selection to accelerate genetic improvement in pepper breeding programs.

Keywords

Capsicum · Genetic resources · Molecular markers · Quantitative trait loci · Marker-assisted selection · Genome-wide association studies · Linkage mapping

5.1 Introduction: *Capsicum* Genera

The hot pepper, a member of the genus *Capsicum* within the *Solanaceae* family, is one of the oldest and most widely utilized vegetables, condiments, and spices worldwide. This genus comprises approximately 32 species, with origins traced to Central and South America as well as Mexico. While the initial domestication of *Capsicum* is primarily associated with these regions, later influences from Southwest Asia and China suggest a multi-regional model of domestication (Kraft et al. 2014). Utilized since ancient times, the plant has served various purposes, including as a vegetable, natural colorant, and traditional medicine. It is distinguished as one of the earliest domesticated plants in the Mesoamerican subcontinent, with archaeological evidence of consumption dating back to approximately 7500 B.C. (Bosland et al. 1996). Over time, multiple species have undergone domestication, resulting in a diverse array of cultivated peppers exhibiting flavors ranging from mild and sweet to hot and intensely pungent. Currently, sweet and pungent peppers are predominantly used as vegetables and hot spices, respectively (Hunziker 2001; Paran and van der Knaap 2007).

Five prominently cultivated species within the genus *Capsicum* include *Capsicum annuum*, *Capsicum chinense*, *Capsicum frutescens*, *Capsicum baccatum*, and *Capsicum pubescens* (Basu and De 2003; Knapp 2007). Morphological characteristics such as fruit size, color, and shape facilitate the differentiation between wild and cultivated peppers (Aguilar-Meléndez et al. 2009). The pungency in *Capsicum* is primarily attributed to five major capsaicinoid compounds: capsaicin (69%), dihydrocapsaicin (22%), nordihydrocapsaicin (7%), homocapsaicin (1%), and homodihydrocapsaicin (1%) (Perry et al. 2007). Capsaicin predominates the capsaicinoid profile in most cultivars, with the exception of *C. baccatum*, where dihydrocapsaicin can constitute up to 51% of the total capsaicinoids (Uarrota et al. 2000). The pungency of various pepper varieties is commonly assessed using the Scoville Heat Units (SHU) scale, which quantifies their spiciness. The standard method for evaluating *Capsicum* pungency was established by Scoville in 1912. Based on this scale, *Capsicum* is classified into five groups: non-pungent paprika (0–700 SHU), mildly pungent (700–3000 SHU), moderately pungent (3000–25,000 SHU), highly pungent (25,000–70,000 SHU), and very highly pungent, exceeding 80,000 SHU. On this scale, the bell pepper is assigned a value of zero.

Whether consumed fresh or dried, hot pepper plays a significant role in enhancing the nutritional quality of the diet (Shetty et al. 2013). The capsaicin content in hot pepper not only contributes to its characteristic flavor but also has been associated with beneficial effects such as the reduction of obesity and diabetes. Additionally, chili is a rich source of vitamins A, B6, and C, as well as antioxidants and essential minerals including calcium, zinc, and iron, which collectively provide protection against dysfunctions of the eyes, skin, and cardiovascular system (Sokona et al. 2013). Extracts from chili are utilized in cosmetic and pharmaceutical industries due to their versatile bioactive properties. Its culinary potential, combined with its aesthetic appeal, also makes hot pepper suitable for cultivation in containers or small-scale gardens.

Researchers are increasingly attracted to *Capsicum* because of its extensive phenotypic diversity, including variations in plant morphology, flower structure, fruit types, color, pungency, and qualitative traits. This diversity renders hot pepper a valuable crop for various applications, particularly in scientific research. The availability of both sweet and hot *Capsicum* types expands its utility as a food source. In temperate regions of Europe and North America, certain pepper varieties are commonly consumed as vegetables, either raw or cooked. Hot pepper significantly enhances the flavor of tropical cuisines, imparting a fiery heat and sharpness to various food preparations. Global interest in chili as a commercial crop is steadily increasing, creating an urgent need to accelerate production to meet rising demand while maximizing its nutritional and medicinal benefits (Herath et al. 2021). Due to limited local production and suboptimal yields, many countries rely on chili imports to satisfy consumption needs. Therefore, the development of high-yielding varieties is a critical priority to boost production. Expansion of cultivation areas largely depends on the availability of such improved varieties. Consequently, numerous breeding programs have been initiated to enhance overall hot pepper production (Bhutia et al. 2015; Nagaraju et al. 2017; Rohini and Lakshmanan 2017).

Molecular breeding within the *Capsicum* genus represents a state-of-the-art approach to improving genetic traits related to yield, disease resistance, and other desirable characteristics (Fig. 5.1). Over the years, considerable efforts have been directed toward elucidating the molecular basis of *Capsicum* species, including studies on genetic diversity, trait mapping, and marker-assisted selection. These molecular breeding initiatives aim to expedite the development of superior cultivars, addressing cultivation challenges and aligning with evolving agricultural and consumer demands. The integration of advanced genomic tools and techniques in *Capsicum* breeding holds great promise for enhancing the resilience, productivity, and overall quality of this agriculturally and economically important crop. This chapter explores key molecular breeding advances within the *Capsicum* genus, underscoring their critical role in shaping the future of pepper cultivation.

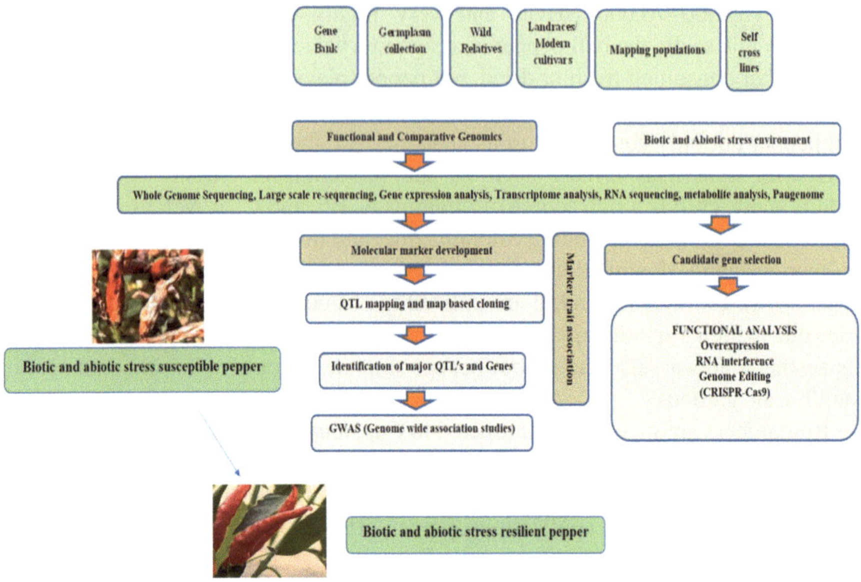

Fig. 5.1 Diagrammatic representation of the improvement of *Capsicum* genus utilizing various advanced molecular tools

5.2 Breeding Objectives of *Capsicum*

Pepper breeding objectives, encompassing both hot and bell pepper varieties, vary considerably based on factors such as the country of cultivation, intended purpose, growing conditions, end use, and regional consumer preferences (Selvakumar et al. 2023). Preferences range from regions favoring hot and pungent peppers to those that prefer sweet varieties. The diseases affecting pepper crops also differ according to the prevailing climatic conditions in each region. Notably, disease resistance breeding remains a primary focus in pepper improvement programs (Paran et al. 2004). The process of breeding for disease resistance begins with the identification of resistant sources and an understanding of their genetic basis, followed by the introgression of resistance genes into elite genotypes. The *Capsicum* genus has substantially benefited from the incorporation of disease and pest resistance traits derived from wild species into cultivated elite cultivars to enhance resilience (Pohronezny 2003). Interspecific hybridization programs aimed at introgressing resistance genes also require careful assessment of crossability levels between species (Mondo et al. 2022).

Currently, significant diseases affecting *Capsicum* include viral pathogens such as Potyviruses (e.g., Potato virus Y, Pepper mottle virus) (Quenouille et al. 2013; Fang et al. 2021), Begomoviruses (Shingote et al. 2022), and Tospoviruses (e.g., Tomato spotted wilt virus) (Oliver and Whitefield 2016). Additionally, fungal diseases like powdery mildew and Phytophthora root rot, bacterial diseases such as bacterial wilt and bacterial spot, as well as pests including whiteflies, thrips, mites, and root-knot nematodes, pose considerable threats to pepper cultivation. The introgression of disease resistance genes from wild germplasm into elite genetic backgrounds significantly contributes to crop improvement by enhancing yield, quality, and stability in pepper production. However, challenges often arise when resistance traits are governed by complex quantitative genetic mechanisms linked to undesirable horticultural and economic traits, complicating breeding efforts. Furthermore, the continual emergence of new pathogen races and strains that overcome existing resistance highlights the critical need for ongoing identification and deployment of novel resistance sources (Barchenger et al. 2018).

Hybrid breeding programs are gaining increased importance to achieve the critical objective of enhancing yield and overall productivity in *Capsicum* cultivation. Emphasis is placed on developing hybrids using male sterility systems, which reduce the time and labor required for hybrid seed production. Both genetic male sterility (GMS) and cytoplasmic male sterility (CMS) systems have been employed in hybrid development; however, the CMS system is more widely utilized. Consequently, the identification of new CMS sources and their maintainer lines, diversification of CMS cytoplasms, and discovery of restorer lines with strong general and specific combining ability have become key objectives. Additionally, incorporating disease resistance genes into CMS lines and restorers remains an essential breeding goal to develop high-performing, resilient hybrids (Meena et al. 2018). Breeding objectives for pepper are closely aligned with market demand and end-use requirements, encompassing both horticultural and biochemical traits. For fresh-market varieties, breeders typically select for attributes such as fruit color at the unripe stage, usually varying shades of green (light, medium, or dark), as well as fruit length, width, and pericarp thickness (Zhang et al. 2023). In addition to these traits, the level of pungency represents a distinctive and commercially important aspect of pepper breeding. Understanding regional consumer preferences for pungency is crucial for developing varieties tailored to specific markets.

Pungency in peppers is primarily attributed to the presence of capsaicinoids, a class of compounds unique to the *Capsicum* genus (Perucka et al. 2001). The two most abundant capsaicinoids are capsaicin and dihydrocapsaicin, together constituting approximately 90% of the total capsaicinoid content. Capsaicin alone typically accounts for about 71% of the total capsaicinoids in most pungent varieties (Kosuge 1970). Capsaicin content is a major determinant of commercial quality in peppers (Kawabata et al. 2006; Hachiya et al. 2007; Ohnuki et al. 2001). Uniquely, *Capsicum* species possess the enzyme capsanthin-capsorubin synthase (CCS), which catalyzes the synthesis of two red pigments, capsanthin and capsorubin (Guzmán et al. 2011). Breeding efforts targeting higher capsanthin content are particularly important for red dried peppers, which are used as whole dried spices or powders, as well as for industrial extracts such as paprika oleoresin, capsaicinoids, and carotenoids (Villa-Rivera et al. 2020). The red color intensity in chili is measured in ASTA (American spice trade association) units, where a higher ASTA value correlates with greater brightness and richness of the final product. Capsanthin content is categorized as low (up to 70 ASTA units), medium (71–100 ASTA units), and high (101–150 ASTA units). While ASTA color influences product brightness, the surface color of the fruit primarily affects its hue.

The development of paprika varieties that produce non-pungent pods with high color value, suitable for oleoresin extraction industries, represents a key objective in pepper breeding. In addition to their conventional nutritional and culinary uses, peppers also have diverse non-food applications, including roles in defence, spiritual practices, and ethnobotanical traditions (Kumar et al. 2014; Meghvansi et al. 2010). Consequently, breeding goals are adapted to address these varied uses and requirements.

For dry peppers, dry matter content is a critical quality trait targeted in breeding programs aimed at producing dried powder and whole dried fruits. These characteristics are particularly important for export markets. From a commercial standpoint in the spice industry, high dry matter content in red hot pepper fruits is desirable; however, there is no positive correlation between dry matter content and capsaicin concentration (Dhall 2008). Additionally, a relatively thin pericarp is preferred, as it facilitates more efficient drying. Fruits with thicker pericarps tend to develop a wrinkled surface and dull appearance after drying, which can reduce their market value. With increasing industrialization, coupled with the risks of crop failure due to climate change and growing domestic and international demand for nutritious and safe foods, breeding efforts are increasingly focused on developing genotypes with enhanced tolerance to high temperatures, drought, and broader adaptability.

Sweet pepper, or bell pepper, scientifically classified as *Capsicum annuum*, lacks the pungent compound capsaicin, resulting in a mild and sweet flavor. The primary objectives in sweet pepper genetic improvement focus on developing varieties with a blocky fruit shape and diverse colors, such as medium to dark green at the unripe stage and red, yellow, or orange at maturity (Sharma et al. 2020). Another key goal is to select and develop breeding lines or cultivars with elevated levels of antioxidants and vitamins, including ascorbic acid (vitamin C), flavonoids (phenolics), and red and yellow/orange carotenoids, which encompass vitamin A precursors like α- and β-carotene and β-cryptoxanthin (Hamed et al. 2019). Breeding efforts also emphasize traits such as high fruit set and yield across varied growing conditions both open-field and protected environments along with tolerance to abiotic stresses like low temperature, water deficit, and salinity (Elkind et al. 2012). Additional focus is placed on improving the stability of carotenoid extracts under long-term storage and protection against photooxidation. While sweet and hot peppers share many common pathogens, sweet pepper is particularly susceptible to diseases such as *Phytophthora* fruit rot, anthracnose, viral infections, powdery mildew, and bacterial wilt under open cultivation. Developing sweet pepper genotypes with broader adaptability is essential, as it is predominantly a cool-season crop; thus, "tropicalization" of sweet pepper is necessary to enable cultivation in nontraditional regions for extended periods (Srivastava et al. 2023). Under protected cultivation, major breeding goals include developing sweet pepper lines with an indeterminate growth habit suitable for training and pruning, a blocky fruit shape, and resistance to root-knot nematodes (Pegard et al. 2005; Sharma et al. 2020).

5.3 Discovery and Utilization of Genetic Resources in *Capsicum*

The utilization of *Capsicum* genetic resources has been highly successful in contemporary plant breeding programs, leading to the development and commercialization of a wide range of sweet and hot pepper cultivars with diverse market applications. These cultivars are specifically bred to exhibit resistance to various biotic and abiotic stresses, frequently incorporating genes sourced from landraces and different cultivated species within the *Capsicum* genus. In modern breeding initiatives, the integration of genetic diversity from *Capsicum* germplasm has facilitated the creation of cultivars that not only meet diverse market preferences but also demonstrate enhanced resilience to a variety of environmental challenges. The incorporation of genetic material from landraces and cultivated species remains a key strategy for introducing valuable traits, such as resistance to pests, diseases, and adverse abiotic conditions. Breeding programs particularly focus on developing cultivars with resistance to multiple biotic stresses (e.g., pests and diseases) and abiotic stresses (e.g., drought, extreme temperatures). The deployment of resistance genes from diverse genetic backgrounds confers broad-spectrum resistance, thereby enhancing the robustness of cultivars against evolving environmental and pathogenic pressures. To provide a comprehensive overview of these breeding efforts, the status of germplasm screening and characterization against selected biotic and abiotic stresses has been meticulously documented and summarized in Table 5.1.

5.4 Development of Molecular Markers and Genetic Maps in *Capsicum* Genera

Molecular markers and genetic maps serve as indispensable tools in modern genetics and genomics, addressing scientific questions that were previously unexplored or insufficiently studied. Their significance spans various critical areas, including the assessment of genetic diversity and germplasm characterization, trait mapping for the identification and tracking of genes associated with specific phenotypes, quantitative trait loci (QTL) mapping, marker-assisted selection, and resistance breeding for both biotic and abiotic stresses. The integration of molecular techniques has markedly enhanced the precision, efficiency, and sustainability of breeding programs, while substantially advancing our understanding of plant genomes. This section provides a comprehensive overview of the development and application of molecular markers and genetic maps within the *Capsicum* genus. The primary focus is on tracing the evolution of methodologies used in the molecular characterization of *Capsicum* species, beginning with early morphological and isozyme marker studies and progressing to the current era of high-throughput genotyping technologies.

Table 5.1 Genetic resources and their utilization for the improvement of *Capsicum* genus

Trait	Genetic resource identified	Genetic resources utilized		Genetics of resistance	Markers identified	References
		From	To			
Anthracnose	*C. baccatum*: PBC-80, PBC 81, PI594137, PI1497985-1, PI260550 *C. chinense*: PBC 932	PBC932	*C. annuum*	Single recessive genes; QTLs	SCAR-Indel SSR-HpmsE032 SNP markers	Gniffe et al. (2013), Kim et al. (2008), Mahasuk et al. (2009), Lin et al. (2007), and Suwor et al. (2015, 2017)
		PBC 80	*C. annuum* using bridging species C. chinense			
Bacterial Wilt	MC-4, PBC1347, PBC066 (Malasian strain) PBC631 (SriLankan strain) PBC473 (Indonesian strain) LS2341			Monogenic recessive Bw1 (major QTL), QTL-*qRRs-10.1*	AFLP marker, SSR-CAMS451, CAPS-CA04g2500 InDel marker- ID10-194305124	Thakur et al. (2019), Mimura et al. (2009), Ha et al. (2019), and Du et al. (2019)
		From	To			
Phytophtora Root Rot	CM334, PI201232, PI201234, Perennial	CM334	*C. annuum*	QTLs: *Phyto-U*, *Phyto 5.2,* and *Pc 5.1*	SSR-ZL6726 linked to *CaPhyto*	Sy et al. (2008), Quirin et al. (2005), Lefebvre and Palloix (1996), Ogundiwin et al. (2005), Mallard et al. (2013), Minamiyama et al. (2006), Kim et al. (2008), and Wang et al. (2016)
		PI201232, PI201234	*C. annuum*			
Viruses						
Potyviruses	C. *annuum:* IHR2451, IHR4503, Yolo Y, Florida VR@, Avelar, CM334, Perennial, NW4 C. chinense: PI159236, PI152225 PI159236			Monogenic recessive: pvr1, pvr2, pvr6 Dominant genes: Pvr7, Pvr4	SNP markers (SNP-H1.5, SNP-H1.6, SNPH1.7, SNPH1.8)	Lee et al. (2013), Naresh et al. (2018), and Venkatesh et al. (2018)

Cucumoviruses	*C. annuum*: Sapporo-oonaga, Nanbuonaga, Bukang, BJ0747, Perennial, IHR2451, IHR4503 *C. baccatum*: Pen-3 PI439381-1-3 *C. frutescens*: PBC688, LS1839-2-4, Tabasco	From	To	Monogenic recessive, partially dominant Single dominant gene: *Cmr1* (Bukang) Polygenic (Perennial) Two major QTLs (BJ0747)	CAPs marker for *Cmr1* SNP-CA02g19570 for *qCmr2.1*	Suzuki et al. (2003), Kang et al. (2010), Yao et al. (2013), Nono-Womdim et al. (1991), Pochard and Daubeze (1989), Lapidot et al. (1997), Chaim et al. (2001), and Naresh et al. (2018)
		Pen-3	Vania (*C. annuum*)			
		PBC688	98-42 (inbred line)			
Begomoviruses	*C. annuum*: DLS-Sel-10, WBC-Sel-5, PBC345, PBC143, PBC1444, PBC149, PBC495, VI012005, GKC-29, BS-35, EC-497636, Kalyanpur Chanchal, S-343 *C. chinense*: BG-3821, Bhut Jhalokia *S. pseudocapsicum*	S-343	CH-27 (hybrid)	Single monogenic resistance (Bhut jhalokia, DLS-sel-10, WBC-Sel-5) Single dominant gene (S-343)	SSR-CA 516044 and PAU-LC-343-1	Anaya-Lopez et al. (2003), Adluri et al. (2017), Srivastava et al. (2015, 2017), Kenyon et al. (2014), Singh et al. (2016), Mathur et al. (2018), Thakur et al. (2019), and Srivastava et al. (2021)
Tospoviruses	*C. chinense*: PI152225, PI59236, PI59234, CNPH 275, PI-15, C00943, ECU-973 *C. baccatum*: PIM26-1			Dominant gene: Tsw	RAPD marker mapped on Ch. 10	Boiteux et al. (1993), Moury et al. (1997), Jahn et al. (2000), and Cebolla-Cornejo et al. (2003)
Tobamoviruses	*C. chinense*: PI 315008, PI 315023, PI 315024, PI 159236, PI 152225, KC667 *C. annuum:* Bruinsma Wonder, Verbeterde Glas, KC780, Nanbu-Ohnaga *C. frutescens* Tabasco *C. chacoense:* PI 260429. *C. baccatum*: PI 439381-1-3			Single Dominant genes: L^1, L^2, L^3, L^4	RFLP- TG36, *L* locus mapped on Chr 11 SCAR marker WA31-1500S	Rast (1988), Hamada et al. (2002), Berzal-Herranz et al. (1995), Garcìa-Neria and Rivera-Bustamante (2011), Tomita et al. (2011), and Sawada et al. (2004)

(continued)

Table 5.1 (continued)

Trait	Genetic resource identified	Genetic resources utilized	Genetics of resistance	Markers identified	References
Root-knot nematode	*C. frutescens*: UFGFR 05; *C. chinense*: UFGCH 24 C. annuum: PI322719, PI201234, and CM334	–	Single dominant gene: N gene Six Me genes (Me1, Me3, Me7, Me4, Mech1 and Mech2)	SNP based CAPS-C2At2g06530 N-SCAR-315 CASSR37	Uncu et al. (2015), Celik et al. (2016), Wang et al. (2009), Djian-Caporalino et al. (2007), Marques et al. (2019), and Barbary et al. (2015)
Insects	P BC145: resistant to thrips, mites, aphids *C. annuum*: AC1979 resistant to thrips, IIHR-243-1-1-15, Musalwadi , Jwala, G-5, Pant C-1, IC342390, I572492, I337281	–	Major QTLs	HpmsE078 and HpmsE031on chr 5 and 6	Gniffke et al. (2013), Maharijaya et al. (2011, 2012, 2015), Borah (1987), Linders et al. (2010), and Rameash et al. (2016)
Drought	*C. chinense*: IHR4502	–	–	–	Naresh et al. (2018)
High temperature	*C. annuum:* Medusa, Thai Hot, Explosive Ember, Treasure Red, AVPP9905, AVPP9823, DLS-161-1, DLS-152-1, DLS-20-11	Introgression of HSPs AVPP0702 to Malaysian chilli variety Kulai	Polygenic	64 QTLs mapped	Usman et al. (2018), Srivastava et al. (2020, 2023), and TS et al. (2023)
Salt tolerance	*C. annuum*: CO-1, K1, Jayanthi, Arka Suphal, EC497636	–	–	–	Balasankar et al. (2017)

5.4.1 Molecular Markers in *Capsicum*

The discovery of molecular markers marked a breakthrough, ushering in a new era of molecular breeding. These advancements have accelerated traditional breeding by enabling the precise identification and manipulation of specific genes linked to economically important traits in the *Capsicum* genus. Molecular markers encompass morphological, biochemical, and isozyme markers; however, DNA-based markers are the most widely utilized due to their high specificity and reproducibility. These markers exploit variations in DNA sequences to distinguish genetic differences. The commonly studied DNA markers include Restriction Fragment Length Polymorphisms (RFLPs), Random Amplified Polymorphic DNA (RAPDs), Cleaved Amplified Polymorphic Sequences (CAPSs), Sequence Characterized Amplified Regions (SCARs), Amplified Fragment Length Polymorphisms (AFLPs), Simple Sequence Repeats (SSRs), and Single-Nucleotide Polymorphisms (SNPs). Genetic mapping remains a pivotal tool for pinpointing gene locations within the genome and identifying QTLs responsible for phenotypic variation. Recently, the construction of genetic maps has been greatly expedited through rapid, genome-wide SNP detection technologies facilitated by next-generation sequencing (NGS) methods such as genotyping-by-sequencing (GBS), restriction site-associated DNA sequencing (RAD-seq), and specific-locus amplified fragment sequencing (SLAF-seq).

5.4.1.1 Early Markers: Isozymes and RFLPs

Early genetic studies in *Capsicum* species primarily utilized isozyme analysis and RFLPs. Isozyme markers are co-dominant and detect variations in amino acid sequences, which are analyzed through protein electrophoresis (Kang et al. 2001). For example, three isozyme markers—*Gpi-2*, *Idh-1*, and *Pgm-2*—were mapped on an interspecific genetic linkage map derived from a cross between *Capsicum annuum* 'NuMex RNaky' and *Capsicum chinense* 'PI159234'. Additionally, *Idh-1* and *Pgm* isozyme markers were used to perform comparative genetic mapping between pepper and tomato, revealing conserved genomic regions (Livingstone et al. 1999). In another interspecific map developed from a cross of *C. annuum* 'Perennial' × 'Yolo Wonder', the isozyme marker *Mnr-1*, corresponding to menadione reductase, was successfully positioned (Lefebvre et al. 2002). Furthermore, Kumar et al. (2014) analyzed genetic diversity among 21 pepper cultivars using electrophoretic patterns of two isozyme markers, peroxidase and polyphenol oxidase. This approach facilitated genotype fingerprinting based on intraspecific variation. Similarly, Mondal et al. (2016) employed peroxidase enzyme gel electrophoresis to assess genetic diversity and categorize *Capsicum annuum* genotypes into distinct groups.

RFLP markers are co-dominant markers known for their high reproducibility and good transferability across species (Botstein et al. 1980). Lefebvre et al. (1993) conducted a study on RFLP variation among different pepper cultivars, analyzing accessions from *Capsicum annuum* var. *annuum* and *C. baccatum* var. *pendulum*. Using restriction enzymes to digest DNA, the study revealed that genetic variation between *C. annuum* and *C. baccatum* was greater than the variation observed among

C. annuum cultivars, demonstrating the effectiveness of RFLP markers in both inter- and intraspecific genetic studies. Similarly, Prince et al. (1995) investigated interspecific genetic variation within the *Capsicum* genus using shared restriction fragments detected by Southern blot analysis, further illustrating the utility of RFLP markers in genetic diversity assessments.

5.4.1.2 Advent of PCR-Based Markers (RAPDs and AFLPs)

RAPD markers are widely used molecular tools in genetic studies, valued for their cost-effectiveness and time efficiency. They are particularly suitable for species with limited genomic information, as they do not require prior knowledge of the genome. RAPD markers have proven effective in phylogenetic analyses, polymorphism detection, and assessment of genetic diversity. For instance, Rodriguez et al. (1999) investigated variation among and within different *Capsicum* species using 110 RAPD markers. This study successfully reclassified some *Capsicum* accessions that were previously misclassified based on morphological traits, highlighting the higher resolving power of RAPD markers. Similarly, Bahrami et al. (2008) utilized RAPD markers to analyze genetic diversity among several *Capsicum* genotypes, clustering them according to polymorphic band patterns using Jaccard's similarity coefficient and UPGMA clustering methods.

The application of RAPD markers extends beyond diversity studies; Ilbi (2003) demonstrated their utility in seed quality control for cultivar identification and hybrid purity testing in *Capsicum annuum*. Additionally, Arnedo et al. (2002) identified RAPD markers linked to the *Pvr4* gene, which confers resistance to Potato Virus Y, facilitating marker-assisted selection in breeding programs. In another example, Arumingtyas et al. (2022) developed a SCAR marker derived from RAPD profiles to detect fast-flowering pepper strains. During RAPD screening with 20 primers, primer OPB6 produced polymorphic bands that could differentiate between early- and late-flowering lines. More recently, Rodrigues et al. (1999) evaluated genetic divergence among multiple pepper accessions using RAPD markers, identifying promising genotypes with high genetic variability suitable for breeding programs. Several studies have utilized Amplified Fragment Length Polymorphism (AFLP) markers to assess the genetic diversity of *Capsicum* species, including works by Guzmán et al. (2005), Thul et al. (2006), Baba et al. (2016), and Cardoso et al. (2018). In addition to diversity analysis, AFLP markers have been applied in disease resistance research. For example, Lee et al. (2013) investigated resistance to hot pepper veinal mottle virus in *Capsicum* and identified new dominant sources of resistance. Their study also reported two AFLP markers linked to this dominant resistance gene, both mapped to chromosome 6.

5.4.1.3 Transition to Sequence-Based Markers (SSRs and SNPs)

SSRs and SNPs offer significant advantages over conventional molecular markers in genetic studies. Their compatibility with modern genomic technologies, high reproducibility, and cost-effectiveness have made SSRs and SNPs pivotal tools for advancing genetic research beyond the limitations of earlier marker systems. From the genomic DNA of *Capsicum annuum*, Minamiyama et al. (2006) constructed

SSR-enriched libraries, sequenced numerous clones, and identified unique SSR markers. More recently, Jaiswal et al. (2020) developed non-coding RNA-based SSR markers distributed across all 12 *Capsicum* chromosomes and successfully genotyped various accessions with these markers. This study underscored the potential of SSR markers in gene and QTL mapping, as well as in the improvement of diverse traits through breeding programs. With the advent of NGS technologies, whole-genome sequences for many crops, including *Capsicum* species, have become available. Using data mining approaches, Zhong et al. (2021) identified SSR markers from *Capsicum frutescens* genome sequences and screened them across different cultivars, revealing substantial genetic diversity within the species.

Comparative genetic analysis with tomato and potato enabled the development of SNP markers in *Capsicum* species, as reported by Yang et al. (2009). The first documented use of SNP markers in pepper for varietal identification was by Jung et al. (2010), demonstrating the potential of SNPs as reliable genetic markers. Garces et al. (2007) identified SNPs associated with pungency within an approximately 307 bp expressed sequence tag derived from placental tissue expression analysis in *Capsicum* fruits. They successfully distinguished SNPs between pungent and non-pungent accessions. Using CAPS and tetra-primer amplification refractory mutation system PCR (ARMS-PCR), they developed a marker linked to pungency. This approach proved to be a faster, more cost-effective, and reliable method for identifying pungent peppers compared to traditional sensory evaluation, facilitating early-stage screening in breeding programs. Advancements in high-throughput sequencing technologies have made whole-genome resequencing feasible, allowing comprehensive identification of allelic variants. Nicolai et al. (2012) sequenced *Capsicum annuum* cv. 'Yolo Wonder', and read mapping facilitated the discovery of numerous SNPs and SSRs. Similarly, Truong et al. (2011) developed gene-based SNP markers linked to the bacterial spot resistance gene *Bs-2* using tetra-primer ARMS-PCR methodology, which were subsequently validated across diverse cultivars.

5.4.2 Construction of Genetic Maps

The construction of a genetic map is a fundamental step in genetic research, allowing for the systematic arrangement of genes along chromosomes. This process involves identifying and mapping specific loci, or genetic markers, based on linkage relationships and recombination frequencies. Genetic maps are crucial tools for elucidating genome organization, aiding gene discovery, and supporting a wide range of genetic studies across diverse species.

5.4.2.1 Linkage Mapping in *Capsicum* Species

The first genetic linkage map for peppers was developed using 84 RFLP markers derived from a shared set of cDNA and single-copy genomic clones. This map was constructed from a population of 46 individuals resulting from a cross between *Capsicum annuum* 'Doux des Landes (CA50)' and *Capsicum chinense* 'CA4'

(Tanksley et al. 1988). Subsequently, Prince et al. (1995) created a molecular map of pepper spanning approximately 720 cM by combining isozyme and RFLP markers. This study also performed a comparative analysis with the tomato genome, revealing insights into their divergence from a common ancestor. Genome evolution within the *Solanaceae* family was attributed to centric fusions and chromosome breakage events.

Kang et al. (2001) constructed a molecular linkage map of pepper using an interspecific F_2 population, incorporating 150 RFLP and 430 AFLP markers. Their work included mapping genes involved in the biosynthesis of carotenoids and capsaicinoids, contributing to the understanding of secondary metabolite pathways in pepper. Rao et al. (2003) performed an advanced backcross QTL analysis in a cross between *Capsicum annuum* and *Capsicum frutescens*, developing a genetic map using RFLP markers within this genome. Paran et al. (2004) developed an integrated genetic linkage map comprising 2262 markers, including AFLP, RFLP, RAPD, isozyme, and morphological markers, based on data from six distinct progenies. Minamiyama et al. (2006) created an SSR-based linkage map for an intraspecific doubled haploid population of *Capsicum annuum*, combining AFLP, CAPS, RAPD, and nine previously reported SSR markers distributed across 13 linkage groups, providing a valuable reference genetic map for *Capsicum* species. Similarly, Kim (2000) constructed a molecular linkage map for hot pepper from an interspecific population of *C. annuum* × *C. chinense*, using RFLP and AFLP markers. This map consisted of 11 large and 5 small linkage groups, covering approximately 1320 cM with an average marker spacing of 7.5 cM. Genes linked to carotenoid and capsaicinoid biosynthesis were also mapped, laying the foundation for genetic studies of secondary metabolites in pepper. More recently, Sugita et al. (2013) developed a high-density SSR-based linkage map of *C. annuum* to facilitate genetic analysis and marker-assisted breeding. This map spans about 2028 cM and comprises 12 linkage groups.

5.4.2.2 Comparative Mapping Studies

In a significant study, a CAPS marker linked to the dominant potyvirus resistance gene *Pvr4* enabled the mapping of both *Pvr4* and *Pvr7* to linkage group 10 on an interspecific pepper map. The findings revealed that *Pvr4*, *Pvr7*, and *Tsw* genes conferring resistance to the tomato spotted wilt virus constitute the first recognized cluster of dominant disease resistance genes in *Capsicum* (Grube et al. 2000). A comparative genomic analysis between the pepper and tomato genomes, using 299 orthologous markers including 263 conserved ortholog set II (COSII) markers, identified at least 19 chromosomal inversions and 6 translocations, along with numerous potential single-gene transpositions. Despite these differences, 35 conserved syntenic segments were found where gene/marker order is highly preserved (Wu et al. 2009). Using a population of 126 F_8 recombinant inbred lines derived from a cross between the Phytophthora root rot-resistant genotype *YCM334* and the susceptible genotype *Tean*, a genetic map covering approximately 2177.5 cM with an average marker spacing of 5.2 cM was constructed (Troung et al. 2010). Lee et al. (2016) developed a genetic map of an interspecific population of *Capsicum baccatum* using SNPs identified through NGS (Next generation sequencing) of parental lines. Comparative analysis with the reference physical map of *C. annuum* suggested that chromosomal translocations may act as genetic barriers between these species. Further comparative mapping between eggplant and pepper confirmed the presence of 10 translocations and 8 inversions. This study identified 151 QTLs in pepper and 212 QTLs in eggplant, including 76 major QTLs affecting key agronomic traits (Rinaldi et al. 2016). Recent genomic studies have reported the sequences of two pepper varieties *viz. C. baccatum* 'PBC81' and *C. chinense* 'PI159236' alongside an improved reference genome of *C. annuum* 'CM334'. These analyses revealed dynamic chromosomal rearrangements, particularly translocations involving chromosomes 3, 5, and 9 in *C. baccatum* compared to the other species, suggesting ongoing speciation and evolutionary processes within the *Capsicum* genus (Kim et al. 2017).

5.5 Quantitative Trait Loci (QTL) Mapping

QTLs are defined as genomic regions associated with quantitative traits such as yield, heat tolerance, quality, or disease resistance where clusters of genes collectively influence a particular phenotype. QTL mapping involves identifying associations between measurable phenotypes and genetic markers. Two primary approaches are used in genetic research for QTL mapping: linkage mapping and association analysis. In plant breeding, markers linked to QTLs enable the identification and selection of genomic regions that control complex traits, thereby enhancing breeding efficiency. Numerous QTL mapping studies in *Capsicum* species have revealed significant QTLs that explain a substantial proportion of phenotypic variance, highlighting the promising potential of marker-assisted selection in *Capsicum* improvement programs (Rani et al. 2023).

5.5.1 Genetic Mapping of Economic Traits in *Capsicum*

Molecular breeding techniques such as QTL mapping, identification of causative genes, and molecular marker development have been extensively applied to economically important traits in *Capsicum* species. Numerous studies have compiled QTL data for a range of plant- and fruit-related traits, including primary stem length, internode length, plant height, branching, and flowering time; key factors influencing overall plant growth. Efforts to map genes controlling these traits have been conducted by several researchers, summarized in Table 5.2 (Barchi et al. 2009; Han et al. 2016; Dwivedi et al. 2015; Yarnes et al. 2013). In particular, QTLs associated with plant height (PHT) and various yield-related traits such as number of fruits per plant (NFP), ten fruits weight (TFW), fruit length (FL), fruit width (FW), total fruit weight (ToFW), and pericarp thickness (PT) have been mapped in intraspecific advanced recombinant inbred line (RIL) populations of pepper. Linkage maps were constructed using SSR, RAPD, and SCAR markers, resulting in the identification of

10 QTLs influencing yield traits across four linkage groups. These stable QTL regions serve as promising targets for fine mapping and the development of molecular markers to improve yield and morphological traits in pepper (Dwivedi et al. 2015).

More recently, further advances in *Capsicum* QTL mapping have been made. For example, a mapping population ($F_{2:3}$) derived from a cross between the wild *Capsicum annuum* accession 'Chiltepin' and the cultivated variety 'Puya' was used to identify genomic regions linked to 19 domestication and agronomic traits. A genetic map comprising 1023 SNP markers distributed over 12 linkage groups spanning 1263.87 cM was constructed. QTL analysis revealed 20 marker-trait associations for 13 phenotypes, including 14 previously known loci and six novel genomic regions. This study highlighted genetic parallels in *Capsicum* domestication with other crop species, characterized by a few loci exerting significant effects, QTL clusters in specific genomic regions, and the presence of recessive alleles selected during domestication (Lopez et al. 2023).

Fruit size and weight QTLs have been identified in numerous studies. A QTL analysis of pepper fruit characteristics was performed using an F_3 population derived from a cross between two *Capsicum annuum* genotypes: the bell-type cultivar 'Maor' and the Indian small-fruited line 'Perennial'. A genetic map constructed with 177 markers facilitated comparative analysis with tomato. Interval mapping with an LOD threshold of 3.0 identified a total of 55 QTLs related to traits such as fruit diameter, weight, pericarp thickness, and pedicel diameter. Notably, chromosome 3 harbored QTLs associated with most of the evaluated traits in this population (Ben Chaim et al. 2001). Among these, two major QTLs for fruit weight, fw2.1 and fw4.2, have been found to be conserved across three *Capsicum* species (*C. annuum*, *C. frutescens*, and *C. chinense*). The fw2.1 locus exhibited the most significant effect in multiple populations (Ben Chaim et al. 2001; Rao et al. 2003; Zygier et al. 2005). Total fruit yield in pepper is influenced by factors such as the NFP and fruit weight. Several studies have reported clustering of QTLs for yield with those for fruit traits, particularly in the same region of chromosome 2 (Ben Chaim et al. 2001; Rao et al. 2003; Zygier et al. 2005; Barchi et al. 2009; Chunthawodtiporn et al. 2018). Similarly, QTLs for pericarp thickness, which is positively correlated with fruit weight, are often co-located with fruit weight QTLs in common genomic regions (Ben Chaim et al. 2001; Yarnes et al. 2013; Naegele et al. 2014).

Parameters most commonly studied for fruit shape include fruit weight, FL, and fruit shape index (length/width ratio). Two major QTLs, fs3.1 and fs10.1, have been identified as key regulators of fruit elongation (Ben Chaim et al. 2001; Ben Chaim et al. 2003; Borovsky and Paran 2011) (Table 5.2). Fruit shape QTLs have been mapped to chromosomes 1, 3, 4, and 10 (Ben Chaim et al. 2001; Ben Chaim et al. 2003; Naegele et al. 2014; Dwivedi et al. 2015; Han et al. 2016). To date, no genes underlying natural fruit shape variation in pepper have been cloned or functionally characterized. The color of ripe pepper fruits is primarily determined by carotenoids, while anthocyanins and chlorophyll dictate the color of immature fruits (Reference). Major QTLs regulating unripe fruit color have been identified on chromosome 10 (Brand et al. 2012; Eggink et al. 2014; Han et al. 2016).

Table 5.2 Catalogue of QTLs discovered for plant growth and fruit characteristics in *Capsicum*

Trait	Type of mapping population used	Method of QTL analysis	No. of QTLs	Major QTLs identified	References
Axis growth speed	Recombinant Inbred Lines (RILs)	Interval Mapping (IM) and Composite Interval Mapping (CIM)	4		Barchi et al. (2009)
Internode growth time			2		
Internode length			3		
Primary axis length			5		
Flowering time			5		
Fruit weight			7		
Fruit length			4	*Frl4.1*	
Fruit diameter			12		
Fruit shape index			7		
Pericarp thickness			8		
Number of locules			9	*NloLG25.1*	
Pedicel length			6		
Internode length	Recombinant Inbred Lines (RILs)	Composite Interval Mapping (CIM)	5		Han et al. (2016)
Main stem length			6		
Plant height			6		
Plant width			2		
Lateral branch number			2		
Fruit weight			6		
Fruit diameter			5	*FD-1*, *FD-3.2*	
Fruit length			6	*FL-3.1*, *FL-3.2*, *FL-3.3*	
Fruit shape index			4	*FS-3.1*, *FS-3.2*	
Immature fruit color			6	*IFC-10.2*	
Fruit position			6		
Plant Height	Recombinant inbred line (RILs)	Composite Interval Mapping (CIM)	1	*Qpht. iivr.5.1*	Dwivedi et al. (2015)
Number of fruits/ plants			1		
Ten fruit weight			1	*Qtfw. iivr-2.1*	
Total fruit weight			3	*Qtofw. iivr-1.1*	
Fruit weight			1	*Qfw.iivr-2.1*	
Pericarp thickness			1	*Qpt.iivr-2.1*	
Fruit length			2	*Qfl.iivr.3.2*	

(continued)

Table 5.2 (continued)

Trait	Type of mapping population used	Method of QTL analysis	No. of QTLs	Major QTLs identified	References
Branching	Recombinant inbred line (RILs)	Composite Interval Mapping (CIM)	6		Yarnes et al. (2013)
Flowering time			8	*2.6, 2.8*	
Pericarp thickness			10	*4.4*	
Pericarp area			8		
Fruit shape index			51	*2.5, 2.6, 2.8, 2.10, 4.4, 11.4*	
Capsaicinoid			12	*4.2, 4.14, 4.15*	
Fruit Weight	Backcross populations	Interval Mapping (IM)	3	*LG1_8*	Eggink et al. (2014)
Fruit length			4	*LG10.1*	
Fruit Width			7	*LG1_8 &LG9*	
Fruit shape index			5	*LG1_8*	
Fruit color ripe			3		
Fruit color unripe			3	*LG10.1*	
Biochemical composition				*LG1_8, LG10.1*	
Fruit Size					Chaim et al. (2001)
Fruit Shape Index					
Fruit Shape Index					Chaim et al. (2003a, b)
Fruit Weight					Chaim et al. (2006)
Capsaicin					
Dihydrocapsaicin					
Nordihydrocapsaicin					
Total Capsaicinoids					
Trichome density			11	*Ptel1, Ptel2, Ptel9*	Kim et al. (2011)
Flowering time			1	Flw2.1	Borovsky et al. (2015)
Fruit shape index			3	*fs1.1, fs10.1*	Borovsky and Paran (2011)
Number of leaves			2	*Nle1.1, Nle12.1, Nle2.1*	Mimura et al. (2010) and Tan et al. (2015)
Fruit weight			3	*fw2.1, fw4.1, fw4.2*	Zygier et al. (2005)
Fruit shape index			3	*fs4.2*	
Fruit weight			16		Nimmakayala et al. (2016a, b)
Capsaicinoids			14		

(continued)

Table 5.2 (continued)

Trait	Type of mapping population used	Method of QTL analysis	No. of QTLs	Major QTLs identified	References
Pericarp thickness			1	*3.1*	Naegele et al. (2014)
Fruit shape index			5	*4.1*	
Fruit firmness			1	*12.1*	
Fruit length			5	*3.1*	Lee et al. (2011)
Capsaicin			3	*Qcap3.1, qcap6.1*	Lee et al. (2016)
Dihydrocapsaicin			2	*Qhdc2.1, qhdc2.2*	Lee et al. (2016)
Capsaicinoid			1	*cap*	Blum et al. (2003)
Metabolites			279		Wahyuni et al. (2014)
Chlorophyll content			2	*Pc8.1*	Brand et al. (2012)
Fruit orientation			1	*Up12.1*	Cheng et al. (2016)
Post-harvest water loss			3	*PWL10.1, PWL10.2*	Popovsky-Sarid et al. (2017)
Pungency	F_2 population	Composite interval mapping (CIM)	24	*Pun2_PJ_Gibbs_3.11*	Yi et al. (2022)
Restorer of fertility				*Rf2*	
Cytoplasmic male sterility	Near-isogenic lines	–	2	*atp6-2*, *$coxII_{708}$*	Kim and Kim (2005)
	F_2, F_3 and F_4 generations	–	1	*orf507*	Gulyas et al. (2010)
	CMS and fertile lines	–	3	*orf507, atp6 and accD-U*	Ji et al. (2014)
Nuclear male sterility	F_2 population	Composite interval mapping (CIM)	1	*ms8*	Bartoszewski et al. (2012)
	F_2 population	Composite interval mapping (CIM)	1	*ms1*	Jeong et al. (2018)
	F_2 population, Near-isogenic lines NILs	Bulk segregant analysis	1	*ms3*	Naresh et al. (2018)

Molecular markers have played a crucial role in selecting for pungency and capsaicinoid content in peppers. Non-pungency is primarily governed by the recessive pun1 gene, which encodes an acyltransferase enzyme essential for capsaicinoid biosynthesis (Stewart et al. 2005). Additional recessive alleles, pun12 and pun13, have been identified in *Capsicum chinense* and *Capsicum frutescens*, respectively (Stewart et al. 2007; Stellari et al. 2010). Loss-of-function mutations in the pAMT gene lead to non-pungent peppers that produce capsinoids, which are capsaicinoid analogs with reduced pungency (Lang et al. 2009; Tanaka et al. 2014). Several studies have focused on developing molecular markers linked to the Pun1 gene to facilitate marker-assisted selection (Garcés-Claver et al. 2007; Truong et al. 2009; Rodriguez-Maza et al. 2012; Wyatt et al. 2012). Blum et al. (2003) performed QTL analysis for capsaicinoid content and mapped a major QTL, cap, on chromosome 7, explaining 34–38% of the phenotypic variation. Ben Chaim et al. (2006) identified six QTLs controlling capsaicinoid levels, including cap3.1, cap4.1, cap4.2, cap7.1, and cap7.2 for capsaicin, as well as dhc4.1, dhc4.2, dhc7.1, and dhc7.2 for dihydrocapsaicin. Paran et al. (2010) confirmed the effect of cap7.1 and discovered a second QTL, cap7.2, on chromosome 7 in a *C. frutescens* × *C. annuum* cross. Additionally, in the *Perennial* × *Maor* cross, cap8.1 on chromosome 8 was identified alongside cap7.1. A genome-wide association study by Reddy et al. (2014) further suggested that capsaicinoid content is strongly influenced by the expression of Pun1.

Initially, due to limitations in molecular biology techniques and the absence of a reference genome, pepper fertility restoration (‘*Rf*’) genes were only roughly mapped. Various studies applied methods such as bulked segregant analysis (BSA), AFLP, and other molecular markers to identify major and minor QTLs associated with fertility restoration on chromosomes 1, 2, 3, 5, and 6 (Zhang et al. 2000; Wang et al. 2004; Kim et al. 2006; Lee et al. 2008; Yang et al. 2010; Jo et al. 2010). With advances in research and the publication of the pepper reference genome, the ‘Rf’ gene was consistently mapped to chromosome 6 in later studies (Qin et al. 2014; Kim et al. 2014). Using comparative genomics, Kompetitive Allele-Specific PCR (KASP) markers, and high-throughput sequencing, researchers identified several candidate genes potentially involved in fertility restoration, including CaPPR6, Rfu, Capana06g002967, Capana06g002969, CaRf, and CaRf032 (Jo et al. 2016; Wu et al. 2019; Cheng et al. 2020; Zhang et al. 2020a, b, c, 2021; Kang et al. 2022). Despite these advances, functional cloning and definitive characterization of the ‘Rf’ gene in pepper CMS lines remain unreported to date. In a related study on microspore embryogenesis in *Capsicum* species, 142 RILs were analyzed to understand the genetic factors influencing this trait. Using a genetic map, QTLs associated with microspore embryogenesis were mapped, revealing that this trait is quantitatively inherited. Seven QTLs were identified across linkage groups LG2, LG6, LG8, LG11, and LG12 (Bai et al. 2023).

5.5.2 QTL/Genes Identified for Resistance to Biotic Stresses Affecting *Capsicum* Genera

Among biotic stresses, *Capsicum* is affected by a wide range of pathogens, including fungi such as *Phytophthora capsici*, *Rhizoctonia solani*, *Verticillium dahliae*, *Colletotrichum scovillei* and *C. truncatum*, *Leveillula taurica*, and various *Fusarium* species; bacteria including *Xanthomonas* spp.; viruses like Tospoviruses (e.g., Tomato spotted wilt orthotospovirus and Impatiens necrotic spot orthotospovirus), Potyviruses (e.g., Potato virus Y, Tobacco etch virus, Pepper mottle virus), Tobamoviruses (e.g., Tobacco mosaic virus, Tomato mosaic virus), Cucumoviruses (e.g., Cucumber mosaic virus); as well as nematodes such as *Meloidogyne* spp. and insect pests including mites, aphids, Lepidoptera, and thrips. Given the growing demand for sustainable agriculture, resistance breeding remains the most practical and eco-friendly approach to developing pepper varieties and hybrids resistant to these devastating pathogens. The recent availability of the whole-genome sequence of *Capsicum annuum* (Qin et al. 2014) has greatly accelerated pepper breeding programs. Molecular tools such as molecular markers and QTL mapping, combined with high-throughput NGS and genotyping technologies, enable the identification of numerous genes and QTLs associated with economically important traits. These discoveries facilitate molecular breeding approaches, including marker-assisted selection and gene pyramiding, to develop disease-resistant cultivars.

Furthermore, several global databases provide extensive collections of *Capsicum* germplasm as sources of resistance genes against various biotic stresses. Major repositories include the Chile Variety Database (Chile Database 2018), the National Plant Germplasm System (NPGS) Germplasm Collection Genebanks of USDA-ARS (Peppers 2018), the World Vegetable Center Database (AVGRIS 2018), the Center for Genetic Resources, the Netherlands (CGN) at Wageningen University (CGN 2018), and the National Bureau of Plant Genetic Resources (India) (PGR Portal 2018). Table 5.3 summarizes the known QTLs/genes and their associated molecular markers conferring resistance to fungal, bacterial, viral, and root-knot nematode diseases in pepper.

5.5.3 QTLs/Genes Identified for Tolerance to Abiotic Stresses Affecting *Capsicum* Genera

Research on identifying QTLs and genes associated with abiotic stress tolerance in the *Capsicum* genus has been crucial for elucidating the genetic mechanisms underlying stress responses in these plants. Abiotic stresses such as drought, salinity, extreme temperatures, and nutrient deficiencies significantly impact *Capsicum* species by affecting their growth, productivity, and overall resilience. Furthermore, climate change is intensifying these threats, posing additional challenges to crop stability and yield. Numerous studies have sought to uncover the genetic components conferring tolerance to these stressors. By employing molecular techniques including genome-wide association studies (GWAS), QTL mapping,

Table 5.3 QTLs/genes and associated markers having a role in conferring biotic stress tolerance in pepper

Diseases		Resistance genes	Chromosomal location of the gene	Molecular markers developed	References
Fungal					
	Anthracnose	QTLs (*CaR12.2* and *CcR9*) pathogen pathotypes: *Colletotrichum acutatum* and *Colletotrichum capsici*	LG 12 (6.1 cM position), LG 9 (10.0 cM position)	SSR	Lee et al. (2010)
		QTLs (*AnRGO5, AnRGT5,AnRGD5, AnRRO5, AnRRT5, AnRRD5*), pathogen pathotypes: *Colletotrichum acutatum*	5	In Del, SSR, SCAR, CAPS	Sun et al. (2015)
		co1, co, co4 and *Co5*	2, 12, and 9	SNP	Mahasuk et al. (2009)
		*Major QTLs-(An8.1, An9.1) Minor QTLs-(An7.3, An7.4,An4.1, An3.1, An3.2),*Pathogen pathotypes: *Colletotrichum acutatum*			Kim et al. (2010)
	Phytophthora blight	*QTL, Pc5.1*	5	SSR, SCAR	Mallard et al. (2013)
		CaPhyto	5	SSR	Wang et al. (2016)
Bacterial	Bacterial wilt	*QTL Bw1 QTL*	1	SSR	Mimura et al. (2009)
qRRs-10.1	10	QTL *qRRs*-10.1	10	InDel and SNP	Du et al. (2019)

	Bacterial leaf spot	*Bs1, Bs2, Bs3, Bs4, bs5, bs6 and Bs7*	*Bs2, bs6* (chromosome 6)	RAPD, SCAR, CAPS	Wai et al. (2015), Jones et al. (2004), Csillery et al. (2004), Tai et al. (1999), Pierre et al. (2000), and Vallejos et al. (2010)
Viral	Chilli leaf curl disease	*ChiLCVD*-resistant gene	6	SSR	Thakur et al. (2019)
	Pepper Yellow leaf curl virus	*QTL peplcv-1, peplcv-7, and peplcv-12*	1,7,12	SSR	Siddique et al. (2022)
	Cucumber Mosaic virus	*Cmr1, cmr2*	2	SNP	Kang et al. (2010)
		qCmr2.1 (CA02g19570)	2	SNP	Guo et al. (2017a, b)
	Potyviruses	*pvr1 (eIF4E), pvr2,pvr3, Pvr4 (CaNBARC322), pvr5, pvr8,pvr6(eIF(iso)4E) ,Pvr7*	*pvr1, pvr2* (4), *Pvr4* (10), *pvr6* (3), *Pvr7* (10)	CAPS, SCAR, SSR, SNP	Kyle and Palloix (1997), Yeam et al. (2005), Ruffel et al. (2002), Dato et al. (2015), Caranta et al. (1999), Andres et al. (2004), Venkatesh et al. (2018), and Arnedo-Andres et al. (2002)

(continued)

Table 5.3 (continued)

Diseases		Resistance genes	Chromosomal location of the gene	Molecular markers developed	References
Nematodes	Root-knot disease	*Me Loci, Me1, Mech2, Me3, Me4, Me 7*	1, 9	CAPS, SCAR, SNP, SSR	Djian-Caporalino et al. (2007), Uncu et al. (2015), Wang et al. (2018), Barbary et al. (2016), Pinar et al. (2016), Guo et al. (2017a, b), and Changkwian et al. (2019)

transcriptomics, and functional genomics, researchers have identified key genomic regions and specific genes linked to abiotic stress tolerance in *Capsicum* species (see Table 5.4). Among the various abiotic stresses, the increasing frequency of drought and elevated temperatures represents imminent threats to global agricultural productivity. For example, Armita et al. (2017) reported that drought stress in *C. frutescens* caused a decline in chlorophyll content and an increase in secondary metabolites, particularly carotenoids, resulting in prolonged disruption of metabolic activity. Notably, *C. chinense*, characterized by higher accumulation of secondary metabolites and enhanced antioxidant activity, demonstrated greater drought resilience compared to *C. annuum* and *C. frutescens*. Drought negatively impacts *Capsicum* plants at all growth stages; however, the vegetative stage is particularly vulnerable, demanding greater research focus relative to flowering or fruiting phases (Okunlola et al. 2017).

Examining the expression of stress-responsive genes such as Ca-DREBLP1 and CaGLIP1 highlights their vital roles in conferring drought resistance in *Capsicum.* Notably, Ca-DREBLP1 expression significantly increases under dehydration and salt stress conditions (Hong and Kim 2005). Similarly, the overexpression of CaGLIP1 in *Arabidopsis* enhances tolerance to drought and glucose stress, suggesting its potential role in stress adaptation. Several genomic regions and candidate genes associated with drought tolerance traits have been identified in *Capsicum* (Table 5.4). These genes are likely involved in regulating water uptake and retention or participate in stress-responsive pathways such as abscisic acid (ABA) signaling and osmotic adjustment. Temperature fluctuations, including both heat spikes and cold stress, detrimentally affect pepper fruit set and development. Genes or genomic regions linked to heat shock proteins, cold-responsive genes, and regulatory elements governing temperature stress responses are key targets for breeding programs. Table 5.4 summarizes major genes implicated in heat tolerance within the *Capsicum* genus. Recent studies exploring the effects of temperature and exposure duration on chili pepper germination and seedling growth revealed a strong correlation where increased temperatures and prolonged exposure significantly reduced shoot length, root length, root number, seedling weight, leaf count, and chlorophyll content. Elevated temperatures during germination markedly impaired seedling development and physiological parameters. Furthermore, QTL mapping for heat tolerance in hot pepper identified 64 QTLs with LOD scores ranging from 2.517 to 11.170, explaining 4.05% to 19.39% of phenotypic variance. Among these, 21 major QTLs accounted for over 10% of the variance and controlled nine morphological, three physiological, and two biochemical traits (TS et al. 2023).

Soil salinity is another major abiotic stress limiting pepper growth and productivity. *Capsicum annuum* is notably salt-sensitive, and salinity adversely affects its development and yield. Butt et al. (2021) evaluated the impact of five NaCl salinity levels on growth, biochemical traits, and yield in two chili genotypes. They observed reductions in fresh and dry weight, relative water content, water use efficiency, leaf osmotic potential, glycine betaine content, photosynthetic rate, transpiration rate, stomatal conductance, and chlorophyll content under salinity stress. Salinity stress also elevates malondialdehyde levels and activates antioxidant enzymes such as

Table 5.4 Summary of genes reported to be involved in abiotic stresses in capsicum

S. No.	Genes identified (name and symbol)	Description (function)	Abiotic stress	References
1	Dehydrin	ROS signaling pathway	Osmotic stress	Chung et al. (2003)
2	UBI-3, β-TUB, and GAPDH	Transportation of ubiquitin from the ubiquitin-activating enzyme (E1) to the ubiquitin-ligase enzyme (E3) and substrate TUB has a role in structural support, a pathway for transport, and force generation in cell division GAPDH is an important enzyme for energy metabolism and the production of ATP and pyruvate through anaerobic glycolysis in the cytoplasm	Heat stress	Wan et al. (2011)
3	Genes of the CCCH Zinc finger proteins (ZFPs) family	Sequence-specific transcription factors involved in the regulation of physiological adversity in abiotic stress	Heat, drought, cold, salt, and osmotic	Maurya et al. (2014)
4	CaDHN3. Dehydrin gene	Regulation of osmotic stress responses through the ROS signaling pathway	Drought and salt stress	Meng et al. (2021) and Jing et al. (2016)
5	BiP genes ((CaBiP1, CaBiP2, and CaBiP3))	*Protective role by alleviating endoplasmic reticulum (ER) stress*	Heat, salinity, osmotic, and drought stress	Wang et al. (2017)
6	Hsp70 and sHsp	Assist a large variety of protein folding processes	Heat stress	Magaji et al. (2018)
7	Genes for CaAQPs (aquaporin proteins)	Plant growth and development regulate plant water homeostasis	Drought Stress	Sahitya et al. (2019)
8	CaMADS, a MADS-box	*Regulation of plant tolerance to extreme conditions*	cold, salt, and osmotic	Chen et al. (2019)
9	Pepper bHLH protein genes	Roles in stress tolerance, reproduction (flower and fruit development), and in the biosynthesis of secondary metabolites	Heat, drought, salt, and cold	Zhang et al. (2020a, b, c)

(continued)

Table 5.4 (continued)

S. No.	Genes identified (name and symbol)	Description (function)	Abiotic stress	References
10	*CaNAC035* NAC Transcription Factor	Regulator of many signaling pathways	Role in tolerance to cold, salt, and drought stress tolerance	Zhang et al. (2020a, b, c)
11	CaChiVI2	*Cytokinin induced F-Box protein* carries heat shock elements, a decisive role in stress tolerance	Heat, drought	Ali et al. (2020)
12	PMEI13, TUB8, EXPA13, XK-1, PME1, CEL5, CSLE1), KASI, LACS2), (CDC2), (BASS1), (SnRK2.10, TINY2 and ERD4) (PSBP-1)	Genes related to cell wall biosynthesis and expansion Wax and fatty acid biosynthesis Cell division Vitamin transport ABA-signaling Photosynthesis	Salt stress	López-Serrano et al. (2021)
13	Heat Shock Proteins and Heat Transcription Factors	Heat stress recovery responses	Heat stress	Wang et al. (2021)
14	β-carotene hydroxylase (*CaCA1*, *CaCA2*), zeaxanthin epoxidase (*CaZEP*), and violaxanthin de-epoxidase (*CaVDE*)	Promotion of endogenous zeaxanthin synthesis	Cold stress	Tang et al. (2021)
15	Transcription factor GAMYB-like, Ethylene-responsive transcription factor RAP2-7, Dof zinc finger protein DOF3.1-like, WRKY transcription factor 21	Gene expression regulation and various metabolic pathways	Heat, cold, osmotic, and salinity stress	Baruah et al. (2021)

superoxide dismutase (SOD), catalase (CAT), and peroxidase (POD) (Abogadallah 2010). The genetic basis of salinity tolerance in *Capsicum* involves genes and QTLs regulating ion homeostasis, osmotic adjustment, ion transport, and antioxidant defense systems that mitigate oxidative damage caused by high salt concentrations. Key genes, including CaDREBLP1, CaRMa1H1, CaKR1, CaOSM1, CaPROX1, and CaPIP2, have been implicated in plant adaptation to salt stress, suggesting their roles in diverse physiological and molecular processes essential for stress tolerance (Maurya et al. 2014).

5.6 Marker-Assisted Selection (MAS) in Pepper Breeding

Marker-assisted selection (MAS) has significantly advanced pepper (*Capsicum* spp.) breeding by providing a precise and efficient method for selecting plants with desirable traits. MAS involves the use of molecular markers that are closely linked to target genes or QTLs, allowing breeders to track important traits throughout the breeding process. One of the key benefits of MAS is the ability to identify and select plants with favorable traits at early developmental stages, which accelerates the breeding cycle. In pepper breeding, MAS has been particularly valuable in several areas: Disease resistance is a major focus, as *Capsicum* species are susceptible to a wide range of pathogens, including bacteria, fungi, and viruses. MAS allows for the accurate selection of plants carrying resistance genes or QTLs, supporting the development of varieties resistant to diseases such as bacterial spot, Phytophthora blight, and various viral infections. Yield improvement is another critical goal. By identifying QTLs associated with high yield, MAS enables the selection of genotypes with greater productivity, which is especially important in regions where pepper is an economically vital crop. Fruit quality traits, such as size, color, flavor, and texture, are also essential for marketability and consumer preference. MAS facilitates the selection of plants with improved quality traits, contributing to the development of varieties that meet commercial and consumer standards. In addition, MAS enables the introgression of beneficial traits from wild relatives into cultivated varieties. This enhances genetic diversity and adaptability, helping plants better withstand climate change, pests, and new disease pressures. Overall, MAS is a powerful tool in modern pepper breeding, enabling the efficient and targeted development of improved cultivars with enhanced resistance, yield, and quality.

MAS has become feasible for traits governed by both major genes and QTLs. The effectiveness of a molecular marker depends on its ability to detect polymorphisms in nucleotide sequences, enabling the distinction between different alleles. These polymorphisms can be identified using various molecular techniques, including RFLP, AFLP, microsatellites or SSR, RAPD, CAPS, single-strand conformation polymorphism (SSCP), and SNPs (Khlestkina 2014). The application of MAS has been demonstrated in breeding new cultivars containing capsinoids. Tanaka et al. (2014) studied the use of DNA markers to identify genotypes with capsinoid biosynthesis genes, which are preferred due to their lower pungency compared to capsaicinoids. MAS has also been employed to develop resistance to potato virus Y (PVY) in *Capsicum annuum*. Local germplasm carrying recessive resistance genes was used to develop PVY-resistant pepper lines, and allele-specific amplification using the ARMS-PCR method enabled the detection of resistant and susceptible alleles in F_1 and F_2 generations (Moodely et al. 2019). More recently, Na Jinda et al. (2023) applied MAS in the backcross breeding of male sterile and maintainer lines. Their objective was to develop a novel maternal (A-line) and maintainer (B-line) pair by introducing male sterility and corresponding maintainer genes from a donor variety into restorer lines (C-lines), thus enabling efficient hybrid seed production. In conclusion, advancements in molecular markers and genetic mapping within the *Capsicum* genus mark significant progress in plant genetics and breeding. These tools have enhanced our understanding of genetic diversity and have provided breeders with powerful methods for the efficient and precise development of improved pepper varieties.

5.7 Genome-Wide Association Studies in *Capsicum* Genera

GWAS is a powerful research method used to identify genetic variations statistically associated with specific traits or diseases. In the *Capsicum* genus, GWAS has become a pivotal tool for understanding the genetic basis of various traits in this phenotypically diverse group of plants. *Capsicum* includes a wide range of pepper species exhibiting extensive variability in traits such as fruit size, color, flavor, and resistance to abiotic and biotic stresses. GWAS in peppers involves scanning the entire genome to detect genetic variants linked to traits of interest. By analyzing large populations along with genotypic data, researchers can identify genomic regions or loci associated with key characteristics, thereby providing valuable insights into the genetic architecture underlying phenotypic diversity. One of the major applications of GWAS in *Capsicum* has been the investigation of capsaicinoid content the compounds responsible for pungency in chili peppers. Kim et al. (2014) identified specific genomic regions involved in capsaicinoid biosynthesis, offering a deeper understanding of the molecular mechanisms controlling this trait.

Similarly, GWAS has been employed to study fruit-related traits such as fruit weight, size, and shape in *Capsicum annuum*. The study by Qin et al. (2014) provided critical genetic information for breeding programs focused on improving fruit quality and yield. Another significant contribution came from Hulse-Kemp et al. (2018), who used GWAS to uncover genomic regions linked to resistance against the pathogen *Phytophthora capsici*. Their research identified candidate genes involved in the plant's disease response, highlighting potential targets for developing disease-resistant pepper varieties. The emergence of advanced sequencing technologies, including NGS and high-throughput genotyping, has greatly enhanced the depth and scope of GWAS in *Capsicum* species, making it a vital approach in modern pepper breeding and genetics research.

GWAS have been extensively applied in *Capsicum* research to identify genetic variants linked to various economically important traits. One notable study conducted by Ro et al. (2023) identified SNPs associated with anthracnose resistance. Using a dataset of 53,518 SNPs, they found 25 and 32 SNPs significantly associated with resistance based on phenotypic data from field and in vitro experiments, respectively. GWAS was also performed on a diverse pepper germplasm collection to discover novel sources of resistance against the highly virulent *Xanthomonas gardneri* strain. This study identified significant SNPs associated with defoliation in response to *X. gardneri* infection (Potnis et al. 2019). The functionally relevant candidate genes identified in this analysis were involved in disease resistance/susceptibility, hormone signaling, and basal resistance mechanisms against multiple pathogens. In another study, GWAS was utilized to pinpoint SNPs associated with resistance to *Phytophthora capsici* (isolate KCP7). A significant SNP, "Chr02-1126," located at 112 Mb on chromosome 2, was found to predict *P. capsici* resistance with 78.5% accuracy. Similarly, the QTL5-1 marker on chromosome 5 predicted resistance with 80.2% accuracy. The combination of these markers could be employed to enhance prediction accuracy for *Phytophthora* blight resistance in pepper genetic resources (Ro et al. 2022). Han et al. (2018) identified five QTL regions consistently associated with capsaicin synthesis located on chromosomes 1, 2, 3, 4, and 10. These regions were validated across RIL populations and multiple environments. Within these regions, five candidate genes involved in capsaicinoid biosynthesis were identified, further strengthening the genetic basis for trait improvement. Overall, GWAS has been pivotal in advancing our understanding of the genetic architecture of key traits in the *Capsicum* genus, including pungency, yield, and resistance to biotic and abiotic stresses. These insights provide indispensable tools for breeding programs aiming to enhance disease resistance, fruit quality, and agronomic performance. In the context of climate change, GWAS will continue to serve as a valuable resource for identifying diverse genotypes and mining novel QTLs and genes for crop improvement.

5.8 Genome Sequencing of *Capsicum* Species and Its Utility

Genome sequencing efforts within the *Capsicum* genus represent a major advancement in deciphering the genetic architecture of pepper species. These initiatives have significantly contributed to understanding trait inheritance and have opened new avenues for targeted crop improvement. The *Capsicum* genus, encompassing diverse pepper species, has been the focus of extensive genomic studies aimed at uncovering genetic diversity, genome structure, and the molecular mechanisms controlling important agronomic traits such as pungency, fruit color, and disease resistance. The sequencing of *Capsicum annuum*, one of the most widely cultivated pepper species, has been particularly transformative. In 2014, Kim et al. reported the complete genome sequences of *C. annuum* accession CM334 and *C. chinense* accession PI159236, with estimated genome sizes of 3.48 Gb and 3.14 Gb, respectively. These assemblies enabled valuable interspecies comparative analyses. The same study also included the resequencing of two cultivated *C. annuum* varieties, Perennial and Dempsey adding further depth to intra-species genomic variation. In parallel, Qin et al. (2014) published the reference genome of Zunla-1 (*C. annuum* L.), a cultivated variety, and its wild progenitor Chiltepin (*C. annuum* var. *glabriusculum*). Their work predicted 35,336 and 34,476 protein-coding genes in Zunla-1 and Chiltepin, respectively, offering a valuable resource for evolutionary and functional genomic studies.

Subsequent advancements included the genome sequencing of *C. baccatum* PBC81 and updated, high-quality assemblies of CM334 and *C. chinense* PI159236 (Kim et al. 2017). Notably, the *C. baccatum* genome showed a substantial expansion due to the proliferation of *Athila* elements, which were found at more than twice the abundance compared to other species. This contributed to species-specific genome size increase in the *C. baccatum* lineage. These genome sequencing projects have facilitated the identification of candidate genes linked to critical traits, including heat level, fruit quality, and resistance to pests and diseases. A deeper understanding of the genetic basis of these traits enables breeders to implement more efficient and precise strategies for the development of improved cultivars, ultimately contributing to sustainable agriculture and food security.

In the genomic exploration of pepper, a total of 16,956 gene families comprising 22,885 genes were identified from 34,447 predicted protein-coding sequences in the study by Kim et al. (2014). Similarly, the genome analysis by Qin et al. (2014) revealed 16,770 gene families encompassing 26,444 genes out of 35,336 predicted protein-coding sequences. Among these, over 2000 genes were classified as transcription factors and transcription regulators, representing approximately 6.25% of the total gene content. These were grouped into 80 distinct families, notably including 73 *WRKY* and 106 *NAC* gene family members, both of which are critical to plant development and stress response pathways. Gene expression analysis of the *NAC* family indicated that approximately 30% of these genes exhibited high expression levels during fruit development, with 13% peaking in developing fruits. Furthermore, 123 genes belonging to the *AP2/ERF* superfamily; key regulators in cellular response, growth, and developmental processes were identified. This superfamily was predominantly composed of the *ERF* subfamily (80%), followed by *AP2* (17%) and *RAV* (1.6%). Additional gene families involved in developmental processes included flowering genes, which influence flowering time and shoot architecture, and cuticle biosynthesis genes that play a protective role against abiotic and biotic stresses by regulating water retention and gas exchange. Pepper genomes also harbor a wide array of functional gene families related to growth, defence, and physiological regulation. These include RNA-binding proteins, auxin response factors (*ARFs*), 867 receptor-like kinase genes, 684 nucleotide-binding site (NBS) genes, and members of the glycoside hydrolase gene family.

One practical application of genome sequencing in the *Capsicum* genus is the acceleration of traditional breeding programs through MAS. Molecular markers linked to specific traits identified via genome sequencing enable breeders to efficiently screen and select plants possessing desirable characteristics. This targeted selection significantly reduces the time and resources required to develop improved pepper varieties with traits such as enhanced pungency, disease resistance, and other agronomically important features. In addition, the insights gained from genome sequencing have catalyzed progress in functional genomics, facilitating a deeper understanding of gene function and regulation. This knowledge supports the precise modification of target genes through genetic engineering, thereby enabling the development of genetically modified (GM) pepper lines with improved attributes, including extended shelf life, enhanced nutritional value, and customized flavor profiles.

In conclusion, genome sequencing efforts in the *Capsicum* genus have transformed the landscape of pepper breeding and cultivation. The practical applications of genomic information span from the refinement of conventional breeding techniques to the advancement of genetic engineering strategies, offering a comprehensive approach to crop improvement. As our understanding of *Capsicum* genomics deepens, the potential to develop resilient, high-yielding, and nutritionally enriched pepper varieties continues to grow, contributing to a more sustainable and productive future for global pepper agriculture.

5.9 Conclusions, Future Trends, and Perspectives

The application of QTL mapping and GWAS in the *Capsicum* genus has significantly enhanced our understanding of the genetic architecture underlying key traits in pepper plants. These molecular approaches have provided critical insights into complex traits such as pungency, fruit quality, disease resistance, and overall productivity. QTL mapping enables the identification of genomic regions associated with specific phenotypic traits, thereby revealing the genetic factors responsible for trait variability within *Capsicum*. This information has proven highly valuable in conventional breeding programs, supporting the development of improved pepper varieties. By integrating QTL mapping into breeding strategies, researchers have accelerated the selection of elite lines, enhancing both the efficiency and precision of crop improvement.

GWAS, in contrast, offers a broader genome-wide perspective, identifying associations between phenotypic traits and genetic markers across diverse populations. This approach has uncovered the polygenic nature of many traits in *Capsicum*, revealing that multiple loci often contribute to the expression of a single trait. The insights gained from GWAS have provided a more comprehensive view of trait architecture, facilitating a more informed and strategic approach to breeding. Looking forward, future trends in QTL and GWAS research in *Capsicum* are expected to be shaped by advancements in high-throughput genotyping technologies and the integration of multi-omics datasets. The adoption of NGS and high-density marker platforms will allow for more precise and exhaustive mapping of traits. Additionally, the incorporation of transcriptomic, epigenomic, and metabolomic data will yield a systems-level understanding of gene regulation and trait expression, enabling a more refined dissection of complex traits. These developments hold great promise for precision breeding, enabling the creation of tailor-made cultivars by targeting specific genomic regions associated with desirable traits. This approach will be vital for addressing the emerging challenges of climate change, evolving pathogens, and shifting consumer demands.

Furthermore, the combination of genomic data with advanced phenotyping technologies will facilitate the discovery of novel traits and deepen our understanding of genotype-environment interactions. Such insights will be instrumental in developing resilient *Capsicum* cultivars that are well-suited to dynamic agro-ecological conditions and contribute to global food security. In conclusion, QTL mapping and GWAS have ushered in a new era of precision breeding in the *Capsicum* genus, offering unprecedented insight into the genetic basis of important traits. The continued evolution of genomics technologies, coupled with multi-omics integration, will pave the way for the development of resilient, high-yielding, and consumer-oriented pepper varieties. As we continue to explore the complexities of the *Capsicum* genome, the synergy between molecular genetics and plant breeding will be central to shaping the future of sustainable pepper cultivation.

References

Abogadallah GM (2010) Antioxidative defense under salt stress. Plant Signal Behav 5(4):369–374. https://doi.org/10.4161/psb.5.4.10873. Epub 2010 Apr 7. PMID: 20118663

Adluri PK, Baldoldiya GM, Nath PD (2017) Screening of Bhut Jolokia (Capsicum chinense Jacq.) germplasm of North East India against chili leaf curl virus. Int J Pure Appl Biosci 5(4):1189–1196

Aguilar-Meléndez A, Morrell PL, Roose ML, Kim SC (2009) Genetic diversity and structure in semiwild and domesticated chiles (Capsicum annuum; Solanaceae) from Mexico. Am J Bot 96(6):1190–1202. https://doi.org/10.3732/ajb.0800155

Ali M, Muhammad I, Ul Haq S, Alam M, Khattak AM, Akhtar K, Ullah H, Khan A, Lu G, Gong ZH (2020) The CaChiVI2 gene of Capsicum annuum L. confers resistance against heat stress and infection of Phytophthora capsici. Front Plant Sci 11:219

Anaya-López JL, Torres-Pacheco I, González-Chavira M, Garzon-Tiznado JA, Pons-Hernandez JL, Guevara-González RG et al (2003) Resistance to geminivirus mixed infections in Mexican wild peppers. HortScience 38(2):251–255

Andrés A, Luis Arteaga M, Gil Ortega R (2004) New genes related to PVY resistance in C. annuum L. 'Serrano Criollo de Morelos-334'. In: Voorrips RE (ed) Proceedings of the XIIth EUCARPIA Meeting on genetics and breeding of Capsicum and eggplant, Noordwijkerhout, The Netherlands, 17–19 May 2004. Publisher Plant Research International, Wageningen, pp 134–138

Armita D, Arumyngtyas EL, Mastuti R (2017) Tolerance level of three genotypes of cayenne pepper (Capsicum frutescens L.) toward drought stress of vegetative phase based on morphological and physiological responses. Intl J Chem Tech Res 10(2):183–192

Arnedo-Andrés M, Gil-Ortega R, Luis-Arteaga M, Hormaza J (2002) Development of RAPD and SCAR markers linked to the Pvr4 locus for resistance to PVY in pepper (Capsicum annuum L.). Theor Appl Genet 105:1067–1074

Arumingtyas EL, Agustina BRE, Kusnadi J (2022) Developing RAPD-derived SCAR (Sequence Characterized Amplified Region) marker for flowering time in chili pepper. In: 7th international conference on biological science (ICBS 2021). Atlantis Press, pp 78–81

AVGRIS. The Asian Vegetable Center (ARVDC) (Taiwan) (2018) Available online: http://www.avrdc.org

Baba VY, Rocha KR, Gomes GP, de Fátima Ruas C, Ruas PM, Rodrigues R, Gonçalves LSA (2016) Genetic diversity of Capsicum chinense accessions based on fruit morphological characterization and AFLP markers. Genet Resour Crop Evol 63(8):1371–1381

Bahrami Rad M, Hassani ME, Mohammadi A, Lessan SH,Ghazi Zade S (2008) Evaluation of genetic diversity in Capsicum spp. as revealed by RAPD markers. In: VI international symposium on in vitro culture and horticultural breeding, vol 829, pp 275–278

Bai J, Chen B, Qi M, Du H, Wang Q, Geng S, Zhang X (2023) QTL mapping and candidate gene analysis of microspore embryogenesis in Capsicum. Euphytica 219:3

Balasankar D, Praneetha S, Armugam T, Manivannan N et al (2017) Assessment of salinity tolerance in chilli (Capsicum annuum L.) genotypes. Int J Chem Stud 5(6):1194–1198

Bartoszewski G, Waszczak C, Gawroński P, Stępień I, Bolibok-Brągoszewska H, Palloix A et al (2012) Mapping of the ms8 male sterility gene in sweet pepper (Capsicum annuum L.) onthe chromosome P4 using PCR-based markers useful for breeding programmes. Euphytica 186(2):453–461

Boiteux LS, de B. Giordano L (1993) Genetic basis of resistance against two Tospovirus species in tomato (Lycopersicon esculentum). Euphytica 71(1):151–154

Borah DC (1987) Bio-ecology of Polyphagotarsonemus latus (Banks) and Scirtothrips dorsalis (Hood) infesting chilli and their natural enemies. Ph. D. Thesis, University of Agricultural Sciences, Dharwad, p 234

Brand A, Borovsky Y, Meir S, Rogachev I, Aharoni A, Paran I (2012) pc8. 1, a major QTL for pigment content in pepper fruit, is associated with variation in plastid compartment size. Planta 235(3):579–588

Barbary A, Djian-Caporalino C, Palloix A, Castagnone-Sereno P (2015) Host genetic resistance to root-knot nematodes, *Meloidogyne* spp.; in *Solanaceae*: from genes to the field. Pest Manag Sci 71:1591–1598

Barbary A, Djian-Caporalino C, Marteu N, Fazari A, Caromel B, Castagnone-Sereno P, Palloix A (2016) Plant genetic background increasing the efficiency and durability of major resistance genes to root-knotnematodes can be resolved into a few resistance QTLs. Front Plant Sci 7:632

Barchenger DW, Lamour KH, Bosland PW (2018) Challenges and strategies for breeding resistance in Capsicum annuum to the multifarious pathogen, Phytophthora capsici. Front Plant Sci 9:628. https://doi.org/10.3389/fpls.2018.00628

Barchi L, Lefebvre V, Sage-Palloix AM, Lanteri S, Palloix A (2009) QTL analysis of plant development and fruit traits in pepper and performance of selective phenotyping. Theor Appl Genet 118(6):1157–1171. https://doi.org/10.1007/s00122-009-0970-0. Epub 2009 Feb 15. PMID: 19219599

Baruah PM, Krishnatreya DB, Bordoloi KS, Gill SJ, Agarwala N (2021) Genome wide identification and characterization of abiotic stress responsive lncRNAs in *Capsicum annuum*. Plant Physiol Biochem 162:221–236

Basu SK, De KA (2003) Capsicum: historical and botanical perspectives. CRC Press

Berzal-Herranz A, De La Cruz A, Tenllado F, Diaz-Ruiz JR, Lopez L, Sanz AI, Vaquero C, Serra MT, Garcia-Luque I (1995) The *Capsicum* L3 gene-mediated resistance against the tobamoviruses is elicited by the coat protein. Virology 209:498–505

Bhutia ND, Seth T, Shende VD, Dutta S, Chattopadhyay A (2015) Estimation of heterosis, dominance effect and genetic control offresh fruit yield, quality and leaf curl disease severity traits of chilli pepper (Capsicum annuum L.). Sci Hortic 182:47–55

Blum EM, Mazourek M, O'Connell J, Curry T, Thorup K, Liu M, Jahn M, Paran I (2003) Molecular mapping of capsaicinoid biosynthesis genes and quantitative trait loci analysis for capsaicinoid content in Capsicum. Theor Appl Genet 108:79–86

Borovsky Y, Paran I (2011) Characterization of fs10.1, a major QTL controlling fruit elongation in Capsicum. Theor Appl Genet 123(4):657–665. https://doi.org/10.1007/s00122-011-1615-7. Epub 2011 May 21. PMID: 21603875

Borovsky Y, Sharma VK, Verbakel H, Paran I (2015) CaAP2 transcription factor is a candidate gene for a flowering repressor and a candidate for controlling natural variation of flowering time in Capsicum annuum. Theor Appl Genet 128(6):1073–1082. https://doi.org/10.1007/s00122-015-2491-3. Epub 2015 Mar 8. PMID: 25748116

Bosland PW, Bailey AL, Iglesias-Olivas J (1996) Capsicum pepper varieties and classification new mexico state university cooperative extension service circular no. 530

Botstein B, White RL, Sholnick M, Davis RW (1980) Construction of a genetic linkage map in man using restriction fragment length polymorphisms. Am J Hum Genet 32:314–331

Butt M, Sattar A, Abbas T, Hussain R, Ijaz M, Sher A, Shahzad U, Ullah S, Brestic M, Zivcak M, Gasparovic K, Aljuaid BS, El-Shehawi AM, Zuan ATK (2021) Morpho-physiological and biochemical attributes of Chili (Capsicum annum L.) genotypes grown under varying salinity levels. PLoS One 4:16(11):e0257893. https://doi.org/10.1371/journal.pone.0257893

Caranta C, Thabuis A, Palloix A (1999) Development of a CAPS marker for the Pvr4 locus: a tool for pyramiding potyvirus resistance genes in pepper. Genome 42:1111–1116

Cardoso R, Ruas CF, Giacomin RM, Ruas PM, Ruas EA et al (2018) Genetic variability in Brazilian Capsicum baccatum germplasm collection assessed by morphological fruit traits and AFLP markers. PLoS One 13(5):e0196468

Cebolla-Cornejo J, Soler S, Gomar B, Soria MD, Nuez F (2003) Screening capsicum germplasm for resistance to tomato spotted wilt virus (TSWV). Ann Appl Biol 143(2):143–152
Celik I, Sogut MA, Özkaynak E, Doganlar S, Frary A (2016) Physical mapping of NBS-coding resistance genes to the *Me*-gene cluster on chromosome P9 reveals markers tightly linked to the N gene for root-knot nematode resistance in pepper. Mol Breed 36:137
CGN. The Centre for Genetic Resources, the Netherlands (CGN) of Wageningen University (2018) Available online: https://www.wur.nl/en/Research-Results/Statutory-research-tasks/Centre-for-Genetic-Resourcesthe-Netherlands-1/Expertise-areas/PlantGenetics
Chaim AB, Paran I, Grube RC, Jahn M, Van Wijk R, Peleman J (2001) QTL mapping of fruit-related traits in pepper (Capsicum annuum). Theor Appl Genet 102:1016–1028
Chaim AB, Borovsky Y, De Jong W, Paran I (2003a) Linkage of the A locus for the presence of anthocyanin and fs10.1, a major fruit-shape QTL in pepper. Theor Appl Genet 106(5):889–894. https://doi.org/10.1007/s00122-002-1132-9. Epub 2002 Oct 30. PMID: 12647064
Chaim AB, Borovsky Y, Rao GU, Tanyolac B, Paran I (2003b) *fs3.1*: a major fruit shape QTL conserved in Capsicum. Genome 46(1):1–9. https://doi.org/10.1139/g02-096. PMID: 12669791
Chaim AB, Borovsky Y, Falise M, Mazourek M, Kang BC, Paran I, Jahn M (2006) QTL analysis for capsaicinoid content in *Capsicum*. Theor Appl Genet 113:1481–1490
Changkwian A, Venkatesh J, Lee JH, Han JW, Kwon JK, Siddique MI et al (2019) Physical localization of the root-knot nematode (Meloidogyne incognita) resistance locus Me7 in pepper (Capsicum annuum). Front Plant Sci 10:886
Cheng J, Qin C, Tang X, Zhou H, Hu Y, Zhao Z et al (2016) Development of a SNP array and its application to genetic mapping and diversity assessment in pepper (capsicum spp.). Sci Rep 6(1):33293
Chen R, Ma J, Luo D, Hou X, Ma F, Zhang Y, Meng Y, Zhang H, Guo W (2019) CaMADS, a MADS-box transcription factor from pepper, plays an important role in the response to cold, salt, and osmotic stress. Plant Sci 280:164–174
Cheng J, Chen Y, Hu Y, Zhou Z, Hu F, Dong J, Chen W, Cui J, Wu Z, Hu K (2020) Fine mapping of *restorer-of-fertility* gene based on high-density genetic mapping and collinearity analysis in pepper (*Capsicum annuum* L.). Theor Appl Genet 133:889–902. https://doi.org/10.1007/s00122-019-03513-y
Chile Database.Resistance Sources from Chile Variety Database (2018) Available online: www.g6csy.net/chile/database.html. Accessed 12 Mar 2020
Chung E, Kim SY, Yi SY, Choi D (2003) *Capsicum annuum* dehydrin, an osmotic-stress gene in hot pepper plants. Mol Cell 15(3):327–332
Chunthawodtiporn J, Hill T, Stoffel K, Van Deynze A (2018) Quantitative trait loci controlling fruit size and other horticultural traits in bell pepper (Capsicum annuum). Plant Genome 11(1):160125
Csillery G, Szarka E, Sardi E, Mityko J, Kapitany J, Nagy B, Szarka J (2004) The unity of plant defense: genetics, breeding and physiology. In: Proceedings of the 12th Eucarpia meeting on genetics and breeding of capsicum and egg-plant, Noordwijkerhout, The Netherlands, 17–19 May 2004, pp 147–153
Di Dato F, Parisi M, Cardi T, Tripodi P (2015) Genetic diversity and assessment of markers linked to resistance and pungency genes in Capsicum germplasm. Euphytica 204:103–119
Djian-Caporalino C, Fazari A, Arguel MJ, Vernie T, Van de Casteele C, Faure I, Brunoud G, Pijarowski L, Palloix A, Lefebvre V et al (2007) Root-knot nematode (*Meloidogyne* spp.) Me resistance genes in pepper (*Capsicum annuum* L.) are clustered on the P9 chromosome. Theor Appl Genet 114:473–476
Du H, Wen C, Zhang X, Xu X, Yang J, Chen B et al (2019) Identification of a major QTL (qRRs-10.1) that confers resistance to Ralstonia solanacearum in pepper (Capsicum annuum) using SLAF-BSA and QTL mapping. Int J Mol Sci 20:5887
Dwivedi N, Kumar R, Paliwal R, Kumar U, Kumar S, Singh M, Singh RK (2015) QTL mapping for important horticultural traits in pepper (Capsicum annuum L.). J Plant Biochem Biotechnol 24:154–160

Dhall RK (2008) Breeding for quality traits in chilli: a review. J Res Punjab Agric Univ 45(3 and 4):156–160

Eggink PM, Tikunov Y, Maliepaard C, Haanstra JPW, De Rooij H et al (2014) Capturing flavors from Capsicum baccatum by introgression in sweet pepper. Theor Appl Genet 127(2):373–390

Elkind Y, Glick E, Jacobson Y, Schor-Fumbarov T (2012) Breeding and agrotechnical modifications of blocky type pepper cultivars adapted for passive protected cultivation under mild winter conditions. In: International CIPA conference 2012 on plasticulture for a green planet, vol 1015, pp 21–27

Fang M, Yu J, Kim KH (2021) Pepper mottle virus and its host interactions: current state of knowledge. Viruses 13(10):1930. doi: https://doi.org/10.3390/v13101930. PMID: 34696360; PMCID: PMC8539092

Garcés-Claver A, Fellman SM, Gil-Ortega R, Jahn M, Arnedo-Andrés MS (2007) Identification, validation and survey of a single nucleotide polymorphism (SNP) associated with pungency in Capsicum spp. Theor Appl Genet 115:907–916

Garcìa-Neria MA, Rivera-Bustamante RF (2011) Characterization of geminivirus resistance in an accession of *Capsicum chinense* Jacq. Mol Plant Microbe 24:172–182

Gniffke PA, Shieh SC, Lin SW, Sheu ZM, Chen JR, Ho FI et al (2013) Pepper research and breeding at AVRDC–the world vegetable Center. In: XV EUCARPIA meeting on genetics and breeding of capsicum and eggplant, pp 305–311

Grube RC, Blauth JR, Arnedo Andres MS, Caranta C, Jahn MK (2000) Identification and comparative mapping of a dominant potyvirus resistance gene cluster in Capsicum. Theor Appl Genet 101:852–859

Gulyas G, Shin Y, Kim H, Lee JS, Hirata Y (2010) Altered transcript reveals an Orf 507 sterility-related gene in chili pepper (Capsicum annuum L.). Plant Mol Biol Report 28(4):605–612

Guo G, Wang S, Liu J, Pan B, Diao W, Ge W, Gao C, Snyder JC (2017a) Rapid identification of QTLs underlying resistance to Cucumber mosaic virus in pepper (Capsicum frutescens). Theor Appl Genet 130(1):41–52. https://doi.org/10.1007/s00122-016-2790-3

Guo X, Yang XH, Yang Y, Mao ZC, Liu F, Ma WQ et al (2017b) Bacterial artificial chromosome library construction of root-knot nematode resistant pepper genotype HDA149 and identification of clones linked to Me3 resistant locus. J Intergr Agric 16:57–64

Guzmán FA, Ayala H, Azurdia C, Duque MC, De Vicente MC (2005) AFLP assessment of genetic diversity of capsicum genetic resources in Guatemala: home gardens as an option for conservation. Crop Sci 45(1):363–370

Guzmán S, Ruiz Gurrola B, Cruz-Zaragoza E, Tufino A, Furetta C, Favalli A, Brown F (2011) Study of the glow curve structure of the minerals separated from black pepper (Piper nigrum L.). Radiat Eff Defects Solids 166(4):288–296

Ha J, Lee D, Ahn YK, Kim ST, Jun TH (2019) Development of SNP based markers associated with bacterial wilt resistance in pepper (*Capsicum annuum*). Int J Agric Biol 22:1431–1143

Hachiya S, Kawabata F, Ohnuki K, Inoue N, Yoneda H, Yazawa S, Fushiki T (2007) Effects of CH-19 sweet, a non-pungent cultivar of red pepper, on sympathetic nervous activity, body temperature, heart rate, and blood pressure in humans. Biosci Biotechnol Biochem 71(3):671–676

Hamada H, Takeuchi S, Morita Y, Sawada H, Kiba A, Hikichi Y (2002) Amino acid changes *in Pepper mild mottle virus* coat protein that affect L3 gene-mediated resistance in pepper. J Gen Plant Pathol 68:155–162

Han K, Jeong HJ, Yang HB, Kang SM, Kwon JK, Kim S, Choi D, Kang BC (2016) An ultra-high-density bin map facilitates high-throughput QTL mapping of horticultural traits in pepper (*Capsicum annuum*). DNA Res 23(2):81–91. https://doi.org/10.1093/dnares/dsv038. Epub 2016 Jan 6. PMID: 26744365; PMCID: PMC4833416

Hamed M, Kalita D, Bartolo ME, Jayanty SS (2019) Capsaicinoids, polyphenols and antioxidant activities of Capsicum annuum: comparative study of the effect of ripening stage and cooking methods. Antioxidants 8(9):364

Han K, Lee HY, Ro NY, Hur OS, Lee JH, Kwon JK, Kang BC (2018) QT mapping and GWAS reveal candidate genes controlling capsaicinoid content in *Capsicum*. Plant Biotechnol J16:1546–1558. https://doi.org/10.1111/pbi.12894

Herath HMSN, Rafii MY, Ismail SI, Nakasha JJ, Ramlee SI (2021) Improvement of important economic traits in chilli through heterosis breeding: a review. J Hortic Sci Biotechnol 96:14–23
Hong JP, Kim WT (2005) Isolation and functional characterization of the *Ca-DREBLP1* gene encoding a dehydration-responsive element binding-factor-like protein 1 in hot pepper (*Capsicum annuum* L. cvPukang). Planta 220:875–888
Hulse-Kemp AM, Maheshwari S, Stoffel K et al (2018) Reference quality assembly of the 3.5-Gb genome of *Capsicum annuum* from a single linked-read library. Hortic Res 5(4):1–11
Hunziker AT (2001) Genera Solanacearum. The genera of solanaceae illustrated, arranged according to a new system. Gantner Verlag, Ruggell, Liechtenstein. Introgression of heat shock protein (Hsp70 and sHsp) genes into the Malaysian elite chilli variety Kulai (*Capsicum annuum* L.) through the application of marker-assisted backcrossing (MAB)
Ilbi H (2003) RAPD markers assisted varietal identification and genetic purity test in pepper, *Capsicum annuum*. Sci Hortic 97:211–218
Jahn M, Paran I, Hoffmann K, Radwanski ER, Livingstone KD, Grube RC et al (2000) Genetic mapping of the Tsw locus for resistance to the Tospovirus tomato spotted wilt virus in capsicum spp. and its relationship to the Sw-5 gene for resistance to the same pathogen in tomato. Mol Plant-Microbe Interact 13(6):673–682
Jaiswal V, Rawoof A, Dubey M, Chhapekar SS, Sharma V, Ramchiary N (2020) Development and characterization of non-coding RNA based simple sequence repeat markers in Capsicum species. Genomics 112:1554–1564
Jeong K, Choi D, Lee J (2018) Fine mapping of the genic male-sterile ms1 gene in Capsicum annuum L. Theor Appl Genet 131:183–191
Ji JJ, Huang W, Yin YX, Li Z, Gong ZH (2014) Development of a SCAR marker for early identification of S-cytoplasm based on mitochondrial SRAP analysis in pepper (*Capsicum annuum* L.). Mol Breed 33:679–690
Jing H, Li C, Ma F, Ma JH, Khan A, Wang X et al (2016) Genome-wide identification, expression diversication of dehydrin gene family and characterization of *CaDHN3* in pepper (*Capsicum annuum* L.). PLoS ONE 11(8):e0161073
Jo YD, Kim YM, Park MN, Yoo JH, Park M, Kim BD, Kang BC (2010) Development and evaluation of broadly applicable markers for restorer-of-fertility in pepper. Mol Breed 25:187–201. https://doi.org/10.1007/s11032-009-9318-3
Jo YD, Ha Y, Lee JH, Park M, Bergsma AC, Choi HI, Goritschnig S, Kloosterman B, van Dijk PJ, Choi D et al (2016) Fine mapping of *Restorer-of-fertility* in pepper (*Capsicum annuum* L.) identified a candidate gene encoding a pentatricopeptide repeat (PPR)-containing protein. Theor Appl Genet 129:2003–2017. https://doi.org/10.1007/s00122-016-2755-6
Jones JB, Lacy GH, Bouzar H, Stall RE, Schaad NW (2004) Reclassification of the xanthomonads associated with bacterial spot disease of tomato and pepper. Syst Appl Microbiol 27:755–762
Jung JK, Park SW, Liu WY, Kang BC (2010) Discovery of single nucleotide polymorphism in Capsicum and SNP markers for cultivar identification. Euphytica 175:91–107
Kang BC, Nahm SH, Huh JH, Yoo HS, Yu JW, Lee MH, Kim BD (2001) An interspecific (Capsicum annuum× C. chinese) F_2 linkage map in pepper using RFLP and AFLP markers. Theor Appl Genet 102:531–539
Kang WH, Hoang N, Yang HB, Kwon JK, Jo SH, Seo JK, Kim KH, Choi D, Kang BC (2010) Molecular mapping and characterization of a single dominant gene controlling CMV resistance in peppers (Capsicum annuum L.). Theor Appl Genet 120:1587–1596
Kang MC, Kang HJ, Jung SY, Lee HY, Kang MY, Jo YD, Kang BC (2022) The *Unstable Restorer-of-fertility* locus in pepper (*Capsicum annuum* L.) is delimited to a genomic region containing PPR genes. Theor Appl Genet 135:1923–1937. https://doi.org/10.1007/s00122-022-04084-1
Kawabata F, Inoue N, Yazawa S, Kawada T, Inoue K, Fushiki T (2006) Effects of CH-19 sweet, a nonpungent cultivar of red pepper, in decreasing the body weight and suppressing body fat accumulation by sympathetic nerve activation in humans. Biosci Biotechnol Biochem 70(12):2824–2835
Kenyon L, Tsai WS, Shih SL, Lee LM (2014) Emergence and diversity of begomoviruses infecting solanaceous crops in east and Southeast Asia. Virus Res 186:104–113

Khlestkina EK (2014) Molecular markers in genetic studies and breeding. Russ J Genet Appl Res 4:236–244
Kim BD (2000) Linkage map construction and molecular genetic approach in Capsicum spp. Kor J Plant Tissue Cult 27:367–373
Kim DS (2005) Development of RAPD and AFLP markers linked to fertility restorer (Rf) gene in chili pepper (*Capsicum annuum* L.). Seoul National University
Kim DS, Kim DH, Yoo JH, Kim BD (2006) Cleaved amplified polymorphic sequence and amplified fragment length polymorphism markers linked to the fertility restorer gene in chili pepper (*Capsicum annuum* L.). Mol Cell 21:135–140
Kim HJ, Nahm SH, Lee HR, Yoon GB et al (2008) BAC-derived markers converted from RFLP linked to Phytophthoracapsici resistance in pepper (Capsicum annuum L.). Theor Appl Genet 118:15–27
Kim S, Kim KT, Kim DH, Yang EY, Cho MC, Jamal A, Chae Y, Pae DH, Oh DG, Hwang JK (2010) Identification of quantitativetrait loci associated with anthracnose resistance in chilipepper (Capsicum spp.). Kor J Hortic Sci Technol 28:1014–1024
Kim HJ, Han JH, Kim S, Lee HR, Shin JS, Kim JH, Cho J, Kim YH, Lee HJ, Kim BD, Choi D (2011) Trichome density of main stem is tightly linked to PepMoV resistance in chili pepper (*Capsicum annuum* L.). Theor Appl Genet 122(6):1051–1058
Kim S, Park M, Yeom SI, Kim YM, Lee JM, Lee HA, Seo E, Choi J, Cheong K, Kim KT et al (2014) Genome sequence of the hot pepper provides insights into the evolution of pungency in *Capsicum* species. Nat Genet 46:270–278. https://doi.org/10.1038/ng.2877
Kim S, Park J, Yeom SI, Kim YM, Seo E et al (2017) New reference genome sequences of hot pepper reveal the massive evolution of plant disease-resistance genes by retroduplication. Genome Biol 18:210
Knapp S (2007) Some like it hot. Science 315(5814):946–947. https://doi.org/10.1126/science.1138308
Kosuge S, Furuta M (1970) Studies on the pungent principle of capsicum: part XIV chemical constitution of the pungent principle. Agric Biol Chem 34(2):248–256
Kraft KH, Brown CH, Nabhan GP, Luedeling E, de Jesús Luna Ruiz J, d'Eeckenbrugge GC, Hijmans RJ, Gepts P (2014) Multiple lines of evidence for the origin of domesticated chili pepper, Capsicum annuum, in Mexico. PNAs 111(17):6165–6170
Kumar OA, Tata SS, Rupavathi T (2014) Evaluation of genetic diversity in 21 cultivars of chili pepper (Capsicum annuum L.) using isozyme markers. Eur J Exp Biol 4:44–49
Kyle MM, Palloix A (1997) Proposed revision of nomenclature for potyvirus resistance genes in Capsicum. Euphytica 97:183–188
Lang Y, Kisaka H, Sugiyama R, Nomura K, Morita A, Watanabe T et al (2009) Functional loss of pAMT results in biosynthesis of capsinoids, capsaicinoid analogs, in Capsicum annuum cv. CH-19 sweet. Plant J 59(6):953–961
Lapidot M, Paran I, Ben-Joseph R, Ben-Harush S, Pilowsky M, Cohen S, Shifriss C (1997) Tolerance to cucumber mosaic virus in pepper: development of advanced breeding lines and evaluation of virus level. Plant Dis 81(2):185–188
Lee J, Yoon JB, Park HG (2008) A CAPS marker associated with the partial restoration of cytoplasmic male sterility in chili pepper (*Capsicum annuum* L.). Mol Breed 21:95–104. https://doi.org/10.1007/s11032-007-9111-0
Lee YR, Yoon JB, Lee J (2016) A SNP-based genetic linkage map of *Capsicum baccatum* and its comparison to the *Capsicum annuum* reference physical map. Mol Breed 36:1–11
Lee J, Hong J, Do JW (2010) Identification of QTLs for resistance to anthracnose to two *Colletotrichum* species in pepper. J Crop Sci Biotechnol 13:227–233. https://doi.org/10.1007/s12892-010-0081-0
Lee HR, An HJ, You YG, Lee J, Kim HJ, Kang BC, Harn CH (2013) Development of a novel codominant molecular marker for chili veinal mottle virus resistance in Capsicum annuum L. Euphytica 193:197–205
Lee MS, Kim CT, Kim IH, Kim Y (2011) Effects of capsaicin on lipid catabolism in 3T3-L1 adipocytes. Phytother Res 25(6):935–939

Lefebvre V, Palloix A, Rives M (1993) Nuclear RFLP between pepper cultivars (Capsicum annuum L.). Euphytica 71:189–199

Lefebvre V, Palloix A (1996) Both epistatic and additive effects of QTLs are involved in polygenic induced resistance to disease: a case study, the interaction pepper – Phytophthora capsici Leonian. Theor Appl Genet 93(4):503–511

Lefebvre V, Pflieger S, Thabuis A, Caranta C, Blattes A, Chauvet JC, Daubeze AM, Palloix A (2002) Towards the saturation of the pepper linkage map by alignment of three intraspecific maps including known-function genes. Genome 46:839–854

Lin WC, Hill BD (2007) Neural network modelling of fruit colour and crop variables to predict harvest dates of greenhouse-grown sweet peppers. Can J Plant Sci 87(1):137–143

Linders EGA, Nicolet JLME, Van Wijk HJ (2010) Insect resistant plant, Vol. PCT/EP2008/055374: Syngenta Participations AG United States

Livingstone KD, Lackney VK, Blauth JR, van Wijk R, Jahn MK (1999) Genome mapping in Capsicum and the evolution of genome structure in the Solanaceae. Genetics 152:1183–1202

Lopez-Moreno H, Basurto-Garduño AC, Torres-Meraz MA et al (2023) Genetic analysis and QTL mapping of domestication-related traits in chili pepper (Capsicum annuum L.). Front Genet 14:1101401

López-Serrano L, Calatayud Á, López-Galarza S et al (2021) Uncovering salt tolerance mechanisms in pepper plants: a physiological and transcriptomic approach. BMC Plant Biol 21:169

Magaji GU, Rafii MY, Martini MY, Yusuff OA, Ismail MR, Gous M (2018) Introgression of heat shock protein (Hsp70 and sHsp) genes into the Malaysian elite chilli variety Kulai (*Capsicum annuum* L.) through the application of marker-assisted backcrossing (MAB). Cell Stress Chaperones 23(2):223–234

Maharijaya A, Vosman B, Steenhuis-Broers G, Harpenas A, Purwito A, Visser RG, Voorrips RE (2011) Screening of pepper accessions for resistance against two thrips species (Frankliniella occidentalis and Thrips parvispinus). Euphytica 177(3):401–410

Maharijaya A, Vosman B, Verstappen F, Steenhuis-Broers G, Mumm R, Purwito A et al (2012) Resistance factors in pepper inhibit larval development of thrips (F rankliniella occidentalis). Entomol Exp Appl 145(1):62–71

Maharijaya A, Vosman B, Steenhuis-Broers G, Pelgrom K, Purwito A, Visser RG, Voorrips RE (2015) QTL mapping of thrips resistance in pepper. Theor Appl Genet 128(10):1945–1956

Mahasuk P, Taylor PWJ, Mongkolporn O (2009) Identification oftwo new genes conferring resistance to Colletotrichumacutatum in Capsicum baccatum. Phytopathology 99:1100–1104. https://doi.org/10.1094/PHYTO-99-9-1100

Mallard S, Cantet S, Massire A, Bachellez A, Ewert S, Lefebvre V (2013) A key QTL cluster is conserved among accessions and exhibits broad-spectrum resistance to Phytophthoracapsici: a valuable locus for pepper breeding. Mol Breed 32:349–364. https://doi.org/10.1007/s11032-013-9875-3

Marques MLS, Chadud JVG, Oliveira MF, Nascimento AR, Rocha MR (2019) Identification of chili pepper genotypes (Capsicum spp.) resistant to Meloidogyne Enterolobii. J Agric Sci 11:165–175

Mathur AK, Sharma V, Mangal M, Srivastava A, Maurya P (2018) Study of gene effects for resistance to leaf curl disease in chilli using generation mean analysis. J Hill Agric 9(4):387–390

Maurya V, Srinivasan R, Nalini E, Ramesh N, Gothandam KM (2014) Annual Res Rev Biol 6(1):66–78. https://doi.org/10.9734/ARRB/2015/14107

Meena OP, Dhaliwal MS, Jindal SK (2018) Development of cytoplasmic male sterile lines in chilli (*Capsicum annuum* L.) and their evaluation across multiple environments. Breed Sci 68(4):404–412. https://doi.org/10.1270/jsbbs.17150. Epub 2018 Aug 25. PMID: 30369814; PMCID: PMC6198902

Meghvansi MK, Siddiqui S, Khan MH, Gupta VK, Vairale MG, Gogoi HK, Singh L (2010) Naga chilli: a potential source of capsaicinoids with broad-spectrum ethnopharmacological applications. J Ethnopharmacol 132(1):1–14

Meng YC, Zhang HF, Pan XX, Chen N et al (2021) CaDHN3, a pepper (*Capsicum annuum* L.) dehydrin gene enhances the tolerance against salt and drought stresses by reducing ROS accumulation. Int J Mol Sci 22(6):3205. https://doi.org/10.3390/ijms2206320

Mimura Y, Kageyama T, Minamiyama Y, Masashi H (2009) QTL analysis for resistance to Ralstonia solanacearum in *Capsicum* accession 'LS2341'. J Jpn Soc Hortic Sci 78:307–313

Mimura Y, Minamiyama Y, Sano H, Hirai M (2010) Mapping for axillary shooting, flowering date, primary axis length, and number of leaves in pepper (*Capsicum annuum*). J Jpn Soc Hortic Sci 79(1):56–63

Minamiyama Y, Tsuro M, Hirai M (2006) An SSR-based linkage map of Capsicum annuum. Mol Breed 18:157–169

Mondal CK, Acharyya P, Saha U (2016) Study on genetic diversity in chilli (*Capsicum annuum*) based on multivariate analysis and isozyme analysis. J Appl Natural Sci 8:1884–1892

Mondo JM, Agre PA, Edemodu A, Asiedu R, Akoroda MO, Asfaw A (2022) Cross compatibility in intraspecific and interspecific hybridization in yam (*Dioscorea spp.*). Sci Rep 12(1):3432. https://doi.org/10.1038/s41598-022-07484-x. PMID: 35236890; PMCID: PMC8891288

Moodley V, Naidoo R, Gubba A, Mafongoya PL (2019) Development of Potato virus Y (PVY) resistant pepper (*Capsicum annuum* L.) lines using marker-assisted selection (MAS). Physiol Mol Plant Pathol 105:96–101

Moury B, Palloix A, Gebre Selassie K, Marchoux G (1997) Hypersensitive resistance to tomato spotted wilt virus in three Capsicum chinense accessions is controlled by a single gene and is overcome by virulent strains. Euphytica 94(1):45–52

Naegele RP, Ashrafi H, Hill TA, Chin-Wo SR, Van Deynze AE, Hausbeck MK (2014) QTL mapping of fruit rot resistance to the plant pathogen *Phytophthora capsici* in a recombinant inbred line *Capsicum annuum* population. Phytopathology 104(5):479–483. https://doi.org/10.1094/PHYTO-05-13-0143-R. PMID: 24168044

Nagaraju MM, Kumary IS, Celine VA, Devi CRS, Manju P (2017) Development of F1 hybrids in chilli (*Capsicum annuum* L.) for dual purpose (Green as well as Dry). Int J Curr Microbiol App Sci 6:84–96

Na Jinda A, Nikornpun M, Jeeatid N, Thumdee S, Thippachote K, Pusadee T, Kumchai J (2023) Marker-assisted selection of male-sterile and maintainer line in chili improvement by backcross breeding. Horticulturae 9(3):357

Naresh P, Lin SW, Lin CY, Wang YW, Schafleitner R, Kilian A, Kumar S (2018) Molecular markers associated to two non-allelic genic male sterility genes in peppers (*Capsicum annuum* L.). Front Plant Sci 9:1343

Nicolaï M, Pisani C, Bouchet JP, Vuylsteke M, Palloix A (2012) Discovery of a large set of SNP and SSR genetic markers by high-throughput sequencing of pepper (*Capsicum annuum*). Genet Mol Res 11:2295–2300

Nimmakayala P, Abburi VL, Saminathan T, Alaparthi SB, Almeida A, Davenport B, Nadimi M, Davidson J, Tonapi K, Yadav L, Malkaram S, Vajja G, Hankins G, Harris R, Park M, Choi D, Stommel J, Reddy UK (2016a) Genome-wide diversity and association mapping for capsaicinoids and fruit weight in Capsicum annuum L. Sci Rep 6:38081. https://doi.org/10.1038/srep38081. PMID: 27901114; PMCID: PMC5128918

Nimmakayala P, Abburi VL, Saminathan T, Almeida A, Davenport B, Davidson J, Reddy CVCM, Hankins G, Ebert A, Choi D, Stommel J, Reddy UK (2016b) Genome-wide divergence and linkage disequilibrium analyses for *Capsicum baccatum* revealed by genome-anchored single nucleotide polymorphisms. Front Plant Sci 7:1646

Nono-Womdim R, Marchoux G, Gebre-Selassie K (1991) Application des méthodes DAS et PAS ELISA pour la détection de plusieurs virus infectant le Piment. Phytopathol Mediterr:14–22

Ogundiwin EA, Berke TF, Massoudi M, Black LL, Huestis G, Choi D et al (2005) Construction of 2 intraspecific linkage maps and identification of resistance QTLs for Phytophthora capsici root-rot and foliar-blight diseases of pepper (Capsicum annuum L.). Genome 48(4):698–711

Ohnuki K, Haramizu S, Oki K, Watanabe T, Yazawa S, Fushiki T (2001) Administration of capsiate, anon-pungent capsaicin analog, promotes energy metabolism and suppresses body fat accumulation in mice. Biosci Biotechnol Biochem 65(12):2735–2740

Okunlola GO, Olatunji OA, Akinwale RO, Tariq A, Adelusi AA (2017) Physiological response of the three most cultivated pepper species (*Capsicum* spp.) in Africa to drought stress imposed at three stages of growth and development. Sci Hortic 224:198–205

Oliver JE, Whitfield AE (2016) The genus *tospovirus*: emerging bunyaviruses that threaten food security. Annual Rev Virol 3(1):101–124

Paran I, van der Knaap E (2007) Genetic and molecular regulation of fruit and plant domestication traits in tomato and pepper. J Exp Bot 58(14):3841–3852. https://doi.org/10.1093/jxb/erm257

Paran I, van der Voort JR, Lefebvre V et al (2004) An integrated genetic linkage map of pepper (Capsicum spp.). Mol Breed 13:251–261

Paran I, Akler T, Borovsky Y (2010) QTLs for capsaicinoids content in *Capsicum*. In: Prohens J, Rodríguez-Burruezo A (eds) Advances in genetics and breeding of capsicum and eggplant. Editorial de la Universitat Politècnica de València, Valencia, pp 273–278

Pegard A, Brizzard G, Fazari A, Soucaze O, Abad P, Djian-Caporalino C (2005) Histological characterization of resistance to different root-knot nematode species related to phenolics accumulation in Capsicum annuum. Phytopathology 95(2):158–165

Perry L, Dickau R, Zarrillo S, Holst I, Pearsall DM et al (2007) Starch fossils and the domestication and dispersal of chili peppers (Capsicum spp. L.) in the Americas. Science 315(5814):986–988. https://doi.org/10.1126/science.1136914

Perucka I, Materska M (2001) Phenylalanine ammonia-lyase and antioxidant activities of lipophilic fraction of fresh pepper fruits Capsicum annum L. Innovative Food Sci Emerg Technol 2(3):189–192

PGR Portal National Bureau of Plant India (2018) Available online: www.nbpgr.ernet.in/pgrportal

Pierre M, Noel L, Lahaye T, Ballvora A, Veuskens J, Ganal M, Bonas U (2000) High-resolution genetic mapping of the pepper resistance locus *Bs3* governing recognition of the *Xanthomonas campestris* pv *vesicatoria* AvrBs3 protein. Theor Appl Genet 101:255–263

Pinar H, Mutlu N, Ozaslandan A, Argun D, Keles D, Canhilal R (2016) Reliability assessment of molecular markers linked to resistance genes against *Meloidogyne* spp. in diverse peppers genotypes. Egypt J Biol Pest Co 26:515–521

Pochard E, Daubèze AM (1989) Progressive construction of a polygenic resistance to cucumber mosaic virus in the pepper. In: Proceedings of the 7th EUCARPIA meeting on genetics and breeding of capsicum and eggplant, pp 187–192

Popovsky-Sarid S, Borovsky Y, Faigenboim A, Parsons EP, Lohrey GT, Alkalai-Tuvia S et al (2017) Genetic and biochemical analysis reveals linked QTLs determining natural variation for fruit post-harvest water loss in pepper (capsicum). Theor Appl Genet 130(2):445–459

Potnis N, Branham SE, Jones JB, Wechter PW (2019) Genome-wide association study of resistance to *Xanthomonas gardneri* in the USDA pepper (*Capsicum*) collection. Phytopathology 109(7):1217–12250. https://doi.org/10.1094/PHYTO-06-18-0211-R

Prince JP, Lackney VK, Angeles C et al (1995) A survey of DNA polymorphism within the genus Capsicum and the fingerprinting of pepper cultivars. Genome 38:224–231

Qin C, Yu C, Shen Y, Fang X, Chen L, Min J, Cheng J, Zhao S, Xu M, Luo Y et al (2014) Whole-genome sequencing of cultivated and wild peppers provides insights into *Capsicum* domestication and specialization. Proc Natl Acad Sci USA 111:5135–5140. https://doi.org/10.1073/pnas.1400975111

Quenouille J, Vassilakos N, Moury B (2013) Potato virus Y. Mol Plant Pathol 14:439–452. https://doi.org/10.1111/mpp.12024

Quirin EA, Ogundiwin EA, Prince JP, Mazourek M, Briggs MO, Chlanda TS et al (2005) Development of sequence characterized amplified region (SCAR) primers for the detection of Phyto. 5.2, a major QTL for resistance to Phytophthora capsici Leon. in pepper. Theor Appl Genet 110(4):605–612

Rameash K, Pandravada SR, Sivaraj N, Balijepalli SB, Chakrabarty SK (2016) Diversity and distribution of Capsicum annuum genotypes resistant to Scirtothrips dorsalis in India: an analysis through geographical information system. J Food Agric Environ 14(1):51–64

Rani K, Kumar M, Razzaq A et al (2023) Recent advances in molecular marker technology for QTL mapping in plants. QTL Map Crop Improv:1–15

Rao GU, Ben Chaim A, Borovsky Y, Paran I (2003) Mapping of yield-related QTLs in pepper in an interspecific cross of Capsicum annuum and C. frutescens. Theor Appl Genet 106:1457–1466
Rast ATB (1988) Pepper *tobamoviruses* and pathotypes used in resistance breeding. Capsicum Newsl 7:20–23
Reddy UK, Almeida A, Abburi VL, Alaparthi SB, Unselt D, Hankins G, Park M, Choi D, Nimmakayala P (2014) Identification of gene-specific polymorphisms and association with capsaicin pathway metabolites in *Capsicum annuum* L. collections. PLoS ONE 9:e86393
Rinaldi R, Deynze AV, Portis E, Rotino GL, Toppino L et al (2016) New insights on eggplant/tomato/pepper synteny and identification of eggplant and pepper orthologous QTL. Front Plant Sci 7:1031
Ro N, Haile M, Hur O, Geum B, Rhee J, Hwang A, Kim B, Lee J, Hahn B-S, Lee J, Kang B-C (2022) Genome-wide association study of resistance to *Phytophthora capsici* in the pepper (Capsicum spp.) collection. Front Plant Sci 13:902464. https://doi.org/10.3389/fpls.2022.902464
Ro N, Haile M, Hur O et al (2023) Genome-wide association study of resistance to anthracnose in pepper (*Capsicum chinense*) germplasm. BMC Plant Biol 23:389. https://doi.org/10.1186/s12870-023-04388-4
Rodriguez JM, Berke T, Engle L, Nienhuis J (1999) Variation among and within Capsicum species revealed by RAPD markers. Theor Appl Genet 99:147–156
Rohini N, Lakshmanan V (2017) Evaluation studies of hot pepper hybrids (Capsicum annuum L.) for yield and quality characters. Electron J Plant Breed 8:643–651
Ruffel S, Dussault MH, Palloix A, Moury B, Bendahmane A, Robaglia C, Caranta C (2002) A natural recessive resistance gene against Potato virus Y in pepper corresponds to the eukaryotic initiation factor 4E (eIF4E). Plant J 32:1067–1075
Sahitya UL, Krishna MSR, Suneetha P (2019) Integrated approaches to study the drought tolerance mechanism in hot pepper (Capsicum annuum L.). Physiol Mol Biol Plants 25(3):637–647. https://doi.org/10.1007/s12298-019-00655-7
Sawada H, Takeuchi S, Hamada H, Kiba A, Matsumoto M, Hikichi Y (2004) A new tobamovirus-resistance gene, L-1a, of sweet pepper (*Capsicum annuum* L.). J Jpn Soc Hortic Sci 73:552–557
Selvakumar R, Chandregowda Manjunathagowda D, Kumar Singh P (2023) Capsicum: breeding prospects and perspectives for higher productivity. Capsicum Curr Trends Perspect. IntechOpen. Available at: https://doi.org/10.5772/intechopen.104739
Sharma VK, Srivastava A, Mangal M (2020) Recent trends in sweet pepper breeding. In: Accelerated plant breeding, volume 2: vegetable crops. Springer International Publishing, Cham, pp 417–444
Shetty PK, Ayyappan S, Swaminathan MS (2013) Climate change and sustainable food security. New Delhi, NIAS/ICAR
Shingote PR, Wasule DL, Parma VS, Holkar SK, Karkute SG, Parlawar ND, Senanayake DMJB (2022) An overview of chili leaf curl disease: molecular mechanisms, impact, challenges, and disease management strategies in Indian subcontinent. Front Microbiol 13:899512. https://doi.org/10.3389/fmicb.2022.899512
Siddique MI, Lee J-H, Ahn J-H, Kusumawardhani MK, Safitri R, Harpenas A et al (2022) Genotyping-by-sequencing-based QTLmapping reveals novel loci for *Pepper yellow leafcurl virus* (PepYLCV) resistance in *Capsicumannuum*. PLoS ONE 17(2):e0264026. https://doi.org/10.1371/journal.pone.0264026
Singh AK, Kushwaha N, Chakraborty S (2016) Synergistic interaction among begomoviruses leads to the suppression of host defense-related gene expression and breakdown of resistance in chilli. Appl Microbiol Biotechnol 100(9):4035–4049
Sokona D, Niamoye YD, Paul NS, Olagorite A, Aminata DN, Kadidiatou GT, Aissata TTR, Sériba K, Daoulé DB (2013) Overview of pepper (Capsicum spp.) breeding in West Africa. Afr J Agric Res 8:1108–1114
Srivastava A, Kumar S, Jaidi M, Raj SK (2015) Characterization of a novel begomovirus associated with yellow mosaic disease of three ornamental species of Jatropha grown in India. Virus Res 201:41–49

Srivastava A, Mangal M, Saritha RK, Kalia P (2017) Screening of chilli pepper (capsicum spp.) lines for resistance to the begomoviruses causing chilli leaf curl disease in India. Crop Prot 100:177–185

Srivastava A, Mangal M, Parihar BR (2020) Development and characterization of heat tolerant genotypes in chilli (Capsicum annuum L.). Ann Hortic 13:34–38. https://doi.org/10.5958/0976-4623.2020.00004.3

Srivastava A, Srivastava S, Gupta T, Rishi N (2021) Next-generation molecular diagnostics for plant viruses. In: Virus diseases of ornamental plants: characterization, identification, diagnosis and management. Springer Singapore, Singapore, pp 397–406

Srivastava A, Singh K, Khar A, Parihar BR, Tomar BS, Mangal M (2022) Morphological, biochemical and molecular insights on responses to heat stress in chilli. Indian J Hortic 79(1):15–22. https://doi.org/10.5958/0974-0112.2022.00003.2

Srivastava A, Tomar BS, Behera TK, Krishna H, Jain PK, Pandey R, Singh B, Gupta R, Mangal M (2023) Genetic analysis of heat tolerance in hot pepper: insights from comprehensive phenotyping and QTL mapping. Front Plant Sci 14:1232800. https://doi.org/10.3389/fpls.2023.1232800

Stellari GM, Mazourek M, Jahn MM (2010) Contrasting modes for loss of pungency between cultivated and wild species of capsicum. Heredity 104(5):460–471

Stewart C Jr, Kang BC, Liu K, Mazourek M, Moore SL, Yoo EY et al (2005) The Pun1 gene for pungency in pepper encodes a putative acyltransferase. Plant J 42(5):675–688

Stewart C Jr, Mazourek M, Stellari GM, O'Connell M, Jahn M (2007) Genetic control of pungency in C. chinense via the Pun1 locus. J Exp Bot 58(5):979–991

Sugita T, Semi Y, Sawada H, Utoyama Y, Hosomi Y, Yoshimoto E et al (2013) Developmentof simple sequence repeat markers and construction of a high-density linkage map of Capsicum annuum. Mol Breed 31(4):909–920

Sun C, Mao SL, Zhang ZH, Palloix A, Wang LH, Zhang BX (2015) Resistances to anthracnose (Colletotrichumacutatum) ofCapsicum mature green and ripe fruit are controlled by amajor dominant cluster of QTLs on chromosome P5. Sci Hortic 181:81–88. https://doi.org/10.1016/j.scienta.2014.10.033

Suwor P, Thummabenjapone P, Sanitchon J, Kumar S, Techawongstien S (2015) Phenotypic and genotypic responses of chili (Capsicum annuum L.) progressive lines with different resistant genes against anthracnose pathogen (Colletotrichum spp.). Eur J Plant Pathol 143(4):725–736

Suwor P, Sanitchon J, Thummabenjapone P, Kumar S, Techawongstien S (2017) Inheritance analysis of anthracnose resistance and marker-assisted selection in introgression populations of chili (Capsicum annuum L.). Sci Hortic 220:20–26

Suzuki K, Kuroda T, Miura Y, Murai J (2003) Screening and field trials of virus resistant sources in capsicum spp. Plant Dis 87(7):779–783

Sy O, Steiner R, Bosland PW (2008) Recombinant inbred line differential identifies race-specific resistance to Phytophthora root rot in Capsicum annuum. Phytopathology 98(8):867–870

Tai T, Dahlbeck D, Stall RE, Peleman J, Staskawicz BJ (1999) High-resolution genetic and physical mapping of the region containing the Bs2 resistance gene of pepper. Theor Appl Genet 99:1201–1206

Tan S, Cheng JW, Zhang L, Qin C, Nong DG, Li WP, Tang X, Wu ZM, Hu KL (2015) Construction of an interspecific genetic map based on InDel and SSR for mapping the QTLs affecting the initiation of flower primordia in pepper (Capsicum spp.). PLoS One 10(3):e0119389. https://doi.org/10.1371/journal.pone.0119389. PMID: 25781878; PMCID: PMC4363154

Tanaka Y, Yoneda H, Hosokawa M, Miwa T, Yazawa S (2014) Application of marker-assisted selection in breeding of a new fresh pepper cultivar (Capsicum annuum) containing capsinoids, low-pungent capsaicinoidanalogs. Sci Hortic 165:242–245

Tang C, Xie J, Jian L, Li J, Zhang J, Wang C, Liang G (2021) Alleviating damage of photosystem and oxidative stress from chilling stress with exogenous zeaxanthin in pepper (*Capsicum annuum* L.) seedlings. Plant Physiol Biochem 162:395–409

Tanksley SD, Bernatzky R, Lapitan NL, Prince JP (1988) Conservation of gene repertoire but not gene order in pepper and tomato. Proc Natl Acad Sci USA 85:6419–6423

Thakur H, Jindal SK, Sharma A, Dhaliwal MS (2019) A monogenic dominant resistance for leaf curl virus disease in chilli pepper (Capsicum annuum L.). Crop Prot 116:115–120

Thul ST, Shasany AK, Darokar MP, Khanuja SP (2006) AFLP analysis for genetic diversity in Capsicum annuum and related species. Nat Prod Commun 1(3):1934578X0600100309

Tomita R, KenTaro S, Hiroyuki M, Sakamoto M, Murai J, Kiba A, Hikichi Y, Suzuki K, Kobayashi K (2011) Genetic basis for the hierarchical interaction between Tobamovirus spp. and L resistance gene alleles from different pepper species. Mol Plant-Microbe Interact 24:108–117

Truong HTH, Kim KT, Kim S, Kim HR, Cho MC, Woo JG (2009) Development of gene-based markers for the Pun1 pungency gene in pepper (capsicum spp.) for marker-assisted selection. Hortic Environ Biotechnol 50(4):358–365

Truong HTH, Kim KT, Kim S, Chae Y, Park JH, Oh DG, Cho MC (2010) Comparative mapping of consensus SSR markers in an intraspecific F_8 recombinant inbred line population in capsicum. Hortic Environ Biotechnol 51:193–206

Truong HTH, Kim KT, Kim S, Cho MC, Kim HR, Woo JG (2011) Development of gene-based markers for the Bs2 bacterial spot resistance gene for marker-assisted selection in pepper (Capsicum spp.). Hortic Environ Biotechnol 52:65–73

Ts A, Srivastava A, Tomar BS, Behera TK, Krishna H, Jain PK, Pandey R, Singh B, Gupta R, Mangal M (2023) Genetic analysis of heat tolerance in hot pepper: insights from comprehensive phenotyping and QTL mapping. Front Plant Sci 14:1232800

Uarrota VG, Maraschin M, de Bairros A, de Pedreschi R (2000) Factors affecting the capsaicinoid profile of hot peppers and biological activity of their non-pungent analogs (capsinoids) present in sweet peppers. Crit Rev Food Sci Nutr 2020:1–17. https://doi.org/10.1080/10408398.2020.1743642

Uncu AT, Celik I, Devran Z, Özkaynak E, Frary A, Frary A, Doganlar S (2015) Development of a SNP-based CAPS assay for the Me1 gene conferring resistance to root knot nematode in pepper. Euphytica 206:393–399

Usman MG, Rafii MY, Martini MY, Yusuff OA, Ismail MR, Miah G (2018) Introgression of heat shock protein (Hsp70 and sHsp) genes into the Malaysian elite chilli variety Kulai (Capsicum annuum L.) through the application of marker-assisted backcrossing (MAB). Cell Stress Chaperones 23(2):223–234

Vallejos CE, Jones V, Stall RE, Jones JB, Minsavage GV, Schultz DC, Rodrigues R, Olsen LE, Mazourek M (2010) Characterization of two recessive genes controlling resistance to all races of bacterial spot in peppers. Theor Appl Genet 121:37–46

Venkatesh J, An J, Kang WH, Jahn M, Kang BC (2018) Fine mapping of the dominant potyvirus resistance gene Pvr7 reveals a relationship with Pvr4 in Capsicum annuum. Phytopathology 108:142–148

Villa-Rivera MG, Ochoa-Alejo N (2020) Chili pepper carotenoids: nutraceutical properties and mechanisms of action. Molecules 25(23):5573

Wahyuni Y, Stahl-Hermes V, Ballester AR, de Vos RC, Voorrips RE, Maharijaya A et al (2014) Genetic mapping of semi-polar metabolites in pepper fruits (capsicum sp.): towards unravelling the molecular regulation of flavonoid quantitative trait loci. Mol Breed 33(3):503–518

Wai KP, Siddique MI, Mo HS, Yoo HJ, Byeon SE, Jegal Y, Mekuriaw AA, Kim BS (2015) Pathotypes of bacterial spot pathogen infecting Capsicum peppers in Korea. Plant Pathol J 31:428–432

Wan H, Yuan W, Ruan M, Ye Q, Wang R, Li Z et al (2011) Identification of reference genes for reverse transcription quantitative real-time PCR normalization in pepper (*Capsicum annuum* L.). Biochem Biophys Res Commun 416:24–30

Wang LH, Zhang BX, Lefebvre V, Huang SW, Daubeze AM, Palloix A (2004) QTL analysis of fertility restoration in cytoplasmic male sterile pepper. Theor Appl Genet 109:1058–1063. https://doi.org/10.1007/s00122-004-1715-8

Wang LH, Gu XH, Hua MY, Mao SL, Zhang ZH, Peng DL, Yun XF, Zhang BX (2009) A SCAR marker linked to the N gene for resistance to root knot nematodes (Meloidogyne spp.) in pepper (*Capsicum annuum* L.). Sci Hortic 122(2):318–322

Wang P, Wang L, Guo J, Yang W, Shen H (2016) Molecular mapping of a gene conferring resistance to *Phytophthora capsici* Leonian race 2 in pepper line PI201234 (*Capsicum annuum* L.). Mol Breed 36:66. https://doi.org/10.1007/s11032-016-0464-0

Wang H, Niu H, Zhai Y, Lu M (2017) Characterization of *BiP* Genes from Pepper (*Capsicum annuum* L.) and the Role of *CaBiP1* in response to endoplasmic reticulum and multiple abiotic stresses. Front Plant Sci 8:1122

Wang X, Fazari A, Cao Y, Zhang Z, Palloix A, Mao S et al (2018) Fine mapping of the root-knot nematode resistance gene Me1 in pepper (*Capsicum annuum* L.) and development of markers tightly linked to Me1. Mol Breed 38:39

Wang F, Yin Y, Yu C, Li N, Shen S, Liu Y, Gao S, Jiao C, Yao M (2021) Transcriptomics analysis of heat stress-induced genes in pepper (*Capsicum annuum* L.) seedlings. Horticulturae 7:339

Wu F, Eannetta NT, Xu Y, Durrett R, Mazourek M et al (2009) A COSII genetic map of the pepper genome provides a detailed picture of synteny with tomato and new insights into recent chromosome evolution in the genus *Capsicum*. Theor Appl Genet 118:1279–1293

Wu L, Wang P, Wang Y, Cheng Q, Lu Q, Liu J, Li T, Ai Y, Yang W, Sun L et al (2019) Genome-wide correlation of 36 agronomic traits in the 287 pepper (*Capsicum*) accessions obtained from the SLAF-seq-Based GWAS. Int J Mol Sci 20:5675. https://doi.org/10.3390/ijms20225675

Wyatt LE, Eannetta NT, Stellari GM, Mazourek M (2012) Development and application of a suite of nonpungency markers for the Pun1 gene in pepper (capsicum spp.). Mol Breed 30(3):1525–1529

Yang HB, Liu WY, Kang WH, Jahn M, Kang BC (2009) Development of SNP markers linked to the L locus in Capsicum spp. by a comparative genetic analysis. Mol Breed 24:433–446

Yang J, Wang W, Shen H (2010) SSR marker mapping and marker-assisted selection of *Rf* gene in pepper. Chin Cucurbits Veg 23:1–5

Yao M, Li N, Wang F, Ye Z (2013) Genetic analysis and identification of QTLs for resistance to cucumber mosaic virus in chili pepper (Capsicum annuum L.). Euphytica 193(2):135–145

Yarnes SC, Ashrafi H, Reyes-Chin-Wo S, Hill TA, Stoffel K, Van Deynze A (2013) Identification of QTLs for capsaicinoids, fruit quality, and plant architecture-related traits in an interspecific Capsicum RIL population. Genome 56(1):61–74

Yeam I, Kang BC, Lindeman W, Frantz JD, Faber N, Jahn MM (2005) Allele-specific CAPS markers based on point mutations in resistance alleles at the pvr1 locus encoding eIF4E in Capsicum. Theor Appl Genet 112:178–186

Yi S, Lee D-G, Back S, Hong J-P, Jang S, Han K, Kang BC (2022) Genetic mapping revealed that the Pun2 gene in Capsicum chacoense encodes a putative aminotransferase. Front Plant Sci 13:1039393. https://doi.org/10.3389/fpls.2022.1039393

Zhang B, Huang S, Yang G, Guo J (2000) Two RAPD markers linked to a major fertility restoration gene in pepper. Euphytica 113:155–161

Zhang H, Ma F, Wang X, Liu S, Saeed UH, Hou X, Zhang Y, Luo D, Meng Y, Zhang W, Abid K, Chen R (2020a) Molecular and functional characterization of *CaNAC035*, an NAC transcription factor from pepper (*Capsicum annuum* L.). Front Plant Sci 11:14

Zhang Z, Chen J, Liang C et al (2020b) Genome-wide identification and characterization of the bHLH transcription factor family in pepper (*Capsicum annuum* L.). Front Genet 11:570156

Zhang Z, Zhu Y, Cao Y, Yu H, Bai R, Zhao H, Zhang B, Wang L (2020c) Fine mapping of the male fertility restoration gene *CaRf032* in *Capsicum annuum* L. Theor Appl Genet 133:1177–1187. https://doi.org/10.1007/s00122-020-03540-0

Zhang ZH, An DL, Cao YC, Yu HL, Zhu YS, Mei YJ, Zhang BX, Wang LH (2021) Development and application of KASP markers associated with *Restorer-of-fertility* gene in *Capsicum annuum* L. Physiol Mol Biol Plants 27:2757–2765. https://doi.org/10.1007/s12298-021-01109-9

Zhang J, Wang C, Wang J, Yang Y, Han K, Bakpa EP, Li J, Lyu J, Yu J, Xie J (2023) Comprehensive fruit quality assessment and identification of aroma-active compounds in green pepper (Capsicum annuum L.). Front Nutr 9:1027605. https://doi.org/10.3389/fnut.2022.1027605

Zhong Y, Cheng Y, Ruan M, Ye Q, Wan R, Yao Z, Wan H (2021) High-throughput SSR marker development and the analysis of genetic diversity in Capsicum frutescens. Horticulturae 7:187

Zygier S, Chaim AB, Efrati A, Kaluzky G, Borovsky Y, Paran I (2005) QTLs mapping for fruit size and shape in chromosomes 2 and 4 in pepper and a comparison of the pepper QTL map with that of tomato. Theor Appl Genet 111(3):437–445. https://doi.org/10.1007/s00122-005-2015-7. Epub 2005 Jun 28. PMID: 15983758

Advances in QTL Mapping and GWAS: Implications for Cucumber Breeding

6

Gayatri B. K. and Shyam Sundar Dey

Abstract

Advancements in genome-wide association studies (GWAS) and quantitative trait locus (QTL) mapping have significantly benefited the field of cucumber breeding. These approaches enable the identification of key genetic factors underlying important agronomic traits. QTL mapping has been instrumental in pinpointing genomic regions associated with disease resistance, yield, and fruit quality. For example, the discovery of QTLs linked to resistance against powdery mildew and downy mildew has facilitated the development of more resilient cucumber varieties. Leveraging the genetic diversity present in diverse germplasm collections, GWAS has further identified candidate genes associated with quantitative traits, thereby enhancing our understanding of trait inheritance. Through GWAS, genetic loci related to fruit size, color, and stress tolerance have been successfully mapped. Consequently, cucumber breeding programs have become more efficient, enabling the production of cultivars that align with consumer preferences and adapt to various environmental conditions. The integration of QTL mapping and GWAS has accelerated the development of improved cucumber varieties by providing a robust framework for marker-assisted selection (MAS) and genomic selection. These genomic tools are revolutionizing cucumber breeding, facilitating higher yields, enhanced disease resistance, and superior fruit quality.

Keywords

Cucumber · QTL analysis · Resequencing · Gene annotation · Candidate gene · GWAS · Quantitative trait loci

G. B. K. · S. S. Dey (✉)
Division of Vegetable Science, ICAR-Indian Agricultural Research Institute, New Delhi, India

S. S. Dey et al. (eds.), *GWAS and QTL Mapping in Horticultural Crops*,
https://doi.org/10.1007/978-981-95-2306-1_6

Abbreviations

CGMMV	Cucumber Green Mottle Mosaic Virus
Cm	Centimorgan
CMV	Cucumber Mosaic Virus
CVYV	Cucumber Vein Yellowing Virus
DM	Downy Mildew
GA	Gibberellic Acid
GWAS	Genome-Wide Association Studies
PM	Powdery Mildew
PRSV	Papaya Ring Spot Virus
QTL	Quantitative Trait Loci
Rils	Recombinant Inbred Lines
SSR	Simple Sequence Repeats
WMV	Watermelon Mosaic Virus

6.1 Introduction

Cucumbers (*Cucumis sativus* L.) are an economically important vegetable crop cultivated worldwide (Zhou et al. 2015a, b, c). The tender fruits are consumed fresh in salads, as pickles, in sweets, and as cooked vegetables. Fresh cucumbers, which are largely composed of water, provide essential nutrients beneficial to human health. The flesh is rich in vitamins A and C, as well as folic acid, while the firm skin contains minerals such as calcium, potassium, and magnesium. Cucumbers are recognized for their anti-inflammatory properties and are traditionally believed to alleviate various health conditions, including jaundice, eczema, constipation, hypertension, and even cancer. They hold a significant position in Ayurvedic medicine and have diverse applications in culinary arts, therapeutics, and cosmetics (Jia and Wang 2021). Beyond their dietary importance, cucumbers serve as a model system for studying plant vascular biology and sex determination (Zhou et al. 2015a, b, c). Cucumber breeding has undergone considerable evolution with the advent of modern technological advancements. Traditional breeding methods, which primarily focused on improving yield, disease resistance, and stress tolerance, have laid the groundwork for contemporary approaches that utilize cutting-edge genomic tools. Progress through conventional breeding is typically slow and time-consuming due to the extended duration required for selecting desirable traits in cultivars. To accelerate this process and address the demands of a growing population, researchers have employed novel molecular and transgenic techniques in cucumber breeding (Kaur and Sharma 2022). The integration of technologies such as quantitative trait loci (QTL) mapping, genome-wide association studies (GWAS), and high-throughput sequencing has revolutionized cucumber breeding by enabling the precise identification and manipulation of genes responsible for key agronomic traits.

6.2 Genetic Diversity

Genetic diversity is essential for the success and long-term sustainability of cucumber breeding programs. It serves as the foundation for selecting and developing improved cultivars with desirable traits such as higher yield, enhanced fruit quality, disease resistance, and tolerance to abiotic stress. Cucumber (*Cucumis sativus* L.) belongs to the diverse *Cucurbitaceae* family. The *Cucumis* genus, which includes cultivated species such as cucumber and melon, along with wild relatives like *C. hystrix* and *C. sativus* var. *hardwickii*, exhibits substantial phenotypic variation. This includes differences in growth habit, fruit skin texture, spine density, fruit size, sex expression, and flesh bitterness. This diversity is distributed across approximately 50 recognized species native to regions in Africa, Asia, and India. Despite the broad phenotypic and geographic diversity, efforts to hybridize cultivated cucumbers with their wild relatives have met with limited success (Weng et al. 2010; Kacar et al. 2012; Lv et al. 2012; Zhang et al. 2012; Qi et al. 2013). One major challenge in cucumber breeding is its relatively low genetic diversity compared to other crops, largely a consequence of domestication bottlenecks and selective breeding focused on a narrow set of agronomic traits. Nevertheless, wild cucumber species and traditional landraces continue to harbor significant genetic variation that can be harnessed for crop improvement. Advances in next-generation sequencing technologies have facilitated more comprehensive analyses of cucumber genetic resources, enabling the discovery of novel alleles and gene variants (Kaur and Sharma 2022). This highlights the critical importance of conserving and studying wild *Cucumis* species to support future breeding efforts. Currently, QTL mapping and GWAS are preferred approaches in cucumber breeding, as they streamline the identification of genomic loci associated with important traits. These loci serve as valuable targets for MAS and the development of improved cucumber varieties.

6.3 QTL Mapping in Cucumber

QTL mapping is a statistical method used to identify genomic regions (loci) associated with specific quantitative traits, such as plant height, fruit yield, or disease resistance. These traits are typically controlled by multiple genes and influenced by environmental factors. QTL mapping has significantly advanced cucumber breeding by offering a genetic roadmap for traits of agronomic importance. It involves establishing associations between phenotypic variation (observable traits) and genotypic variation (genetic markers). With the advent of high-throughput sequencing technologies, researchers can now rapidly identify loci linked to target traits, facilitating the dissection of complex traits into simpler genetic components. This enables more precise and efficient breeding through approaches such as linkage analysis and marker-assisted selection (MAS).

For instance, Qi et al. (2013) identified five QTLs—*fl1.1, fl3.1, fl4.1, fl4.2, and fl6.1*—associated with fruit length by analyzing resequencing data from 115 cucumber accessions. Subsequently, Sheng et al. (2019) and Pan et al. (2020) contributed nine additional stable QTLs—*FS1.3, FS1.4, FS2.3, FS4.2, FS4.3, FS5.2, FS5.3, FS6.3,* and *FS7.2*—which further refined the regulatory model of cucumber fruit shape and other key plant architecture traits (Li et al. 2008). These studies provide critical insights into the genetic architecture of complex traits. A more recent advancement, QTL-seq, integrates traditional QTL mapping with next-generation sequencing technologies to expedite the discovery of trait-associated loci. This method involves sequencing the genomes of two phenotypically contrasting groups and analyzing differences in allele frequency to identify genomic regions associated with the trait of interest. The identified QTLs for various cucumber traits are summarized in Table 6.1.

6.4 GWAS in Cucumber Breeding

GWAS offer high-resolution insights into the genetic architecture of traits across diverse populations. By leveraging the natural genetic variation present in germplasm collections, GWAS enables the identification of candidate genes associated with QTLs for important agronomic traits in cucumber, including disease resistance, abiotic stress tolerance, fruit quality, and yield. Through the association of genetic variants with phenotypic traits, GWAS facilitates the discovery of loci and candidate genes that can be functionally validated and integrated into breeding programs.

The advent of high-throughput sequencing technologies has provided access to vast genomic datasets, allowing for comprehensive genotyping and precise phenotyping. This data-driven approach accelerates breeding cycles, improves selection accuracy, and broadens the genetic base of cultivated cucumber varieties.

In cucumber, several key loci have been identified through GWAS. Bo et al. (2019a, b) reported a novel locus, *fsd6.1*, associated with fruit spines, and two loci, *qgf5.1* and *qgf3.1*, linked to green fruit flesh. Shang et al. (2014) identified the Bi gene responsible for bitterness in a study involving 115 cucumber germplasm accessions. In the context of disease resistance, Wang et al. (2018a, b) and Liu et al. (2020) evaluated 395 cucumber germplasm resources for downy mildew resistance and identified 18 associated loci, with five candidate genes predicted at major-effect loci. For powdery mildew resistance, Jia et al. (2021) analyzed 231 core germplasm (CG) accessions and identified 12 resistance-associated loci, predicting 63 candidate genes. Zhang et al. (2018a, b) identified *Csa5M622830*, a gene encoding a GATA transcription factor, as a potential candidate conferring powdery mildew resistance derived from *Cucumis hystrix*. In a Korean breeding line, *CsGy5G015660*, encoding a leucine-rich repeat receptor-like serine/threonine-protein kinase (RPK2), was proposed as a causative gene for resistance (Zhang et al. 2021a, b). In terms of abiotic stress tolerance, several studies have reported significant loci: Wang et al. (2019a, b) identified four loci associated with low-temperature resistance at the seedling stage; Zhang et al. (2019) found five loci related to high-temperature tolerance during seed germination; and Wei et al. (2019) reported seven loci associated with heat tolerance in seedlings, predicting 67 candidate genes at the *gHII4.1* and-v*gHII4.2* loci. The novel genes identified for various cucumber traits through GWAS are summarized in Table 6.2.

Table 6.1 List of QTLs identified for various traits in cucumber

Sl no.	Trait	QTL	Location	Source	References
I. Plant architecture					
1	Determinate	*De*	Chromosome 6 (1.4 cM)	Gy7 × H-19 RIL	Weng et al. (2010)
2	Little leaf	*Ll*	Chromosome 6 (4.2 cM)	Gy7 × H-19 RIL	Weng et al. (2010)
3	Compact plant	*Cp*	Chromosome 4 (2.1 cM) Chromosome 4 (2.1 cM) Chromosome 4 (2.0 cM)	USDA North Central Regional Plant Introduction Station	Li et al. (2011)
4	Plant height	*qPL3* *qPL4*	Chromosome 3 Chromosome 4	$F_{2:3}$ C18 (determinate) and C19 (indeterminate)	Kalvandi et al. (2018)
II. Fruit and flower-related traits					
5	Femaleness	*F* locus *sex1.1* *sex1.2*	Chromosome 1 5 cM and 0.8 cM from *F* locus	RILs and F_2 G421 (gynoecious) × H19 (monoecious)	Fazio et al. (2003)
6	Sub-gynoecious	*sg3.1* linked to SSR13466 *sg6.1* and *sg6.2* SSR01308 and SSR02123 respectively	Chromosome 3, 6	F_2 240-1-2-2-3-1 (gynoecious) × 3- 5-1-3-2-1-1-1-1-2 (monoecious)	Bu et al. (2016)
7	Green flesh	*qgf5.1(Csa5G021320)* *qgf3.1 (Csa3G177380)*	Chromosome 5, 3	F_2 one BC_1P_1, BC_1P_2 9110Gt (P_1) × PI183967 (P_2)	Bo et al. (2019)
8	β-carotene (*Ore* gene)	*ec3.1* *CsBCH1 (CsGy3G017310.1)*	Chromosome 3	RIL-F_9 CC3 × SWCC8	Bo et al. (2012) and Qi et al. (2013)
9	Fruit length	*fl3.1*, *fl5.1*, *fl6.1*	Chromosome 3, 5, 6	F_2 and $F_{2:3}$ WI7200 × WI7167P	Bo et al. (2015) and Pan et al. (2017)

(continued)

Table 6.1 (continued)

Sl no.	Trait	QTL	Location	Source	References
III. Traits related to disease resistance					
10	Downy mildew resistance *(dm)*	*dm1.1*	Chromosome 1 SSR marker to *dm1.1* was 9.7 cM	F_2 and $F_{2:3}$ families of Inbred line K 8 (R) × K18 (S)	Bai et al. (2008)
11		*dm4.1* *dm5.1*	Chromosome 4 Chromosome 5	243 $F_{2:3}$ Cross between DM-resistant inbred line WI7120 (PI 330628) and susceptible '9930'	Wang et al. (2016)
12		*dm2.2*	Chromosome 2	F2-derived F3 ($F_{2:3}$) TH118FLM (resistant to DM) and WMEJ (susceptible to DM)	Win et al. (2017)
13		*Dm4.1* *Dm1.1* *Dm5.2*	Chromosome 4 Chromosome 1 Chromosome 5	183 F2-derived F3 ($F_{2:3}$) PI 197088 (resistant) and Changchunmici (susceptible)	Li et al. (2018)
14		*dm5.1, dm5.2, dm5.3*	Chromosome 5	$F_{2:3}$ families of PI 197088 (R) × Coolgree (S)	Wang et al. (2018a, b)
15		*dm-1*	Chromosome 1	RILs of Gy14 (R, derived from PI 197087) × 9930 and $F_{2:3}$ families of WTF23 (R, derived from PI 197087) × True Lemon	Wang et al. (2019a, b)
16		*dmG7.1*	Chromosome 7	Core germplasm	Liu et al. (2020)
17		*dm1.1* *dm5.1* *dm5.2* *dm5.3* *dm6.1*	Chromosome 1 2128970- 6490330 Chromosome 5 6608865- 20656139 Chromosome 5 20656139- 22056501 Chromosome 6 25127215- 25127410	F_2 and $F_{2:3}$ families derived from the cross K8 × K18	Zhang et al. (2013)

18	Powdery mildew	*pm5.1, pm5.2, pm5.3, pm6.1*	Chromosome 5, 6	F_2 and $F_{2:3}$ families of K8 (R) × K18 (S)	Zhang et al. (2011)
19		*pm5.2* *pm6.1*	Chromosome 5 Chromosome 6	F_1, F_2, and F2-derived F3	Zhang et al. (2021a, b)
20		*pm1.1, pm2.1, pm4.1, pm6.1*	Chromosome 1, 2, 4, 6	RILs ($F_{6:7}$ family) of Inbred line S94 (R) × S06(S)	Liu et al. (2008)
21	Gummy stem blight	SSR02697, SSR06632	Chromosome 4, 6	ILs derived from *C. Hystrix C. Sativus*	Lou et al. (2013)
22		*gsb3.1, gsb3.2, gsb3.3, gsb4.1, gsb5.1, gsb6.1*	Chromosome 3, 4, 5, 6	RILs derived from PI 183967 931	Liu et al. (2017)
23		*gsb-s1.1, gsb-s2.1, gsbs6.1, gsb-s6.2,*	Chromosome 1, 2, 6	RILs derived from PI 183967 931	Zhang et al. (2017)
24		*Gsb1*	Chromosome 1	RILs derived from IL77 8419	Zhang et al. (2018a, b)

Table 6.2 List of candidate genes identified for different traits using GWAS in cucumber

Sl no.	Trait	QTL	Location	Population	Putative gene	Gene annotation	References
I.	Genes identified for plant architecture, growth, and development						
1	Pre-harvest sprouting in cucumber	*qPHS4.1*	Chromosome 4 0.53 Mb	F_2 Q12 (North China fresh market cucumber, PHS resistant, P1) and P60 (North China fresh market cucumber, PHS susceptible, P2)	*Csa4G622760* *Csa4G622800*	Chalcone isomerase-like protein Peptide methionine sulfoxide reductase msrB	Cao et al. (2021)
2	Stem girth (*gSD*)	*gSD1.1*	Chromosome 1: 15,337,811bp		*CsaV3_1G028310*	Associated with antibiotic biosynthesis	Yang et al. (2022)
3		*gSD2.1*	Chromosome 2: 3,445,551		*CsaV3_2G006960*	Proteins associated with TPR related to mitochondrial RNA processing	
4		*gSD3.1*	Chromosome 3: 7,785,079		*CsaV3_3G009560*	LITTLE ZIPPER protein	
5		*gSD5.2*	Chromosome 5: 25,600,905		*CsaV3_5G031320*	encoded a kinase function in chloroplast	
6		*gSD6.1*	Chromosome 6: 17,507,945		*CsaV3_6G031260*	associated with gibberellin	

II	Genes identified for flowers and flowering						
7	Ovary- and Fruit-Related Traits	*qOHN4.1*	chromosome 4 114-kb	155 $F_{7:8}$ RIL CCMC and IL52		Predicted 13 genes in the candidate region of ovary hypanthium neck (*qOHN4.1*) locus according to 9930 v3.0 draft genome	Wang et al. (2023)
III	Genes identified for fruits						
8	fruit size/ shape (FS)	*FS5.2*	Chromosome 5 95.5 kb	NIL-derived segregating populations (BC1 to BC4, BC4F2) WI7167 and WI7237		A homolog of the Arabidopsis CRABS CLAW (CsCRC)	Pan et al. (2022)
9	Fruit length	*Fl4.1*	Chromosome 4	RIL (151) "HW" and "XTMC"		*Csa4G337340* encoding an auxin efflux carrier family protein	Xing et al. (2023)
10	Pericarp color	*CsPC1*	Chromosome 3 94-kb	F_2, BC_4F_2 G35 (light-green pericarp color) and Q51 (dark-green pericarp color)		*Csa3G912920* which encodes a GATA transcription factor (*CsGATA1*)	Huang et al. (2022)

(continued)

Table 6.2 (continued)

Sl no.	Trait	QTL	Location	Population	Putative gene	Gene annotation	References
IV	Genes identified for disease resistance						
11	Downy mildew resistance *(dm)*	*DM4.1.2*	Chromosome4:15766975	NILs and $F_{2:3}$ families of PI 197088 (R) × HS279 (S)	*CsLRK10L2* *Csa4M410850*	Receptor-like kinase; contains IPR011009 (Protein kinase-like domain)	Berg et al. (2020)
12		*dmG4.1*	Chromosome.4:5250-5350kb	Core germplasm	*Csa4G064680*	an eukaryotic translation initiation factor 5A	Liu et al. (2020)
13		*dmG5.2*	Chromosome.5:22883182-22313023		*Csa5G606470*	a WRKY transcription factor 2–4	
14		*dmG7.1*	Chromosome.7:34900-134900		*Csa7G004020*	–	
15	Downy Mildew Resistance and Powdery Mildew Resistance	*pm/dm5.2*	24,738,421	RILs changchunmici (CCMC) × IL52	*Csa5M622830.1*	GATA transcriptional factor gene	Zhang et al. (2018a, b)
16	Powdery mildew resistance	SNP (s5_16047445)	Chromosome.5 : 16,045,687-16,048,001 +	264 accessions, which included the local wild varieties and the breeding lines,	*Csa5G453160*	Short-chain dehydrogenase/ reductase	Lee et al. (2020)
17		SNP (s5_16623037)	Chromosome.5 : 16,621,524-16,631,782		*Csa5G471070*	Anaphase-promoting complex subunit	
18		*pm5.2, pm6.1*	Chromosome.5 Chromosome.6	F_2 population of PM-R × PM-S	*CsGy5G015660*	–	Zhang et al. (2021a, b)

19	*Cucumber Mosaic Virus* resistance	*cmv6.1* (SSR9-56–SSR11-177)	Chromosome6 8,004,868 to 9,628,895 bp	RIL 65Gx 02245	*Csa6M133680*	a bZIP domain-containing transcriptional activator	Shi et al. (2018)
20	*Cucumber Vein Yellowing Virus* (CVYV) resistance		Chromosome05:5,088,092-8,208,448 bp	CE0754 × CE0749 F_1, F_2, and $F_{2:3}$ plants were inoculated mechanically with CVYV-AILM	*CsCvy-1*	–	Martínez-García et al. (2004) and Pujol et al. (2019)
21	*Cucumber Mosaic Virus*	*cmv6.1*	Chromosome 6 1624.0 kb	RILs 65G × 02245		*Csa6M133680* bZIP transcription factor family protein	Shi et al. (2018)

(continued)

Table 6.2 (continued)

Sl no.	Trait	QTL	Location	Population	Putative gene	Gene annotation	References
V	Genes identified for abiotic stress tolerance						
22	Heat tolerance	*qHT3.2*	Chromosome.3: 31,699,402–33,960,084 bp	Recombinant inbred lines (RILs) from 65G × 02245	*Csa3G822410* *Csa3G824920* *Csa3G824940* *Csa3G823060* *Csa3G825010* *Csa3G824990* *Csa3G822440* *Csa3G825020* *Csa3G824850*	HSP20-likechaperone NBS-containing resistance like protein TIR-NBS-LRR resistance protein Calmodulin-like protein 1 Calmodulin-likeProtein NAC domain protein AP2-like ethylene-responsive transcription factor Dof zinc finger protein MYB transcription factor	Dong et al. (2020)
23		*qHT1.1*	Chromosome1: 987,005–990,966bp	931 × 99281 F_2 and RIL population	*Csa1G004990*	RING finger protein	Liu et al. (2021)
24	Low-temperature tolerance	*qLTG1.1* (SSR11340)	Chromosome1: 14,855,251–15,424,716 bp	65G × 02245 F_9 RILs	*Csa1M408720*	encodes the DELLA protein	Song et al. (2018)
25		*gLTG1.2*	Chromosome1: 29,185,530-29,524,806bp	151 accessions from different geographical origins	*CsaV3_1G044080*	encoded a pentatricopeptide repeat-containing protein (CsPPR)	Li et al. (2023)

26	Salt tolerance	*gST2.1*	Chromosome 2: 23,409,202~23,509,202 bp	220 accessions	*CsaV3_2G035120*	encoding an isocitrate lyase	Liu et al. (2022)
27		*gST3.1*, *gST3.2*	Chromosome.3: 21,496,729~21,596,729 bp		*CsaV3_3G023710*	encoding an ankyrin repeat family protein	
28		*gST4.2*	Chromosome.4: 23,423,770~23,523,770 bp		*CsaV3_4G033150*	encoding a phosphoenolpyruvate carboxylase protein	

Implications of QTL Mapping and GWAS A list of QTLs and genes identified in cucumbers for different traits is presented in Tables 6.1 and 6.2.

6.5 Plant Architecture

6.5.1 Plant Height

Typically, cucumber cultivars exhibit an indeterminate growth habit, characterized by tall, vining plants with numerous lateral branches and long internodes. This growth pattern is well-suited for trellis-supported production systems. However, compact or dwarf-growing cucumber varieties offer significant advantages, particularly for mechanized harvesting, by improving efficiency and reducing labor costs (Cramer and Wehner 2000). The manipulation of plant architecture, especially traits related to reduced plant height, plays a critical role in modern plant breeding. In many crops, plant height is a key yield-related trait. Dwarf, semi-dwarf, or bushy growth habits are often associated with shorter internodes and stronger stems, which enable higher planting densities and thus greater productivity per unit area (Hedden 2003; Teng et al. 2013). The most prominent examples of this concept are the "Green Revolution" genes, which affect the biosynthesis and signaling pathways of the plant hormone gibberellic acid (GA). Notable among these are the *Rht* (reduced height) genes in wheat (Peng et al. 1999) and the *sd1* (semi-dwarf) genes in rice (Monna et al. 2002; Sasaki et al. 2002; Spielmeyer et al. 2002).

The compact (cp) gene was initially identified in two plant introduction (PI) lines, PI308915 and PI308916 (Kauffman and Lower 1976), and was later mapped to a 220-kb genomic region on cucumber chromosome 4 (Li et al. 2011). A chemically induced compact mutant, **cp-2**, characterized by shorter internodes and smaller seeds, was described by Kubicki et al. (1986). Another compact cucumber mutant, described by Crienen et al. (2009), differs from both *cp* and *cp-2* and displays semi-dominant inheritance. A recessive gene, **scp**, responsible for an ultra-compact phenotype, causes significantly reduced stem length, absence of branching, and wrinkled, dark-green leaves (Niemirowicz-Szczytt et al. 1996). Dwarfism and determinate growth have also been linked to a loss-of-function mutation in the cyclin-dependent protein kinase inhibitor gene (*CsSMR1*) (Li et al. 2022). Several other compact or super-compact cucumber mutants have been reported. These include super compact-1 (*scp-1*), which encodes the cytochrome P450 protein *CsCYP85A1* (Wang et al. 2017); super compact-2 (*scp-2*), which encodes the steroid 5-alpha-reductase *CsDET2* (Hou et al. 2017); and the compact plant architecture *(cpa)* mutant, involving the 7-dehydrocholesterol reductase gene (Zhang et al. 2021a, b). These studies highlight the complexity of genetic control over plant height and architecture in cucumber, with the brassinosteroid (BR) biosynthesis and signaling pathway playing a central role.

6.5.2 Stem Diameter

The stem plays a critical role in providing structural support and facilitating the transport of water, nutrients, and photosynthates. Research has demonstrated a positive correlation between stem diameter and yield in cucumbers. In mini cucumber cultivars, increased stem diameter has been shown to enhance yield potential (Yang and Wang 2016; Cao et al. 2013). Additionally, blue LED light has been reported to significantly promote stem thickening. Biological fertilizers have also been found to positively influence stem diameter and other yield-related traits (Tomar and Khajanji 2009). A similar positive association between stem diameter and yield was observed by Kamil et al. (2015). Given the limitations of available arable land, developing cucumber cultivars with thicker stems is essential to improving productivity. Therefore, identifying the key genes and molecular pathways that regulate stem morphology is crucial for the targeted genetic improvement of cucumber.

6.6 Genes Controlling Quality Traits

Fruit quality in cucumber is regulated by complex networks involving genetic, biochemical, metabolic, and physiological processes, particularly those active during fruit ripening. These quality attributes are critical for domestication, commercial value, and breeding objectives (Qi et al. 2013). Cucumber exhibits a wide range of fruit sizes, shapes, and colors, which are controlled by various genes and QTLs. The QTLs and candidate genes associated with different fruit quality traits are summarized in Tables 6.1 and 6.2.

6.6.1 Fruit Size and Shape

Wenzel et al. (1995) were the first to identify five QTLs associated with marketable fruit length, three QTLs for fruit diameter, and four QTLs for fruit shape index using an F_2:$_3$ population. Zhao et al. (2019) identified the fruitfull (CsFUL1) gene, which plays a role in fruit elongation. Short fruit mutants (*sf1*) exhibit increased degradation and elevated expression of the *ACS2* gene, which encodes a cucurbit-specific RING-type E3 ligase involved in ethylene biosynthesis (Xin et al. 2019). A short fruit neck is a highly desirable trait due to its contribution to a more uniform appearance and the production of non-bitter fruits, whereas long-necked fruits are often associated with bitterness. Several studies have also identified QTLs regulating fruit peduncle length (Song et al. 2016; Yuan et al. 2008).

6.6.2 Fruit Flesh Color

Fruit flesh color is a key quality trait influenced by the composition and concentration of chlorophyll and carotenoid pigments, which are often used by consumers to assess freshness and flavor. Breeders prioritize the development of cucumber varieties with appealing flesh colors to enhance marketability. Variation in pigment content gives rise to diverse flesh colors in cucurbit fruits, including red, orange, yellow, green, light green, and white. Chlorophyll and its derivatives, such as chlorophyllin, have been shown to exhibit protective effects against carcinogens like aflatoxins in animals. Carotenoids—including β-carotene, α-carotene, and lutein—serve as precursors to vitamin A and play vital roles in human health, particularly in maintaining vision. While white flesh is predominant in cultivated cucumber varieties, orange flesh has been observed in the semi-wild Xishuangbanna cucumber (*Cucumis sativus* L. var. *xishuangbannanesis* Qi et Yuan), which exhibits high levels of β-carotene accumulation due to a mutation in the *CsBCH* gene (Qi et al. 2013; Bo et al. 2012). The first report of orange flesh in semi-wild cucumber species due to β-carotene synthesis was made by Qi et al. (1983). Genetic studies have shown that two recessive genes govern the orange mesocarp, while a single recessive gene controls orange pigmentation in the endocarp, which has been mapped to chromosome 3 (Cuevas et al. 2010).

Another variety, PI200815 from Myanmar, exhibits yellow flesh at maturity, with the trait governed by a single recessive gene (*yf*), which was preliminarily mapped by Lu et al. (2015). Earlier, Kooistra (1971) identified two genes responsible for variations in flesh color, including orange, yellow, dingy white, and intense white. In contrast, green flesh is a critical trait in modern cucumber breeding, particularly because it is present throughout fruit development. Green flesh color (*gf*) is controlled by two loci and results from chlorophyll accumulation. Unlike orange and yellow pigmentation, which appears only at the mature stage, green flesh contributes to the market value of immature cucumbers and is therefore more desirable in commercial breeding programs.

6.6.3 Fruit Pericarp Color

Pericarp color is a critical trait in the horticultural industry, as it significantly influences consumer preference and exhibits extensive phenotypic variation that is valuable for breeding programs. Numerous QTLs and genes associated with pericarp color have been identified and cloned in various crops. In cucumber, the white pericarp trait is controlled by the locus *w* on chromosome 3, which harbors a single gene, *Csa3G904140* (*APRR2*) (Liu et al. 2015). This gene encodes a nuclear-localized transcription factor that modulates pericarp color by reducing chlorophyll and chloroplast content (Cortleven and Schmulling 2015; Jiao et al. 2017). Another gene, *Csa7G051430*, identified via bulked segregant analysis sequencing (BSA-seq) from a cross between a light-green pericarp mutant (*lgp*) and wild-type cucumber, is homologous to *Arabidopsis ARC5* and plays a vital role in chloroplast division (Gao et al. 2003; Zhou et al. 2015a, b, c). Similarly, *Csa6G133820*, mapped through the light-green leaf and pericarp mutant *M218*, encodes a Ycf54-like protein essential for chlorophyll biosynthesis and is designated *CsYcf54* (Bollivar et al. 2014; Lun et al. 2015). Additionally, *Csa2G352940* (*CsMYB36*), which encodes the

MYB36 transcription factor, regulates yellow-green peel coloration in cucumber (Hao et al. 2018). Li et al. (2013) demonstrated that a single dominant gene (*B*), located on chromosome 4, controls black spine and orange fruit color by encoding an R2R3-MYB transcription factor. The genetic mechanism underlying green pericarp color in cucumber remains unclear. Further research focusing on inheritance patterns and the identification of candidate genes associated with green pericarp color will provide valuable insights for breeding programs.

6.7 Flowering and Sex Expression

6.7.1 Flowering Time

In cucumber, early yield is influenced by the flowering period and the position of the female flower. Miao et al. (2011) identified QTLs for the first female flower-bearing node in recombinant inbred lines (RILs) derived from crosses between monoecious (ff) and gynoecious (FF) lines, detecting three major-effect QTLs. Various studies have reported the genetic basis of the first female flower trait, which is closely related to flowering time and sex expression. Tan et al. (2015) identified the *m-1* allele controlling bisexual flowering in andromonoecious lines, resulting from a 14-bp deletion in the *CsACS2* gene. This deletion produces a truncated, loss-of-function 1-aminocyclopropane-1-carboxylic acid synthase protein, associated with elongated fruit shape and perfect flowers. Sheng et al. (2020) reported a major QTL, *FT1.1*, regulating flowering time, along with a minor QTL, *FT6.3*, controlling photoperiod-sensitive flowering time in cultivated cucumber. In segregating populations, four QTLs (*Sgy3.1, Sgy4.1, Sgy6.1,* and *Sgy6.1*) were identified for female flower traits, with the major QTL *Sgy3.1* explaining 54.6% of phenotypic variation on chromosome 3 (Bu et al. 2016). Win et al. (2019) identified three QTLs (*sg1.2, sg3.1,* and *sg1.1*) associated with gynoecy in cucumber, reporting two major QTLs conferring femaleness.

6.7.2 Sex Expression

Although sex expression in cucumber is a complex genetic trait, it is essential for the genetic improvement of the crop (Dhall et al. 2023). Cucumber plants can produce three distinct flower types: staminate (male), pistillate (female), and hermaphrodite (bisexual or perfect). Early studies on cucumber sex determination identified three key genes—*F* (femaleness), *m* (andromonoecy), and *a* (androecy)—all belonging to the 1-aminocyclopropane-1-carboxylic acid synthase (ACS) gene family: *CsACS1G*, *CsACS2*, *CsACS11*, and *CsACO2*, corresponding to *F*, *m*, *a*, and *a-1*, respectively (Kubicki 1969; Pierce and Wehner 1990; Trebitsh et al. 1997; Mibus and Tatlioglu 2004; Li et al. 2009). Consequently, the different sex expression phenotypes in cucumber can be classified as monoecious (*MMffAA*), gynoecious

(*MMFFA/a*), subgynoecious (*MMFfA/a*), andromonoecious (*mmffAA*), hermaphroditic (*mmFFA/a*), and androecious (*MMffaa*/*mmffaa*) (Li et al. 2020).

6.7.3 Parthenocarpy

Parthenocarpy is widespread in cucumber germplasm resources, allowing fruit set without pollination, which is particularly important under protected cultivation conditions. Parthenocarpy can be induced by exogenous application of plant growth regulators (PGRs) (Fu et al. 2008; Serrani et al. 2010), developed through trait introgression via backcrossing (Sun et al. 2006a), or manipulated by transgenic expression of regulatory genes. For instance, Yin et al. (2006) demonstrated that overexpression of the *DEFH9-IaaM* gene can stimulate parthenocarpy in cucumber. The trait is controlled by an incompletely dominant gene, *Pc* (Pike and Peterson 1969). Four major QTLs associated with parthenocarpy, exhibiting significant epistasis and genotype-by-environment interactions, have been identified in processing cucumber (Sun et al. 2006b). A major QTL, *Parth2-1*, located on chromosome 2, was identified along with two candidate genes linked to parthenocarpy (Zhee et al. 2015). Lietzow et al. (2016) discovered seven QTLs conferring parthenocarpy, with four (parth5.1, parth7.1, parth6.1, and parth6.2) showing equal distribution. Wu et al. (2016) identified one major-effect QTL, *Parth2.1*, among seven QTLs mapped on chromosomes 1, 2, 3, 5, and 7.

Previous studies indicate that the hormonal regulation of fruit set involves complex and sometimes conflicting interactions (Wu et al. 2016; Li et al. 2017; Su et al. 2021). Recent research has highlighted several major loci (*parth2.1, parth5.1, parth7.1, parth6.1,* and *parth6.2*) and candidate genes such as *CsARF19*, *CsWD40*, and *CsEIN1* (Wu et al. 2016; Lietzow et al. 2016; Gou et al. 2022). However, the comprehensive molecular network underlying parthenocarpy remains to be fully elucidated, and a global understanding of the parthenocarpic process in cucumber is yet to be achieved.

6.7.4 Disease Resistance

Given the complexity of the regulatory mechanisms underlying disease resistance, the identification and dissection of resistance QTLs to facilitate the detection of resistance or susceptibility genes from inbred lines are fundamentally important for cucumber breeding programs. Despite extensive research and the accumulation of considerable data, the molecular defense mechanisms involved in disease resistance remain insufficiently elucidated. Consequently, only a limited number of findings have been successfully translated into practical applications in MAS to date.

6.7.5 Powdery Mildew

Powdery mildew (PM) is a foliar disease that typically manifests during the middle to late growth stages but can occur throughout the entire growing season. Two fungal pathogens, *Podosphaera xanthii* (formerly *Sphaerotheca fuliginea*) and *Golovinomyces cichoracearum* (previously *Erysiphe cichoracearum*), are the primary causal agents of cucumber PM, with *P. xanthii* being the predominant pathogen in most countries (Kooistra 1968). Early infection appears as small, white, powdery fungal colonies on the upper leaf surface, which gradually coalesce. As the disease progresses, affected leaf areas become chlorotic, shrink, and eventually lose photosynthetic capacity, leading to premature leaf senescence. This reduction in foliage and photosynthetic efficiency ultimately decreases yield. The development of resistant cultivars remains the most economical and effective strategy for controlling PM in cucumber. PM resistance was first identified in Asian germplasm during the 1940s and 1950s. Notable resistant lines include PR37 from Chinese germplasm, Yomaki (PI 288,238), and cv. Natsufushinari from Japan, as well as PI 197087, cv. Bangalore, and Burma lines (PI 200,815 and PI 200,818) from India. Resistance in these accessions and older cultivars has been characterized as incomplete and inherited as a polygenic and/or recessive trait (Kooistra 1968; Ishii et al. 2001; Panstruga and Schulze-Lefert 2002; Pérez-García et al. 2009). Xu et al. (2016) fine-mapped a dominantly inherited major QTL for PM resistance to a 41.1-kb region and identified two cysteine-rich receptor-like kinase genes (*Csa1M064780* and *Csa1M064790*) as candidate genes from the Jin5-508-derived SSSL0.7 line.

PM resistance in cucumber is controlled by multiple genes, making the underlying mechanism complex. The gene *CsMLO1* has been identified and functionally confirmed through transgenic complementation. However, some cucumber lines homozygous recessive for the disease-resistant allele (*csmlo1/csmlo1*) still exhibit susceptibility to PM (Nie et al. 2015), indicating that *CsMLO1* is necessary but not sufficient for complete PM resistance. Therefore, identifying and cloning additional PM resistance genes is critical for breeding resistant cucumber cultivars. Using QTL mapping, Sun et al. (2024) identified *PM5.2* as a major-effect QTL responsible for PM resistance. Fine mapping in backcross populations localized *PM5.2* to an 18.1-kb region, with segregation predominantly associated with the *PM5.2* loci. Through map-based cloning, identification of natural variants, and transgenic complementation, *Csa5M471600*—designated *CsPM5.2*—was validated as a PM resistance gene. *CsPM5.2* encodes a phosphate transporter-like protein, *PHO1; H3*, which is broadly expressed and upregulated upon PM infection. Full PM resistance in cucumber requires the combined action of *CsPM5.2* and *cspm5.1* (*csmlo1*). Transcriptome analyses suggest that the salicylic acid (SA) signaling pathway plays a pivotal role in *CsPM5.2*-mediated resistance. These findings enhance our understanding of PM resistance mechanisms and offer valuable targets for developing resistant cucumber cultivars.

6.7.6 Downy Mildew

Cucumber production is severely threatened by downy mildew (DM), caused by the obligate biotrophic oomycete *Pseudoperonospora cubensis* [(Berkeley and M. A. Curtis) Rostovzev] (Lebeda and Urban 2004; Colucci et al. 2006). The predominant symptoms include chlorotic or yellow lesions on the upper leaf surface, typically bounded by prominent leaf veins. Symptom development is highly influenced by environmental conditions, favoring damp and humid environments, as the pathogen requires moisture to germinate and initiate new infections. Under conducive conditions, chlorotic lesions rapidly become necrotic, and sporulation appears as downy brownish-gray growth on the undersides of infected leaves. The disease eventually leads to defoliation and significant yield losses due to the loss of photosynthetically active tissue (Liu et al. 2020). The first DM-resistant cucumber cultivar, Palmetto, derived from the PR line PR40, was released in 1948 (Criswell et al. 2010). However, its resistance was overcome within two years (Epps and Barnes 1952), prompting the search for new resistance sources, including the Indian accession PI 197087 (Barnes and Epps 1954). DM resistance in PI 197087 is controlled by a recessive gene, *dm-1*, identified as a mutant allele of the *Staygreen* (*Sgr*) gene (Wang et al. 2019a, b). The *dm-1* gene, located on chromosome 5, effectively controlled DM for nearly five decades until new pathogen strains emerged in 2004, overcoming the resistance (Cohen et al. 2015).

More recently, the QTL *DM4.1*, identified in PI 197088, has been dissected into three subQTLs, each contributing to distinct resistance components: chlorosis, sporulation, and necrosis. A cucumber receptor-like kinase gene, *CsLRK10L2*, is the likely causal gene for subQTL *DM4.1.2*, which mediates resistance by reducing sporulation (Berg et al. 2020). Additionally, a loss-of-function mutation in the cucumber amino acid permease gene *CsAAP2A* underlies resistance conferred by subQTL *DM4.1.3* by limiting DM-induced chlorosis (Berg 2020). QTL regions on chromosomes 4 and 5 have been repeatedly identified across multiple studies and cucumber accessions, indicating their major and stable contributions to DM resistance. However, it remains unclear whether these overlapping QTL regions harbor the same causal genes or represent different alleles of key resistance genes such as *Sgr*, *CsLRK10L2*, and *CsAAP2A*. To date, numerous QTLs and candidate genes associated with polygenic DM resistance have been reported across all cucumber chromosomes (Tables 6.1 and 6.2).

6.8 Gummy Stem Blight

Gummy stem blight (GSB) is a serious disease caused by *Didymella bryoniae* (Auersw.) Rehm that occurs in both open field and greenhouse environments under warm and humid conditions (Van Steekelenburg 1982). The disease affects all above-ground and vegetative parts of the cucumber plant, including leaves, petioles, vines, stems, tendrils, pedicels, flowers, peduncles, fruit, and seeds. Initially, the asexual reproductive stage (anamorph) of the causal ascomycete was named

Ascochyta Cucumis Fautrey & Roum., while the sexual reproductive stage (teleomorph) was identified as *Didymella bryoniae* (Auersw.) Rehm (Chiu and Walker 1949; Sherf and MacNab 1986). Lou et al. (2013) identified two QTLs associated with GSB resistance: one located on chromosome 4 within a 12 cM interval spanning 3.569 Mbp, and another on chromosome 6 within an 11 cM interval covering 1.299 Mbp. These QTLs provide a foundation for gene cloning and elucidating the molecular mechanisms underlying GSB resistance. Zhang et al. (2017) reported that GSB resistance is controlled by three pairs of additive epistatic major genes and identified five QTLs (gsb-s1.1, gsb-s2.1, gsb-s6.1, gsb-s6.2, and gsb-s6.3) linked to resistance. Between SSR04083 and SSR02940, 117 candidate genes related to GSB resistance were predicted. Liu et al. (2017) detected six QTLs conferring GSB resistance in cucumber. Furthermore, Hu et al. (2018) identified a candidate GSB resistance gene on pseudo-chromosome 4 and developed a high-density genetic map comprising 12,932 recombination bin markers covering 1818 cM. Collectively, these studies contribute significantly to the fine mapping of major QTLs involved in disease resistance and facilitate the development of resistant cucumber cultivars.

6.8.1 Viruses

Cucumbers are susceptible to infection by at least 59 virus species (Loebenstein and Lecoq 2012), and increasing globalization facilitates the spread of new viral strains and species. Viruses pose one of the most challenging biological problems in crop production. Cucumber mosaic virus (CMV), a member of the genus *Cucumovirus* in the family *Bromoviridae* (Edwardson and Christie 1992), has a worldwide distribution and an exceptionally broad host range. It was first reported in cucumbers exhibiting mosaic symptoms in 1934 (Price 1934). CMV can infect cucumber plants throughout their growth cycle, causing symptoms such as pronounced leaf mosaic, leaf distortion, stunted growth, reduced flower production, and fruit lesions, leading to yield losses of 10–20%. Given CMV's wide host range, multiple insect vectors, and lack of effective chemical controls, breeding for genetic resistance remains the most viable management strategy (Lin et al. 2012). Reports on the inheritance of CMV resistance are inconsistent, ranging from dominant and recessive to monogenic and polygenic controls. Wang (2010) identified 17 QTLs (cmv1 to cmv17) controlling CMV resistance in "F-3" cucumber lines, located on chromosomes 3, 4, 5, and 6. Among these, cmv17 was the major locus, explaining 67.3% of phenotypic variance, and was tightly linked (0.5 cM) to the anti-CMV SCAR marker cmv-200. Further analysis of RILs revealed a QTL on chromosome 6, cmv6.1, spanning 1624 kb and encompassing nine disease resistance-related genes. Comparative genomic analysis between resistant and susceptible lines identified four single-base substitutions within the coding sequence and two in the 3′-UTR of *Csa6M133680*, suggesting it as a candidate gene for CMV resistance, while the other eight genes showed identical exonic sequences (Shi et al. 2018).

Cucumber vein yellowing virus (CVYV), an *Ipomovirus* in the family *Potyviridae*, is transmitted semi-persistently by the whitefly *Bemisia tabaci*. It induces severe vein clearing, chlorosis, and necrosis in melon (*Cucumis melo* L.) and cucumber (*Cucumis sativus* L.). Resistance QTLs for zucchini yellow mosaic virus (ZYMV), papaya ringspot virus (PRSV), and watermelon mosaic virus (WMV) have been mapped on chromosome 6 (Tian et al. 2015, 2016; Ramírez-Madera and Havey 2017). Silencing of the RNA-dependent RNA polymerase 1b (*RDR1b*) gene on chromosome 5 confers resistance to CMV, CVYV, and cucumber green mottle mosaic virus (CGMMV) (Leibman et al. 2011, 2018). Moreover, CRISPR/Cas9-mediated modification of the eukaryotic initiation factor 4E (*EIF4E*) gene on chromosome 1 conferred CMV resistance (Chandrasekaran et al. 2016). While monogenic resistance genes have been mapped and introduced into elite cucumber germplasm (Martín-Hernández and Picó 2021), they are often overcome by evolving viral strains. Polygenic resistance is considered more durable (Quenouille et al. 2013). However, genetic studies to date have surveyed only a fraction of cucumber diversity and have not fully characterized the genetic architecture of virus resistance. GWAS utilizing large, diverse panels could address this gap by linking genetic variants to phenotypic variation. Nevertheless, GWAS have been underutilized for mapping plant-virus resistance due to methodological challenges (Monnot et al. 2021).

6.9 Other Diseases

Fusarium Wilt (FW) is a major soil-borne disease in cucumber caused by *Fusarium oxysporum* f. sp. *cucumerinum*, resulting in significant yield losses. Understanding the genetics of FW resistance is critical for identifying and developing resistant sources (Pierce and Wehner 1990). Resistance to races 1 and 2 of FW is controlled by a single dominant gene in the Wis248 line. Additionally, resistance in cucumber lines WI 2757, SMR18, and Wis248 was found to be allelic (Vakalounakis 1995). Several QTLs linked to FW resistance, including *Foc2.1*, *qFoc2.1*, *qFoc6.1*, *qFoc6.2*, and *Foc4*, have been mapped to chromosomes 2, 4, and 6 (Wang 2005; De Milliano et al. 2012; Zhang et al. 2014; Zhou et al. 2015a). Fruit rot, caused by *Phytophthora capsici*, is a serious disease in field cucumbers characterized by small water-soaked lesions that expand to cover the entire fruit, leading to yield reductions (Hausbeck and Lamour 2004). Resistant sources have been identified in U.S. cucumber collections, such as PI 109483, PI 178884, and PI 214049 (Colle et al. 2014). Cucumber Scab, caused by *Cladosporium cucumerinum*, is particularly problematic in cool, foggy climates. Severe infections can cause seedling death and reduce fruit quality. Resistance is governed by a single dominant gene, *Ccu*, which is mapped on chromosome 2 (Bailey and Burgess 1934; Kang et al. 2011). Bacterial wilt, caused by *Erwinia tracheiphila*, is one of the most destructive and common diseases affecting cucumbers. It is transmitted by the striped cucumber beetle and spotted cucumber beetle. Resistance in the PI 200818 cucumber line is controlled by a single dominant gene (Iezzoni and Peterson 1980).

6.10 Abiotic Stress Tolerance

Abiotic stresses such as drought, salinity, and nutrient deficiencies can significantly affect cucumber growth, yield, and quality (Zhang et al. 2009; Amer et al. 2009; Wan Shafiin et al. 2020; Jang et al. 2021). To elucidate the genetic basis of abiotic stress tolerance in cucumber, various molecular approaches have been applied, including QTL mapping, QTL-seq analysis, and GWAS. For instance, recent research identified QTLs associated with salt tolerance on chromosomes 2 and 6, with the QTL on chromosome 2 accounting for up to 22% of the phenotypic variation. These regions harbor candidate genes involved in ion homeostasis, osmotic adjustment, and antioxidant defense, which represent promising targets for MAS or genetic engineering aimed at enhancing cucumber resilience to salt stress (Mousavi-Derazmahalleh et al. 2018).

References

Amer KH, Midan SA, Hatfield JL (2009) Effect of deficit irrigation and fertilization on cucumber. Wiley 101(6):1556–1564

Bai Z, Yuan X, Cai R, Liu L, He H et al (2008) QTL analysis of downy mildew resistance in cucumber. Prog Nat Sci 18:706–710

Bailey RM, Burgess IM (1934) Breeding cucumbers resistant to scab. Proc Am Soc Hortic Sci 32:474

Barnes W, Epps W (1954) An unreported type of resistance to cucumber downy mildew. Plant Dis Rep 38:620

Berg JA (2020) Cucumber mildew resistance: identification of cucumber genes involved in susceptibility and resistance to Powdery and Downy Mildew. Ph.D. thesis, Wageningen University & Research, Wageningen

Berg JA, Hermans FW, Beenders F, Lou L, Vriezen WH et al (2020) Analysis of QTL *DM4.1* for downy mildew resistance in cucumber reveals multiple subQTL: a novel RLK as candidate gene for the most important subQTL. Front Plant Sci 11:1601

Bo KL, Song H, Shen J, Qian CT, Staub JE et al (2012) Inheritance and mapping of the ore gene controlling the quantity of β-carotene in cucumber (*Cucumis sativus* L.) endocarp. Mol Breed 30(1):335–344

Bo K, Ma Z, Chen J, Weng Y (2015) Molecular mapping reveals structural rearrangements and quantitative trait loci underlying traits with local adaptation in semi-wild Xishuangbanna cucumber (*Cucumis sativus* L. var. *xishuangbannanesis* Qi et Yuan). Theor Appl Genet 128(1):25–39

Bo KBK, Wei Shuang WS, Wang WeiPing WW, Miao Han MH, Dong Shao Y (2019) QTL mapping and genome-wide association study reveal two novel loci associated with green flesh color in cucumber

Bo K, Miao H, Wang M, Xie X, Song Z et al (2019a) Novel loci *fsd6.1* and *Csgl3* regulate ultra-high fruit spine density in cucumber. Theor Appl Genet 132:27–40

Bo K, Wei S, Wang W, Miao H, Dong S et al (2019b) QTL mapping and genome-wide association study reveal two novel loci associated with green flesh color in cucumber. BMC Plant Biol 19:243

Bollivar D, Braumann I, Berendt K, Gough SP, Hansson M (2014) The Ycf54 protein is part of the membrane component of Mg-protoporphyrin IX monomethyl ester cyclase from barley (*Hordeum vulgare* L.). FEBS J 281(10):2377–2386

Bu F, Chen H, Shi Q, Zhou Q, Gao D et al (2016) A major quantitative trait locus conferring subgynoecy in cucumber. Theor Appl Genet 129:97–104

Cao G, Zhang GB, Yu JH, Ma YX (2013) Effects of different LED light qualities on cucumber seedling growth and chlorophyll fluorescence parameters. Sci Agric Sin 46:1297–1304

Cao M, Li S, Deng Q, Wang H, Yang R (2021) Identification of a major-effect QTL associated with pre-harvest sprouting in cucumber (*Cucumis sativus* L.) using the QTL-seq method. BMC Genomics 22:249

Chandrasekaran J, Brumin M, Wolf D, Leibman D, Klap C et al (2016) Development of broad virus resistance in non-transgenic cucumber using CRISPR/Cas9 technology. Mol Plant Pathol 17(7):1140–1153

Chiu WF, Walker JC (1949) Physiology and pathogenicity of the cucurbit black-rot fungus. J Agric Res 78:589–615

Cohen Y, Van den Langenberg KM, Wehner TC, Ojiambo PS, Hausbeck M et al (2015) Resurgence of *Pseudoperonospora cubensis*: the causal agent of cucurbit Downy Mildew. Phytopathology 105:998–1012

Colle M, Straley EN, Makela SB, Hammar SA, Grumet R (2014) Screening the cucumber plant introduction collection for young fruit resistance to Phytophthora capsici. Hortic Sci 49:244–249

Colucci SJ, Wehner TC, Holmes GJ (2006) The downy mildew epidemic of 2004 and 2005 in the eastern United States. In: Proceedings of the Cucurbitaceae Asheville NC USA 17–21, pp 403–410

Cortleven A, Schmülling T (2015) Regulation of chloroplast development and function by cytokinin. J Exp Bot 66:4999–5013

Cramer CS, Wehner TC (2000) Path analysis of the correlation between fruit number and plant traits of cucumber populations. HortScience 35(4):708–711

Crienen J, Reuling G, Segers B, van de Wal M (2009) New cucumber plants with a compact growth habit [patent] International publication number WO; vol (2009) (059777):A1

Criswell AD, Call AD, Wehner TC (2010) Genetic control of downy mildew resistance in cucumber—a review. Cucurbit Genet Coop Rep 33:13–16

Cuevas HE, Song H, Staub JE, Simon PW (2010) Inheritance of beta-caroteneassociated flesh color in cucumber (*Cucumis sativus* L.) fruit. Euphytica 171(3):301–311

De Milliano MJK, Folkertsma RT, van Paassen MQM, De Vries JS, Sela MB (2012) Fusarium resistant cucumber plants. US patent US 2012/0066790 A1

Dhall RK, Kaur H, Manchanda P, Eshanee (2023) Genetics and marker assisted breeding for sex expression in cucumber. Front Genet 14:1180083

Dong Song Z, Shuang W, Liu Y, Li C (2020) Identification of quantitative trait loci controlling high-temperature tolerance in cucumber (*Cucumis sativus* L.) Seedlings. Plants 9(9):1155

Edwardson JR, Christie RG (1992) CRC handbook of viruses infecting legumes. CRC Press, p 505

Epps W, Barnes W (1952) The increased susceptibility of the Palmetto cucumber to downy mildew in South Carolin A Plant. Dis Rep 36:14–15

Fazio G, Staub JE, Stevens MR (2003) Genetic mapping and QTL analysis of horticultural traits in cucumber (*Cucumis sativus* L.) using recombinant inbred lines. Theor Appl Genet 107:864–874

Fu FQ, Mao WH, Shi K, Zhou YH, Asami T (2008) A role of brassinosteroids in early fruit development in cucumber. J Exp Bot 59:2299–2308

Gao H, Kadirjan-Kalbach D, Froehlich JE, Osteryoung KW (2003) ARC5 a cytosolic dynamin-like protein from plants is part of the chloroplast division machinery. Proc Natl Acad Sci USA 100(7):4328–4333

Gou C, Zhu P, Meng Y, Yang F, Xu Y (2022) Evaluation and genetic analysis of parthenocarpic germplasms in cucumber. Gene 13(2):225

Hao N, Du Y, Li H, Wang C, Gong S et al (2018) CsMYB36 is involved in the formation of yellow green peel in cucumber (*Cucumis sativus* L.). Theor Appl Genet 131(8):1659–1669

Hausbeck MK, Lamour KH (2004) Phytophthora capsici on vegetable crops: research progress and management challenges. Plant Dis 88:1292–1303. https://doi.org/10.1094/PDIS.2004.88.12.1292

Hedden P (2003) The genes of the Green Revolution. Trends Genet 19(1):5–9
Hou S, Niu H, Tao Q, Wang S, Gong Z et al (2017) A mutant in the *CsDET2* gene leads to a systemic brassinosteriod deficiency and super compact phenotype in cucumber (*Cucumis sativus* L.). Theor Appl Genet 130(8):1693–1703
Hu Z, Deng G, Mou H, Xu Y, Chen L et al (2018) A re-sequencing-based ultra-dense genetic map reveals a gummy stem blight resistance-associated gene in *Cucumis melo*. DNA Res 25:1–10
Huang H, Yang Q, Zhang L, Kong W, Wang H et al (2022) Genome-wide association analysis reveals a novel QTL *CsPC1* for pericarp color in cucumber. BMC Genomics 23:383
Iezzoni AF, Peterson CE (1980) Linkage of bacterial wilt resistance and sex expression in cucumber. HortScience 15:257–258
Ishii H, Fraaije BA, Sugiyama T, Noguchi K, Nishimura K et al (2001) Occurrence and molecular characterization of strobilurin resistance in cucumber Powdery Mildew and Downy Mildew. Phytopathology 91:1166–1171
Jang SJ, Park HH, Kuk YI (2021) Application of various extracts enhances the growth and yield of cucumber (*Cucumis sativus* L.) without compromising the biochemical content. Multidiscip Digit Publish Inst 11(3):505–505
Jia H, Wang H (2021) Introductory Chapter: Studies on cucumber. In: Cucumber economic values and its cultivation and breeding. IntechOpen
Jia HX, Li XX, Song JP, Lin YE, Zhang XH et al (2021) Genome-wide association study of powdery mildew resistance of cucumber core germplasms
Jiao J, Liu H, Liu J, Cui M, Xu J et al (2017) Identifcation and functional characterization of APRR2 controlling green immature fruit color in cucumber (*Cucumis sativus* L.). Plant Growth Regul 83(2):233–243
Kacar YA, Simsek O, Solmaz I, Sari N, Mendi YY (2012) Genetic diversity among melon accessions (*Cucumis melo*) from Turkey based on SSR markers. Genet Mol Res 11:4622–4631
Kalvand H, Olfati JA, Samizadeh H, Vafaee Y (2018) Quantitative trait locus analysis of plant height–related traits in cucumber using F2: 3 population. Agric Conspec Sci 83(2):155–159
Kamil SS, Sarkawt AA, Ismael AH, Pshtiwan HA (2015) Effect of bio-fertilizer and chemical fertilizer on growth and yield in cucumber (Cucumis sativus) in green house condition. Pak J Biol Sci 18:129–134
Kang H, Weng Y, Yang Y, Zhang Z, Zhang S et al (2011) Fine genetic mapping localizes cucumber scab resistance gene Ccu into an R gene cluster. Theor Appl Genet 122:795–803
Kauffman C, Lower R (1976) Inheritance of an extreme dwarf plant type in the cucumber. J Am Sci Hortic Sci 101:150–151
Kaur M, Sharma P (2022) Recent advances in cucumber (*Cucumis sativus* L.). J Hortic Sci Biotechnol 97(1):3–23
Kooistra E (1968) Powdery mildew resistance in cucumber. Euphytica 17:236–244
Kooistra E (1971) Inheritance of fruit flesh and skin colours in powdery mildew resistant cucumbers (*Cucumis sativus* L.). Euphytica 20(4):521–523
Kubicki B (1969) Investigation of sex determination in cucumber (*Cucumis sativus* L). Genet Pol 10:123–143
Kubicki B, Sołtysiak U, Korzeniewska A (1986) Induced mutation in cucumber (*Cucumis sativus* L.). V: Compact type of growth. Genet Pol 27:289–298
Lebeda A, Urban J (2004) Distribution harmfulness and pathogenic variability of cucurbit downy mildew in the Czech Republic. Acta Fytotech Zootech 7:170–173
Lee HY, Kim JG, Kang BC, Song K (2020) Assessment of the genetic diversity of the breeding lines and a genome wide association study of three horticultural traits using worldwide cucumber (*Cucumis spp.*). Germplasm Collection. Agronomy 10:1736
Leibman D, Wolf D, Saharan V, Zelcer A, Arazi T et al (2011) A high level of transgenic viral small RNA is associated with broad potyvirus resistance in cucurbits. Mol Plant-Microbe Interact 24(10):1220–1123
Leibman D, Kravchik M, Wolf D, Haviv S, Weissberg M et al (2018) Differential expression of cucumber RNA-dependent RNA polymerase 1 genes during antiviral defence and resistance. Mol Plant Pathol 19(2):300–312

Li XZ, Yuan XJ, Jiang S, Pan J, Deng SL et al (2008) Detecting QTLs for plant architecture traits in cucumber (*Cucumis sativus* L.). Jpn Soc Breed 58(4):453–460
Li Z, Huang S, Liu S, Pan J, Zhang Z et al (2009) Molecular isolation of the M gene suggests that a conserved-residue conversion induces the formation of bisexual flowers in cucumber plants. Genetics 182(4):1381–1138
Li Y, Yang L, Pathak M, He X, Weng Y (2011) Fine genetic mapping of *cp*: a recessive gene for compact (dwarf) plant architecture in cucumber *Cucumis sativus* L. Theor Appl Genet 123(6):973–983
Li Y, Wen C, Weng Y (2013) Fine mapping of the pleiotropic locus *B* for black spine and orange mature fruit color in cucumber identifies a 50 kb region containing a R2R3-MYB transcription factor. Theor Appl Genet 126:2187–2196
Li, J, Xu, J, Guo, Q. W, Wu, Z, Zhang, T et al (2017). Proteomic insight into fruit set of cucumber (Cucumis sativus L.) suggests the cues of hormone-independent parthenocarpy. BMC Genomics, 18, 1-18.
Li L, He H, Zou Z, Li Y (2018) QTL analysis for downy mildew resistance in cucumber inbred line PI 197088. Plant Dis 102:1240–1245
Li Z, Han Y, Niu H, Wang Y, Jiang B et al (2020) Gynoecy instability in cucumber (*Cucumis sativus* L.) is due to unequal crossover at the copy number variation-dependent Femaleness (F) locus. Hortic Res 7:32
Li S, Zhang Q, Zhang H, Wang J, Sun J et al (2022) Deletion of a cyclindependent protein kinase inhibitor *CsSMR1* leads to dwarf and determinate growth in cucumber (*Cucumis sativus* L.). Theor Appl Genet 135(3):915–927
Li C, Dong S, Beckles DM, Liu X, Guan J et al (2023) GWAS reveals novel loci and identifies a pentatricopeptide repeat-containing protein *(CsPPR)* that improves low temperature germination in cucumber. Front Plant Sci 14:1116214
Lietzow CD, Zhu H, Pandey S, Havey MJ, Weng Y (2016) QTL mapping of parthenocarpic fruit set in North American processing cucumber. Theor App Genet 129:2387–2401
Lin CY, Ku HM, Chiang YH, Ho HY, Yu TA, Jan FJ (2012) Development of transgenic watermelon resistant to *Cucumber mosaic virus* and *Watermelon mosaic virus* by using a single chimeric transgene construct. Transgenic Res 21(5):983–993
Liu L, Yuan X, Cai R, Pan J, Zhu L (2008) Quantitative trait loci for resistance to powdery mildew in cucumber under seedling spray inoculation and leaf disc infection. J Phytopathol 156:691–697
Liu H, Meng H, Pan Y, Liang X, Jiao J et al (2015) Fine genetic mapping of the white immature fruit color gene w to a 33.0-kb region in cucumber (*Cucumis sativus* L.). Theor Appl Genet 128(12):2375–2385
Liu S, Shi YX, Miao H, Wang M, Li B (2017) Genetic analysis and QTL mapping of resistance to gummy stem blight in *Cucumis sativus* seedling stage. Plant Dis 101:1145–1152
Liu X, Lu H, Liu P, Miao H, Bai Y et al (2020) Identification of novel loci and candidate genes for cucumber downy mildew resistance using GWAS. Plants (Basel) 9(12):1659
Liu Y, Dong S, Wei S, Han M, Bo K et al (2021) QTL mapping of heat tolerance in cucumber (*Cucumis sativus* L.) at adult stage. Plants (Basel) 10(2):324
Liu D, Dong S, Miao H et al (2022) A large-scale genomic association analysis identifies the candidate genes regulating salt tolerance in cucumber (*Cucumis sativus* L.) seedlings. Int J Mol Sci 23(15):8260
Loebenstein G, Lecoq H (eds) (2012) Viruses and virus diseases of vegetables in the Mediterranean basin, vol 84. Academic
Lou LN, Wang HY, Qian CT, Liu J, Bai YL (2013) Genetic mapping of gummy stem blight (*Didymella bryoniae*) resistance genes in *Cucumis sativus-hystrix* introgression lines. Euphytica 192:359–369
Lu HW, Miao H, Tian GL, Wehner TC, Gu XF, Zhang SP (2015) Molecular mapping and candidate gene analysis for yellow fruit flesh in cucumber. Mol Breed 35(2):64
Lun Y, Wang X, Zhang C, Yang L, Gao D et al (2015) A CsYcf54 variant conferring light green coloration in cucumber. Euphytica 208(3):509–517

Lv J, Qi J, Shi Q, Shen D, Zhang S et al (2012). Genetic diversity and population structure of cucumber (*Cucumis sativus* L.)

Martínez-García B, Marco CF, Goytia E, López-abella D, María S et al (2004) Development and use of detection methods specific for *Cucumber vein yellowing virus* (CVYV). Eur J Plant Pathol 110(8):811–821

Martín-Hernández AM, Picó B (2021) Natural resistances to viruses in cucurbits. Agronomy 11:1–29

Miao H, Gu XF, Zhang SP, Zhang ZH, Huang SW et al (2011) Mapping QTLs for fruit-associated traits. Sci Agric Sin 44:5031–5040

Mibus H, Tatlioglu T (2004) Molecular characterization and isolation of the F/f gene for femaleness in cucumber (*Cucumis sativus* L.). Theor Appl Genet 109:1669–1676

Monna L, Kitazawa N, Yoshino R, Suzuki J, Masuda H et al (2002) Positional cloning of rice semi dwarfing gene *sd-1*: rice "Green Revolution Gene" encodes a mutant enzyme involved in gibberellin synthesis. DNA Res 9(1):11–17

Monnot S, Desaint H, Mary-Huard T, Moreau L, Schurdi-Levraud V, Boissot N (2021) Deciphering the genetic architecture of plant virus resistance by GWAS, state of the art and potential advances. Cells 10(11):3080

Mousavi-Derazmahalleh M, Bayer PE, Hane JK, Valliyodan B, Nguyen HT (2018) Adapting legume crops to climate change using genomic approaches. Wiley-Blackwell 42(1):6–19

Nie J, He H, Peng J, Yang X, Bie B et al (2015) Identification and fine mapping of pm5. 1: a recessive gene for powdery mildew resistance in cucumber (*Cucumis sativus* L.). Mol Breed 35:1–13

Niemirowicz-Szczytt K, Rucinska M, Korzeniewska (1996) A An induced mutation in cucumber (*Cucumis sativus* L.): super compact. Cucurbit Genet Coop Rep 19:1–3

Pan Y, Qu S, Bo K, Gao M, Haider KR et al (2017) QTL mapping of domestication diversifying selection related traits in round-fruited semi-wild Xishuangbanna cucumber (*Cucumis sativus* L. var *xishuangbannanesis*). Theor Appl Genet 130(7):1531–1548

Pan Y, Wang Y, McGregor C, Liu S, Luan F, Gao M et al (2020) Genetic architecture of fruit size and shape variation in cucurbits: a comparative perspective. Theor Appl Genet 133:1–21. https://doi.org/10.1007/s00122-019-03481-3

Pan Y, Chen B, Qiao L, Chen F, Zhao J et al (2022) Phenotypic characterization and fine mapping of a major-effect fruit shape QTL *FS5.2* in Cucumber *Cucumis sativus* L with Near-Isogenic Line-Derived Segregating Populations. Int J Mol Sci 23(21):13384

Panstruga R, Schulze-Lefert P (2002) Live and let live: insights into powdery mildew disease and resistance. Mol Plant Pathol 3(6):495–502

Peng J, Richards DE, Hartley NM, Murphy GP, Devos KM et al (1999) 'Green revolution' genes encode mutant gibberellin response modulators. Nature 400(6741):256–261

Pérez-García A, Romero D, Fernández-Ortuño D, López-Ruiz F, De Vicente A, Tores JA (2009) The powdery mildew fungus Podosphaera fusca (synonym Podosphaera xanthii) a constant threat to cucurbits. Mol Plant Pathol 10:153–160

Pierce LK, Wehner TC (1990) Review of genes and linkage groups in cucumber. HortSci 25(6):605–615

Pike LM, Peterson CE (1969) Inheritance of parthenocarpy in the cucumber (*Cucumis sativus* L.). Euphytica 18:101–105

Price WC (1934) Isolation and study of some yellow strains of cucumber mosaic. Phytopathology 24:743–761

Pujol M, Alexiou KG, Fontaine AS, Mayor P, Miras M et al (2019) Mapping cucumber vein yellowing virus resistance in cucumber (*Cucumis sativus* L.) by using BSA-seq analysis. Front Plant Sci 10:1583

Qi C, Yuan Z, Li Y (1983) A new type of cucumber—*Cucumis sativus L. var. xishuangbannanesis* [sic]. Acta Hortic Sin 10:259–264

Qi J, Liu X, Shen D, Miao H, Xie B et al (2013) A genomic variation map provides insights into the genetic basis of cucumber domestication and diversity. Nat Genet 45:1510–1515

Quenouille J, Montarry J, Palloix A et al (2013) Farther slower stronger: how the plant genetic background protects a major resistance gene from breakdown. Mol Plant Pathol 14:109–118

Ramírez-Madera AO, Havey MJ (2017) Different haplotypes encode the same protein for independent sources of zucchini yellow mosaic virus resistance in cucumber. HortScience 52:1040–1042

Sasaki A, Ashikari M, Ueguchi-Tanaka M, Itoh H, Nishimura A et al (2002) Green revolution: a mutant gibberellin-synthesis gene in rice. Nature 416(6882):701–702

Serrani JC, Carrera E, Ruiz-Rivero O, Gallego-Giraldo L, Peres LEP, Garcia-Martinez JL (2010) Inhibition of auxin transport from the ovary or from the apical shoot induces parthenocarpic fruit-set in tomato mediated by gibberellins. Plant Physiol 153:851–862

Shang Y, Ma Y, Zhou Y, Zhang H, Duan L et al (2014) Biosynthesis regulation and domestication of bitterness in cucumber. Science 346:1084–1088

Sheng Y, Pan Y, Li Y, Yang L, Weng Y, Flachowsky H (2019) Quantitative trait loci for fruit size and flowering time-related traits under domestication and diversifying selection in cucumber (Cucumis sativus). Plant Breed 139:176–191. https://doi.org/10.1111/pbr.12754

Sheng Y, Pan Y, Li Y, Yang L, Weng Y et al (2020) Quantitative trait loci for fruit size and flowering time-related traits under domestication and diversifying selection in cucumber (*Cucumis sativus*). Plant Breed 139:176–191

Sherf AF, Mac Nab AA (1986) Vegetable diseases and their control, 2nd edn. Wiley, New York

Shi L, Yang Y, Xie Q, Miao H, Bo K, Song Z et al (2018) Inheritance and QTL mapping of cucumber mosaic virus resistance in cucumber (*Cucumis Sativus* L.). PLoS ONE 13(7):e0200571

Song ZC, Miao H, Zhang S, Wang Y, Zhang SP, Gu XF (2016) Genetic analysis and QTL mapping of fruit peduncle length in cucumber (*Cucumis sativus* L.). PLoS One 11(12):e0167845

Song Z, Wang W, Shi L, Zhang S, Xie Q et al (2018) Identification of QTLs controlling low-temperature tolerance during the germination stage in cucumber (*Cucumis sativus* L.). Plant Breed 137(4):629–637

Spielmeyer W, Ellis MH, Chandler PM (2002) Semidwarf (*sd-1*) "green revolution" rice contains a defective gibberellin 20-oxidase gene. Proc Natl Acad Sci USA 99(13):9043–9048

Su L, Rahat S, Ren N, Kojima M, Takebayashi Y et al (2021) Cytokinin and auxin modulate cucumber parthenocarpy fruit development. Sci Hortic 282:110026

Sun Z, Lower RL, Staub JE (2006a) Analysis of generation means and components of variance for parthenocarpy in cucumber (*Cucumis sativus* L.). Plant Breed 125:277–280

Sun Z, Staub JE, Chung SM, Lower RL (2006b) Identification and comparative analysis of quantitative trait loci associated with parthenocarpy in processing cucumber. Plant Breed 125:281–287

Sun J, Nie J, Xiao T, Guo C, Lv D et al (2024) *CsPM5.2* a phosphate transporter protein-like gene promotes powdery mildew resistance in cucumber. Plant J 117:1487–1502

Tan J, Tao Q, Niu H, Zhang Z, Li D et al (2015) A novel allele of monoecious (*m*) locus is responsible for elongated fruit shape and perfect flowers in cucumber (*Cucumis sativus* L.). Theor Appl Genet 128:2483–2493

Teng F, Zhai L, Liu R, Bai W, Wang L et al (2013) ZmGA3ox2 a candidate gene for a major QTL *qPH3.1* for plant height in maize. Plant J 73(3):405–416

Tian G, Yang Y, Zhang S, Miao H, Lu H et al (2015) Genetic analysis and gene mapping of papaya ring spot virus resistance in cucumber. Mol Breed 35:1–9

Tian G, Miao H, Yang Y, Zhou J, Lu H et al (2016) Genetic analysis and fine mapping of Watermelon mosaic virus resistance gene in cucumber. Mol Breed 36:1–11

Tomar GS, Khajanji SN (2009) Effect of organic manuring and mineral fertilizer on the growth yield and economics of soybean [*Glycine max* (L.) Merrill]. Int J Agric Sci 5:590–594

Trebitsh T, Staub JE, O'Neill SD (1997) Identification of a 1- aminocyclopropane-1-carboxylic acid synthase gene linked to the Female (*F*) locus that enhances female sex expression in cucumber. Plant Physiol 113(3):987–995

Vakalounakis DJ (1995) Allelism of the Fcu-1 and Foc genes conferring resistance to fusarium wilt in cucumber. Eur J Plant Pathol 102:855–858

Van Steekelenburg NAM (1982) Factors influencing external fruit rot of cucumber caused by Didymella bryoniae. Eur J Plant Pathol 88:47–56

Wan Shafiin WN, Ablah NL, Nur Fatihah HN, Alam MA, Ma'arup R et al (2020) Breeding strategies for enhancing nutrient content and quality in Cucurbitaceae: a review. Int J Veg Sci 27:415–438

Wang YJ (2005) Studies on molecular marker of Fusarium wilt resistance related gene in cucumber (*Cucumis sativus* L.). China Northwest A&F University, pp 1–55

Wang JH (2010) Construction of a genetic linkage map and QTL analysis of the anti-CMV trait in cucumber (*Cucumis sativus* L.). Dissertation, Northwest A&F University

Wang Y, VandenLangenberg K, Wehner TC, Kraan PA, Suelmann JZ et al (2016) QTL mapping for downy mildew resistance in cucumber inbred line WI7120 (PI 330628). Theor Appl Genet 129:1493–1505

Wang H, Li W, Qin Y, Pan Y, Wang X et al (2017) The cytochrome P450 gene CsCYP85A1 is a putative candidate for Super Compact-1 (*Scp-1*) plant architecture mutation in cucumber (*Cucumis sativus* L.). Front Plant Sci 8:266

Wang X, Bao K, Reddy UK, Bai Y, Hammar SA et al (2018a) The USDA cucumber (*Cucumis sativus* L.) collection: Genetic diversity population structure genome-wide association studies and core collection development. Hortic Res 5:64

Wang Y, VandenLangenberg K, Wen C, Wehner TC, Weng Y (2018b) QTL mapping of downy and powdery mildew resistances in PI 197088 cucumber with genotyping-by-sequencing in RIL population. Theor Appl Genet 131:597–611

Wang Y, Tan J, Wu Z, VandenLangenberg K, Wehner TC et al (2019a) STAYGREEN, STAY HEALTHY: a loss-of-susceptibility mutation in the STAYGREEN gene provides durable, broad-spectrum disease resistances for over 50 years of US cucumber production. New Phytol 221(1):415–430

Wang WP, Song ZC, Bo KL, Dong SY, Wei S et al (2019b) Evaluation and genome-wide association study (GWAS) of low-temperature tolerance at seedling stage in cucumber core germplasm. J Plant Genet Resour 20:1606–1612

Wang Y, Fang Y, Ning S, Xia L, Zhan J et al (2023) QTL mapping for ovary- and fruit-related traits in *Cucumis sativus-C. hystrix* introgression line IL52. Genes (Basel) 14(6):1133

Wei S, Zhang S, Bo KL, Wang WP, Miao H et al (2019) Evaluation and genome-wide association study (GWAS) of seedling thermotolerance in cucumber core germplasm. J Plant Genet Resour 20:1223–1231

Weng Y, Johnson S, Staub JE, Huang S (2010) An extended intervarietal microsatellite linkage map of cucumber *Cucumis sativus* L. Hortic Sci 45(6):882–886

Wenzel G, Kennard WC, Havey MJ (1995) Quantitative trait analysis of fruit quality in cucumber: QTL detection confirmation and comparison with mating-design variation. Theoret Appl Genetics 91:53–61

Win KT, Vegas J, Zhang C, Song K, Lee S (2017) QTL mapping for downy mildew resistance in cucumber via bulked segregant analysis using next-generation sequencing and conventional methods. Theor Appl Genet 130:199–211

Win KT, Zhang C, Silva RR, Lee JH, Kim Y et al (2019) Identification of quantitative trait loci governing subgynoecy in cucumber. Theor Appl Genet 132:1505–1521

Wu Z, Zhang T, Li L, Xu J, Qin X et al (2016). Identification of a stable major-effect QTL (Parth 2.1) controlling parthenocarpy in cucumber and associated candidate gene analysis via whole genome re-sequencing. BMC Plant Biol 16:182

Xin TX, Zhang Z, Li S, Zhang S, Li Q et al (2019) Genetic regulation of ethylene dosage for cucumber fruit elongation. Plant Cell 31:1063–1076

Xing Y, Cao Y, Ma Y, Wang F, Xin S (2023) QTL mapping and transcriptomic analysis of fruit length in cucumber. Front Plant Sci 14:1208675

Xu X, Yu T, Xu R, Shi Y, Lin X (2016) Fine mapping of a dominantly inherited powdery mildew resistance major-effect QTL Pm1. 1 in cucumber identifies a 41.1 kb region containing two tandemly arrayed cysteine-rich receptor-like protein kinase genes. Theor Appl Genet 129:507–516

Yang S, Wang LP (2016) Analysis on heterosis and combining ability of six mini-cucumber inbred lines. J Changjiang Veg 20:44–49

Yang Y, Dong S, Miao H, Liu X, Dai Z et al (2022) Genome-wide association studies reveal candidate genes related to stem diameter in cucumber (*Cucumis sativus* L.). Gene 13(6):1095

Yin Z, Malinowski R, Ziolkowska A, Sommer H, Plcader W et al (2006) The DefH9-iaaMcontaining construct efficiently induces parthenocarpy in cucumber. Cell Mol Biol Lett 11:279–290

Yuan XJ, Pan JS, Cai R, Guan Y, Liu LZ (2008) Genetic mapping and QTL analysis of fruit and flower related traits in cucumber (*Cucumis sativus* L.) using recombinant inbred lines. Euphytica 164:473–491

Zhang Z, Liu ZQ, Zhang Y, Shu Q (2009) Effect of Cu stress on photosynthesis and nutrient absorption of grafted and ungrafted cucumber seedling. Science Press 17(1):135–139

Zhang S, Liu M, Miao H, Zhang S, Yang Y et al (2011) QTL mapping of resistance genes to powdery mildew in cucumber (*Cucumis sativus* L.). Sci Agric Sin 44:3584–3593

Zhang C, Pratap AS, Natarajan S, Pugalenhdhi L, Kikuchi S (2012) Evaluation of morphological and molecular diversity among South Asian germplasms of *Cucumis sativus* and *C. melo*. Int Scholar Res Netw Agron 134134

Zhang S, Liu M, Miao H, Zhang S, Yang Y et al (2013) Chromosomal mapping and QTL analysis of resistance to downy mildew in *Cucumis sativus*. Plant Dis 97:245–251

Zhang SP, Miao H, Yang YH, Xie BY, Wang Y et al (2014) A major quantitative trait locus conferring resistance to Fusarium wilt was detected in cucumber by using recombinant inbred lines. Mol Breed 34:1805–1815

Zhang S, Liu S, Miao H, Shi Y, Wang M et al (2017) Inheritance and QTL mapping of resistance to gummy stem blight in cucumber stem. Mol Breed 37:49

Zhang K, Wang X, Zhu W, Qin X, Xu J et al (2018a) Complete resistance to powdery mildew and partial resistance to downy mildew in a *Cucumis hystrix* introgression line of cucumber were controlled by a co-localized locus. Theor Appl Genet 131:2229–2243

Zhang X, Xu J, Li J, Lou QF, Chen JF (2018b) QTL mapping and identification of candidate gene for resistance to gummy stem blight in *Cucumis sativus/hystrix* introgression line 'IL77'. Acta Hortic Sin 45:2141e2152

Zhang S, Miao H, Song ZC, Bo KL, Wang Y et al (2019) Evaluation and genome-wide association study (GWAS) of thermotolerance for cucumber at the germination stage. J Plant Genet Resour 20:335–346

Zhang M, Song M, Cheng F, Yang Z, Davoudi M et al (2021a) Identification of a putative candidate gene encoding 7-dehydrocholesterol reductase involved in brassinosteroids biosynthesis for compact plant architecture in Cucumber (*Cucumis sativus* L.). Theor Appl Genet 134(7):2023–2034

Zhang C, Anarjan MB, Win KT, Begum S, Lee S (2021b) QTL-seq analysis of powdery mildew resistance in a Korean cucumber inbred line. Theor Appl Genet 134:435–451

Zhao JY, Jiang L, Che G, Pan YP, Li YQ et al (2019) A functional allele of CsFUL1 regulates fruit length through repressing CsSUP and inhibiting auxin transport in cucumber. Plant Cell 31:1289–1307

Zhee W, Lei L, Ting Z, Ting-lin Z, Ji L et al (2015) QTL mapping for parthenocarpy in cucumber. Sci Agric Sin 48:112–119

Zhou Q, Wang S, Hu B, Chen H, Zhang Z, Huang S (2015a) An ACCUMULATION AND REPLICATION OF CHLOROPLASTS 5 gene mutation confers light green peel in cucumber. J Integr Plant Biol 57(11):936–942

Zhou HM, Dong CJ, Zhang HY, Ren Y, Guo SG, Yang WC (2015b) SSR molecular markers and location of the gene Foc-4 linked to the resistance to Fusarium Wilt of cucumber. Mol Plant Breed 13:1980–1986

Zhou Q, Miao H, Li S, Zhang S, Wang Y et al (2015c) A sequencing-based linkage map of cucumber. Elsevier BV 8(6):961–963

GWAS and QTL Mapping in Muskmelon and Watermelon

7

Raman Selvakumar

Abstract

Two economically significant Cucurbitaceae commodities that are cultivated globally are watermelon, *Citrullus lanatus* (Thunb.) Matsum., and melon (*Cucumis melo* L.). The quality of melon and watermelon can be significantly assessed by considering the following characteristics: fruit shape, fruit epidermis pattern, fruit skin color, fruit flesh, fruit aroma, and soluble solid content (SSC). The flavor level of a melon has a substantial impact on its quality and consumer appeal. Fruit yield increase is one of the primary objectives of both public and private watermelon breeding programs. Fruit yield, average fruit weight, and fruit count are significant determinants of the marketability and profitability of the watermelon produce industry. Polygenic in nature and characterized by moderate to low heritability, fruit yield is determined by a combination of interdependent component characteristics such as vine architecture, plant density (number of plants per area), fruit count per plant, and fruit weight. The interaction between environmental factors and the physiological and morphological attributes of the plant determines the quantity of each of these yield component traits. The development of consumer-appealing fruit quality and tolerance to biotic stress in watermelon and other cucurbit crops has been the subject of protracted and iterative selection efforts. Utilizing reliable phenotypic data, high-density single-nucleotide polymorphism (SNP) markers, and historical meiotic events, the GWAS method attempts to identify marker-trait associations. The discovery of genomic regions that govern fruit yield and maturation time will be advantageous for trait introgressions assisted by markers. Utilize GWAS to examine the genomic regions (cantaloupe and melon candidate genes and significant SNPs) that underlie the observed phenotypic variation. The GWAS and quantitative trait

R. Selvakumar (✉)
Centre for Protected Cultivation Technology, ICAR-Indian Agricultural Research Institute, New Delhi, India

S. S. Dey et al. (eds.), *GWAS and QTL Mapping in Horticultural Crops*,
https://doi.org/10.1007/978-981-95-2306-1_7

loci (QTL) methodology was effectively employed to ascertain marker-trait associations through the utilization of SNP markers obtained from whole-genome resequencing.

Keywords

Watermelons · Muskmelons · Genotyping · Quality traits · Quantitative trait loci · Genome-wide association studies · Trait discovery

7.1 Introduction

The Cucurbitaceae family, which has its origins in Asia, includes economically significant species such as watermelon (*Citrullus lanatus* [Thumb.] Matsum, Nakai), squash (*Cucurbita maxima* Dutch), pumpkin (*Cucurbita moschata* Dutch), zucchini (*Cucurbita pepo* L.), cucumber (*Cucumis sativus* L.), kiwano (*Cucumis metuliferus* Mey. ex Naud), and the West Indian gherkin (*Cucumis anguria* L.). Melon is classified as a diploid species with 12 chromosomes ($2n = 24$) and a genome size of 450 Mbp (Arumuganathan and Earle 1991). C. melo is an annual herbaceous plant that produces succulent fruit consisting of three ovarian carpels, of which the placenta provides the palatable portion. Melons exhibit remarkable sensitivity to irradiation from light and temperature for the purposes of development, blossoming, and fruit production. By virtue of favorable sunlight and temperature conditions that extend from 24 to 35 °C, these organisms thrive in tropical and temperate zone greenhouses and open fields. Melons, when fully mature, comprise a considerable quantity of sugar and are commonly consumed in their raw state as an appetizer or dessert or blended into chilled bouillon or salad. On the contrary, certain varieties of melons are harvested at an early stage of ripeness and possess an astringent quality; they are then consumed raw, broiled, or preserved. Melon seeds are not typically consumed. Unusual melons are utilized for ornamental purposes in Japan and numerous other countries. Developed nations ingest the fruit of the sweet melon more frequently on account of its high market value and the fact that seed companies developed the majority of its varieties. Nonsweet melons, which exert a reduced economic influence, are predominantly sold at conventional local markets.

7.2 Melon Origin and Sub-specific Classification

Conventional taxonomic classifications (Kirkbride 1993) link *C. melo* to the African species *C. hirsutus*, *C. humifructus*, and *C. sagittatus*, which suggests the plant originated in Africa. Recent molecular phylogenies that combine chloroplast and nuclear gene regions and employ a more extensive sampling of species indicate that charcoal melon is most closely related to an Australian/Asian clade (Kocyan et al. 2007; Renner et al. 2007; Schaefer et al. 2009). This implies that the origin of *C. melo* might have been in Asia, followed by its subsequent migration to Africa. It is not uncommon to encounter uncultivated melons in East and West Africa, as well as India (McCormack and Staub 1993; Rubatzky and Yamaguchi 1997; McCreight and Bemis 1976). Similarly, wild mangoes are renowned throughout Central Asia. Bates and Robinson (1995) have additionally posited that autonomous domestications may have occurred in Asia and Africa.

Irrespective of morphological and physiological characteristics, melons have experienced an extensive period of diversification, resulting in their current state of notable diversity. From the Mediterranean Sea to Eastern Asia, the principal and secondary centers of genetic diversity are located. The categorization of the observed diversity within the species has been the focus of an extensive amount of effort. Recent classifications identify two subspecies based on the hairiness of the ovary: subspecies melo (distinguished by its long hairs) and subspecies agrestis (distinguished by its short hairs) (Jeffrey 1980, 2005). Wild forms and cultigens are both included in both subspecies, according to Stepansky et al. (1999). Munger and Robinson (1991) initially introduced a classification system that gained widespread recognition and was characterized by its simplicity. Robinson and Decker-Walters (1997) subsequently revised this system. It is composed of six cultivar groupings, which are differentiated predominantly on the basis of fruit attributes and sex classification. For example, cantalupensis, inodorus, flexuosus, and dudaim are all members of ssp. agrestis, whereas momordica and conomon are constituents of cantalupensis. Pitrat (2008) classified a total of 15 botanical varieties, including but not limited to *cantalupensis, reticulatus, adana, chandalak, ameri, inodorus, chate, flexuosus, dudaim, momordica, conomon, chinensis, makuwa, acidulus,* and *tigis*.

Cantaloupes, also known as muskmelons, of the group cantaloupensis are known to generate both andromonoecious and monoecious fruits that vary in size from medium to large, are ovoid to flat in shape, and may be sutured, ribbed, or warted on occasion. These fruits feature delectable orange and green flesh, a netted or unnetted exterior, fragrant attributes, and climacteric behavior. This collective comprises botanical varieties cantalupensis and reticulatus. It harbors a multitude of economically valuable cultivars that have their origins in Europe, Asia, and America. Notable examples of such cultivars include 'Cantaloupes,' 'Galia,' 'Charentais,' 'Western Shippers,' 'Hale's Best Jumbo,' and 'Dulce.' Furthermore, the botanical varieties Adana, Chandalak, and Ameri are included in this group. These fruits are both andromonoecious and monoecious, with a medium girth. They are climacteric in nature and possess a faint scent; their flesh is pale orange, green, or white in hue. The edges may occasionally be fringed or subtly corrugated.

Inodorus group (winter melons): Andromonoecious, substantial fruits that vary in shape from round to elliptic may be subtly adorned with a net, rib, or wrinkle, contain white or green flesh that is palatable, lack climacteric properties, and emit negligible to no aroma. In addition to Western and Central Asia, market classes such as 'Piel de sapo,' 'Rochet,' 'Tendral,' 'Amarillo,' 'Kirkagac,' and 'Hassanbey' are prevalent in Eastern and Southern Europe. 'Honeydew' is the common name for these groups on a global scale (Group flexuosus, snake melons): Fruits are typically monoecious, elliptical to extremely elongated, with furrowed or corrugated surfaces; flesh is pale orange to light orange in color; devoid of sugar and aroma;

climacteric; immature specimens are consumed similarly to cucumbers. ‘Carosello,’ ‘Alficoz,’ ‘Fakouss,’ ‘Silka,’ and ‘Kakri,’ among others, are cultivated varieties of var. chate (less elongated) and var. flexuosus that are indigenous to the Near East, Southern Europe, and Northern Africa.

Group dudaim, alternatively referred to as mango melons, are diminutive, spherical, andromonoecious fruits. Their climacteric, desiccated flesh is devoid of sugar and is typically orange, red, or brown in pigmentation. A potent aroma is also emanating from the tissues. This category comprises ornamental oriental cultivars, such as ‘Queen Anne’s’ pocket melon. As members of the Momordica group, snap melons are monoecious and produce fruits varying in length from flat to elongated. The flesh is pale in color, possesses a negligible fragrance, and has a low sugar content. Particularly climacteric are the fruits. At maturity, the fruit erupts in flames, which serves as an emblematic characteristic. Momordica cultigens are cultivated in India; some examples include ‘MR-1,’ ‘PI 124112,’ ‘PI 414723,’ and ‘Kharbuja.’ Furthermore, this particular variety can be discovered in India. It produces fruits that are oblong to elliptic in shape, consisting solely of firm white flesh that is bereft of any sugary or aromatic qualities. Group conomon, alternatively referred to as marinating melons: Conomons are distinguished by their elongated, monomorphic shape, rigid flesh that is either white or green in color, minimal to no aroma, and, above all else, the absence of climacteric qualities. The vast majority are ingested as vegetables. var. *Conomon* and the Far East Asian varieties of var. *chinensis* and makuwa are included in this group. These two cultivars yield fruits that vary in shape from pear to flat, featuring flesh that is pale orange or green in color and may exhibit sutures and ribs. Furthermore, their sugar content is minimal, and they emit a subtle aroma. Certain substances are climacteric or, in an intermediate stage, display climacteric characteristics. Ginsen makuwa, Freeman’s cucumber, Songwhan Charmi, and Shiro Uri Okayama are renowned cultigens within the Cokonon group. Esteras et al. (2009) found that genomic analyses indicate that these three varieties are similar.

Pitrat et al. (2000) introduced the term “tibish,” which they defined as an andromonoecious botanical variety distinguished by its nonclimacteric nature, solid white flesh devoid of sugar and aroma, and ovoid shape, which had not been described in previous classifications. Its composition includes varieties cultivated in Sudan. The aforementioned classified variety is referred to as the melo subspecies. Contrarily, molecular research classifies it as subspecies *agrestis*, which includes numerous wild and cultivated varieties that have their origins in Asia and Africa (Esteras et al. 2009). Var. *agrestis* is a prevalent name for melons that are grown in the open in Asia and Africa. These organisms lack climacteric traits and are monoecious. They are identified by their inedible, small fruits (less than 5 cm in length), which contain flesh that is pale green or translucent in color and is devoid of any sugar or aroma.

Multiple studies have determined that a considerable amount of the genetic diversity within this species is specific to *S. agrestis* (Stepansky et al. 1999; Akashi et al. 2002; Monforte et al. 2003; Dhillon et al. 2007, 2009; Fergany et al. 2010). The disease resistance of *cantalupensis* and *inodorus* cultivars has been enhanced by this variability; nevertheless, it has not been applied to fruit quality breeding processes. Hence, for the purpose of improving the fruit quality of 'Occidental melon' cultivars, this germplasm could be considered a genetic diversity reservoir that is underutilized.

7.3 In Vitro Culture

Organogenesis from cotyledons, hypocotyls, roots, leaves, protoplasts, and shoot meristems is capable of directly or indirectly regenerating entire melon plants. In addition, somatic embryogenesis, which typically originates from cotyledonary tissue, is possible. Given the substantial genetic variation observed among melon cultivars, it is critical to develop individualized protocols that account for the genotypes and source tissue of the melon (see Nunez-Palenius et al. 2008 for a comprehensive analysis). Breeders have the ability to propagate uncommon or intriguing genotypes vegetatively through the use of these fundamental in vitro culture techniques. Furthermore, it is possible to decrease the time span between generations by means of embryo rescue. The generation of double haploids and melon haploids through chromosome duplication with colchicine has also been accomplished by inducing in situ gynogenetic haploid parthenogenesis via pollination with gamma or X-ray-irradiated pollen (Sauton and Dumas de Vaulx 1987). The initial protocol has been successfully implemented and improved in numerous melon cultivars (Cuny and Roudot 1991; Cuny et al. 1993; Yetisir and Sari 2003; Nunez-Palenius et al. 2006; Lim and Earle 2009). Breeders commonly generate double-haploid genotypes to acquire inbred lines and expeditiously correct desirable alleles. Double-haploid lines have been utilized in foundational melon research to map genetic regions and analyze QTL (Monforte et al. 2004; Gonzalo et al. 2005).

7.4 Molecular Markers and Linkage map

Compared to other significant crop species, the development of molecular marker technology and linkage maps for melons transpired at a relatively later stage. In the early 1990s, restriction fragment length polymorphisms (RFLPs) were first documented by researchers (Baudracco-Arnas and Pitrat 1996; Oliver et al. 2001). In the years that followed, simple sequence repeats (SSRs) and, more recently, SNPs (Deleu et al. 2009), emerged as well (Danin-Poleg et al. 2001; Ritschel et al. 2004; Gonzalo et al. 2005; Fernandez-Silva et al. 2008; Fukino et al. 2008). The issue pertaining to the availability of markers in melon genetic analysis has been resolved with the recent release of an expressed sequence tag (EST) database by the International Cucumis Genomics Initiative (ICuGI). This database contains thousands of novel SSR and SNP markers and more than 100,000 ESTs. (It can be found at www.icugi.org.) The initial molecular map was developed by Baudracco-Arnas and Pitrat (1996) primarily using RFLPs and morphological characteristics. Nevertheless, this map was insufficient in its coverage of all 12 melon chromosomes. The first diagrams depicting the entire set of 12 chromosomes were published by Oliver et al. (2001) and Périn et al. (2002a). Furthermore, SNPs and SSRs have been utilized to generate dense linkage maps (Deleu et al. 2009; Gonzalo et al. 2005; Fukino et al. 2008; Fernandez-Silva et al. 2008). The absence of standardized terminology for linkage groups (LGs) impeded the capacity to compare disparate linkage maps. Adopted in 2005 by the international community was the nomenclature that Périn et al. (2002a) had proposed. The Roman numerals that comprise the Melon LG sequence span from LG I to XII. By integrating over 1500 markers and incorporating the preponderance of published maps, the ICuGI has produced an integrated map. Over the last 15 years, a considerable cohort of scientists has made significant contributions to the dissemination of an extensive collection of genetic maps, sequences, and markers pertaining to the melon community. This quantity is on par with that of other well-known fruit commodities.

7.5 Sex Expression

Melon plants possess the capacity to manifest andromonoecious or monoecious flowering patterns. Male and female blooms coexist on monoecious vegetation. Completely hermaphrodite accessions and gynoecious lines (producing only female blossoms) have been developed. Andromonoecy is notably advanced in modern melon cultivars and landraces, whereas monoecy predominates in primordial, untamed melons. Therefore, it is probable that andromonoecy originated as a strategy to increase productivity via forced pollination after the domestication of melons. In this context, two genes are involved: gene a governs the presence or absence of stamens on female flowers (Poole and Grimball 1939); gene g specifies whether or not two varieties of flowers can coexist on a single plant (Kenigsbuch and Cohen 1990). Both genomes have been cloned as of late. The gene encodes *CmACS-7*, which is a 1-aminoacyclopropane-1-carboxylic acid synthase (Boualem et al. 2008). As a result of *CMACS-7* inactivation, developing blossoms exhibit the andromonoecious phenotype. Gynoecious outcomes, such as the production of exclusively female flowers and the absence of CmWIPI expression in monoecious plants, are the consequence of transposon insertion-induced epigenetic modifications in the promoter region of the transcription factor CmWIPI (Martin et al. 2009). Monoecious cultivars have been attempted to be introduced into andromonoecious cultivars in an effort to expedite the production of F_1 commercial hybrids. However, excessive elongation of fruits in commercial production, which can be attributed to the absence of filaments in female flowers, is generally considered undesirable (Monforte et al. 2004; Abdelmohsin and Pitrat 2008).

7.6 Fruit Quality

The multifaceted nature of melon fruit quality includes characteristics such as dimensions, shape, hue, consistency, fragrance, satiety, nutritional composition, and postharvest longevity. Every single one of these characteristics is governed by a distinct gene and is subject to the impact of environmental factors. The response to selection in conventional breeding is generally minimal as a result of the complex genetic control and low genetic variability in the gene pools employed (Grandillo et al. 1999). The classification of a melon is predominantly ascertained by its variety. Proximus, acidulus, and conomon melons, which are immature, find application in culinary preparations, salads, and marinating. On the contrary, mature cantalupensis and indorus melons are frequently incorporated into fruit salads and served as dessert fruits owing to their gustatory appeal and substantial sugar content, both of which are highly regarded by consumers. Even within the realm of succulent cantaloupe, variations in desired quality attributes can be discerned. To indicate the optimal stage of maturation, cantaloupes should possess an aromatic quality, demonstrate a significant sugar content, and endure a color transition from green to yellow. In contrast, specific inodorus cultivars (e.g., the Spanish ‘Piel de Sapo’) should not experience any alteration in coloration or the development of a fragrance throughout the maturation process. On the other hand, they ought to have an extended shelf life and a greater sugar content in comparison to cantaloupes, which would likely counterbalance the reduced aromas.

7.7 QTL for Fruit Morphology

Product quality is significantly influenced by the morphology of commercial melons, as consumers hold specific expectations regarding morphologies that are indicative of particular cultivar varieties (Stepansky et al. 1999; Monforte et al. 2005). The fruit size of flexuosus accessions can vary from a few tenths of grams in wild melons to several kilograms.

There exists significant diversity with regard to fruit morphology, as evidenced by the oblate, round, ovate, elliptical, and extremely elongated fruits produced by various cultivars or accessions (Stepansky et al. 1999; Dhillon et al. 2007, 2009). Economically significant cultivars of Galia and cantaloupes are round or ovoid in shape and of a medium size (approximately 1 kg). ‘Piel de Sapo’ cultivars, on the other hand, have an elliptic shape and range in weight from 1.5 to 3 kg. Current breeding objectives encompass the maintenance of round or oval morphologies, the prevention of elongated form development, and the pursuit of moderate to small proportions (limited to 1 kg or less in weight). The model posited by Gillaspy et al. (1993) delineates the stages of melon fruit development as follows: anthesis, which follows the development of the ovary, signifies the beginning of the first stage; cellular division is the predominant mechanism of growth during the second stage; cell elongation induces growth in the third stage; and biochemical and physical transformations, which ultimately determine the final quality attributes (e.g., sugar and n), commence in the final stage. In the same way that variations in cell division rate during early fruit development stages likely explain the differences in fruit size among melon cultivars (Higashi et al. 1999), as demonstrated in tomatoes (Frary et al. 2000; Cong et al. 2002). Complex interrelationships among polygenic factors determine the dimensions of melon specimens. Quantitative trait loci (QTL) for this specific trait have been documented in multiple publications (Monforte et al. 2004; Zalapa et al. 2007; Paris et al. 2008). The direct transmission of these QTL into breeding programs is not feasible due to their origin in experimental populations. Prior to anthesis, the initial morphology of the fruit is determined (Périn et al. 2002b; Monforte et al. 2005; Eduardo et al. 2007). The fruit's form is established during the initial stages of development. It was established by Moreno et al. (2008a) that the morphological characteristics of the flower bud can be ascertained within two days of its emergence. Divergences in fruit longitudinal axis growth seem to account for the majority of the observed variation, in contrast to fluctuations along the equatorial axis (Périn et al. 2002b; Monforte et al. 2005).

Furthermore, complex regulations govern the morphology of melon products. Quarterly Taylor Reports (Périn et al. 2002b; Monforte et al. 2004; Paris et al. 2008) have identified QTL. Parrion et al. (2002b) and Monforte et al. (2004) identified four QTLs at identical positions in two distinct investigations. According to this result, the influence of QTLs on this specific attribute may be relatively consistent. As a result, fruit morphological modification could potentially be genetically modified via QTL analysis. Furthermore, the fruit morphology is influenced by the pleiotropic effects of two primary genes, namely a (as mentioned in the previous section) and p. The gene p controls the variation in carpel number; while a small number of accessions, PI161375, possess five carpels, the majority of melon genotypes have three. Fruit with a rounded morphology is characterized by the presence of five carpels, as opposed to fruit with three carpels (Périn et al. 2002b; Monforte et al. 2004; Eduardo et al. 2007; Paris et al. 2008). In recent times, Eduardo et al. (2007) presented a line genomic library that facilitates introgression. This library was compiled through a cross between the 'Piel de Sapo' cultivar from Spain and the Korean accession PI 161375 (Eduardo et al. 2005).

7.8 QTL for External Color

The external appearance of melon fruit is determined by the color of the epidermis (e.g., white, green, or yellow), the structure (e.g., smooth, furrowed, or fringed), and the presence or absence of longitudinal sutures or ribs. The intricate, multi-layered rind color may consist of a background hue and secondary hues arranged in a variety of patterns; it may also be speckled (spots <0.5 cm), spotted (spots >0.5 cm), maculate, striped (bands form from peduncle to blossom scar), or streaked. An identifiable characteristic for the consumer is ascribed to the combination of secondary design elements, tints, and texture (such as wrinkles, webbing, ribs, or warts) present in each cultivar. Specific varieties of cantalupensis are frequently embellished with interwoven ribs or sutures, which impart an exterior gray coloration to the upper surface and a green backdrop: Cantaloupes originating from the United States are characterized by dense netting, whereas those from the 'Piel de Sapo' variety exhibit spots in shades of dark green and yellow, furrowing and yellow in color, and a smooth and white appearance. Honeydews, on the other hand, are yellow in color and devoid of any spots. Mapping has been performed on significant genes associated with several of these traits: The researchers identified a severe epicarp (st) on LG III, rind marks (mt-2) on LG II, and rind sutures (s-2) on LG XI (Danin-Poleg et al. 2002). Monforte et al. (2004) posit that the modulation of rind pigmentation is subject to a multifaceted interplay of epistatic interactions that encompass over two genes. The gene accountable for the yellow coloration of the backdrop in 'Piel de Sapo' was sequenced by the LG X (Monforte et al. 2004; Eduardo et al. 2007).

7.9 QTL for Flesh Color

The edible flesh envelops the seeds, which are contained within the melon's central cavity. Cavity morphology is an essential quality attribute; consumers have a preference for cavities that are compact and free from any empty spaces between carpels. Present-day melons exhibit an internal structure that is comparatively compact, distinguished by a significant quantity of edible flesh and a dense encasement of seeds situated in the fruit's central region. On the contrary, there is considerable variation in cavity size across different melon germplasm. This includes compact modern fruits as well as those that contain edible flesh but a substantial portion of empty cavity; wild melons, for instance, contain edible flesh despite having a thin endocarp and a seed-filled interior. A significant gene associated with LG II (empty cavity, Ec) controls the dimensions of cavities (Périn et al. 2002a). The genetic underpinnings that control the vast majority of variations in tissue thickness remain unknown (Flores et al. 2001). Melon flesh pigmentation is a property that is not influenced by ethylene. The flesh may be salmon, orange, white, or green in color. The flesh acquires a green tint due to the accumulation of chlorophyll, while a yellow hue is associated with the accumulation of transviolaxanthin and lutein (Rodrigo et al. 2008). Furthermore, it has been observed that the flesh of pale orange melons contains colorless precursors of phytofluene and γ-carotene (Watanabe et al. 1991). Due to the accumulation of carotene, which imparts an orange coloration, orange-fleshed varieties gain additional nutritional value. The regulation of flesh pigmentation is accomplished through epistatic interactions between two principal genes (Clayberg 1992): green flesh (gf; Hughes 1948) and white flesh (wf; Iman et al. 1972). Flesh becomes orange upon the combination of *wf* +→/− −*gf* + and green upon the combination of *wfwf*/*gf* +−. Obtaining molecular markers that are linked to the mapped genes *gf* and *wf* on LG VIII and IX, respectively, has been demonstrated to be feasible (Monforte et al. 2004; Eduardo et al. 2007; Périn et al. 2002a; Fukino et al. 2008). The variation in flesh color intensity observed in diverse melon germplasm, which spans from dark orange or salmon to dark green, suggests that additional genes may exert an influence on the activity of these primary genes. Eduardo et al. (2007) and Monforte et al. (2004) linked QRTL associated with flesh pigmentation intensity to LG III, VI, and VII.

7.10 QTL for Nutrition

In addition to vitamins C and E, folic acid, α-tocopherol, niacin, B6, and carotenoids (lutein, zeaxanthin, and β-carotene), dietary fiber, and minerals (potassium, calcium, iron, magnesium, phosphorus, and zinc), melon flesh also contains additional nutrients that are advantageous to human health. The determining factor is the quantity of β-carotene present in cultivars characterized by orange flesh. Prior research has demonstrated that the concentrations of ascorbic acid and β-carotene, which are predominantly associated with the color and intensity of flesh, differ among germplasm (Burger et al. 2006; Dhillon et al. 2007). Variations have also been observed in other carotenoids. Cuevas et al. (2008) state that numerous QTL regulate the concentration of β-carotene; however, the presence of β-carotene in meat is determined by the gene wf (see previous discourse). In the realm of vitamins and antioxidants, two or three QTLs associated with ascorbic acid levels have been identified by a number of authors (Sinclair et al. 2004, 2006; Obando et al. 2009b; Park et al. 2009). To fully comprehend how genetics regulates nutritional components and how to apply this understanding to practical breeding initiatives, however, further investigation is necessary.

7.11 QTL for Organic Acids and Sugars

As a maturity indicator, total soluble solids (TSSs) in melons are exceptionally dependent on soluble organic acids, soluble sugars, and soluble pectin. Primarily due to sucrose, one of the most significant intrinsic characteristics of melons is their sweetness. Burger et al. (2002) have identified a recessive gene responsible for regulating sucrose accumulation in sweet melon varieties, but this gene is absent in nonsweet varieties. There are multiple QTLs that govern the variability observed in sucrose accumulation (Monforte et al. 2004; Sinclair et al. 2006; Eduardo et al. 2007; Obando-Ulloa et al. 2008, 2009b; Park et al. 2009). However, integrating QTL information into breeding programs remains difficult due to the fact that this trait frequently demonstrates substantial genotype–environment interaction.

As stated by Burger et al. (2002), the acidity of mature melon is also regulated by a single gene. Although specific varieties of nonsweet melons, namely acidulus and flexuosus, possess a high acidity level, sweet melons generally exhibit a moderate acidity. Citric acid is frequently the predominant organic acid identified in delicious melons. Subsequently, succinic, tartaric, L-glutamic, fumaric, isocitric, ascorbic, and so forth are present. The QTL for particular organic acids were identified by Obando et al. (2008); nevertheless, further comprehensive investigation is necessary to successfully modify the acid composition and develop cultivars that offer unique added values.

7.12 QTL for Aroma

Melon aroma volatiles are produced through a variety of metabolic pathways, which include the metabolism of terpenoids, amino acids, phenols, and fatty acids (Pech et al. 2008b). In contrast to decarboxylation or direct aldehyde production, the degradation of amino acids into aroma volatiles in melon fruit tissues may begin via the transamination pathway. Compared to nonclimacteric melons, climacteric melons exhibited greater activity of amino acid transaminase, an enzyme derived from two genes. The transaminases identified in the flesh of mature melon fruits converted L-isoleucine, L-leucine, L-valine, L-methionine, or L-phenylalanine into their respective a-keto acids, using a-ketoglutarate as the amine acceptor (Gonda et al. 2010). Climacteric melons differ significantly in aroma from nonclimacteric varieties.

The predominant components of the aroma in climacteric melons are acetate and nonacetate esters and alcohols. In contrast, nonclimacteric varieties are characterized by the existence of terpenes, aldehydes, and organic acids, albeit in lesser quantities (Shalit et al. 2001). These differentiations among cultivar varieties are the cause of the distinctive aroma of climacteric melons. Other substances discovered in fragrance profiles, according to Obando-Ulloa et al. (2008, 2009a, 2010); Shalit et al. (2001), are norisoprenes, aliphatic molecules, ketones, and sulfur-derived compounds. Obando-Ulloa et al. (2008) establish a connection between the pleiotropic effects of QTL that regulate climacteric behavior and variations in scent profiles. Obando et al. (2010) additionally identified four aroma QTL in LGs IV, VIII, and XI that are not dependent on ethylene. The synthesis of octanal, (Z, Z)-3,6-nonadienal, and 3-hydroxy-2,4,4-trimethylpentyl 2-methylpropanoate is associated with these QTLs.

7.13 QTL for Postharvest Quality

Melons are an ideal system to study the genetic regulation underlying climacteric and nonclimacteric ripening, as they display a wide variety of maturation characteristics, and varieties can be climacteric or nonclimacteric. The climacteric maturation process, experienced by various cultivars of *cantalupensis*, is characterized by a reduction in shelf life, an increase in respiration rate, and autocatalytic ethylene production. Nonclimacteric fruit maturation, characterized by the absence of respiration and an increase in ethylene levels, and fruit with an extended shelf life are distinguishing features of inodorus cultivars (Lurie and Klein 1989; Kays and Paull 2004; Flores et al. 2001, 2002; Obando-Ulloa et al. 2009c). Melon demonstrates the concurrent presence of maturation pathways that are both dependent and independent of ethylene. The regulation of rind color changes, softening of the majority of fruit parts, susceptibility to chilling injury, and biosynthesis of aromatic compounds are all facets of postharvest shelf life that are influenced by ethylene (Ayub et al. 1996; Flores et al. 2001, 2002; El-Sharkawy et al. 2005; Fernández-Trujillo et al. 2008; Obando-Ulloa et al. 2008, 2009a, c; Pech et al. 2008a). Ayub et al. (1996) and

Pech et al. (2008a) list additional processes that transpire in the absence of ethylene, namely fruit portion softening, acidity loss, and pulp coloration. Developing cultivars that retain the desirable qualities of climacteric varieties, including aroma and rind color, while also having an extended shelf life, is a fundamental objective of breeding. The implementation of strategies to reduce ethylene biosynthesis has resulted in the extension of product shelf life. However, this approach has also been associated with the degradation of surface pigments and fragrances (Ayub et al. 1996), which severely restricts the commercial feasibility of these methods.

In comparison to nonclimatic ripening, climatic ripening is characterized by genetic dominance (Périn et al. 2002c). Determining the location of two prominent genes—*Al-3* and *Al-4*—on LG VIII and IX, ethylene synthesis is regulated. Additional studies (Périn et al. 2002c; Moreno et al. 2008b) have identified additional QTL that control the production of climacteric LGs I, II, III, and XI in hybrids containing nonclimacteric genotypes. These findings indicate that this trait is subject to a complex genetic regulation algorithm.

Melon consumers consider texture to be an essential quality. Conversely, cantaloupes necessitate a marginally finer texture (Ferrer et al. 2007). Specific inodorus cultivars, such as 'Piel de Sapo,' exhibit a preference for a brittle texture. Genetic control research has been impeded, probably be damned, by the intricacy of the attributes. Recent studies (Moreno et al. 2008b; Obando et al. 2008; Paris et al. 2008) have identified candidate genes for flesh firmness among the QTL associated with whole fruit rigidity, juiciness, and flesh firmness. There is a notable risk of physiological disorders developing during storage, especially in international markets where melon fruits are transported to different continents (e.g., a portion of the melons consumed in Europe during the winter are produced in South America). Ostenation, shriveling, flavor loss, oxidation, senescence spots, disintegrating skin, greasiness, overripeness, and frost injury are a few of the issues that can be identified (Fernández-Trujillo et al. 2007, 2008; Martínez et al. 2009; Obando-Ulloa et al. 2009a). Merely a handful of studies (Fernández-Trujillo et al. 2007, 2008) have attempted to ascertain the QTL governing these characteristics via genetic investigation.

7.14 Genomics for Melon Quality

Recently, despite the considerable importance of the melon crop, there has been a scarcity of genetic technologies that can be utilized to improve produce quality. An improvement in melon cultivars could be facilitated by a greater comprehension of the genes that regulate crucial fruit qualities such as acidity, sugar concentration, maturation, and aroma. Diverse genomic data from cucurbit species could be consolidated with the establishment of the ICuGI in 2006. ICuGI database entries in March 2010 revealed that fruit tissues originating from diverse melon varieties and at various phases of development accounted for around 25% of the 127,106 melon ESTs, or 23,762 unigenes. González-Ibeas et al. (2007) report that a fraction of these EST sequences were obtained through the initiatives of national genomic consortiums. Furthermore, as a result of advancements in next-generation sequencing technology (RNA-seq), numerous laboratories have already generated and will shortly publish additional EST sequences from various plant tissues, including fruit. Existing in the ICuGI EST database are gene orthologs. ICuGI EST, a database particularly abundant in sequences from fruit tissue, already contains orthologs of genes associated with the principal metabolic mechanisms that affect fruit quality. Numerous novel methodologies will become available subsequent to the most recent sequencing of the melon genome (Garcia-Mas and Rodriguez-Concepcion 2016), enabling the identification of candidate genes and the localization of quantitative trait loci associated with fruit quality attributes (Portnoy et al. 2008). Mining the EST database enabled the identification of novel sesquiterpene synthases involved in the production of volatile sesquiterpenes in melon rinds, which are responsible for the distinct fragrances of various melon varieties. Ezura and Owino (2008) and Moreno et al. (2008b) report that the melon EST database also contains the vast majority of genes identified as being involved in cell wall disintegration, perception, response, and ethylene production. To date, two melon microarrays have been documented: the first (www.icugi.org), which contained 3000 unigenes primarily derived from fruit tissues, and the more recent one (Mascarell-Creus et al. 2009), which is enriched with ESTs from fruit tissues and contains 17,000 unigenes. (Saladié et al. 2014) Through hybridization of fruit tissue cDNAs from two climacteric and two nonclimacteric melon varieties at four stages of fruit development, a group of genes that are differentially expressed in climacteric melon varieties was identified. A comprehension of fruit maturation processes would be beneficial in the development of melon cultivars that possess an extended shelf life (Pech et al. 2008a). In recent years, platforms for functional genomics research into melon have utilized targeting-induced local lesions in genomes (TILLING). Bennahmane et al. (unpublished) identify the genetic background of Charentais as the location of the other platform, whereas Gonzalez-Ibeas et al. (2007) identify the genetic background of inodorus. Boualem et al. (2008) and Martin et al. (2009) report that the second TILLING platform facilitated the recent identification of mutants in the melon sex determination genes *a* and *g*. It is expected that the enormous quantity of genetic data that will be generated in the coming years will be beneficial in melon breeding to improve produce quality, as cucurbit species have just entered the "-omics" era.

7.15 Future Directions

Melon cultivars grown commercially on a large scale are mainly classed as cantalupensis, inodorus, and reticulatus. The other melon botanical groups show considerable variety in fruit characteristics, which might be leveraged to develop new melon cultivars with unique added values or to solve current concerns. The primary objectives of modern melon cultivars have been greater yield and disease resistance. Because of the significant cost, time, and resource requirements of quality-oriented breeding programs, only a few multinational corporations have fully adopted them. One of the typical unsolved issues is breaking the physiological link between scent creation in different climates and ethylene production, which results in a limited shelf life. Aroma synthesis has significantly lowered as a consequence of all methods to restrict ethylene accumulation, whether via genetic engineering or natural variation. Breeders are expected to continue functioning in this sector. Breeders will gain from new information on the genetic and molecular regulation of melon fruit ripening in both climacteric and nonclimacteric varieties, but the process will not be easy. We're learning more about how genetics affects fruit form, color, and rind traits such as sutures and spots. Major genes and their stable QTL have been found in genomic maps, enabling marker-assisted selection to modify them. Melon fruits will soon be able to be modified in terms of size, shape, flesh or rind color, and rind structure. Finally, modern breeders should strive to supply clients with a broader range of external and interior appearances. Furthermore, the significance of sensory aspects is growing. Sugar accumulation control is probably still elusive because it is difficult to pick fruit at the ripening stage that most influences sugar content and because external variables strongly alter this trait's expression. On the other hand, genetics can more readily regulate acidity, and changing acidity may change the flavor of melon fruit. Fruit fragrance alteration is getting new opportunities thanks to recent discoveries about the genetic and molecular control of volatile component production. Research on the nutritional makeup of melon flesh is required to understand management and subsequent modification of the flesh. Some progress has been achieved in creating cultivars with high concentrations of β-carotene by changing the color of the flesh, but the genetic regulation of other nutrients, such as ascorbic acid, is still unknown (Burger et al. 2002, 2006).

Other unresolved issues include defining the senescence point and harvest maturity of melons, lacking nondestructive analysis to measure quality repeatedly over time with the same individual melons, lacking reliable and well-defined methodologies, lacking information on the breeding perspective's kinetics of quality, and variation between and sometimes within the same individual (Tijskens et al. 2003, 2009; Hertog et al. 2007). The aforementioned will enhance our comprehension of the dynamics of fruit quality attributes in diverse pre- and postharvest scenarios, together with their correlations with QTL, and provide novel perspectives on the biological diversity present in a harvested batch at different biological levels.

Genomic and biotechnological tools are currently available for melon research. Improvements are still needed to transformation procedures to increase their overall efficacy and suitability for a larger range of cultivars. A few of the several genomics tools that have been created recently include mutant populations, TILLING platforms, dense genetic maps, EST databases, molecular markers, and microarrays for gene expression. The cucumber genome sequence that is now available will shortly be supplemented by further genome sequences (Huang et al. 2009).

We are ready to enter the postgenomics era, when genetic information and phenotypes—that is, conventional or modern as metabolomics data—must be combined. Thanks to sophisticated genomics methods and the vast variability found in melon, breeders will have almost endless possibilities when it comes to developing new melon cultivars.

7.16 Watermelon

The genus Citrullus is composed of seven species, including *C. mucosospermus* (which is responsible for producing egusi melon), *C. amarus* (citron), *C. colocynthis*, *C. ecirrhosus*, C. *rehmii*, and *C. naudinianus* (Chomicki and Renner 2015). *Citrullus lanatus*, a non-bitter desert cantaloupe, is one of these seven species. A widely held notion that persisted for the majority of the twentieth century was that watermelon originated in Southern Africa from *C. amarus* (Bailey 1930). As more genomic evidence emerged in the 2010s, the closest relative of watermelon and the likely site of domestication shifted from the citron melons of Southern Africa to *C. mucosospermus* of Western Africa (Guo et al. 2013). However, new research indicates that the region presently known as Sudan in northeastern Africa is the most likely location of the origin site (Renner et al. 2021). Archaeological findings and the recent discovery of Kordofan melon (*C. lanatus* susp. *cordophanus*) watermelons with white flesh (Chomicki and Renner 2015; Renner et al. 2021). China is estimated to have contributed 60% of the global cantaloupe production that surpassed 100 million tons in 2019. Turkey, in comparison to China, occupies a considerably smaller market share as the second-largest producer, contributing only 3.8% to the overall global production. Further significant participants in worldwide production include the United States, India, Brazil, Algeria, Iran, and Russia, each of which contributes between 1.6% and 2.5% (FAOSTAT 2020).

Whole or precut segments or cut portions of watermelon fruit are frequently consumed (Wehner 2008). Apart from their application in the production of edible seeds, the interior and rind of the fruit can also be candied, preserved, juiced, or used to make preserves (Wehner 2008). Particular cultivars are utilized in the production of palatable seeds (Zhang and Jiang 1990; Gusmini et al. 2004). There are several critical factors that dictate the significance of watermelon quality attributes, the majority of which are linked to consumer inclination. Within the realm of desserts, the preservation of an appetizing and highly sought-after product for customers is contingent upon internal fruit quality attributes such as sugar content, aroma, color, and texture (Wehner 2008). The inclination of consumers in different regions of the world toward specific phenotypes is impacted by regional differences in fruit sizes and cuticle patterns (Wu et al. 2019; Guo et al. 2013). The characteristic of rind thickness has gained increasing significance in the watermelon industry as a result of the expansion of cultivators' markets brought about by globalization. Watermelons characterized by substantial rind thickness exhibit enhanced resistance to transport-related deterioration. Nevertheless, this advantage must be evaluated in light of consumer inclination and fruit dimensions, as rinds that constitute an inconsequential proportion of the fruit are more favored (Wehner 2008; Wu et al. 2019). The preservation of gene inventories dates back to 1944 (Wehner 2012). Research has been dedicated to examining the Mendelian inheritance of fruit properties in watermelon for more than a century (Porter 1937; Porter et al. 1940; Weetman 1937; Poole and Grimball 1945; McKay 1936; Poole 1944). Through the application of genomics and molecular biology, it has become possible to discern the loci, genes, and allelic variation associated with watermelon fruit quality attributes and other horticulturally significant qualities that have influenced the course of domestication and selection. The watermelon's genome size (~425 Mbp) is relatively modest and diploid, which contributed to the swift adoption of novel genomic tools (Arumuganathan and Earle 1991). The Cucurbit Genomics Database (CuGenDB) presently contains two high-quality draft genomes (Guo et al. 2013, 2019; Wu et al. 2019; Zheng et al. 2018). In addition, CuGenDB includes transcriptome and comparative genomics tools, such as SyntenyViewer, which are specifically designed for the analysis of watermelon (Sandlin et al. 2012; Ren et al. 2014; Pan et al. 2020; Zheng et al. 2018). In order to ascertain critical genes and loci influencing the quality of watermelon fruit, researchers have made extensive use of comparative genomics, genetic mapping, and quantitative trait locus-seq (Branham et al. 2017; Wang et al. 2019; Dou et al. 2018b; Park et al. 2016). The breeding initiatives for this essential horticultural crop are influenced by consumer preference, which is determined by these loci and genes.

7.17 GWAS and QTL Mapping in *Citrullus lanatus*

7.17.1 Fruit Bitterness

Cucurbitacin, a disagreeable characteristic found in palatable watermelon, is a cucurbit crop-derived acrid compound. Knowledge of the genetic regulation of cucurbitacin is of the utmost importance for watermelon breeding programs. In order to ascertain the correlation between bitterness and cucurbitacin, a multitude of studies have conducted comparisons between cultivated watermelon and its crop wild relatives (CWRs). Cucurbitacin is detected in diverse structural configurations throughout the stems, fruits, leaves, and roots of Citrullus (Zhou et al. 2016; Kim et al. 2018). The compounds cucurbitacin E-2-O-glucoside (CuEGlu), cucurbitacin B (CuB), and cucurbitacin C (CuC) are among these. The principal acrid component identified in the fruit of Citrullus is CuE, also known as elaterinide (Chambliss et al. 1968; Matsuo et al. 1999). The synthesis of cucurbitacin, which is primarily encoded in the Bi gene, controls the bitterness of Citrullus fruits (Chambliss et al. 1968; Navot et al. 1990). On the contrary, the expression of the suppressor of bitterness gene, which is recessive in nature, governs the presence or absence of bitterness in watermelon fruit (Chambliss et al. 1968; Robinson et al. 1976). Prior research has classified the *Bi* gene located on chromosome 6 of the watermelon genome as an oxidosqualene cyclase (OSC; Cla007080) gene (Davidovich-Rikanati et al. 2015; Zhou et al. 2016). OSC genes are responsible for the initial phase of cucurbitacin biosynthesis (Shibuya et al. 2004). Specific genes, namely *Bl* (bitter_leaf), *Br* (bitter_root), and *Bt* (bitter_fruit), are associated with cucurbitacin accumulation in leaves, roots, and fruits, respectively (Shang et al. 2014; Zhou et al. 2016). The distribution of cucurbitacin accumulation across the plant is heterogeneous. Zhou et al. (2016), Li et al. (2018), and Zhang et al. (2018) identified the *Bt* gene as a basic helix–loop–helix transcription factor (bHLH; Cla011508) located on chromosome 1. According to the findings of Zhou et al. (2016) and Li et al. (2018), a truncated protein associated with non-bitter fruit is the consequence of a second exon single-base-pair mutation (C382T).

7.18 Sugar Content and Composition

7.18.1 Soluble Solids Content (SSC)

The watermelon plant is widely recognized for its ability to yield delectable and lustrous fruit, characterized by a Brix score that spans from 10 to 13. Guo et al. (2019) report that the SSC of the fruit of other Citrullus species is considerably lower; *C. colocynthis, C. amarus,* and *C. mucosospermus* have Brix values of 1.6, 3.4, and 3.4, respectively. The association between SSC and various QTLs has been documented in multiple studies (Fall et al. 2019; Liu et al. 2015; Ren et al. 2014; Sandlin et al. 2012), indicating that this characteristic is highly heritable.

Sandlin et al. (2012) identified the four QTLs associated with Brix on chromosomes 1, 2, and 7. The QTLs in question were brix1 (PVE ~ 8%), brix2.1 (PVE ~ 21%), brix9.2 (12% PVE), and brix7 (7% PVE), which were synonymous. While QBRX2.1 was identified amidst a background of *C. mucosospermus* and *C. lanatus*, the latter was found to contain *Qbrx1, Qbrx2.2,* and *Qbrx7* (Klondike Black Seeded < New Hampshire Midget). In a *C. lanatus* × *C. amarus* population, Ren et al. (2014) identified additional QTLs on chromosomes 2, 6, (Qbrix6; PVE ~ 8%), and 8, specifically *QBRX2.2* (PVE ~ 28%) and *Qbrix2.3* (syn. *Qbrix2.1*; PVE ~ 22%). *Qfruc2.1* (PVE ~ 23%) and *Qsuc2.1* (PVE ~ 10%), which are notable QTLs associated with fructose and sucrose, were also detected in the same population as *Qbrix2.3* (Ren et al. 2014). According to Ren et al. (2018), the observed effect of *QBRIX2.1* may be attributed to changes in the expression of the Tonoplast Sugar Transporter gene (ClTST2; Cla000264). This effect is characterized by an SNP in the promoter region of the gene. It is worth mentioning that erythema development is also correlated with *ClTST2* overexpression (Ren et al. 2018). The authors of the study conducted by Guo et al. (2019) established that Qbrix6 is sensitive to selection throughout the entire domestication process. Nevertheless, the locus in question displayed selective pressure throughout both the domestication and enhancement of watermelon. The occurrence of sugar accumulation selection during domestication was established by Guo et al. (2019) in relation to the *ClAGA2* alkaline a-galactosidase gene situated on chromosome 4. The sugar content of watermelon fruit is influenced by the variability in gene expression of *ClAGA2*, a critical constituent in the hydrolysis process of raffinose to sucrose and galactose (Ren et al. 2021). Two single-nucleotide polymorphisms (SNPs) in the *ClAGA2* promoter, which interact with the nuclear transcription factor Y subunit C transcription factor, are responsible for the variation in expression (Ren et al. 2021). Sugars will eventually be extracted. Transporter 3 (ClSWEET3; Cla97C01G000640), situated on chromosome 1, is associated with sugar storage in vacuoles. For sugar accumulation in watermelons, *ClTST2* and *ClAGA2* are crucial genes (Ren et al. 2021).

7.19 Sugar Composition

The sugar composition of fruit has a substantial impact on its flavor, which subsequently influences the preferences and tastes of consumers (Brueckner et al. 2007; Kyriacou and Rouphael 2018; Evans 2008). Sucrose, fructose, and glucose are the principal sugar components identified in fully developed watermelon fruit. Significant variation exists among genotypes in the proportions of these sugars (Yativ et al. 2010; Yoo et al. 2012; Pardo et al. 1997). Ren et al. (2014) identified five QTLs that exhibited a correlation with fructose content in a population of *C. lanatus* × *C. amarus* (97103 × PI 296341-FR). Among these QTLs, three were associated with glucose content and five with fructose content. The QTLs corresponding to fructose content were identified on chromosomes 2, 6, 21, and 8% by weight: *Qfru2.1*, PVE ~ 23%; *Qfru2.2*, PVE ~ 25%; and Qfru2.3, PVE ~ 28%. On chromosome 6 was the only QTL associated with glucose content (*Qglu6*, PVE ~ 23%). Two QTLs pertaining to sucrose content (*Qsuc2.1*, PVE ~ 1%; *Qsuc2.2*, PVE ~ 14%) were identified on chromosome 2, while one was identified on chromosome 5 (*Qsuc5*, PVE ~ 11%). Qbrx2.3 exhibited colocalization with Qfru2.1 and *Qsuc2.1*, while *Qbrx2.2* and *Qsuc2.2*, respectively, demonstrated colocalization with Qfru2.3 and Qbrx2.1. The glucose (*Qclu6*) and fructose (*Qfru6*) content QTLs, akin to *Qbrix6*, are colocalized on chromosome 6, albeit within a separate genomic region. The association between the accumulation of sucrose in the interspecific 97103 < PI 296341-FR population and the truncation of the vacuolar sugar transporter (ClVST1) via an SNP is documented (Ren et al. 2021).

Within the *C. lanatus* background, QTLs were identified for fruit glucose content (Qglu5.1, PVE ~ 13%; Qglu5.2, PVE ~ 10%; Qglu5.3, PVE ~ 12%; Qglu8.1, PVE ~ 17%; Qglu8.2, PVE ~ 15%), fructose content (Qfru5.1, PVE ~ 18%; Qfru5.2, PVE ~ 18%), and sucrose content (Qsur5.2, PVE ~ 25%; Qsur8.1, PVE ~ 9%). A QTL governing the fructose, glucose, and sucrose content of the flesh's edge portion was identified by Cheng et al. (2016) on chromosomes 10 and 1: Qfru10.1 syn. FCE10.1, PVE ~ 11.8%; Qglu1 syn. SCE 1.1, PVE ~ 7%; and Qsuc1 syn. GCE, PVE ~ 6.5%. The differentiation observed in interspecific populations from the elite *C. lanatus* background and the QTLs in the former implies that separate areas underwent selective pressures during domestication and subsequent refinement.

7.20 Fruit Shape and Size

The proportions and configuration of crops are consequential with respect to both consumer preference and transportation logistics. It is not unusual for specific sizes and shapes to be favored by regions; in fact, current trends favor fruits that are smaller and more angular (Wehner 2008). Consistent with other cucurbits, the fruit's form is largely determined before anthesis; thus, there is a robust correlation between the shape of the ovary and the shape of the fruit (McKay 1936; Pan et al. 2020; Weetman 1937; Legendre et al. 2020). The O gene has been identified as the regulator of fruit morphology; the variants OO, Oo, and oo are associated with fruits that are spherical, oval, and elongated, respectively (Poole and Grimball 1945; Tanaka et al. 1995; Weetman 1937). A total of 50 QTLs associated with the size and morphology of watermelons have been discovered (Pan et al. 2020). Due to the relationship between fruit morphology (FSI), fruit length (FL), and fruit diameter (FD), QTLs associated with these variables colocalize frequently (FSI = FL/FD). A main QTL for FSI (Qfsi3.1) is located on chromosome 3 of the watermelon genome. This QTL is stable in a wide range of genetic contexts (Dou et al. 2018b; Liu et al. 2014, 2016; Sandlin et al. 2012). The candidate gene for Qfsi3.1, hypothesized to be the location of the O gene, was determined to be Cla011257, a member of the SUN family (ClSUN25-26-27a) (Dou et al. 2018b; Pan et al. 2020; Kim et al. 2015; Legendre et al. 2020). The wild-type allele (cultivar 97103) is linked to round fruit, whereas an SNP and a 159-base pair deletion in the third exon of the gene are associated with elongated and blocky fruit, respectively (Dou et al. 2018b; Kim et al. 2015; Legendre et al. 2020; Pan et al. 2020). Extensive documentation exists regarding the influence of the SUN gene, an IQ Domain (IQD) constituent, on the morphology of tomato fruits. This influence is mediated by the gene's control over fruit elongation in the early stages of development, following pollination and fertilization (van der Knaap and Tanksley 2001). While the cell patterns are already determined prior to anthesis, the gene in question primarily influences the fruit morphology of tomatoes in the early stages of fruit development (Wu et al. 2011; Xiao et al. 2008). Conversely, variations in the morphology of watermelon are ascribed to genetic mutations occurring in the coding regions of ClSUN25-26-27a. Conversely, the elongation observed in tomato fruits can be ascribed to the duplication of genes responsible for the overexpression of SUN (Xiao et al. 2008). In populations derived from *C. amarus*, *C. lanatus*, and *C. mucosospermus*, Sandlin et al. (2012) also identified a significant QTL on chromosome 2 (Qfd2.2; Qfwt2.2) that was associated with fruit weight (FD). Qfl4.1, situated on chromosome 4, and Qfw2.1 and Qfd8.1, situated on chromosomes 2 and 8, respectively, are additional QTLs that have been identified in individuals with diverse genetic backgrounds (Cheng et al. 2016; Liu et al. 2016; Lu et al. 2016; Pan et al. 2020; Reddy et al. 2015; Ren et al. 2014; Sandlin et al. 2012). The genes responsible for these QTL are currently unknown. Previous studies have identified a pleiotropic effect of sex expression on the morphology of watermelon fruits (Loy 2006; Poole and Grimball 1945; Rosa 1928). Distillate flowers have a fruit morphology characterized by greater elongation when compared to hermaphroditic flowers. Chromosome 3 (Qand3) was found to contain a significant QTL associated with sex expression; the candidate gene was identified as an ACS (CitACS4; 1-aminocyclopropane-1-carbo xylate synthase) gene (Cla011230) (Prothro et al. 2013). It has been established that ACS genes play a crucial role in the expression of cucurbit sex (Boualem et al. 2008, 2009, 2016). Previous studies have provided evidence that fruit morphology and romonoecy are both impacted by a missense mutation in CitACS4 (Aguado et al. 2018; Manzano et al. 2016).

7.21 Flesh Color

A substantial variety of flesh pigments has been observed in cantaloupe. One epistatic interaction between *Wf* (short for white flesh) and *B* (short for yellow flesh; b) has been identified as one of numerous genes and interactions that regulate the pigmentation of flesh (Henderson 1992; Robinson et al. 1976; Shimotsuma 1963). *YCrl* predominates over orange flesh (*yO*), whereas flesh colors other than salmon yellow (*y*) and coral red (*YCrl*) are recessive to *C* (canary yellow), with the exception of white flesh (*Wf*) (Henderson et al. 1998; Henderson 1989; Poole 1944; Porter 1937). Conversely, previous studies have identified an inhibitor to *C (i-C)* that induces red flesh in the presence of the *C* allele via epistatic interaction (Poole 1944; Henderson et al. 1998). Furthermore, a higher prevalence of scarlet red flesh (*YScr*) was observed in comparison to coral red flesh (*YCrl*), according to a study by Gusmini and Wehner (2006). The accumulation of unique carotenoids is responsible for the variation in flesh color observed in watermelons (Fang et al. 2020; Perkins-Veazie et al. 2006; Subburaj et al. 2019; Sun et al. 2018; Zhao et al. 2013). The principal carotenoid present in red-fleshed watermelon is lycopene. In contrast, previous studies have identified α-carotene, pro-lycopene, β-carotene, and phytoene as the carotenoids that exert the greatest impact on the flesh color of orange-fleshed watermelon (Fang et al. 2020; Subburaj et al. 2019; Tadmor et al. 2005). Primary carotenoids that accumulate in the flesh of fruits with yellow flesh include lutein, phytoene, vioxanthin, and b-carotene (Fang et al. 2020; Subburaj et al. 2019).

A substantial QTL associated with flesh color was discovered on chromosome 4 in a population that was produced by crossing LSW-177 (which produces crimson flesh) and Cream of Saskatchewan (COS), which produces pale yellow flesh) (Liu et al. 2015). A QTL (*Qlcyb4.1*) and *Qflc4.1* were observed colocalized, and this colocalization was associated with the watermelon genome location of the lycopene β-cyclase (LCYB, Cla005011) gene and 92% of the PVE for lycopene content in this population. Ansa Nan et al. (2016); Wang et al. (2019); Subburaj et al. (2019); Liu et al. (2015). LCYB converts lycopene to b-carotene through the utilization of γ-carotene (Giuliano et al. 2008). Watermelon germplasm contains two SNPs—G676T and G1305C—as well as two amino acid substitutions (V226F and K435N) that encode for LCYB. According to Bang et al. (2007), Wang et al. (2019), and Zhang et al. (2020), individuals carrying the *V226/K435* genotype have crimson flesh, while those carrying the *F226/N435* and *F226/K435* genotypes have yellow and white flesh, respectively. These mutations lead to differential accumulation of LCYB protein, as opposed to differential RNA expression (Zhang et al. 2020). Branham et al. (2017) established an epistatic interaction between *Qflc1.1* and *Qlcyb4.1*. The researchers identified a primary QTL (*Qflc1.1*, syn. qFC.1) located on chromosome 1 that was linked to the accumulation of b-carotene in watermelon fruit from a population that differentiated between orange and yellow flesh. Phytoene synthase 1 (PSY1; Cla009122), the rate-limiting enzyme in the carotenoid pathway, is postulated to be the candidate gene underlying *Qflc1.1*. A correlation has been observed between the concentrations of carotenoids present in the flesh of watermelons and the levels of *PSY1* expression (Fang et al. 2020; Lv et al. 2015; Sun et al. 2018). In addition, two non-synonymous mutations (N133D and K148E) in Cla009122 have been associated with orange-fleshed fruit (Fang et al. 2020). The QTL (Qflc6.2) associated with the YScr gene was identified as residing in a 40 Kb region of chromosome 6. Consequently, three potential candidate genes were identified by Li et al. (2020): Cla018769, Cla018770, and Cla018771. Subsequent research has identified additional minor QTLs on chromosomes 3 (*Qflc3.1*), 6 (*Qflc6.1*), and 11 (*Qflc 11.1*) that are associated with the flesh color of watermelons (Fall et al. 2019; Liu et al. 2015; Subburaj et al. 2019).

7.22 Rind Pattern

Watermelon fruit rinds can be either solid or variegated (patterned exterior). The visual presentation of the rind has a substantial impact on consumer preference. For example, a considerable proportion of consumers prefer fruits with variegated rinds to those with uniform rinds (Lou and Wehner 2016). The stripes may vary in hue and width, with the background being represented by the lighter hue, and the darker hue typically signifying the stripe.

The patterns display a range of edge characteristics, including diffuse and pointed. Specific stripes exhibit a pronounced visual differentiation, while others are more understated, interspersed, or variegated (UPOV 2013; Gusmini and Wehner 2006). Variegated patterns, reticulated netting (veins), and ground patches are additional manifestations of patterns that can be observed on the rind of watermelons (Weetman 1937; Lou and Wehner 2016; UPOV 2013).

Additional unique patterns, such as the yellow epidermis of Royal Golden and the yellow regions of Moon and Star, have been recorded by Barham (1956) and Poole (1944).

Numerous patterns' qualitative inheritance has been identified by scientists investigating the genetic regulation of watermelon rind patterns since the 1930s (Barham 1956; Gusmini and Wehner 2006; Kumar and Wehner 2011; Lou and Wehner 2016; Poole 1944; Porter 1937; Weetman 1937). According to the model, patterned rind (*gs*) and uniform dark green rind (*G*) are less prevalent than gray rind (*g-1* and *g-2*) and patterned rind (*G*). The presence of a dominant allele at either locus confers dark green epidermis, while intermittent stripes (ins) are recessive in nature (Gusmini and Wehner 2006). The yellow-colored ground spot (*Yb*) and the Moon and Stars pattern, which comprises both large and small yellow spots on the fruit and foliage, are dominant (*Sp*) and recessive (*go*), respectively (Poole 1944; Rhodes 1986; Gusmini and Wehner 2006). Royal Golden, on the other hand, has a recessive yellow rind (Barham 1956; Robinson et al. 1976). Yang et al. (2015) presented an alternative paradigm in their recent study wherein they employed three loci—*Dgo* (Depth), *S* (Stripe), and *D* (Dominant golden color)—instead of concentrating three alleles at a single locus. This model focuses on the locus *Dgo*, which regulates the background color (yellow > green) and is differentiated from the recessive go (green > yellow) that was previously delineated. The *D* locus governs the deep (dark) epidermis color, which is predominant in comparison to the standard color locus. The presence or absence of foreground stripes is determined by the *S* locus; however, their presence is preferred over their absence. The loci *Dgo*, *D*, and *S* were mapped to chromosomes 4, 8, and 6, respectively, and Park et al. (2016) devised markers to aid in the selection of loci. Studies employing GWAS have verified the positions of these loci. Cla97C04G068530 (2-phytyl-1,4-beta-naphthoquin one methyltransferase), Cla97C08G161570 (2-phytyl-1,4-beta-naphthoquinone methyltransferase), and Cla97C06G126710 (WD40 repeat) have been proposed as candidate genes for Dgo, D, and S, respectively (Guo et al. 2019; Wu et al. 2019). Furthermore, Dou et al. (2018a) established that chromosome 4 harbors the predominant yellow epidermis pigment. However, they proposed as a candidate gene a gene whose function is unknown (Cla002755). Li et al. (2019) identified a rise in the expression level of ClCG08G017810, which is a homolog of Cla97C08G161570 in the genome of Charleston Gray, in the black rind. The authors devised a CAPS marker in order to discern the non-synonymous SNP located in the fifth exon of the gene that is linked to the observed phenotype. The mapping of the Citrullus lanatus Stripe Pattern (ClSP) on chromosome 6 unveiled a degree of overlap with the previously identified S locus; nevertheless, the exact characteristics of the locus that both represent are presently unknown (Yue et al. 2021). Oren et al. (2019) identified a QTL located on chromosome 9 that exhibited a significant association with dark rind color, as opposed to pallid rind color. An SNP in ClCG09G012330 (ClAPRR2; Arabidopsis pseudo-response regulator2-like) is linked to the phenotype; this SNP causes a 16-bp insertion in the mRNA of the dark rind parent and an alternative splice junction. CmAPRR2, a homolog of ClAPRR2 (Melo3C003375), is further responsible for the melon rind's vibrant coloration. It is worth mentioning that *CmAPRR2* is additionally implicated in the process of pigment accumulation (carotenoids) within the flesh of melons (Oren et al. 2019). Liu et al. (2021) have observed that the epidermis and leaves of the heirloom cultivar Moon and Stars bear noticeable yellow patches, varying in size from small to large. The regulation of these areas is carried out by two complementary loci situated on chromosomes 1 (*Sp1*) and 8 (*Sp2*). Rhodes (1986) arrived at the conclusion that the trait is governed by a solitary gene. The present result, however, challenges the conclusions of Rhodes (1986), potentially due to the utilization of unique genetic backgrounds in the two investigations (Liu et al. 2021).

7.23 Flesh Firmness

Quality and sensory perception of cantaloupe flesh are substantially impacted by the degree of flesh firmness. Fruits containing robust flesh may undergo a detrimental effect on flavor as a result of the diminished amount of liquid present; conversely, fruits with delicate flesh might necessitate an extended period of time to recuperate from storage (Harker et al. 2003; Kyriacou and Rouphael 2018; Risse et al. 1990). The firmness of tissue is regulated by metabolic processes, including the synthesis and degradation of cell wall components (Gao et al. 2020). A variety of factors contribute to the softening of flesh pulp during maturation; modifications in cellular composition, such as protopectin, cellulose, pectin, and hemicellulose, are the primary causes (Brummell and Harpster 2001; Sun et al. 2020; Anees et al. 2021).

The QTL (*Qffi6.1*), which governs the firmness of flesh in *C. amarus*, is located on chromosome 6 of the watermelon genome (Gao 2018; Gao et al. 2016; Juarez et al. 2013). Cla018816, a xyloglucan endotransglucosylase/hydrolase (XTH) gene that demonstrates distinct expression patterns in firm and soft-fleshed near-isogenic lines, is the presumed candidate gene for Qffi6.1 (Anees et al. 2021). Cell wall modification occurs during fruit maturation (Matsui et al. 2005; Munoz-Bertomeu et al. 2013). Through transcriptome analysis, Anees et al. (2021) identified a number of critical genes associated with the firmness of watermelon flesh, including cellulose synthase (*Cla012351*), ethylene response element transcription factor 1 (*Cla004120*), galactosyltransferase (*Cla006648*), pectinesterase gene (*Cla004251*), and cellulose synthase (*Cla012351*). The precise role of the ethylene-responsive element within watermelon fruits is unknown; however, it may have a suggestive effect on the regulation of cell wall genes, specifically those linked to pectin (Anees et al. 2021). Sun et al. (2020) identified QTLs associated with tissue firmness on chromosomes 2 (*Qffi2.1*; PVE ~ 14) and 8 (*Qffi8.1*; PVE ~ 24) in *C. mucosospermus*. A DUF579 family member (Cla016033) and a MADS-box transcription factor (Cla012507) were identified as the most likely candidate genes for *Qffi2.1* and *Qffi8.1*, respectively, on the basis of differential transcription observed during fruit maturation in genotypes with firm and soft-fleshed fruits. Sun et al. (2020) propose that an 11-bp insertion (GGCAATCCAAA) in the promoter region of Cla016033 in the progenitor *C. mucosospermus* is the causative mutation associated with firm flesh.

7.24 Rind Thickness and Toughness

The qualities of the internal epidermis, including its firmness and thickness, are crucial factors that influence consumer preference and the capacity of the container. Thick rinds provide watermelon fruit with the necessary resilience to withstand transportation, thereby allowing cultivators to serve a global clientele. However, this tendency toward dense rinds is restrained by consumer preference, which stipulates that the fruit's exterior should comprise a relatively small portion of the whole (Wehner 2008). Given the current consumer preference for watermelons that have the appearance of iceboxes, breeders are required to find a middle ground between shippability and the quality of the fruit inside, as perceived by consumers. A significant positive correlation ($r = 0.38–0.71$) was observed between characteristics of fruit size and the thickness of the rind. Furthermore, two separate populations exhibited a principal QTL situated on chromosome 2 (*Qrth2*), which exhibited a correlation with rind thickness and colocalized with QTLs corresponding to FD (*Qfd2.2*) and FW (Sandlin et al. 2012; Yang et al. 2021). Nonetheless, enormous fruits with thin rinds and tiny fruits with heavy rinds are not uncommon. Additionally, Sandlin et al. (2012) have identified on chromosomes 5, 6, and 8 additional minor QTLs associated with rind thickness. Preliminary investigations determined that the thickness of the rind was not affected by the toughness of the rind (E), with the latter bearing more significance than the former (Poole 1944; Porter 1937). Fruits that possess explosive rinds rupture or divide when sliced, which destroys their suitability for transportation (Wehner 2012). Extensive use has been made of explosive epidermis in the breeding of nonharvestable pollinizers, which facilitates the pulverizing of small, unharvested watermelons (Wehner 2012). Epidermis toughness is frequently defined as the degree to which the epidermis resists fracturing and/or becoming rigid. A range of methodologies can be employed to evaluate it, including knife cutting, the "thumb" test, and texture analyzer utilization (Yang et al. 2021; Wehner 2008; Liao and Hu 2020). A notable QTL associated with the resilience of the epidermis is located at position *Q210.1* on chromosome 10 of the watermelon genome (Liao and Hu 2020). It was postulated that the ethylene-responsive transcription factor 4 (*ClERF4*) gene deletion, consisting of 11 base pairs, was the mutated sequence responsible for Qrto10.1. The QTLs associated with rind rigidity and rind hardness were identified by Yang et al. (2021) on chromosomes 9 (*Qrto9*; PVE = 5.4) and 10 (*Qrto10.2*; PVE = 49.1). In contrast to *Qrto10.1*, *Qrto10.2* is not redundant with *Qrto*. In the latter study, a positive correlation was observed between rind thickness and both rind hardness and rind rigidity.

7.25 Fruit Quality

The significant differences in fruit characteristics—such as sweetness, bitterness, size, and flesh color—observable in the CWRs of cultivated and wild watermelon provide compelling proof of the pervasive influence that breeding, domestication, and selection pressure have exerted. Edible watermelon, as opposed to its CWR counterparts, is characterized by an exceptionally limited genetic diversity that stems from a genetic bottleneck that took place during the process of domestication (Guo et al. 2013, 2019; Levi et al. 2001). Although this phenomenon is observed in most cultivated plants, it is especially conspicuous in the case of watermelon (Hardigan et al. 2017). When crops are diversified or domesticated, selection efforts frequently target genes associated with fruit quality traits (Doebley et al. 2006).

Cla011508 specifies that the genotypes of *C. lanatus* harboring the non-bitter alleles at the Qbt1 gene remain unchanged. On the contrary, the bitter allele-containing accessions of *C. amarus* and *C. colocynths* remain fixed, whereas the segregation of *C. mucosospermus* alleles at the locus indicates that selection took place in the nascent phases of domestication (Guo et al. 2019; Renner et al. 2021). The diminished nucleotide diversity observed in the fruit sweetness-related *ClAGA2* and *ClSWEET3* genes of *C. lanatus* and *C. mucosospermus* indicates that fruit sugar content potentially played a role in the selection process during early domestication (Guo et al. 2019; Ren et al. 2021). According to Guo et al. (2019), the region encompassing *QBRX2.1* (ClTST2) seemed to have been selected for both domestication and enhancement, while Qbrix6 was chosen during the process of domestication. During the enhancement phase, it seemed that Qfru10.1, which was identified in the background of *C. lanatus*, was selected (Cheng et al. 2016; Guo et al. 2019). The vivid pigments that are distinctive of sweet watermelon are absent in the pale flesh of *C. mucosospermus* and *C. amarus*. PSY1 (Cla009122) and LCYB (Cla005011) variants, which are linked to flesh color, were chosen during domestication (Guo et al. 2019). The early domestication of this characteristic in *C. lanatus* subsp. *cordophanus* was demonstrated by the absence of the *V226F* mutation in *LCYB* and the presence of white flesh (Renner et al. 2021). One common goal of domestication selection is to increase the amount of fruit that can be digested (Doebley et al. 2006). During domestication and enhancement, *Qfsi3.1* and *Qfwt2.1*, which were associated with the morphology and size of the fruit, were selectively targeted (Ren et al. 2014).

7.26 Future Perspectives

Genes and allelic variation associated with significant watermelon fruit quality phenotypes, which have been the subject of research for nearly a century, have been identified as a result of genomics advancements. Enhanced knowledge of the molecular mechanisms that govern these characteristics will facilitate the development of more effective and focused selection strategies, thereby promoting breeding efficiency for watermelon consumer-relevant traits.

References

Abdelmohsin ME, Pitrat M (2008) Pleiotropic effect of sex expression on fruit shape in melon. In: Pitrat M (ed) Cucurbitaceae 2008: Proceedings of the IXth Eucarpia Meeting on Genetics and Breeding of Cucurbitaceae. Avignon, Paris, pp 551–555

Aguado E, García A, Manzano S et al (2018) The sexdetermining gene CitACS4 is a pleiotropic regulator of flower and fruit development in watermelon (Citrullus lanatus). Plant Reprod 31(4):411–426. https://doi.org/10.1007/s00497-018-0346-1

Akashi Y, Fukuda N, Wako T et al (2002) Genetic variation and phylogenetic relationships in East and South Asian melons, *Cucumis melo* L., based on the analysis of five isozymes. Euphytica 125(3):385–396

Anees M, Gao L, Umer MJ et al (2021) Identification of key gene networks associated with cell wall components leading to flesh firmness in watermelon. Front Plant Sci 12(1164):630243. https://doi.org/10.3389/fpls.2021.630243

Arumuganathan K, Earle ED (1991) Nuclear DNA content of some important plant species. Plant Mol Biol Report 9(3):208–219

Ayub R, Guis M, BenAmor M et al (1996) Expression of ACC oxidase antisense gene inhibits ripening of cantaloupe melon fruits. Nat Biotechnol 14(7):862–866

Bailey LH (1930) Three discussions in cucurbitaceae. Gentes Herb 2:175–186

Bang H, Kim S, Leskovar D, King S (2007) Development of a codominant CAPS marker for allelic selection between canary yellow and red watermelon based on SNP in lycopene b-cyclase (LCYB) gene. Mol Breed 20(1):63–72. https://doi.org/10.1007/s11032-006-9076-4

Barham WS (1956) A study of the Royal Golden watermelon with emphasis on the inheritance of the chlorotic condition characteristic of this variety. Proc Am Soc Hortic Sci 67:487–489

Bates DM, Robinson RW (1995) Cucumbers, melons and water-melons. In: Smartt J, Simmonds NW (eds) Evolution of crop plants, 2nd edn. Longman, Essex, pp 89–96

Baudracco-Arnas S, Pitrat M (1996) A genetic map of melon (*Cucumis melo* L.) with RFLP, RAPD, isozyme, disease resistance and morphological markers. Theor Appl Genet 93(1–2):57–64

Boualem A, Fergany M, Fernandez R et al (2008) A conserved mutation in an ethylene biosynthesis enzyme leads to andromonoecy in melons. Science 321(5890):836–838. https://doi.org/10.1126/science1159023

Boualem A, Troadec C, Kovalski I et al (2009) A conserved ethylene biosynthesis enzyme leads to andromonoecy in two Cucumis species. PLoS ONE 4(7):e6144

Boualem A, Lemhemdi A, Sari MA et al (2016) The andromonoecious sex determination gene predates the separation of *Cucumis* and *Citrullus* genera. PLoS ONE 11:e0155444

Branham S, Vexler L, Meir A et al (2017) Genetic mapping of a major codominant QTL associated with b-carotene accumulation in watermelon. Mol Breed 37(12):146. https://doi.org/10.1007/s11032-017-0747-0

Brueckner B, Schonhof I, Schroedter R et al (2007) Improved flavour acceptability of cherry tomatoes. Target group: children. Food Qual Prefer 18(1):152–160. https://doi.org/10.1016/j.foodqual.2005.09.011

Brummell DA, Harpster MH (2001) Cell wall metabolism in fruit softening and quality and its manipulation in transgenic plants. Plant Mol Biol 47:311–340

Burger Y, Sa'ar U, Katzir N et al (2002) A single recessive gene for sucrose accumulation in *Cucumis melo* fruit. J Am Soc Hortic Sci 127(6):938–943

Burger Y, Sa'ar U, Paris HS et al (2006) Genetic variability for valuable fruit quality traits in *Cucumis melo*. Isr J Plant Sci 54(3):233–242

Chambliss OL, Erickson HT, Jones CM (1968) Genetic control of bitterness in watermelon fruits. Proc Am Soc Hortic Sci 93:539–546

Cheng Y, Luan F, Wang X et al (2016) Construction of a genetic linkage map of watermelon (Citrullus lanatus) using CAPS and SSR markers and QTL analysis for fruit quality traits. Sci Hortic 202:25–31. https://doi.org/10.1016/j.scienta.2016.01.004

Chomicki G, Renner SS (2015) Watermelon origin solved with molecular phylogenetics including Linnaean material: another example of museomics. New Phytol 205(2):526–532. https://doi.org/10.1111/nph.13163

Clayberg CD (1992) Interaction and linkage test of flesh colour genes in *Cucumis melo* L. Cucurbit Genet Coop Rpt 15:53

Cong B, Liu JP, Tanksley SD (2002) Natural alleles at a tomato fruit size quantitative trait locus differ by heterochronic regulatory mutations. Proc Natl Acad Sci USA 99(21):13606–13611

Cuevas HE, Staub JE, Simon PW et al (2008) Mapping of genetic loci that regulate quantity of beta-carotene in fruit of US Western Shipping melon (*Cucumis melo* L.). Theor Appl Genet 117(8):1345–1359

Cuny F, Roudot AC (1991) *In vitro* pollen growth and melon pollen germination (*Cucumis melo* L.) after gamma-irradiation. Environ Exp Bot 31(3):277–283

Cuny F, Grotte M, Devaulx RD et al (1993) Effects of gamma-irradiation of pollen on parthenogenetic haploid production in muskmelon (*Cucumis melo* L.). Environ Exp Bot 33(2):301–310

Danin-Poleg Y, Reis N, Tzuri G et al (2001) Development and characterization of microsatellite markers in *Cucumis*. Theor Appl Genet 102(1):61–72

Danin-Poleg Y, Tadmor Y, Tzuri G et al (2002) Construction of a genetic map of melon with molecular markers and horticultural traits, and localization of genes associated with ZYMV resistance. Euphytica 125(3):373–384

Davidovich-Rikanati R, Shalev L, Baranes N et al (2015) Recombinant yeast as a functional tool for understanding bitterness and cucurbitacin biosynthesis in watermelon (Citrullus spp.). Yeast 32(1):103–114

Deleu W, Esteras C, Roig C et al (2009) A set of EST-SNPs for map saturation and cultivar identification in melon. BioMed Central Plant Biol 9:90

Dhillon NPS, Ranjana R, Singh K et al (2007) Diversity among landraces of Indian snapmelon (*Cucumis melo* var. *momordica*). Genet Resour Crop Evol 54(6):1267–1283

Dhillon NPS, Singh J, Fergany M et al (2009) Phenotypic and molecular diversity among landraces of snapmelon (*Cucumis melo* var. *momordica*) adapted to the hot and humid tropics of eastern India. Plant Genet Resour Character Util 7(3):291–300

Doebley JF, Gaut BS, Smith BD (2006) The molecular genetics of crop domestication. Cell 127(7):1309–1321. https://doi.org/10.1016/j.cell.2006.12.006

Dou J, Lu X, Ali A et al (2018a) Genetic mapping reveals a marker for yellow skin in watermelon (Citrullus lanatus L.). PLoS ONE 13(9):e0200617. https://doi.org/10.1371/journal.pone.0200617

Dou J, Zhao S, Lu X et al (2018b) Genetic mapping reveals a candidate gene (ClFS1) for fruit shape in watermelon (Citrullus lanatus L.). Theor Appl Genet 131(4):947–958. https://doi.org/10.1007/s00122-018-3050-5

Eduardo I, Arus P, Monforte AJ (2005) Development of a genomic library of near isogenic lines (NILs) in melon (*Cucumis melo* L.) from the exotic accession PI161375. Theor Appl Genet 112(1):139–148

Eduardo I, Arus P, Monforte AJ et al (2007) Estimating the genetic architecture of fruit quality traits in melon using a genomic library of near isogenic lines. J Am Soc Hortic Sci 132(1):80–89

El-Sharkawy I, Manriquez D, Flores FB et al (2005) Functional characterization of a melon alcohol acyl-transferase gene family involved in the biosynthesis of ester volatiles. Identification of the crucial role of a threonine residue for enzyme activity. Plant Mol Biol 59(2):345–362

Esteras C, Lunn J, Sulpice R et al (2009) Phenotyping a highly diverse core collection of melons to be screened using EcoTILLING. In: 8th Plant GEM Congress Lisbon. Plant Genomics European Meeting, p 214, abstract S8P16. Lisbon

Evans CB (2008) Consumer preferences for watermelons: a conjoint analysis. Auburn University

Ezura H, Owino WO (2008) Melon, an alternative model plant for elucidating fruit ripening. Plant Sci 175(1–2):121–129

Fall LA, Perkins-Veazie P, Ma G et al (2019) QTLs associated with flesh quality traits in an elite _ elite watermelon population. Euphytica 215(2):30. https://doi.org/10.1007/s10681-0192356-y

Fang X, Liu S, Gao P et al (2020) Expression of ClPAP and ClPSY1 in watermelon correlates with chromoplast differentiation, carotenoid accumulation, and flesh color formation. Sci Hortic 270:109437. https://doi.org/10.1016/j.scienta.2020.109437

FAOSTAT (2020) Food and agriculture organization of the united nations, crop production. http://www.fao.org/faostat/en/#data/QC

Fergany M, Balvir K, Monforte AJ et al (2010) Variation in melon (*Cucumis melo*) landraces adapted to the humid tropics of southern India. Genet Resour Crop Evol. https://doi.org/10.1007/s10722-010-9564-6

Fernandez-Silva I, Eduardo I, Blanca J et al (2008) Bin mapping of genomic and EST-derived SSRs in melon (*Cucumis melo* L.). Theor Appl Genet 118(1):139–150

Fernandez-Trujillo JP, Obando J, Martinez JA et al (2007) Mapping fruit susceptibility to postharvest physiological disorders and decay using a collection of near-isogenic lines of melon. J Am Soc Hortic Sci 132(5):739–748

Fernandez-Trujillo JP, Obando-Ulloa JM, Martinez JA et al (2008) Climacteric or non-climacteric behavior in melon fruit 2. Linking climacteric pattern and main postharvest disorders and decay in a set of near-isogenic lines. Postharvest Biol Technol 50(2–3):125–134

Ferrer MJ, Obando J, Moreno E et al (2007) Differences in sensory quality associated with climacteric or non-climacteric behavior by using near-isogenic lines. In: Artes A (ed) V Congreso Iberoamericano de Tecnología Postcosecha y Agroexportaciones. Cartagena, pp 504–514

Flores F, Ben Amor M, Jones B et al (2001) The use of ethylene-suppressed lines to assess differential sensitivity to ethylene of the various ripening pathways in Cantaloupe melons. Physiologia Plantarum 113:128–133

Flores F, El Yahyaoui F, de Billerbeck G et al (2002) Role of ethylene in the biosynthetic pathway of aliphatic ester aroma volatiles in Charentais Cantaloupe melons. J Exp Bot 53(367):201–206

Frary A, Nesbitt TC, Grandillo S et al (2000) *fw2.2*: a quantitative trait locus key to the evolution of tomato fruit size. Science 289(5476):85–88

Fukino N, Ohara T, Monforte AJ et al (2008) Identification of QTLs for resistance to powdery mildew and SSR markers diagnostic for powdery mildew resistance genes in melon (*Cucumis melo* L.). Theor Appl Genet 118(1):165–175
Gao L (2018) Transcriptome analysis and fine mapping of major genes controlling flesh firmness and sour flesh in watermelon. Chinese Academy of Agricultural Sciences, Beijing
Gao L, Zhao SJ, Lu XQ et al (2016) Linkage analysis on flesh firmness of watermelon [Citrullus lanatus (Thunb.) Mansf.] by SSR molecular marker. J Plant Genet Resour 17:866–870
Gao Y, Guo Y, Su Z et al (2020) Transcriptome analysis of genes related to fruit texture in watermelon. Sci Hortic 262:109075. https://doi.org/10.1016/j.scienta.2019.109075
Gillaspy G, Bendavid H, Gruissem W (1993) Fruits—a developmental perspective. Plant Cell 5(10):1439–1451
Giuliano G, Tavazza R, Diretto G et al (2008) Metabolic engineering of carotenoid biosynthesis in plants. Trends Biotechnol 26(3):139–145. https://doi.org/10.1016/j.tibtech.2007.12.003
Gonda I, Bar E, Portnoy V et al (2010) Branched-chain and aromatic amino acid catabolism into aroma volatiles in *Cucumis melo* L. fruit. J Exp Bot 61(4):1111–1123
Gonzalez-Ibeas D, Blanca J, Roig C, González-To M, Picó B, Truniger V, Gómez P, Deleu W, Caño-Delgado A, Arús P, Nuez F, Garcia-Mas J, Puigdomènech P, Aranda MA (2007) MELOGEN: an EST database for melon functional genomics. BMC Genomics 8:306. https://doi.org/10.1186/1471-2164-8-306
Gonzalo MJ, Oliver M, Garcia-Mas J et al (2005) Simple-sequence repeat markers used in merging linkage maps of melon (*Cucumis melo* L.). Theor Appl Genet 110(5):802–811
Garcia-Mas J, Rodriguez-Concepcion M (2016) The carrot genome sequence brings colors out of the dark. Nat Genet 48(6):589–590. https://doi.org/10.1038/ng.3574
Grandillo S, Zamir D, Tanksley SD (1999) Genetic improvement of processing tomatoes: a 20 years perspective. Euphytica 110(2):85–97
Guo S, Zhang J, Sun H et al (2013) The draft genome of watermelon (Citrullus lanatus) and resequencing of 20 diverse accessions. Nat Genet 45:51–58
Guo S, Zhao S, Sun H et al (2019) Resequencing of 414 cultivated and wild watermelon accessions identifies selection for fruit quality traits. Nat Genet 51(11):1616–1623. https://doi.org/10.1038/s41588-019-0518-4
Gusmini G, Wehner TC (2006) Qualitative inheritance of rind pattern and flesh color in watermelon. J Heredity 97:177–185
Gusmini G, Wehner TC, Jarret RL (2004) Inheritance of egusi seed type in watermelon. J Hered 95(3):268–270. https://doi.org/10.1093/jhered/esh031
Hardigan MA, Laimbeer FPE, Newton L et al (2017) Genome diversity of tuber-bearing Solanum uncovers complex evolutionary history and targets of domestication in the cultivated potato. Proc Natl Acad Sci 114(46):E9999–E10008. https://doi.org/10.1073/pnas.1714380114
Harker FR, Gunson FA, Jaeger SR (2003) The case for fruit quality: an interpretive review of consumer attitudes, and preferences for apples. Postharvest Biol Technol 28:333–347
Henderson WR (1989) Inheritance of orange flesh color in watermelon. Cucurbit Genet Coop Rpt 12:59–63
Henderson WR (1992) Corrigenda to the 1991 Watermelon gene list (CGC 14:129–137). Cucurbit Genet Coop Rpt 15:110
Henderson WR, Scott GH, Wehner TC (1998) Interaction of flesh color genes in watermelon. J Heredity 89:50–53
Hertog MLATM, Lammertyn J, De Ketelaere B et al (2007) Managing quality variance in the postharvest food chain. Trends Food Sci Technol 18(6):320–332
Higashi K, Hosoya K, Ezura H (1999) Histological analysis of fruit development between two melon (*Cucumis melo* L. *reticulatus*) genotypes setting a different size of fruit. J Exp Bot 50(339):1593–1597
Huang SW, Li RQ, Zhang Z et al (2009) The genome of the cucumber, *Cucumis sativus* L. Nat Genet 41(12):1275–1281
Hughes MN (1948) The inheritance of two characters of *Cucumis melo* and their interrelationship. Proc Am Soc Hortic Sci 52:399–402

Iman MK, Abo-Bakr MA, Hanna HY (1972) Inheritance of some economic characters in crosses between sweet melon and snake cucumber. I. Inheritance of qualitative characters. *Assiut.* J Agric Sci 3(1):363–380

Jeffrey C (1980) A review of the *Cucurbitaceae*. Bot J Linn Soc 81:233–247

Jeffrey C (2005) A new system of *Cucurbitaceae*. Botanicheskij Zhurnal 90:332–335

Juarez B, King JJ, Bachlava E et al (2013) Methods and compositions for watermelon firmness, Patent Application Publication. Patent No: WO2013033611A1, Pub.No.: US 2013/0055466A1

Kays SJ, Paull RE (2004) Postharvest biology. Exon Press, Athens

Kenigsbuch D, Cohen Y (1990) The inheritance of gynoecy in muskmelon. Genome 33(3):317–320

Kim KH, Hwang JH, Han DY et al (2015) Major quantitative trait loci and putative candidate genes for powdery mildew resistance and fruit related traits revealed by an intraspecific genetic map for watermelon (Citrullus lanatus var. lanatus). PLoS ONE 10:e0145665

Kim Y-C, Choi D, Zhang C et al (2018) Profiling cucurbitacins from diverse watermelons (Citrullus spp.). Hortic Environ Biotechnol 59(4):557–566

Kirkbride JH Jr (1993) Biosystematic monograph of the Genus Cucumis (Cucurbitaceae). Parkway Publishers, Boone

Kocyan A, Zhang LB, Schaeffer H et al (2007) A multi-locus chloroplast phylogeny for the *Cucurbitaceae* and its implications for character evolution and classification. Mol Phylogenet Evol 44(2):553–577

Kumar R, Wehner TC (2011) Discovery of second gene for solid dark green versus light green rind pattern in watermelon. J Heredity 102(4):489–493

Kyriacou MC, Rouphael Y (2018) Towards a new definition of quality for fresh fruits and vegetables. Sci Hortic 234:463–469. https://doi.org/10.1016/j.scienta.2017.09.046

Legendre R, Kuzy J, McGregor C (2020) Markers for selection of three alleles of ClSUN25-26-27a (Cla011257) associated with fruit shape in watermelon. Mol Breed 40(2):19. https://doi.org/10.1007/s11032-020-1104-2

Levi A, Thomas CE, Wehner TC et al (2001) Low genetic diversity indicates the need to broaden the genetic base of cultivated watermelon. HortScience 36(6):1096–1101

Li B, Lu X, Dou J et al (2018) Construction of a high density genetic map and mapping of fruit traits in watermelon (Citrullus lanatus L.) based on whole genome resequencing. Int J Mol Sci 19:3268

Li B, Zhao S, Dou J et al (2019) Genetic mapping and development of molecular markers for a candidate gene locus controlling rind color in watermelon. Theor Appl Genet 132(10):2741–2753. https://doi.org/10.1007/s00122-019-03384-3

Li N, Shang J, Wang J et al (2020) Discovery of the genomic region and candidate genes of the Scarlet Red Flesh Color (Yscr) locus in watermelon (Citrullus Lanatus L.). Front Plant Sci 11(116). https://doi.org/10.3389/fpls.2020.00116

Liao N, Hu Z, Li Yet al (2020) Ethylene-responsive factor 4 is associated with the desirable rind hardness trait conferring cracking resistance in fresh fruits of watermelon. Plant Biotechnol J 18(4):1066–1077. https://doi.org/10.1111/pbi.13276

Lim W, Earle ED (2009) Enhanced recovery of doubled haploid lines from parthenogenetic plants of melon (*Cucumis melo* L.). Plant Cell Tissue Org Cult 98(3):351–356

Liu CQ, Gao P, Luan FS (2014) A linkage map for watermelon and QTL analysis for fruit related traits. Sci Agric Sin 47:2814–2829

Liu S, Gao P, Wang X et al (2015) Mapping of quantitative trait loci for lycopene content and fruit traits in Citrullus lanatus. Euphytica 202(3):411–426. https://doi.org/10.1007/s10681-014-1308-9

Liu S, Gao P, Zhu QL et al (2016) Development of cleaved amplified polymorphic sequence markers and a CAPS-based genetic linkage map in watermelon (Citrullus lanatus [Thunb.] Matsum. and Nakai) constructed using wholegenome re-sequencing data. Breed Sci 66:244–259

Liu D, Sun D, Liang J et al (2021) Characterization and bulk segregant analysis of 'moon and star' appearance in watermelon. Sci Hortic 285:110140. https://doi.org/10.1016/j.scienta.2021.110140

Lou L, Wehner TC (2016) Qualitative inheritance of external fruit traits in watermelon. HortScience 51(5):487
Loy BJ (2006) Fruit size in melon in monoecious and andromonoecious isolines. Cucurbit Genet Coop Rpt 2005–2006(28–29):12–13
Lu BY, Zhou HW, Chen X et al (2016) QTL analysis of fruit traits in watermelon. J Fruit Sci 33:1206–1218
Lurie S, Klein JD (1989) Cyanide metabolism in relation to ethylene production and cyanide insensitive respiration in climacteric and non-climacteric fruits. J Plant Physiol 135:518–521
Lv P, Li N, Liu H et al (2015) Changes in carotenoid profiles and in the expression pattern of the genes in carotenoid metabolisms during fruit development and ripening in four watermelon cultivars. Food Chem 174:52–59. https://doi.org/10.1016/j.foodchem.2014.11.022
Manzano S, Aguado E, Martínez C et al (2016) The ethylene biosynthesis gene CitACS4 regulates monoecy/andromonoecy in watermelon (Citrullus lanatus). PLoS ONE 11(5):e0154362. https://doi.org/10.1371/journal.pone.0154362
Martin A, Troadec C, Boualem A et al (2009) A transposon-induced epigenetic change leads to sex determination in melon. Nature 461(7267):1135–1138
Martinez JA, Jowkar MM, Obando-Ulloa JM et al (2009) Uncommon disorders and decay in near-isogenic lines of melon and reference cultivars. Horticult Bras (Brasilia) 27(4):506–515
Mascarell-Creus A, Canizares J, Vilarrasa-Blasi J et al (2009) An oligo-based microarray offers novel transcriptomic approaches for the analysis of pathogen resistance and fruit quality traits in melon (*Cucumis melo* L.). BMC Genomics 10(1471):467
Matsui A, Yokoyama R, Seki M et al (2005) AtXTH27 plays an essential role in cell wall modification during the development of tracheary elements. Plant J 42:525–534
Matsuo K, DeMilo AB, Schroder RFW et al (1999) Rapid high-performance liquid chromatography method to quantitate elaterinide in juice and reconstituted residues from a bitter mutant of Hawkesbury watermelon. J Agric Food Chem 47:2755–2759
McCormack JH, Staub JE (1993) Uncultivated and wild species of Cucumis. CGC Report 16:85–90
McCreight JD, Bemis WP (1976) Occurrence of wild Cucumis species in Africa and India. HortScience 11(3):222–224
McKay JW (1936) Factor interaction in Citrullus: seedcoat color, fruit shape and markings show evidence of Mendelian inheritance in watermelon crosses. J Hered 27:110–112
Monforte AJ, Garcia-Mas J, Arus P (2003) Genetic variability in melon based on microsatellite variation. Plant Breed 122(2):153–157
Monforte AJ, Oliver M, Gonzalo MJ et al (2004) Identification of quantitative trait loci involved in fruit quality traits in melon (*Cucumis melo* L.). Theor Appl Genet 108(4):750–758
Monforte AJ, Eduardo I, Abad S et al (2005) Inheritance mode of fruit traits in melon: heterosis for fruit shape and its correlation with genetic distance. Euphytica 144:31–38
Moreno E, Fernandez-Silva I, Eduardo I et al (2008a) Agronomical, genetical and developmental characterization of fs6.4: a quantitative trait locus controlling melon fruit shape. In: Pitrat M (ed) Cucurbitaceae 2008: Proceedings of the Ixth Eucarpia Meeting on Genetics and Breeding of Cucurbitaceae. Avignon, Paris, pp 101–108
Moreno E, Obando JM, Dos-Santos N et al (2008b) Candidate genes and QTLs for fruit ripening and softening in melon. Theor Appl Genet 116(4):589–602
Munger HM, Robinson RW (1991) Nomenclature of *Cucumis melo* L. Cucurbit Genet Coop Rpt 14:43–44
Muñoz-Bertomeu J, Miedes E, Lorences EP (2013) Expression of xyloglucan endotransglucosylase/hydrolase (XTH) genes and XET activity in ethylene treated apple and tomato fruits. J Plant Physiol 170(13):1194–1201. https://doi.org/10.1016/j.jplph.2013.03.015
Nan W, Shi L, Peng G et al (2016) Developmental changes in gene expression drive accumulation of lycopene and b-Carotene in watermelon. J Am Soc Hortic Sci 141(5):434–443. https://doi.org/10.21273/JASHS03690-16
Navot N, Sarfatti M, Zamir D (1990) Linkage relationships of genes affecting bitterness and flash color in watermelon. J Heredity 81:162–165

Nunez-Palenius HG, Klee HJ, Cantliffe DJ (2006) Embryo-rescue culture of the 'Galia' muskmelon (*Cucumis melo* L. var. *reticulatus* Ser.) male parental line. Plant Cell Tissue Org Cult 85(3):345–352

Nunez-Palenius HG, Gomez-Lim M, Ochoa-Alejo N et al (2008) Melon fruits: genetic diversity, physiology, and biotechnology features. Crit Rev Biotechnol 28(1):13–55

Obando J, Fernandez-Trujillo JP, Martinez JA et al (2008) Identification of melon fruit quality quantitative trait loci using near-isogenic lines. J Am Soc Hortic Sci 133(1):139–151

Obando-Ulloa JM, Moreno E, Garcia-Mas J et al (2008) Climacteric or non-climacteric behavior in melon fruit 1. Aroma volatiles. Postharvest Biol Technol 49(1):27–37

Obando-Ulloa JM, Nicolai B, Lammertyn J et al (2009a) Aroma volatiles associated with the senescence of climacteric or non-climacteric melon fruit. Postharvest Biol Technol 52(2):146–155

Obando-Ulloa JM, Eduardo I, Monforte AJ et al (2009b) Identification of QTLs related to sugar and organic acid composition in melon using near-isogenic lines. Sci Hortic 121(4):425–433

Obando-Ulloa JM, Jowkar MM, Moreno E et al (2009c) Discrimination of climacteric and non-climacteric fruit at harvest and at senescence stage by quality traits. J Sci Food Agric 89(10):1743–1753

Obando-Ulloa JM, Ruiz J, Monforte AJ et al (2010) Aroma profile of a collection of near-isogenic lines of melon. Food Chem 118(3):815–822

Oliver M, Garcia-Mas J, Cardus M et al (2001) Construction of a reference linkage map for melon. Genome 44(5):836–845

Oren E, Tzuri G, Vexler L et al (2019) The multi-allelic APRR2 gene is associated with fruit pigment accumulation in melon and watermelon. J Exp Bot 70(15):3781–3794. https://doi.org/10.1093/jxb/erz182

Pan Y, Wang Y, McGregor C et al (2020) Genetic architecture of fruit size and shape variation in cucurbits: a comparative perspective. Theor Appl Genet 133(1):1–21. https://doi.org/10.1007/s00122-019-03481-3

Pardo JE, Gomez R, Tardaguila J et al (1997) Quality evaluation of watermelon varieties (Citrullus vulgaris S.). J Food Qual 20:547–557

Paris MK, Zalapa JE, McCreight JD et al (2008) Genetic dissection of fruit quality components in melon (*Cucumis melo* L.) using a RIL population derived from exotic x elite US Western Shipping germplasm. Mol Breed 22(3):405–419

Park SO, Hwang HY, Crosby KM (2009) A genetic linkage map including loci for male sterility, sugars, and ascorbic acid in melon. J Am Soc Hortic Sci 134(1):67–76

Park S, Kim K-T, Kang S-C et al (2016) Rapid and practical molecular marker development for rind traits in watermelon. Hortic Environ Biotechnol 57:385–391

Pech JC, Bouzayen M, Latche A (2008a) Climacteric fruit ripening: Ethylene-dependent and independent regulation of ripening pathways in melon fruit. Plant Sci 175(1–2):114–120

Pech JC, Latche A, van der Rest B (2008b) Genes involved in the biosynthesis of aroma volatiles in fruit and vegetables and biotechnological applications. In: Bruckner B, Wyllie SG (eds) Fruit and vegetable flavour: recent advances and future prospects. Woodhead Publishers, Cambridge, pp 254–271

Perin C, Hagen LS, De Conto V et al (2002a) A reference map of *Cucumis melo* based on two recombinant inbred line populations. Theor Appl Genet 104(6–7):1017–1034

Perin C, Hagen LS, Giovinazzo N et al (2002b) Genetic control of fruit shape acts prior to anthesis in melon (*Cucumis melo* L.). Mol Gen Genomics 266(6):933–941

Perin C, Gomez-Jimenez M, Hagen L et al (2002c) Molecular and genetic characterization of a non-climacteric phenotype in melon reveals two loci conferring altered ethylene response in fruit. Plant Physiol 129(1):300–309

Perkins-Veazie P, Collins JK, Davis AR et al (2006) Carotenoid content of 50 watermelon cultivars. J Agric Food Chem 54:2593–2597

Pitrat M (2008) Melon (*Cucumis melo* L.). In: Prohens J, Nuez F (eds) Handbook of plant breeding. vegetables, vol I. Springer, New York, pp 283–315

Pitrat M, Hanelt P, Hammer K (2000) Some comments on infraspecific classification of cultivar of melon. Acta Hortic 510:29–36

Poole CF (1944) Genetics of cultivated cucurbits. J Hered 35:122–128

Poole CF, Grimball PC (1939) Inheritance of new sex forms in *Cucumis melo* L. J Hered 30(1):21–25

Poole CF, Grimball PC (1945) Interaction of sex, shape, and weight genes in watermelon. J Agric Res 71:533–552

Porter DR (1937) Inheritance of certain fruit and seed characters in watermelons. Hilgardia 10:489–509

Porter DR, Bission CS, Allinger HW (1940) Factors affecting the total soluble solids, reducing sugars, and sucrose in watermelons. Hilgardia 13:31–66

Portnoy V, Benyamini Y, Bar E et al (2008) The molecular and biochemical basis for varietal variation in sesquiterpene content in melon (*Cucumis melo*, L.) rinds. Plant Mol Biol 66(6):647–661

Prothro J, Abdel-Haleem H, Bachlava E al (2013) Quantitative trait loci associated with sex expression in an inter-subspecific watermelon population. J Am Soc Hortic Sci 138(2):125–130

Reddy UK, Abburi L, Abburi VL et al (2015) A genomewide scan of selective sweeps and association mapping of fruit traits using microsatellite markers in watermelon. J Hered 106(2):166–176. https://doi.org/10.1093/jhered/esu077

Ren Y, McGregor C, Zhang Y et al (2014) An integratedgenetic map based on four mapping populations and quantitative trait loci associated with economically important traits in watermelon (Citrullus lanatus). BMC Plant Biol 14(1):33. https://doi.org/10.1186/1471-2229-14-33

Ren Y, Guo S, Zhang J et al (2018) A tonoplast sugar transporter underlies a sugar accumulation QTL in watermelon. Plant Physiol 176(1):836–850. https://doi.org/10.1104/pp.17.01290

Ren Y, Li M, Guo S et al (2021) Evolutionary gain of oligosaccharide hydrolysis and sugar transport enhanced carbohydrate partitioning in sweet watermelon fruits. Plant Cell:koab055. https://doi.org/10.1093/plcell/koab055

Renner SS, Schaefer H, Kocyan A (2007) Phylogenetics of *Cucumis* (*Cucurbitaceae*): Cucumber (*C. sativus*) belongs in an Asian/Australian clade far from melon (*C. melo*). BioMed Central Evol Biol 7:58

Renner SS, Wu S, Pérez-Escobar OA et al (2021) A chromosome-level genome of a Kordofan melon illuminates the origin of domesticated watermelons. Proc Natl Acad Sci 118(23):e2101486118. https://doi.org/10.1073/pnas.2101486118

Rhodes B (1986) Gene affecting foliage color in watermelon. J Hered 77:134–135

Risse LA, Brecht JK, Sargent SA et al (1990) Storage characteristics of small watermelon cultivars. J Am Soc Hortic Sci 115:440–443

Ritschel PS, Lins TCL, Tristan RL et al (2004) Development of microsatellite markers from an enriched genomic library for genetic analysis of melon (*Cucumis melo* L.). *BioMed Central.* Plant Biol 4:9

Robinson RW, Decker-Walters DS (1997) *Cucurbits*, crop production science in horticulture, 6. CAB International, New York

Robinson RW, Munger HM, Whitaker TW et al (1976) Genes of the cucurbitaceae. HortScience 11:554–568

Rodrigo MJ, Zacarias L, Deleu W et al (2008) Flesh carotenoid profile in climacteric near-isogenic lines: Of melon obtained from non-climacteric parentals: Implications for fruit design. Acta Hortic 858:43–49

Rosa JT (1928) The inheritance of flower types in Cucumis and Citrullus. Hilgardia 3:233–250

Rubatzky VE, Yamaguchi M (1997) World vegetable principles, production and nutritive values, 2nd edn. Chapman Hall, New York

Saladié M, Wright LP, Garcia-Mas J, Rodríguez-Concepción M, Phillips MA (2014) The 2-C-methylerythritol 4-phosphate pathway in melon is regulated by specialized isoforms for the first and last steps. J Exp Bot 65(17):5077–5092. https://doi.org/10.1093/jxb/eru275

Sandlin KC, Prothro JM, Heesacker AF et al (2012) Comparative mapping in watermelon [Citrullus lanatus (Thunb.) Matsum. et Nakai]. Theor Appl Genet 125:1603–1618

Sauton A, Dumas de Devaulx R (1987) Induction of gynogenetic haploid plants in muskmelon (*Cucumis melo* L.) by use of irradiated pollen. Agronomie 7(2):141–148

Schaefer H, Heibl C, Renner SS (2009) Gourds afloat: a dated phylogeny reveals an Asian origin of the gourd family (*Cucurbitaceae*) and numerous oversea dispersal events. Proc R Soc B Biol Sci 276(1658):843–851

Shalit M, Katzir N, Tadmor Y et al (2001) Acetyl-CoA: alcohol acetyltransferase activity and aroma formation in ripening melon fruits. J Agric Food Chem 49(2):794–799

Shang Y, Ma Y, Zhou Y et al (2014) Biosynthesis, regulation, and domestication of bitterness in cucumber. Science 346(6213):1084–1088. https://doi.org/10.1126/science.1259215

Shibuya M, Adachi S, Ebizuka Y (2004) Cucurbitadienol synthase, the first committed enzyme for cucurbitacin biosynthesis, is a distinct enzyme from cycloartenol synthase for phytosterol biosynthesis. Tetrahedron Lett 60:6995–7003

Shimotsuma M (1963) Cytogenetical studies in the genus Citrullus. VII. Inheritance of several characters in watermelons. Jpn J Breed 13:235–240

Sinclair JW, Park SO, Crosby KM (2004) Identification of QTL affecting vitamin C in melon. Subtrop Plant Sci 56:10–15

Sinclair JW, Park SO, Lester GE et al (2006) Identification and confirmation of RAPD markers and andromonoecious associated with quantitative trait loci for sugars in melon. J Am Soc Hortic Sci 131(6):360–371

Stepansky A, Kovalski I, Perl-Treves R (1999) Intraspecific classification of melons (*Cucumis melo* L.) in view of their phenotypic and molecular variation. Plant Syst Evol 217(3–4):313–332

Subburaj S, Lee K, Jeon Y et al (2019) Whole genome resequencing of watermelons to identify single nucleotide polymorphisms related to flesh color and lycopene content. PLoS ONE 14(10):e0223441. https://doi.org/10.1371/journal.pone.0223441

Sun T, Yuan H, Cao H et al (2018) Carotenoid metabolism in plants: the role of plastids. Mol Plant 11(1):58–74. https://doi.org/10.1016/j.molp.2017.09.010

Sun L, Zhang Y, Cui H et al (2020) Linkage mapping and comparative transcriptome analysis of firmness in watermelon (Citrullus lanatus). Front Plant Sci 11(831). https://doi.org/10.3389/fpls.2020.00831

Tadmor Y, King S, Levi A et al (2005) Comparative fruit colouration in watermelon and tomato. Food Res Int 38(8):837–841. https://doi.org/10.1016/j.foodres.2004.07.011

Tanaka T, Wimol S, Mizutani T (1995) Inheritance of fruit shape and seed size of watermelon. J Jpn Soc Hortic Sci 64:543–548

Tijskens LMM, Konopacki P, Simcˇicˇ M (2003) Biological variance, burden or benefit? Postharvest Biol Technol 27(1):15–25

Tijskens LMM, Dos-Santos N, Jowkar MM et al (2009) Postharvest firmness behaviour of near-isogenic lines of melon. Postharvest Biol Technol 51(3):320–326

UPOV (2013) Guidelines for the conduct of tests for distinctness, uniformity and stability. International Union for the Protection of New Varieties—Watermelon TG/142/5

van der Knaap E, Tanksley SD (2001) Identification and characterization of a novel locus controlling early fruit development in tomato. Theor Appl Genet 103(2):353–358. https://doi.org/10.1007/s001220100623

Wang C, Qiao A, Fang X et al (2019) Fine mapping of lycopene content and flesh color related gene and development of molecular marker–sssisted selection for flesh color in watermelon (Citrullus lanatus). Front. Plant Sci 10(1240). https://doi.org/10.3389/fpls.2019.01240

Watanabe K, Saito T, Hirota S et al (1991) Carotenoid pigments in orange, light orange and white flesh colored fruits of melon (*Cucumis melo* L.). J Jpn Soc Food Sci Technol 38(2):153–159

Weetman LM (1937) Inheritance and correlation of shape, size and color in the watermelon Citrullus vulgaris

Wehner T (2008) Watermelon. In: Prohens J, Nuez F (eds) Vegetables I: asteraceae, brassicaceae, chenopodicaceae, and cucurbitaceae, vol 2. Springer, New York, pp 381–418

Wehner T (2012) 2012 Gene list for watermelon. The cucurbit genetics cooperative. Accessed 9 June 2021

Whitaker TW, Bemis WP (1976) Cucurbits, Cucumis, Citrullus, Cucurbita, Lagenaria (Cucurbitaceae). In: Smartt J, Simmonds NW (eds) Evolution of crop plants, 2nd edn. Longman, Essex, pp 64–69

Wu S, Xiao H, Cabrera A et al (2011) SUN regulates vegetative and reproductive organ shape by changing cell division patterns. Plant Physiol 157:1175–1186

Wu S, Wang X, Reddy U et al (2019) Genome of 'Charleston Gray', the principal American watermelon cultivar, and genetic characterization of 1,365 accessions in the U.S. National Plant Germplasm System watermelon collection. Plant Biotechnol J 17:2246–2258. https://doi.org/10.1111/pbi.13136

Xiao H, Jiang N, Schaffner E et al (2008) A retrotransposon-mediated gene duplication underlies morphological variation of tomato fruit. Science 319:1527–1530

Yang H, Park S, Park Y-H et al (2015) Linkage analysis of the three loci determining rind color and stripe pattern in watermelon. Korean J Hortic Sci Technol 33:559–565

Yang T, Amanullah S, Pan J et al (2021) Identification of putative genetic regions for watermelon rind hardness and related traits by BSA-seq and QTL mapping. Euphytica 217:19. https://doi.org/10.1007/s10681-020-02758-9

Yativ M, Harary I, Wolf S (2010) Sucrose accumulation in watermelon fruits: genetic variation and biochemical analysis. J Plant Physiol 167(8):589–596. https://doi.org/10.1016/j.jplph.2009.11.009

Yetisir H, Sari N (2003) A new method for haploid muskmelon (*Cucumis melo* L.) dihaploidization. Sci Hortic 98(3):277–283

Yoo KS, Bang H, Lee EJ et al (2012) Variation of carotenoid, sugar, and ascorbic acid concentrations in watermelon genotypes and genetic analysis. Hortic Environ Biotechnol 53(6):552–560. https://doi.org/10.1007/s13580-012-0014-6

Yue Z, Ma R, Cheng D et al (2021) Candidate gene analysis of watermelon stripe pattern locus ClSP ongoing recombination suppression. Theor Appl Genet. https://doi.org/10.1007/s00122-021-03891-2

Zalapa JE, Staub JE, McCreight JD et al (2007) Detection of QTL for yield-related traits using recombinant inbred lines derived from exotic and elite US Western Shipping melon germplasm. Theor Appl Genet 114(7):1185–1201

Zhang XP, Jiang Y (1990) Edible seed watermelons (Citrullus lanatus (Thunb.) Matsum. , Nakai) in northwest China. Cucurbit Genet Coop Rpt 13:40–42

Zhang Z, Zhang Y, Sun L et al (2018) Construction of a genetic map for Citrullus lanatus based on CAPS markers and mapping of three qualitative traits. Sci Hortic 233:532–538. https://doi.org/10.1016/j.scienta.2017.10.029

Zhang J, Sun H, Guo S et al (2020) Decreased protein abundance of Lycopene b-Cyclase contributes to red flesh in domesticated watermelon. Plant Physiol 183(3):1171–1183. https://doi.org/10.1104/pp.19.01409

Zhao W, Lv P, Gu H (2013) Studies on carotenoids in watermelon flesh. Agric Sci 4:13–20

Zheng Y, Wu S, Bai Y et al (2018) Cucurbit Genomics Database (CuGenDB): a central portal for comparative and functional genomics of cucurbit crops. Nucleic Acids Res 47(D1):D1128–D1136. https://doi.org/10.1093/nar/gky944

Zhou Y, Ma Y, Zeng J et al (2016) Convergence and divergence of bitterness biosynthesis and regulation in Cucurbitaceae. Nat Plants 2:16183. https://doi.org/10.1038/nplants.2016.183

GWAS and QTL Mapping in Cauliflower (*Brassica oleracea* var. *botrytis* L.)

8

Srija Priyadarsini, Saurabh Singh, Alok Nandi,
Shyam Sundar Dey, and Reeta Bhatia Dey

Abstract

Deciphering the genetic basis of complicated agronomic, yield, biotic and abiotic stress resistance, and quality traits is the principal pursuit of genetics. In the post-genomics era, state-of-the-art techniques like QTL-seq, genotyping by sequencing, and genome-wide association studies (GWAS) have accelerated the identification of QTLs associated with complex traits in crop plants. GWAS utilizes the presence of high recombination in natural populations used for association studies. These innovative tools have also been proved effective in *Brassica oleracea* vegetables, and numerous sets of QTLs by employing GWAS have been determined. Among the cole group of vegetables, cauliflower (*Brassica oleracea* var. *botrytis* L.) predominates in India and globally, being an essential contributor in balanced human diet. The most economic part in cauliflower in curd and understanding of curd complexities based on genomics approaches is of

S. Priyadarsini
Department of Vegetable Science, Odisha University of Agriculture and Technology (OUAT), Bhubaneswar, Odisha, India

S. Singh (✉)
Department of Vegetable Science, Rani Lakshmi Bai Central Agricultural University, Jhansi, Uttar Pradesh, India

A. Nandi
Institute of Agricultural Sciences, Siksha 'O' Anusandhan (Deemed to be University), Bhubaneswar, Odisha, India

S. S. Dey
Division of Vegetable Science, ICAR-Indian Agricultural Research Institute, New Delhi, India

R. Bhatia Dey
Division of Floriculture and Landscaping, ICAR-Indian Agricultural Research Institute, New Delhi, India

S. S. Dey et al. (eds.), *GWAS and QTL Mapping in Horticultural Crops*,
https://doi.org/10.1007/978-981-95-2306-1_8

utmost importance for revealing the QTLs regulating curd architecture in cauliflower. GWAS and QTL analysis have been successfully applied in understanding curd-related traits and curd architecture in cauliflower. Different sets of mapping populations like doubled haploid, biparental mapping populations, and natural populations have been proved effective in QTL analysis studies in cauliflower. But as compared to other *Brassica oleracea* vegetables, limited research has been carried out in determining the QTLs and genetic basis of various quality, biotic stress resistance, and abiotic stress-resistance traits in cauliflower. In this context, the GWAS and QTL-seq can be potentially applied in elucidating complex traits in cauliflower. This chapter summarizes the up-to-date progress made in GWAS and QTL mapping of different traits in cauliflower.

Keywords

QTL analysis · Cauliflower · Curd traits · Association studies · Mapping · GBS

8.1 Introduction

The principal pursuit of genetics is the elucidation of the genetic basis of complex traits. The QTL mapping is a potential technique to elucidate the genetic control of complex traits (Li et al. 2023). Quantitative trait locus (QTL) mapping combines the phenotypic and genotypic information to study the genetic basis of complex traits. It allows the scientists, researchers, plant breeders to link certain complex phenotypic traits to specific locations on chromosomes (Miles and Wayne 2008). For majority of the crops, improvement in quality, plant architecture, high yield, and stress resistance are core breeding objectives which are having complex inheritance governed by multiple genes (Zhang et al. 2023). In such cases, the QTLs governing promising traits can be mapped and exploited in selection breeding programs effectively. To detect the QTLs controlling these traits, researchers utilize different techniques like mapping QTLs based on linkage analysis, BSA-seq, and genome-wide association studies (GWAS). Traditionally, QTL mapping relies on bi-parental mapping populations by using the linkage analysis approach. However, in the NGS era, GWAS is a powerful genomics tool to determine the QTLs underlying complex quantitative traits (Hamblin et al. 2011). The association mapping (AM) or linkage disequilibrium (LD) mapping determines the association of linked markers and phenotypic traits in the natural populations and identifies controlling QTLs. In simple terms the aim of GWAS is to determine the association between allelic frequency and governing traits. Recently, owing to their efficiency and suitability, both GWAS and genomic selection (GS) based on genome-wide marker data are gaining popularity among plant breeders for predicting phenotypes (Hamblin et al. 2011). The beauty of GWAS is that it does not require the construction of a specific population and can effectively detect the genetic architecture of complex traits (Alseekh et al. 2021). GWAS utilizes the presence of high recombination in natural populations used for association studies. Over the last decade, GWAS has been utilized effectively across the vegetable crops to map QTLs controlling various complex traits such as mapping QTLs for fruit traits in tomato (Kim et al. 2021), QTLs for carrot root traits (Brainard et al. 2022), QTLs for curd architecture in cauliflower (Zhao et al. 2020), physiological and agronomic traits in potato (Alvarez-Morezuelas et al. 2023) and QTLs for agro-morphological traits in eggplant (Ro et al. 2022). As compared to standard bi-parental mapping population, the GWAS enables the better mapping resolution by detecting relationships between genotypic data based on molecular markers and desirable phenotypic traits (Xu et al. 2017; Ro et al. 2022). In QTL mapping studies, genotyping-by-sequencing (GBS) is effective in producing SNP markers, which can be useful in the determination of alleles linked with target traits by employing GWAS and QTL mapping approach (Sallam et al. 2016). In the era of genomics, the availability of reference genomes of array of crops and genome resequencing data has facilitated the utilization of GWAS and integration of omics tools in understanding the genetic basis of complex traits. With the advancements in NGS approaches and reduction in cost of genotyping, the GWAS and QTL mapping have demonstrated the genetic basis of complex phenotypes across the vegetable crops, including *Brassica oleracea* vegetables (Zhao et al. 2020).

Among the cole group of vegetables, cauliflower (*Brassica oleracea* var. *botrytis* L.) predominates in India and globally, being an essential contributor to a balanced human diet (Thakur et al. 2015; Singh et al. 2021a, 2023). Cauliflower is a rich source of dietary minerals, vitamin C, vitamin A, glucosinolates as secondary metabolites, carotenoids, anthocyanins, and other phyto-nutraceutical compounds having antioxidant potential (Singh et al. 2018a, b). Due to the worldwide value of cauliflower as an important component of healthy, nutritious diet, much progress has been made in developing high-yielding hybrids in cauliflower by exploiting genetic mechanisms of sporophytic self-incompatibility and cytoplasmic male sterility (Sehgal and Singh 2018; Singh et al. 2019a, b, 2021b). With the development of high-temperature-tolerant hybrids, cauliflower cultivation is expanding in non-traditional areas. However, the productivity is still not at the satisfactory level owing to the thermosensitive nature of curd development in cauliflower (Matschegewski et al. 2015). The curd initiation and development in cauliflower is temperature dependent, but a set of genes regulates curd formation in cauliflower viz. *BoCAL1*, *BoAP1-a*, *CCE1*, and *BoREM1* (Duclos and Bjorkman 2008; Li et al. 2017). Understanding of curd complexities based on genomics approaches is of utmost importance for revealing the QTLs regulating curd architecture in cauliflower. The important traits in cauliflower are curd solidity, curd color, curd structure, curd size, resistance to bacterial, fungal, and viral diseases, plant architecture traits, plant types, leaf traits, and yield traits. Recently, Thorwarth et al. (2018) conducted a study involving 174 gene bank accessions of cauliflower for curd-related traits and identified 24 significant associations by employing GWAS approach. To stimulate the cauliflower genetic improvement, several researchers have identified genomic loci for complex agronomic traits. In QTL mapping doubled haploid (DH) populations are playing a significant role across the vegetable crops, especially in cross-pollinated crops. In this context recently, 20 QTLs were identified in cauliflower for curd-related traits by using two different DH populations (Zhao et al. 2020). The

QTLs for stalk length (qSL.C6–1 and qSL.C6–2) and curd solidity (qCS.C6–1 and qCS.C6–2) were also mapped in this study (Zhao et al. 2020). Based on GWAS and QTL mapping studies, the *BoVRN2* and *BoFLC2* genes were determined as regulators of floral transition in cauliflower (Matschegewski et al. 2015). GBS-GWAS has been successfully utilized in the crop plants for identification of candidate genes controlling complex phenotypes and facilitating crop breeding programs in the era of climate change. Much progress has been made in elucidating the genetic basis of complex agronomic traits in cauliflower, including curd architecture, plant architecture, curd color, biotic and abiotic stresses. This chapter summarizes the up-to-date information and provides critical literature analysis regarding exploitation of GWAS in determining major QTLs for different plant, curd, disease, and abiotic stress resistance traits in cauliflower.

8.2 Curd Architecture and Related Traits

The cauliflower (*Brassica oleracea* var. *botrytis*, $2n = 2\times = 18$) curd is an economic organ and is actually a prefloral fleshy apical meristem (Natalini et al. 2021). The curd initiation and flower development in cauliflower is a temperature-regulated phenomenon. Curd is the most significant yield determinant trait in cauliflower, where vernalization response is also an important feature. The wide range of cultivars with obligate and facultative chilling requirements have been developed in the past decades (Thorwarth et al. 2018). Recently, GWAS have been utilized successfully in cauliflower to dissect the genetic basis of important agronomic, phenological, reproductive, and yield traits (Table 8.1). Among the curd traits, curd architecture is an important player in regulating the curd morphology (Zhao et al. 2020). The wide variation is present in cauliflower with respect to curd morphology like loose curd cauliflower and compact curd cauliflowers. Successful attempts have been made in determining the genes/QTLs underlying differences in different curd phenotypes. The QTL analysis based on DH mapping populations has identified QTLs for stalk length and curd solidity in cauliflower (Zhao et al. 2020). The curd solidity in cauliflower is a potent trait that relies on the inside architecture of the curd. The solid, compact curds have short stalk length and vice versa in loose curd cultivars (Gu et al. 2014). The co-localization of QTLs governing curd size and earliness has been demonstrated, which is attributed to imposed selection pressure for pyramiding these phenotypic traits from different sources (Lan and Paterson 2000). The curd architecture is a complex trait comprising an array of related parameters determining curd morphology. Recently, Zhang et al. (2023) determined the genetic basis of four different traits in cauliflower viz. main stem height, purplish curd, external leaf wing, and curd weight. In this context, a set of 220 core accessions of loose curd cauliflower were subjected to whole genome resequencing and GWAS-based QTL mapping. The QTL regions regulating temperature-mediated curd initiation have been depicted in cauliflower on chromosomes C06 and C09 (Hasan et al. 2016). In the era of NGS, genome-based predictions for time to curd initiation in cauliflower have been made feasible. To investigate the genetic basis of temperature-mediated

Table 8.1 Status of QTL mapping in cauliflower

Trait	Mapping population	Technique	Molecular markers	QTLs/genes	References
Stalk length	Two DH populations IZ population (IL4305 × ZN198) ID population (IL4305 × DL3203–61)	QTL mapping	SSR, SRAP, SNP	qSL.C6–1, qSL.C6–2	Zhao et al. (2020)
Curd solidity	Two DH populations IZ population (IL4305 × ZN198) ID population (IL4305 × DL3203–61)	QTL mapping	SSR, SRAP, SNP	qCS.C6–1 and qCS.C6–2	Zhao et al. (2020)
Temperature-dependent curd initiation	DH population	QTL mapping	SNP	One major QTL on C06	Hasan et al. (2016)
Prediction of curd induction time (slope and temperature response functions)	265 DH lines	Genome wide QTL mapping	20k SNP array	5 QTLs for slope. 2 QTLs for temperature-response functions.	Rosen et al. (2018)
Leaf appearance rate in curds	265 DH lines	Genome wide QTL mapping	20k SNP array	10 QTLs on	Rosen et al. (2018)
Weight of single curd	220 inbred lines of loose curd cauliflower	GWAS and RNA-seq	SNPs	*BOB01G136670* candidate gene	Zhang et al. (2023)
Green curd color	F_2 mapping population (Stovepipe × ACX800)	QTL-seq	CAPS based SNPs	Two QTLs: *Gr5.1*, *Gr7.1*	Tan et al. (2020)
Black rot	F_2 mapping population (Pusa Himjyoti × BR-161)	Molecular mapping	RAPD, ISSR, SSR	*Xca1bo*	Saha et al. (2014)
Downy mildew	F_2 mapping population (Pusa Himjyoti × BR-2)	Molecular mapping	RAPD, ISSR	*Ppa3*	Singh et al. (2012)
Downy mildew	RILs mapping population from Pusa Sharad × DMR-2-0-7	Molecular mapping	SSR	*Ppa207*	Saha et al. (2020)

curding initiation in cauliflower, a GWAS based study involving 111 commercial lines of cauliflower was conducted by Matschegewski et al. (2015) and they determined 18 QTLs localized on seven different chromosomes (O1, O2, O3, O4, O6, O8, and O9) within the genomic regions of flowering genes. Recently, based on GWASs and RNA-seq analyses using 220 inbred lines of loose curd cauliflower, Zhang et al. (2023) designated *BOB01G136670* as a candidate gene controlling the weight of a single curd in cauliflower. For purplish curd, Zhang et al. (2023) determined three protein-coding genes (*BOB02G088210, BOB02G088220,* and *BOB02G088870*). In the Indian cauliflower, Rakshita et al. (2021) reported 12 QTLs regulating curd formation and development in Indian cauliflower. The five QTLs were linked with curd length, width, days to 50% curd maturity, and MCW (marketable curd weight), while the remaining seven QTLs were associated with CL (curd length), CW (curd width), days to 50% curd harvest, gross plant weight, LL (leaf length), net curd weight, and leaf numbers per plant.

8.3 Curd Color

The genetic architecture and morphology of cauliflower curd are quite complex. In general, the cauliflower genotypes with white colour curds are predominant in markets globally. But spontaneous mutations in cauliflower have introduced green, orange, and purple color curd (Chiu et al. 2010; Tan et al. 2020; Welsch et al. 2020), where '*Gr*', '*Pr*', and '*Or*' genes have been reported to determine green, purple, and orange color curds, respectively. The *Or* mutant confers high accumulation of β-carotene in cauliflower curds, while *Gr* and *Pr* mutations are characterized by chlorophyll and anthocyanin accumulation, respectively (Chiu et al. 2010; Tan et al. 2020; Welsch et al. 2020). The green curd development in cauliflower is attributed to ectopic generation of chloroplast in curd tissues, which is under the genetic control of single semi-dominant gene (Zhou et al. 2011). To identify the genomic regions governing complex traits, rapid, and cost-effective determination of QTLs, the QTL-seq is an effective technique (Takagi et al. 2013). This process is further facilitated by identification of SNPs between the parental genotypes. Based on this approach by whole genome resequencing and SNP analysis, Tan et al. (2020) identified two QTLs on chromosome 5 and chromosome 7 determining green color curds in cauliflower (Table 8.1). The QTL located on chromosome 5 was suggested as a major QTL and a minor QTL on chromosome 7.

8.4 Biotic Stress Resistance

Cauliflower production is limited by various fungal, bacterial, and viral pathogens globally. Of which black rot (*Xanthomonas campestris* pv. *campestris*), downy mildew (*Hyaloperonospora parasitica*), and stalk rot are some of the most devastating in cauliflower (Singh et al. 2018c). The continuous search for new resistant sources, identification of linked molecular markers with resistance loci, and QTL mapping

Table 8.2 Genetic resources for resistance to major diseases in cauliflower

Disease	Genetic resource
Black rot (*Xanthomonas campestris* pv. *campestris*)	SN-445, MGS-2-3, Pua Kea, S-12, Selection 1614, BR-161, BR-207, PN-1, DC-79, Pusa Snowball K-1, Pusa Shubhra
Downy mildew (*Hyaloperonospora parasitica*)	BR-2, 1-6-1-2, CC, CCm, CCm-6, DMR-1-0-6, DMR-2-0-7, Pusa Kartik Shankar, Pusa Hybrid-2
Alternaria leaf spot (*A. brassicicola*)	IIHR 142, MGS-2-3, Pusa Kea
Stalk rot (*Sclerotinia sclerotiorum*)	KN 81, KT-25, *B. villosa*

are instrumental in stimulating breeding for disease resistance. The genetic resources for major diseases in cauliflower (Table 8.2) have been exploited for mapping genes/QTLs to accelerate resistance breeding. The resistance to black rot is under the control of single dominant gene in the cauliflower resistant sources, SN 445, BR-161, and BR-207 (Saha et al. 2014). Different *R genes/QTLs* determining black rot resistance have been reported in cauliflower and related *Brassica oleracea* crops (Table 8.1) (Singh et al. 2018c). In this context, the black rot resistance gene '*Xca1bo*' identified in Indian cauliflower and was mapped on chromosome 3 (Table 8.1) (Saha et al. 2014). The resistance to downy mildew is reported to be under the control of a single dominant gene in cauliflower, two independent dominant genes, and 3 to 4 complementary dominant genes in *Brassica oleracea* (Saha et al. 2014). The downy mildew resistance in cauliflower is associated with QTLs at different developmental stages of plant growth (Shaw et al. 2021). The resistance to downy mildew in Indian cauliflower is governed by a monogenic dominant gene '*Ppa3*' (Table 8.1) (Singh et al. 2012). Recently, a new downy mildew resistance gene '*Ppa207*' was mapped on chromosome 2 at 4.8 cM interval by using RILs as a mapping population by Saha et al. (2020). The two SSR markers, BoGMS0486 and BoGMS0900, flanked the resistance gene at a distance of 3.6 and 1.2 cM, respectively, from *Ppa207*. In Chinese cabbage four major QTLs, viz. *sBrDM8, yBrDM8, rBrDM8,* and *hBrDM8,* have been reported on chromosome A08 to regulate the resistance to downy mildew (Yu et al. 2016). These results indicate the role of QTLs in controlling resistance to downy mildew. In the same fashion, the QTLs for resistance to downy mildew can be traced by using different sets of mapping populations in cauliflower for facilitating disease-resistance breeding.

8.5 Abiotic Stress Resistance

In the era of climate change, abiotic stresses such as cold, heat, salinity, and drought have aggravated their drastic effects on crop growth and yield (Bulgari et al. 2019). Global warming is a real peril for the thermosensitive crops like cauliflower, attributed to alterations in optimal temperatures required for optimum growth and development (Saeed et al. 2023). Thus, there is an urgent need for understanding the genetic basis of tolerance to abiotic stresses and developing climate-resilient cultivars to feed the billions. Genomic revolution have increased our understanding of complex traits across the crops during the past two decades. The more rapid use of genomic tools like GBS, QTL-seq, and GWAS in the identification of SNPs and determining underlying QTLs for various abiotic stresses in crop plants have accelerated plant breeding for developing resistant varieties (Roorkiwal et al. 2020). The population's linkage disequilibrium (LD) and SNPs in the genome are major components of GWAS polymorphisms (SNPs) in the genome. The true characterization of crop wild relatives (CWRs) is an essential component in the era of climate change to facilitate crop breeding against various abiotic stresses (Raj and Nadarajah 2023). A high level of genetic diversity is preserved in the CWRs, as they have lived in close association with natural environments for several years. Heat stress in cauliflower and related cole crops causes a wide array of undesirable characteristics for marketing like leafiness, bracting, riceyness, loose curd, reduction in curd size and weight, decrease in quality and yield. The major breeding obstacle for heat stress in *Brassica oleracea* is lack of research for evaluation of new genetic resources and understanding genetics for heat and other abiotic stresses. In an attempt to identify QTLs for heat stress tolerance in related cole vegetable, broccoli, Branham et al. (2017) identified QTLs for heat tolerance in broccoli by employing a GBS approach. Different sets of QTLs (QHT_C02, QHT_C03, QHT_C05, and QHT_C09) for heat stress were identified by using DH mapping population. This study would pave the way for initiating breeding for heat stress in cauliflower and determining the genetic basis for heat tolerance. Much progress has been made in determining the genetic basis of abiotic stress tolerance in related *Brassica species*, but less efforts have been made in case of *Brassica oleracea* vegetables. For instance, Limbalkar et al. (2023) reported 29 QTLs for drought stress in introgressions lines (ILs) of *Brassica juncea* carrying genomic segments of *B. carinata* by using genotyping-by-sequencing (GBS) approach. Among 29 QTls, *qWUE.7B.1* QTL (with flanking markers 7B_30548045 and 7B_870706) was responsible for enhancing water use efficiency under moisture stress conditions. In the era of genomics, GWAS has been successfully used in the identification of QTLs for abiotic stresses across the Brassica genus (Chen et al. 2022). The extensive studies are needed for dissecting the genetic basis of abiotic stress tolerance in cauliflower globally, as very less reports are there particularly on cauliflower. The availability of GWAS, QTL-seq, and GBS techniques in identification of QTLs for various complex traits would be instrumental in accelerating the cauliflower breeding for climate resilience.

8.6 Conclusion

QTL mapping is a potential technique that enables plant scientists to unravel the genetic basis of complicated traits in crop plants. Additionally, GWAS has been proved another vital technique involved in QTL research based on natural populations to detect the genetic architecture of complex characters. Recently, with the generation of high-density SNPs (single-nucleotide polymorphism) BeadCap based on GBS and whole-genome resequencing of gene bank accessions, GWAS has been effectively utilized to map the candidate genes and underlying QTLs to facilitate plant breeding programs. In case of cauliflower, curd architecture, curd size, and morphology are one of the important traits determining curd yield. Some of the results reported in related *Brassica oleracea* vegetables regarding QTLs for disease resistance, abiotic stress resistance, and secondary metabolites lacks replication in cauliflower. The reports on cauliflower are not substantial to provide robust conclusions regarding multi-faceted applications of GWAS and QTL mapping. Further, the GWAS-derived transcriptome and metabolome-wide association studies can be potential tools to narrow down the knowledge gaps regarding the genetic basis of complex physiological, phenotypic, biochemical, and stress resistance traits in cauliflower (Ferguson et al. 2021; Tello and Ibanez 2023). Additionally, meta-GWAS has evolved as a vital tool for dissecting complex traits to strengthen the use of single-marker GWAS and facilitate the identification of potential and stable genomic loci spanning space and time while eliminating false positives. The application of these potential tools can be instrumental in cauliflower breeding in genomics era.

References

Alseekh S, Kostova D, Bulut M, Fernie AR (2021) Genome-wide association studies: assessing trait characteristics in model and crop plants. Cell Mol Life Sci 78:5743–5754. https://doi.org/10.1007/s00018-021-03868-w

Alvarez-Morezuelas A, Barandalla L, Ritter E, Ruiz de Galarreta JI (2023) Genome-wide association study of agronomic and physiological traits related to drought tolerance in potato. Plants (Basel) 12(4):734

Brainard SH, Ellison SL, Simon PW, Dawson JC, Goldman IL (2022) Genetic characterization of carrot root shape and size using genome-wide association analysis and genomic-estimated breeding values. Theor Appl Genet 135:605–622. https://doi.org/10.1007/s00122-021-03988-8

Branham SE, Stansell ZJ, Couillard DM, Farnham MW (2017) Quantitative trait loci mapping of heat tolerance in broccoli (*Brassica oleracea var. italica*) using genotyping-by-sequencing. Theor Appl Genet 130:529–538. https://doi.org/10.1007/s00122-016-2832-x

Bulgari R, Franzoni G, Ferrante A (2019) Biostimulants application in horticultural crops under abiotic stress conditions. Agronomy 9(6):309. https://doi.org/10.3390/agronomy9060306

Chen S, Hayward A, Dey SS, Choudhary M, Witt Hmon KP, Inturrisi FC, Dolatabadian A, Neik TX, Yang H, Siddique KHM et al (2022) Quantitative trait loci for heat stress tolerance in *Brassica rapa* L. are distributed across the genome and occur in diverse genetic groups, flowering phenologies and morphotypes. Gene 13(2):296. https://doi.org/10.3390/genes13020296

Chiu LW, Zhou X, Burke S, Wu X, Prior RL, Li L (2010) The purple cauliflower arises from activation of a MYB transcription factor. Plant Physiol 154(3):1470–1480. https://doi.org/10.1104/pp.110.164160

Duclos DV, Björkman T (2008) Meristem identity gene expression during curd proliferation and flower initiation in *Brassica oleracea*. J Exp Bot 59(2):421–433. https://doi.org/10.1093/jxb/erm327

Ferguson JN, Fernandes SB, Monier B, Miller ND, Allen D, Dmitrieva A, Schmuker P, Lozano R, Valluru R, Buckler ES, Gore MA, Brown PJ, Spalding EP, Leakey ADB (2021) Machine learning-enabled phenotyping for GWAS and TWAS of WUE traits in 869 field-grown sorghum accessions. Plant Physiol 187(3):1481–1500. https://doi.org/10.1093/plphys/kiab346

Gu H, Zhao Z, Sheng X, Yu H, Wang J (2014) Efficient doubled haploid production in microspore culture of loose-curd cauliflower (*Brassica oleracea* var. *botrytis*). Euphytica 195(3):467–475

Hamblin M, Buckler ES, Jannink JL (2011) Population genetics of genomics-based crop improvement methods. Trends Genet 27:98–106. https://doi.org/10.1016/j.tig.2010.12.003

Hasan Y, Briggs W, Matschegewski C, Ordon F, Stützel H, Zetzsche H, Groen S, Uptmoor R (2016) Quantitative trait loci controlling leaf appearance and curd initiation of cauliflower in relation to temperature. Theor Appl Genet 129:1273–1288

Kim M, Nguyen TTP, Ahn JH, Kim GJ, Sim SC (2021) Genome-wide association study identifies QTL for eight fruit traits in cultivated tomato (*Solanum lycopersicum* L.). Hortic Res 8:203. https://doi.org/10.1038/s41438-021-00638-4

Lan TH, Paterson AH (2000) Comparative mapping of quantitative trait loci sculpting the curd of Brassica oleracea. Genetics 155(4):1927

Li H, Liu Q, Zhang Q, Qin E, Jin C, Wang Y, Wu M, Shen G, Chen C, Song W, Wang C (2017) Curd development associated gene (CDAG1) in cauliflower (*Brassica oleracea* L. var. *botrytis*) could result in enlarged organ size and increased biomass. Plant Sci 254:82–94. https://doi.org/10.1016/j.plantsci.2016.10.009

Li F, Liu Z, Chen H, Wu J, Cai X, Wang H, Wang X, Liang J (2023) QTL mapping of leaf-related traits using a high-density bin map in Brassica rapa. Horticulturae 9(4):433. https://doi.org/10.3390/horticulturae9040433

Limbalkar OM, Vasisth P, Singh G et al (2023) Dissection of QTLs conferring drought tolerance in *B. carinata* derived *B. juncea* introgression lines. BMC Plant Biol 23:664. https://doi.org/10.1186/s12870-023-04614-z

Matschegewski C, Zetzsche H, Hasan Y, Leibeguth L, Briggs W, Ordon F, Uptmoor R (2015) Genetic variation of temperature-regulated curd induction in caulifower: elucidation of foral transition by genome-wide association mapping and gene expression analysis. Front Plant Sci 6:720. https://doi.org/10.3389/fpls.2015.00720

Miles C, Wayne M (2008) Quantitative trait locus (QTL) analysis. Nat Educ 1(1):208

Natalini A, Acciarri N, Cardi T (2021) Breeding for nutritional and organoleptic quality in vegetable crops: the case of tomato and cauliflower. Agriculture 11(7):606. https://doi.org/10.3390/agriculture11070606

Raj SRG, Nadarajah K (2023) QTL and candidate genes: techniques and advancement in abiotic stress resistance breeding of major cereals. Int J Mol Sci 24(1):6. https://doi.org/10.3390/ijms24010006

Rakshita KN, Singh S, Verma VK, Sharma BB, Saini N, Iquebal MA, Behera TK (2021) Understanding population structure and detection of QTLs for curding-related traits in Indian cauliflower by genotyping by sequencing analysis. Funct Integr Genomics 21:679–693. https://doi.org/10.1007/s10142-021-00811-x

Ro N, Haile M, Kim B, Cho GT, Lee J, Lee YJ, Hyun DY (2022) Genome-wide association study for agro-morphological traits in eggplant core collection. Plants (Basel) 11(19):2627

Roorkiwal M, Bharadwaj C, Barmukh R, Dixit GP, Thudi M, Gaur PM et al (2020) Integrating genomics for chickpea improvement: achievements and opportunities. Theor Appl Genet 133:1703–1720. https://doi.org/10.1007/s00122-020-03584-2

Rosen A, Hasan Y, Briggs W, Uptmoor R (2018) Genome-based prediction of time to curd induction in cauliflower. Front Plant Sci 9:78. https://doi.org/10.3389/fpls.2018.00078

Saeed F, Chaudhry UK, Raza A, Charagh S, Bakhsh A, Bohra A, Ali S, Chitikineni A, Saeed Y, Visser RGF, Siddique KHM, Varshney RK (2023) Developing future heat-resilient vegetable crops. Funct Integr Genomics 23(1):47. https://doi.org/10.1007/s10142-023-00967-8

Saha P, Kalia P, Sonah H, Sharma TR (2014) Molecular mapping of black rot resistance locus *Xca1bo* on chromosome 3 in Indian cauliflower (*Brassica oleracea* var. *botrytis* L.). Plant Breed 133:268–274. https://doi.org/10.1111/pbr.12152

Saha P, Ghoshal C, Ray S, Saha ND, Srivastava M, Kalia P, Tomar BS (2020) Genetic analysis of downy mildew resistance and identification of molecular markers linked to resistance gene *Ppa207* on chromosome 2 in cauliflower. Euphytica 216:183. https://doi.org/10.1007/s10681-020-02696-6

Sallam A, Arbaoui M, El-Esawi M, Abshire N, Martsch R (2016) Identification and verification of QTL associated with frost tolerance using linkage mapping and GWAS in winter faba bean. Front Plant Sci 7:1098

Sehgal N, Singh S (2018) Progress on deciphering the molecular aspects of cell-to-cell communication in Brassica self-incompatibility response. 3. Biotech 8(8):347. https://doi.org/10.1007/s13205-018-1372-2

Shaw RK, Shen Y, Zhao Z, Sheng X, Wang J, Yu H, Gu H (2021) Molecular breeding strategy and challenges towards improvement of downy mildew resistance in cauliflower (*Brassica oleracea* var. *botrytis* L.). Front Plant Sci 12:667757. https://doi.org/10.3389/fpls.2021.667757

Singh S, Sharma SR, Kalia P, Deshmukh R, Kumar V, Sharma P, Sharma TR (2012) Molecular mapping of the downy mildew resistance gene *Ppa3* in cauliflower (*Brassica oleracea* var. *botrytis* L.). J Hortic Sci Biotechnol 87:137–143. https://doi.org/10.1080/14620316.2012.11512844

Singh S, Bhatia R, Kumar R, Sharma K, Dash S, Dey SS (2018a) Cytoplasmic male sterile and doubled haploid lines with desirable combining ability enhances the concentration of important antioxidant attributes in *Brassica oleracea*. Euphytica 214:207. https://doi.org/10.1007/s10681-018-2291-3

Singh S, Singh R, Thakur P, Kumar R (2018b) Phytochemicals, functionality and breeding for enrichment of cole vegetables (*Brassica oleracea* L.). In: Petropoulos SA, Ferreira ICFR, Barros L (eds) Phytochemicals in vegetables: a valuable sourcce of bioactive compounds. Bentham Science Publishers, Sharjah, pp 256–295

Singh S, Dey SS, Bhatia R, Batley J, Kumar R (2018c) Molecular breeding for resistance to black rot [*Xanthomonas campestris* pv. *campestris* (Pammel) Dowson] in Brassicas: recent advances. Euphytica 214:196. https://doi.org/10.1007/s10681-018-2275-3

Singh S, Dey SS, Bhatia R, Kumar R, Behera TK (2019a) Current understanding of male sterility systems in vegetable Brassicas and their exploitation in hybrid breeding. Plant Reprod 32:231–256. https://doi.org/10.1007/s00497-019-00371-y

Singh S, Dey SS, Bhatia R, Kumar R, Sharma K, Behera TK (2019b) Heterosis and combining ability in cytoplasmic male sterile and doubled haploid based *Brassica oleracea* progenies and prediction of heterosis using microsatellites. PLoS One 14:e0210772. https://doi.org/10.1371/journal.pone.0210772

Singh S, Bhatia R, Kumar R, Behera TK, Kumari K, Pramanik A, Ghemeray H, Sharma K, Bhattacharya RC, Dey SS (2021a) Elucidating mitochondrial DNA markers of ogura-based CMS lines in Indian cauliflowers (*Brassica oleracea* var. *botrytis* l.) and their floral abnormalities due to diversity in cytonuclear interactions. Front Plant Sci 12:631489. https://doi.org/10.3389/fpls.2021.631489

Singh S, Bhatia R, Kumar R, Das A, Ghemeray H, Behera TK, Dey SS (2021b) Characterization and genetic analysis of OguCMS and doubled haploid based large genetic arsenal of Indian cauliflowers (*Brassica oleracea* var. *botrytis* L.) for morphological, reproductive and seed yield traits revealed their breeding potential. Genet Resour Crop Evol 68:1603–1623. https://doi.org/10.1007/s10722-020-01090-4

Singh S, Dey SS, Bhatia R, Kumar R (2023) Organelle genome-based markers in deciphering mito-nuclear conflict displaying floral malformations and distinguishing cytoplasm types in snowball cauliflower. Acta Hortic 1362:507–512. https://doi.org/10.17660/ActaHortic.2023.1362.68

Takagi H, Abe A, Yoshida K, Kosugi S, Natsume S, Mitsuoka C, Uemura A, Utsushi H, Tamiru M, Takuno S, Innan H, Cano LM, Kamoun S, Terauchi R (2013) QTL-seq: rapid mapping of quantitative trait loci in rice by whole genome resequencing of DNA from two bulked populations. Plant J 74:174–183

Tan H, Wang X, Fei Z, Li H, Tadmor Y, Mazourek M, Li L (2020) Genetic mapping of green curd gene Gr in cauliflower. Theor Appl Genet 133(1):353–364. https://doi.org/10.1007/s00122-019-03466-2

Tello J, Ibanez J (2023) Review: Status and prospects of association mapping in grapevine. Plant Sci 327:111539. https://doi.org/10.1016/j.plantsci.2022.111539

Thakur P, Vidyasagar, Singh S (2015) Evaluation of cytoplasmic male sterile (CMS) progenies and maintainer lines for yield and horticultural traits in cabbage (*Brassica oleracea* var. *capitata* L.). SABRAO J Breed Genet 47:29–39

Thorwarth P, Yousef E, Schmid K (2018) Genomic prediction and association mapping of curd-related traits in gene bank accessions of cauliflower. G3 Genes Genomes Genetics 8:707–718. https://doi.org/10.1534/g3.117.300199

Welsch R, Zhou X, Koschmieder J, Schlossarek T, Yuan H, Sun T, Li L (2020) Characterization of cauliflower OR mutant variants. Front Plant Sci 10:1716. https://doi.org/10.3389/fpls.2019.01716

Xu Y, Li P, Yang Z, Xu C (2017) Genetic mapping of quantitative trait loci in crops. Crop J 5:175–184

Yu S, Su T, Zhi S, Zhang F, Wang W, Zhang D, Zhao X, Yu Y (2016) Construction of a sequence-based bin map and mapping of QTLs for downy mildew resistance at four developmental stages in Chinese cabbage (*Brassica rapa* L. ssp. *pekinensis*). Mol Breed 36:44. https://doi.org/10.1007/s11032-016-0467-x

Zhang X, Wen Z, Jiang H, Niu G, Liu L, Yao X, Sun D, Shan X (2023) Identification of loci for four important agronomic traits in loose-curd cauliflower based on genome-wide association studies. Horticulturae 9(9):970. https://doi.org/10.3390/horticulturae9090970

Zhao ZQ, Sheng XG, Yu HF, Wang JS, Shen YS, Gu HH (2020) Identification of QTLs associated with curd architecture in cauliflower. BMC Plant Biol 20(1):177. https://doi.org/10.1186/s12870-020-02377-5

Zhou X, Fei Z, Thannhauser TW, Li L (2011) Transcriptome analysis of ectopic chloroplast development in green curd cauliflower (*Brassica oleracea* L. var. *botrytis*). BMC Plant Biol 11:169

9 GWAS and QTL Mapping in Cabbage (*Brassica oleracea* var. *capitata* L.)

Srija Priyadarsini, Saurabh Singh, Alok Nandi,
Reeta Bhatia Dey, and Shyam Sundar Dey

Abstract

Cabbage is an important member of the cole group of vegetables belonging to the family Brassicaceae. In cabbage, head characteristics and related head traits, agronomic, disease resistance, and abiotic stress resistance traits are the prime targets of cabbage breeders. Unravelling the genetic basis of these complex quantitative and qualitative traits is the principal objective of genetics. Much progress has been made in revealing the QTLs underlying different complex traits in cabbage. In the era of genomics, state-of-the-art tools such as association mapping, QTL-seq, and genotyping by sequencing have accelerated the identification of QTLs linked with complicated traits in crop species. These innovative tools have also been proved effective in cole vegetables, and their progress has been highlighted in this chapter. Among the cole group of vegetables, cabbage is a crucial component in a balanced human diet. GWAS and QTL analysis have

S. Priyadarsini
Department of Vegetable Science, Odisha University of Agriculture and Technology (OUAT), Bhubaneswar, Odisha, India

S. Singh (✉)
Department of Vegetable Science, Rani Lakshmi Bai Central Agricultural University, Jhansi, Uttar Pradesh, India

A. Nandi
Institute of Agricultural Sciences, Siksha 'O' Anusandhan (Deemed to be University), Bhubaneswar, Odisha, India

R. Bhatia Dey
Division of Floriculture and Landscaping, ICAR-Indian Agricultural Research Institute, New Delhi, India

S. S. Dey
Division of Vegetable Science, ICAR-Indian Agricultural Research Institute, New Delhi, India

S. S. Dey et al. (eds.), *GWAS and QTL Mapping in Horticultural Crops*,
https://doi.org/10.1007/978-981-95-2306-1_9

been exploited commercially in cole crops in pursuit of determining the genetics of complex traits. This chapter provides a comprehensive view of the status of QTL mapping in cabbage in the era of genomics.

Keywords

Cabbage · QTL-seq · Head traits · Abiotic stresses · Genome-wide association studies · Molecular mapping

9.1 Introduction

The cole group of vegetables represents a group of six morphotypes (cauliflower, cabbage, broccoli, kale, brussel's sprout, knol khol) evolved from a common ancestor, *Brassica oleracea* var. *sylvestris* (Fig. 9.1) (Mabry et al. 2021). Among these, cabbage (*Brassica oleracea* L. var. *capitata*) is an economically important vegetable used for human consumption globally (Singh and Vidyasagar 2012, 2015; Thakur et al. 2015). The compact leafy heads are edible organs of the cabbage plants where different head shapes, like round, oblate, conical, flat head, and blunt pointed, are available according to genotype (Zhang et al. 2016). Cabbage group vegetables are reservoirs of vitamins, minerals, antioxidant compounds, carotenoids, anthocyanins, and secondary metabolites, which are having health promoting properties (Singh et al. 2018a, b). Indole-3-carbinol present in cabbage and related crucifers is a potential compound for combatting cancer (Katz et al. 2018).

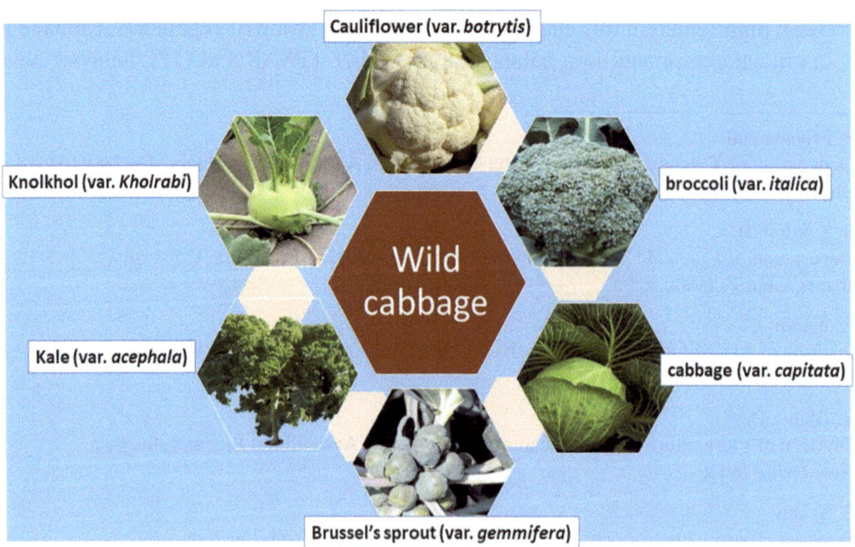

Fig. 9.1 Six brassica morphotypes evolved from a common ancestor, wild cabbage (*B. oleracea* var. *sylvestris*)

In addition to the economic and nutritional value of cabbage, it also represents as a model plant for understanding genome evolution, as the genome of cabbage is CC. The CC genome is one of the three basic diploid genomic *Brassica* species in Nagaharu U's triangle. The genome sequence of cabbage is also available, which facilitates the identification of genes regulating various traits in cabbage (Liu et al. 2014). The most important agronomic traits in cabbage are plant height, size, maturity period, and yield, which are under the control of polygenic traits (Lv et al. 2016). Head shape is also an important agronomic trait, which is related to head vertical diameter, head transverse diameter and head shape index, and ratio of polar to equatorial diameter (Zhang et al. 2016). Head splitting is another important trait which determines the yield and quality of cabbage heads (Su et al. 2015). Cabbage production is also limited by different bacterial, fungal, and viral diseases (Singh et al. 2018c). The economical and health benefits of cabbage have pushed cabbage breeders to develop high-yielding hybrids in cabbage (Sehgal and Singh 2018; Singh et al. 2023a). In this context, the availability of genetic mechanisms of sporophytic self-incompatibility and cytoplasmic male sterility has been proved instrumental in cabbage hybrid breeding globally (Singh et al. 2019a, b, 2021a, b). The protection of plant varieties is also an important component in crop improvement programs. In this context, distinctness, uniformity, and stability (DUS) characterization is a vital component of plant variety protection (PVP) for safeguarding breeder's and farmer's rights (Jo et al. 2022). During the early stages, the SSR markers were predominantly used in cultivar identification (Louarn et al. 2007). But there are some constraints in the use of SSR markers like profound use of dinucleotide loci, difficulty in distinction among neighboring alleles due to their closeness. Relative to SSR, SNPs offer numerous advantages including genome-wide abundance, low mutation rate, and high reproducibility (Mammadov et al. 2012). The advances in NGS techniques have facilitated the genome-wide discovery of SNPs, which is a booster for plant breeding.

The segregation analysis and QTL mapping are potential approaches to understand the genetic control of complicated polygenic traits (Su et al. 2015; Li et al. 2023a, b). QTL mapping helps the plant breeders to associate certain complex agronomic and related plant traits to specific locations on the genome (Miles and Wayne 2008). The core breeding objectives in major crops are genetic improvement in quality traits, plant architecture, high-yield potential, and resistance to abiotic and biotic stresses, which are having quantitative inheritance (Zhang et al. 2023). In this context, the determination of QTLs regulating these complex traits is of utmost importance with the help of advanced techniques like mapping QTLs based on linkage analysis using bi-parental populations, BSA-seq, and genome-wide association studies (GWAS). With the advancements in genomics tools, GWAS (genome-wide association studies) has become a powerful technique to identify the QTLs regulating complex polygenic traits (Hamblin et al. 2011). Tapping of QTLs in the natural populations and determination of association between linked DNA markers and complex plant traits, the association mapping (AM) or linkage disequilibrium (LD) have been proved instrumental. Recently, due to their efficiency and efficacy, both GWAS and genomic selection (GS) based on genome-wide marker data are gaining worldwide popularity among plant breeders for predicting plant phenotypes (Hamblin et al. 2011). GWAS avoids the need of constructing a specific population for determining the genetic architecture of quantitative traits (Alseekh et al. 2021). The association mapping based on GWAS utilizes availability of high recombination in natural populations. With the advancements in genomics, GWAS has been effectively used in mapping complex traits across the vegetables such as tomato (Kim et al. 2021), carrot (Brainard et al. 2022), cauliflower (Zhao et al. 2020), potato (Alvarez-Morezuelas et al. 2023) and eggplant (Ro et al. 2022). As compared to standard bi-parental mapping population, the GWAS enables the better mapping resolution by detecting relationships between genotypic data based on molecular markers and desirable phenotypic traits (Xu et al. 2017; Ro et al. 2022). Recently, genotyping-by-sequencing (GBS) has accelerated the identification and generation of SNP markers, which are further stimulating the applications of GWAS and QTL mapping approach (Sallam et al. 2016). The NGS techniques have facilitated the determination of reference genomes of crops, including *Brassica oleracea* vegetables, which in turn have made the rapid applicability of GWAS in mapping QTLs for complicated traits.

In cabbage, a linkage map has been developed by using a different set of molecular markers. GBS-GWAS has been successfully utilized in the crop plants for identification of candidate genes controlling complex phenotypes and facilitating crop breeding programs in the era of climate change. Much progress has been made in elucidating the genetic basis of complex agronomic traits in cabbage, and QTLs for head splitting resistance (HSR), disease resistance, and abiotic stress resistance have been mapped using a different set of mapping populations (Lee et al. 2015; Su et al. 2015; Lv et al. 2016). This chapter summarizes the status of GWAS, GBS, QTL-seq, and other genomics tools in cabbage breeding for determining the genetic basis of complex polygenic traits.

9.2 Mapping QTLs for Agronomic Traits

Cabbage is one of the potential nutritional and food value crops among the cole group of vegetables globally. Understanding of the genetic basis of complex agronomic traits in the era of genomics has been made feasible with the application of GWAS, GBS, QTL-seq, and associated genomics techniques. The potent agronomic traits in cabbage are head traits (head compactness, head splitting, head shape, head size, and head weight), plant height, days to maturity, head quality, leaf waxiness, and marketable yield (Lv et al. 2016; Wang et al. 2023). In this context, Lv et al. (2016) identified 144 QTLs for 24 agronomic traits in cabbage (*Brassica oleracea* var. *capitata*) using a doubled haploid mapping population (Table 9.1). Of which the major QTLs regulating different agronomic traits were "*Ph 3.1*" regulating plant height with an LOD value of 28.2, QTL "*Ll 3.2*" for leaf length with an LOD score of 13.95 and QTL for head transverse diameter "*Htd 3.2*" with an LOD score of 9.49. In this study, for the four different plant type related characters, a set of 25 QTLs were located on five different chromosomes (C01, C03, C05, C06, and C08).

Table 9.1 Major QTLs governing complex traits in cabbage

Trait	Mapping population	Technique	Molecular markers	QTLs/genes	References
Agronomic traits	Doubled haploid (DH from cross of 01-20 × 96-100-308)	Linkage analysis based meta-QTL technique	InDel and SSR markers	*Ph 3.1* (plant height) *Ll 3.2* (leaf length) *Htd 3.2* (Head transverse diameter) *Cw 9.1* and *Cw 9.2* (core width)	Lv et al. (2016)
Head splitting resistance (HSR)	Doubled haploid (DH from cross of 79–156 × 96-100)	Polygenic inheritance analysis and QTL mapping	SSR InDel markers	*Hsr 3.2* *Hsr 4.2* *Hsr 9.2*	Su et al. (2015)
Head splitting resistance (HSR)	F_2 mapping population (derived from cross of 748 × 747)	Linkage and QTL analysis	SSR and IP markers	*SPL-2-1*, *SPL-2-2* (on chromosome 2) and *SPL-4-1* (on chromosome 4) *SPL-6-1*, *SPL-6-2* (on chromosome 6)	Pang et al. (2015)
Black rot resistance	F_2, $F_{2:3}$ mapping populations from the cross of C1184 and C1234	Whole genome resequencing and QTL mapping	SNP, dCAPS, InDel	*BRQTL-C1_1*, *BRQTL-C1_2*, *BRQTL-C3*, *BRQTL-C6*	Lee et al. (2015)
Black rot	F_2 mapping population derived from cross of Early Fuji and a broccoli line	Linkage analysis-based QTL mapping	EST-SNPs	*QTL-1*, *QTL-2*, and *QTL-3*	Kifuji et al. (2013)

(continued)

Table 9.1 (continued)

Trait	Mapping population	Technique	Molecular markers	QTLs/genes	References
Black rot race 1	F_2 mapping population derived from cross of cabbage inbred lines SC31 and BR155	GBS-based linkage mapping of QTLs	SNPs	*qCaBR-C2-1, qCaBR-C5, qCaBR-C2-2, qCaBR-C6-1, qCaBR-C7, qCaBR-C2-3, qCaBR-C6-2*	Lu et al. (2023)
Club root	F_2 mapping population derived from cross of cabbage inbred lines 263 and GZ87	QTL-seq, RNA-seq based QTL mapping	SNPs	*qCRc7-2, qCRc7-3, and qCRc7-4*	Ce et al. (2021)
Club root	BC1/BC1S1 mapping population derived from ECD11 (resistant line) and DH line T010000DH3	GBS	SNPs	*Rcr_C03-1, Rcr_C08-1, Minor QTLs: Rcr_C01-1, Rcr_C03-2, Rcr_C03-3, Rcr_C04-1 and Rcr_C08-2*	Karim and Yu (2023)
Flowering time	BC_1P_1, BC_1P_2, F_2, and $F_{2:3}$ populations from cross of 195” (late-flowering) and “93219” (early flowering)	Whole-genome QTL-seq strategy	SNPs, InDel markers	*Ef2.1*	Shu et al. (2018)
Glucosinolates (sinigrin, glucobrassicin, 4-methoxyglucobrassicin	F_2 mapping population derived from cross of cabbage inbred lines 747 and 748	QTL mapping	SSR	*qSIN, qGBS, q4MGBS*	Oh et al. (2018)

For plant type 1, five QTLs were reported viz. *Pt3.1, Pt 5.1, Pt 5.2, Pt 6.2,* and *Pt 8.1*. Likewise, five different QTLs regulating plant type II were identified as *Pt 5.3, Pt 6.1, Pt 6.2, pt 6.3,* and *Pt 8.1*, while 9 QTLs were reported to regulate plant diameter (*Pd 1.1f, Pd 1.2, Pd 1.3, Pd 3.1, Pd 3.2, Pd 5.1, Pd 6.1, and Pd 8.1*) (Lv et al. 2016). The plant height was reported to be controlled by six QTLs viz. *Ph 1.1, Ph 1.2, Ph 1.3, Ph 3.1,* and *Ph 3.1.* A total of 64 QTLs regulating 10 different leaf-related traits were mapped on all 9 chromosomes of cabbage. The robust QTLs for leaf color included *LcL 8.1* and *LcL 8.2.* The QTL analysis revealed that the genetic factors controlling leaf-related characters were located on three different chromosomes, C02, C03, and C09 of *Brassica oleracea.* The study also revealed that all 9 chromosomes harbored 55 QTLs for head-related traits in cabbage. The robust QTLs for core width in cabbage were *Cw 9.1, Cw 9.2* and for head solidity the robust QTL is *Hs2.1.* Further, the resistance to head splitting in cabbage is one of the important agronomic traits determining the yield and quality of cabbage heads (Su et al. 2015). The resistance to head splitting in cabbage is a complicated trait under the genetic control of multiple genes. In this regard, by using DH-based mapping population, a set of 9 QTLs regulating head splitting resistance (HSR) in cabbage were identified on four different Brassica *oleracea* chromosomes (C3, C4, C7, and C9) (Su et al. 2015). Of these, the three major QTLs, *Hsr 3.2, Hsr 4.2,* and *Hsr 9.2,* exhibited large effects (Table 9.1). These QTLs would be instrumental in marker-assisted selection (MAS) advancement in cabbage breeding for developing head splitting resistant varieties. Similarly, Pang et al. (2015) identified six QTLs for head splitting resistance in cabbage by using F_2 mapping population. The physiological basis of head-splitting in cabbage is not well understood, but different responsive factors have been mentioned in various reports. The responsive factors for head splitting are viz. excess application of water during later stages of growth, heavy irrigation or heavy rain after a long dry spell, which is attributed to brittle tissue and poor cell flexibility, and delay in harvesting (Wang et al. 2023). All these factors are related to genetic features of cabbage and are genotype dependent. In the post-genomics era, much progress has been made in the identification of SNPs and understanding the genetic basis of complex traits, facilitating smart plant breeding (Singh et al. 2023b). Flowering is also an important agronomic character in cabbage for hybrid breeding. Thus, understanding the genetic basis of flowering is essential to facilitate hybrid breeding in *Brassica oleracea* vegetables. In line with this, Shu et al. (2018) reported a major QTL for flowering time in broccoli × cabbage populations (Table 9.1) on chromosome C02. A candidate gene, "*Bol024659*" identified in the QTL region was designated as "*BolGRF6,*" which is a major gene for regulating flowering (Shu et al. 2018).

9.3 QTL Mapping of Biotic Stress Resistance in Cabbage

The availability of draft genome sequences of *Brassica oleracea* vegetables has accelerated the identification of genes/QTLs regulating biotic stress resistance in cole vegetables. Cabbage production is limited by bacterial, fungal, and viral pathogens in combination with insect pests like diamond back moth, aphids, cabbage butterfly, and cabbage webworm. Among the biotic stresses, black rot, caused by *Xanthomonas campestris* pv. *campestris* (*Xcc*), is one of the devastating bacterial menaces causing huge yield loss in cabbage and related crucifer vegetables (Singh et al. 2018c). Detailed progress regarding black rot resistance genes and QTLs has been reviewed by Singh et al. (2018c). Mapping QTLs for black rot resistance in *Brassica oleracea* has been a matter of interest for many years. The races R1 and R4 are the most widespread and the most virulent races of the black rot pathogen, and QTLs governing resistance to races 1 and 4 in *Brassica* vegetables have been mapped (Singh et al. 2018c). Deciphering the genetic basis of black rot resistance is essential for breeding resistant varieties. The polygenic control of resistance and evolution of new races of the black rot pathogen are major obstacles in black rot resistance breeding (Lee et al. 2015). Several QTLs for resistance to *Xcc* have been mapped across the *Brassica oleracea* vegetables. Lee et al. (2015) identified an array of SNPs and mapped QTLs for black rot resistance in cabbage based on whole genome resequencing and the QTL mapping approach. QTLs for black rot resistance were mapped on chromosome numbers C01, C03, and C06 (Table 9.1) (Lee et al. 2015). Previously, Kifuji et al. (2013) reported three QTLs for black rot resistance using 161 EST-SNP markers (Table 9.1). Recently, Lu et al. (2023) identified seven QTLs for resistance to race 1 of *Xcc,* usingGBS (genotyping-by-sequencing) approach for identification of SNPs and genotyping of F_2 mapping population (Table 9.1).

Fusarium wilt in cabbage (*Fusarium oxysporum* f. sp. *conglutinans*) is another devastating disease-causing loss in cabbage production globally. A monogenic dominant gene "*FOC1*" has been identified again in fusarium wilt in *B. oleracea* (Liu et al. 2017). Clubroot disease (*Plasmodiophora brassicae*) is one of the major threats to *Brassica* crops globally, including cabbage, Chinese cabbage, and oilseed rape (Ce et al. 2021). Based on QTL-seq, three QTLs were reported on chromosome number C07 against clubroot in cabbage (Table 9.1) (Ce et al. 2021). Further, two candidate genes, *Bol037115* and *Bol042270,* were identified from these three QTLs using RNA-seq and qRT-PCR analysis (Ce et al. 2021). Previously, Peng et al. (2018) identified two QTLs, DIC.I-1, and DIC.II-1 on chromosome C08 in the resistant inbred line "GZ87" of cabbage against race 4 of the club root pathogen. Recently, Karim and Yu (2023) reported two major and five minor QTLs for clubroot resistance in cabbage (Table 9.1). The major QTLs were located on chromosomes C03 and C08, respectively, while the five minor QTLs were detected on chromosomes C01, C03, C04, and C08 (Karim and Yu 2023).

9.4 Quality Traits

Brassica vegetables are storehouses of secondary metabolites, glucosinolates, which have anticancer and human health-promoting properties. In addition, the presence of secondary metabolites is also related to plant defense against biotic and abiotic stresses (Singh et al. 2018b). The glucosinolate accumulation in crops is genotype, environment, and development stage dependent (Velasco et al. 2008). A better understanding of the genetics of beneficial glucosinolate accumulation is required for the improvement of useful glucosinolates in crop plants. Some efforts have been made in the past to understand the genetic basis of glucosinolate accumulation, distribution, and biosynthesis in *Brassicaceae* crops. In this context, Sotelo et al. (2014) identified metabolic QTLs for glucosinolate accumulation in leaves, flower buds, and seeds of *Brassica oleracea* using a doubled haploid mapping population. Later on, Oh et al. (2018) reported three QTLs (qSIN on chromosome C01, qGBS on chromosome C08, and q4MGBS on chromosome C08) linked with sinigrin, glucobrassicin, and 4-methoxyglucobrassicin in cabbage based on the F_2 mapping population. Previously, Sotelo et al. (2014) located sinigrin on chromosome C09 and C05. Thus, there exists a wide diversity with respect to genomic regions harboring QTLs for glucosinolates.

9.5 Abiotic Stress Resistance

All the life forms on earth are facing the challenges of anthropogenic climate change for their existence, which has aggravated the ill effects of drought, heat, cold, salinity, and floods on humanity and agricultural production (Bulgari et al. 2019; Singh et al. 2024). *Brassica* vegetables being a group of cool season vegetables are more sensitive to heat stress. Under high temperature stress conditions, cabbage plant growth is reduced, the symptoms of wilting, yellowing, and bolting occur (Li et al. 2023a, b). The activity of superoxide dismutase (SOD), catalase (CAT), and peroxidase (POD) increases under heat stress conditions in *Brassica oleracea.* Different indicators like water-use efficiency (WUE), drought susceptibility index (DSI), leaf wilting index (LWI), and relative vigour index (RVI) are important components in breeding and research activities for abiotic stresses (Zhang et al. 2014). Thus, understanding the genetic basis of tolerance to abiotic stresses and developing climate-resilient cultivars is one of the prime objectives of 21st vegetable breeding. In the past decades, the advances in NGS tools have accelerated the breeding for complex traits, including abiotic stresses across the crop plants. The more rapid use of genomic tools like GBS, QTL-seq, and GWAS in the identification of SNPs and the determination of underlying QTLs for various abiotic stresses in crop plants have accelerated plant breeding for developing resistant varieties (Singh et al. 2024; Chen et al. 2022). The major breeding obstacle for heat stress in *Brassica oleracea* is lack of research for evaluation of new genetic resources and understanding genetics for heat and other abiotic stresses. The QTLs for heat stress have been reported in related cole vegetables like broccoli by Branham et al. (2017) who detected different set of QTLs viz. QHT_C02, QHT_C03, QHT_C05, and QHT_C09 for heat stress. Currently, the GWAS and QTL-seq-based approaches are vital tools for determining the genetic basis of abiotic stresses, which can be exploited efficiently in cabbage and related cole vegetables. Much progress has been made in determining the genetic basis of abiotic stress tolerance in related *Brassica species*, but fewer reports are available in the cole group of vegetables. In the era of genomics, GWAS has been successfully used in the identification of QTLs for abiotic stresses across the Brassica genus (Chen et al. 2022). The extensive studies are needed for dissecting the genetic basis of abiotic stress in cabbage and related crucifers. In the NGS era, the availability of reference genomes of crop plants and the advent of genomics approaches like GWAS, QTL-seq, RNA-seq have successfully explored the genetics of complex traits in the plant kingdom. The applicability of these genomic tools can be effective in enhancing the breeding of cabbage and related cole vegetables in the era of climate change.

9.6 Conclusion

Brassica oleracea vegetables are an abundant source of phytochemicals, minerals, vitamins, and antioxidant compounds. Understanding the genetic basis of complex agronomic, yield, and stress-related traits would be instrumental in enhancing the genetic improvement of cole vegetables. Traditionally, QTL mapping has enabled the crop breeders to decipher the genetics of polygenic traits in vegetable crops. In recent times, GWAS based on exploiting the recombination frequencies of natural populations has stimulated the identification of QTLs underlying complicated traits. The whole genome-based identification of SNPs, QTL-seq, GBS, and GWAS tools are the choice of plant scientists to unravel complex trait genetics in crop plants globally. The reports in cabbage are not substantial to provide robust conclusions regarding genetic basis of complex traits using GWAS and linkage analysis-based QTL mapping. Additionally, the GWAS-derived transcriptome and metabolome-wide association studies are vital to narrow down the knowledge gaps regarding the genetic basis of complex physiological, phenotypic, biochemical, and stress resistance traits in cabbage (Tello and Ibanez 2023). The QTL mapping and GWAS have now become essential technique for candidate gene identification and elucidating genetics of complicated traits. The application of these potential tools can be effective in accelerating cabbage improvement in the genomics era.

References

Alseekh S, Kostova D, Bulut M, Fernie AR (2021) Genome-wide association studies: assessing trait characteristics in model and crop plants. Cell Mol Life Sci 78:5743–5754. https://doi.org/10.1007/s00018-021-03868-w

Alvarez-Morezuelas A, Barandalla L, Ritter E, Ruiz de Galarreta JI (2023) Genome-wide association study of agronomic and physiological traits related to drought tolerance in potato. Plants (Basel) 12(4):734

Brainard SH, Ellison SL, Simon PW, Dawson JC, Goldman IL (2022) Genetic characterization of carrot root shape and size using genome-wide association analysis and genomic-estimated breeding values. Theor Appl Genet 135:605–622. https://doi.org/10.1007/s00122-021-03988-8

Branham SE, Stansell ZJ, Couillard DM, Farnham MW (2017) Quantitative trait loci mapping of heat tolerance in broccoli (*Brassica oleracea var. italica*) using genotyping-by-sequencing. Theor Appl Genet 130:529–538. https://doi.org/10.1007/s00122-016-2832-x

Bulgari R, Franzoni G, Ferrante A (2019) Biostimulants application in horticultural crops under abiotic stress conditions. Agronomy 9(6):309. https://doi.org/10.3390/agronomy9060306

Ce F, Mei J, He H, Zhao Y, Hu W, Yu F, Li Q, Ren X, Si J, Song H, Qian W (2021) Identification of candidate genes for clubroot-resistance in *Brassica oleracea* using quantitative trait loci-sequencing. Front Plant Sci 12:703520. https://doi.org/10.3389/fpls.2021.703520

Chen S, Hayward A, Dey SS, Choudhary M, Witt Hmon KP, Inturrisi FC, Dolatabadian A, Neik TX, Yang H, Siddique KHM, Batley J, Cowling WA (2022) Quantitative trait loci for heat stress tolerance in *Brassica rapa* l. are distributed across the genome and occur in diverse genetic groups, flowering phenologies and morphotypes. Genes (Basel) 13(2):296

Hamblin M, Buckler ES, Jannink JL (2011) Population genetics of genomics-based crop improvement methods. Trends Genet 27:98–106. https://doi.org/10.1016/j.tig.2010.12.003

Jo J, Kang MY, Kim KS, Youk HR, Shim EJ, Kim H, Park JS, Sim SC, Yu BC, Jung JK (2022) Genome-wide analysis-based single nucleotide polymorphism marker sets to identify diverse genotypes in cabbage cultivars (*Brassica oleracea* var. *capitata*). Sci Rep 12(1):20030. https://doi.org/10.1038/s41598-022-24477-y

Karim MM, Yu F (2023) Identification of QTLs for resistance to 10 pathotypes of *Plasmodiophora brassicae* in *Brassica oleracea* cultivar ECD11 through genotyping-by-sequencing. Theor Appl Genet 136:249. https://doi.org/10.1007/s00122-023-04483-y

Katz E, Nisani S, Chamovitz DA (2018) Indole-3-carbinol: a plant hormone combatting cancer. F1000Res 7: F1000 Faculty Rev-689

Kifuji Y, Hanzawa H, Terasawa Y, Ashutosh, Nishio T (2013) QTL analysis of black rot resistance in cabbage using newly developed EST-SNP markers. Euphytica 190:289–295. https://doi.org/10.1007/s10681-012-0847-1

Kim M, Nguyen TTP, Ahn JH, Kim GJ, Sim SC (2021) Genome-wide association study identifies QTL for eight fruit traits in cultivated tomato (*Solanum lycopersicum* L.). Hortic Res 8:203. https://doi.org/10.1038/s41438-021-00638-4

Lee J, Izzah NK, Jayakodi M et al (2015) Genome-wide SNP identification and QTL mapping for black rot resistance in cabbage. BMC Plant Biol 15:32. https://doi.org/10.1186/s12870-015-0424-6

Li F, Liu Z, Chen H, Wu J, Cai X, Wang H, Wang X, Liang J (2023a) QTL mapping of leaf-related traits using a high-density bin map in *Brassica rapa*. Horticulturae 9(4):433. https://doi.org/10.3390/horticulturae9040433

Li Q, Zhang G, Zhao Y, Gu L, Wang Y, Yu X, Abdullah S (2023b) Hub gene identification and heat-stress-related transcriptional regulation mechanism in cabbage (*Brassica oleracea* L.). *Horticulturae* 9(9):977. https://doi.org/10.3390/horticulturae9090977

Liu S, Liu Y, Yang X, Tong C, Edwards D, Parkin IA et al (2014) The *Brassica oleracea* genome reveals the asymmetrical evolution of polyploid genomes. Nat Commun 5:3930

Liu X, Han F, Kong C, Fang Z, Yang L, Zhang Y, Zhuang M, Liu Y, Li Z, Lv H (2017) Rapid introgression of the fusarium wilt resistance gene into an elite cabbage line through the combined application of a microspore culture, genome background analysis, and disease resistance-specific marker assisted foreground selection. Front Plant Sci 8:354. https://doi.org/10.3389/fpls.2017.00354

Louarn S, Torp AM, Holme IB, Andersen SB, Jensen BD (2007) Database derived microsatellite markers (SSRs) for cultivar differentiation in *Brassica oleracea*. Genet Resour Crop Evol 54:1717–1725

Lu L, Choi SR, Lim YP, Kang SY, Yi SY (2023) A GBS-based genetic linkage map and quantitative trait loci (QTL) associated with resistance to *Xanthomonas campestris* pv. *campestris race* 1 identified in Brassica oleracea. Front. Plant Sci 14:1205681. https://doi.org/10.3389/fpls.2023.1205681

Lv H, Wang Q, Liu X, Han F, Fang Z, Yang L, Zhuang M, Liu Y, Li Z, Zhang Y (2016) Whole-genome mapping reveals novel QTL clusters associated with main agronomic traits of cabbage (*Brassica oleracea* var. *capitata* L.). Front Plant Sci 7:989. https://doi.org/10.3389/fpls.2016.00989

Mabry ME, Turner-Hissong SD, Gallagher EY, McAlvay AC, An H, Edger PP, Moore JD, Pink DAC, Teakle GR, Stevens CJ, Barker G, Labate J, Fuller DQ, Allaby RG, Beissinger T, Decker JE, Gore MA, Pires JC (2021) The evolutionary history of wild, domesticated, and feral *Brassica oleracea* (Brassicaceae). Mol Biol Evol 8(10):4419–4434. https://doi.org/10.1093/molbev/msab183

Mammadov J, Aggarwal R, Buyyarapu R, Kumpatla S (2012) SNP markers and their impact on plant breeding. Int J Plant Genom 2012:728398

Miles C, Wayne M (2008) Quantitative trait locus (QTL) analysis. Nat Educ 1(1):208

Oh SH, Choi SR, Pang W, Rameneni JJ, Yi SY, Kim MS, Im SB, Lim YP (2018) Identification of glucosinolate-associated QTLs in cabbage (*Brassica oleracea* L. var. *capitata* L.). Korean J Agric Sci 45:1–8

Pang W, Li X, Choi SR et al (2015) Mapping QTLs of resistance to head splitting in cabbage (*Brassica oleracea* L. var. *capitata* L.). Mol Breed 35:126. https://doi.org/10.1007/s11032-015-0318-1

Peng L, Zhou L, Li Q, Wei D, Ren X, Song H, Mei J, Si J, Qian W (2018) Identification of quantitative trait loci for clubroot resistance in *Brassica oleracea* with the use of Brassica SNP microarray. Front Plant Sci 9:822. https://doi.org/10.3389/fpls.2018.00822

Ro N, Haile M, Kim B, Cho GT, Lee J, Lee YJ, Hyun DY (2022) Genome-wide association study for agro-morphological traits in eggplant core collection. Plants (Basel) 11(19):2627

Sallam A, Arbaoui M, El-Esawi M, Abshire N, Martsch R (2016) Identification and verification of QTL associated with frost tolerance using linkage mapping and GWAS in winter faba bean. Front Plant Sci 7:1098

Sehgal N, Singh S (2018) Progress on deciphering the molecular aspects of cell-to-cell communication in Brassica self-incompatibility response. 3. Biotech 8(8):347. https://doi.org/10.1007/s13205-018-1372-2

Shu J, Liu Y, Zhang L, Li Z, Fang Z, Yang L, Zhuang M, Zhang Y, Lv H (2018) QTL-seq for rapid identification of candidate genes for flowering time in broccoli × cabbage. Theor Appl Genet 131(4):917–928. https://doi.org/10.1007/s00122-017-3047-5

Singh S, Vidyasagar (2012) Effect of common salt (NaCl) sprays to overcome the self-incompatibility in the s-allele lines of *Brassica oleracea* var. *capitata* L. SABRAO J Breed Genet 44:339–348

Singh S, Vidyasagar (2015) Assessment of strength of self-incompatibility in s-allele lines of cabbage (*Brassica oleracea* var. *capitata* L.). Indian J Ecol 42:259–261

Singh S, Bhatia R, Kumar R, Sharma K, Dash S, Dey SS (2018a) Cytoplasmic male sterile and doubled haploid lines with desirable combining ability enhances the concentration of important antioxidant attributes in *Brassica oleracea*. Euphytica 214:207. https://doi.org/10.1007/s10681-018-2291-3

Singh S, Singh R, Thakur P, Kumar R (2018b) Phytochemicals, functionality and breeding for enrichment of cole vegetables (*Brassica oleracea* L.). In: Petropoulos SA, Ferreira ICFR, Barros L (eds) Phytochemicals in vegetables: a valuable sourcce of bioactive compounds. Bentham Science Publishers, Sharjah, pp 256–295

Singh S, Dey SS, Bhatia R, Batley J, Kumar R (2018c) Molecular breeding for resistance to black rot [*Xanthomonas campestris* pv. *campestris* (Pammel) Dowson] in Brassicas: recent advances. Euphytica 214:196. https://doi.org/10.1007/s10681-018-2275-3

Singh S, Dey SS, Bhatia R, Kumar R, Behera TK (2019a) Current understanding of male sterility systems in vegetable Brassicas and their exploitation in hybrid breeding. Plant Reprod 32:231–256. https://doi.org/10.1007/s00497-019-00371-y

Singh S, Dey SS, Bhatia R, Kumar R, Sharma K, Behera TK (2019b) Heterosis and combining ability in cytoplasmic male sterile and doubled haploid based *Brassica oleracea* progenies and prediction of heterosis using microsatellites. PLoS One 14:e0210772. https://doi.org/10.1371/journal.pone.0210772

Singh S, Bhatia R, Kumar R, Behera TK, Kumari K, Pramanik A, Ghemeray H, Sharma K, Bhattacharya RC, Dey SS (2021a) Elucidating mitochondrial DNA markers of ogura-based CMS lines in Indian cauliflowers (*Brassica oleracea* var. *botrytis* l.) and their floral abnormalities due to diversity in cytonuclear interactions. Front Plant Sci 12:631489. https://doi.org/10.3389/fpls.2021.631489

Singh S, Bhatia R, Kumar R, Das A, Ghemeray H, Behera TK, Dey SS (2021b) Characterization and genetic analysis of OguCMS and doubled haploid based large genetic arsenal of Indian cauliflowers (*Brassica oleracea* var. *botrytis* L.) for morphological, reproductive and seed yield traits revealed their breeding potential. Genet Resour Crop Evol 68:1603–1623. https://doi.org/10.1007/s10722-020-01090-4

Singh S, Dey SS, Bhatia R, Kumar R (2023a) Organelle genome-based markers in deciphering mito-nuclear conflict displaying floral malformations and distinguishing cytoplasm types in snowball cauliflower. Acta Hortic 1362:507–512. https://doi.org/10.17660/ActaHortic.2023.1362.68

Singh S, Singh R, Sharma D, Sharma SK, Dey SS, Bhatia R, Ghemeray H, Kumar R (2023b) Insight into carrot carotenoids in post-genomic world for higher nutrition. In: Singh S, Sharma D, Sharma SK, Singh R (eds) Smart plant breeding for vegetable crops in post-genomics era. Springer, Singapore, pp 367–382. https://doi.org/10.1007/978-981-19-5367-5_15

Singh S, Singh R, Priyadarsini S, Ola AL (2024) Genomics empowering conservation action and improvement of celery in the face of climate change. Planta 259:42. https://doi.org/10.1007/s00425-023-04321-x

Sotelo T, Soengas P, Velasco P, Rodríguez VM, Cartea ME (2014) Identification of metabolic QTLs and candidate genes for glucosinolate synthesis in *Brassica oleracea* leaves, seeds and flower buds. PLoS One 9(3):e91428. https://doi.org/10.1371/journal.pone.0091428

Su Y, Liu Y, Li Z, Fang Z, Yang L, Zhuang M, Zhang Y (2015) QTL analysis of head splitting resistance in cabbage (*Brassica oleracea* L. var. *capitata*) using SSR and InDel makers based on whole-genome re-sequencing. PLoS One 10(9):e0138073. https://doi.org/10.1371/journal.pone.0138073

Tello J, Ibanez J (2023) Review: Status and prospects of association mapping in grapevine. Plant Sci 327:111539. https://doi.org/10.1016/j.plantsci.2022.111539

Thakur P, Vidyasagar, Singh S (2015) Evaluation of cytoplasmic male sterile (CMS) progenies and maintainer lines for yield and horticultural traits in cabbage (Brassica oleracea var. capitata L.). SABRAO J Breed Genet 47:29–39

Velasco P, Soengas P, Vilar M, Elena Cartea M, del Rio M (2008) Comparison of glucosinolate profiles in leaf and seed tissues of different Brassica napus crops. J Am Soc Hortic Sci 133:551–558

Wang Y, Li Q, Zhang G, Gu L, Zhao Y, Zhou L, Dong Y, Dong H, Song X (2023) Mechanism of tolerance to head-splitting of cabbage (*Brassica oleracea* L. var. *capitata* L.): a review of current knowledge and future directions. Horticulturae 9(2):251. https://doi.org/10.3390/horticulturae9020251

Xu Y, Li P, Yang Z, Xu C (2017) Genetic mapping of quantitative trait loci in crops. Crop J 5:175–184

Zhang X, Lu G, Long W, Zou X, Li F, Nishio T (2014) Recent progress in drought and salt tolerance studies in Brassica crops. Breed Sci 64(1):60–73

Zhang X, Su Y, Liu Y, Fang Z, Yang L, Zhuang M, Zhang Y, Li Z, Lv H (2016) Genetic analysis and QTL mapping of traits related to head shape in cabbage (*Brassica oleracea* var. *capitata* L.). Sci Hortic 207:82–88. https://doi.org/10.1016/j.scienta.2016.05.015

Zhang X, Wen Z, Jiang H, Niu G, Liu L, Yao X, Sun D, Shan X (2023) Identification of loci for four important agronomic traits in loose-curd cauliflower based on genome-wide association studies. Horticulturae 9(9):970. https://doi.org/10.3390/horticulturae9090970

Zhao ZQ, Sheng XG, Yu HF, Wang JS, Shen YS, Gu HH (2020) Identification of QTLs associated with curd architecture in cauliflower. BMC Plant Biol 20(1):177. https://doi.org/10.1186/s12870-020-02377-5

Advances in GWAS and QTL Mapping in Gourd Vegetables

10

Shreyas Aradhya C. S., K. K. Gangopadhyay, and Shyam Sundar Dey

Abstract

Gourd vegetables are a significant vegetable group known for their medicinal properties and useful fibers. They require a variety of traits, including yield and quality components. The focus is on producing varieties with strong resistance to pathogens, physiological disorders, and stress. Contemporary cultivars have undergone introgression processes to integrate these traits, aiming to provide high-yielding, disease-resistant varieties that meet consumer preferences for quality and nutritional value. Research has been conducted to create a genetic map and pinpoint quantitative trait loci (QTL) in gourd vegetables. QTL mapping is a crucial research process that identifies genes associated with quantitative traits using molecular markers. It involves meticulous crossbreeding planning and precise phenotyping. The primary objectives of QTL mapping include exploring the number of genes influencing a trait, determining gene locations, and assessing the impact of gene dosage on trait variation. Genetic mapping, the initial step in map-based cloning, facilitates DNA-based marker-assisted selection (MAS) and investigates the linkage between genes of interest. QTL mapping is useful for pinpointing genetic regions that co-segregate with the target trait within breeding populations. This methodology is applicable to diverse population types, including F2 populations, double-haploid populations, and families of backcross or recombinant inbred lines. Various markers, such as restriction fragment length polymorphism (RFLP), amplified fragment length polymorphism (AFLP), microsatellite or simple sequence repeat (SSR), and single-nucleotide polymorphism (SNP), can be used in QTL mapping. This chapter aims to explore QTL mapping and GWS analysis, focusing on their potential applications in gourd crops. It aims to provide a comprehensive understanding of QTLs and

S. Aradhya C. S. · K. K. Gangopadhyay · S. S. Dey (✉)
Division of Vegetable Science, ICAR-Indian Agricultural Research Institute, New Delhi, India

S. S. Dey et al. (eds.), *GWAS and QTL Mapping in Horticultural Crops*,
https://doi.org/10.1007/978-981-95-2306-1_10

genome-wide association (GWAS), aiding researchers in identifying areas for further investigation and enhancing their knowledge in this field.

Keywords

Gourd vegetables · Germplasm · Quantitative trait loci · Single-nucleotide polymorphism · Double-haploid · Next-generation sequencing

10.1 Introduction

Gourd vegetables are the most important group of vegetables consumed for their medicinal properties and useful fibers, and therefore have high demand for a broad range of traits, including both characteristics related to yield and components associated with quality. Furthermore, there is a considerable focus on creating varieties that exhibit strong resistance to pathogens, physiological disorders, and various forms of stress. In response to these requirements, contemporary cultivars have undergone introgression processes to integrate these sought-after traits. This coordinated endeavor seeks to offer growers varieties that are both high-yielding and resistant to diseases, meeting consumer preferences for superior quality and nutritional value at the same time. Considerable research has been conducted to create a genetic map and pinpoint quantitative trait loci (QTL) in gourd vegetables (Rao et al. 2021).

QTL mapping is the process of identifying genes associated with quantitative traits through the use of molecular markers, constituting a fundamental research endeavor that demands meticulous crossbreeding planning and precise phenotyping. A pivotal breakthrough in understanding quantitative traits and enabling the selection of QTLs was initiated with the advent of DNA or molecular markers in the 1980s. The primary objectives of QTL mapping encompass exploring the number of genes influencing a trait, determining gene locations, and assessing the impact of gene dosage on trait variation. Genetic mapping, the initial step in map-based cloning, serves various purposes, such as facilitating DNA-based marker-assisted selection (MAS) and investigating the linkage between genes of interest. Notably, a prevalent observation is that numerous commercially important traits in crops exhibit a quantitative nature, influenced by multiple polygenes. Through the utilization of molecular markers and robust statistical approaches, researchers can precisely identify chromosome locations housing genes responsible for regulating these complex or quantitative traits. QTL mapping has emerged as a valuable approach for pinpointing genetic regions that co-segregate with the target trait within breeding populations. This methodology is versatile and applicable to diverse population types, including F2 populations, double-haploid (DH) populations, as well as families of backcross or recombinant inbred lines (RILs). Various types of markers, such as restriction fragment length polymorphism (RFLP), amplified fragment length polymorphism (AFLP), microsatellite or simple sequence repeat (SSR), and single-nucleotide polymorphism (SNP) markers, can be employed in this process (Alqudah et al. 2020). Mapping studies have extensively documented QTLs across various crop plants, elucidating their involvement in a range of traits like yield, quality, disease and insect resistance, tolerance to abiotic stress, and adaptation to environmental conditions.

In this chapter, our objective is to thoroughly explore QTL mapping and GWS analysis, with a specific emphasis on their potential applications in the realm of gourd crops. We believe that the perspectives presented in this chapter will provide a thorough understanding of QTLs and GWAS, helping researchers identify potential areas for further investigation and enhance their knowledge in this particular domain (Table 10.1).

Table 10.1 QTLs associated with quantitative traits in bitter gourd

Traits	Parent	Genes/QTL	Location of chromosome	Mapping population	References
Node at the first pistillate flower	'DBGy-201' (PVGy-201) × 'Pusa Do Mousami' (PDM)	*qNPF5-F3*	5	F2:3	Rao et al. (2018)
		qNPF9-F3	9	F2:3	
		qNPF14-F3	14	F2:3	
Days to first pistillate flower appearance		*qNPF9-F2*	9	F2	
		qNPF14-F2	14	F2	
		qDPF3-F3	3	F2:3	
		qDPF5a-F3	5	F2:3	
		qDPF5b-F3	5	F2:3	
		qDPF14-F3	14	F2:3	
		qDPF16a-F3	16	F2:3	
		qDPF16b-F3	16	F2:3	
		qDPF3a-F2	3	F2	
		qDPF3b-F2	3	F2	
Sex ratio (male: female)		*qSR9a-F3*	9	F2:3	
		qSR9b-F3	9	F2:3	
		qSR13-F3	13	F2:3	
		qSR14a-F3	14	F2:3	
		qSR14b-F3	14	F2:3	
		qSR14a-F2	14	F2	
		qSR14b-F2	14	F2	
		qSR16a-F2	16	F2	
		qSR16b-F2	16	F2	
Gynoecy	K44 × Dali-11	*gy1.1*	1	F2	Cui et al. (2018)
		gy1.2	1	F2	
The first female flower node		*fffn1.1*	1	F2	
		fffn1.2	1	F2	
Fruit wart		*Fwa*	4	F2	
Width of fruit ridge		*Wr*	4	F2	
White fruit color		*W*	10	F2	
Hue angle		*H °*	10	F2	
Lightness variable		*L*	10	F2	

(continued)

Table 10.1 (continued)

Traits	Parent	Genes/QTL	Location of chromosome	Mapping population	References
Fruit length (cm)	'DBGy-201' × 'Pusa Do Mausami'	*qFL1* *qFL5* *qFL14*	1 5 14	F2:3 F2:3 F2:3	Rao et al. (2021)
Fruit diameter (cm)		*qFD1* *qFD3* *qFD13* *qFD15* *qFD16* *qFD20*	1 3 13 15 16 20	F2:3 F2:3 F2:3 F2:3 F2:3 F2:3	
Fruit weight (g)		*qFW1*	1	F2:3	
		qFW15 *qFW20*	15 20	F2:3 F2:3	
Fruit flesh thickness (mm)		*qFT14*	14	F2:3	
Number of fruits per plant		*qFN5* *qFN14* *qFN20*	5 14 20	F2:3 F2:3 F2:3	
Yield per plant (g)		*qYD1* *qYD15* *qYD20*	1 15 20	F2:3 F2:3 F2:3	

10.2 Genetic Diversity and Distribution of Cucurbits

Cucurbits play a crucial role in global agriculture, holding significant economic importance and thriving in diverse climates, from tropical to sub-temperate zones. Key cucurbit-growing nations include China, Turkey, India, and Iran. Notably, bitter gourd (*Momordica charantia*) stands out as a vital crop in various regions such as India, the Philippines, Malaysia, China, Australia, Africa, the Middle East, Latin America, and the Caribbean. Its origins trace back to domestication in Eastern India or Southern China. The global expansion of cucurbits has led to the development of diverse genetic resources adapted to a wide range of environmental conditions. Recognizing and understanding the existing genetic diversity among cucurbits is crucial for optimizing initiatives related to their collection and conservation. Additionally, this knowledge supports plant breeders worldwide in improving crops such as melon, cucumber, watermelon, and squash by integrating traits from their wild relatives.

The Cucurbitaceae family includes some of the most economically significant crops, serving various human purposes like food, containers, musical instruments, and medicine for more than 12,000 years (Whitaker and Davis 1962). Initial descriptions by Jeffrey (1990a, b) and Robinson and Decker-Walters (1997) identified 118 genera and approximately 825 species within the family. Later, classifications by Jeffrey (2005) and Jeffrey and De Wilde (2006) indicate an expanded scope, encompassing 130 genera and around 800 species. This family is divided into two subfamilies: Nhandiroboideae, also known as Zanonioideae (Jeffrey 1990b), with 19 genera and approximately 60 species lacking economic value, and Cucurbitoideae, with 111 genera and 740 species. However, according to another study, Cucurbitaceae consists of over 965 species grouped into 95 genera (Christenhusz and Byng 2016).

The subfamily Cucurbitoideae includes significant tribes and genera within cucurbits, such as Benincaseae, Luffeae, Cucurbiteae, Sicyeae, Joliffieae, Scizopeponeae, Trichosantheae, Bryonieae, Herpetospermae, and Coniandreae (Jeffrey 1990b, 2005; Rubatzky and Yamaguchi 1997). Genera like *Cucumis, Cucurbita*, and *Citrullus*, containing species like cucumber, melon, watermelon, and squash, are among the most widely cultivated crops globally. Additionally, noteworthy cucurbits of local or regional economic importance include *Lagenaria, Momordica, Benincasa, Luffa*, and *Sechium*. The diversification of cucurbits worldwide has resulted in a substantial number of genetic resources adapted to various environmental and growing conditions across different regions. Recognizing the existing genetic diversity is essential for optimizing collection and conservation programs and supporting ongoing efforts by plant breeders to enhance melon, cucumber, watermelon, and squash through the utilization of wild relatives.

Various cucurbit species exhibit distinctive distribution patterns. Sponge gourd (*Luffa cylindrica*) boasts wild populations spanning Southern Central Asia to northeastern Australia and the South Pacific, while its domesticated variety is cultivated in Asia, Africa, and tropical America. Ridge gourd (*Luffa acutangula*) is primarily grown in Southeast Asia and other tropical countries, with concentrated diversity in specific regions of India, including the Indo-Gangetic plains, northeastern areas, northwestern Himalayas, the Western and Eastern Ghats, and sporadically in tribal-dominant belts of Central India. Bottle gourd and ash gourd diversity is prevalent in Northeast India, with a similar concentration found in the Eastern peninsular tract for *Luffa* species.

The *Momordica* genus, comprising 60 species globally, includes seven species (*Momordica charantia, M. balsamina, M. dioica, M. cochinchinensis, M. tuberosa, M. subangulata, M. macrophylla*) found in the Indian subcontinent. Notable species within this genus include *M. dioica* Roxb. ex Wild., *M. cochinchinensis* Spreng., *M. tuberosa* (Roxb) Cogm. (also known as *M. cymbalaria*), *M. balsamina* L., and *M. cabriei*. Among them, *M. dioica* Roxb. ex Wild. and *M. cochinchinensis* Spreng. are edible and extensively cultivated. Other wild species such as *M. subangulata* Bl., *M. denudata*, and *M. macrophylla* also exist (Behera et al. 2023).

In India, bitter gourd is classified into two botanical forms based on fruit characteristics. The first, *M. charantia var. charantia*, bears large fusiform fruits with triangular tubercles and non-tapering ends. The second, *M. charantia var. muricata* (Wild), bears small, round fruits with tubercles and tapering ends. Additionally, Yang and Walters (1992) categorized bitter gourd into three horticultural groups: the small-fruited type (10–20 cm long, dark green, and very bitter), the long-fruited type (30–60 cm long, light green, and slightly bitter), and the triangular-fruited type (9–12 cm long, light to dark green with moderate to strong bitterness). Various accessions/lines with unique horticultural traits from different regions in India have been preserved and registered in the national germplasm collections of the ICAR

National Bureau of Plant Genetic Resources, New Delhi. These collections are valuable for potential utilization in future breeding programs (Behera et al. 2023).

The *Lagenaria* genus encompasses the domesticated bottle gourd and five wild species, namely *L. breviflora* (Benth.) Roberty, *L. abyssinica* (Hook f.) Jeffrey, *L. rufa* (Gilg.) Jeffrey, *L. sphaerica* (Sonder) Naudin, and *L. guineensis* (G. Don) Jeffrey (Whitaker 1971; Jeffrey 1980). Morphologically, the cultivated bottle gourd displays considerable diversity in comparison to its wild counterparts, particularly in quantitative fruit traits (Morimoto et al. 2005). Intra-specific variations have been reported between *L. siceraria* and *L. sphaerica*, as well as *L. abyssinica* and *L. breviflora* (Decker-Walters et al. 2001). Genotypic variation is also prevalent within *L. sphaerica*. However, the phylogenetic relationships among genotypes of *L. abyssinica* and *L. breviflora* remain inconclusive and require resolution through diagnostic molecular markers (Morimoto et al. 2005).

Wild species of bottle gourd could serve as valuable sources of phytochemical compounds, particularly cucurbitacin, with potential applications in the pharmaceutical industry. Additionally, there is limited information available on the nutritional compositions of wild bottle gourd species. Exploring and comparing macronutrient, micronutrient, amino acid, oil, and fatty acid contents between cultivated and wild species is essential for guiding breeding strategies to enhance the nutritional profile of the domesticated bottle gourd. This knowledge gap underscores the potential for breeding efforts to develop nutritionally enhanced genotypes within the cultivated bottle gourd.

10.3 Biotic Stress

10.3.1 Disease Resistance

Developing cultivars with resistance or tolerance to biotic stress through traditional breeding approaches is challenging and time-consuming. This difficulty arises from the rapid changes and mutations in strains, races, and pathotypes, allowing them to overcome resistance quickly. The presence of linkage drag issues further complicates conventional breeding, as it involves the undesired combination of multiple genes with the desired ones, hindering the achievement of both yield potential and stress tolerance (Wang et al. 2015). Recent advancements in molecular biology and genomics have enabled the identification of key resistant genes and quantitative trait loci (QTLs) associated with significant biotic stress. Additionally, progress in marker technologies has streamlined the process of breeding for biotic stress tolerance, allowing for a more efficient completion of the breeding programs (Table 10.2).

Fusarium wilt is a serious disease of cucurbit crops, including cultivated *Luffa* species (*Luffa aegyptica, Luffa acutangula*), causing a considerable amount of reduction in yield and quality. Recently, 63 *Luffa* accessions from the World Vegetable Center genebank were evaluated for resistance to an aggressive isolate of *Fusarium oxysporum* f. *FoCu*- 1 (Fsp-66). And at the second level of screening, these accessions were evaluated for resistance to 2 more isolates, FoCu-1 and

Table 10.2 QTLs associated with morphological and biotic stress-related traits in *Luffa*

Traits	Parent	Gene/QTL	Location of chromosomes	Mapping population	References
Resistance to cucumber mosaic virus	P1–21 × P2–16 (advanced inbred lines of Sponge gourd)	2 QTLs	1 and 4	F2	Lou et al. (2019)
Maturity-related traits	S1174 (*Luffa acutangula*) × P93075 (*Luffa cylindrica*)	12 QTLs	1, 7, 9, 10, 11, and 13	BC1	Liu et al. (2022)
Flower-size-related traits		3 QTLs	5 and 7	BC1	
Plant-growth-related traits		9 QTLs	5, 7, 8, 9, 10, and 11	BC1	
Fruit-related traits		16 QTLs (2 QTLs have greater effect)	Across 8 chromosomes (2 major QTLs on 7 and 5)	BC1	
Internode diameter		*qID1* and *qID2*	9	BC1	

FoM-6. Out of these 63 accessions, 14 accessions exhibited resistance against Fsp-66 in visual screening, and 11 were confirmed to be resistant against isolate Fsp-66. In addition, 13 accessions showed high resistance against isolates FoCu-1 and FoM-6 (Bindal et al. 2023). As mentioned here, initially, the identification, assessment, and integration of resistance genes were conducted via phenotypic selection and traditional breeding approaches. Presently, a significant portion of disease-resistant breeding in gourds still relies on similar procedures. Nonetheless, the application of traditional methods has posed difficulties in effectively transferring resistance from wild species to cultivated gourds, thereby restricting the incorporation of resistance into numerous elite lines.

Currently, limited information is available regarding quantitative trait loci (QTLs) governing both qualitative and quantitative traits, as well as resistance to biotic stresses in bottle gourd. The prevalence of plant diseases poses a significant challenge in bottle gourd production (Mashilo et al. 2017). Recent advances in bottle gourd genome sequencing have allowed for the mapping of specific loci associated with disease resistance. For instance, a dominant monogenic locus (Prs) that confers resistance to papaya ringspot virus has been successfully mapped (Wu et al. 2017a, b). Additionally, multiple genes such as LsWD, LsGAPDH, LsH3, LsP4H, and LsCYP linked to resistance against cucumber green mottle mosaic virus affecting bottle gourd have been identified and mapped (Zhang et al. 2018). Notably, a newly identified gene, HG_GLEAN_10011803, is considered a major contributor to resistance against Fusarium wilt in bottle gourd (Yanwei et al. 2021). High-density genetic maps, developed through various molecular markers including ISSR, SSR, and SNPs, have allowed marker-trait associations using genome-wide association (GWAS). Despite the identification of bottle gourd genotypes resistant to economically important diseases, such as zucchini yellow mosaic virus, watermelon mosaic virus, squash mosaic virus, tomato ringspot virus, tobacco mosaic virus, cucumber green mottle mosaic virus, and powdery mildew, QTL controlling disease resistance remains relatively understudied. This knowledge gap hinders marker-assisted breeding for biotic stress resistance in bottle gourd.

Powdery mildew, a significant foliar fungal disease in bitter gourd caused by *Podosphaera xanthii*, has a detrimental impact on production by causing premature foliage loss and ultimately reducing growth and yield (Keinath and DuBose 2004). In an effort to address this issue, World Veg, Taiwan, introduced five powdery mildew-resistant bitter gourd lines (THMC 113, THMC 143, THMC 153, THMC 167, and THMC 170) in 2018. The resistance in these lines is controlled by two independent, recessive genes (Dhillon et al. 2018, 2019). Bitter gourd distortion mosaic virus poses another serious threat, with infestation levels ranging from 41% to 95% (Rabindranath and Pillai 1986; Hollingsworth et al. 1997). The resistance to BDMV is determined by polygenes with non-additive gene action, and their expression is highly influenced by environmental factors (Arunachalam 2002). Additionally, Tomato leaf curl New Delhi virus (ToLCNDV), a bipartite Begomovirus of the genus Begomovirus, family Geminiviridae, leads to severe epidemics causing losses in bitter gourd cultivation. The genetics of several other biotic stresses in bitter gourd have not been extensively studied, and there is a need for further research to understand the durability of resistance and identify new sources for application in resistance breeding programs.

10.3.2 Insect Resistance

Despite previous research on host-plant resistance to the red pumpkin beetle in bottle gourd and ridge gourd, many of these studies lack information on the heritability of resistance. Knowledge of heritability is crucial for expediting the breeding process. In the case of bitter gourd, specifically Pant Karela, a selection from inbreds of indigenous germplasm at Pantnagar has been developed and is identified as highly resistant to the red pumpkin beetle. This development took place at GBPUAT, Pantnagar, Uttarakhand. When breeding for resistance to a particular insect species, it is prudent to ensure that the transfer of genes for susceptibility to other economically significant pests is avoided. In situations where resistance and susceptibility are controlled by separate genes (no linkage), breeding can effectively eliminate enhanced susceptibility.

10.4 Abiotic Stress Resistance

Recently, there has been a development and standardization of genomic/genetic resources as well as plant transformation protocols specifically tailored for cucurbits (Nanasato et al. 2013; Sun et al. 2017; Montero-Pau et al. 2017, 2018). The wild relatives of cucurbits have been characterized, contributing to their potential utilization in breeding and other initiatives related to crop improvement (Holdsworth et al. 2016). Notably, many wild relatives exhibit multiple stress tolerance capacities, encompassing both biotic and abiotic stresses. However, not many examples are there in gourd vegetables in particular (Table 10.2). Certain varieties of gourds that demonstrate resistance to abiotic stress, particularly heat, include:

1. *Lagenaria siceraria*: Thar Samridhi
2. *Luffa acutangula*: Thar Kami
3. *Luffa cylindrica*: Thar Tapish

10.5 Agronomic and Fruit Quality Traits

(a) *Luffa sp.*, or sponge gourd/ridge gourd, is targeted for breeding improvements in both quality and agronomic traits to meet producer and consumer demands. Quality enhancements focus on visual appeal and taste through improvements in fruit quality, size, and shape. Nutritional quality is also a priority, aligning with health-conscious consumer preferences. Agronomically, the breeding program emphasizes increasing yield potential, adaptability to diverse environments, and resilience to stress factors. Post-harvest attributes, including storability and transportability, are considered for sustained commercial viability. The overarching goal is to strike a balance between superior agronomic performance and enhanced quality, promoting environmentally sustainable cultivation and ensuring *Luffa* sp.'s continued significance in global agriculture and nutrition (Table 10.3).

Table 10.3 QTLs associated with fruit quality and abiotic stress (heat)-related traits in bottle gourd

Traits	Parent	Gene/ QTL	Location of chromosomes	Mapping population	References
Fruit bitterness	Hangzhou Gourd × Puxian Gourd	*QBt.1* *QBt.2*	2 9	F2 F2	Wu et al. (2019)
Heat tolerance	L1 (P17) × L6 (P23)	*qHT1.1* *qHT2.1* *qHT2.2* *qHT5.1* *qHT6.1* *qHT7.1* *qHT8.1*	1 2 2 5 6 7 8	F2 F2 F2 F2 F2 F2 F2	Song et al. (2020)

Over the past decades, the primary focus of *Luffa* breeding programs has centered on traditional methods, with limited attention given to genetic and molecular research. Notably, Wu et al. (2016) conducted ground breaking work by constructing an interspecific genetic linkage map between *Luffa acutangula* and *Luffa cylindrica*. This map, based on 177 expressed sequence tag-derived simple sequence repeat (EST-SSR) markers, revealed 14 linkage groups spanning 1436.12 cM, with six putative QTLs associated with reproductive isolation traits identified (Wu et al. 2016). In a subsequent study, Lou et al. (2020) advanced genetic research by constructing a high-density genetic map using the genetic background of *Luffa cylindrica*. Utilizing 3701 polymorphic markers through specific-locus amplified fragment sequencing (SLAF-seq), the map covered 13 linkage groups and a total distance of 1518.56 cM. This genetic map played a crucial role in predicting an LTR Copia-type retrotransposon as a candidate gene responsible for CMV resistance. Importantly, the identification and characterization of a key candidate gene, LacCRWN3, controlling internode diameter emerged as a significant outcome. The differential expression profiles associated with sequence variations in promoter regions were identified as the main causative factors for stem thickness divergence between the thin-stem parental line S1174 and the thick-stem parental line P93075. (Lou et al. 2020).

The crucial quantitative trait locus (QTL), qID2, located on chromosome 9, significantly influences internode diameter, explaining 11% of phenotypic variation within a 23 kb distance. LacCRWN3 (Lac09g006860), a gene encoding a nuclear lamina protein crucial for nuclear morphology control, is exclusively housed within qID2. Noteworthy distinctions in promoter sequences between the parental alleles of LacCRWN3 correlate with a distinctive higher expression level in the thick-stemmed P93075 compared to the thin-stemmed S1174. This QTL and the differential expression of LacCRWN3 offer key insights into the genetic basis of internode diameter variation, aiding targeted breeding strategies for *Luffa* sp (Lou et al. 2020). These studies collectively underscore the significance of QTL analysis in *Luffa* breeding, providing valuable insights into the genetic basis of important traits, paving the way for more targeted and efficient breeding strategies in the future.

(b) *Bottle gourd*

Young and immature bottle gourd fruits are highly favored in semi-urban and urban areas as fruit vegetables. To advance bottle gourd cultivation, collaborative research efforts involving plant breeders, agronomists, and geneticists are essential for knowledge and germplasm sharing. The focus should be on designing, developing, and releasing adaptable bottle gourd cultivars with versatile product profiles, including quality and quantity of leaves, fruit, fodder, seed, and nutritional compositions for various value chains in the food and feed industry. Fruit shape is a critical trait influencing crop usability, ranging from club-shaped to oval, and understanding the gene action controlling fruit shape is crucial for creating unique and attractive fruits that cater to diverse market opportunities. Oval and pear shapes are controlled by a combination of dominant and recessive genes, while round shapes are influenced by two recessive genes, and dominant genes dictate long fruit shapes. The variability in fruit neck shape and length also provides opportunities for ornamental purposes, with two genes controlling complementary gene actions. Fruit color is another essential trait influencing market value, with smooth or corrugated textures contributing to the overall horticultural diversity and breeding potential of bottle gourd. Two quantitative trait loci (QTLs), specifically *QBt.*1 and *QBt.*2, were identified as regulators of bitter taste in bottle gourd, according to the findings of Wu et al. (2019). Thirty-seven genes responsible for governing plant growth and development have been discovered in the bottle gourd.

Bottle gourd exhibits extensive phenotypic variation in agronomic traits, encompassing characteristics such as flowering time, plant height, the count of main and lateral branches per plant, fruit weight, and the number of fruits per plant, as documented by Mashilo et al. in 2015 and 2017. The variation in fruit shape and size directly influences fruit yield. Numerous studies have highlighted the significance of direct selection for agronomic traits such as the number of female flowers per plant, the number of branches per plant, fruit width, and plant height in increasing the potential fruit yield in bottle gourd. Surprisingly, there remains limited information to date regarding quantitative trait loci (QTLs) that control agronomic traits in bottle gourd (Narayan et al. 1996; Behera et al. 2015; Yao et al. 2015).

(c) *Bitter gourd*

The breeding objective for bitter gourd prioritizes improvements in fruit size, shape, and resistance attributes. Additional goals include the early appearance of pistillate flowers for early fruiting, development of non-bitter cultivars with medicinal benefits, promotion of a high female-to-male sex ratio for increased yield, and adherence to international export standards. Desired characteristics encompass specific fruit color, firmness, shape, and flesh thickness. Nutritional enhancements and compliance with export criteria are integral aspects of the breeding program.

In a study, a comprehensive linkage map of *Momordica charantia* (bitter gourd) was established using genotyping-by-sequencing (GBS) technology. The mapping population consisted of individuals resulting from the cross between 'DBGy-201' and 'Pusa Do Mausami'. A total F2:3 of 2013 high-quality single-nucleotide polymorphisms (SNPs) were successfully mapped onto 20 linkage groups (LGs), spanning a cumulative distance of 2329.2 centimorgans (CM) with an average genetic distance of 1.16 CM. The research focused on the quantitative trait loci (QTL) analysis of six key yield-contributing traits: fruit length, fruit diameter, fruit weight, fruit flesh thickness, number of fruits per plant, and yield per plant. Utilizing composite interval mapping (CIM), a total of 19 QTLs were identified, with 9 QTLs exhibiting a logarithm of odds (LOD) score greater than 3. Notably, among these QTLs, 12 were associated with 'Pusa Do Mausami' and demonstrated a negative additive effect, indicating a reduction in trait scores with an increase in the allele from this parent. Conversely, the remaining 7 QTLs from 'DBGy-201' exhibited a positive additive effect, indicating an increase in trait scores with the allele from this parent. The phenotypic variation (R2%) explained by these QTLs ranged from 0.09% for fruit flesh thickness on LG 14–32.65% for fruit diameter on LG 16, revealing six major QTLs in total (Rao et al. 2021). This insightful research, conducted by Rao et al. in 2021, contributes valuable information to the understanding of the genetic basis of important traits in bitter gourd.

10.6 Genome-Wide Association Studies (GWAS)

The advent of whole-genome sequencing for various gourd vegetables, such as ridge gourd, sponge gourd, ash gourd, and bitter gourd, has facilitated the annotation of numerous coding genes. Notably, research has focused on identifying genes associated with key traits in these vegetables, spanning diverse characteristics like pathogen resistance, fruit size, mass, color, texture, length, shape, ripening behavior, bitterness, flavor and aroma, sex determination, and tendrils (Huang et al. 2009a, b; Zhang et al. 2019; Guo et al. 2019). Through population analysis and genome-wide association studies (GWAS) involving diverse accessions of these species, researchers have pinpointed several candidate genes responsible for desirable fruit and vegetable traits (Zhang et al. 2019). This wealth of information aids in formulating effective breeding strategies and contributes to the development of high-quality, resilient elite cultivars of gourd vegetables (Garcia-Mas et al. 2012; Zhao et al. 2019) (Table 10.4).

The utilization of genome-wide association studies (GWAS) stands out as a potent and efficient method for establishing associations between the genome and phenotypes, as well as for pinpointing causative loci or genes. In GWAS, the fundamental approach involves evaluating the correlation between each genetic marker and a specific phenotype that has been assessed across unrelated individuals within a diverse collection (unrelated individuals referring to those distantly related and displaying heterogeneity) (Huang and Han 2014). The robust efficacy of GWAS in unraveling the intricacies of complex traits has been substantiated, and ongoing advancements, such as large populations and high-throughput sequencing technology, are anticipated to enhance the efficiency of identifying causative loci or genes for quantitative traits. The success of high-resolution mapping in GWAS can be attributed to historical recombination events and the increased allele diversity incorporated in the analysis. Within association mapping populations, historical recombination accumulated over generations leads to the rapid decay of linkage

Table 10.4 QTLs associated with biotic stress and fruit quality traits through GWAS in bottle gourd

Traits	Mapping population	QTLs	References
Disease resistance			
Fusarium wilt	89 inbred lines	10 SNPs	Li et al. (2021)
Fruit quality trait			
Free glutamate content (Umami taste)	139 inbred lines	17 SNPs	Wu et al. (2017a, b)

disequilibrium (LD), enhancing resolution for association analysis. In contrast, family-based populations, especially doubled haploid (DH) populations with a limited number of recombination events, often result in lower mapping resolution and wider recombination values for locus pairs, leading to larger linkage blocks and increased LD (Rafalski 2010).

In the world of genome-wide association studies (GWAS) aimed at understanding traits, several key factors play a crucial role in success. First and foremost is phenotypic variation. We need to look closely at raw trait data, finding and removing unusual data points using tools like boxplots, which are helpful visual aids. Methods like BLUP and BLUE help improve the trait data by considering how genetics interacts with the environment across different places or years. Second, the population size being studied is also very important for getting meaningful results. A population of 100 to 500 individuals is necessary to have enough statistical power (Kumar et al. 2012). This allows us to find connections with traits more effectively, especially when dealing with rare variations. Third, understanding the details of the population structure is a must. Statistical methods carefully calculate how individuals are related, taking into account factors like mixtures and historical backgrounds. The choice of the population, based on things like location or growth habits, introduces specific genetic differences, influencing the following association analysis. Reliable methods, such as STRUCTURE and EIGENSTRAT using principal component analysis (PCA), help accurately define and control the population structure (Pritchard et al. 2000). Fourth, considering allele frequency is also crucial. Finding rare gene versions can be tricky, but it is essential for understanding natural differences and contributing useful insights to breeding and genetics (Soto-Cerda and Cloutier 2012). Fifth, when exploring how genes are linked, we need to pay special attention to linkage disequilibrium (LD). The LD serves as a measure for identifying the distance between genetic loci, a crucial factor in determining the optimal number of markers for a comprehensive genome scan. In other words, a high LD value indicates that fewer markers are necessary to effectively cover the entire genome (Myles et al. 2009; Semagn et al. 2010). Ignoring LD could lead to incorrect findings by including both important and unimportant gene versions in subsequent analyses. Calculating LD from the beginning is vital, with measures like r^2 helping to understand the connection between genes. SNPs with an r^2 value above 0.2 are particularly important, helping identify genes responsible for traits (Flint-Garcia et al. 2003). These methods help us make sense of genetic connections and contribute to our understanding of traits in a meaningful way.

Steps for performing a successful GWAS experiment (Alqudah et al. 2020):

Step I: The focus is on phenotyping, where all genetic types are assessed for a specific trait or set of traits based on the study's objectives. Achieving precise phenotyping is crucial for identifying associations between genotypes and phenotypes. It is emphasized that the phenotyping process should be replicated across different conditions, locations, and/or years to ensure the reliability of the results.

Step II: Genotyping is conducted on the same set of individuals that underwent phenotyping. This involves the use of DNA molecular markers, with genotyping-by-sequencing (GBS) being a commonly employed method due to its cost-effectiveness and ability to generate numerous single-nucleotide polymorphism (SNP) markers covering the entire genome. The GBS-generated SNPs undergo filtering based on criteria such as missing data, heterozygosity, and minor allele frequency.

Step III: Population structure is examined to determine the appropriate genome-wide association study (GWAS) model. Two commonly used statistical models are the general linear model (GLM) and the mixed linear model (MLM). GLM is suitable for populations without a discernible population structure, while MLM, incorporating factors like kinship and principal components, is more appropriate for populations with a clear structure.

Step IV: Phenotypic and genotypic data are integrated using dedicated software like TASSEL. This integration allows for the detection of alleles associated with specific traits. The stepwise procedure ensures a comprehensive approach, from phenotyping and genotyping to statistical modeling and data integration, facilitating the identification of genotype-phenotype associations in the study.

10.7 Tools for Conducting Genome-Wide Association Studies (GWAS)

I. *Trait Analysis by Association, Evolution, and Linkage (TASSEL):*

- TASSEL is a widely used software for GWAS in plants. It offers powerful statistical methods, including general linear model (GLM) and mixed linear model (MLM), for association analysis (Bradbury et al. 2007).
- TASSEL can analyze population structure using kinship and principal component analysis (PCA), and it incorporates linkage disequilibrium (LD) analysis. The latest version, TASSEL 5.0, can analyze genetic diversity, perform SNP calling from genotyping-by-sequencing (GBS) data, and include visualization tools like scatter plots, Manhattan plots, and heat maps (Pasam et al. 2012).

II. *GenStat for Windows Edition:*

- GenStat is another statistical software for marker-trait association analysis in genetically diverse populations using bi-allelic and multi-allelic markers.
- GWAS can be conducted with GLM or MLM models, incorporating population structure correction through PCA or kinship. GenStat allows defining the significance threshold and can determine and visualize LD decay, essential for determining the required number of markers for GWAS (Alqudah et al. 2014).

III. *PLINK:*

- PLINK is free software for studying large datasets of phenotypes and genotypes. It performs analyses for population stratification, basic association tests, meta-analyses, gene-based tests, and epistasis screening.
- PLINK produces graphical images such as Manhattan plots, Q-Q plots, and multidimensional scaling for population structure, along with presenting results in tables (Renteria et al. 2013).

IV. *R Statistical Environment (GAPIT—Genome Association and Prediction Integrated Tool):*

- GAPIT, an R package, is a powerful tool for GWAS and genomic selection. It handles large datasets efficiently and includes statistical methods like MLM, P3D, and EMMA (Lipka et al. 2012).
- Results can be visualized through Manhattan plots, QQ plots, and tables with various information like *p*-values, minor allele frequency, and phenotypic variance explained (Storey and Tibshirani 2003).

10.8 Genome Sequencing in Gourd Vegetables

Sequencing offers a crucial understanding of an organism's genetic composition, while re-sequencing provides valuable information about genetic variability and evolutionary patterns. Additionally, re-sequencing aids in the detection of mutations linked to diseases or particular traits (Schuster 2008).

Researchers have gained valuable insights into the pivotal genes and proteins influencing crucial aspects of crop development, opening up numerous opportunities for enhancing the traits of chosen plants (Le et al. 2016). The advent of next-generation sequencing technologies has greatly enhanced this cognitive process by providing a cost-effective and efficient analysis of DNA or RNA at high-throughput rates (Hu et al. 2021). Sequencing not only sheds light on the genetics of organisms, enabling specific gene identification, but also deepens our understanding of gene function and interactions. This encompasses the capacity to interpret intricate genetic information, decode regulatory elements, and pinpoint the genetic underpinnings of traits, including those that are complex in nature (Auffray et al. 2003).

First-generation sequencing involves the utilization of chemicals or chain-terminating nucleotides to cleave bases within a DNA molecule, followed by manual separation of fragments through electrophoresis (Akintunde et al. 2023). In contrast, second-generation sequencing, which emerged in 2005, addresses the limitations of the first generation by generating millions of short reads simultaneously, significantly expediting the sequencing process compared to the first generation. Third-generation sequencing, known as long-read sequencing, produces substantially longer reads than second-generation sequencing. Despite allowing direct sequencing of single DNA molecules, third-generation technologies have higher error rates, complicating downstream genome assembly and data analysis. Ongoing development is expected to bring further improvements to third-generation sequencing technologies (Heather and Chain 2016; Meslier et al. 2022).

The advent of next-generation sequencing (NGS) technology has facilitated whole-genome sequencing, unveiling previously undetected variants, mutations, and structural changes. NGS has also made whole-genome sequencing of RNA transcriptomes (RNA-Seq) feasible, providing insights into gene expression, transcriptome mapping, and alternative splicing events (Metzker 2010). Third-generation sequencing technologies, such as those by Pacific Biosciences (PacBio) and Oxford Nanopore Technologies (ONT), have played a pivotal role in moving away from amplification-based NGS methods (Kchouk et al. 2017). NGS methods have become prevalent, enabling rapid and cost-effective sequencing of large genome portions or entire genomes, leading to the creation of reference genomes. In epigenetics research, NGS is a powerful tool providing a high-resolution, genome-wide view of epigenetic changes. For example, whole-genome bisulfite sequencing (WGBS), an NGS-based method, is widely used to map DNA methylation in plant genomes, influencing gene expression (Lister and Ecker 2009). As NGS technology evolves and computational tools improve, plant epigenetics knowledge will likely expand, offering innovative applications in agriculture and plant biology. Moreover, genome-wide association studies (GWAS) utilize sequenced genomes to correlate specific genetic changes with observable traits. These studies have detected genomic variants associated with agronomic or biochemical phenotypes, facilitating gene cloning, accelerated crop breeding through marker-assisted selection (MAS), and genetic engineering applications (Cortes et al. 2021).

Plant genomes exhibit inherent complexity, marked by factors such as polyploidy, extensive repetitive sequences, and epigenetic modifications (Michael and VanBuren 2020; Leitch and Leitch 2008; Bennetzen and Wang 2014). Transposable elements (TEs), particularly retrotransposons with long terminal repeats, play a significant role in shaping the size and complexity of plant genomes by changing their positions (Schnable et al. 2009; Lisch 2013). Additionally, complex regulatory networks, encompassing cis-regulatory elements, transcription factor binding sites, and epigenetic markers, add another layer of complexity, enabling adaptation and phenotypic plasticity (Varshney et al. 2019). Advanced sequencing technologies facilitate the complex annotation of genomes, providing insights for targeted crop improvement. In 2009, a milestone was achieved with the sequencing of the first genome within the Cucurbitaceae family, cucumber (*Cucumis sativus*). Since then, 87 plant varieties and cultivars within this family have been described globally (Huang et al. 2009a, b). Notably, the genomes of gourd vegetables were documented in later years.

Subsequently, bitter gourd (*Momordica charantia*) genomes, including the reference genome OHB3-1, were sequenced in 2017, revealing a genome size of 285.6 Mb and 45,859 predicted genes (Urasaki et al. 2017). Further research by different groups provided additional insights into bitter gourd genomes, including variations in genome size and the number of protein-coding genes (Matsumura et al. 2020). Other gourds, such as *Lagenaria siceraria* (bottle gourd), were also studied, revealing genome sizes around 297.9 Mb and approximately 23,000 predicted genes (Xu et al. 2021).

Advancements in genome cognition techniques led to the identification of ridge gourd (*Luffa acutangula*), silver-seed gourds (*Cucurbita argyrosperma subsp. sororia* or *C. argyrosperma subsp. argyrosperma*), and snake gourd (*Trichosanthes anguina*) genomes (Pootakham et al. 2021; Barrera-Redondo et al. 2021; Barrera-Redondo et al. 2019). The snake gourd, with a genome size of 919.8 Mb, stands out as one of the largest genomes in the Cucurbitaceae family (Ma et al. 2020). More recent genome identifications (2019–2023) include tropical vines like sponge gourd (*Luffa cylindrica*) with varying genome sizes. The designated genome for loofah is structured across 13 chromosomes, measuring 656.2 Mb according to the first reference (Wu et al. 2020), while the second publication on this species reported a slightly larger size of 669.7 Mb (Zhang et al. 2020).

Finally, the genome sequences of wax gourd (*Benincasa hispida*) and its cultivars B227 and PF3 were discovered in 2019 and 2023. The wax gourd genomes exhibited sizes of 913.0 Mb and 975.6 Mb for B227 and pf3, respectively, making them among the largest genomes in the Cucurbitaceae family (Xie et al. 2019; Luo et al. 2023).

10.9 Challenges and Future Perspective

QTL mapping in gourd vegetables and genome-wide association studies (GWAS) encounter complex challenges rooted in the genetic and phenotypic intricacies of these crops. Gourd vegetables, characterized by significant genetic diversity and polygenic traits, pose difficulties in accurately identifying and mapping QTL. Phenotypic variability influenced by environmental factors adds further complexity, necessitating meticulous experimental design. Limited genomic resources in some gourd vegetables, lacking comprehensive reference genomes and markers, present additional obstacles. In overcoming these challenges, researchers must leverage advanced genomic tools, employ rigorous experimental methodologies, and consider the impact of environmental factors to enhance the precision and reliability of QTL mapping in gourd vegetables.

In GWAS, the challenge of synthetic associations arises, where noncausal SNPs are erroneously deemed more significant than true causal variants. This can lead to misleading outcomes, especially when significant peaks are located in distinct LD blocks, complicating accurate gene identification. Synthetic associations become more problematic in cases of allelic heterogeneity and rare causal alleles, making their detection challenging. Despite the ongoing challenge of determining an appropriate significance threshold, GWAS holds great promise in various applications, including breeding, genetic mapping, and gene editing. Its advantages, such as higher resolution and a broader genetic base, make it a powerful tool in advancing plant breeding and genomic analysis, providing comprehensive insights into the genetic basis of traits.

References

Akintunde O, Tucker T, Carabetta VJ (2023) The evolution of next-generation sequencing technologies. arXiv 2023, arXiv:2305.08724v1

Alqudah AM, Sharma R, Pasam RK, Graner A, Kilian B, Schnurbusch T (2014) Genetic dissection of photoperiod response based on GWAS of pre-anthesis phase duration in spring barley. PLoS ONE 9(11):e113120

Alqudah AM, Sallam A, Stephen Baenziger P, Börner A (2020) GWAS: fast-forwarding gene identification and characterization in temperate Cereals: lessons from Barley—a review. J Adv Res 4(22):119–135

Arunachalam P (2002) Breeding for resistance to distortion mosaic virus in bitter gourd (*Momordica charantia* L.). Ph. D. Thesis, Faculty of Agriculture, Kerala Agriculture University, Vellanikara

Auffray C, Imbeaud S, Roux-Rouquié M, Hood L (2003) From functional genomics to systems biology: concepts and practices. C R Biol 326:879–892

Barrera-Redondo J, Ibarra-Laclette E, Vázquez-Lobo A, Gutiérrez-Guerrero YT, Sánchez de la Vega G, Piñero D, Montes Hernández S, Lira-Saade R, Eguiarte LE (2019) The genome of *Cucurbita argyrosperma* (silver-seed gourd) reveals faster rates of protein-coding gene and long noncoding RNA turnover and neofunctionalization within *Cucurbita*. Mol Plant 12:506–520

Barrera-Redondo J, Sánchez-de la Vega G, Aguirre-Liguori JA, Castellanos-Morales G, Gutiérrez-Guerrero YT, Aguirre Dugua X, Aguirre-Planter E, Tenaillon MI, Lira-Saade R, Eguiarte LE (2021) The domestication of *Cucurbita argyrosperma* as revealed by the genome of its wild relative. Hortic Res 8:109

Behera T, Sirohi P, Pal A, Singh I (2015) 'Pusa Santushti': a new high yielding bottle gourd variety. Int J Veg Sci 21:323–328. https://doi.org/10.1080/19315260.2013.876565

Behera TK, Bhardwaj DR, Gautam KK (2023) Bitter gourd: breeding and genomics. Vegetab Sci 50(spl):189–207

Bennetzen JL, Wang H (2014) The contributions of transposable elements to the structure, function, and evolution of plant genomes. Annu Rev Plant Biol 65:505–530

Bindal S, Sheu ZM, Kenyon L, Taher D, Rakha M (2023) Novel sources of resistance to fusarium wilt in *Luffa* species. Front Plant Sci 14:1116006. https://doi.org/10.3389/fpls.2023.1116006. PMID: 37360710; PMCID: PMC10288365

Bradbury PJ, Zhang Z, Kroon DE, Casstevens TM, Ramdoss Y, Buckler ES (2007) TASSEL: software for association mapping of complex traits in diverse samples. Bioinformatics 23(19):2633–2635

Christenhusz MJM, Byng JW (2016) The number of known plants species in the world and its annual increase. Phytotaxa 261:201–217

Cortes LT, Zhang Z, Yu J (2021) Status and prospects of genome-wide association studies in plants. Plant Genome 14:e20077

Cui J, Luo S, Niu Y, Huang R, Wen Q, Su J, Miao N, He W, Dong Z, Cheng J, Hu K (2018) A RAD-based genetic map for anchoring scaffold sequences and identifying QTLs in bitter gourd (*Momordica charantia*). Front Plant Sci 9. https://doi.org/10.3389/fpls.2018.00477

Decker-Walters DS, Staub JE, Lopez-Sese A, Nakata E (2001) Diversity in landraces and cultivars of bottle gourd (*Lagenaria siceraria*; Cucurbitaceae) as assessed by random amplified polymorphic DNA. Genet Resour Crop Evol 48:369–380

Dhillon NPS, Sanguansil S, Srimat S, Schafleitner R, Manjunath B, Agarwal P, Xiang Q, Masud MAT, Myint T, Hanh NT (2018) Cucurbit powdery mildew-resistant bitter gourd breeding lines reveal four races of Podosphaera xanthii in Asia. HortScience 48:1078–1089

Dhillon NPS, Sanguansil S, Srimat S, Laenoi S, Schafleitner R, Pitrat M, McCreight JD (2019) Inheritance of resistance to cucurbit powdery mildew in bitter gourd. HortScience 54:1013–1016

Flint-Garcia SA, Thornsberry JM, Buckler ES (2003) Structure of linkage disequilibrium in plants. Annu Rev Plant Biol 54(1):357–374

Gangadhara Rao P, Behera TK, Gaikwad AB, Munshi AD, Jat GS, Boopalakrishnan G (2018) Mapping and QTL analysis of gynoecy and earliness in bitter gourd (*Momordica charantia*

L.) using genotyping-by-sequencing (GBS) technology. Front Plant Sci 9:1555. https://doi.org/10.3389/fpls.2018.01555
Garcia-Mas J, Benjak A, Sanseverino W et al (2012) The genome of melon (*Cucumis melo* L.). Proc Natl Acad Sci USA 109:11872–11877
Guo S, Zhao S, Sun H et al (2019) Resequencing of 414 cultivated and wild watermelon accessions identifies selection for fruit quality traits. Nat Genet 51:1616–1623
Heather JM, Chain B (2016) The sequence of sequencers: the history of sequencing DNA. Genomics 107:1–8
Holdsworth WL, LaPlant KE, Bell DC, Jahn MM, Mazourek M (2016) Cultivar-based introgression mapping reveals wild species-derived Pm-0, the major powdery mildew resistance locus in squash. PLoS One 11:e0167715. https://doi.org/10.1371/journal.pone.0167715
Hollingsworth R, Vagalo M, Tsatsia F (1997) Biology of melon fly with special reference to the Solomon Islands. In: Allwood AJ, Drew RAI (eds) Management of fruit flies in the Pacific. Proceedings of the Australian Country Indian Agriculture Research, vol 76, pp 140–144
Hu T, Chitnis N, Monos D, Dinh A (2021) Next-generation sequencing technologies: an overview. Hum Immunol 82:801–811
Huang X, Han B (2014) Natural variations and genome-wide association studies in crop plants. Annu Rev Plant Biol 65(1):531–551
Huang S, Li R, Zhang Z et al (2009a) The genome of the cucumber, *Cucumis sativus* L. Nat Genet 41:1275–1281
Huang S, Li R, Zhang Z, Li L, Gu X, Fan W, Lucas WJ, Wang X, Xie B, Ni P et al (2009b) The Genome of the cucumber, Cucumis sativus L. Nat Genet 41:1275–1281
Jeffrey C (1980) A review of the *Cucurbitaceae*. Bot J Linn Soc 81(3):233–247
Jeffrey C (1990a) Systematics of the Cucurbitaceae: an overview. In: Bates DM, Robinson RW, Jeffrey C (eds) Biology and utilization of the Cucurbitaceae. Cornell University Press, Ithaca, pp 3–9
Jeffrey C (1990b) An outline classification of the Cucurbitaceae (Appendix). In: Bates DM, Robinson RW, Jeffrey C (eds) Biology and utilization of the Cucurbitaceae. Cornell University Press, Ithaca, pp 449–463
Jeffrey C (2005) A new system of Cucurbitaceae. Bot Zhurn 90:332–335
Jeffrey C, De Wilde WJJO (2006) A review of the subtribe Thladianthinae (Cucurbitaceae). Bot Zhurn 91:766–776
Kchouk M, Gibrat J-F, Elloumi M (2017) Generations of sequencing technologies: from first to next generation. Biol Med 9:395
Keinath AP, DuBose B (2004) Evaluation of fungicides for prevention and management of powdery mildew on watermelon. Crop Prot 23:35–42
Kumar J, Pratap A, Solanki RK, Gupta DS, Goyal A, Chaturvedi SK et al (2012) Genomic resources for improving food legume crops. J Agric Sci 150(3):289–318
Le DT, Chu HD, Le NQ (2016) Improving nutritional quality of plant proteins through genetic engineering. Curr Genomics 17:220–229
Leitch AR, Leitch IJ (2008) Genomic plasticity and the diversity of polyploid plants. Science 320:481–483
Li Y, Wang Y, Wu X, Wang J, Wu X, Wang B, Lu Z, Li G (2021) Novel genomic regions of fusarium wilt resistance in bottle gourd [*Lagenaria siceraria* (Mol.) standl.] discovered in genome-wide association study. Front Plant Sci 12:650157
Lipka AE, Tian F, Wang Q, Peiffer J, Li M, Bradbury PJ et al (2012) GAPIT: genome association and prediction integrated tool. Bioinformatics 28(18):2397–2399
Lisch D (2013) How important are transposons for plant evolution? Nat Rev Genet 14:49–61
Lister R, Ecker JR (2009) Finding the fifth base: Genome-wide sequencing of cytosine methylation. Genome Res 19:959–966
Liu L, Gan Y, Luo J, Li J, Zheng X, Gong H, Liu X, Deng L, Zhao G, Wu H (2022) QTL mapping reveals candidate genes for main agronomic traits in *Luffa* based on a high-resolution genetic map. Front Plant Sci 13. https://doi.org/10.3389/fpls.2022.1069618

Lou L, Liu Z, Liu X, Su X (2019) Construction of a high-density genetic linkage map and QTL analysis of resistance to cucumber mosaic virus in *Luffa cylindrica* (L.) Roem. based on specific length amplified fragment sequencing. Res Square. https://doi.org/10.21203/rs.2.15852/v2

Lou L, Su X, Liu X, Liu Z (2020) Construction of a high-density genetic linkage map and identification of gene controlling resistance to cucumber mosaic virus in Luffa cylindrica (L.) Roem. based onspecific-length amplified fragment sequencing. Mol Biol Rep 47:5831–5841. https://doi.org/10.1007/s11033-020-05652-8

Luo W, Yan J, Luo S, Liu W, Xie D, Jiang B (2023) A chromosome-level reference genome of the wax gourd (*Benincasa hispida*). Sci Data 10:78

Ma L, Wang Q, Mu J, Fu A, Wen C, Zhao X, Gao L, Li J, Shi K, Wang Y et al (2020) The genome and transcriptome analysis of snake gourd provide insights into its evolution and fruit development and ripening. Hortic Res 7:199

Mashilo J, Shimelis H, Odindo A (2015) Genetic diversity of bottle gourd (*Lagenaria siceraria* (Molina) Standl.) landraces of South Africa assessed by morphological traits and simple sequence repeat markers. SA J Plant Soil 33:113–124. https://doi.org/10.1080/02571862.2015.1090024

Mashilo J, Shimelis H, Odindo A (2017) Phenotypic and genotypic characterization of bottle gourd [*Lagenaria siceraria* (Molina) Standl.] and implications for breeding: a review. Sci Hortic 222:136–144. https://doi.org/10.1016/j.scienta.2017.05.020

Matsumura H, Hsiao M-C, Lin Y-P, Toyoda A, Taniai N, Tarora K, Urasaki N, Anand SS, Dhillon NPS, Schafleitner R et al (2020) Long-read bitter gourd (*Momordica charantia*) genome and the genomic architecture of non-classic domestication. Proc Natl Acad Sci USA 117:14543–14551

Meslier V, Quinquis B, Da Silva K, Oñate FP, Pons N, Roume H, Podar M, Almeida M (2022) Benchmarking second and third-generation sequencing platforms for microbial metagenomics. Sci Data 9:694

Metzker ML (2010) Sequencing technologies—the next generation. Nat Rev Genet 11:31–46

Michael TP, VanBuren R (2020) Building near-complete plant genomes. Curr Opin Plant Biol 54:26–33

Montero-Pau J, Blanca J, Esteras C, Martínez-Pérez EM, Gómez P, Monforte AJ et al (2017) An SNP-based saturated genetic map and QTL analysis of fruit-related traits in zucchini using genotyping- by-sequencing. BMC Genomics 18:3439. https://doi.org/10.1186/s12864-016-3439-y

Montero-Pau J, Blanca J, Bombarely A, Ziarsolo P, Esteras C, Martí-Gómez C et al (2018) De novo assembly of the zucchini genome reveals a wholegenome duplication associated with the origin of the cucurbita genus. Plant Biotechnol J 16:1161–1171. https://doi.org/10.1111/pbi.12860

Morimoto Y, Maundu P, Fujimaki H, Morishima H (2005) Diversity of landraces of the white flowered gourd (*Lagenaria siceraria*) and its wild relatives in Kenya: fruit and seed morphology. Genet Resour Crop Evol 52(6):737–747

Myles S, Peiffer J, Brown PJ, Ersoz ES, Zhang Z, Costich DE et al (2009) Association mapping: critical considerations shift from genotyping to experimental design. Plant Cell 21(8):2194–2202

Nanasato Y, Okuzaki A, Tabei Y (2013) Improving the transformation efficiency of cucurbita species: factors and strategy for practical application. Plant Biotechnol 30:287–294. https://doi.org/10.5511/plantbiotechnology.13.0331a

Narayan R, Singh SP, Sharma DK, Rastogi KB (1996) Genetic variability and selection parameters in bottle gourd. Indian J Hort 53:53–58

Pasam RK, Sharma R, Malosetti M, van Eeuwijk FA, Haseneyer G, Kilian B et al (2012) Genome-wide association studies for agronomical traits in a world-wide spring barley collection. BMC Plant Biol 12:16

Pootakham W, Sonthirod C, Naktang C, Nawae W, Yoocha T, Kongkachana W, Sangsrakru D, Jomchai N, U-thoomporn S, Sheedy JR et al (2021) De novo assemblies of *Luffa acutangula* and *Luffa cylindrica* genomes reveal an expansion associated with substantial accumulation of transposable elements. Mol Ecol Resour 21:212–225

Pritchard JK, Stephens M, Donnelly P (2000) Inference of population structure using multilocus genotype data. Genetics 155(2):945–959

Rabindranath K, Pillai KS (1986) Control of fruit fly of bitter gourd using synthetic pyrethroids. Entomon 11:269–272
Rafalski JA (2010) Association genetics in crop improvement. Curr Opin Plant Biol 13(2):174–180
Rao PG, Behera TK, Gaikwad AB, Munshi AD, Srivastava A, Boopalakrishnan G, Vinod. (2021) Genetic analysis and QTL mapping of yield and fruit traits in bitter gourd (*Momordica charantia* L.). Sci Rep 11(1). https://doi.org/10.1038/s41598-021-83548-8
Renteria ME, Cortes A, Medland SE (2013) Using PLINK for genome-wide association studies (GWAS) and data analysis. In: Gondro C, van der Werf J, Hayes B (eds) Genome-wide association studies and genomic prediction. Humana Press, Totowa, pp 193–213
Robinson RW, Decker-Walters DS (1997) Cucurbits. CAB International, Wallingford, p 226
Rubatzky VE, Yamaguchi M (1997) World vegetable principles, production and nutritive values, 2nd edn. Chapman and Hall/International Thompson Publications, New York, p 853
Schnable PS, Ware D, Fulton RS, Stein JC, Wei FS, Pasternak S, Liang CZ, Zhang JW, Fulton L, Graves TA et al (2009) The B73 maize genome: Complexity, diversity, and dynamics. Science 326:1112–1115
Schuster SC (2008) Next-generation sequencing transforms today's biology. Nature 200:16–18
Semagn K, Bjornstad A, Xu YB (2010) The genetic dissection of quantitative traits in crops. Electron J Biotechnol 13(5)
Song H, Huang Y, Gu B (2020) QTL-Seq identifies quantitative trait loci of relative electrical conductivity associated with heat tolerance in bottle gourd (*Lagenaria siceraria*). PLoS ONE 15(11):e0227663. https://doi.org/10.1371/journal.pone.0227663
Soto-Cerda BJ, Cloutier S (2012) Association mapping in plant genomes. In: Genetic diversity in plants. InTech
Storey JD, Tibshirani R (2003) Statistical significance for genomewide studies. PNAS 100(16):9440–9445
Sun H, Wu S, Zhang G, Jiao C, Guo S, Ren Y (2017) Karyotype stability and unbiased fractionation in the paleo allotetraploid cucurbita genomes. Mol Plant 10:1293–1306. https://doi.org/10.1016/j.molp.2017.09.003
Urasaki N, Takagi H, Natsume S, Uemura A, Taniai N, Miyagi N, Fukushima M, Suzuki S, Tarora K, Tamaki M et al (2017) Draft genome sequence of bitter gourd (*Momordica charantia*), a vegetable and medicinal plant in tropical and subtropical regions. DNA Res 24:51–58
Varshney RK, Pandey MK, Bohra A, Singh VK, Thudi M, Saxena RK (2019) Toward the sequence-based breeding in legumes in the post-genome sequencing era. Theor Appl Genet 132:797–816
Wang W, Zhang Y, Xu C, Ren J, Liu X, Black K, Ren H (2015) Cucumber ECERIFERUM1 (CsCER1), which influences the cuticle properties and drought tolerance of cucumber, plays a key role in VLC alkanes biosynthesis. Plant Mol Biol 87(3):219–233
Whitaker TW, Davis GN (1962) Cucurbits: botany, cultivation and utilization. World Crops Books, London/New York, p 249
Whitaker TW, Cutler HC (1971) Pre-historic cucurbits from the Valley of Oaxaca. Econ Bot 25(2):123–127
Wu H, He X, Gong H, Luo S, Li M, Chen J, Zhang C, Yu T, Huang W, Luo J (2016) Genetic linkage map construction and QTL analysis of two interspecific reproductive isolation traits in Sponge Gourd. Front Plant Sci 7. https://doi.org/10.3389/fpls.2016.00980
Wu X, Xu P, Wu X, Wang B, Lu Z, Li G (2017a) Development of insertion and deletion markers for bottle gourd based on restriction site-associated DNA sequencing data. Hortic Plant J 3:13–16. https://doi.org/10.1016/j.hpj.2017.01.01
Wu X, Xu P, Wu X, Wang B, Lu Z, Li G (2017b) Genome-wide association analysis of free glutamate content, a key factor conferring umami taste in the bottle gourd [*Lagenaria Siceraria* (mol.) standl.]. Sci Hortic 225:795–801. https://doi.org/10.1016/j.scienta.2017.08.015
Wu X, Wu X, Wang Y, Wang B, Lu Z, Xu P, Li G (2019) Molecular genetic mapping of two complementary genes underpinning fruit bitterness in the bottle gourd (*Lagenaria Siceraria* [mol.] standl.). Front Plant Sci 10. https://doi.org/10.3389/fpls.2019.01493
Wu H, Zhao G, Gong H, Li J, Luo C, He X, Luo S, Zheng X, Liu X, Guo J et al (2020) A high-quality sponge gourd (*Luffa cylindrica*) genome. Hortic Res 7:128

Xie D, Xu Y, Wang J, Liu W, Zhou Q, Luo S, Huang W, He X, Li Q, Peng Q et al (2019) The wax gourd genomes offer insights into the genetic diversity and ancestral Cucurbit karyotype. Nat Commun 10:5158

Xu P, Wang Y, Sun F, Wu R, Du H, Wang Y, Jiang L, Wu X, Wu X, Yang L et al (2021) Long-read genome assembly and genetic architecture of fruit shape in the bottle gourd. Plant J 107:956–968

Yang SL, Walters TW (1992) Ethnobotany and the economic role of the Cucurbitaceae of China. Econ Bot 46:349–367

Yanwei L, Ying W, Xinyi W, Jian W, Xiaohua W, Baogen W et al (2021) Novel genomic regions of Fusarium wilt resistance in bottle gourd [*Lagenaria siceraria* (Mol.) Standl.] discovered in genome-wide association study. Front Plant Sci 12:650157. https://doi.org/10.3389/fpls.2021.650157

Yao KAG, Koffi KK, Ondo-Azi SA, Baudoin JP, Zoro BIA (2015) Seed yield component identification and analysis for exploiting recombinative heterosis in bottle gourd. Int J Veg Sci 21:441–453. https://doi.org/10.1080/19315260.2014.895791

Zhang C, Zheng H, Wu X, Xu H, Han K, Peng J et al (2018) Genomicwide identification of new reference genes for RT-qPCR normalization in CGMMV-infected *Lagenaria siceraria*. Front Plant Sci 11:177. https://doi.org/10.3389/fpls.2020.00177

Zhang H, Li X, Yu H et al (2019) A high-quality melon genome assembly provides insights into genetic basis of fruit trait improvement. iScience 2(22):16–27

Zhang T, Ren X, Zhang Z, Ming Y, Yang Z, Hu J, Li S, Wang Y, Sun S, Sun K et al (2020) Long-read sequencing and de novo assembly of the *Luffa cylindrica* (L.). Roem Genome Mol Ecol Resour 20:511–519

Zhao G, Lian Q, Zhang Z et al (2019) A comprehensive genome variation map of melon identifies multiple domestication events and loci influencing agronomic traits. Nat Genet 51:1607–1615

GWAS and QTL Mapping in Carrot (*Daucus carota* L.)

11

Raman Selvakumar

Abstract

Carrots (*Daucus carota* L.), a root vegetable that is significant during the cool season, are cultivated and consumed worldwide. Moreover, in addition to imparting a noticeable flavour to consumers, the storage root is abundant in nutritional fibre, carotenoids, vitamins, and anthocyanins. Molecular breeding techniques, whether traditional or contemporary, have been instrumental in enhancing the yield and quality of crops in temperate, subtropical, and tropical areas worldwide. This chapter provides a concise overview of the existing genetic resources; advancements in carrot breeding with respect to significant biotic and abiotic stresses, yield, and quality; genomic resources, genome-wide association studies, and quantitative trait loci (QTL) mapping. Furthermore, this paper addresses the obstacles and potential progress associated with tropical carrot breeding in terms of enhancing critical characteristics, including but not limited to beta-carotene content, obtuse root morphology, and scar-free smooth roots.

Keywords

Carrot · Mapping population · Quality traits · Genome-wide association studies · Quantitative trait loci · Trait discovery

11.1 Introduction

Carrots (*Daucus carota* L.; Apiaceae; $2n = 2X = 18$) are a root vegetable known for their remarkable nutrient density, which confers on them a wide array of health advantages and nutritional qualities. Due to its historical importance and inclusion

R. Selvakumar (✉)
Centre for Protected Cultivation Technology, ICAR-Indian Agricultural Research Institute, New Delhi, India

S. S. Dey et al. (eds.), *GWAS and QTL Mapping in Horticultural Crops*,
https://doi.org/10.1007/978-981-95-2306-1_11

in the diet, the term 'carotenoid' is commonly utilised in reference to a remedy for night blindness. Carrots are a rapidly maturing agricultural product that delivers a higher concentration of nutrients per unit of area and time compared to other commodities. Carrots are a source of carotenoids with a high yield per unit and can be ingested raw or prepared. A vegetable commodity of considerable financial value, it is both cultivated and consumed on a global scale, including in Indonesia, China, and the United States. The food item in question is rich in vitamin A (6500–8258 μg/100 g) and can easily fulfil the body's need for β-carotene. Carrots are regarded as more advantageous than alternative vegetables due to their pleasant taste, ease of digestion, and higher levels of provitamin A, other carotenoids, and fibre. The introduction of 'baby carrots' or conveniently packaged 'cut & peel' carrots represents an endeavour to enhance the reputation of carrots as the preeminent vegetable nibble food in Western countries. Furthermore, the company has broadened its clientele by offering exquisite pastries (such as halwa and gajrela), pickles, vegetable preparations, and beverages (whether pure or blended). Through genetic analysis, the diversity of carrot germplasm in terms of tolerance and adaptability to biotic stresses has also been exhaustively documented. Enhancing resistance to biotic and abiotic stresses, improved quality, increased yield, and enhanced adaptability constitute a fundamental requirement for augmenting domestic carrot production and utilising it as a preventive measure against micronutrient deficiencies, specifically vitamin A deficiency. A nutrient analysis conducted on 100 g of carrot roots revealed the following components: 86 g of moisture, 10.6 g of carbohydrates, 0.9 g of protein, 0.2 g of fat, 1890 mg beta-carotene, 3.0 mg vitamin C, 0.04 mg thiamine, 0.02 mg riboflavin, 80 mg calcium, and 2.2 mg iron. Furthermore, carrots are categorised as temperate, Western, or Asian (tropical) in origin. The orange hue exhibited by the roots of European carrots can be attributed to the elevated concentration of provitamin A carotenoids, particularly β-carotene. Moreover, when compared to carrots from the East, their mean sugar content is elevated by 18%. This group appears more developed and demonstrates greater adaptability to commercial production; the processing industry prefers roots that are cylindrical or attenuated cylindrical in shape. The likelihood of bolting in plants is diminished when the pubescence of the foliage is reduced (Rubatzky et al. 1999). The carrots are also distinguished by the orange pigmentation of their roots, which is predominantly attributed to the presence of provitamin A carotenoids (beta-carotene). European carrots are biennial plants that solely fruit and produce seeds in temperate regions. This is attributed to their temperature preferences, which span from 5 to 8 °C for a duration of 40–60 days, while they transition from the vegetative to reproductive phases. Eastern carrots are commonly cultivated in Asia, where particular cultivars produce roots that exhibit characteristics such as increased density, reduced length, narrow conical shape, and a propensity for branching. Moreover, their price is greater than that of Western carrots. Their glabrous foliage and early blooming period often render them more prone to displaying primitive characteristics (Rubatzky et al. 1999). Asiatic varieties are adapted to benign growing conditions, have a gentler texture, are less aromatic, have a moderate flavour, and bloom more swiftly. Annual plants of the Asiatic variety have broad centres, extensive roots, and erect summits, all of which contribute to their increased productivity. Provitamin A carotenoids are scarce in their roots, which are coloured ivory (lutein), red (lycopene), or purple (anthocyanin). Asiatic carrot roots demonstrate an increased concentration of phenolic compounds and are more efficient in eliminating free radicals. Currently, carrot breeding places considerable emphasis on the development of varieties and hybrids that exhibit future-oriented characteristics and satisfy consumer preferences. The aforementioned qualities may encompass visually pleasing tints, elevated moisture and sugar content, self-sustaining cores, antioxidant properties, and a greater quantity of essential minerals in comparison to existing carrot cultivars. In the carrot breeding industry, traditional desirable qualities such as consistent root diameter and shape, compact, self-coloured tissue featuring slender centres, crack-free roots, delayed blossoming, finer texture, and substantial root production remain highly relevant. In addition to tolerance to heat and late bolters, carrot breeding places emphasis on the development of resistance against biotic stresses, including root knot nematode, Alternaria blight, Cercospora leaf blight, root fissures, and crown rot. In recent years, carrot genetics has also incorporated temperature tolerance, coloured carrots (including red, cream, black, purple, orange, and multicoloured cultivars), and speciality carrots (including infant carrots, cut-and-peel carrots and multicoloured carrots). In light of these considerations, the present chapter analyses the progression and prospective trajectories of carrot breeding via quantitative trait loci (QTL) mapping and genome-wide association studies.

11.2 Molecular Marker Application in Carrot Improvement

The origins of the carrot extend across numerous global regions, contributing to its notable genetic diversity. Carrot genetic germplasms were rigorously analysed and classified by employing morphological and phenotypic distinctions. In accordance with the findings of a genetic molecular marker, cultivated cultivars, putative ancestors, and wild relatives display a substantial amount of phenotypic variation. Regarding genetic and molecular marker studies in populations of self-pollinated heterozygous carrot plants, Westphal and Wricke (1991) and Schulz et al. (1994) have conducted research. The investigation of open-pollinated cultivars, cultivated carrots, wild carrots, and land races has been the subject of extensive genetic and gene dispersal research. The experiments conducted by Santos and Simon (2002) have illustrated how these combinations can be utilised to evaluate genetic diversity and distinguish progenitor donors. A number of molecular marker studies were examined in an effort to determine the relationship between carrot diversity and said diversity. An assortment of taxonomic relationships between geographical groups and carrot germplasm has been elucidated by means of molecular investigations employing random molecular markers, including isozymes (St. Pierre and Bayer 1991; St. Pierre et al. 1990), random-amplified polymorphic DNA (RAPD) (Grzebelus et al. 2001; Nakajima et al. 1998, 1999, 2001), and amplified-fragment length polymorphism (AFLP) (Ichikawa et al. 1987, 1989).

The *atpA* antigen was generated, and the protein product and mitochondrial fragment were utilised to conduct molecular analysis of the cytoplasm of fertile, petaloid, and brown anthers. This analysis unveiled distinct patterns. Innovative CMS systems were produced as a consequence of these interspecific and subspecific crosses (Nothnagel 1992). Vivek and Simon (1999a, b) found that the *Rs* gene and the AFLP marker P3B30XA were found to have a tenuous association. A linkage map was constructed with 8 linkage groups averaging 13.1 cM distance, using 16 RAPD markers, 6 isolated restriction fragment length polymorphism (RFLP) probes obtained from PCR fragments, 10 isozyme loci, and 14 RFLPs (Schulz et al. 1993). A 109-point molecular linkage map was generated for the phenotypic loci (*P1*, *Y2*, and *Rs*) of carrot (*Daucus carota* L. ssp. *sativus*). This map was based on 6 RFLPs, 2 RAPDs, 96 AFLPs, and 2 selective SAMPL markers, exhibiting polymorphisms of 36%, 20%, and 42%, respectively. Furthermore, the saturated map consisted of a clearance of 4.9 cM and covered a total area of 534.4 ft. The marker sizes for AFLP P6B15 and AFLP-P1B34 were noted by to be 1.7 cM and 2.2 cM, respectively, with Rs at 8.1 cM. Scheulz et al. (1994) utilised 1926 markers distributed across 5–8 linkage groups to construct the molecular map, in addition to four self-ed (S1) carrot plants and phenotypic loci, isozymes, RFLPs, and RAPDs. The restriction patterns observed in the mitochondrial genomes of three male-fertile carrot lines and six petaloid male-sterile carrot lines were indicative of specific geographic origins, according to the research conducted by Bowes and Wolyn (1998). PHENOtype (phen1/phen1) of annual flowering carrot was mapped through the combination of morphological markers and RAPD, RFLP, and isozyme markers. The *phen1/phen1* genetic composition was characterised and described in a study conducted by Schulz et al. (1994). The specimen possessed distinctive roots, exhibited small, coiled, dark green leaves, and displayed annual flowering that was consistent and did not require vernalisation.

A research investigation carried out by Nakajima et al. (1998) utilised RAPD and AFLP to discern and differentiate among five cultivars/inbred lines and six accessions representing six native species and carrot (*D. carota* ssp. *sativus*) of *Daucus carota*. The interconnections between diversity and eight native and cultivated species were investigated in a study conducted by Vivek and Simon (1999b). In contrast to the study by Sáenz Lán (1981) of the conventional classification of five sections, they identified that these species are belonging to four distinct sections. Utilising AFLP markers, Nakajima et al. (1998) established that two Japanese carrot cultivars exhibited a higher degree of genetic similarity to wild carrots than to their European counterparts. The hypothesis that four species of wild carrots could be categorised into a single subgroup was examined by Bradeen et al. (2002) through the utilisation of AFLP and ISSR markers. The six predominant RAPD markers identified by Nakajima et al. (1999) were used to distinguish between cytoplasmic sterile and male-fertile carrot lines could be distinguished. In addition, the terminal sequences of these markers were cloned and characterised.

By means of linkage group C, which is genetically unlinked to Y2 and P1, but distant from an AFLP marker, Vivek and Simon (1999a) mapped the reducing sugar phenotype (Rs). Simon, 1996 specified that the marker measure for Rs was 8.1 cM. The successful transmission of the Rs gene through marker-assisted selection was documented by Yau et al. (2005). The resistance of the root knot nematode gene was found to be significantly correlated with the size of an RAPD marker, which was determined to be 6.5 cM by Boiteux et al. (2000). According to a study by Shim and Jørgensen (2000), AFLP polymorphic markers were successfully utilised to determine the geographical origin of wild carrot populations and to purify strains. In order to differentiate genetic heterogeneity within and between inbred lines of carrots, Grzebelus et al. (2001) employed eight AFLP markers. Furthermore, the markers were successfully employed to assess the integrity of eight F1 hybrids that originated from four inbred lines, which comprised male-sterile lines and carrot maintainer lines. Robison and Wolyn (2002) successfully mapped more than 20 genes and pseudogenes onto the genome of CMS carrot through the utilisation of mitochondrial gene fragments of carrot derived from alien plant species. Furthermore, a distance of 255 kilobases was identified between circular subgenomic DNA molecules known as cox1, cox2, and atp8, which were found to be reversibly located on these molecules and were linked to repeated mitochondrial genome regions. Distinguished two unrelated carrot F2 populations that were dissimilar in appearance and size along the same linkage group by employing the AFLP marker.

An effective analysis of the genetic diversity of 124 carrot accessions was conducted using a closed-linked SCAR marker (*Y2*), nematode resistance (*Mj-1*), and a chloroplast-specific conditioning gene (CSC). Additionally, AFLP, ISSR, and two SCAR markers (*Y2* and *Mj-1*) were evaluated. This research distinguished feral accessions from cultivated ones by employing unstructured, high-molecular-level analysis. Bach et al. (2002) subsequently conducted research to differentiate between the fertile and barren cytoplasms of male carrot germplasm by employing RFLP markers. A variety of seventeen markers were found to be applicable, as demonstrated by their methodology. Furthermore, they first identified tightly linked RFLP markers that were associated with six conserved genes in the male-fertile line (K831B) and the petaloid male-sterile line (K826A). Linke et al. (2003) identified a significant correlation between MADS box genes, which are homologous to GLOBOSA and DEFICIENCS, and cytoplasmic male sterility markers. In regard to the locus accountable for disease control, Westphal and Wricke (1997) and Vivek and Simon (1999a) formulated two molecular maps. The nematode resistance locus (*M. hapla*) was determined using two hundred microsatellite markers RCAPD, AFLP, RAMP, and AFLP, all of which were derived from *D. carota* subsp. *azoricus* (Ali 2007). To investigate the genetic variability of forty indigenous lines of subtropical carrot, Jhang et al. (2010) employed sixteen simple sequence repeats (SSR), ten URPs, sixteen ISSRs, and ten URPs. In contrast, the amplification of the SSR markers was limited to 100–300 bp, whereas the ISSR, RAPD, and URP markers utilised 200–3500 bp. Furthermore, they described how ISSR and URPs were effectively implemented to characterise the genetic diversity of tropical carrots.

11.3 Mapping Populations

The most critical stage in the construction of linkage maps utilising molecular markers is the establishment of mapping populations. Critical determinants of mapping project success include the generational sequences employed for molecular marker analysis and phenotypic evaluation, the magnitude of the mapping populations, and the selection of parents for the crossing programme (Young 1994). Santos and Simon (2002b, 2004) detailed the methodologies utilised in the development and propagation of carrot mapping populations, in addition to the provenance of the mapping populations and the particular characteristics that qualify them for mapping purposes. Frequently employed in the construction of linkage maps for crop species are segregating populations, including backcrosses and F2 populations.

In the context of genetic mapping, a higher level of discernible DNA sequence polymorphism between parental individuals is adequate (Helentjaris 1987). Aside from a limited number of allogamous species, however, sufficient polymorphism will result from crosses involving unrelated individuals to enable mapping. The identification of DNA polymorphism in crop species that inbreed naturally presents a challenge, owing to the minimal degree of sequence variation. The requirement for sufficient DNA sequence polymorphism in certain narrow-based crossings may impede the utilisation of DNA markers (Young 1994). Probes based on minisatellites (Dallas 1988), or simple repeated tetra-nucleotide motifs (Weising et al. 1989), or electrophoresis systems capable of segregating DNA molecules with a single base pair change are instances of contemporary technologies that provide enhanced methods for identifying polymorphisms among closely related individuals. Given the substantial variation observed at the DNA sequence level, it is likely that these sequences will ultimately function as useful markers for narrow-base crosses in mapping (Winter et al. 1999; Choumane et al. 2000).

11.4 Selection of Population Segregations

The identification of suitable progenitor individuals for the purpose of generating genetic populations is an essential component in the construction of the linkage map. A diverse range of genetic population types exists. The most straightforward populations are backcross populations and F_2 populations derived from authentic F_1 hybrids. The construction of these populations for the majority of plant species is straightforward; nevertheless, the number of potential parent combinations may be limited, particularly in wide crosses, due to the sterility of the F_1 hybrid. One notable drawback of F_2 and backcross populations is their transient characteristics. Seeds acquired from these individuals via selfing will not generate authentic progeny. The process of phenotyping individuals with S_1 genotypes using progeny from the advanced generation is delineated by Westphal and Wricke (1991). A method was suggested by Schulz et al. (1994) and Westphal and Wricke (1997), which involves the use of S1 plants obtained from a single F_1 plant selfing in order to create and maintain a carrot mapping population. Multiple seasons guarantee access to an 'immortal' mapping population through the use of this technique. F_2 populations were generated using analogous inbred lineages; in addition, they present a multitude of benefits in terms of linkage mapping compared to S_1s (Vivek and Simon 1999b; Santos and Simon 2004). Durable mapping resources can be established through the utilisation of inbred populations composed of recombinant inbred lines (RILs) acquired from individual F2 plants (Burr et al. 1988; Burr and Burr 1991).

11.5 Map Construction

Molecular markers employed in map construction can be broadly categorised into two groups: those based on biochemical agents, such as isozymes, and those based on DNA. There is a hypothesis that allelic variation at isozyme loci is selectively neutral, allowing for the accumulation of gene mutations at isozyme loci, which manifest as altered protein mobilities when detected by starch or polyacrylamide electrophoretic gels. The presence of codominance in isozyme data is a characteristic that allows for the differentiation between heterozygous and homozygous individuals of either parental class. The limited number of loci that can be definitively assessed significantly restricts the applicability of isozymes in the construction of linkage maps in plants. Conversely, the number of DNA markers is virtually infinite, and their application in the development of linkage maps is primarily limited by the researcher's ability to detect polymorphisms.

Regarding outcrossing species, such as carrots, this has not demonstrated itself to be a significant barrier. In carrots, maps have been constructed utilising a variety of DNA-based marker systems, including restriction fragment length polymorphism (RFLP), random-amplified polymorphic DNA (RAPD), and amplified-fragment length polymorphism (AFLP). The RFLP procedure, which was introduced by Botstein et al. in 1980, operates on the principle of DNA–DNA hybridisation. By capitalising on the specificity of a restriction enzyme with respect to its designated DNA sequence, this methodology attains outstanding accuracy and uniformity. RFLP data generally illustrate the presence of a predominance effect. However, the RFLP method is associated with significant effort, expense, and duration due to the requirement of segregating and manipulating genomic DNA samples on the microgram scale. The PCR-based RAPD technique, initially proposed by Williams et al. in 1990, is characterised by its technological simplicity, low cost, and rapidity. This method solely requires nanogram quantities of genomic DNA. Conversely, RAPD markers demonstrate preponderance in natural systems, which undermines the ability to discern the genetic composition of heterozygotes.

Several authors have raised concerns regarding the reliability and reproducibility of RAPD markers when used for lineage mapping and assessment (Echt et al. 1992; Riedy et al. 1992; Heun and Helentjaris 1993; Thormann et al. 1994; Hallden et al. 1996). But it is important to acknowledge that a considerable number of plant species have utilised RAPD markers in a successful manner to construct linkage maps (Kennard et al. 1994; Mudge et al. 1996; Sondur et al. 1996; Bucci et al. 1997; Davis and Yu 1997; Jiang et al. 1997; King et al. 1998; Boiteux et al. 2000).

The documentation of AFLP markers was initiated by Vos et al. in 1995. The methodology integrates the speed and simplicity of PCR detection with the reliability and precision of restriction digestion. The extensive application of the method in the construction of linkage maps in plants has provided substantial evidence of its robustness (Ballvora et al. 1995; Becker et al. 1995; Meksem et al. 1995; Mackill et al. 1996; Bendahmane et al. 1997; Brigneti et al. 1997; Li et al. 1998; Vuylsteke et al. 1999; Bradeen et al. 2001). Similar to RAPD data, AFLP data exhibit a preponderant characteristic.

11.6 Gene Mapping

Schulz et al. (1994) mapped carrots using a variety of structural markers, such as phen1, RFLP, RAPD, and isozyme markers. The phen1/phen1 feature was classified as 'complex' by the scientists due to the presence of tiny, dark green, coiled leaves and a preference for annual flowering without vernalisation, despite the apparent normality of root development. Phenotype-based gene mapping has not been previously implemented in carrots, notwithstanding the apparent lack of association between phen1 and carrot cultivation. Following this, it was determined that carrots contain numerous cultivation-critical genes. Male sterility, disease resistance, and root characteristics are governed by these genes. Carrot roots are rich in carotenoids, including carotene and anthocyanins, which give rise to the distinctive orange/yellow and purple hues observed on the roots. The synthesis of beta-carotene in carrot roots is controlled by an intricate genetic code comprising a multitude of activator and inhibitor genes. One of the genes implicated in the inheritance of carrot xylem core colour was identified as *Y2* by Kust (1970). Carrot core carotene accumulation is inhibited by the *Y2* locus. *Y2/y2* individuals are characterised by an orange core containing significant amounts of α-, β-, and γ-carotene. Conversely, *Y2*-individuals, whose genetic background determines their colour, have a yellow or colourless core devoid of carotene (Buishand and Gabelman 1979).

Bradeen and Simon (1998) examined the separation of primary colours in 103 F_2 offspring resulting from the cross B9304 × YC7262. Bulk segregant analysis (MSA; Michelmore et al. 1991) was the technique utilised. As an additional step beyond F_2 localisation, six AFLP markers were identified by the researchers in proximity to and surrounding the Y_2 locus. The variation in distance between the markers and the gene spanned from 3.8 to 15.8 crystal units. Then, utilising the identical F_2 mapping population, Vivek and Simon (1999a) identified a solitary AFLP marker situated 2.2 cM away from the Y_2 locus. They claim that the locus was situated at one extremity of linkage group B. The accumulation of carrot phloem anthocyanin is regulated at the P1 locus. The prevalence of purple (*P1*) anthocyanin is greater than that of non-purple (*P1*) anthocyanin. The inheritance patterns of *P1* and *Y2* traits in F_2 and BC populations descended from were investigated by Simon (1996).

Information extracted from the eastern carrot was utilised to demonstrate the absence of a correlation between the two loci. Consequently, Vivek and Simon (1999a) established a connection between P1 and linkage group A by employing

AFLP markers positioned 1.7 and 8.1 cM from the gene adjacent to P1, omitting the need for *Y2*. Carrot root comprises over 95% pure sugars and up to 60% total carbohydrates. These are composed primarily of sucrose, fructose, and glucose, which are reducing sugars. Mostly determined genetically by sugar concentration is the palatability of carrots. Both the quantity and composition of sugars are crucial. Based on the findings of Freeman and Simon's 1983 study, the sugar structure of carrot inbred lines exhibited considerable heterogeneity, despite the relatively constant total sugar content.

While certain inbred lines contained reducing sugars comprising 51–78% of total sugars, others contained only 10–12% (Freeman and Simon 1983). Moreover, it was ascertained by the researchers that the quantity of sucrose is controlled by a solitary gene designated Rs. Although *RS-rs* genotypes store the majority of sucrose, *Rs* genotypes store the majority of reducing carbohydrates, including glucose and fructose. It was demonstrated by Freeman and Simon (1983) that the *Rs* protein might influence the production or activity of invertase. Invertase is the name of the enzyme capable of converting sucrose to fructose and glucose. The prediction was validated upon the discovery that the rs allele is a non-functional, naturally occurring mutant of a carrot invertase isozyme (Yau and Simon 2003). This type of allele is incapable of producing an enzyme that functions. Vivek and Simon (1999a) discovered that Rs was associated with linkage group C at one end, 8.1 cM away from the AFLP marker. Consistent with inheritance data, Rs has no genetic relationship with *Y2* or *P1*, according to mapping results (Simon 1996).

Information pertaining to resistance against the rootknot worm (Meloidogyne javanica) is encoded in the *Mj-1* gene. The *Mj-1* variant is a descendant of the Brasilia variety of Brazilian carrot. The inheritance pattern is uniparental and comprises either a single dominant locus or two closely related loci that are spaced 10–20 cM apart (Simon et al. 2000). Boiteux et al. (2000) applied BSA to a sample of 442 participants classified as F_2. Through a cross between the susceptible parent B6274 and the resistant parent Brasilia-1252, the population was generated.

Using 1000 RAPD primers to screen resistant and sensitive F_2 bulks, 63 fragments believed to be associated with *Mj-1* were identified. In an additional F_2 investigation (Boiteux et al. 2000), four RAPD fragments adjacent to and connected to *Mj-1* were identified and mapped by the authors. The resistance values and markers varied between 0.8 and 5.7 cm. Each RAPD marker was associated with resistance throughout the coupling phase. This indicates that the presence of the RAPD fragment is associated with the *Mj-1* resistance gene.

From the genomic DNA of carrots, Robison and Wolyn (2002) used PCR primers to amplify mitochondrial gene fragments. The primers were constructed using duplicate mitochondrial genes that were identified in multiple plant species. Subsequently, the fragments were employed to assemble a physical map of the 255-kilobase CMS carrot mitochondrial genome, on which more than 20 genes and pseudogenes were located. There is evidence linking three genes—*cox1, cox2*, and *atp8*—to repetitive regions within the genome. These regions are composed of subgenomic circular DNA molecules that are present in both orientations, albeit with potentially non-stoichiometric quantities. As of now, neither the economic significance of these genes nor their potential involvement in male sterility is known. Significant genes, such as those governing root colouration and resistance to diseases, have been identified and investigated at the DNA level; furthermore, they play a crucial role in the field of agriculture (Simon 2000). Including these and other genes in the developing molecular maps of carrots appears to be one of the primary objectives.

11.7 Nuclear Linkage Map

Analogous to a multitude of other botanical species, the primary molecular linkage map for carrot was artificially synthesised, using only spectroscopic isozyme markers (Westphal and Wricke 1991). BC_1 populations were generated by Westphal and Wricke (1991) using self-pollinated S1 carrot lines. Linkage analysis and band variation analysis were performed on ten unique isozyme systems. Evidence of locus duplication was provided by the authors for three distinct systems, namely phosphoglucomutase (PGM), phosphoglucose isomerase (PGI), and 6-phosphogluconic dehydrogenase (6-PGD). The presence of duplication in 6-PGD and PGM was confirmed via segregation analysis. In all, 12 loci were assigned to 4 linkage groups. Given the difficulties inherent in identifying appropriate isozyme markers, Westphal and Wricke (1991) deduced that DNA-based markers, including RFLP and RAPD markers, might offer greater potential for constructing comprehensive linkage maps for carrot. Nevertheless, the report of isozyme linkages by the authors functioned as the primary linkage map for carrot and laid the groundwork for subsequent efforts in the field of mapping.

Subsequently, Schulz et al. (1993) integrated RFLP, RAPD, and isozyme markers in order to develop a carrot linkage map. In total, 4 S_1 populations were analysed; 2 of these contained a minimum of 70 individuals each and were derived from the carrot cultivars 'Finaê', 'Primatoê', and 'Gustoê'. A segregation analysis was performed on 28 RAPD markers, 14 RFLP markers, and 10 isozyme markers. The authors employed six RAPD fragments that exhibited monomorphism in at least one mapping population as RFLP probes. In aggregate, 58 molecular markers were mapped. The loci were categorised into eight linkage groups across the mapping populations, which was insufficient for a comprehensive carrot map, which would have required nine linkage groups. Three markers remained unlinked thereafter.

The linkage groups exhibited a range of compositions, consisting of two to nine markers, and their lengths varied from 1.7 cM (two markers) to 78.6 cM (seven markers). Mean distance between markers is estimated to be 13.1 cM. Subsequently, Westphal and Wricke (1997) expanded the map to include approximately 200 markers, including isozyme, RFLP, RAPD, microsatellite, and AFLP markers, by employing the same mapping populations. The final map was composed of nine principal linkage groups, with 'several' markers remaining unmapped.

Documented by Vivek and Simon (1999a), the carrot linkage map comprised 106 molecular markers. There were a total of 96 markers present, including two microsatellite-derived markers, two RAMPs, and six RFLPs. The authors utilised a mapping population comprising 103 F_2 individuals. This population was formed by self-pollination involving a solitary F_1 offspring resulting from the cross between B9304 and YC7262. A total of eleven linkage groups, each comprising five to fifteen markers, were established using a conservative LOD of 4.0. Seventeen additional markers were connected via linkages consisting of two to three markers, while twelve markers remained unlinked.

The mean distance between mapped markers was 4.9 cM, and the length of the resulting molecular map was 524.1 cM. The authors estimate that the carrot genome spans a total of 900 centimetres (Vivek and Simon 1999a). In order to generate the most current and comprehensive linkage map for carrot, Santos and Simon (2004) combined marker data from two separate F2 populations. The procedure of map merging entails the integration of marker data generated by separate populations in order to form a unified and consistent map (Stam 1993). The process of merging requires the creation of a framework consisting of identifiers that are universally shared by both populations. Following this, framework-specific identifiers pertaining to a particular population are statistically inserted. Upon completion of this process for each marker that represents a population, a unified and harmonious map is generated. Successful map merging has been achieved for a number of plant species (Sebastian et al. 2000; Bradeen et al. 2001; Paran et al. 2004). Santos and Simon (2004) employed F_2 populations of carrots obtained from a variety of germplasm sources. The preliminary crossing occurred between an inbred carrot line (B493) and a wild carrot (*D. carota* ssp. carota; QAL). In the second cross, a high carotene strain (HCM) was bred with Brasilia, an open-pollinated Brazilian cultivar. AFLP data were generated for each of these populations by employing two restriction enzyme combinations: *EcoRI/MseI* and *PstI/MseI*. A limited number of codominant markers were also identified for both populations via PCR (Santos and Simon 2004). An imperative component of efficient map integration entails the detection of markers that are universally present across all mapping populations. As a result of the complexity of the banding patterns generated by this technique and the prevalence of the majority of AFLP markers, it is essential to verify that AFLP markers of similar sizes originating from separate populations are identical by descent. Santos and Simon (2002b) determined in a previous investigation that AFLP markers of comparable sizes and originating from the same F_2 mapping populations demonstrated a DNA sequence identity similarity surpassing 91%, with a minimum of 84%. To facilitate the integration of maps, a framework for mapping was developed utilising two PCR markers and 28 AFLP markers that were shared by both mapping populations. Following that, identifiers that were specific to each population were integrated into the framework. In total, six linkage groups could be consolidated. With a range of 18–30 markers, the length of the merged linkage groups varied between 117.6 cM (27 markers) and 68.9 cM (20 markers) in composition. The average distance between markers on the merged linkage groups was 3.75 cm. The inability of three additional linkage groups from each population to merge was due to the absence of shared markers (Santos and Simon 2004). At this time, linkage maps for *Daucus* species other than *D. carota* have not been documented. The Apiaceae family encompasses a wide variety of cultivated species, such as carrot, parsnip, fennel, parsley, cumin, coriander, dill, and anise. Linkage mapping among the aforementioned species has been documented solely for celery (Huestis et al. 1993; Yang and Quiros 1995). Celery has been mapped using RFLP, RAPD, isozyme, and morphological markers; nevertheless, the lack of shared markers between the carrot and celery maps precludes the utilisation of previously developed maps to compare the mapping of these species.

11.8 Assessment of Marker-Based Linkage Mapping Systems

A variety of marker systems have been implemented to map relationships using carrots. Each marking system possesses both benefits and drawbacks. In their study, Vivek and Simon (1999a) examined the efficacy of RFLP, RAPD, and AFLP markers as they pertain to the development of carrot linkage maps. Their F_2 mapping population was generated by phenotypically distinct progenitor genotypes. The sole source of one progenitor, B9304, was germplasm derived from Western carrots. Provided context: Material from eastern carrots was incorporated into the second progenitor, YC7262 (a full-sibling of the carrot inbred B7262; Simon et al. 1997). However, out of the 300 genomic DNA clones that were assessed as RFLP probes, a mere thirty (10%) exhibited conspicuous scorability and polymorphism across the genotypes of their progenitors. Additionally, in a study employing 164 RAPD primers, the authors documented a 20% polymorphism rate (33 primers) between parental genotypes; nevertheless, accurate scoring was only possible for two of the RAPD markers (Vivek and Simon 1999a). Schulz et al. (1993) reported that a self-pollinating mapping population derived from a *D. carota* cultivar exhibited a 33% observed polymorphism rate (20/60 RAPD primers). Potentially explaining the observed discrepancies in the polymorphism rates of RAPD markers utilised in these two investigations are genetic background variations. In conclusion, Vivek and Simon (1999a) identified 164 polymorphic fragments from 404 AFLP bands generated by seven primer combinations, representing an observed polymorphism rate of 42%. AFLP is thus more than twice as effective as RFLP and RAPD, according to the scientists, at producing markers for the development of carrot linkage maps. In their 1998 study, Nakajima et al. consistently demonstrated that while both AFLP and RAPD markers proved valuable for phylogenetic analysis in Daucus, the AFLP method produced over four times as many functional markers per reaction. For carrot-optimised RAPD (Boiteux et al. 1999) and AFLP (Briard et al. 2000), published protocols are available.

11.9 Linkage Mapping in Carrots

The utilisation of a primary genetic map, which consists of easily scored polymorphism marker loci dispersed throughout a genome, is essential for conducting exhaustive genetic investigations on any given organism. Genetic investigations, including but not limited to positional gene cloning, marker-assisted selection, and quantitative trait mapping, rely heavily on saturated linkage maps. Throughout history, the development of linkage maps of this kind was limited to organisms that were the subject of extensive scientific investigation, such as yeast, fruit flies, or bacteria, wherein a multitude of observable mutations functioned as genetic markers. Since the early 1980s, this limitation has been overcome due to the widespread adoption of molecular marker techniques that facilitate the observation of DNA-level polymorphisms. Linkage mapping entails the organisation of marker loci (and quantitative trait loci), the calculation of their relative distances, and the allocation of these components to their corresponding linkage groups based on the recombination values obtained from every conceivable combination of three and two points. The groundbreaking molecular marker-based linkage map of the human genome, developed by Botstein et al. in 1980, acted as an impetus for the development of similar maps in subsequent organisms. Researchers successfully produced genetic linkage maps for carrot through the meticulous application of genetic markers (Schulz et al. 1994; Bradeen et al. 1997; Niemann et al. 1997; Westphal and Wricke 1989, 1991, 1997; Vivek and Simon 1999a, b; Boiteux et al. 2000; Le Clerc et al. 2003; Santos and Simon 2004; Just et al. 2007; Bradeen and Simon 2007). On these maps are depicted isomers, RFLPs, RAPD, microsatellites, SNPs, ISSRs, SAMPLs, and morphological markers. Comparisons are unattainable on the majority of these maps due to the lack of common markers. To ascertain six linkage groups, two populations of Brasilia × HCM BA and B493 × QAL were utilised in conjunction with two codominant markers and 28 sequenced dominant AFLP (Santos and Simon 2004). As demonstrated by the most researchers, in Carrot (Vivek and Simon 1999a; Niemann et al. 1997; Boiteux et al. 2000, 2004; Yau and Simon 2003), the utility of PCR-based markers in distantly related populations is limited by their small number. The preliminary molecular linkage map of carrot was generated by employing an isozyme marker in backcross populations derived from self-pollinating (S1) carrot lines (Westphal and Wricke 1991). Westphal and Wricke postulated in their 1991 article that RFLP and RAPD markers could potentially yield more accurate linkage map constructions in carrot. Following this, the linkage map was completed by Schulz et al. (1993) through the utilisation of four S1 populations that originated in the Finaê, Primatoê, and Gustoê. Seventy individuals composed these populations, which were represented by fourteen RFLP, twenty-eight RAPD, and ten isozyme markers. Each of the eight linkage groups was separated by 13.1cM as a result of the analysis. After that, Westphal and Wricke (1997) extended the linkage map with an additional 200 markers, including AFLP, microsatellites, isozymes, RFLP, and RAPD, utilising the identical mapping populations. The RAPD and AFLP marker systems were utilised to evaluate the genetic diversity of Daucus carota in an efficient manner. This finding illustrated the ability to differentiate cultivated carrots from their wild counterparts (Nakajima et al. 1998). A 109-point linkage map was generated for the phenotypic loci (*P1*, *Y2*, and *Rs*) of carrot by employing 6RFLP, 2RAPDs, 96AFLPs, and 2SAMPL. The corresponding polymorphism yields for RFLP probes, RAPD primers, and AFLP primers were 36%, 20%, and 42%, respectively. The phenotypic attributes of purple pigmentation (*P1*), white root (*Y2*), and reducing sugar (*Rs*) exhibited significant associations with AFLP markers P6B15,

P1B34, and P3B30XA, respectively, at distances of 1.7 cM, 2.2 cM, and 8.1 cM (Vivek and Simon 1999b).

Bradeen and Simon (1998) developed six AFLP markers that exhibited a strong association with the *Y2* locus and spanned 6.6 cM. In total, 264 base pairs were employed in the process of cloning and sequencing the markers. A 2 kb fragment was sequenced by employing a sequenced AFLP marker and the *Y2Y2* allele parent. This sequence was used to characterise the *Y1Y1* parent, which served as the inverse PCR template. The outcome was a 310 bp codominant marker for the *Y2* allele, which could be used to identify the *Y2* allele parent (172 bp for the *Y2* maker and 138 bp for the indel). Using the 103F_2 mapping population of B9304 × YC7262, the linkage map was produced. The mapping procedure employed a comprehensive set of 106 markers, comprising 96 AFLPs, 6 RFLPs, 2 RAPDs, and 2 microsatellite-derived markers. The eleven resulting linkage groups comprised five to fifteen markers, with an average distance of 4.9 cM and a total length of 524.1 cM. The estimated extent of the carrot genome, as indicated by this map, was 900 cM (Vivek and Simon 1999a).

At a distance of 8.1cM, the reducing sugar (Rs) type was mapped into the C group linkage using the AFLP marker. Conversely, the Rs locus demonstrated a marginal correlation with *Y2* and *P1* (Simon 1996; Vivek and Simon 1999a). Santos and Simon (2002b) demonstrated that 84% of AFLP markers of comparable size produced from the same F2 mapping populations exhibited over 91% DNA sequence identity overlap. To facilitate the integration of maps, a mapping framework was developed using 2 PCR markers and 28 AFLP markers shared by both mapping populations. Following that, identifiers that were specific to each population were integrated into the framework. The linkage map was generated from F_2 mapping populations of B493 × QAL and Brasilia × HCM using the AFLP markers in combination with *EcoRI/MseI* and *PstI/MseI* restriction enzymes. However, PCR-based codominant markers were also utilised in these two populations. Santos and Simon (2004) reported that the merged linkage group demonstrated a variation in distance from 118.6 cM (27 markers) to 68.9 cM (20 markers), with an average distance of 3.75 cM.

In their study, Just et al. (2007) identified twenty-two carotenoid genes in the carotenoid biosynthetic pathway across eight of the nine linkage groups in the carrot genetic linkage originating from the population of orange × white carrot. Furthermore, twenty-four putative genes and two co-localised quantitative trait loci (QTL) associated with carotenoid accumulation were identified through the utilisation of sequence data derived from carrot and various other plant species.

The genetic linkage map was constructed utilising F_2 mapping populations of wild carrot QAL (*D. carota* subsp. *carota*) and B493 (*D. carota* subsp. *sativus*). Gene-specific size, single-nucleotide polymorphism, DcMaster, and Krak transposon insertion sites were added to the exhaustive genetic map after their designation as 250 AFLP markers. At this time, there are 381 markers on this saturated genetic map, and the genome sizes of QAL and B493 were 1242.0 cM and 1474.7 cM, respectively (Santos and Simon 2002; Just et al. 2007; Grzebelus et al. 2007; Grzebelus and Simon 2009). A total of 440 GSSRs, ESSRs, and BSSRs were recently developed with the intention of conducting linkage mapping, assessing the genetic diversity of the carrot family, and determining the transferability of markers throughout the Apiaceae lineage. The formation of nine linkage groups was achieved through the utilisation of these broadly distributed SSR markers (Cavagnaro et al. 2011).

11.10 Organelle Physical Maps

Both the production of hybrid seeds and the propagation of carrots are highly reliant on CMS. Carrot male sterility is governed by genes located in the mitochondrial genome. Conversely, components present in the nuclear genome have the potential to restore male fertility. Considering the interplay between nuclear and organellar components in regulating the expression of sterility, this system is suitably designated a genic CMS system. Borner (1995) investigated the genetics of carrot CMS. The mitochondrial genome of higher plants is dynamically arranged and may comprise numerous subgenome circular DNA molecules, as opposed to being composed of a single enormous master chromosome (Folkerts and Hanson 1991; Yamato et al. 1992; Janska and Mackenzie 1993; Kubo et al. 2000). Bowes and Wolyn (1998) examined restriction patterns in the mitochondrial genomes of six petaloid male-sterile carrot lines and three male-fertile maintainer lines originating from three distinct regions. Despite the petaloid male-sterile lines originating from locations that were over 1500 km apart, minimal variability was observed among them. On the contrary, they observed considerable mitochondrial diversity among male-fertile varieties, which aligns with the notion that the mitochondrial genome architecture is dynamic. The scientists arrived at the conclusion that petaloid CMS lineages were most likely generated by a shared rearrangement of the mitochondrial genome.

The physical mapping of the mitochondrial genome of the petaloid male-sterile carrot line W259A was conducted by Robison and Wolyn (2002). The arrangement of cosmid fragments was accomplished through the utilisation of restriction site analysis to construct a map. The finalised map is approximately 255 kilobytes in size. The investigators identified three significant regions of repetitive genomic DNA, ranging in size from 0.53 to 2.0 kb. Frequently, higher plant species possess repetitive regions within their mitochondrial DNA. It is postulated that recombination occurring across repeat regions could induce stoichiometric variations in distinct regions of the genome and enable the reversible synthesis of subgenomic circular DNA molecules. By consistently employing Southern hybridisation, Robison and Wolyn (2002) demonstrated that the three repetitive regions present in the mitochondrial genome of the CMS carrot experience reorganisation in living organisms. By elucidating the manner in which the CMS carrot mitochondrial genome is structured into subgenomic circular DNA molecules, they arrived at the deduction that the CMS mitochondrial genome's overall configuration precludes its existence as a solitary large circular chromosome. Also included are stoichiometric variations in the composition of the plant mitochondrial genome. CMS has been associated with stoichiometric variations in specific mitochondrial genome regions in several plant species. Several studies have established a link between male sterility in carrots and the modification or expression of mitochondrial genes (Szklarczyk et al. 2000; Nakajima et al. 2001; Rurek et al. 2001). However, no specific gene or region of the CMS carrot mitochondria has been conclusively linked to male sterility.

11.11 QTL Mapping in Carrots

The utilisation and advancement of molecular marker technologies to identify and delineate quantitative trait loci has become a crucial approach in the investigation of the genetic and phenotypic foundations of complex traits (Edwards et al. 1987, 1991; Paterson et al. 1988). By establishing a connection between particular genetic elements and a complex trait, research can focus on the autonomous functionality of each locus, unaffected by the possible impact of other segregating loci that could introduce confounding variables.

Bradeen and Simon (1998) identified six AFLP markers associated with the Y2 locus at distances of 2.8 cM and 15.8 cM from 103 F_2 populations in a cross between B9304 and YC7262. Bulk segregant analysis was employed to partition these populations according to core pigmentation, as outlined in the methodology of Michelmore et al. (1991). Subsequently, Vivek and Simon (1999a) employed this population to discern a single AFLP marker located at the Y2 locus and assign it to linkage group B, which had a dimension of 2.2cM. Moreover, through segregation analysis, horticulturally significant QTLs have been identified.

According to Santos and Simon (2002a), the 160F2 Brasilia HCM population was employed to map a comprehensive set of 287 AFLP markers. These markers were distinguished by a medium orange characteristic line and a high carotene line, respectively. All of the evaluated characteristics were discernible when these markers were present; they were associated with α and β-carotene, lycopene, and the γ-carotene and phytoene QTL conditioning of the precursor. For β-carotene, five QTLs were identified, while for -carotene, three QTLs, one QTL, and -carotene were identified, respectively. The precursors carotene and phytoene explained 40%, 20%, 7.2%, 16.3%, and 28%, respectively, of the total phenotypic variation, while the QTLs of Brasilia and HCM exhibited 3.7% to 13.2% of the total variation. Twenty key quantitative trait loci (QTL) have been identified as regulators of the carotenoid content in orange carrots (Santos and Simon 2002a). In 1979, Buishand and Gabelman identified clusters of genes and quantitative trait loci (QTLs). Among these, two primary genes (Y and Y2) were found to be associated with the carotenoid biosynthetic pathway, which was common to the orange carrot (yyy2y2) and white carrot (YYY2Y2) populations. These populations were distinguished by their respective pale orange and yellow pigmentation. Santos and Simon (2002a) established that these markers were also intimately associated with the Y2 gene, whereas Bradeen and Simon (1998) identified SNP-based markers that map to this region. They considered ZDS2 and ZEP to be possible candidate genes for the *Y2*

QTL. Yellow (xanthophyll) and orange (α- and β-carotene) phenotypes were found to be affected by *Y2* (Buishand and Gabelman 1979). The origin of the Y2 gene was traced back to yellow segregants present in the B493 × QAL population, as was disclosed. Moreover, Y2's mapping to this substantial region suggests that it may account for a considerable proportion, if not the entirety, of the quantitative trait locus effect that Santos and Simon (2004) have documented. The identification of quantitative trait loci (QTLs) pertaining to total carotenoids was carried out by Santos and Simon (2002b). The concentration of carotenoids was impacted by these quantitative trait loci (QTLs), which accounted for 37.7%, 44.2%, and the accumulation of α-carotene, lycopene, and phytoene, respectively, of the overall phenotypic variance. Just et al. (2007) reported that for eight of the nine linkage groups, gene-specific codominant polymorphisms were identified in 22 genes involved in the biosynthesis and metabolism of carotenoids. A primary quantitative trait locus (QTL) has been identified in the F2 population of P50006 and HCM A.C., regulating β-carotene, total carotene, and lycopene accumulation, using sequence-related amplified polymorphism (SRAP) markers. This QTL explains 12.79%, 12.87%, and 14.61% of the overall phenotypic variations, respectively. The SRAP marker exhibited a strong binding at a distance of 502.9 cM across nine linkage groups, with an average interval of 5.5 cM. These three quantitative trait loci displayed genetic variability that was explained by additive genetic variance. Furthermore, a set of epistasis quantitative trait loci (QTL) governing the accumulation of lycopene and β-carotene contributed 6.5% and 15.1%, respectively, to the overall variation in phenotype. To increase the levels of carotenoids and lycopene in carrots, these SRAP markers linked to these QTLs may be utilised in selection or QTL pyramiding (Cheng-Gang et al. 2010).

Explorations utilising the genetic data that was included in the review will provide valuable insights into various aspects of the carrot. Future research will concentrate on carrot colour as a natural food colouring source, carrot colour genetics, the development of a mapping population, the application of various markers, linkage and QTL mapping analysis in carrot, and other related topics. A comprehension of carrot genetics and molecular breeding would expedite the development of carrot hybrids in the near future that are exceptionally rich in vitamins and minerals.

11.12 Mapping for Carotenoid Content

Carrots are rich in beta-carotene, a highly effective antioxidant, and are therefore among the most prolific sources of vitamin A (6500–8258 g/100 g). This is due to the high concentration of β-carotene in carrot cultivars (Sun et al. 2009), which accounts for approximately 44–79% of the total carotenoids present in carrots. Carotene, by virtue of its provitamin A activity, is a vital nutrient. However, there are differences in the accession parameters of carrots, such as Selection-141 (52 mg/100 g), HC-1 (3.11 mg/100 g), HCM (17,206/100 g FW), and Violette Turque (318 g/100 g FW) (Nicolle et al. 2004). Although the carrot has been credited with coining the term 'carotenoids' since ancient times, little research has been undertaken to determine its nutritional value. Additional investigation into the genetic composition of carrot roots revealed their complex attributes, which included genes responsible for both the regulation of specific pigments and the inhibition of pigment growth (Kust 1970; Bushland and Gabelman 1979). Carrots display a preponderance of the white hue, with intermediate yellow and orange tints also present. The yellow pigmentation appears to be monogenic in nature. The orange × white cross involves three contributing factors, while the purple-rooted plants generated by the progeny of the orange × purple cross demonstrate the interaction of complementary genes. The orange (W93) carrot crossed with the red (KHN, KOS, KDA) carrots implicated at least two alleles, demonstrating orange's superiority over red. Each carrot cultivar possesses recessive alleles that confer an increase in carotene content. Deep orange (ABcc), orange (aaBcc), and yellow-orange (AAbbcc) roots comprise a 3:1:1 proportion of the overall length of *D. carota* var. *hollandicus* cv 'Nantes 3'. Aabbcc is found in genotypes that are colourless or translucent. In *D. carota* var. *boisseri*, the heterozygous genotype Abbc is denoted by an 8:3:4:1 ratio of purple to orange to yellow to white. The orange hue can be attributed to C, the yellow to A, and the purple to B. The recessive yy gene is accountable for the orange colouration, whereas the y genotype governs the development of white or orange-white roots (Buishand and Gabelman 1979). The suppression of orange pigment synthesis in roots is attributed to the activity of the dominant genes Y, Y1, and Y2. Kust documented in 1970 five gene systems that govern the pigmentation of xylem and phloem. The predominance of light colour over dark colour is supported by a polygenic control of carrot root colour that outlined. An attempt was made to increase the total amount of carotene in two populations, which produced the subsequent outcomes: Population A, which experienced two iterations of positive selection for varying levels of carotene content (from high to low), acquired an annual increase of 21 μg in total carotene (on a fresh weight basis). On the other hand, Population B exhibited an annual increase of 16 μg (fresh weight basis) in carotene levels.

Goldman and Breitbach (1996) found that the presence of a single recessive gene called rp is responsible for a 92% decrease in the overall carotenoid content of rprp genotypes compared to those that do not possess the rp gene. Early in the developmental process, the scientists observed a correlation between the rp gene and white foliage; they anticipate that this correlation will be indispensable in elucidating the carotenoid pathway. Conversely, Koch and Goldman (2005) observed 96% reductions in α and β-carotene levels and 23–43% reductions in α-tocopherol when compared to near-isogenic lines. A simultaneous development of α-tocopherol and carotenoids is supported by the positive correlations observed between α-tocopherol and β-carotene ($r = 0.65$; p 0.001) and α-carotene and α-tocopherol ($r = 0.52$; p 0.001), respectively, as demonstrated by this study. Just et al. (2009) ascribed a significant portion of the variance in carotenoid accumulation in carrot root to the *Y* and *Y2* loci, which were discovered on linkage groups 2 and 5, respectively. These specific genes function as potential loci for the enzymes carotene hydroxylase, zeaxanthin epoxidase, and carotenoid dioxygenase. As of now, a considerable number of quantitative trait loci (QTL) encoding provitamin A, α and β-carotenoids, γ-carotene, lycopene, and precursors in the carotene pathway have been identified in carrots. Santos and Simon (2002) identified the quantitative trait loci (QTLs) that are linked to both total carotenoids and five-component carotenoids in carrots in their previous investigation. The aforementioned constituents consist of lycopene, phytoene, β-carotene, α-carotene, and γ-carotene. Additionally, it has been disclosed by Just et al. (2007) and Santos and Simon (2002) that a significant proportion of the structural genes associated with the carotenoid pathway have been identified. Bradeen and Simon (1998) analysed AFLP markers in the neighbouring *Y2* region by employing bulked segregant analysis and F_2 mapping. Between 15.8 and 3.8 cm of separation distinguished the markers from the gene. Simon Vivek and Simon (1999a, b) detected an additional AFLP marker at a distance of 2.2 cm from the *Y2* locus. The ability to detect carotenoids in leaf tissue during the early stages of development, as Yau et al. (2005) illustrated in relation to the sugar content of leaves in a plant, may eliminate the necessity for mature root formation. The detection of additional codominant markers and the swift advancement of cartography may potentially streamline the implementation of these genetic tools in marker-based breeding.

11.13 Mapping for Anthocyanin Content

Anthocyanins are a class of water-soluble pigmented phytonutrients that are potent antioxidants. This pigment gives higher plants their characteristic purple, red, and blue tints. It provides defence against oxidative stress, inflammation, cardiovascular disease, certain types of cancer, and numerous age-related ailments. On the other hand, the abundance and bioavailability of dietary anthocyanins influence the health benefits they provide. The excretion and bioavailability of anthocyanins and other polyphenols are substantially influenced by their chemical composition, which comprises the type of sugar molecules, the presence of phenolic aglycone, and the degree of acylation. Cyanidin compounds constitute the majority of the composition of a carrot. Structural genes—which encode the enzymes essential for the direct synthesis of anthocyanin pigments—and regulatory genes—which regulate the transcription of structural genes—are the two types of genes involved in anthocyanin biosynthesis. The researchers assert that an extensive array of model organ isms, including petunia, snapdragon, and maize, possess a substantial quantity of anthocyanin-producing genes that have been successfully identified and published. Structural and regulatory genes exhibit substantial variation across different species as well. Inconsistent cross-species transmission of information regarding the genetic regulation of flavonoid biosynthesis may result from this phenomenon.

The regulation of anthocyanin synthesis in carrot roots is attributed to a solitary dominant gene known as *P1*. Several anthocyanin biosynthesis genes, including five structural genes (*F3H, FLS1, PAL3, LDOX2*, and *UFGT*) and three regulatory genes (*DcEFR1, DcMYB3*, and *DcMYB5*), were genetically mapped in a recent study by Yildiz et al. (2013). Nevertheless, no substantial association with chromosome 3 was detected; solely the anthocyanin biosynthesis genes *F3H* and *FLS1* exhibited a correlation with the gene *P1*. The research findings indicated a clear association between the accumulation of transcripts and anthocyanins, despite the differential expression of five key genes (*CHS1, DFR1, LDOX2, F3H* and *PAL3*) in orange carrots, solid purple carrots, and purple orange carrots. Although genetic factors have been associated with variations in the abundance of anthocyanin colour in carrot roots, the underlying molecular and metabolic mechanisms remain elusive. Furthermore, the mechanism by which anthocyanin production in organs and tissues other than taproot carrots is inherited remains obscure. Simon (1996) hypothesised a relationship between *P1* and *P2* after identifying a straightforward dominant locus, denoted as *P2*, which regulates the preference for purple or green pigmentation in petioles. A potential gene for *P3* has been identified in *DcMYB7* located on chromosome 3. Additionally, genes responsible for the accumulation of carrot anthocyanin *P1, P2,* and *P3* have been identified (Cavagnaro et al. 2014; Xu et al. 2019). As stated in Wild carrots, on the other hand, have a short tap root that complicates harvesting, whereas cultivated carrots have a broad and deep root. Suggest that the cult gene, located on chromosome 2, modulates the root diameter. *DcAHLc1*, an enzyme that oversees organ development, could potentially serve as a candidate for this gene. Furthermore, investigations utilising genome-wide association studies (GWAS) have been conducted with carrots. A genome-wide association study (GWAS) by Keilwagen et al. (2017) validated the existence of a correlation between terpenoid carrot taste volatiles and the genes implicated in terpenoid synthesis. Identified the Or gene in a comprehensive set of carrots characterised by distinct root carotenoid-based tints through the implementation of a genome-wide association analysis (GWAS). The transformation of proplastids to chromoplasts occurs *via* the *Or* gene.

In order to facilitate genome assembly, precise quantitative trait locus mapping, marker-assisted breeding, gene identification, comparative synteny analysis, precise quantitative trait locus mapping, and high-density linkage mapping based on relevant sequence-based markers such as SNPs and SSRs are required. In a few instances, the accuracy of the carrot maps was enhanced through the incorporation of restriction fragment length polymorphism (RFLP) or codominant SSR markers. Grzebelus et al. (2014) found in a recent study that carrots possess the highest map resolution at 1.1 cM. The absence of established Diversity Arrays Technology (DArT) markers poses a challenge in the identification of potential genes and the execution of fundamental comparison map analyses. Cavagnaro et al. (2014) developed a high-resolution gene-derived SNP-based linkage map of carrots, including 894 markers spanning 635.1 cM and a map resolution of 1.3 cM. In order to estimate total root anthocyanin (RTPE) and map 15 major quantitative trait loci (QTL) associated with anthocyanin compounds, six chromosomes were utilised. Chromosome 3 (RTPE) harboured eight significant phenotypic QTLs, in addition to those governing anthocyanin glycosides and root total pigment estimates, among others. A solitary dominant gene that regulates anthocyanin acylation within a solitary cell was identified. Anthocyanin pigment in carrots is regulated by a minimum of three distinct genes, contrary to previous findings, as demonstrated by comparison mapping with two additional purple carrot populations. The pigmentation of purple petioles and roots is controlled by a single dominant gene. The researchers observed the identical genetic regulation for purple pigmentation in a distinct genetic background as they had previously identified. They discovered that P1 and additional loci were associated with the anthocyanin pigmentation of the root and petiole in three distinct carrot backgrounds. To study the production of anthocyanins in carrots, Xu et al. (2014a, b) identified six new structural genes: *CHS1, CA4H1, 4CL1, F3H1, F3'H1, PAL3/PAL4, DFR1,* and *LDOX1/LDOX2*. Light-dependent anthocyanin synthesis may be inhibited by mutations in the carrot *F3H1, F3'H1, CHS1, CHI1, DFR1,* and *LDOX1/LDOX2* genes, which are present in orange and yellow carrots. Numerous structural genes have been associated with the accumulation and loss of anthocyanin in the taproots of purple and non-purple carrots. Gene inactivation in regulator genes is one possible explanation for non-purple carrot taproot gene inactivation.

11.14 Mapping of Root Sugar Content

Carrot sweetness is an indispensable characteristic for its applications as a beverage, confection ingredient, and pool (halwa) construction material. Additionally, the refining of carrots is dependent on their sugar content. While inferential by definition, empirical evidence does not support the correlation between this factor and the texture and flavour of the roots. When evaluating the GCA effects of total sucrose, Karkleliene et al. (2005) determined that the CMS line NS 744 exhibited the highest efficacy in relation to this attribute. Recently, it was discovered that the rs allele, which produces no active enzymes, is a naturally occurring knockout mutant for the carrot invertase isozyme (Yau and Simon, 2003). Vivek and Simon (1999a, b) utilised an AFLP marker to identify *Rs* at the uppermost point of linkage group C, specifically at 8.1 cM. Based on mapping investigations and inheritance data, it has been determined that *Y2* and *P1* do not share a genetic relationship with Rs (Simon, 1996). The Rs gene was subjected to marker-assisted selection (Yau et al. 2005). *Susy1*Dc1* and *Susy1Dc2* are sucrose synthase genes found in carrots. Surprisingly, dissimilar are the expression patterns of these two sucrose synthase alleles. The expression of Susy1Dc1 is prevalent in the developing seeds, roots, leaves, and flowers of carrots, while Susy1Dc2 is typically observed in the flowers. Despite the apparent criticality of sucrose synthase in plant growth-related metabolic processes, numerous studies indicate that its impact on sucrose partitioning in carrots is negligible. By employing a candidate gene approach, Yau and Simon (2003) demonstrated the direct participation of soluble acid invertase in sucrose partitioning and aroma synthesis in carrot roots. The accumulation of fructose and glucose, which are simple hexose carbohydrates, occurs in carrot roots as a result of the dominant Rs allele. Individuals who carry the *RS/RS* homozygosity accumulate sucrose in their root systems. The Rs locus is of significant importance in the regulation of carrot root's sugar content and aroma profile. Yau and Simon (2003) determined the molecular basis of the Rs locus using the carrot inbred B4367 by establishing isogenic lines *rs/rs* and *Rs/Rs* that were close by. Transcripts for several enzymes involved in the sucrose/simple sugar pathway were detected via RT-PCR, including those of sucrose synthase, extracellular acid invertase, and two isoforms of soluble acid invertase (isoform I and II). A reliable correlation was identified between the genetically characterised Rs locus and the aberrant soluble acid invertase isoform II allele. This implies that the transcription of soluble acid invertase isoform II occurs at the Rs locus.

11.15 GWAS for Biotic Stress

In carrots, leaf spot (*Cercospora apii* Fres.), leaf blight (*Alternaria radicina*), powdery mildew (*Erysiphe polygoni* DC), and soft rot (*Erwinia carotovora* (Jones) Holland) are commonly occurring diseases in India and cause considerable reduction in the yield unless chemical sprays are used to control them. Breeding efforts have been reviewed by Stein and Nothnagel (1995). In recent years, the major focus has been on the use of molecular markers for rapid and accurate breeding of resistant varieties. Using a mix of biparental and multiparental linked population studies, Le Clerc et al. (2015) detected 11 quantitative trait loci (QTLs) in carrot, which confer resistance to leaf blight (*Alternaria dauci*). The heritability of populations PC2 (75%) and PC3 (78%) was strong, showing that environmental factors had little influence on carrot resistance to *A. dauci*; however, the instability of QTL over time may be related to shifting environmental conditions. *Meloidogyne incognita* is a significant carrot pest in warmer regions of the globe, inflicting direct economic damage to the crop by galling and forking of the carrot roots, leaving an infected carrot unmarketable. Given the high cost and toxicity of soil-applied nematicides, genetic resistance presents an appropriate approach for RKN management. The *Mj-1* locus in carrot was found in the line 'Br1252' of a 'Brasilia' cultivar, and it confers resistance to *M. javanica* as a monogenic dominant trait (Simon et al. 2000). Simon et al. (2000) also discovered partial resistance to *M. incognita* in Br1252 and its derivatives, which could not be explained only by *Mj-1*. Recent mapping of a second *M. javanica* resistance gene, *Mj-2*, in PI 652188 from China revealed that it is distantly related to *Mj-1* on chromosome 8 (Ali et al. 2014a, b). Pearsons et al. (2015) mapped the genetic resistance to Meloidogyne incognita in carrots from three distinct regions: Syria (HM), Europe (SFF), and South America (Br1091). They created two F_2 mapping populations, (Br1091 × HM1) and (SFF × HM2), using these parents, as well as a segregating population obtained from self-pollination of an HM plant (HM3). Four QTLs associated with resistance were identified in Br1091 × HM1, three in SFF × HM2, and three in HM3. The consensus genetic map of three populations identified five non-overlapping QTLs for *M. incognita* resistance, one on carrot chromosomes 1, 2, 4, 8, and 9. In all three groups, one QTL was found in the same area of chromosome 8 as *Mj-1*, which confers resistance to *M. javanica*.

11.16 GWAS for Abiotic Stress Tolerance

The *PR-5* pathogenesis-related protein family includes osmotin and osmotin-like proteins, which are induced in a range of plant species in response to biotic and abiotic stresses. Ali et al. (2014a, b) employed *Agrobacterium*-mediated transformation to transform carrots with the tobacco osmotin gene, which results in a protein that lacks the vacuolar sorting motif and has a stretches of 20 different amino acid at the C-terminal end, according to their findings. All transformants wilted at a slower rate than wild-type plants under drought stress and recovered more rapidly after the drought was relieved. Transformants showed reduced hydrogen peroxide generation, lipid peroxidation, and electrolyte leakage under drought stress, as well as enhanced leaf water content. Stressors in the environment have a significant impact on plant growth and development. Because of these difficulties, higher plants have developed a diverse range of complicated molecular systems to adapt with them. It is important for different stages of developmental stages of growth in response to environmental challenges to have transcription factors with a basic helix-loop-helix (*bHLH*) structure. Chen et al. (2015) found 146 *DcbHLH TF*s in carrots using a genomic and transcriptome database. Using the former Arabidopsis thaliana classification approach, the *DcbHLH TF*s were divided into 17 subfamilies. There are 109 *DcbHLH* proteins related to DNA, according to several sequence alignments (*52 DcbHLHs* were G-box binders and 83 proteins were E-box binders). The number of *bHLH TF*s found in plants, fungi, and metazoans increased over this period. Two carrot cultivars, Junchuanhong and Kurodagosun, were found to express eight *DcbHLH* genes from subfamily 15 in three tissues as well as in response to four abiotic stimuli.

The transcription factors of the *AP2/ERF* family influence a number of physiological functions in plants, including plant development and stress response. There are 267 potential AP2/ERF factors in the carrot genome. Based on amino acid sequence similarity, the *AP2/ERF* proteins were phylogenetically classified into five subclasses. The *AP2/ERF* domain seems to have increased through time, based on the distribution and comparative genomics of *AP2/ERF* components in plants. There are 60 orthologous and 145 co-orthologous *AP2/ERF* gene pairings in carrot and Arabidopsis, as well as an orthologous gene interaction network. Abiotic stressor expression patterns are seen in eight *AP2/ERF* family genes from each subfamily (*DREB, ERF, AP2,* and *RAV*) (Li et al. 2015a, b).

11.17 Conclusion and Future Directions

It is indisputable that carrot linkage maps will continue to undergo development for the foreseeable future. By increasing the spacing between markers, the resulting maps will exhibit greater density and saturation, thereby accurately representing a larger proportion of the carrot genome. Genes of agricultural significance will continue to be localised. In the near future, gene mapping may target genes that control male sterility (including nuclear restorers of fertility in male-sterile cytoplasms), root phenotypes (e.g. Rf). At present, endeavors are underway to establish connections between structural genes involved in pigment biosynthetic pathways and principal colour genes, as well as quantitative trait loci (QTL) in carrots, via candidate gene analysis (Just 2004). By increasing the resolution of the mitochondrial genome in petaloid male-sterile carrots, it might be possible to discern prospective genes responsible for this particular trait. There is a strong likelihood that the mapping of the mitochondrial genome will also occur for male-fertile carrot lines, facilitating a comparative examination with the genome of male-sterile lines. The extension of marker assisted selection (MAS) integration into carrot breeding programmes could be facilitated through the development of converted markers that are linked to agriculturally significant traits. MAS is especially suitable for quantifying attributes that are costly, difficult, or time-intensive. There is growing interest in the linkage maps of Apiaceae species, which could potentially lead to comparative mapping efforts involving the nuclear genome of carrot. Engaging in this undertaking may enhance our understanding of the evolutionary lineage of agriculturally significant genes and the comprehensive genome organisation that underlies this family. Positional cloning of carrots is an imminent development. Considerable insert libraries have been established for the species. Positional cloning candidates may include genes that control the biosynthesis of carotenoids and anthocyanins, in addition to the root knot nematode resistance gene *Mj-1*. The ongoing utilisation of comparative transcriptome analyses in carrots shows potential in accelerating the discovery agriculturally significant genes. The development of genomics resources is likely to occur, encompassing microarrays, in situ hybridisation methodologies, transposon-tagging systems, and expressed sequence tag libraries.

References

Ali G, Rather AC, Ishfaq AS, Dar A, Wani S, Khan MN (2007) Gene action for grain yield and its attributes in maize (Zea mays L.). Int J Agric Sci 3(2):278

Ali A, Mathews WC, Cavagnaro PF, Iorizzo M, Roberts PA, Simon PW (2014a) Inheritance and mapping of *Mj-2*, a new source of root-knot nematode (*Meloidogyne javanica*) resistancein carrot. J Hered 105(2):288–291

Ali A, Rathore K, Crosby K (2014b) Overexpression of a tobacco osmotin gene in carrot (*Daucus carota* L.) enhances drought tolerance. In Vitro Cell Dev Biol - Plant 50:299–306

Bach IC, Olesen A, Simon PW (2002) PCR-based markers to differentiate the mitochondrial genome of petaloid and malefertile carrot (*Daucus carota* L.). Euphytica 127:353–365

Ballvora A, Hesselbach J, Niewohner J, Leister D, Salamini F, Gebhardt C (1995) Marker enrichment and high-resolution map of the segment of potato chromosome VII harbouring the nematode resistance gene Gro1. Mol Gen Genet 249:82–90

Becker J, Vos P, Kuiper M, Salamini F, Heun M (1995) Combined mapping of AFLP and RFLP markers in barley. Mol Gen Genet 249:65–73

Bendahmane A, Kanyuka K, Baulcombe DC (1997) Highresolution genetical and physical mapping of the Rx gene for extreme resistance to potato virus × in tetraploid potato. Theor Appl Genet 95:153–162

Boiteux LS, Fonseca MEN, Simon PW (1999) Effects of plant tissue and DNA purification method on randomly amplified polymorphicDNA-based genetic fingerprinting analysis in carrot. J Am Soc Hortic Sci 124:32–38

Boiteux LS, Belter JG, Roberts PA, Simon PW (2000) RAPD linkage map of the genomic region encompassing the rootknot nematode (*Meloidogyne javanica)* resistance locus in carrot. Theor Appl Genet 100:439–446

Boiteux LS, Hyman JR, Bach IC, Fonseca MEN, Matthews WC, Roberts PA, Simon PW (2004) Employment of flanking codominant STS markers to estimate allelic substitution effects of a nematode resistance locus in carrot. Euphytica 136:37–44

Borner T, Linke B, Nothnagel T, Scheike R, Schulz B, Steinborn R, Brennicke A, Stein M, Wricke G (1995) Inheritance of nuclear and cytoplasmic factors affecting male sterility in *Daucus carota*. Adv Plant Breed 18:111–122

Botstein D, White RL, Skolnick M, Davis RW (1980) Construction of a genetic linkage map in man using restriction fragment length polymorphisms. Am J Hum Genet 32:314–331

Bowes CE, Wolyn DJ (1998) Phylogenetic relationships among fertile and petaloid male-sterile accessions of carrot, *Daucus carota* L. Theor Appl Genet 96:928–932

Bradeen JM, Simon PW (1998) Conversion of an AFLP fragment linked to the carrot Y2 locus to a simple, codominant, PCR based marker form. Theor Appl Genet 97:960–967

Bradeen JM, Staub JE, Wye C, Antonise R, Peleman J (2001) Towards an expanded and integrated linkage map of cucumber (*Cucumis sativus* L.). Genome 44:111–119

Bradeen JM, Bach IC, Briard M, le Clerc V, Grzebelus D, Senalik DA, Simon PW (2002) Molecular diversity analysis of cultivated carrot (*Daucus carota* L.) and wild *Daucus* populations reveals a genetically nonstructured composition. J Am Soc Hortic Sci 127:383–391

Bradeen JM, Simon PW (2007) Carrot. In Kole C. Berlin (ed) Genome mapping and molecular breeding in plants. Springer-Verlag, pp 161–184

Bradeen JM, Vivek BS, Simon PW (1997) Detailed genetic mapping of the Y2 carotenoid locus in carrot. J Appl Genet 38A:28–32

Briard M, le Clerc V, Grzebelus D, Senalik D, Simon PW (2000) Modified protocols for rapid carrot genomic DNA extraction and AFLP analysis using silver stain or radioisotopes. Plant Mol Biol Report 18:235–241

Brigneti G, GarciaMas J, Baulcombe DC (1997) Molecular mapping of the potato virus Y resistance gene Rysto in potato. Theor Appl Genet 94:198–203

Bucci G, Kubisiak TL, Nance WL, Menozzi P (1997) A population consensus, partial linkage map of Picea abies Karst. based on RAPD markers. Theor Appl Genet 95:643–654

Buishand JG, Gabelman WH (1979) Investigations on the inheritance of color and carotenoid content in phloemand xylem of carrot roots (*Daucus carota* L.). Euphytica 28:611–632

Burr B, Burr FA (1991) Recombinant inbreds for molecular mapping in maize: theoretical and practical considerations. Trends Genet 7(2):55–60

Burr B, Burr FA, Thompson KH, Albertsen MC, Stuber CW (1988) Gene mapping with recombinant inbreds in maize. Genetics 118(3):519–526

Cavagnaro PF, Chung SM, Manin S et al (2011) Microsatellite isolation and marker development in carrot - genomic distribution, linkage mapping, genetic diversity analysis and marker transferability across Apiaceae. BMC Genomics 12:386. https://doi.org/10.1186/1471-2164-12-386

Cavagnaro PF, Iorizzo M, Yildiz M, Senalik D, Parsons J, Ellison S, Simon PW (2014) A gene-derived SNP-based high resolution linkage map of carrot including the location of QTL conditioning root and leaf anthocyanins pigmentation. BMC Genomics 15:1118

Chen Y-Y, Li M-Y, Xue-Jun W, Huang Y, Ma J, Xiong A-S (2015) Genome-wide analysis of basic helix2loop2helix family transcription factors and their role in responses to abiotic stress in carrot. Mol Breed 35:125

Cheng-Gang OU, Deng, BT, Bao SY, Zhao ZW, Hu H, Zhuang FY, Mao SM(2010) QTL mapping for contents of main carotenes and lycopene in carrot (Daucus carota L.). Hereditas 32(12): 1290–1295

Choumane W, Winter P, Weigand F, Kahl G (2000) Conservation and variability of sequence-tagged microsatellite sites (STMSs) from chickpea [Cicer arietinum(sic) L.] within the genus Cicer Theor Appl Genet101:269–278

Dallas JF (1988) Detection of DNA fingerprints of cultivated rice by hybridization with a human minisatellite DNA probe. Proc Natl Acad Sci USA 85:6831–6835

Davis TM, Yu H (1997) A linkage map of the diploid strawberry, Fragaria vesca. J Hered 88:215–221
Echt CS, Erdahl LA, McCoy TJ (1992) Genetic segregation of random amplified polymorphic DNA in diploid cultivated alfalfa. Genome 35:84–87
Edwards K, Johnstone C, Thompson C (1991) A simple and rapid method for the preparation of plant genomic DNA for PCR analysis. Nucleic Acids Res 19(6):1349. https://doi.org/10.1093/nar/19.6.1349
Edwards MD, Stuber CW, Wendel JF (1987) Molecular marker facilitated investigations of quantitative trait loci in maize. I. Numbers, genomic distribution and types of gene action. Genetics 115:113–125
Folkerts O, Hanson MR (1991) The male sterility-associated *pcf* gene and the normal *atp9-1* gene in petunia are located on different mitochondrial DNA molecules. Genetics 129:885–895
Freeman RE, Simon PW (1983) Evidence for simple genetic control of sugar type in carrot (*Daucus carota* L.). J Am Soc Hortic Sci 108:50–54
Goldman IL, Breitbach DN (1996) Inheritance of a recessive character controlling reduced carotenoid pigmentation in carrot (Daucus carota L.). J Hered 87(5):380–382
Grzebelus D, Senalik D, Jagosz B, Simon PW, Michalik B (2001) The use of AFLP markers for the identification of carrot breeding lines and F1 hybrids. Plant Breed 120:526–528
Grzebelus D, Jagosz B, Simon PW (2007) The DcMaster transposon display maps polymorphic insertion sites in the carrot (Daucus carota L.) genome. Gene 390:67–74
Grzebelus D, Simon PW (2009) Diversity of DcMaster-like elements of the PIF/Harbinger super family in the carrot genome. Genetica 135:347–353
Grzebelus D, Iorizzo M, Senalik D et al (2014) Diversity, genetic mapping, and signatures of domestication in the carrot (Daucus carota L.) genome, as revealed by Diversity Arrays Technology (DArT) markers. Mol Breed 33(3):625–637
Hallden C, Hansen M, Nilsson NO, Hjerdin A (1996) Competition as a source of errors in RAPD analysis. Theor Appl Genet 93:1188–1192
Helentjaris T (1987) A genetic linkage map for maize based on RFLPs. Trends Genet 3:217–221
Heun M, Helentjaris T (1993) Inheritance of RAPDs in F1 hybrids of corn. Theor Appl Genet 85:961–968
Huestis GM, McGrath JM, Quiros CF (1993) Development of genetic markers in celery based on restriction fragment length polymorphisms. Theor Appl Genet 85:889–896
Ichikawa H, Tanno-Suenaga L, Immamura J (1987) Selection of Daucus cybrids based on metabolic complementation between X-irradiated D. capillifolius and iodoacetamidetreated D. carota by somatic cell fusion. Theor Appl Genet 74:746–752
Ichikawa H, Tanno-Suenaga L, Immamura J (1989) Mitochondrial genome diversity among cultivars of Daucus carota (ssp. sativus) and their wild relatives. Theor Appl Genet 77:39–43
Janska H, Mackenzie SA (1993) Unusual mitochondrial genome organization in cytoplasmic male-sterile common bean and the nature of cytoplasmic reversion to fertility. Genetics 135:869–879
Jhang T, Kaur M, Kalia P, Sharma TR (2010) Efficiency of different marker systems for molecular characterization of subtropical carrot germplasm. J Agri Sci 148:171–181
Jiang C, Lewis ME, Sink KC (1997) Combined RAPD and RFLP molecular linkage map of asparagus. Genome 40:69–76
Just B (2004) Geneticmapping of carotenoid pathway structural genes and major gene QTLs for carotenoid accumulation in wild and domesticated carrot (*Daucus carota* L.). Dissertation, University of Wisconsin, Madison
Just BJ, Santos CAF, Fonseca MEN, Boiteux LS, Oloizia BB, Simon PW (2007) Carotenoid biosynthesis structural genes in carrot (*Daucus carota*): isolation, sequence-characterization, single nucleotide polymorphism (SNP) markers and genome mapping. Theor Appl Genet 114:693–704
Just BJ, Santos CAF, Yandell BS, Simon PW (2009) Major QTL for carrot color are positionally associated with carotenoid biosynthetic genes and interact epistatically in a domesticated x wild carrot cross. Theor Appl Genet 119:1155–1169

Karkleliene R, Bobinas E, Staniene G (2005) Combining ability of morphological traits andbiochemical parameters in carrot (*Daucus sativus*Röhl.) CMS lines. Biologija 3:15–18

Keilwagen J, et al (2017) The terpene synthase gene family of carrot (Daucus carota L.). Front Plant Sci 8:1930. https://doi.org/10.3389/fpls.2017.01930

Kennard WC, Poetter K, Dijkhuizen A, Meglic V, Staub JE, Havey MJ (1994) Linkages among RFLP, RAPD, isozyme, diseaseresistance, andmorphologicalmarkers in narrowandwide crosses of cucumber. Theor Appl Genet 89:42–48

King JJ, Bradeen JM, Bark O, McCallum JA, Havey MJ (1998) A low-density genetic map of onion reveals a role for tandem duplication in the evolution of an extremely large diploid genome. Theor Appl Genet 96:52–62

Koch T, Goldman IL (2005) Relationship of carotenoids and tocopherols in a sample of carrot root-color accessions and carrot germplasm carrying Rp and rp alleles. J Agri Food Chem 53:325–331

Kubo T, Nishizawa S, Sugawara A, Itchoda N, Estiati A, Mikami T (2000) The complete nucleotide sequence of the mitochondrial genome of sugar beet (*Beta vulgaris* L.) reveals a novel gene for tRNACys(GCA). Nucleic Acids Res 28:2571–2576

Kust AF (1970) Inheritance and differential formation of color and associated pigments in xymlem and phloem of carrot, *Daucus carota*. L. Dissertation, University of Wisconsin, Madison

le Clerc V, Briard M, Granger J, Delettre J (2003) Genebank biodiversity assessments regarding optimal sample size and seed harvesting techniques for the regeneration of carrot accessions. Biodivers Conserv 12:2227–2236

Le Clerc V, Marques S, Suel A, Huet S, Hamama L, Voisine L, Auperpin E, Jourdan M, Barrot L, Prieur R, Briard M (2015) QTL mapping of carrot resistance to leaf blight with connected populations: stability across years and consequences for breeding. Theor Appl Genet. https://doi.org/10.1007/s00122-015-2576-z

Li X, van Eck HJ, Rouppe van der Voort JNAM, Huigen DJ, Stam P, Jacobsen E (1998) Autotetraploids and genetic mapping using common AFLP markers: the R2 allele conferring resistance to *Phytophthora infestans* mapped on potato chromosome 4. Theor Appl Genet 96:1121–1128

Li Q, Li J, Liu S, Huang J, Lin H, Wang K et al (2015a) A comparative proteomic analysis of the buds and the young expanding leaves of the tea plant (*Camellia sinensis* L.). Int J Mol Sci 1:14007–14038

Li M-Y, Xu Z-S, Huang Y, Tian C, Wang F, Xiong A-S (2015b) Genome-wide analysis of AP2/ERF transcription factors in carrot (*Daucus carota* L.) reveals evolution and expression profiles under abiotic stress. Mol Gen Genomics. https://doi.org/10.1007/s00438-015-1061-3

Linke B, Nothnagel T, Borner T (2003) Flower development in carrot CMS plants: mitochondria affect the expression of MADS box genes homologous to GLOBOSA and DEFICIENS. Plant J 34:27–37

Mackill DJ, Zhang Z, Redona ED, Colowit PM (1996) Level of polymorphism and genetic mapping of AFLP markers in rice. Genome 39:969–977

Meksem K, Leister D, Peleman J, Zabeau M, Salamini F, Gebhardt C (1995) A high-resolution map of the vicinity of the R1 locus on chromosome V of potato based on RFLP and AFLP markers. Mol Gen Genet 249:74–81

Michelmore RW, Paran I, Kesseli RV (1991) Identification of markers linked to disease-resistance genes by bulked segregant analysis: a rapid method to detect markers in specific genomic regions by using segregating populations. Proc Natl Acad Sci USA 88:9828–9832

Mudge J, Andersen WR, Kehrer RL, Fairbanks DJ (1996) A RAPD genetic map of Saccharum officinarum. Crop Sci 36:1362–1366

Nakajima Y, Oeda K, Yamamoto T (1998) Characterization of genetic diversity of nuclear and mitochondrial genomes in *Daucus* varieties by RAPD and AFLP. Plant Cell Rep 17:848–853

Nakajima Y, Yamamoto T, Muranaka T, Oeda K (1999) Genetic variation of petaloid male-sterile cytoplasm of carrots revealed by sequence-tagged sites (STSs). Theor Appl Genet 99:837–843

Nicolle C, Simon G, Rock E, Amouroux P, Remesy C (2004) Genetic variability influences carotenoid, vitamin, phenolic, and mineral content in white, yellow, purple, orange, and dark-orange carrot cultivars. J Am Soc Hort Sci 129:523–529

Nakajima Y, Yamamoto T, Muranaka T, Oeda K (2001) A novel *orfB*-related gene of carrot mitochondrial genomes that is associated with homeotic cytoplasmic male sterility (CMS). Plant Mol Biol 46:99–107

Niemann M, Westphal L, Wricke G (1997) Analysis of microsatellite markers in carrot (Daucus carota L. sativus). J Appl Genet 38A:20–27

Nothnagel T (1992) Results in the development of alloplasmic carrots (Daucus carota var. sativusHoffm). Plant Breed 109:67–74

Paran I, Rouppe van der Voort J, Lefebvre V, Jahn M, Landry L, van Schriek M, Tanyolac B, Caranta C, Chaim AB, Livingstone K, Palloix A, Peleman J (2004) An integrated genetic map of pepper (*Capsicum* spp.). Mol Breed 13:251–261

Parsons J, Matthews W, Iorizzo M, Roberts P, Simon P (2015) *Meloidogyne incognita* nematode resistance QTL in carrot. Mol Breed 35:114. https://doi.org/10.1007/s11032-015-0309-2

Paterson AH, Lander, ES, Hewitt JD, Peterson S, Lincoln SE, Tanksley SD (1988) Resolution of quantitative traits into Mendelian factors using a complete RFLP linkage map. Nature 335(6192):721–726. https://doi.org/10.1038/335721a0

Pierre MDS, Bayer RJ (1991) The impact of domestication on the genetic variability in the orange carrot, cultivated *Daucus carota* ssp. *sativus* and the genetic homogeneity of various cultivars. Theo Appl Genet 82:249–253

Plunkett GM, Soltis DE, Soltis PS (1997) Clarification of the relationship between Apiaceae and Araliaceae based on *matK* and *rbcL* sequence data. Am J Bot 84:565–580

Riedy MF, Hamilton WJ, Aquadro CF (1992) Excess of nonparental bands in offspring fromknown primate pedigrees assayed using RAPD PCR. Nucleic Acids Res 20:918

Robison MM, Wolyn DJ (2002) Complex organization of the mitochondrial genome of petaloid CMS carrot. Mol Genet Genomics 268:232–239

Rubatzky VE, Quiros CF, Simon PW (1999) Carrots and related vegetable umbelliferae. Crop production science in horticultural series: 10. CABI Publishers, New York

Rurek M, Szklarczyk M, Adamczyk N, Michalik B, Augustyniak H (2001) Differences in editing of mitochondrial *nad3* transcripts from CMA and fertile carrots. Acta Biochim Polonica 48:711–717

Saenz Lain C (1981) Research on *Daucus* L. (*Umbelliferae*). Anal Jardin Bot Madrid 37:481–533

Santos CAF, Simon PW (2002) QTL analyses reveal clustered loci for accumulation of major provitamin A carotenes and lycopene in carrot roots. Mol Gen Genomics 268:122–129

Santos CAF, Simon PW (2002a) QTL analyses reveal clustered loci for accumulation ofmajor provitamin A carotenes and lycopene in carrot roots. Mol Genet Genomics 268:122–129

Santos CAF, Simon PW (2002b) Some AFLP amplicons are highly conserved DNA sequences mapping to the same linkage groups in two F2 populations of carrot. Genet Mol Biol 25:195–201

Santos CAF, Simon PW (2004) Merging carrot linkage groups based on conserved dominant AFLP markers in F2 populations. J Am Soc Hortic Sci 129:211–217

Schulz B, Westphal L, Wricke G (1993) Linkage groups of isozymes, RFLP and RAPD markers in carrot (*Daucus carota* L. *sativus*). Euphytica 74:67–76

Schulz B, Westphal L, Wricke G (1994) Linkage groups of isozymes, RFLP and RAPD markers in carrot (*Daucus carota* L. *sativus*). Euphytica 74:67–76

Sebastian RL, Howell EC, King GJ, Marshall DF, Kearsey MJ (2000) An integrated AFLP and RFLP Brassica oleracea linkage map from two morphologically distinct doubled-haploid mapping populations. Theor Appl Genet 100:75–81

Shim SI, Jørgensen RB (2000) Genetic structure in cultivated and wild carrots (Daucus carota L.) revealed by AFLP analysis. Theor Appl Genet 101:227–233

Simon PW (1996) Inheritance and expression of purple and yellow storage root colour in carrot. J Hered 87(1):63–66

Simon PW (2000) Domestication, historical development, and modern breeding of carrot. Plant Breed Rev 19:157–190

Simon PW, Rubatzky VE, Bassett MJ, Strandberg JO, White JM (1997) B7262, purple carrot inbred. HortScience 32:146–147
Simon PW, Matthews WC, Roberts PA (2000) Evidence for simply inherited dominant resistance to *Meloidogyne javanica* in carrot. Theor Appl Genet 100:735–742
Sondur SN, Manshardt RM, Stiles JI (1996) A genetic linkage map of papaya based on randomly amplified polymorphic DNA markers. Theor Appl Genet 93:547–553
St. Pierre MD, Bayer RJ, Weis IM (1990) An isozyme-based assessment of the genetic variability within the *Dau cus carota complex* (*Apiaceae:* Caucalideae). Can J Bot 68:2449–2457
Stam P (1993) Construction of integrated genetic linkage maps by means of a new computer package: JoinMap. Plant J 3:739–744
Stein M, Nothnagel T (1995) Some remarks on carrot breeding (*Daucus carota sativus* Hoffm.). Plant Breed 114:1–11
Sun T, Simon PW, Tanumihardjo SA (2009) Antioxidant phytochemicals and antioxidant capacity of biofortified carrots (Daucus carota L.) of various colors. J Agric Food Chem 57:4142–4147
Szklarczyk M, Oczkowski M, Augustyniak H, Borner T, Linke B, Michalik B (2000) Organisation and expression of mitochondrial atp9 genes from CMS and fertile carrots. Theor Appl Genet 100:263–270
Thormann CE, Ferreira ME, Camargo LEA, Tivang JG, Osborn TC (1994) ComparisonofRFLP and RAPD markers for estimating genetic relationships within and among cruiferous species. Theor Appl Genet 88:973–980
Vivek BS, Simon PW (1999a) Linkage relationships among molecularmarkers and storage root traits of carrot (Daucus carota L. ssp. sativus). Theor Appl Genet 99:58–64
Vivek BS, Simon PW (1999b) Phylogeny and relationships in Daucus based on restriction fragment length polymorphisms (RFLPs) of the chloroplast and mitochondrial genomes. Euphytica 105:183–189
Vuylsteke M, Mank R, Antonise R, Bastiaans E, Senior ML, Stuber CW, Melchinger AE, Lubbersted T, Xia XC, Stam P, Zabeau M, Kuiper M (1999) Two high-density AFLP linkage maps of Zeamays L.: analysis of distribution of AFLP markers. Theor Appl Genet 99:921–935
Weising K, Weigand F, Driesel A, Kahl T, Zischler H, Epplen JT (1989) Polymorphic GATAIGACA repeats in plant genomes. Nucl Acids Res 17:10–28
Westphal L, Wricke G (1991) Genetic and linkage analysis of isozyme loci in Daucus carota L. Euphytica 56:259–267
Westphal L, Wricke G (1989) Genetic analysis of DIA, GOT and PGI isozyme loci in Daucus carota L. sativus. Plant Breed 102:51–57
Westphal L, Wricke G (1997) Construction of a linkage map of Daucus carota L. sativus and its application for the mapping of disease resistance and restorer genes. J Appl Genet 38A:13–19
Winter P, Pfaff T, Udapa SM, Hiittel B, Sharma PC, Sahi S, Arreguin-Espinoza R, Weigand F, Muehlbauer FJ, Kahl G (1999) Characterization and mapping of sequence-tagged microsatellite sites in thechickpea (Cicer areitinurn L.) genome. Mol Gen Genet 262:90–101
Xu ZS, Tan HW, Wang F, Hou XL, Xiong AS (2014a). CarrotDB: a genomic and transcriptomic database for carrot. Database 2014:bau096
Xu Z-S, Huang Y, Wang F, Song X, Wang G-L, Xiong A-S (2014b) Transcript profiling of structural genes involved in cyanidin-based anthocyanin biosynthesis between purple and non-purple carrot (*Daucus carota* L.) cultivars reveals distinct patterns. BMC Plant Biol 14:262
Xu ZS, Feng K, Xiong AS (2019) CRISPR/Cas9-mediated multiply targeted mutagenesis in orange and purple carrot plants. Mol Biotechnol 61(3):191–199
Yamato K, Ogura Y, Kanegae T, Yamada Y, Ohyama K (1992) Mitochondrial genome structure of rice suspension culture from cytoplasmic male-sterile line (A-58CMS): reappraisal of the master circle. Theor Appl Genet 83:279–288
Yang XF, Quiros CF (1995) Construction of a genetic linkage mapin celery using DNA-basedmarkers. Genome 38:36–44
Yau Y, Simon PW (2003) A 2.5-kb insert eliminates acid soluble invertase isozyme II transcript in carrot (Daucus carota L.) roots, causing high sucrose accumulation. Plant Mol Biol 53:151–162

Yau Y, Santos K, Simon PW (2005) Molecular tagging and selection for sugar type in carrot roots with codominant, PCR-based markers. Mol Breed 16:1–10

Yildiz M, Willis DK, Cavagnaro PF, Iorizzo M, Abak K, Simon PW (2013) Expression and mapping of anthocyanin biosynthesis genes in carrot. Theor Appl Genet 126:1689–1702

Young ND (1994) Constructing a plant genetic linkage map with DNA markers. In: Phillips RI, Vasil IK (eds) DNA-based markers in plants, pp 39–57

QTL Mapping in Onion and Garlic

12

Priyanka Anand, Karanveer Kaur, Neha Verma, and Jiffinvir Khosa

Abstract

Onion and garlic are major *Allium* vegetables grown worldwide and valued for their culinary, nutritional, and health benefits. However, genetic and molecular improvements in these crops have been challenging due to their complex genetic makeup characterized by high inbreeding depression, biennial life cycles, outcrossing nature, along with large and complex genomes. Understanding recent advancements in *Allium* molecular genetics is vital for advancing production, quality, and disease resistance. This chapter provides an in-depth review of *Allium* crop genetics with a particular focus on quantitative trait loci (QTL) mapping of key agronomic traits, highlights the existing challenges, and explores future opportunities for advancing breeding strategies. Utilization of advanced genomic technologies like genetic mapping, RNA sequencing, and third-generation long-read sequencing, along with new population genomics strategies offers exciting opportunities for accelerating genetic gains in *Allium* crops.

Keywords

Onion · Garlic · Mapping population · Single-nucleotide polymorphism · Quantitative trait loci · Flavour · Recombinant inbred lines

P. Anand · K. Kaur · N. Verma · J. Khosa (✉)
Department of Vegetable Science, Punjab Agricultural University, Ludhiana, Punjab, India
e-mail: jiffinvir@pau.edu

S. S. Dey et al. (eds.), *GWAS and QTL Mapping in Horticultural Crops*,
https://doi.org/10.1007/978-981-95-2306-1_12

12.1 Introduction

Allium vegetables, a diverse group belonging to family Alliaceae, include species such as onion, garlic, leeks, and chives. They are known for their aromatic attributes arising from the enzymatic breakdown of flavour precursors, notably S-alk(en)yl cysteine sulfoxides (Jones 2004; Randle 1992). These compounds release sulfur compounds and bioactives with preventive and therapeutic properties against various human ailments (Liguori et al. 2017; Bisen and Emerald 2016). Since domestication, traditional breeding techniques have been employed to enhance onion production, quality, and resistance to biotic stresses, enabling cultivation across diverse regions (Khosa et al. 2016a). However, this process is often slow and primarily focused on plant phenotype, which can be influenced by external factors, leading to genotype × environment interactions that complicate the accurate prediction of desired traits (Bharti and Chimata 2019; Lema 2018).

The current understanding of genetic and genomic makeup in case of onion and garlic is inadequate in comparison to other crops (Havey 2018). There is a requirement to improve *Allium* breeding programmes through the integrated use of genomics technologies in order to meet consumer expectations in the upcoming years (McCallum et al. 2007a, b; Havey 2018). But challenges such as large genome size, biennial life cycle, high inbreeding depression, and small research communities have impeded molecular tool development. To advance onion and garlic improvement programmes, breeders must focus on accelerating the process by utilizing genetic diversity to enhance quality, yield, stress tolerance, and adaptability to changing climates, incorporating new insights, and modern molecular tools (Khosa et al. 2016a). The advancement of genomic resources, including homozygous inbred populations, molecular markers, and saturated linkage maps, enables the identification of the genetic basis of phenotypes and improves breeding efficiency by means of marker-assisted selection (Havey 2018; Malik et al. 2021). With advancements in next-generation sequencing, exciting opportunities are emerging for accelerated onion breeding programmes, benefiting both farmers and consumers. A collaborative effort produced the first de novo onion genome assembly with 10.6 Gb DNA, while recent chromosome-level assemblies for Welsh onion, garlic, and onion offer crucial genetic insights for *Allium* crop breeding (Finkers et al. 2021; Hao et al. 2023). This chapter provides an overview of the current state of knowledge in *Allium* crops (onion and garlic) genetics of various traits, highlighting challenges and opportunities for future advancements in breeding strategies.

12.2 Mapping Populations

In agriculture, different kinds of populations such as F_2 generations, recombinant inbred lines (RILs), near-isogenic lines (NILs), and diversity sets have been used for mapping different traits (Collard et al. 2005). Onion faces challenges in developing inbred populations due to severe inbreeding depression. Typically, F_2 families, sometimes derived as doubled haploids or mostly through inbreeding, have been used for onion genetics, segregating for economically important traits in diverse long- and short-day inbreds. Generating large experimental stocks without inbreeding is achieved by developing multiple large F_2 families through mass pollination of clonal hybrid bulbs from wide crosses to doubled haploid lines (Baldwin et al. 2014). An alternative strategy for onion and garlic researchers involves utilizing a diverse panel or core collection of germplasm for association mapping purposes (Jia et al. 2023). Employing association mapping and genomic selection strategies is crucial for a better understanding and improvement of polygenic traits, as demonstrated in other crop plants for enhancing genetic gains (Varshney et al. 2014).

12.3 Gene/QTL Mapping

Significant progress has been achieved in generating genetic maps through both intra- and interspecific crosses over the last decade, particularly in a small number of *Allium* species. The first genetic map of onions was developed, relying on co-dominant molecular markers within the F_2 generation (BYG15-23 × AC43) (King et al. 1998a, b). Subsequent studies investigated the construction of onion genetic linkage maps, using next-generation sequencing (NGS) technology (Baldwin et al. 2012; Choi et al. 2020; Duangjit et al. 2013; Jo et al. 2017; Labate et al. 2020). These maps shed light on the genomic structure, sequencing, and linkage/recombination processes of the organisms studied. This data assists breeders in estimating recombination outcomes for various breeding procedures, including hybridization and self-fertilization. Furthermore, knowing the physical location of their genes of interest helps plant biotechnologists isolate and transfer genes through genetic transformation. Notably, over the past decade, key genes in the *Allium* genus, particularly in onion, Japanese bunching onion, and the wild species *Allium roylei* Stearn. have been effectively traced to specific chromosomal locations. These include factors such as male sterility (S/T/N cytotype, *Ms/ms* locus) (Gökçe et al. 2002), fertility restoration (*Rf/rf*) (Yamashita et al. 2005), and quality attributes such as pungency (sulfur content) (Masamura et al. 2012; McCallum et al. 2007a, b), dry matter content (McCallum et al. 2006), antioxidant/medicinal potential (quercetin levels) (Tsukazaki et al. 2011), bulb pigments (anthocyanins/flavonoids) (Masuzaki et al. 2006), adaptive traits (McCallum et al. 2016), and biotic resistance (Scholten et al. 2016). These findings now allow for the quick production of elite cultivars adapted for specific agronomic or industrial purposes using marker-assisted breeding and genetic transformation. Enhanced understanding of inheritance patterns and trait relationships has reduced onion breeders' reliance on potentially misleading phenotypic assessments. Although ongoing efforts to map genes, markers, and quantitative trait loci (QTLs) on genomes aid in constructing genetic maps, the task is challenging for *Allium* species due to their extensive and repetitive genomes, making it difficult to establish clear linkages and correlations, unlike many other common vegetable and cereal crops.

NGS platforms, such as SSRs and SNPs, are increasingly widely used for marker discovery. Baldwin et al. (2012) used NGS to discover 921 SSRs from a 6.6-Mbp genomic sequence to build advanced genomic SSR markers for a double haploid population. Onion whole-genome sequencing approaches have produced a large number of SNPs, with frequent detection aided by sophisticated genotyping technology and twofold haploid techniques (Finkers et al. 2021; Kim et al. 2015a, b; Shigyo et al. 2019). Validated SNPs in extensive onion germplasm were mapped onto the existing linkage map (Duangjit et al. 2013). High-throughput genotyping methods such as genotyping-by-sequencing (DArTseq) and double digest restriction site-associated DNA sequencing (ddRAD-seq) have been used for diversity analysis and SNP discovery in garlic and Korean short-day onions, respectively (Egea et al. 2017; Lee et al. 2018), assisting in the development of precise linkage maps and the management of genetic resources for onions and related *Alliums*.

A joint initiative involving Wageningen University and Research and private seed companies in the Netherlands led to the initial de novo assembly of the onion genome, revealing 10.6 Gb of DNA with abundant ancient repeats alongside a majority of eukaryotic core genes (Finkers et al. 2021). Recently, researchers have made a significant contribution to *Allium* crop breeding by overcoming the hurdle of lacking high-quality reference genomes. This is achieved by creating chromosome-level genome assemblies for three crucial *Allium* crops: Welsh onion, garlic, and onion (Hao et al. 2023). These genomes are large, measuring 11.17 Gb, 15.52 Gb, and 15.78 Gb, respectively. The remarkable contig N50 values for Welsh onion, garlic, and onion reflect the quality and consistency of the assembled genomic sequences. These genome assemblies, crucial for future research and breeding in *Allium* crops, establish a connection by providing valuable genetic information to breeders for planning crossbreeding programmes and understanding the genetics of key traits in *Alliums*.

12.3.1 Male Sterility and Fertility Restoration

Allium hybrid breeding programmes rely on cytoplasmic male sterility (CMS), highlighting the need for improved genetic resources and techniques to help in cytotypes and nuclear restorer loci identification. CMS exhibits spontaneous rearrangements of mitochondrial genomic sequences, which result from emergence of novel open reading frames (ORFs) within mitochondrial chimeric mutants. These mutations characterized by their variable sequence, function, and structure resulting in shifting of sub-stoichiometric (Hanson and Bentolila 2004). Therefore, pollen production is disrupted and rendering the plant unable to produce functional pollen grains. However, nuclear encoded restorer of fertility (*Rf*) genes can restore pollen fertility in male sterile cytoplasm by modulating the CMS ORFs expression via post-transcriptional or post-translational modifications (Tang et al. 2017). The CMS was first utilized in the middle of twentieth century in onion for the development of F_1 hybrids (McCallum et al. 2007a, b). Hybrid onion seeds are produced utilizing different cytotypes (CMS-S, CMS-R, and CMS-T) (Havey and Kim 2021). S cytoplasm is the most commonly used cytoplasm worldwide including India (Jones and Clarke 1943). In CMS, a three-line system for hybrid seed production is required, viz., male sterile (A-line; S*msms*), maintainer (B-line; N*msms*), and pollinator (C/R-line; N*msms*, N*Msms*, S*Msms*, and S/N*MsMs*) lines (Jones and Clarke 1943). However, due to onion biennial generation time, classical crossing, and high inbreeding depression, segregation analysis takes years (>10 years) to establish cytoplasm (S/N) and genotype at *Ms* (*msms/Msms/MsMs*). Marker-assisted selection can significantly accelerate this process by identifying the cytotype and fertility restoration locus at the seedling stage (Von Kohn et al. 2013; Havey and Kohn 2017). Researchers reported different markers to identify the cytotypes (*N/S/T*) and fertility restorer gene(s) worldwide in onion. Initially, the use of RFLP profiling of chloroplast and mitochondrial DNA classified the cytoplasm of onion into two groups, viz., M (N and T cytoplasm) and S (S cytoplasm) groups (Havey 1993). The N and T cytoplasm were found to be autoplasmic and possibly introgressed or originating from *Allium vavilovii*, whereas the S cytoplasm is likely to be derived from interspecific cross. However, due to high cost and low polymorphism, RFLP was not employed for further studies (McCallum et al. 2007a, b; Khosa et al. 2016a). Later, Havey suggested that 'Pran' (*Allium* × *cornutum*), a triploid viviparous onion, might be the source of S cytoplasm, while T cytoplasm was speculated to have been acquired from Japanese or Dutch populations (Havey 2000).

Organelle mutation (associated with S cytoplasm) molecular genetic analysis led to the development of PCR-based assay for differentiating normal and sterile cytotype based on chloroplast and mitochondria (Havey 1995; Sato 1998; Engelke et al. 2003). Kim et al. (2009) also developed another PCR marker based on *orf725* and *cox1* relative copy numbers, which enables the differentiation of S/N/T cytoplasm using single PCR marker (Kim et al. 2009). The chloroplast genome sequencing and annotation led to the identification of *accD* marker that targets small insertions and deletions (*indels*) in the *accD* gene to distinguish N and S cytoplasms (Von Kohn et al. 2013). Advancement in sequencing technologies resulted in identification of N and T cytotype variants, viz., Cytotype X and Y (Kim et al. 2015a, b; Kim et al. 2019). Further research in cytotypes of diverse geographic regions has led to the identification of new variants of T cytoplasm (T, T-like, and Y; Kim et al. 2019). In 2021, Havey and Kim proposed four different cytotypes (S, R, T, and N), of which S and R cytoplasms being widely used, while T cytoplasm has limited commercial utility. Various new markers, viz., PCR based (Kim et al. 2009; Von Kohn et al. 2013), SNP (Havey 2024), and HRM (Khrustaleva et al. 2023; Ahn and Kim 2023) linked to the cytoplasm were identified and utilized due to their wide genomic distribution, high reproducibility, multi-allelic nature and ease to score. Novel forms of cytoplasmic male sterility (CMS) have emerged in onion by incorporating cytoplasm from *Allium galanthum*, leading to unique male-sterility characteristics among onion populations (Havey 1999). The exploitation of *Allium roylei* to develop new male sterile cytoplasmic lines was successful, and *A. roylei* can be used as a new source of male sterility (Vu et al. 2011). A recently identified SNP specific to *Allium galanthum* cytoplasm (Gal-CMS) has led to the development of a precise KASP assay capable of distinguishing Gal-CMS from other cytotypes (Havey 2024).

Cytoplasm identification is the first step in male sterile line isolation, followed by identification of fertility restoration (*Ms*) locus in onion OP populations for the isolation of maintainer (N*msms*) as well as restorer of fertility (N*MsMs*/N*Msms*) lines. The male fertility of CMS-S, CMS-R, and CMS-T restored by single *Rf* locus (*Ms*; Jones and Clark 1943), interaction among three nuclear loci (Schweisguth) and *Ms* locus (Kim 2014; Kim et al. 2015a, b), respectively. In *Allium* genetics, detailed and high-resolution restorer loci mapping stands out as an exception, as most mapping studies exhibit low resolution and power. The *Ms* locus was initially mapped in BYG15-23 × Alisa Craig population to interval of 28 cM flanked by markers RFLP (AOB210-H3-6.6/5.5 and API65-E5-10.0) on chromosome 2 (King et al. 1998a, b; Martin et al. 2005). However, subsequent improvement in test-cross families phenotyping as well as marker coverage resulted in change of linkage order and a marker located at 0.9 cM from the *Ms* locus (Gökçe et al. 2002). Subsequently, co-dominant CAPS markers, viz., OPT/PsaO (Bang et al. 2013) and *jnurf05/jnurf17* (Park et al. 2013) and SCAR markers (DNF-566 AND RNS 357) (Yang et al. 2013; Kim 2014) that co-segregate with *Ms* locus were identified. Markers PsaO/OPT developed from AGF136 and AOB272 linked to *Ms* locus at 6.4 and 1.5 cM distance, respectively (Bang et al. 2011). Furthermore, Park and coworkers in 2013 identified CAPS markers (*jnurf05* and *jnurf17*) closely linked to *Ms* locus, with *jnurf05* estimated to be only 0.05 cM distant (Park et al. 2013). These markers can be utilized to identify the linkage disequilibrium of *Ms* locus-specific alleles in segregating families. The genetic markers linked to *Ms* locus have been identified and mapped on chromosome 2 of onion. Havey in 2013 screened near isogenic lines with 930 SNPs and identified three linked SNP on chromosome 2 which showed strong linkage disequilibrium with locus *Ms* in both open pollination (OP) and inbred onion populations. Two SCAR markers were developed by Yang et al. (2013) based on hiTAIL-PCR and AFLP to identify *Ms* locus. Among 14 PCR markers, *jnurf13* and *jnurf12* were developed, with *jnurf13* showing linkage disequilibrium with the *Ms* locus in diverse genetic background (Kim 2014). In addition, *Ms* locus has been tagged and new markers based on PCR, viz., AcPMS1 and *AcSKP1* were developed (Huo et al. 2015; Kim et al. 2015a, b). A combined BSA and RNA-Seq techniques identified a molecular marker (*AcPMS1*) linked with *PMS1* candidate gene which involved in DNA mismatch repair and govern fertility restoration (Kim et al. 2015a, b). Identification of multiplex PCR-based co-dominant marker (*AcSKP1*) consisting four compatible primers, viz., FD898, FU898, SD628, and SU628, facilitates the differentiation of various alleles in OP population and showed LD with *Ms* locus (Huo et al. 2015). Utilization of these markers for MAS of *Ms* locus in Brazilian and Indian onion germplasm showing a difference between the molecular marker targeting the *Ms* locus and male fertility phenotype (Khar and Saini 2016). Later, the Tyramide FISH method was used to develop the physical map of *Ms* region of chromosome 2 by developing the intron spanning probes from the SNP marker identified by Duangjit et al. (2013). The physical map showed that markers were scattered in proximal centromere of chromosome 2 in the long arm region of lower recombination (Khrustaleva et al. 2016). This elucidates that the indiscrepancy of the linked markers observed to detect the *Ms* locus in majority of onion populations (Khar and Saini 2016; Ferreira and Santos 2018) may be due to the lower recombination event, possibility of another fertility restoration locus, and challenges in transferring these markers in out breeding crop having poorly characterized genome (Havey and Kohn 2017). However, Havey and Kohn (2017) corroborated that marker, viz., *AcPMS1* and *jnurf*13 accurately predicted *Ms* locus in North American onion populations, with *AcPMS1* performing better amongst two. Additionally, Kim and Kim et al. (2021) identified a unique *AcPMS1* allele in the onion breeding line ORI854, which is scarcely found within onion germplasm. However recently, a new *Rf* locus (*Ms2*) identified via utilizing the BSA-based RNA-Seq approach at chromosome 2, showing reduced transcription of orf725 transcripts in male fertile F_2 individuals, unlike locus *Ms* which did not effect orf725 transcription. Therefore, *Ms2* male fertility mechanism differs from *Ms* locus at the transcription level. Utilization of the identified PCR markers at preliminary stage can accelerate the process of identification of maintainer and male sterile lines. These markers are especially important in regions where hybrids of onion are still not widely commercialized. With the availability of reference onion genome (Finkers et al. 2021), extensive study on male sterility systems across diverse geographic locations may be useful for developing markers which shows linkage disequilibrium with locus *Ms*.

12.3.2 Quality Characteristics

Onions, known for their culinary appeal and health benefits, are primarily made up of water (~90%), carbohydrates (7–9%), dietary fibre (1–2%), and essential nutrients such as vitamin C, B6, folate, and potassium. Furthermore, onions contain a variety of beneficial compounds, including flavonoids, fructans, fructooligosaccharides (FOS), allicin, and thiosulfate, which have been linked to a variety of health benefits, including antioxidant, anticarcinogenic, antimicrobial, antiplatelet, antithrombotic, and cardioprotective effects (Suleria et al. 2015). These chemicals, as well as the bulb's carbohydrate profile (which includes glucose, fructose, sucrose, and fructans) and sulfur content, have a substantial impact on its viability for a variety of applications. Understanding quantitative trait loci (QTLs), pinpointing the associated genes influencing variation in these beneficial compounds, and elucidating their molecular pathways can equip plant breeders with the information required to develop cultivars abundant in crucial nutraceutical characteristics.

12.3.2.1 Onion Flavour and Pungency

The pungency in onion bulbs is due to sulfur compounds and pungent flavour results from the enzymatic conversion of trans-S-1-propenyl-L-cysteine sulfoxide (*PRENCSO*) to onion lachrymatory factor (*LF*) through successive reactions catalysed by lachrymatory factor synthase (*LFS*) (Imai et al. 2002; McCallum et al. 2007a). These compounds affect the intensity of onion flavour, which can be measured either by the production of pyruvic acid or the presence of the lachrymatory factor (*LF*) (Griffiths et al. 2002). Organosulfur compounds of *Alliums* contribute to their flavour and other bioactivities (Block 2010). Various researchers found that the levels of organic sulfur compounds depend on the regulation of genes in the S assimilation pathway in *Allium* (McCallum et al. 2007a, b; Sun et al. 2012). A molecular marker map targeting sulfur assimilation genes identified genomic regions influencing onion pungency in the 'W202A' × 'Texas Grano 438' cross (McCallum et al. 2007a, b). Key genes such as plastidic ferredoxin-sulfite reductase (*SiR*) and plastidic ATP sulfurylase (*ATPS*) on chromosome 3 were closely linked (1–2 cM). QTL analysis linked *ATPS* and *SiR* loci with bulb pungency, explaining 30–50% of genetic variation. These markers offer a practical way to select for lower pungency without affecting bulb solids content. Numerous studies highlighted environmental factors, including location, temperature, and sulfur availability during bulb production, impacting overall pungency or the levels of flavour precursors (Bedford 1984; Coolong and Randle 2003; Lancaster et al. 1998; Platenius and Knott 1935; Randle 1992; Vavrina and Smittle 1993). Major QTL for pseudostem pungency had been mapped on chromosome 2a in bunching onion (Tsukazaki et al. 2011).

The production of flavour precursors is still poorly known (Jones 2004), limiting the use of candidate gene techniques. The study of onion genetics reveals important chromosomal areas that affect pungency. Several candidate genes related to sulfur assimilation were revealed including serine acetyl-transferase (McManus et al. 2005), ATP sulfurylase (ATPS), and sulfite reductase (McCallum et al. 2007a, b) by QTL analysis. ATPS and *SiR* loci on chromosome 3 were found to have significant relationships with bulb pungency, presenting possible markers for selecting low-pungency onions (McCallum et al. 2007a, b). Several candidate genes related to sulfur assimilation were revealed in genus *Allium* (Brewster 2008; Kamenetsky et al. 2015; McManus et al. 2005, 2012; Sun et al. 2016).

LFS enzyme responsible for tearing action on chromosome 5A was initially mapped using alien monosomic addition lines and later confirmed through an interspecific cross of *A. cepa* × *A. roylei* located on the onion's chromosome 5 (Masamura et al. 2012). Co-segregation with chromosome 5 markers was identified via heteroduplex analysis of *LFS* amplicons. Development of onion EST-based markers, including the SNP marker 'ACP052', further linked to the *LFS* marker. Screening of two BAC clones harbouring the *LFS* gene revealed eight positive clones, indicating at least two *LFS* loci. Fluorescence in situ hybridization (FISH) karyotype analysis mapped these clones to the proximal region of chromosome 5's long arm, with specific positions noted for clones 2E8/10 and 4F10/155. Two distinct haplotypes, 'haplotype 1' (4F10/155) and 'haplotype 2' (2E8/10), were identified through double haploid analysis, showing variation in the upper flanking region of the *LFS* gene. This confirms the presence of at least two potential loci near the dry matter QTL on chromosome 5. During the development of flavour in *Allium* crops, the considerable duplication and differentiation of lachrymatory factor synthase and alliinase appear as significant evolutionary events (Edwards et al. 2022). 'Tear-less' onions were developed by anti-sense downregulation of *LFS* (Eady et al. 2008). An advantage of these genetically engineered 'non-tear-inducing' onions was the increased presence of secondary sulfur compounds known for their health-promoting properties (Aoyagi et al. 2011; Eady et al. 2008; Thomson et al. 2013). Low tearing action of onion were also achieved by mutagen treatment (Kato et al. 2016).

Flavonoid molecular regulation in bulb onions is not as thoroughly investigated as in other crops, despite their importance in diets and economies. RNA-Seq data has been utilized to pinpoint essential genes involved in the synthesis of flavonoids (Schwinn et al. 2016). In *Allium fistulosum*, transcriptome study for the biosynthesis pathway of flavonoids resulting in presence of unigene responsible for flavonoid biosynthesis on chromosome 5A (Abdelrahman et al. 2019). Qin et al. (2023) explored S-Alk(en)ylcysteine sulfoxides (CSOs) dynamics during onion bulb expansion which is an important contributor to onion flavour revealed that particularly S-methylcysteine sulfoxide (MCSO) increased during bulb expansion stage, while S-propylcysteine sulfoxide (PeCSO) and S-allylcysteine sulfoxide (ACSO) remained unchanged. Key genes (*AcSULTRs, AcCSs, AcATPS, AcSAT, AcGL, AcGS, AcGGT, AcFMO*) in CSOs biosynthesis were identified through transcriptomic analysis, providing insights into organosulfur compound regulation in onions.

12.3.2.2 Carbohydrate Content

Sucrose, glucose, fructose, and fructans are the most prevalent free sugar components in onion bulbs, forming the major part (65–80%) of the dry matter content (Darbyshire and Henry 1979). Onion bulb quality and storage are greatly influenced by carbohydrate composition, starting at 5% in fresh salads to more than 20% in dehydrated onions (Jones and Mann 1964; Moshfegh et al. 1999). Fructans, known for their health benefits, are linked to reduced colorectal cancer risk, diarrhoea inhibition, constipation relief, and osteoporosis prevention. Researchers are investigating QTLs affecting onion bulb carbohydrates. Higher fructan levels correlated with increased SSC, pungency, bulb dormancy, and onion-induced antiplatelet activity (OIAA). The QTL analysis for SSC, DM%, pungency, and OIAA was conducted by Galmarini et al. (2001), with the creation of the BYG15-23 × AC43 linkage map marking a crucial step, enabling the first quantitative analysis of loci impacting bulb composition. Linkage group E (chromosome 5) exhibited significant association with all traits. The identification of QTLs related to onion bulb characteristics is significantly influenced by the duration of storage, highlighting its crucial role in breeding programmes aimed at enhancing desirable traits. Further, phenotyping of bulb carbohydrate in a subset of F_3 families confirmed the chromosome 3 locus impact on bulb sucrose levels (Havey et al. 2004). The presence of sucrose in high-solid onions modulates fructan synthesis, which affects soluble carbohydrate accumulation and its association with thiosulfinates, pungency, and OIAA. QTL analysis of carbohydrate composition in onion populations identified a significant dominant gene, *Frc*, on chromosome 8, influencing notable variations in fructan content in onion populations (McCallum et al. 2006). In the F2:3 families resulting from the 'W202A' × 'Texas Grano 438' cross, markers ACM171 and ACM235 were notably linked to DM%. The recombination estimates for ACM235 and ACM033 markers were approximately 10.5 ± 5 cM and 16.5 ± 5 cM, respectively, indicating close linkage across various crosses. The location of the QTL on chromosome 8 was verified using the 'BYG15-23' × 'AC43' mapping population. This QTL is positioned between the SSR marker ACM033 and SNP marker ACABE58, elucidating 50.3% of the phenotypic variance and impacting onion carbohydrate levels. Additionally, the mapping of genes responsible for fructan biosynthesis and carbohydrate metabolism regulation on chromosomes 6 and 8, respectively, highlighting their important roles in onion carbohydrate metabolism (Fujishima et al. 2005; Hang et al. 2004; Vijn and Smeekens 1999). Earlier research identified QTLs for sucrose and fructose but not for glucose, and these QTLs were located on low-density linkage maps with uncertain positions (Havey et al. 2004; Khosa et al. 2016a; McCallum et al. 2006). Havey et al. (2004) demonstrated a sucrose-associated QTL on linkage group D using a more reliable linkage map. Nevertheless, this map was also low density and partially divided. Raines et al. (2009) studied genetic effects on DM% and soluble carbohydrates in segregating families from 'BYG15-23' (high soluble solids) and 'AC43' (low soluble solids) crosses. They found codominance on chromosomes 5 and 8 with markers API66, API92, AOB236, ACM033, and ACABE58. API92 and ACM033 significantly affected all traits, influencing dry weight, fructan, and sugar levels. Chromosome 5 (API66 to API47) was related to increased DM%, while chromosome 8 impacted DM% and fructan. Specific regions on chromosomes 3, 5, and 8 played key roles in carbohydrate levels. ACM033 and ACABE58 were closely linked to a major fructan QTL on chromosome 8, crucial for marker-assisted selection (MAS) in breeding. Lee et al. (2021) introduced the first SNP-based onion genetic linkage map through GBS analysis, and identified four loci (*qSC4.1*, *qGC5.1*, *qFC5.1*, *qTSC5.1*) associated with sucrose, glucose, fructose, and total sugar content.

RNA-seq study of carbohydrate metabolism in onion was carried out by Zhang et al. (2016) and revealed peak expression of sucrose transporter genes during early swelling, implicating their role at this stage. Sucrose hydrolysis involved sucrose synthase, cell wall invertase, and invertase, with glucose and fructose increasing, and sucrose decreasing from 15 to 40 days after swelling, suggesting a period of rapid expansion. The relationship between carbohydrate metabolism and the process of flowering was studied, revealing an expected association between onion bolting and bulb sucrose levels through genetic mapping (Baldwin et al. 2014). Chaturvedi et al. (2018) revealed that in 'Shani' cultivar of garlic, preplanting storage at 2 °C enhances carbohydrate metabolism genes, elevating sucrose and fructose in internal buds and storage leaves compared to 33 °C. This influences *FT2* expression, affecting meristem termination and bulb development timing after replanting. Studies on carbohydrate metabolism in onion and garlic highlight the complex interaction of genes, enzymes, temperature, and genetics, affecting bulb growth and flowering.

12.3.2.3 Bulb Colour

The colour of the onion bulb is an important quality characteristic, with variations such as white, yellow, red, pink, chartreuse, and gold. Early genetic research showed that onion colour is determined by at least five loci in its genes, known as loci (*I*, *C*, *G*, *L*, and *R*) (Clarke et al. 1944; El-Shafie and Davis 1967). As per Khar et al. (2008) and Masuzaki et al. (2006), chromosomes 4 and 7 harbour the *L* and *R* loci, while chromosome 6 carries the *C* locus. Tsukazaki et al. (2017) identified and characterized quantitative trait loci (QTLs) and a major fraction of the genes associated to these QTLs within the anthocyanin pathway in onions. The initial studies focussed on identifying the loci responsible for the distinctive red scale colour in shallots (Shigyo et al. 1997). Kim et al. (2004a, b) cloned various genes (*CHS* (chalcone synthase), *F3H*, *DFR* (dihydroflavonol 4-reductase), *ANS* (anthocyanidin synthase), and *FLS* (flavonol synthase) and studied gene expression, where only *DFR* gene was expressed in red onions. The red colour of onions is controlled by at least three loci, according to Khar et al. (2008), and concluded that yellow onions had mutations in the genes that cause red bulb pigmentation. Kim et al. (2005a) compared gene expression in differently coloured onions, revealing reduced *CHS* gene transcription in white onions (associated with the *C* locus) and exclusive *DFR* gene transcription in red onions. In case of pink onions, which are controlled by a single recessive gene, F_3 pink onions produced from crosses between yellow and dark red cultivars have shown reduced ANS gene expression. Follow-up studies proposed that mutations in the cis-acting element likely cause reduced *ANS* gene expression; additionally, sequencing showed a 390-bp insertion in the promoter region of the pink allele (Kim et al. 2004a, b). Khar et al. (2008) found Kim et al. (2005b) markers ineffective, suggesting multiple mutations causing the absence of red colour in onions during domestication. Kim's subsequent studies highlighted diverse mutations in *DFR* and *ANS* genes affecting bulb colour. Song et al. (2014) reported six novel inactive and three active alleles of the *DFRA* gene in diverse onion germplasm, viz., *DFR-APD*, *DFRA5'DEL*, *DFR-AGT* and *DFR-A2AT*; *DFRAK48N* caused an amino acid change from lysine to asparagine at NADPH binding site and *DFR-ATTA* allele with insertion of a leucine near the catalytic site.

The provisionally named *P* locus, associated with certain alleles, was later identified as the L locus through co-segregation analysis in yellow × yellow crosses (Kim et al. 2005b). Sequencing of *ANS* gene revealed changes, like a point mutation altering an amino acid and a 3 kb insertion in the promoter, confirming that the *L* locus is the *ANS* gene. Another *ANS* gene variant (*ANS-h1*) was found, showing a range of colours from pink to dark red in a group of onions, suggesting other factors influence *ANS-h1* expression (Kim et al. 2006). Kim et al. (2016a, b) identified two new non-functional alleles of the *ANS* gene; ANS^{PS}, which has a 4-bp insertion in exon 1 leading to a premature stop codon, and another allele ANS^{S188L}, with a critical amino acid change from serine to leucine. Baek et al. (2017) utilized RNA-seq and bulk segregant analyses on yellow and F_2 white plants to identify a marker associated with the *C* locus, determining white bulb colour in onions. Among their findings, 97 genes, including those previously linked to the flavonoid pathway, exhibited over fivefold expression in the yellow bulk.

To understand onion bulb colour genetics, Havey (2020) bred chartreuse onion bulbs with those of yellow bulbs. The resulting inheritance pattern was consistent with Mendelian genetics predictions. This research of specific onion family with both colours revealed 380 SNPs, some of which were previously unknown. This allowed for the precise identification of the location of the gene for bulb colour (*G* locus) on chromosome 7. A specific SNP marker closely linked to the *G* locus was identified, facilitating the integration of the recessive chartreuse bulb trait into other onion populations for commercial purposes. Next, using genotyping-by-sequencing, researchers developed a SNP-based genetic map in an F_2 onion population to identify quantitative trait loci (QTLs) related to anthocyanin content (Choi et al. 2020). The study revealed a major QTL named *qAS7.1* for bulb colour on chromosome 7, exhibiting an 87.61% phenotypic variation. Additionally, two significant QTLs, *qAC4.1* and *qAC4.2*, were identified on chromosome 4, with R2 values of 19.43% and 26.28%, respectively, associated with anthocyanin content variations. Further research identified candidate gene, *AcLAC12*, linked to the *G* locus (Jeon et al. 2022). This gene produced a protein similar to laccases, enzymes that control the production of quercetin and anthocyanin, resulting in the chartreuse bulb colour. Additionally, studies on the interplay between the *G* and *R* loci revealed that the chartreuse bulb colour only appeared when the *R* locus was recessive, emphasizing the crucial role of inactivating *DFR-A* (dihydroflavonol reductase-A) in this process. A linkage map revealed *G* and *R* loci linkage on chromosome 7 (19.5 cM apart). *AcLAC12A* emerged as the primary laccase gene in onion bulbs, crucial for colouration among eight identified laccase-coding genes. This discovery significantly enhances our understanding of onion genetics and provides valuable information for breeding efforts focused on creating new onion varieties with specific desired colour traits.

12.3.3 Adaptive Traits

Bulbing and flowering traits are vital for both yield and seed production as they hold significant importance for farmers in case of *Allium sp.* Variable requirement for vernalization during flowering is an interesting adaptive characteristic in different *Allium* crops (Brewster 2008). In agricultural plants, the photoperiodic and vernalization mechanisms that control the development of bulbing and flowering are thoroughly researched (Andrés and Coupland 2012).

Mapping QTLs for premature bolting led to the discovery of the *AcBlt1* QTL on chromosome 1. This finding was obtained through a broad cross involving the long-day (LD) doubled-haploid (DH) onion variety 'CUDH2150' and Indian landrace 'Nasik Red' (Baldwin et al. 2014). Two reproducible QTL were detected by Wako et al. (2016) on chromosomes 1 and 2 for bolting time in bunching onion. Transcriptomic analysis revealed conserved candidate genes, notably the *FT*-like family, playing a critical role in onion flowering and bulb formation, as demonstrated by Lee et al. (2013). In Arabidopsis, after winter vernalization and the arrival of longer days in spring, *FLOWERING LOCUS T* (FT) is produced in the leaves and then transported to the shoot apical meristem. At this point, it forms a complex known as the florigen activation complex (FAC) by interacting with 14-3-3 and *FLOWERING LOCUS D* (FD). The FAC plays a pivotal role in initiating the transcriptional activation of various meristem identity genes, which are crucial for transitioning from vegetative growth to the reproductive stage (Abe et al. 2005; Greb et al. 2007; Nakamichi 2011; Yoo et al. 2005). Before the onset of winter, FLOWERING LOCUS C (FLC) suppresses the transcription of FT, thereby preventing premature flowering. However, prolonged exposure to cold temperatures gradually reduces FLC expression. Nonetheless, sustained epigenetic modifications are required for continuous FLC silencing (Whittaker and Dean 2017). Once the vernalization requirement is met and spring days lengthen, the circadian clock genes *GIGANTEA (GI)* and *FLAVIN BINDING KELCH REPEAT F-BOX PROTEIN 1 (FKF1)* form a complex that stabilizes *CONSTANS (CO)*, which subsequently activates the expression of FT, initiating the flowering process (Fornara et al. 2010; Putterill et al. 1995; Sawa et al. 2007; Song et al. 2013).

It has been established that *AcFT4* and *AcFT1* play opposing roles in controlling bulbing. *AcFT4* suppresses bulb formation in immature or mature onion plants cultivated under non-inductive short-day (SD) photoperiods (Lee et al. 2013). However, when the plants are matureand are exposed to inductive long-day (LD) photoperiods, the expression of *AcFT4* declines, causing the activation of another gene, *AcFT1*, which stimulates bulb development. Beyond these *FT* genes, many circadian clock genes in onions have been found, including *AcGI*, *AcFKF-1*, and three *AcCOLs*, but their precise role in bulb formation is unknown (Harun-Ar-Rashid and Thomas 2020; Taylor et al. 2010). Following bulb formation, mature bulbs are harvested and replanted for seed production in the subsequent year. During this period, exposure to cold temperatures triggers the transcriptional activation of *AcFT2*, facilitating flowering (Lee et al. 2013).

Floral organ identity genes, predominantly MADS box genes, play a key role in flower development across various plant species. These genes, grouped into classes A, B, C, D, and E, contribute to the diversity of floral structures in the plant kingdom, including *Alliums* (Thomson et al. 2017). In bulb onion, flower development genes were identified, including MADS box genes (Khosa et al. 2016b). Phylogenetic analysis aligns these onion transcripts with the ABCDE model, revealing distinct clades for B, C, D, and separate clades for E class genes. The highly expressed MADS-box genes in floral organs underscore their significance in bulb onion flower development (Khosa et al. 2016b). Moreover, it has been found that onions have many orthologs of rice genes, which are essential for male fertility and flowering. Continued research aims to identify critical adaption genes in various *Allium cepa* genotypes, using genomic and physiological analysis to learn more about its evolutionary history. RNA-SEQ technologies have also accelerated our understanding of essential gene families and adaptation in garlic (Kamenetsky et al. 2015). Transcriptome study of dormant and growing garlic stem apices revealed approximately 20,000 upregulated unigenes during the sprouting phase (Sun et al. 2012). Sun et al. (2013) identified candidate genes (*ENHYDROUS*, *DAG1*, *DAM*, *DTH8*, etc.) involved in shoot apex dormancy and sprouting in garlic.

12.3.4 Biotic Stress

Onion is susceptible to various diseases and insect pests, which compromise both bulb yield and quality as well as affecting storage life, seed yield, and quality. Majorly purple blotch (PB), *Stemphylium* blight (SB), Pink root (PR), *Fusarium* basal rot (FBR), downy mildew (DM), grey mould, *Iris yellow spot virus* (IYSV), etc. are impacting onion crop. Research efforts have predominantly focused on downy mildew (DM) and *Fusarium* basal rot (FBR) in temperate conditions, while on purple blotch in tropical regions (Khar et al. 2022). Diverse onion germplasm has been identified as a resilient source against these diseases (Scholten et al. 2007; Cramer 2000; Galvan et al. 2008; Taylor et al. 2013; Arias et al. 2020). Wild relative *A. royeli* Stearn showed complete resistance to DM and utilizing backcross breeding and interspecific hybridization, a dominant gene (*Pd*) was introgress for this resistance into a bulb onion (Van der Meer and De Vries 1990; Kofoet et al. 1990). A RAPD/SCAR marker linked to DM resistance mediated by locus *Pd* located at chromosome 3 and positioned at 2.6 cM in *A. cepa* × *A. roylei* interspecific cross (Kik et al. 1997; Van Heusden et al. 2000). However, the marker reliability decreased over time, as positive PCR product associated with *A. roylei* fragment became apparent in susceptible onion accessions (Scholten et al. 2007). In a research patent by Harrewijn et al. (2012) on 'Identification of locus governing DM resistance', the resistance locus (*Pd*) was located on chromosome 3 long arm in *A. cepa*. This locus (*Pd*) is identical to the genomic segment imparting resistance to DM in *A. roylei*. Subsequently, a reliable PCR marker linked to DM resistance was developed by Kim et al. (2016a, b) by using the data from linkage map (Duangjit et al. 2013) and RNA-Seq (Kim et al. 2015a, b). A robust PCR marker DMR1 associated with *Pd1* DM-resistant gene was developed in resistant cv Santero from Europe and *A. roylei* accession PI 59299. A high-resolution map of *A. royeli* fragment consisting *Pd* locus of ~27.6 cM length was identified at the end of chromosome 3, and it is suggested that marker positioned at boundary could be utilized as reliable recombinant selection marker (Cho et al. 2021). Further, the homozygous introgression lines with DM resistance were developed successfully using DMR1 marker linked to *Pd* locus and GISH in a short breeding time.

In India, purple blotch severity varies between 2.5% and 85%, causing yield losses in bulb as well as seed crops (Tripathy et al. 2013; Veeraghanti et al. 2017). The mechanism of PB resistance in onion is controlled by additive and non-additive genes (Evoor et al. 2010) as well as single dominant gene (Abubakar and Ado 2008). Two sets of SSR (*matK* and *rbcL*) markers were assessed for MAS of locus *s1/s2* on chromosome 8 that governs PB resistance, and resistance was shown to be dominant (Chakraborty et al. 2015). The first report on QTL mapping revealed single dominant gene *ApR1* was responsible for PB resistance in onion. BSA of segregating population derived from Arka Kalyan (resistant) × Agrifound Rose (susceptible) cross identified molecular marker linked to PB resistance. Among 288 primers, 59 primers differentiate the parental population and these primers were further utilized for mapping studies. Molecular mapping identified two markers AcSSR7 and ApR-450 closely linked to *ApR1* locus at a distance of 1.3 and 1.1 cM, respectively (Chand et al. 2018). Further, RIL population was used to fine-map the resistant locus with SNP marker (Sahoo et al. 2023) and to identify closely linked KASP markers (ApRsnip14 and ApRsnip25) to *ApR1* locus. The identification of QTLs linked to PB resistance gene opens up new opportunities to improve breeding strategy to enhance PB resistance in onion. Pink root (PR) is another soil-borne disease impacting onion crops and caused by *Pyrenochaeta terrestris* fungus.

Resistance against this disease is governed by a recessive gene (Jones and Perry 1956; Nicholas et al. 1965). Two resistant sources (PR-1 and PR-2) were used by Marzu et al. (2018) for developing segregating mapping families, which revealed that PR resistance from distinct sources mapped on same chromosome 4 region at LODs of 10.3 and 12.5 and explained 35% and 54% of phenotypic variation in PR1 and PR2 families, respectively. The QTL linked to PR resistance was discovered near the SNP marker (isotig26045_1046) which showed positive additive effect and indicating that with each additional W446B allele, survival rate of PR increased by 24%. In a patent authored by Black et al. (2022), mapping of PR resistance in short-day onion population identified two loci (NQ0257277 and NQ0258453) on chromosome 4 which showed an additive effect that increased the survival of seedling by 32%.

Onion resistance to *Fusarium* isolates has been explored across *Allium* species and diverse germplasm, revealing various genetic control from single gene to two genes or polygenic control has been identified (Cramer 2000). Researchers identified partially dominant genes, viz., *Foc1* and *Foc2* which show additive effects for resistance to FBR in long-day inbred cross (FBR resistant × FBR susceptible). The possible involvement of cytoplasmic or nuclear gene in FBR resistance was also suggested by Tsutsui (1986) and Krueger (1986). Early investigation on genetic basis of FBR resistance identified a preliminary QTL on chromosome 8 from *A. fistulosum* (Galvan et al. 2008). Further, a worldwide onion association panel was genotyped using SNP marker identified by Duangjit et al. (2013) and three markers, viz., I30594_1021 (1.1 cM), c00676_1004 (149.8 cM), and i34519_442 (30.2 cM) linked to FBR resistance on chromosome 8, 1, and 6, respectively along with two other unmapped SNP markers (Taylor et al. 2019). A monosomic addition lines set was used in another study and identified a steroidal saponin from shallot on chromosome 2 which plays a function in FBR defence (Abdelrahman et al. 2017). Using the FBR-resistant selection W440, Straley et al. (2021) identified three QTL on chromosome 2 and two on chromosome 4 which segregate independently in B5351 × W440 F_2 population. The markers isotig38484_281 and isotig44683_192 on chromosome 2B and chromosome 4A, respectively, were most significantly correlated with the FBR resistance, while the marker, viz., isotig31106_505 on chromosome 4C was linked with susceptibility to FBR. The FOC isolates diversity may hinder the FBR-resistant germplasm identification and subsequent genetic research. The regions reported by Taylor et al. (2019) and Straley et al. (2021) were not significantly associated with FBR resistance indicating that there may be several QTLs conferring FBR resistance in onion. However, the identified SNPs could be utilized in future research aimed at understanding the basic mechanism of resistance to FBR.

Various other studies also reported SNP/HRM markers linked to resistance against *Botrytis* leaf blight (Scholten et al. 2016) and grey mould (Kim et al. 2021) in onion. A region of QTL on chromosome 6, initially derived from *A. roylei*, was responsible for *Botrytis squamosa* resistance (Scholten et al. 2016). For breeding onion resistant to grey mould, SCAR marker was developed which amplify specific band only in resistant lines. Furthermore, 320 transcripts were identified in resistant line through gene expression studies, while 1316 transcripts were highly expressed in susceptible lines. The resistance-governing transcripts were used to develop SNP markers, and SNP-3 HRM marker present in *AKR* gene was identified to select grey mould-resistant onions (Kim et al. 2021). AKR gene showed higher expression in resistant line than in susceptible line. Thus markers, viz., SNP-3 HRM and SCAR-OPAN1 can be utilized to breed grey mould-resistant lines. Onion germplasm was also screened by Lee et al. (2022) using in-vitro culture and RAPD marker and identified Asia-23 and Asia-24 as resistant and susceptible genotypes, respectively. Expression analysis of genes, viz., *JAR1* (Jasmonate Resistant 1), *COI 1* (Coronatine Insensitive protein 1), and *MYC2* (transcription factor) among resistant and susceptible groups revealed SNP variation between Asia-23 and Asia-24 lines (Lee et al. 2022). Resistance to moderately resistant sources has been reported against *Stemphyllum* disease (Pathak et al. 2001; Sharma and Sain 2003; Behera et al. 2013; Dangi et al. 2019; Chandel et al. 2022), but the genetic mechanism underlying the resistance is unknown.

Alliums are susceptible to various insect pests, among which thrips being a predominant global pest, causing damage via reducing the photosynthetic activity, serving as an entry point for pathogens and acting as a vector to spread IYSV (Khosa and Dhatt 2020). Onion plants characterized as light green leaf colour with different type and amount of epicuticular waxes experience significantly less feeding damage by thrips than waxy foliage. Few tolerant germplasm from different countries have been identified presenting valuable potential for utilization in breeding programme. Specifically, glossy phenotypes from OGL and White Persian show substantially less feeding damage from thrips compared to waxy onions (Jones et al. 1934; Cramer et al. 2014; Damon et al. 2014). The segregating mapping population was developed from waxy and semi glossy onion and genotype with SNP markers. The regions identified on chromosome 5 (decarbonylation pathway of epicuticular wax biosynthesis) were responsible for the amount of epicuticular wax on leaves of onion, while chromosome 2 was linked with primary fatty alcohols (acyl reduction pathways). SNP markers from both chromosomes will be useful for marker-based breeding to develop thrips-resistant onion cultivars (Damon and Havey 2013). Using a glossy White Persian phenotype, Munaiz and Havey (2020) map the glossy phenotype on chromosome 8 (*glwp* locus). The onion cultivar Odourless Greenleaf (OGL) shows recessive gene resistance to thrips and possesses unique wax profile with 28 or fewer carbons. Further, mapping studies of two segregating families (B9885*OGL and OGL*B8667) identified a specific single region on chromosome 6 that governs both unique wax profile and glossy foliage of OGL (Havey et al. 2021).

12.4 Conclusion

Over the past 30 years, considerable advancements have expedited onion breeding and genetic advancement. The development of molecular markers allows the researchers to map the traits of economic importance. Furthermore, linked markers have been used in breeding programmes for marker-assisted selection especially for male sterility and fertility restoration. In future, there is a need to develop tightly linked markers for other traits as well. The utilization of RNA-seq in onion and garlic led to the discovery of genes involved in different pathways such as photoperiodic and vernalization. The availability of chromosome-level genome assemblies of onion, garlic, and bunching onion will open up the new avenues for genomic-assisted breeding in these species.

References

Abdelrahman M, Hirata S, Sawada Y, Hirai MY, Sato S, Hirakawa H, Mine Y, Tanaka K, Shigyo M (2019) Widely targeted metabolome and transcriptome landscapes of *Allium fistulosum-A. cepa* chromosome addition lines revealed a flavonoid hot spot on chromosome 5A. Sci Rep 9:3541. https://doi.org/10.1038/s41598-019-39856-1

Abe M, Kobayashi Y, Yamamoto S, Daimon Y, Yamaguchi A, Ikeda Y, Ichinoki H, Notaguchi M, Goto K, Araki T (2005) FD, a bZIP protein mediating signals from the floral pathway integrator FT at the shoot apex. Science 309:1052–1056. https://doi.org/10.1126/science.1115983

Abubakar L, Ado SG (2008) Heterosis of purple blotch (*Alternaria porri* (Ellis) Cif.) resistance, yield and earliness in tropical onions (*Allium cepa* L.). Euphytica 164:63–74. https://doi.org/10.1007/s10681-007-9639-4

Ahn W, Kim S (2023) Identification of a candidate gene responsible for male sterility conferred by CMS-T cytoplasm in onion (*Allium cepa* L.) and development of molecular markers for detection of CMS-T cytoplasm. Euphytica 219(2):28. https://doi.org/10.1007/s10681-023-03158-5

Andrés F, Coupland G (2012) The genetic basis of flowering responses to seasonal cues. Nat Rev Genet 13:627–639. https://doi.org/10.1038/nrg3291

Aoyagi M, Kamoi T, Kato M, Sasako H, Tsuge N, Imai S (2011) Structure and bioactivity of thiosulfinates resulting from suppression of lachrymatory factor synthase in onion. J Agric Food Chem 59:10893–10900. https://doi.org/10.1021/jf202446q

Arias M, Curbelo N, Rabelino PG, Vicente E, Giménez G, Galván GA (2020) Inheritance of resistance against '*Peronospora destructor*' in onion cv.'Regia'. Aust J Crop Sci 14:1999–2009

Baek G, Kim C-W, Kim S (2017) Development of a molecular marker tightly linked to the C locus conferring a white bulb color in onion (*Allium cepa* L.) using bulked segregant analysis and RNA-Seq. Mol Breed 37:94. https://doi.org/10.1007/s11032-017-0697-6

Baldwin S, Revanna R, Thomson S, Pither-Joyce M, Wright K, Crowhurst R, Fiers M, Chen L, Macknight R, McCallum JA (2012) A toolkit for bulk PCR-based marker design from next-generation sequence data: application for development of a framework linkage map in bulb onion (*Allium cepa* L.). BMC Genomics 13:637. https://doi.org/10.1186/1471-2164-13-637

Baldwin S, Revanna R, Pither-Joyce M, Shaw M, Wright K, Thomson S, Moya L, Lee R, Macknight R, McCallum J (2014) Genetic analyses of bolting in bulb onion (*Allium cepa* L.). Theor Appl Genet 127:535–547. https://doi.org/10.1007/s00122-013-2232-4

Bang H, Cho DY, Yoo KS, Yoon MK, Patil BS, Kim S (2011) Development of simple PCR-based markers linked to the Ms locus, a restorer-of-fertility gene in onion (*Allium cepa* L.). Euphytica 179:439–449. https://doi.org/10.1007/s10681-010-0342-5

Bang H, Kim S, Park SO, Yoo KS, Patil BS (2013) Development of a codominant CAPS marker linked to the Ms locus controlling fertility restoration in onion (*Allium cepa* L.). Sci Hortic 153:42–49. https://doi.org/10.1016/j.scienta.2013.01.020

Bedford LV (1984) Dry matter and pungency tests on British grown onions

Behera S, Santra PALASH, Chattopadhyay S, Das SRIKANTA, Maity TK (2013) Variation in onion varieties for reaction to natural infection of *Alternaria porri* (Ellis) Ciff. and *Stemphylium vesicarium* (Wallr.). The. Bioscan 8(3):759–761. https://doi.org/10.13140/RG.2.2.21107.20003

Bharti G, Chimata MK (2019) Review on new plant breeding techniques. Int J Sci Res 8:723–730

Bisen S, Emerald M (2016) Nutritional and therapeutic potential of garlic and onion (*Allium* sp.). CNF 12:190–199. https://doi.org/10.2174/1573401312666160608121954

Black L, Chan EKF, Colcol JF, Jones R, Kramer C, Xiang W, Seminis Vegetable Seeds Inc (2022) QTLs conferring resistance to fusarium basal rot, pink root and complementary pinks in onions. U.S. Patent 11,457,582

Block (2010) Allium botany and cultivation, ancient and mdern. In: Garlic and other alliums: the lore and the science, pp 1–32

Brewster JL (2008) Onions and other vegetable alliums. CABI

Chakraborty M, Quddus T, Rahman S, Azad MAK, Hossain MR, Hoque A, Saha MK, Haque MA (2015) Molecular characterization of selected mutant lines of onion (*Allium cepa* L.) against purple leaf blotch disease using SSR markers. Am J Exp Agric 8:261–267. https://doi.org/10.9734/AJEA/2015/17504

Chand SK, Nanda S, Joshi RK (2018) Genetics and molecular mapping of a novel purple blotch-resistant gene ApR1 in onion (*Allium cepa* L.) using STS and SSR markers. Mol Breed 38(9):109. https://doi.org/10.1007/s11032-018-0864-4

Chandel R, Kamil D, Singh S, Kumar A, Patel R, Verma P, Zimik M, Khar A (2022) Screening of short-day onions for resistance to Stemphylium leaf blight in the seed-to-bulb stage (stage I) and bulb-to-seed stage (stage II). Front Plant Sci 13:1063685. https://doi.org/10.3389/fpls.2022.1063685

Chaturvedi AK, Rohkin Shalom S, Faigenboim-Doron A, Teper-Bamnolker P, Salam BB, Daus, A, Kamenetsky R, Eshel, D (2018) Differential carbohydrate gene expression during preplanting temperature treatments controls meristem termination and bulbing in garlic. Environ Exp Bot 150:280–291. https://doi.org/10.1016/j.envexpbot.2018.01.013

Cho Y, Kim B, Lee J, Kim S (2021) Construction of a high-resolution linkage map and chromosomal localization of the loci determining major qualitative traits in onion (*Allium cepa* L.). Euphytica 217:1–12. https://doi.org/10.1007/s10681-020-02746-z

Choi Y, Kim S, Lee J (2020) Construction of an onion (Allium cepa L.) genetic linkage map using genotyping-by-sequencing analysis with a reference gene set and identification of QTLs controlling anthocyanin synthesis and content. Plants (Basel) 9:616. https://doi.org/10.3390/plants9050616

Clarke AE, Jones HA, Little T (1944) Inheritance of bulb color in the onion. Genetics 29:569

Collard BCY, Jahufer MZZ, Brouwer JB, Pang ECK (2005) An introduction to markers, quantitative trait loci (QTL) mapping and marker-assisted selection for crop improvement: The basic concepts. Euphytica 142:169–196. https://doi.org/10.1007/s10681-005-1681-5

Coolong TW, Randle WM (2003) Temperature influences flavor intensity and quality in 'Granex 33'onion. J Am Soc Hortic Sci 128:176–181

Cramer CS (2000) Breeding and genetics of Fusarium basal rot resistance in onion. Euphytica 115:159–166. https://doi.org/10.1023/A:1004071907642

Cramer CS, Singh N, Kamal N, Pappu HR (2014) Screening onion plant introduction accessions for tolerance to onion thrips and iris yellow spot. HortScience 49(10):1253–1261. https://doi.org/10.21273/HORTSCI.49.10.1253

Damon S, Havey MJ (2013) Reduced epicuticular waxes are associated with insect tolerance. In: Conference of the American Society of Agronomy. Available at: https://scisoc.confex.com/crops/2013am/webprogram/Paper81114.html. Accessed 20 Aug 2014

Damon SJ, Groves RL, Havey MJ (2014) Variation for epicuticular waxes on onion foliage and impacts on numbers of onion thrips. J Am Soc Hortic Sci 139(4):495–501. https://doi.org/10.21273/JASHS.139.4.495

Dangi R, Sinha P, Islam S, Gupta A, Kumar A, Rajput LS, Kamil D, Khar A (2019) Screening of onion accessions for Stemphylium blight resistance under artificially inoculated field experiments. Australas Plant Pathol 48:375–384. https://doi.org/10.1007/s13313-019-00639-x

Darbyshire B, Henry RJ (1979) The association of fructans with high percentage dry weight in onion cultivars suitable for dehydrating. J Sci Food Agric 30:1035–1038. https://doi.org/10.1002/jsfa.2740301103

Duangjit J, Bohanec B, Chan AP, Town CD, Havey MJ (2013) Transcriptome sequencing to produce SNP-based genetic maps of onion. Theor Appl Genet 126:2093–2101. https://doi.org/10.1007/s00122-013-2121-x

Eady CC, Kamoi T, Kato M, Porter NG, Davis S, Shaw M, Kamoi A, Imai S (2008) Silencing onion lachrymatory factor synthase causes a significant change in the sulfur secondary metabolite profile. Plant Physiol 147:2096–2106

Edwards S, Vallance S, Montgomery R (2022) Interstitial silences in sociotechnical change: the case of genetically modified 'tearless' onions. Sci Cult 31:212–234. https://doi.org/10.1080/09505431.2021.2019214

Egea LA, Mérida-García R, Kilian A, Hernandez P, Dorado G (2017) Assessment of genetic diversity and structure of large garlic (*Allium sativum*) germplasm bank, by diversity arrays technology "genotyping-by-sequencing" platform (DArTseq). Front Genet 8. https://doi.org/10.3389/fgene.2017.00098

El-Shafie MW, Davis GN (1967) Inheritance of bulb color in the onion (*Allium cepa* L.). Hilgardia 38:607–622. https://doi.org/10.3733/hilg.v38n17p607

Engelke T, Terefe D, Tatlioglu T (2003) A PCR-based marker system monitoring CMS-(S), CMS-(T) and (N)-cytoplasm in the onion (*Allium cepa* L.). Theor Appl Genet 107:162–167. https://doi.org/10.1007/s00122-003-1230-3

Evoor S, Veeregowda R, Ganeshan G (2010) Genetics of resistance to purple blotch disease in onion (*Allium cepa* L.). Karnataka J Agric Sci 20:810–812

Ferreira RR, Santos CAF (2018) Partial success of marker-assisted selection of 'A' and 'B' onion lines in Brazilian germplasm. Sci Hortic 242:110–115. https://doi.org/10.1016/j.scienta.2018.08.002

Finkers R, van Kaauwen M, Ament K, Burger-Meijer K, Egging R, Huits H, Kodde L, Kroon L, Shigyo M, Sato S, Vosman B (2021) Insights from the first genome assembly of Onion (*Allium cepa*). G3 11(9):jkab243. https://doi.org/10.1093/g3journal/jkab243

Fornara F, de Montaigu A, Coupland G (2010) SnapShot: control of flowering in Arabidopsis. Cell 141:550–550

Fujishima M, Sakai H, Ueno K, Takahashi N, Onodera S, Benkeblia N, Shiomi N (2005) Purification and characterization of a fructosyltransferase from onion bulbs and its key role in the synthesis of fructo-oligosaccharides in vivo. New Phytol 165:513–524. https://doi.org/10.1111/j.1469-8137.2004.01231.x

Galmarini CR, Goldman IL, Havey MJ (2001) Genetic analyses of correlated solids, flavor, and health-enhancing traits in onion (*Allium cepa* L.). Mol Gen Genomics 265:543–551. https://doi.org/10.1007/s004380100445

Galván GA, Koning-Boucoiran CF, Koopman WJ, Burger-Meijer K, González PH, Waalwijk C, Kik C, Scholten OE (2008) Genetic variation among Fusarium isolates from onion, and resistance to Fusarium basal rot in related *Allium* species. Eur J Plant Pathol 121:499–512. https://doi.org/10.1007/s10658-008-9270-9

Gökçe AF, McCallum J, Sato Y, Havey MJ (2002) Molecular tagging of the Ms locus in onion. J Am Soc Hortic Sci 127(4):576–582. https://doi.org/10.21273/JASHS.127.4.576

Greb T, Mylne JS, Crevillen P, Geraldo N, An H, Gendall AR, Dean C (2007) The PHD finger protein VRN5 functions in the epigenetic silencing of *Arabidopsis* FLC. Curr Biol 17:73–78

Griffiths G, Trueman L, Crowther T, Thomas B, Smith B (2002) Onions—a global benefit to health. Phytother Res 16:603–615. https://doi.org/10.1002/ptr.1222

Hang TTM, Shigyo M, Yaguchi S, Yamauchi N, Tashiro Y (2004) Effect of single alien chromosome from shallot (*Allium cepa* L. Aggregatum group) on carbohydrate production in leaf blade of bunching onion (*A. fistulosum* L.). Genes Genet Syst 79:345–350. https://doi.org/10.1266/ggs.79.345

Hanson MR, Bentolila S (2004) Interactions of mitochondrial and nuclear genes that affect male gametophyte development. Plant Cell 16(suppl_1):S154–S169. https://doi.org/10.1105/tpc.015966

Hao F, Liu X, Zhou B, Tian Z, Zhou L, Zong H, Qi J, He J, Zhang Y, Zeng P, Li Q, Wang K, Xia K, Guo X, Li L, Shao W, Zhang B, Li S, Yang H, Hui L, Chen W, Peng L, Liu F, Rong Z-Q, Peng Y, Zhu W, McCallum JA, Li Z, Xu X, Yang H, Macknight RC, Wang W, Cai J (2023) Chromosome-level genomes of three key Allium crops and their trait evolution. Nat Genet 55:1976–1986. https://doi.org/10.1038/s41588-023-01546-0

Harrewijn JL, Hoogenboom JPH, Nickerson Zwaan BV (2012) Resistance to downy mildew of onion caused by the fungus Peronospora destructor. U.S. Patent 8,119,856

Harun-Ar-Rashid M, Thomas B (2020) Diurnal expression of Arabidopsis gene homologs during daylength-regulated bulb formation in onion (Allium cepa L.)

Havey MJ (1993) A putative donor of S-cytoplasm and its distribution among open-pollinated populations of onion. Theor Appl Genet 86:128–134. https://doi.org/10.1007/BF00223817

Havey MJ (1995) Identification of cytoplasms using the polymerase chain reaction to aid in the extraction of maintainer lines from open-pollinated populations of onion. Theor Appl Genet 90:263–268. https://doi.org/10.1007/BF00222212

Havey MJ (1999) Seed yield, floral morphology and lack of male fertility restoration of male sterile onion (*Allium cepa*) populations possessing the cytoplasm of *Allium galanthum*. J Amer Soc Hort Sci 124:626–629

Havey MJ (2000) Diversity among male-sterility-inducing and male-fertile cytoplasms of onion. Theor Appl Genet 101:778–782. https://doi.org/10.1007/s001220051543

Havey MJ (2018) Onion breeding. Plant Breed Rev 42:39–85. https://doi.org/10.1002/9781119521358.ch2

Havey MJ (2020) Genetic mapping of chartreuse bulb color in onion. J Am Soc Hortic Sci 145:110–119. https://doi.org/10.21273/JASHS04861-20

Havey MJ (2024) A single nucleotide polymorphism unique to the galanthum-CMS of onion. HortScience 59(1):8–10. https://doi.org/10.21273/HORTSCI17412-23

Havey MJ, Kim S (2021) Molecular marker characterization of commercially used cytoplasmic male sterilities in onion. J Am Soc Hortic Sci 146(5):351–355. https://doi.org/10.21273/JASHS05083-21

Havey MJ, von Kohn C (2017) Efficacy of molecular markers Jnurf13 and AcPms1 for prediction of genotypes at the nuclear Ms locus in three open-pollinated populations of onion from North America. HortScience 52:1052–1053. https://doi.org/10.21273/HORTSCI11954-17

Havey MJ, Galmarini CR, Gökçe AF, Henson C (2004) QTL affecting soluble carbohydrate concentrations in stored onion bulbs and their association with flavor and health-enhancing attributes. Genome 47:463–468. https://doi.org/10.1139/g04-005

Havey MJ, Hunsaker DJ, Munaiz ED (2021) Genetic analysis of the unique epicuticular wax profile of 'odourless greenleaf' onion. J Am Soc Hortic Sci 146(2):118–124. https://doi.org/10.21273/JASHS05024-20

Huo YM, Liu BJ, Yang YY, Miao J, Gao LM, Kong SP, Wang ZB, Kitano H, Wu X (2015) AcSKP1, a multiplex PCR-based co-dominant marker in complete linkage disequilibrium with the male-fertility restoration (Ms) locus, and its application in open-pollinated populations of onion. Euphytica 204:711–722. https://doi.org/10.1007/s10681-015-1374-7

Imai S, Tsuge N, Tomotake M, Nagatome Y, Sawada H, Nagata T, Kumagai H (2002) An onion enzyme that makes the eyes water. Nature 419:685–685. https://doi.org/10.1038/419685a

Jeon S, Han J, Kim C-W, Kim J-G, Moon J-H, Kim S (2022) Identification of a candidate gene responsible for the G locus determining chartreuse bulb color in onion (Allium cepa L.) using bulked segregant RNA-Seq. Theor Appl Genet 135:1025–1036. https://doi.org/10.1007/s00122-021-04016-5

Jia H, Zhao Q, Song J, Zhang X, Yang W, Du Z, Zhu Y, Wang H (2023) Large-scale population structure and genetic architecture of agronomic traits of garlic. Hortic Res 10(4):uhad034. https://doi.org/10.1093/hr/uhad034

Jo J, Purushotham PM, Han K, Lee H-R, Nah G, Kang B-C (2017) Development of a genetic map for onion (Allium cepa L.) using reference-free genotyping-by-sequencing and SNP assays. Front Plant Sci 8:1606. https://doi.org/10.3389/fpls.2017.01606

Jones MG (2004) Biosynthesis of the flavour precursors of onion and garlic. J Exp Bot 55:1903–1918. https://doi.org/10.1093/jxb/erh138

Jones HA, Clarke A (1943) Inheritance of male sterility in the onion and the production of hybrid seed. Proc Am Soc Hortic Sci 43:189–194

Jones HA, Mann LK (1964) Onions and their allies. Soil Sci 98:68

Jones HA, Perry BA (1956) Inheritance of resistance to pink root in the onion. J Hered 47(1):33–34

Jones HA, Bailey SF, Emsweller SL (1934) Thrips resistance in the onion. Hilgardia 8:215–232. https://doi.org/10.3733/hilg.v08n07p213

Kamenetsky R, Faigenboim A, Shemesh Mayer E, Ben Michael T, Gershberg C, Kimhi S, Esquira I, Rohkin Shalom S, Eshel D, Rabinowitch HD, Sherman A (2015) Integrated transcriptome catalogue and organ-specific profiling of gene expression in fertile garlic (Allium sativum L.). BMC Genomics 16:12. https://doi.org/10.1186/s12864-015-1212-2

Kato M, Masamura N, Shono J, Okamoto D, Abe T, Imai S (2016) Production and characterization of tearless and non-pungent onion. Sci Rep 6:23779

Khar A, Saini N (2016) Limitations of PCR-based molecular markers to identify male-sterile and maintainer plants from Indian onion (Allium cepa L.) populations. Plant Breed 135(4):519–524. https://doi.org/10.1111/pbr.12373

Khar A, Jakse J, Havey MJ (2008) Segregations for onion bulb colors reveal that red is controlled by at least three loci. J Am Soc Hortic Sci 133:42–47

Khar A, Galván GA, Singh H (2022) Allium breeding against biotic stresses. In: Genomic designing for biotic stress resistant vegetable crops. Springer, Cham, pp 233–259. https://doi.org/10.1007/978-3-030-97785-6_6

Khosa JS, Dhatt AS (2020) Improvement of onion through accelerated approaches. In: Accelerated plant breeding, vol 2: vegetable crops, pp 57–75. https://doi.org/10.1007/978-3-030-47298-6_3

Khosa JS, McCallum J, Dhatt AS, Macknight RC (2016a) Enhancing onion breeding using molecular tools. Plant Breed 135:9–20. https://doi.org/10.1111/pbr.12330

Khosa JS, Lee R, Bräuning S, Lord J, Pither-Joyce M, McCallum J, Macknight RC (2016b) Doubled haploid 'CUDH2107'as a reference for bulb onion (Allium cepa L.) research: development of a transcriptome catalogue and identification of transcripts associated with male fertility. PloS One 11:e0166568. https://doi.org/10.1371/journal.pone.0166568

Khrustaleva L, Jiang J, Havey MJ (2016) High-resolution tyramide-FISH mapping of markers tightly linked to the male-fertility restoration (Ms) locus of onion. Theor Appl Genet 129:535–545. https://doi.org/10.1007/s00122-015-2646-2

Khrustaleva L, Nzeha M, Ermolaev A, Nikitina E, Romanov V (2023) Two-step identification of N-, S-, R-and T-cytoplasm types in onion breeding lines using high-resolution melting (HRM)-based markers. Int J Mol Sci 24(2):1605. https://doi.org/10.3390/ijms24021605

Kik C, Buiteveld J, Verbeek WHJ (1997) Biotechnological aspects of onion breeding. Acta Hortic 433:291–298. https://doi.org/10.17660/ActaHortic.1997.433.30

Kim S (2014) A codominant molecular marker in linkage disequilibrium with a restorer-of-fertility gene (Ms) and its application in reevaluation of inheritance of fertility restoration in onions. Mol Breed 34:769–778. https://doi.org/10.1007/s11032-014-0073-8

Kim S, Binzel ML, Yoo KS, Park S, Pike LM (2004a) Pink (P), a new locus responsible for a pink trait in onions (Allium cepa) resulting from natural mutations of anthocyanidin synthase. Mol Gen Genomics 272:18–27. https://doi.org/10.1007/s00438-004-1041-5

Kim S, Binzel ML, Park S, Yoo K-S, Pike LM (2004b) Inactivation of DFR (Dihydroflavonol 4-reductase) gene transcription results in blockage of anthocyanin production in yellow onions (*Allium cepa*). Mol Breed 14:253–263. https://doi.org/10.1023/B:MOLB.0000047770.92977.04

Kim S, Jones R, Yoo K-S, Pike LM (2005a) The L locus, one of complementary genes required for anthocyanin production in onions (Allium cepa), encodes anthocyanidin synthase. Theor Appl Genet 111:120–127. https://doi.org/10.1007/s00122-005-2000-1

Kim S, Yoo KS, Pike LM (2005b) Development of a PCR-based marker utilizing a deletion mutation in the dihydroflavonol 4-reductase (DFR) gene responsible for the lack of anthocyanin production in yellow onions (*Allium cepa*). Theor Appl Genet 110:588–595. https://doi.org/10.1007/s00122-004-1882-7

Kim S, Bang H, Yoo K-S, Pike LM (2006) Identification of the fourth allele of the ANS (anthocyanidin synthase) gene and its effect on red color intensity in onions (Allium cepa). Euphytica 149:45–51. https://doi.org/10.1007/s10681-005-9053-8

Kim S, Lee ET, Cho DY, Han T, Bang H, Patil BS, Ahn YK, Yoon MK (2009) Identification of a novel chimeric gene, orf725, and its use in development of a molecular marker for distinguishing among three cytoplasm types in onion (Allium cepa L.). Theor Appl Genet 118:433–441. https://doi.org/10.1007/s00122-008-0909-x

Kim S, Kim CW, Park M, Choi D (2015a) Identification of candidate genes associated with fertility restoration of cytoplasmic male-sterility in onion (*Allium cepa* L.) using a combination of bulked segregant analysis and RNA-seq. Theor Appl Genet 128:2289–2299. https://doi.org/10.1007/s00122-015-2584-z

Kim S, Kim MS, Kim YM, Yeom SI, Cheong K, Kim KT, Jeon J, Kim S, Kim DS, Sohn SH, Lee YH (2015b) Integrative structural annotation of de novo RNA-Seq provides an accurate reference gene set of the enormous genome of the onion (*Allium cepa* L.). DNA Res 22(1):19–27. https://doi.org/10.1093/dnares/dsu035

Kim E-Y, Kim C-W, Kim S (2016a) Identification of two novel mutant ANS alleles responsible for inactivation of anthocyanidin synthase and failure of anthocyanin production in onion (Allium cepa L.). Euphytica 212:427–437. https://doi.org/10.1007/s10681-016-1774-3

Kim S, Kim CW, Choi MS, Kim S (2016b) Development of a simple PCR marker tagging the Allium roylei fragment harboring resistance to downy mildew (*Peronospora destructor*) in onion (*Allium cepa* L.). Euphytica 208:561–569. https://doi.org/10.1007/s10681-015-1601-2

Kim B, Kim CW, Kim S (2019) Inheritance of fertility restoration of male-sterility conferred by cytotype Y and identification of instability of male fertility phenotypes in onion (Allium cepa L.). J Hortic Sci Biotechnol 94(3):341–348. https://doi.org/10.1080/14620316.2018.1512383

Kim SJ, Park JS, Park T, Lee HM, Choi JR, Park YD (2021) Development of molecular markers associated with resistance to gray mold disease in onion (Allium cepa L.) through RAPD-PCR and transcriptome analysis. Horticulturae 7(11):436. https://doi.org/10.3390/horticulturae7110436

King JJ, Bradeen JM, Havey MJ (1998a) Variability for restriction fragment-length polymorphisms (RFLPs) and relationships among elite commercial inbred and virtual hybrid onion populations. J Am Soc Hortic Sci 123(6):1034–1037. https://doi.org/10.21273/jashs.123.6.1034

King JJ, Bradeen JM, Bark O, McCallum JA, Havey MJ (1998b) A low-density genetic map of onion reveals a role for tandem duplication in the evolution of an extremely large diploid genome. Theor Appl Genet 96:52–62. https://doi.org/10.1007/s001220050708

Kofoet A, Kik C, Wietsma WA, De Vries JN (1990) Inheritance of resistance to downy mildew (Peronospora destructor [Berk.] Casp.) from Allium roylei Stearn in the backcross *Allium cepa* L.×(*A. roylei*× *A. cepa*). Plant Breed 105(2):144–149. https://doi.org/10.1111/j.1439-0523.1990.tb00467.x

Krueger SK (1986) Resistance to Fusarium basal rot in onions (*Allium cepa* L.). University of Wisconsin—Madison. https://doi.org/10.1007/BF02979850

Labate JA, Glaubitz JC, Havey MJ (2020) Genotyping by sequencing for SNP marker development in onion. Genome 63:607–613. https://doi.org/10.1139/gen-2020-0011

Lancaster JE, Shaw ML, Randle WM (1998) Differential hydrolysis of alk(en)yl cysteine sulphoxides by alliinase in onion macerates: flavour implications. J Sci Food Agric 78:367–372. https://doi.org/10.1002/(SICI)1097-0010(199811)78:3<367::AID-JSFA127>3.0.CO;2-N

Lee H, Baldwin S, Kenel F, McCallum J, Macknight R (2013) FLOWERING LOCUS T genes control onion bulb formation and flowering. Nat Commun 4:2884. https://doi.org/10.1038/ncomms3884

Lee J-H, Natarajan S, Biswas MK, Shirasawa K, Isobe S, Kim H-T, Park J-I, Seong C-N, Nou I-S (2018) SNP discovery of Korean short day onion inbred lines using double digest restriction site-associated DNA sequencing. PLoS One 13:e0201229. https://doi.org/10.1371/journal.pone.0201229

Lee Y-R, Kim CW, Han J, Choi HJ, Han K, Lee ES, Kim D-S, Lee J, Siddique MI, Lee H-E (2021) Genotyping-by-sequencing derived genetic linkage map and quantitative trait loci for sugar content in onion (Allium cepa L.). Plants (Basel) 10:2267. https://doi.org/10.3390/plants10112267

Lee HM, Park JS, Kim SJ, Kim SG, Park YD (2022) Using transcriptome analysis to explore gray mold resistance-related genes in onion (Allium cepa L.). Gene 13(3):542. https://doi.org/10.3390/genes13030542

Lema M (2018) Marker assisted selection in comparison to conventional plant breeding. Agric Res Technol 14:555914

Liguori L, Califano R, Albanese D, Raimo F, Crescitelli A, Di Matteo M (2017) Chemical composition and antioxidant properties of five white onion (*Allium cepa* L.) landraces. J Food Qual 2017:1–9. https://doi.org/10.1155/2017/6873651

Malik G, Dhatt AS, Malik AA (2021) A review of genetic understanding and amelioration of edible allium species. Food Rev Int 37:415-446. https://doi.org/10.1080/87559129.2019.1709202

Martin WJ, McCallum J, Shigyo M, Jakse J, Kuhl JC, Yamane N, Pither-Joyce M, Gokce AF, Sink KC, Town CD, Havey MJ (2005) Genetic mapping of expressed sequences in onion and in silico comparisons with rice show scant colinearity. Mol Gen Genomics 274:197–204. https://doi.org/10.1007/s00438-005-0007-6

Marzu JC, Straley E, Havey MJ (2018) Genetic analyses and mapping of pink-root resistance in onion. J Am Soc Hortic Sci 143(6):503–507. https://doi.org/10.21273/JASHS04509-18

Masamura N, McCallum J, Khrustaleva L, Kenel F, Pither-Joyce M, Shono J, Suzuki G, Mukai Y, Yamauchi N, Shigyo M (2012) Chromosomal organization and sequence diversity of genes encoding lachrymatory factor synthase

Masuzaki S, Shigyo M, Yamauchi N (2006) Complete assignment of structural genes involved in flavonoid biosynthesis influencing bulb color to individual chromosomes of the shallot (Allium cepa L.). Genes Genet Syst 81:255–263. https://doi.org/10.1266/ggs.81.255

McCallum J, Clarke A, Pither-Joyce M, Shaw M, Butler R, Brash D, Scheffer J, Sims I, Van Heusden S, Shigyo M, Havey MJ (2006) Genetic mapping of a major gene affecting onion bulb fructan content. Theor Appl Genet 112:958–967. https://doi.org/10.1007/s00122-005-0199-5

McCallum J, Pither-Joyce M, Shaw M, Kenel F, Davis S, Butler R, Scheffer J, Jakse J, Havey MJ (2007a) Genetic mapping of sulfur assimilation genes reveals a QTL for onion bulb pungency. Theor Appl Genet 114:815–822. https://doi.org/10.1007/s00122-006-0479-8

McCallum J, Wu X, Shannon JG, Sleper DA, Nguyen HT (2007b) Genome mapping and molecular breeding in plants, vol 5. Chittaranjan K (ed). Springer, Berlin/Heidelberg, pp 21–53

McCallum J, Baldwin S, Thomson S, Pither-Joyce M, Kenel F, Lee R, Khosa JS, Macknight R (2016) Molecular genetics analysis of onion (*Allium cepa* L.) adaptive physiology of bulb. Acta Hortic:71–76. https://doi.org/10.17660/ActaHortic.2016.1110.11

McManus MT, Leung S, Lambert A, Scott RW, Pither-Joyce M, Chen B, McCallum J (2005) Molecular and biochemical characterisation of a serine acetyltransferase of onion, (*Allium cepa* L.). Phytochemistry 66:1407–1416

McManus MT, Joshi S, Searle B, Pither-Joyce M, Shaw M, Leung S, Albert N, Shigyo M, Jakse J, Havey MJ (2012) Genotypic variation in sulfur assimilation and metabolism of onion (*Allium cepa* L.) III. Characterization of sulfite reductase. Phytochemistry 83:34–42

Moshfegh AJ, Friday JE, Goldman JP, Ahuja JKC (1999) Presence of inulin and oligofructose in the diets of Americans. J Nutr 129:1407S–1411S

Munaiz ED, Havey MJ (2020) Genetic analyses of epicuticular waxes associated with the glossy foliage of 'White Persian' onion. J Am Soc Hortic Sci 145(1):67–72. https://doi.org/10.21273/JASHS04840-19

Nakamichi N (2011) Molecular mechanisms underlying the Arabidopsis circadian clock. Plant Cell Physiol 52:1709–1718. https://doi.org/10.1093/pcp/pcr118

Nicholas CG, Gabelman WH, Larson RH, Walker JC (1965) The expression and inheritance of resistance to pink root in onion seedlings. Phytopathology 55:752–756

Park J, Bang H, Cho DY, Yoon MK, Patil BS, Kim S (2013) Construction of high-resolution linkage map of the Ms locus, a restorer-of-fertility gene in onion (Allium cepa L.). Euphytica 192:267–278. https://doi.org/10.1007/s10681-012-0851-5

Pathak CS, Black LL, Cherng SJ, Wang TC, Ko SS (2001) Breeding onions for stemphylium leaf blight resistance. Acta Hortic 555:77–81. https://doi.org/10.17660/ActaHortic.2001.555.7

Platenius H, Knott JE (1935) The pungency of the onion bulb as influenced by the stage of development of the plant. Proc Am Soc Hort Sci:481–483

Putterill J, Robson F, Lee K, Simon R, Coupland G (1995) The CONSTANS gene of Arabidopsis promotes flowering and encodes a protein showing similarities to zinc finger transcription factors. Cell 80:847–857

Qin L, Ma H, Zhang X, Zhang Z, Wang Y (2023) Metabolomics and transcriptomics analyses provides insights into S-alk(en)yl cysteine sulfoxides (CSOs) accumulation in onion (*Allium cepa*). Sci Hortic 310:111727. https://doi.org/10.1016/j.scienta.2022.111727

Raines S, Henson C, Havey MJ (2009) Genetic analyses of soluble carbohydrate concentrations in onion bulbs. J Am Soc Hortic Sci 134:618–623

Randle WM (1992) Onion germplasm interacts with sulfur fertility for plant sulfur utilization and bulb pungency. Euphytica 59:151–156. https://doi.org/10.1007/BF00041267

Sahoo J, Mahanty B, Mishra R, Joshi RK (2023) Development of SNP markers linked to purple blotch resistance for marker-assisted selection in onion (*Allium cepa* L.) breeding. 3 Biotech 13(5):137. https://doi.org/10.1007/s13205-023-03562-7

Sato Y (1998) PCR amplification of CMS-specific mitochondrial nucleotide sequences to identify cytoplasmic genotypes of onion (*Allium cepa* L.). Theor Appl Genet 96(3):367–370. https://doi.org/10.1007/s001220050750

Sawa M, Nusinow DA, Kay SA, Imaizumi T (2007) FKF1 and GIGANTEA complex formation is required for day-length measurement in arabidopsis. Science 318:261–265. https://doi.org/10.1126/science.1146994

Scholten OE, Van Heusden AW, Khrustaleva LI, Burger-Meijer K, Mank RA, Antonise RGC, Harrewijn JL, Van Haecke W, Oost EH, Peters RJ, Kik C (2007) The long and winding road leading to the successful introgression of downy mildew resistance into onion. Euphytica 156:345–353. https://doi.org/10.1007/s10681-007-9383-9

Scholten OE, van Kaauwen MP, Shahin A, Hendrickx PM, Keizer LP, Burger K, van Heusden AW, van der Linden CG, Vosman B (2016) SNP-markers in Allium species to facilitate introgression breeding in onion. BMC Plant Biol 16:1–10. https://doi.org/10.1186/s12870-016-0879-0

Schwinn KE, Ngo H, Kenel F, Brummell DA, Albert NW, McCallum JA, Pither-Joyce M, Crowhurst RN, Eady C, Davies KM (2016) The onion (Allium cepa L.) R2R3-MYB gene MYB1 regulates anthocyanin biosynthesis. Front Plant Sci 7:1865

Sharma SN, Sain RS (2003) RO-1 an improved onion variety for the warmer areas of Rajasthan. Indian J Genet Plant Breed 63(03):281–282

Shigyo M, Iino M, Isshiki S, Tashiro Y (1997) Morphological characteristics of a series of alien monosomic addition lines of Japanese bunching onion (Allium fistulosum L.) with extra chromosomes from shallot (A. cepa L. Aggregatum group). Genes Genet Syst 72:181–186. https://doi.org/10.1266/ggs.72.181

Shigyo M, Fujito S, Sato S (2019) Toward repositioning Allium vegetable crops as global commodities: opportunities and challenges for future application of omics and phytomedomics platforms to crop improvement. Acta Hortic:1–8. https://doi.org/10.17660/ActaHortic.2019.1251.1

Song YH, Ito S, Imaizumi T (2013) Flowering time regulation: photoperiod-and temperature-sensing in leaves. Trends Plant Sci 18:575–583

Song S, Kim C-W, Moon JS, Kim S (2014) At least nine independent natural mutations of the DFR-A gene are responsible for appearance of yellow onions (Allium cepa L.) from red progenitors. Mol Breed 33:173–186. https://doi.org/10.1007/s11032-013-9942-9

Straley E, Colcol Marzu J, Havey MJ (2021) Genetic analyses of resistance to fusarium basal rot in onion. Horticulturae 7(12):538. https://doi.org/10.3390/horticulturae7120538

Suleria HAR, Butt MS, Anjum FM, Saeed F, Khalid N (2015) Onion: nature protection against physiological threats. Crit Rev Food Sci Nutr 55:50–66. https://doi.org/10.1080/10408398.2011.646364

Sun X, Zhou S, Meng F, Liu S (2012) De novo assembly and characterization of the garlic (Allium sativum) bud transcriptome by Illumina sequencing. Plant Cell Rep 31:1823–1828. https://doi.org/10.1007/s00299-012-1295-z

Sun XD, Ma GQ, Cheng B, Li H, Liu SQ (2013) Identification of differentially expressed genes in shoot apex of garlic (*Allium sativum* L.) using Illumina sequencing. J Plant Stud 2:136–148. https://doi.org/10.5539/jps.v2n2p136

Sun X-D, Yu X-H, Zhou S-M, Liu S-Q (2016) De novo assembly and characterization of the Welsh onion (*Allium fistulosum* L.) transcriptome using Illumina technology. Mol Gen Genomics 291:647–659. https://doi.org/10.1007/s00438-015-1131-6

Tang H, Xie Y, Liu YG, Chen L (2017) Advances in understanding the molecular mechanisms of cytoplasmic male sterility and restoration in rice. Plant Reprod 30:179–184. https://doi.org/10.1007/s00497-017-0308-z

Taylor A, Massiah AJ, Thomas B (2010) Conservation of Arabidopsis thaliana photoperiodic flowering time genes in onion (Allium cepa L.). Plant Cell Physiol 51:1638–1647

Taylor A, Vagany V, Barbara DJ, Thomas B, Pink DAC, Jones JE, Clarkson JP (2013) Identification of differential resistance to six Fusarium oxysporum f. sp. cepae isolates in commercial onion cultivars through the development of a rapid seedling assay. Plant Pathol 62(1):103–111. https://doi.org/10.1111/j.1365-3059.2012.02624.x

Taylor A, Teakle GR, Walley PG, Finch-Savage WE, Jackson AC, Jones JE, Hand P, Thomas B, Havey MJ, Pink DAC, Clarkson JP (2019) Assembly and characterisation of a unique onion diversity set identifies resistance to Fusarium basal rot and improved seedling vigour. Theor Appl Genet 132:3245–3264. https://doi.org/10.1007/s00122-019-03422-0

Thomson SJ, Rippon P, Butts C, Olsen S, Shaw M, Joyce NI, Eady CC (2013) Inhibition of platelet activation by lachrymatory factor synthase (LFS)-silenced (Tearless) onion juice. J Agric Food Chem 61:10574–10581. https://doi.org/10.1021/jf4030213

Thomson B, Zheng B, Wellmer F (2017) Floral organogenesis: when knowing your ABCs is not enough. Plant Physiol 173:56–64

Tripathy P, Priyadarshini A, Das SK, Sahoo BB, Dash DK (2013) Evaluation of onion (*Allium cepa* L.) genotypes for tolerance to thrips (*Thrips tabaci* L.) and purple blotch [*Alternaria porri* (Ellis) Ciferri]. Int J Bio-Resour Stress Manag 4(Dec, 4):561–564. https://doi.org/10.13140/RG.2.2.11126.63042

Tsukazaki H, Yamashita K-i, Yaguchi S, Yamashita K, Hagihara T, Shigyo M, Kojima A, Wako T (2011) Direct determination of the chromosomal location of bunching onion and bulb onion markers using bunching onion–shallot monosomic additions and allotriploid-bunching onion single alien deletions. Theor Appl Genet 122:501–510. https://doi.org/10.1007/s00122-010-1464-9

Tsukazaki H, Yaguchi S, Yamashita K, Wako T (2017) QTL analysis of morphological traits and pseudostem pigmentation in bunching onion (*Allium fistulosum*). Euphytica 213:152. https://doi.org/10.1007/s10681-017-1944-y

Tsutsui K (1986) Inheritance of resistance to fusarium basal rot in onions (*Allium cepa* L.). University of Wisconsin–Madison

Van der Meer QP, De Vries JN (1990) An interspecific cross between *Allium roylei* Stearn and *Allium cepa* L., and its backcross to *A. cepa*. Euphytica 47:29–31. https://doi.org/10.1007/BF00040359

Van Heusden AW, Van Ooijen JW, Vrielink-van Ginkel R, Verbeek WHJ, Wietsma WA, Kik C (2000) A genetic map of an interspecific cross in *Allium* based on amplified fragment length polymorphism (AFLP TM) markers. Theor Appl Genet 100:118–126. https://doi.org/10.1007/s001220050017

Varshney RK, Terauchi R, McCouch SR (2014) Harvesting the promising fruits of genomics: applying genome sequencing technologies to crop breeding. PLoS Biol 12:e1001883. https://doi.org/10.1371/journal.pbio.1001883

Vavrina CS, Smittle DA (1993) Evaluating sweet onion cultivars for sugar concentrations and pungency. HortScience 28:804–806

Veeraghanti KS, Naik BG, Hegde KT (2017) Management of Purple blotch disease of Onion under field condition. J Pharmacogn Phytochem 6(6):1768–1769

Vijn I, Smeekens S (1999) Fructan: more than a reserve carbohydrate? Plant Physiol 120:351–360

Von Kohn C, Kiełkowska A, Havey MJ (2013) Sequencing and annotation of the chloroplast DNAs and identification of polymorphisms distinguishing normal male-fertile and male-sterile cytoplasms of onion. Genome 56(12):737–742. https://doi.org/10.1139/gen-2013-0182

Vu HQ, Iwata M, Yamauchi N, Shigyo M (2011) Production of novel alloplasmic male sterile lines in *Allium cepa* harbouring the cytoplasm from *Allium roylei*. Plant Breed 130:469–475. https://doi.org/10.1111/j.1439-0523.2011.01855.x

Wako T, Tsukazaki H, Yaguchi S, Yamashita K, Ito S, Shigyo M (2016) Mapping of quantitative trait loci for bolting time in bunching onion (*Allium fistulosum* L.). Euphytica 209:537–546. https://doi.org/10.1007/s10681-016-1686-2

Whittaker C, Dean C (2017) The FLC locus: a platform for discoveries in epigenetics and adaptation. Annu Rev Cell Dev Biol 33:555–575. https://doi.org/10.1146/annurev-cellbio-100616-060546

Yamashita K, Takatori Y, Tashiro Y (2005) Chromosomal location of a pollen fertility-restoring gene, Rf, for CMS in Japanese bunching onion (*Allium fistulosum* L.) possessing the cytoplasm of *A. galanthum* Kar. et Kir. revealed by genomic in situ hybridization. Theor Appl Genet 111:15–22. https://doi.org/10.1007/s00122-005-1941-8

Yang YY, Huo YM, Miao J, Liu BJ, Kong SP, Gao LM, Liu C, Wang ZB, Tahara Y, Kitano H, Wu X (2013) Identification of two SCAR markers co-segregated with the dominant Ms and recessive ms alleles in onion (Allium cepa L.). Euphytica 190:267–277. https://doi.org/10.1007/s10681-012-0842-6

Yoo SK, Chung KS, Kim J, Lee JH, Hong SM, Yoo SJ, Yoo SY, Lee JS, Ahn JH (2005) Constans activates suppressor of overexpression of constans 1 through Flowering Locus T to promote flowering in Arabidopsis. Plant Physiol 139:770–778

Zhang C, Zhang H, Zhan Z, Liu B, Chen Z, Liang Y (2016) Transcriptome analysis of sucrose metabolism during bulb swelling and development in onion (*Allium cepa* L.). Front Plant Sci 7:1425. https://doi.org/10.3389/fpls.2016.01425

Advances in GWAS and QTL Mapping in Garden Pea

13

Brij Bihari Sharma, Akanksha Singh, and Vinod Kumar Sharma

Abstract

Pea (*Pisum sativum*), a vital cool-season pulse crop, is cultivated in over 100 countries across approximately 12 million hectares globally. The garden pea, an important annual leguminous vegetable, belongs to the family Fabaceae—the third-largest family of flowering plants, comprising over 450 genera and approximately 1200 species. The primary centre of origin of pea is believed to be Central Asia, with the Near East recognized as a secondary centre of diversity. India ranks as the second-largest producer of green peas, following China. Genome-wide association studies (GWAS), which focus on single-nucleotide polymorphisms (SNPs), enable researchers to analyse hundreds to thousands of SNPs across large populations simultaneously. Both GWAS and quantitative trait loci (QTL) mapping are critical tools in garden pea breeding, offering valuable insights into the genetic architecture of key agronomic traits such as yield, disease resistance, and nutritional quality. By identifying specific genomic regions associated with these traits, breeders can more efficiently select for desirable characteristics, thereby accelerating the development of improved cultivars. GWAS capitalizes on the natural genetic variation present within pea populations to detect associations between genetic markers and phenotypic traits, while QTL mapping aids in pinpointing genomic regions that govern quantitative traits. Collectively, these approaches contribute to the development of superior pea

B. B. Sharma (✉) · A. Singh
Department of Vegetable Science, College of Horticulture and Forestry, Rani Lakshmi Bai Central Agricultural University, Jhansi, Uttar Pradesh, India

V. K. Sharma
Division of Germplasm Evaluation, ICAR- National Bureau of Plant Genetic Resources, New Delhi, India

S. S. Dey et al. (eds.), *GWAS and QTL Mapping in Horticultural Crops*,
https://doi.org/10.1007/978-981-95-2306-1_13

varieties capable of addressing agricultural challenges and meeting consumer demands, ultimately supporting sustainable crop production and global food security.

Keywords

Garden pea · Single-nucleotide polymorphisms · Genome-wide association studies · Quantitative trait loci · Biotic and abiotic stresses

13.1 Introduction

Pea (*Pisum sativum* L., $2n = 14$) is a significant cool-season pulse crop cultivated in more than 100 countries across over 12 million hectares globally (FAOSTAT 2016; http://www.fao.org/faostat/en/#data/QC). The garden pea (*Pisum sativum var. hortense* L.) is an important annual leguminous vegetable crop. It belongs to the Fabaceae family, the third-largest family of flowering plants, comprising over 450 genera and approximately 1200 species (Stagnari et al. 2017). There are six recognized species within the *Pisum* genus: *P. sativum* (garden pea), *P. arvense* (field pea), *P. elatius* (Mediterranean pea), *P. abyssinicum* (Abyssinian pea), *P. humile* (dwarf pea), and *P. fulvum* (red-yellow pea). The primary centre of origin of pea is believed to be Central Asia, with the Near East considered a secondary centre of origin (Mohan et al. 2013). India ranks as the second-largest producer of green peas in the world, following China. The annual global production is approximately 14.5 million tons for green peas and 22 million tons for dry pea seeds. Within India, the top ten contributing states to garden pea production are Uttar Pradesh, Madhya Pradesh, Punjab, Jharkhand, Himachal Pradesh, West Bengal, Chhattisgarh, Haryana, Uttarakhand, and Manipur (Handbook of Horticulture 2021). Globally, the area under pea cultivation and overall production have steadily increased over time. In 1970, global pea production was approximately 4.7 million tons, with a harvested area of around 764 thousand hectares. By 2021, production had risen to 20 million tons, with the harvested area expanding to 2.5 million hectares (FAOSTAT 2021). According to recent data given by NHB shows that pea occupies 549 thousand ha area with 5680 thousand MT production in Indian condition in 2021-22. Uttar Pradesh alone produces 49% of total production of pea in India. (Anonymous 2022).

Genome-wide association studies (GWAS) are a powerful genetic research approach aimed at identifying genetic variants associated with specific traits or diseases. The term “genome-wide” refers to the comprehensive nature of the analysis, where a large number of genetic markers across the entire genome are examined to detect statistically significant associations. GWAS employs a systematic strategy to scan the genome—primarily focusing on variations known as single-nucleotide polymorphisms (SNPs)—to uncover correlations between genetic variants and phenotypic traits or disease susceptibility. These studies are typically conducted by comparing the genetic data of individuals with a particular trait or disease to those without it. In a typical GWAS, researchers collect DNA samples from a large number of individuals and analyse them to detect SNPs, which are single-base variations (A, T, C, or G) at specific positions in the genome. Statistical analyses are then performed to identify SNPs that are more frequently present in individuals with the trait or disease of interest. If a specific variant is found to be significantly more common in affected individuals, it is considered a potential genetic marker for that condition. The results of GWAS are often validated through replication studies using independent populations to ensure that the associations are robust and not due to random chance. Beyond human medicine, GWAS has important applications in agriculture, where it is used to identify genes associated with desirable traits in crops and livestock. This knowledge supports the development of more resilient, productive, and high-quality agricultural systems by enabling marker-assisted selection (MAS) and genomic breeding strategies. In essence, GWAS serves as a genetic "detective tool", offering valuable insights into the complex relationships between genotype and phenotype across a wide range of organisms and traits.

13.2 Quantitative Trait Loci (QTL)

Quantitative trait loci (QTL) are specific genomic regions associated with the variation of quantitative traits—traits that exhibit a continuous range of phenotypes and are influenced by the combined effects of multiple genes and environmental factors. QTL analysis involves the use of statistical methods to identify and map the chromosomal locations where these genetic factors contribute to observed trait variation. This approach enables researchers to dissect the complex genetic architecture underlying traits such as disease resistance, physiological characteristics, and behavioural or developmental traits. QTL analysis is a cornerstone of marker-assisted breeding (MAB), as it allows plant breeders to more efficiently identify and select for desirable traits. By pinpointing genomic regions associated with important agronomic traits, QTL analysis helps in understanding the genetic basis of trait inheritance. Once QTLs are identified, molecular markers linked to these loci can be developed. These markers act as proxies for the presence of favourable alleles associated with the traits of interest.

Using these markers, breeders can screen plant populations for the desired alleles without the need to wait for full phenotypic expression, thereby significantly accelerating the breeding process. This is especially valuable for complex traits governed by multiple genes, where conventional breeding methods are often slower and less precise. Marker-assisted selection (MAS), facilitated by QTL mapping, enables more precise and targeted breeding by focusing only on the genomic regions linked to desirable traits. This reduces the risk of inadvertently selecting for undesirable characteristics and enhances the efficiency of developing superior crop varieties. In summary, QTL analysis provides foundational knowledge for marker-assisted breeding, empowering breeders to make informed, data-driven decisions. This accelerates the development of improved crop varieties with enhanced traits, contributing to increased productivity, resilience, and overall agricultural sustainability.

Quantitative trait loci (QTLs) play a pivotal role in gene discovery by providing insights into genomic regions associated with specific traits. QTL analysis enables researchers to identify regions of the genome linked to the expression of particular phenotypes. These regions often harbour candidate genes that may influence the trait of interest. Once a QTL is identified, researchers can narrow the focus to specific genes within that genomic region that are likely responsible for the observed phenotypic variation. Functional analyses of these candidate genes are then conducted to elucidate their roles and mechanisms in trait expression. This includes investigating their molecular functions, biological pathways, and gene–gene or gene–environment interactions. While QTL mapping provides a valuable starting point for gene discovery, further validation is essential. Experimental approaches such as gene expression profiling, genetic transformation, or (Clustered regularly interspaced short palindromic repeats and CRISPR-associated protein 9) CRISPR/Cas-mediated gene editing can be employed to confirm the involvement of specific genes in controlling the trait.

QTL studies significantly contribute to our understanding of the genetic architecture underlying complex, polygenic traits. This knowledge is particularly useful in dissecting traits influenced by multiple genes with minor effects. Furthermore, molecular markers associated with identified QTLs can be developed for use in marker-assisted selection (MAS), thereby facilitating the efficient screening and selection of individuals possessing desirable genetic attributes. In addition, QTL information can be utilized in comparative genomics to identify conserved genetic loci across species, enabling the discovery of shared mechanisms governing similar traits in diverse organisms. In essence, QTL mapping serves as a foundational tool for gene discovery, directing researchers to genomic regions most likely to contain functionally relevant genes. This insight is instrumental in enhancing our understanding of the genetic basis of traits and in developing precise strategies for breeding and genetic improvement. QTL introgression involves the transfer of specific genomic regions associated with quantitative traits from a donor population or species into a recipient population. The goal is to introduce beneficial alleles associated with traits of interest—such as yield, stress tolerance, or disease resistance—into the genetic background of the target population. QTL introgression presents numerous possibilities for crop improvement and genetic enhancement by combining traditional breeding approaches with modern molecular tools.

The QTL introgression process begins with the identification of a donor population that possesses desirable QTLs associated with traits of interest. A second population, known as the recipient, is selected as the target for introgression. QTLs linked to the target traits are first identified in the donor population through QTL mapping, which involves associating specific genomic regions with the expression of quantitative traits. Once these QTLs are identified, molecular markers closely linked to the target QTLs are developed. These markers serve as tools to track the presence and inheritance of the desired genomic regions during the introgression process. Crossbreeding is then carried out between individuals from the donor and recipient populations. Using marker-assisted selection (MAS), individuals carrying the desired QTLs from the donor genome are selectively bred into the genetic background of the recipient population. To ensure the stability of the introgressed traits and to minimize the introduction of unwanted donor genetic material, backcrossing is commonly employed. This involves crossing hybrid offspring back to the recipient parent over multiple generations, while continuing to select for individuals carrying the target QTLs using molecular markers. Following backcrossing, phenotypic evaluation is conducted to confirm that the introgressed QTLs confer the expected traits in the recipient background. This step ensures that the desired traits are not only genetically present but also functionally expressed in the recipient population.

Possibilities and Applications

1. *Crop Improvement:* QTL introgression is extensively utilized in crop improvement programs to transfer valuable traits—such as disease resistance, yield enhancement, and stress tolerance—from wild or exotic germplasm into cultivated varieties.
2. *Conservation Biology:* QTL introgression can be employed in conservation strategies to enhance the genetic diversity and adaptive capacity of endangered populations by introducing advantageous traits from closely related populations.
3. *Research and Development:* QTL introgression offers a powerful approach for researchers to elucidate the genetic basis of complex traits and to develop improved varieties or breeds possessing specific desirable characteristics.

While QTL introgression presents promising opportunities for trait enhancement, it necessitates meticulous planning, continuous monitoring, and rigorous validation to ensure the successful transfer and stable expression of the desired traits within the recipient population.

(I) *Agronomic traits*
(II) *Flowering-related traits*

Earliness

Earliness in crop plants plays a crucial role in mitigating heat-induced drought stress and addressing the challenges posed by global warming, which is expected to intensify in the near future. Moreover, early maturing varieties can extend production into northern regions or higher altitudes characterized by shorter growing seasons, thereby expanding arable land and contributing to global food security. The primary objective of the present study was to identify quantitative trait loci (QTLs) associated with super-earliness and key agro-morphological traits in a recombinant inbred line (RIL) population derived from an interspecific cross. A total of 114 RILs, developed through single-seed descent from a cross between *Pisum sativum* L. and *P. fulvum* Sibth. et Sm., were evaluated for this purpose. A genetic linkage map was constructed using 44 simple sequence repeat (SSR) markers distributed across seven chromosomes, spanning a total genetic distance of 262.6 centiMorgans (cM) (Sari et al. 2023).

Days to Flowering

Two QTLs associated with flowering time were identified on linkage group 2 (LGII). The first QTL, designated FLOR2.1, exhibited a logarithm of odds (LOD) score of 3.6. The marker AA205 was the closest to this QTL, which explained 14% of the phenotypic variance with a LOD of 3.2. The second QTL, FLOR2.2, also accounted for 14% of the phenotypic variance and had a LOD score of 3.2. This QTL was closely linked to the marker AA372.1. Both QTLs showed dominance/additive (d/a) ratios of 1.06 and 0.71, respectively, indicating dominant and partially dominant modes of gene action. Together, these QTLs explained 28% of the phenotypic variance for flowering time. Additionally, the Ppd and Lf loci were mapped on LGII, while the Hr and Dne loci were localized on linkage group 3 (LGIII; Sari et al. 2023).

Plant Height

Only one QTL associated with plant height was identified, located on linkage group 3 (LGIII) with a LOD score of 4.35. The QTL, designated PH3.1, was flanked by markers AD73 and AD270, and explained 16% of the phenotypic variation. The dominance/additive (d/a) ratio for PH3.1 was 1.65, indicating an overdominance mode of gene action (Sari et al. 2023).

(III) *Fruiting-related and yield-related traits*

Pods per Plant

Six QTLs associated with the number of pods per plant, a key yield component, were identified. Three QTLs—NP2.1, NP2.2, and NP2.3—were mapped on linkage group 2 (LII2), while one QTL each was located on linkage group 4 (LGIV; NP4.1), linkage group 5 (LGV; NP5.1), and linkage group 6 (LGVI; NP6.1). The LOD scores for NP2.1, NP2.2, and NP2.3 were 4.3, 3.5, and 3.6, respectively. NP2.1 and NP2.2 explained 16% and 13% of the phenotypic variation, respectively, while NP2.3, NP4.1, and NP5.1 accounted for 13%, 14%, and 14% of the variation, respectively. The QTL NP6.1 explained 17% of the phenotypic variance. Collectively, these six QTLs accounted for 87% of the variation in the number of pods per plant. Gene action analysis indicated overdominance for NP2.1, NP2.2, NP5.1, and NP6.1 (Sari et al. 2023).

Pod Length

Only one QTL for pod length, designated PL4.1, was identified. This QTL explained 12% of the phenotypic variation for pod length (Sari et al. 2023).

Table 13.1 QTL mapped for harvest index in garden pea

Characters	Mapped QTL	Linkage group	References
Yield attributing traits			
Days to flowering	*FLO2.1, FLO2.2*	LGII	Sari et al. (2023)
Plant height	*PH3.1*	LGIII	Sari et al. (2023)
Pods per plant	*NP2.1, NP2.2, NP2.3*	LGII	Sari et al. (2023)
	NP4.1	LGIV	
	NP5.1	LGV	
Seeds per pod	*NS6.1*	LGVI	Sari et al. (2023)
Pod length	*PL4.1*	LGIV	Sari et al. (2023)
Harvest index	*HX3.1*	LGIII	Sari et al. (2023)
	HX5.1	LGV	
	HX6.1	LGVI	

Harvest Index

Three QTLs associated with harvest index were identified: HX3.1, HX5.1, and HX6.1. The QTL HX3.1 explained 13% of the phenotypic variation, while HX5.1 and HX6.1, located on linkage group 6 (LGVI), each accounted for 12% of the phenotypic variation for this trait (Sari et al. 2023) (Table 13.1).

(IV) *Biotic stresses*

***Fusarium* Root Rot**

Fusarium root rot, caused by a complex of *Fusarium* species, is a major disease affecting field pea (*Pisum sativum*). Developing genetic resistance remains the most promising strategy for disease management; however, no pea germplasm has been identified with complete resistance to root rot. The objective of this study was to identify quantitative trait loci (QTL) conferring partial resistance to root rot and wilting caused by five fungal isolates representing *Fusarium solani*, *F. avenaceum*, *F. acuminatum*, *F. proliferatum*, and *F. graminearum*. Evaluation of the root rot-tolerant cultivar “00-2067” and the susceptible cultivar “Reward” was conducted with all five isolates. Significant differences ($p < 0.001$) in mean root rot severity were observed between the two cultivars when inoculated with *F. avenaceum* (F4A) and *F. graminearum* (FG2). Consequently, QTL analysis focused on the F8 recombinant inbred line (RIL) population derived from the cross “Reward” × “00-2067”, which was inoculated in the greenhouse four times with only F4A and FG2 isolates. Genotyping of the parental lines and F8 population was performed using 13,200 single-nucleotide polymorphisms (SNPs) and 222 simple sequence repeat (SSR) markers. Significant genotypic effects ($p < 0.05$) and high heritability estimates (79–92.1%) were observed for disease severity, vigour, and plant height following inoculation with both isolates. Significant correlations among these traits suggested that a substantial proportion of the genetic variance was transmitted from parents to progeny.

No significant QTL (LOD > 3) were detected in the RILs inoculated with F4A. However, in the RILs inoculated with FG2, five QTLs associated with root rot severity and three QTLs each for vigour and plant height were identified. The most stable QTL for plant height, Hgt-Ps3.1, was located on chromosome 5/linkage group III. Two stable QTLs conferring partial resistance to FG2, Fg-Ps4.1 and Fg-Ps4.2, were mapped to a 15.1-cM and 11.2-cM genomic region, respectively, on chromosome 4/linkage group IV. The most stable QTL for vigour, Vig-Ps4.1, was also found within this region. Furthermore, 25 major and moderate-effect digenic epistatic interactions were detected, underscoring the complexity of genetic control for these traits. The identified region on chromosome 4/linkage group IV represents a promising target for resistance breeding and marker-assisted selection.

Fusarium root rot is a common biotic constraint limiting pea yields, and genetic resistance represents the most practical approach for improving pea production (Tables 13.2 and 13.3). This study aimed to identify quantitative trait loci (QTL) controlling partial genetic resistance to *Fusarium solani* (Mart.) Sacc. f.sp. *pisi* (F.R. Jones) W.C. Snyder & H.N. Hans (Fsp), the causal agent of *Fusarium* root rot. A recombinant inbred line (RIL) population was evaluated in a *Fusarium* root rot field disease nursery over three consecutive years. Composite interval mapping was

Table 13.2 QTL mapped for *Fusarium* root rot in garden pea

I.*Fusarium* root rot	*Fg-Ps3.1, Fg-Ps3.2, Vig-Ps3.1, Vig-Ps3.2, Hgt-Ps3.1*	Chrom5/LGIII	Wu et al. (2022)
	Fg-Ps4.1, Fg-Ps4.2, Vig-Ps4.1	Chrom4/LGIV	
	Fg-Ps5.1	Chrom3/LGV	
	Hgt-Ps7.1, Hgt-Ps7.2	Chrom7/LGVII	Coyne et al. (2015)
	Fsp-Ps2.1	II	
	Fsp-Ps3.1	III	
	Fsp-Ps4.1	IV	
	Fsp-Ps6.1	VI	
	Fsp-Ps6.1	VII	

Table 13.3 QTL mapped for *Fusarium* wilt in garden pea

Fusarium wilt, race 1	*Fw*	IV	Dirlewanger et al. (1994)
	Fw	III	McClendon et al. (2002), Okubara et al. (2005), Loridon et al. (2005), Kwon et al. (2013), and Jain et al. (2015)
	Fw	I, II, III, V, VI, VII	Cheng et al. (2015)
Fusarium wilt, race 2	*Fnw*	–	McPhee et al. (2012)
	Fnw 4.1, Fnw 3.1, Fnw 3.2	–	McPhee et al. (2012)
Fusarium wilt, Race5	*Fwf*	II	Okubara et al. (2005) and Coyne et al. (2019)

performed using the mean disease severity scores across the three growing seasons to detect QTLs. Five QTLs were identified, including one stable QTL detected consistently in all 3 years. The multiyear QTL, Fsp-Ps2.1, explained a substantial proportion of the phenotypic variance, ranging from 22.1% to 72.2%. Another QTL, Fsp-Ps6.1, contributed 17.3% of the phenotypic variance. Additional QTLs detected in single years are noteworthy as they colocate with previously reported *Fusarium* root rot resistance QTLs in pea. The QTLs Fsp-Ps2.1, Fsp-Ps3.1, Fsp-Ps4.1, and Fsp-Ps7.1 are flanked by codominant simple sequence repeat (SSR) markers, making them promising candidates for marker-assisted selection in breeding programs aimed at enhancing partial resistance to *F. solani* in pea (Wu et al. 2022).

13.3 Candidate Genes

The QTLs associated with partial resistance to *Fusarium graminearum* located on chromosome 5/linkage group III and chromosome 4/linkage group VI were found to flank 4 and 74 candidate genes, respectively. Among the 74 genes on chromosome 4/LGVI, 15 were implicated in plant defence mechanisms. These genes are involved in various biological functions including UDP (Uridine diphosphate) formation and transport, integral membrane protein components, histone-lysine N-methyl transferase activity, phospholipid transport, actin cytoskeleton dynamics, calcium-ion binding, methyl transferase activity, ubiquitin-conjugating enzyme/ RWD (RING finger-containing proteins, WD repeat-containing proteins, and DEAD (DEXD)-like helicases) domain functions, AP-4 adaptor complex formation, oxidoreductase activity, acyl group transfer, hydrolase activity, G protein-coupled receptor signalling, as well as protein phosphorylation and proteolysis. Notably, some genes participate directly in pathways related to plant defence. For example, *Psat4g125440* is involved in cellulose biosynthesis, which contributes to cell wall reinforcement. Additionally, *Psat4g111280*, *Psat4g110800*, *Psat4g108480*, and *Psat4g102720* are associated with protein ubiquitination, a key regulatory mechanism in plant immune responses.

Powdery Mildew

Pea is one of the most important and productive cool season pulse crops cultivated worldwide. Biotic stresses constitute significant constraints to realizing the full-yield potential of pea, necessitating dedicated research and development efforts. The integration of omics resources and advanced breeding technologies is essential to accelerate the development of high-yielding, multiple stress-tolerant, and resistant varieties in a timely manner. Recently, the pea research community has made notable progress in both conventional and molecular breeding approaches to enhance genetic gain. Numerous quantitative trait loci (QTLs) and molecular markers linked to genes conferring resistance to major diseases—such as *Fusarium* wilt, *Fusarium* root rot, powdery mildew, *Ascochyta* blight, rust, common root rot, broomrape, pea enation, and pea seed-borne mosaic virus—have been identified and are available for marker-assisted selection.

Advanced genomic tools, including comprehensive genetic maps and reliable linked DNA markers, offer great potential for the effective introgression of resistance genes from diverse sources, thereby accelerating genetic improvement in pea. This chapter summarizes recent achievements in the development of genetic and genomic resources, the inheritance patterns of genes controlling various biotic stress responses, characterization of pathogen-related genes, mapping of genes and QTLs, as well as advances in transcriptomics and proteomics. Moreover, emerging breeding strategies—such as transgenics, genome editing, genomic selection, epigenetic breeding, and speed breeding—hold considerable promise to revolutionize pea breeding. The strategic integration of conventional and modern omics-enabled approaches is expected to significantly augment genetic gain and expedite the development of biotic stress-resistant cultivars, thereby sustaining pea production under changing climatic conditions. This chapter consolidates, in a single platform, the research accomplishments to date in pea improvement concerning major biotic stresses and outlines future directions to enhance pea productivity through advanced genomic tools and technologies (Parihar et al. 2022) (Table 13.4).

Pea Weevil

Pea seed yield and quality losses caused by pea weevil (*Bruchus pisorum*) worldwide underscore the urgent need to develop resistant cultivars. However, this effort is hindered by the scarcity of resistance sources and limited understanding of the genetic inheritance of resistance. While some sources of partial resistance exist, they are predominantly found in wild relatives or nonelite germplasm, necessitating extensive crossing and rigorous selection. This process is further complicated by environmental influences on the insect's life cycle and infestation levels. The availability of molecular markers is crucial to facilitate efficient selection for resistance.

Table 13.4 QTL mapped for Powdery mildew in garden pea

Powdery mildew	*er1*	VI	Dirlewanger et al. (1994), Timmerman et al. (1994), Rakshit (1997), Tiwari et al. (1998), Janila and Sharma (2004), Ek et al. (2005), Loridon et al. (2005), Tonguç and Weeden (2010), Pereira et al. (2010), Nisar and Ghafoor (2011), Srivastava et al. (2012), Sun et al. (2015), Wang et al. (2015), Ma et al. (2017), and Sun et al. (2019)
	er3	IV	Fondevilla et al. (2012)
	er2	III	Katoch et al. (2010)
	er1-5	VI	Pavan et al. (2011)
	er1-1, er1-4, er1-2, er1-3, er1-5	VI	Pavan et al. (2013)
	er1-7	VI	Sun et al. (2016)
	er1-6	VI	Sun et al. (2016)
	er1-8, er1-9	VI	Sun et al. (2016)

Furthermore, identifying QTLs associated with pea resistance offers the added advantage of combining multiple QTLs conferring resistance to a particular pest or disease, or even different diseases, within a single cultivar.

A recently developed integrated genetic linkage map, constructed from an advanced recombinant inbred line (RIL) population (F8:9) derived from an interspecific cross between *P. sativum* ssp. *syriacum* and *P. sativum* ssp. *sativum*, was genotyped using DArTseq technology. This map comprises seven linkage groups (LGs) spanning a total length of 2503.6 cM, with an average marker density of 2.61 markers per cM and an average intermarker distance of 1.49 cM. A total of 6447 DArTseq-derived markers were polymorphic across a panel of 108 RILs, exhibiting a robust average polymorphic information content (PIC) value of 0.44 (ranging from 0.34 to 0.51), indicating their high informativeness and utility for marker-assisted selection in pea breeding programs (Table 13.5).

Aschochyta Blight

Ascochyta blight of pea (*Pisum sativum* L.) is a globally significant disease caused by *Didymella pinodes*, *Phoma medicaginis* var. *pinodella*, and, in South Australia, *P. koolunga*. This study aimed to identify candidate genes linked to quantitative trait loci (QTL) conferring field resistance to Ascochyta blight and to examine the role of epistatic interactions (Table 13.6).

Table 13.5 QTL mapped for pea weevil in garden pea

Pea weevil	*BpSI.I*	I	Aznar-Fernández et al. (2020)
	BpSI.II	II	
	BpSI.III, BpLD.I	IV	
	ApI, ApII, ApIII, ApIV.1, ApIV.2, ApV	I, II, III, IV, and V	Barilli et al. (2020)

Table 13.6 QTL mapped for *Aschochyta* blight in garden pea

I.I.I.I.I.I.I. *Aschochyta* blight	*Asc1.1*	I	Timmerman-Vaughan et al. (2016)
	Asc2.1	II	
	Asc3.1, Asc3.2	III	
	Asc4.2, Asc4.3	IV	
	Asc5.1	V	
	Asc7.1, Asc7.2, Asc7.3	VII	
	RGA-G3A, RGA2.97, PsPRP4A, Peachi21, PsMnSOD, DRR230-b, PsDof1, pea beta glucan, and DRR49a, QTL mpIII-4 Asc2.1, Asc4.2, Asc4.3, and Asc7.1 QTL, 14 candidate genes abI-IV-1, abI-IV-2, abI-IV-3, abI-IV-4, abIII1, abVII-1, abI-IV-5, abIII-2, abVII-2 abI-IV-2.1 and abI-IV-2.2 3 MTAs	LG2, 3, 7, 4, 6 – LG1, 2, 3, 4, 7 – –	Jha et al. (2023)

Starting with 101 defence-related genes from published literature, synteny with *Medicago truncatula* was employed to refine candidates for mapping. Fourteen pea candidate genes were mapped in two QTL populations, A26 × Rovar and A88 × Rovar. Four candidate genes defined QTL peaks or intervals for Asc2.1, Asc4.2, Asc4.3, and Asc7.1, with three additional genes located within 1.0 LOD confidence intervals. Epistatic interactions, including QTL × background marker and background marker × background marker effects, contributed significantly to the disease resistance phenotypes in both populations. Each population showed five pairwise interactions surpassing a 5% false discovery rate threshold, with two candidate genes involved in significant interactions. Three genomic regions harboured markers participating in multiple epistatic interactions. This study identifies defence-related candidate genes valuable for marker-assisted selection and highlights the complex genetic architecture of Ascochyta blight resistance, which includes nonadditive epistatic effects important for breeders to consider (Gail et al. 2016).

Bruchid

Pea (*Pisum sativum* L.) is an important legume crop for food and feed, but it is threatened by bruchid pests (*Callosobruchus* spp.) in the field and during storage. This study identified a major quantitative trait locus (QTL) controlling seed resistance to *C. chinensis* and *C. maculatus* using F_2 populations derived from a cross between the resistant cultivar PWY19 and the susceptible PHM22. QTL analysis across two F_2 populations grown in different environments consistently detected a single major QTL, qPsBr2.1, on linkage group 2 between markers 18339 and PSSR202109. This QTL explained 50.91–70.94% of the resistance variation, depending on the environment and bruchid species. Fine mapping localized qPsBr2.1 to a 1.07 Mb region on chromosome 2 (chr2LG1), containing seven annotated genes. Among these, *Psat2g026280* (designated *PsXI*), encoding a xylanase inhibitor, emerged as a candidate gene for bruchid resistance. PCR and sequence analyses revealed an insertion of unknown length in an intron of *PsXI* in PWY19, leading to variation in the open reading frame. Additionally, the subcellular localization of *PsXI* differed between the resistant PWY19 and susceptible PHM22. Together, these results suggest that *PsXI*, encoding a xylanase inhibitor, is responsible for the bruchid resistance observed in the field pea PWY19 (Yan et al. 2023) (Table 13.7).

Table 13.7 QTL mapped for bruchid in garden pea

Bruchid	*qPsBr2.1, qPsBr2.2*	2	Yan et al. (2023)

(V) *Abiotic stress*

Table 13.8 QTL mapped for frost stress tolerance in garden pea

Frost stress tolerance	*WFD 3.2*	LGII	Beji et al. (2020)
	WFD 5.1	LGI	
	WFD 6.1	LGVII	

Frost

Frost is a major limiting abiotic stress for the winter pea crop (*Pisum sativum* L.), and identifying the genetic determinants of frost tolerance is critical for breeding varieties adapted to cold northern regions. Quantitative trait loci (QTLs) associated with frost tolerance have previously been identified using biparental mapping populations, providing an initial overview of the genomic regions controlling this trait. The recent development of high-throughput genotyping tools for pea now enables genome-wide association studies (GWAS) to capture greater allelic diversity across large genetic resource collections and to refine the localization of causal polymorphisms due to increased marker density. In this study, a GWAS was conducted on a panel of 365 pea accessions. Phenotyping involved scoring frost damage (FD) both in the field and under controlled conditions. The association mapping panel was genotyped using an Illumina Infinium® BeadChip, generating data for 11,366 single-nucleotide polymorphism (SNP) markers.

Results from the GWAS identified 62 SNPs significantly associated with frost tolerance, distributed across six of the seven pea linkage groups (LGs). These findings confirmed three QTLs previously mapped in multiple environments on LGIII, LGV, and LGVI using biparental populations. Additionally, one novel locus on LGII was discovered, along with two loci on LGI and LGVII that had been detected in only a single environment before. Fifty candidate genes corresponding to annotated significant SNPs, or SNPs in strong linkage disequilibrium with them, were identified underlying the frost damage (FD)-related loci detected by GWAS. Moreover, the analyses allowed for the definition of favourable haplotypes for these FD-related loci and the identification of corresponding accessions within the mapping panel. In conclusion, this study successfully identified FD-related loci, corresponding favourable marker haplotypes, and representative pea accessions that may be utilized in winter pea breeding programs. Among the highlighted candidate genes, particular attention is drawn to the presence of C-repeat binding factors (CBFs), which are potential genetic determinants of the frost tolerance locus on LGVI. (Beji et al. 2020) (see Table 13.8).

13.4 Heat Stress

Heat stress (HS) negatively impacts pea productivity by impairing photosynthesis, shortening the life cycle, causing abortion of flowers, pods, and ovules, ultimately leading to yield loss. To minimize yield reduction and stabilize production, heat resistance is a desirable trait (Endale G. Tafesse). Environmental stresses broadly limit pea productivity. To elucidate the genetic basis of heat resistance, a genome-wide association study (GWAS) was performed on six stress-responsive physiological and agronomic traits in pea, aiming to identify genetic loci associated with these traits. A panel of 135 genetically diverse pea accessions, representing major pea-growing regions worldwide, was phenotyped in field trials conducted across five environments under ambient (control) and heat stress conditions. Statistical analysis of the phenotypic data revealed significant effects of genotype (G), environment (E), and their interaction (G × E) for all traits examined.

For association analysis, a set of 16,877 high-quality, known single-nucleotide polymorphisms (SNPs) was employed to detect marker-trait associations (MTAs). Thirty-two MTAs were identified consistently across at least three environments and were significantly associated with stress resistance traits: six MTAs for chlorophyll concentration measured by a soil–plant analysis development (SPAD) meter; two each for photochemical reflectance index and canopy temperature; seven for reproductive stem length; six for internode length; and nine for pod number. Within 15 kb of these markers, 48 candidate genes were identified. The identified markers and candidate genes offer promising targets for marker-assisted selection to facilitate the development of heat-resistant pea cultivars.

In the era of climate change, the overall productivity of pea (*Pisum sativum* L.) is increasingly threatened by various abiotic stresses, notably heat stress (HS). HS causes substantial yield losses by adversely affecting multiple traits in pea. A reduction in pod yield ranging from 11.1% to 17.5% has been reported with mean daily temperature increases between 1.4 and 2.2 °C. High-temperature stress (30.5–33 °C), particularly during the reproductive phase, is known to drastically reduce both seed yield and germination. Exposure to HS during germination and the early vegetative stage leads to poor emergence, stunted growth, and detrimental effects on physiological functions. To survive and complete their life cycle under HS, plants employ various defence strategies, including heat escape, avoidance, and tolerance mechanisms. However, the threshold temperatures for pea and its specific responses to heat stress remain inconsistent and are not yet clearly defined. Traditional breeding efforts have identified traits such as semileaflessness (afila), upright growth habit, lodging tolerance, lower canopy temperature, and small seed size as valuable for enhancing adaptation to HS in pea. Screening diverse germplasm, including landraces, under targeted environmental conditions remains a straightforward approach to identify HS-tolerant genotypes. Precise phenotyping using modern phenomics tools can further improve breeding efficiency. Integration of next-generation sequencing (NGS) data enables the identification of candidate genes underlying HS tolerance. Moreover, genomic selection, genome-wide association studies (GWAS), and marker-assisted selection (MAS) provide powerful strategies for developing HS-tolerant pea cultivars. Additionally, the development of transgenic lines represents a potential alternative approach to enhance HS tolerance. This chapter comprehensively addresses the mechanisms of HS tolerance in pea, screening methodologies, advances in omics technologies, and the future challenges involved in developing HS-tolerant genotypes (Devi et al. 2023) (see Table 13.9).

Table 13.9 QTL mapped for heat stress tolerance in garden pea

Traits	Loci (No)	Genomic location		*Genes or markers*	References
SPAD value/chlorophyll concentration	06	LGIII	7–13	*Psat5g221440, Psat5g224400, Psat5g224360, Psat5g224280, Psat5g299080, Psat5g299040, Psat5g301440, Psat5g301400, Psat5g303880, Psat5g303840, Psat5g303800*, and *Psat5g303760*	Tafesse et al. (2020)
Photochemical reflectance index	02	LGII and LGVII	9	*Psat6g234040, Psat6g234000*, and *Psat7g148080*	Tafesse et al. (2020)
Canopy temperature	02	LGIII and LGIV	6	*Psat4g203800, Psat4g203760, Psat5g169800*, and *Psat5g169760*	Tafesse et al. (2020)
Reproductive stem length	07	LGV, LGIV, LGIII, and LGVII	4–6	*Psat3g006600, Psat3g006560, Psat4g020520, Psat5g299080, Psat5g299040, Psat5g303680, Psat7g013080, Psat7g013040, Psat7g015240, Psat7g015200, Psat7g015160, Psat7g057080*, and *Psat7g057040*	Tafesse et al. (2020)
Pod number	09	LGI, LGV, LGIII, and One locus on nonchromosomal scaffold	7-10	*Psat2g060680, Psat2g144160, Psat2g155280, Psat2g157440, Psat2g166600, Psat2g166560, Psat2g166520, Psat2g005000, Psat2g004960, Psat3g111000, Psat3g110960*, and *Psat5g270480*	Tafesse et al. (2020)
Internode length	06	LGIV, LGIII, LGII, and LGVII	6–7	*Psat4g039600, Psat4g047680, Psat4g047640, Psat4g047600, Psat5g299080, Psat5g299040, Psat6g211160*, and *Psat7g120120*	Tafesse et al. (2020)

(continued)

Table 13.9 (continued)

Traits	Loci (No)	Genomic location		*Genes or markers*	References
Lamina wax	04	LGVI, LGIV, LGII, and LG7	–	*Psat1g139360, Psat4g112480*, and *Psat7g076840*	Tafesse et al. (2021)
Petiole wax	03	LGIV, LGVII Uscaffold03717_87257	–	*Psat4g011120, Psat7g186040,* and *Psat0s3717g0080*	Tafesse et al. (2021)
Stem thickness	03	LGVII, LGII, and Uscaffold03985_59708	–	*Psat7g071920, Psat7g072040, Psat7g208760,* and *Psat0s3985g0040*	Tafesse et al. (2021)
Flowering duration	02	LGV and LGIII	–	*Psat3g006600* and *Psat5g140600*	Tafesse et al. (2021)
Normalized difference vegetation index (NDVI)	01	LGII	–	*Psat6g028080* and *Psat6g028120*	Tafesse et al. (2021)
Normalized pigment and chlorophyll index (NPCI)	02	LGIII and LGII	–	*Psat5g299040* and *Psat6g231000*	Tafesse et al. (2021)

(VI) *Processing and Quality traits*

Edible pea/snow pea: Pea is an important grain and vegetable crop with a large genome of 4.45 Gb. Pod softness is a key quality trait in peas; however, the genes controlling this trait remain unknown.From the breeder's perspective, a desired pod-use cultivar should be featured by softer and tender pod walls, slower pod fibrosis process for prolonged harvesting period, and longer shelf life, without compromising the pod's nutritious value and plant environmental adaptivity Zhang et al. (2022) conducted sequence variant analysis on a panel of pea accessions. qRT-PCR analysis in immature pods of the parent lines suggested that Psat1g150000, which encodes a pectate lyase superfamily protein, is a candidate gene associated with pod softness. These results provide a solid foundation for cloning the gene responsible for pod softness and for using marker-assisted selection to improve this trait. KASP molecular markers, such as KASP-10 to KASP-14, were developed and identified to be tightly linked with PsPS1. QTL scanning showing that PsPS1 was strongly linked to marker KaSP-10 (site 297000666 bp)–KaSP-14 (site 303001889 bp).

Protein Content
Pea (*Pisum sativum* L.) is one of the most important grain legume crops grown in temperate regions worldwide. Enhancing pea yield remains a critical objective in breeding programs. Nine interconnected recombinant inbred line (RIL) populations were evaluated across nine environments at INRAE Dijon, France, and genotyped using the GenoPea 13.2K SNP array. Each population was assessed in two to four environments. A multipopulation quantitative trait loci (QTL) analysis was conducted for seed weight per plant (SW), seed number per plant (SN), thousand seed weight (TSW), and seed protein content (SPC). The identified QTLs were projected onto a multipopulation consensus map, followed by a meta-analysis. This approach revealed 17 QTLs for SW, 16 QTLs for SN, 35 QTLs for TSW, and 21 QTLs for SPC, providing insights into the relationships among these traits. These QTLs were consolidated into 27 metaQTLs, several of which exhibited narrow confidence intervals of less than 2 cM, encompassing fewer than 100 candidate genes. The refined precision of these metaQTLs and the identification of potential candidate genes facilitate their application in marker-assisted selection and lay the groundwork for map-based cloning of causal polymorphisms (Klein et al. 2020).

13.5 Candidate Genes

The meta-analysis approach enabled more confident localization of QTL regions associated with seed productivity and quality traits. In several cases, the number of genes underlying the metaQTL confidence intervals was relatively narrow. Specifically, the confidence intervals of metaQTLs mQTL3.4, mQTL5.3, mQTL1.5, mQTL7.2, mQTL3.1, and mQTL1.1 included 17, 51, 56, 126, 135, and 141 genes, respectively. The mQTL3.4 and mQTL1.5 regions encompass the *Le* and *Afla* loci,

Table 13.10 QTL mapped for protein contents in garden pea

Protein content	mQTL1.1, mQTL1.2, mQTL1.3, mQTL1.4, mQTL1.5	LG1	Klein et al. (2020)
	mQTL2.1, mQTL2.2, mQTL2.3	LG2	
	mQTL3.1, mQTL3.2, mQTL3.3, mQTL3.4	LG3	
	mQTL4.1, mQTL4.2, mQTL4.3, mQTL4.4, mQTL4.5	LG4	
	mQTL5.1, mQTL5.2, mQTL5.3	LG5	
	mQTL6.1, mQTL6.2, mQTL6.3	LG6	
	mQTL7.1, mQTL7.2, mQTL7.3, mQTL7.4	LG7	

previously described as controlling multiple traits in pea. The *Le* gene corresponds to *Psat5g299720* and encodes a 2OG-Fe(II) oxygenase involved in gibberellin biosynthesis. Selection for short plants in populations 3, 5, and 9 in 2011 did not prevent detection of this QTL because this gene also segregates in populations 6 and 8. This region also contains *Psat5g299400*, a gene belonging to the AUX/IAA family, which is putatively involved in early auxin response. The mQTL1.5 region harbours several genes encoding transcription factors expressed in flowers (*Psat2g173120* and *Psat2g173160*) and seeds (*Psat2g173360, Psat2g173520,* and *Psat2g173880*), as well as genes encoding an aminotransferase (*Psat2g172800*) and a malate transporter (*Psat2g173480*), all of which represent promising candidate genes for this metaQTL. Within the mQTL5.3 regions, two genes encode phosphatidylethanolamine-binding proteins, with expression peaking in the upper leaves as revealed by the Pea Gene Atlas. In soybean and *Arabidopsis thaliana*, these proteins are known to regulate flowering time, plant architecture, and seed germination.

The mQTL1.1 region includes the *AGPS2* locus, encoding adenosine diphosphate (ADP)-glucose pyrophosphorylase (*Psat2g005160*), previously reported to associate with seed size QTL in pea, as well as a gene encoding a subtilase (*Psat2g005680*) expressed in flowers and pods. This subtilase gene is distinct from the one identified by D'Erfurth et al. (*Psat0s1712g0120*). The mQTL3.1 region contains a gene encoding phosphoenolpyruvate carboxylase (*Psat5g031640*), expressed in above-ground vegetative and reproductive tissues, which could impact carbon assimilation and partitioning in the plant. Additionally, this region includes two tandem genes (*Psat7g127600* and *Psat7g127680*) predicted to encode Kelch motif proteins. Notably, Kelch motif-containing serine/threonine protein phosphatases have been associated with seed size QTL in rice. This study identified several robust metaQTLs associated with seed yield and seed protein content in pea and proposed candidate genes underlying these loci. This valuable knowledge enhances marker-assisted breeding efforts by highlighting the positions and functions of genes that may harbour causal polymorphisms for QTL in pea (see Table 13.10).

13.6 Future Prospectus

The future prospects for garden peas, like many other crops, will likely be shaped by a combination of technological advancements, evolving consumer preferences, and global environmental factors. Key trends and opportunities include:

Nutritional Awareness: As health consciousness rises, demand for plant-based proteins is increasing. Peas are a valuable source of protein, fibre, and essential vitamins, which could enhance their popularity among consumers seeking nutritious foods.

Sustainable Agriculture: With growing emphasis on sustainable and eco-friendly farming practices, peas' nitrogen-fixing ability makes them an important crop for improving soil fertility and supporting sustainable cropping systems.

Plant-Based Diets: The surge in plant-based diets and alternative protein sources is driving demand for pea protein, which is increasingly used as an ingredient in a variety of plant-based food products.

Breeding Innovations: Advances in breeding technologies, including genomics and marker-assisted selection, are expected to produce new pea varieties with improved disease resistance, yield potential, and adaptability to diverse environmental conditions.

Processed Pea Products: The food industry's innovation is likely to lead to novel processed pea products, such as healthy snacks, plant-based meat alternatives, and functional pea-derived ingredients tailored to evolving consumer preferences.

Global Market Expansion: Ongoing developments in global trade, including trade agreements and improved infrastructure, may open new markets and expand export opportunities for peas.

Climate Resilience: While climate change poses challenges, peas' preference for cooler climates could make new regions suitable for cultivation. Breeding for climate-resilient varieties will be essential to sustain productivity under changing environmental conditions.

Overall, monitoring technological progress, consumer trends, and agricultural policies will be critical for accurately assessing the future of garden pea cultivation. These factors, combined with favourable environmental conditions, will significantly influence the trajectory of pea production and utilization.

References

Anonymous (2022) Indian horticulture data base. Published by National Horticulture Board, Gurgaon

Aznar-Fernández T, Barilli E, Cobos MJ, Kilian A, Carling J, Rubiales D (2020) Identification of quantitative trait loci (QTL) controlling resistance to pea weevil (Bruchus pisorum) in a high-density integrated DArTseq SNP-based genetic map of pea. Sci Rep 10(1):33

Barilli E, CarrilloPerdomo E, Cobos MJ, Kilian A, Carling J, Rubiales D (2020) Identification of potential candidate genes controlling pea aphid tolerance in a *Pisum fulvum* high density integrated DArTseq SNP based genetic map. Pest Manag Sci 76:1731–1742. https://doi.org/10.1002/ps.5696

Beji S, Fontaine V, Devaux R, Thomas M, Negro SS, Bahrman N, Lejeune-Hénaut I (2020) Genome-wide association study identifies favorable SNP alleles and candidate genes for frost tolerance in pea. BMC Genomics 21(1): 536

Cheng P, Holdsworth W, Ma Y, Coyne CJ, Mazourek M, Grusak MA et al (2015) Association mapping of agronomic and quality traits in USDA pea single-plant collection. Mol Breed 35:1–13

Coyne CJ, Pilet-Nayel ML, McGee RJ, Porter LD, Smýkal P, Grünwald NJ (2015) Identification of QTL controlling high levels of partial resistance to *Fusarium solani* f. sp. pisi in pea. Plant Breed 134:446–453. https://doi.org/10.1111/pbr.12287

Coyne CJ, Porter LD, Boutet G, Ma Y, McGee RJ, Lesné A et al (2019) Confirmation of Fusarium root rot resistance QTL Fsp-Ps 2.1 of pea under controlled conditions. BMC Plant Biol 19:98. https://doi.org/10.1186/s12870-019-1699

Devi J, Sagar V, Mishra GP, Jha PK, Gupta N, Dubey RK, Prasad PV (2023) Heat stress tolerance in peas (Pisum sativum L.): current status and way forward. Front Plant Sci 13:1108276

Dirlewanger E, Isaac P, Ranade S, Belajouza M, Cousin R, Devienne D (1994) Restriction fragment length polymorphism analysis of loci associated with disease resistance genes and developmental traits in *Pisum sativum (L.)*. Theor Appl Genet 88:17–27. https://doi.org/10.1007/BF00222388

Ek M, Eklund M, Von Post R, Dayteg C, Henriksson T, Weibull P et al (2005) Microsatellite markers for powdery mildew resistance in pea (Pisum sativum L.). Hereditas 142:86–91. https://doi.org/10.1111/j.1601-5223.2005.01906.x

FAOSTAT F (2016) Food and agriculture organization of the United Nations (FAO). http://www.fao.org/faostat/en/-data/QC

FAOSTAT (2021) Food and agriculture organization corporate statistical database. http://www.fao.org/faostat/en/#data/QC

Fondevilla S, Rubiales D (2012) Powdery mildew control in pea: a review. Agron Sustain Dev 32(2):401–409

Jain S, Weeden NF, Kumar A, Chittem K, McPhee K (2015) Functional codominant marker for selecting the Fw gene conferring resistance to Fusarium wilt race 1 in pea. Crop Sci 55:2639–2646. https://doi.org/10.2135/cropsci2015.02.0102

Janila P, Sharma B (2004) RAPD and SCAR markers for powdery mildew resistance gene er in pea. Plant Breed 123:271–274. https://doi.org/10.1111/j.1439-0523.2004.00963.x

Jha UC, Nayyar H, Sharma KD, Jha R, Thudi M, Bakır M, Siddique KH (2023) Chickpea diseases: breeding and "omics" approaches for designing next-generation disease-resistant chickpea cultivar. In: Diseases in legume crops: next generation breeding approaches for resistant legume crops. Springer Nature, Singapore, pp 1–30

Katoch V, Sharma S, Pathania S, Banayal K, Sharma SK, Rathour R (2010) Molecular mapping of pea powdery mildew resistance gene er2 to pea linkage group III. Mol Breed 25:229–237. https://doi.org/10.1007/s11032-009-9322-7

Klein A, Houtin H, Rond-Coissieux C et al (2020) Meta-analysis of QTL reveals the genetic control of yield-related traits and seed protein content in pea. Sci Rep 10:15925. https://doi.org/10.1038/s41598-020-72548-9

Kwon SJ, Smýkal P, Hu J, Wang M, Kim SJ, McGee RJ, Coyne CJ (2013) User-friendly markers linked to Fusarium wilt race 1 resistance Fw gene for marker-assisted selection in pea. Plant Breed 132(6):642–648

Loridon K, McPhee KE, Morin J, Dubreuil P, Pilet-Nayel ML, Aubert G et al (2005) Microsatellite marker polymorphism and mapping in pea (Pisum sativum L.). Theor Appl Genet 111:1022–1031. https://doi.org/10.1007/s00122-005-0014-3

Ma Y, Coyne CJ, Main D, et al (2017) Development and validation of breeder-friendly KASPar markers for er1, a powdery mildew resistance gene in pea (Pisum sativum L.). Mol Breed 37:151

McPhee KE, Inglis DA, Gundersen B, Coyne CJ (2012) Mapping QTL for Fusarium wilt race 2 partial resistance in pea (Pisum sativum). Plant Breed 131(2):300–306

McClendon MT, Inglis DA, McPhee KE, Coyne CJ (2002) DNA markers for fusarium wilt race 1 resistance gene in pea. J Am Soc Hortic Sci 127:602–607. https://doi.org/10.21273/jashs.127.4.602

Mohan N, Aghora TS, Wani MA, Divya B (2013) Garden pea improvement in India. J Hortic Sci 8(2):125–164

Nisar M, Ghafoor A (2011) Linkage of a RAPD marker with powdery mildew resistance er-1 gene in *Pisum sativum* L. Russ J Genet 47:300–304. https://doi.org/10.1134/s102279541102013x

Okubara PA, Keller KE, McClendon MT, Inglis DA, McPhee KE, Coyne CJ (2005) Y15_999 Fw, a dominant SCAR marker linked to the *Fusarium wilt race* 1 (Fw) resistance gene in pea. Pisum Genet 37:30–33

Parihar AK, Kumar J, Gupta DS, Lamichaney A, Naik SJ, Singh AK, Toklu F (2022) Genomics enabled breeding strategies for major biotic stresses in pea (Pisum sativum L.). Front Plant Sci 13:861191

Pavan S, Schiavulli A, Appiano M, Marcotrigiano AR, Cillo F, Visser RG et al (2011) Pea powdery mildew er1 resistance is associated to loss-of-function mutations at a MLO homologous locus. Theor Appl Gen 123:1425–1431. https://doi.org/10.1007/s00122-011-1677-6

Pavan S, Schiavulli A, Appiano M, Miacola C, Visser RGF, Bai YL et al (2013) Identification of a complete set of functional markers for the selection of er1 powdery mildew resistance in *Pisum sativum* L. Mol Breed 31:247–253. https://doi.org/10.1007/s11032-012-9781-0

Pereira G, Marques C, Ribeiro R, Formiga S, Dâmaso M, Sousa MT et al (2010) Identification of DNA markers linked to an induced mutated gene conferring resistance to powdery mildew in pea *(Pisum sativum* L.). Euphytica 171:327–335

Rakshit S (1997) Biochemical and molecular analyses of powdery mildew resistance in pea(Pisum sativum L.). PhD dissertation. Indian Agricultural Research Institute, New Delhi

Sari H, Eker T, Tosun HS, Mutlu N, Celik I, Toker C (2023) Mapping QTLs for super-earliness and agro-morphological traits in RILs population derived from interspecific crosses between Pisum sativum × P. fulvum. Curr Issues Mol Biol 45(1):663–676

Srivastava RK, Mishra SK, Singh AK, Mohapatra T (2012) Development of a coupling-phase SCAR marker linked to the powdery mildew resistance gene 'er1' in pea (Pisum sativum L.). Euphytica 186(3):855–866

Stagnari F, Maggio A, Galieni A et al (2017) Multiple benefits of legumes for agriculture sustainability: an overview. Chem Biol Technol Agric 4:2. https://doi.org/10.1186/s40538-016-0085-1

Sun S, Wang Z, Fu H, Duan C, Wang X, Zhu Z (2015) Resistance to powdery mildew in the pea cultivar Xucai 1 is conferred by the gene er1. Crop J 3:489–499. https://doi.org/10.1016/j.cj.2015.07.006

Sun S, Fu H, Wang Z, Duan C, Zong X, Zhu Z (2016) Discovery of a novel er1 allele conferring powdery mildew resistance in Chinese pea (*Pisum sativum* L.) landraces. PLoS One 11:e0147624. https://doi.org/10.1371/journal.pone.0147624

Sun S, Deng D, Duan C, Zong X, Xu D, He Y, Zhu Z (2019) Two novel er1 alleles conferring powdery mildew (*Erysiphe pisi*) resistance identified in a worldwide collection of pea (*Pisum sativum* L.) germplasms. Int J Mol Sci 20(20):5071. https://doi.org/10.3390/ijms20205071

Tafesse EG, Gali KK, Lachagari VBR, Bueckert R, Warkentin TD (2020) Genome-wide association mapping for heat stress responsive traits in field pea. Int J Mol Sci 21:2043. https://doi.org/10.3390/ijms21062043

Tafesse EG, Gali KK, Lachagari VBR, Bueckert R, Warkentin TD (2021) Genome-wide association mapping for heat and drought adaptive traits in pea. Genes 12(12):1897. https://doi.org/10.3390/genes12121897

Timmerman GM, Frew TJ, Weeden NF, et al (1994) Linkage analysis of er-1, a recessive Pisum sativum gene forresistance to powdery mildew fungus (Erysiphe pisi D.C.). Theor Appl Genet 88:1050–1055

Timmerman-Vaughan GM, Moya L, Frew TJ, Murray SR, Crowhurst R (2016) Ascochyta blight disease of pea (Pisum sativum L.): defence-related candidate genes associated with QTL regions and identification of epistatic QTL. Theor Appl Genet 129(5):879–896

Tiwari KR, Penner GA, Warkentin TD (1998) Identification of coupling and repulsion phase RAPD markers for powdery mildew resistance gene er-1 in pea. Genome 41:440–444. https://doi.org/10.1139/g98-014

Tonguç M, Weeden NF (2010) Identification and mapping of molecular markers linked to er1 gene in pea. J Plant Mol Biol Biotechnol 1:1–5
Wang Z, Fu H, Sun S, Duan C, Wu X, Yang X et al (2015) Identification of powdery mildew resistance gene in pea line X9002. Acta Agron Sin 41:515–523. https://doi.org/10.3724/sp.j.1006.2015.00515
Wu L, Fredua-Agyeman R, Strelkov SE, Chang K-F, Hwang S-F (2022) Identification of quantitative trait loci associated with partial resistance to fusarium root rot and wilt caused by Fusarium graminearum in field pea. Front Plant Sci 12:784593. https://doi.org/10.3389/fpls.2021.784593
Yan J, Chen J, Lin Y, Yuan X, Somta P, Zhang Y, Chen X (2023) Mapping of quantitative trait locus reveals PsXI gene encoding xylanase inhibitor as the candidate gene for bruchid (Callosobruchus spp.) resistance in pea (Pisum sativum L.). Front Plant Sci 14:1057577
Zhang P, Wang Y, Sun T et al (2022) Fine mapping PsPS1, a gene controlling pod softness that defines market type in pea (Pisum sativum). Plant Breed 141:418–28

GWAS and QTL Mapping in Radish, Carrot and Beetroot

14

N. Pradeepkumara, Bichhinna Maitri Rout, Saheb Pal, L. C. Sushmitha, Reeta Bhatia Dey, A. D. Munshi, T. K. Behera, and Shyam Sundar Dey

Abstract

Modern techniques like quantitative trait locus (QTL) sequencing (QTL-seq), genotyping by sequencing, and genome-wide association studies (GWAS) have expedited the identification of QTLs associated with complex traits in crop plants. GWAS capitalizes on the high recombination rates present in natural populations used for association studies. Carrot, radish, and beetroot are key vegetable root crops, globally significant for their contributions to balanced human diets. The economically valuable fleshy solid roots of these crops underscore the importance of understanding root complexities through genomic approaches to uncover QTLs governing root architecture. GWAS and QTL analysis have been effectively employed to understand root-related traits and pigmentation-related genes. Various mapping populations, such as doubled haploids (DHs), biparental mapping populations, and natural populations, have proven effective in QTL analysis studies and candidate gene identification. Despite their importance, limited research has focused on identifying QTLs and elucidating the genetic basis

N. Pradeepkumara · S. S. Dey (✉)
ICAR-Central Institute of Temperate Horticulture, Regional Station, Dirang, Arunachal Pradesh, India

B. M. Rout · L. C. Sushmitha · A. D. Munshi
ICAR-Division of Vegetable Science, Indian Agricultural Research Institute, New Delhi, India

S. Pal
ICAR-Indian Agricultural Research Institute, Hazaribagh, Jharkhand, India

R. Bhatia Dey
Division of Floriculture and Landscaping, ICAR-Indian Agricultural Research Institute, New Delhi, India

T. K. Behera
ICAR-Indian Institute of Horticulture Research, Bengaluru, Karnataka, India

S. S. Dey et al. (eds.), *GWAS and QTL Mapping in Horticultural Crops*,
https://doi.org/10.1007/978-981-95-2306-1_14

of quality traits, biotic and abiotic stress resistance in carrot, radish, and beetroot. In this regard, GWAS and QTL-seq hold promise for unraveling complex traits in these root crops. This chapter provides a summary of the latest advancements in GWAS and QTL mapping for different traits in root crops.

Keywords

Genome-wide association studies · Genotyping by sequencing · Candidate gens · Mapping populations · Linkage mapping

14.1 Introduction

Vegetable breeders and researchers have a primary focus on increasing yield, improving quality, and enhancing resistance to biotic and abiotic stresses to ensure global food security and combat malnutrition. Public-sector crop breeding programs play a crucial role in achieving this goal (Rikkerink et al. 2007). With vegetable production estimated at 212.91 million tons in 2022–2023 (NHB-2022–23), vegetable breeding significantly contributes to meeting the growing demand for food. However, various challenges, including biotic and abiotic factors, pose obstacles to genetic improvement, potentially impacting both the quantity and quality of vegetables, especially during preharvest and postharvest stages (Rikkerink et al. 2007). Traditional phenotypic selection methods are limited in their effectiveness due to the significant influence of environmental factors on phenotype, making it challenging to select suitable breeding materials (Fedoroff et al. 2010). Since the 1980s, technological advancements, such as high-throughput sequencing and molecular mapping, have revolutionized vegetable breeding. These advancements enable the study of genetic variation, identification and tagging of genetic loci, development of stress-resistant cultivars, improvement of yield, enhancement of health-promoting compounds, and utilization of marker-trait information for marker-assisted selection (MAS). This technological revolution empowers breeders to manipulate plant genomes and utilize molecular markers effectively in breeding programs.

Advanced genetic designs (such as biparental mapping), agrobiotechnology/omics applications, and cutting-edge breeding methods (including traditional quantitative trait locus [QTL] mapping, association mapping, RNA sequencing, bulked segregant analysis, and genomics) are expected to address the challenges faced by future breeders. In vegetable crops, yield and quality are economically vital, with most quantitative traits primarily governed by multiallelic or polygenic genes. These traits are significantly influenced by environmental factors (Khan et al. 2012). A key objective in quantitative genetics and molecular genomics is to unveil the genetic underpinnings of evolutionarily significant traits (Slate et al. 2010; Stapley et al. 2010; Wellenreuther and Hansson 2016; Hoban et al. 2016). However, achieving this goal is challenging because many of these traits likely have a polygenic background. They are encoded by numerous major and minor effective genes, each with a small individual effect. Consequently, they are difficult to genomically characterize using traditional QTL mapping methods (Falconer and Mackay 1996; Mackay et al. 2009; Rockman 2012; Santure et al. 2015; Wood and Carden 2014). Two common mapping techniques, linkage mapping/QTL mapping and association mapping, are widely employed to identify genetic variations linked to economic traits in vegetable crops (Zhu et al. 2008; Sukumaran et al. 2016) (Table 14.1). These approaches rely on functional polymorphisms and molecular markers to pinpoint quantitative trait loci (QTLs) associated with yield and quality (Devi et al. 2022).

Quantitative trait locus (*QTL*) mapping, a statistical method, facilitates the connection between multiple genotypes and phenotypes (Lynch 1998; Falconer and Mackay 1996). Molecular markers such as single-nucleotide polymorphisms (SNPs) and amplified fragment length polymorphisms (AFLPs) are utilized to link mapped QTLs with phenotypic data (Fan et al. 2016; Mora et al. 2016; Gahlaut

Table 14.1 Differences between traditional QTL mapping and advanced QTL mapping

Characters	Traditional QTL mapping	Association mapping
Effect on QTL range	Only effective on large to moderate, not much effective on minor or small effective QTLs	Effective for identifying quantitative trait loci (QTLs) with significantly smaller effect sizes compared to those detected through linkage mapping
Smaller allele frequencies: effectiveness	More reliable	Not much effective
Working effectiveness on allele number detected	One to two alleles	All the alleles detected
Accuracy on QTLs detected	Both confirmation and validation are necessary	Often confirmation is done by replication studies
Populations utilized for mapping purposes	Generated through the crossing of selected parents	Natural populations, breeding materials, germplasm lines, and lines resulting from multiple crosses
Recombination events exploited	Recombination events occur after the crosses are performed	All recombination events that have taken place since the linkage disequilibrium (LD) was established
Markers identified to be linked to QTLs or genes	A few to several centimorgans (cM) distant from the gene/QTL	Significantly closer than those identified through linkage mapping
Mapping principle	The frequency of recombination between the loci	The linkage disequilibrium (LD) existing between the loci
Structural characteristics of the population	Reduced through controlled breeding crosses	Minimized by estimation of Q or P and its use in AM
Viability across various species	Achievable in annual and biennial species, while not viable in perennial species	Viable in annual, biennial, and perennial species
The quantity of markers required to encompass the entire genome	Low (102) to moderate (103)	Ranging from high (105 for small genomes) to very high (109 for large genomes)

2019). QTL mapping offers the advantage of mapping trait variation across various populations including F_2 populations, doubled haploids (DHs), near-isogenic lines (NILs), backcross populations, chromosomal deletion and substitution lines, nested association mapping (NAM), multiparent advanced generation intercross (MAGIC) populations, and recombinant inbred lines (RIL) populations. However, it is limited to mapping only RIL populations and measuring allelic variation among parents at an F_2 hybrid due to the unavailability of dense linkage maps. On the other hand, genome-wide association studies (GWAS) overcome the limitations of QTL mapping (Korte and Farlow 2013). GWAS utilizes linkage disequilibrium (LD) to identify associations between genetic variants and traits across the entire genome, offering a broader perspective on genetic variation associated with economic traits in vegetable crops. Genome-wide association studies (GWAS), also known as whole-genome association studies, analyze a comprehensive collection of genetic variations across the entire genome to identify any associations with specific traits (Korte and Farlow 2013). GWAS leverage linkage disequilibrium, which are population-level associations between genetic markers and causative mutations. These associations occur because small segments of chromosomes in the current population share ancestry, leading to identical marker alleles or haplotypes within these segments. Consequently, if a quantitative trait locus (QTL) exists within a chromosome segment, it will also carry identical QTL alleles.

Various statistical methodologies exploit these associations, with the simplest being the genome-wide association test using single-marker regression. GWAS typically focus on the connections between traits and single-nucleotide polymorphisms (SNPs), such as comparing DNA from diverse plant species concerning different phenotypic traits. Despite being hypothesis free, GWAS can generate hypotheses and are utilized for analyzing association results, replicating findings, interpreting results, and guiding further investigations based on specific designs, genotyping technologies, and statistical principles (Bush and Jason 2012). Genome-wide association studies (GWAS) based on single-nucleotide polymorphisms (SNPs) generated by genotyping by sequencing (GBS) have demonstrated effectiveness in identifying new alleles influencing morpho-agronomic and fruit quality-related traits in horticultural crops. For instance, QTLs associated with carotenoid pathway products like lycopene and provitamin A carotenes α- and β-carotene were investigated in two unrelated F_2 carrot populations (Santos and Simon 2002). Similarly, a QTL analysis for clubroot resistance (CR) in radish was conducted using an F_2 population (Kamei et al. 2010).

An illustrative example in chili peppers is the work by Nimmakayala et al. (2016), which aimed to identify SNP markers associated with capsaicinoid content and fruit weight traits in *Capsicum annuum* L. Similarly, Pavan et al. (2017) employed a similar approach to conduct preliminary GWAS, identifying SNP/trait associations for flowering time and seed-related traits in a collection of 72 accessions of *Cucumis melo* from Apulia. Additional GWAS for fruit firmness were carried out in a larger collection of melon from Asia and the Western hemisphere (Nimmakayala et al. 2016). Compared to biparental linkage mapping, GWAS offers advantages such as high resolution, cost-efficiency, and the absence of the need to create a mapping population. Therefore, GWAS is preferred for high-resolution mapping experiments (Table 14.1).

Genome-wide association studies (GWAS) rely on three fundamental elements: (i) a comprehensive marker catalogue derived from genome-wide assessment of diversity among individuals in a population; (ii) understanding of population structure and allele frequency spectrum; and (iii) robust phenotypic data for each individual in the population (Fig. 14.1). GWAS also necessitate a large panel of markers, preferably including thousands of single-nucleotide polymorphisms (SNPs), and a substantial sample size, ranging from hundreds to thousands of individuals, to encompass maximal variability for the traits under investigation and ensure adequate statistical power. Population sampling is a critical step in GWAS and should encompass accessions collected from diverse locations, potentially hierarchically structured. Population structure poses a significant source of confounding in GWAS (Korte and Farlow 2013). Therefore, a thorough characterization of population structure is essential to mitigate the risk of false positive (type I error) or false negative (type II error) associations between SNPs and traits.

Statistical analysis to identify significant genotype-phenotype associations involves examining each SNP independently, considering multiple testing. Typically, only markers that meet the Bonferroni-corrected p-value threshold ($\alpha = 0.05$) are deemed significantly associated with phenotypic observations. Alternative approaches to Bonferroni correction have been proposed for establishing significance in GWAS, including the utilization of false discovery rate (FDR) or permutation testing. While genome-wide association studies (GWAS) have proven to be a powerful tool for investigating genotype-phenotype relationships, they still have several limitations. First, rare alleles may go undetected. Second, alleles with small effects on the phenotype can be challenging to identify. Last, the most significant single-nucleotide polymorphisms (SNPs) identified through GWAS may not always represent the true causative or predictive factors for a given trait.

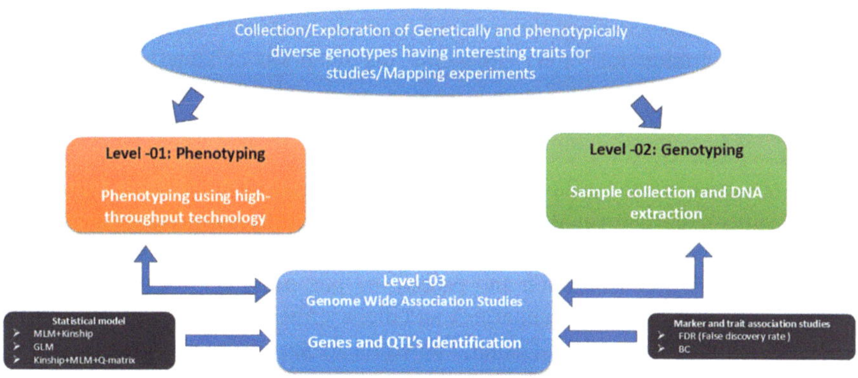

Fig. 14.1 Pictorial representation of genome-wide association study (GWAS) in root vegetables crops

It is highly probable that genomic selection (GS) strategies will supersede GWAS in the near future (Bhat et al. 2016a, b). Despite relying on similar high-throughput genotyping and phenotyping of unrelated individuals in a population, known as the training population (TP), GS has distinct features. The TP data is used to predict genomic-estimated breeding values (GEBV) for a different set of individuals, known as the breeding population (BP), which has been genotyped with dense marker coverage across the genome but lacks phenotypic information. This approach significantly reduces the time and expense associated with phenotyping since phenotypic data is only required in the initial phase to enhance the accuracy of the prediction model. While genomic selection has been predominantly applied in cereals, emerging applications in vegetables, particularly tomato, have been reported (Duangjit et al. 2016; Yamamoto et al. 2020). These studies on tomato assessed the potential of the GS approach to enhance fruit quality traits.

In GWAS, single-nucleotide polymorphisms (SNPs) and simple sequence repeats (SSRs) are predominantly utilized (Cai et al. 2018). Selecting an appropriate population for analysis is crucial to effectively detect candidate genes through GWAS techniques. To mitigate potential false positives resulting from population stratification or structure, it is essential to choose a GWAS population with low stratification (Begum et al. 2015; Yano et al. 2016). However, opting for such a population may limit the sample size, consequently reducing the detection power of the GWAS (Korte and Farlow 2013). Thus, when designing a suitable GWAS population, one must consider the trade-off between population stratification and sample size. A mixed population obtained by blending subpopulations with diverse genetic backgrounds could also be employed in GWAS. One advantage of using such a mixed population is the potential improvement in detecting causal variants through increased sample size. Among vegetable crops, root vegetables are characterized by their prominent, fleshy underground structures, which may include roots, tuberous roots, or hypocotyls with taproots forming below them. Major root vegetables belong to families such as Umbelliferae (carrot), Brassicaceae (radish and turnip), and Chenopodiaceae (beet). These crops exhibit variable food value, processing qualities, export potential, and long storage life, contributing significantly to their nutritional importance. They thrive in regions with cool night temperatures, which slow the respiration rate, thus enhancing the retention of stored carbohydrates (Table 14.2).

Carrot

Carrot (*Daucus carota* L.; $2n = 2x = 18$), belonging to the Umbelliferae family, is a cool-weather crop cultivated worldwide primarily for its edible storage taproots, making it the most significant among root crops. Carrots are extensively grown for both fresh market consumption and processing purposes. They are renowned for being an excellent source of vitamin A and dietary fiber. Black carrots are particularly used in the preparation of Kanji, known for its appetizing properties.

Carrots exhibit two distinct groups: tropical and temperate. Tropical varieties are typically annual, whereas temperate types are biennial. Carrots display a wide range of color variations, ranging from black and red to yellow and orange. Additionally,

Table 14.2 Details of different germplasms/accessions characterized for different resistance in carrot (*Daucus carota* L.)

Trait	Species/germplasm/variety	References
Resistance to *Alternaria* blight	Brasilia	Boiteux et al. (1993)
Resistance to cavity spot	'Redca' and 'Nandor'	Sweet et al. (1989)
Resistance to cavity spot	'Panther', 'E0792', 'Caropride', 'Fannia', and 'Navajo'	Benard and Punja (1995)
Resistance to cavity spot	'Purple Turkey'	Cooper et al. (2004)
Partial resistance to cavity spot	'Six Pak'	McDonald (1994)
Resistance to *Alternaria* blight and Fusarium wilt	*Daucus carota* subsp. *maximus* (Desf.) Ball.; D. broteri Ten.; *D. halophilus* brot.; *D. maximus* Desf.; *D. setifolius* Desf.; *D. carota* Roth.; *D. carota* L.; *D. muricatus* (L.) L.; *D. setifolius* Desf.	Leunov et al. (2021)
Resistance to *Phytophthora* root rot	Imperator II	Stelfox and Henry (1978)
Resistance to hawthorn-carrot aphid	'Afro F1', 'Kongo F1', 'Napa F1', and 'Nipomo F1'	Pobożniak et al. (2021)
Highly resistance		
Resistance to crown rot	'Royal Chantenay' and 'Scarlet Nantes'	Howard and Williams (1976)
Resistance to black rot (*Alternaria radicinia*)	'Panther' and 'Caropak'	Pryor et al. (2000)
Tolerance to *Fusarium avenaceum*	'Nantskaya 4' (transgenic carrot plants overexpressed a single-gene coding for a thaumatin II protein)	Sidorova and Miroshnichenko (2013)
Partial resistance to bacterial blight (*Xanthomonas hortorum* pv. *carotae*)	PI418967, PI432905, and PI432906	Christianson et al. (2015)
Resistance to powdery mildew	*D. c.* subsp. *dentatus*, *D. siculus*, and *D. carota* cultivar 'Bauers Kieler Rote'	Bonnet (1983)
Partial resistance to powdery mildew	' Scarlet Nantes'	Takaichi and Oeda (2000)
Resistance to motley dwarf	CVC-14	Watson and Falk (1994)
Tolerance to motley dwarf	'Kurnella Strongtop' and 'Western Red'	Tomlinson (1965)
Aster yellows—least incidence (average 15.3% infection)	'Scarlet Nantes', 'Royal Chantenay', and 'Gold King'	Gabelman et al. (1994)
Resistance to *M. incognita* and *M. javanica*	'Brasilia'	Huang et al. (2015)
Tolerance to *M. incognita*	'DR-333'	Siddiqui et al. (2011)
Resistance to thrips	'Ingot'	Leiss et al. (2013)

(continued)

Table 14.2 (continued)

Trait	Species/germplasm/variety	References
Partial resistance to carrot fly	'Flyaway'	Ellis (1999) and Simlat et al. (2013)
Resistance to carrot fly	*D. capiltifolius*	Ellis et al. (1993)
	D. c. hispanicus	Frese and Weber (1984)
Tolerance to carrot fly	*D. c. azoricus*	Kraus (1992)
Least susceptible to aphid	Berlikum	Lamb (1953)
Resistance to *M. incognita*	'Br1091', 'Homs', 'Scarlet Fancy', and 'Favourite'	Parsons et al. (2015)
Resistance to tip burn	'Scarlet Nantes'	Tibbitts et al. (1983)

carrots possess medicinal properties, as evidenced by the antibacterial properties found in essential oils extracted from their roots. Carrot seed oil, in particular, has been historically used in treating kidney ailments and dropsy (Chopra 1933; Kirtikar and Basu 1935).

Temperate cultivars of carrots are cultivated as annuals for their fleshy roots but are true biennials requiring 2 years for flowering and seed production. It is believed that carrots originated as a natural variation of the Eurasian wildflower Queen Anne's lace in Afghanistan, particularly in the region where the Himalayan and Hindukush mountains meet. The domestication of carrots is thought to have occurred in Afghanistan and adjacent regions of Russia, Iran, India, Pakistan, and Anatolia (Kirtikar and Basu 1935).

Radish

Radish ($2n = 2x = 18$, *Raphanus sativus* L.), considered one of the earliest domesticated vegetables, traces its origins back to China. The original form of radish likely resembled the modern green and red-fleshed Chinese radishes that remain popular across Asia today. Radish is believed to have originated in both Europe and Asia and has a long history of cultivation, particularly in Egypt. It was introduced to England and France during the early sixteenth century. Unlike some vegetables, radish does not have a wild counterpart; instead, it is thought to have evolved from *R. raphanistrum*, a weed widely found in Europe. European varieties of radish tend to be small in size, mild in flavor, early to mature, and high in quality. These older varieties are true biennials, requiring a period of cold treatment, known as vernalization, to induce flowering (bolting) and seed production. Radish is a significant root vegetable crop within the Brassicaceae family, cultivated globally for its carbohydrates, nutrients, phytochemicals, and dietary fiber, all of which contribute to human nutrition.

Beetroot

Beetroot, scientifically classified as *Beta vulgaris subsp. vulgaris* L. ($2n = 2x = 18$) and belonging to the Chenopodiaceae family, is renowned for its remarkably high antioxidant content, placing it among the top 10 antioxidant-rich vegetables. Its abundance of betalains, phenolic acids, saponins, alkaloids, steroids/triterpenes, catechins, and flavonoids provides robust antioxidant defense against oxidative damage caused by reactive oxygen species (ROS) or free radicals (Pink 1993). These compounds play a crucial role in safeguarding cell metabolism, mitigating DNA damage, and preserving the integrity of lipids and proteins.

Beetroot serves as a valuable dietary addition, offering a plethora of nutrients, minerals, and vitamins, including vitamin B complex and folic acid. Moreover, it contains various bioactive compounds with diverse medicinal properties. Beetroot has been found to exhibit analgesic, hepatoprotective, antimicrobial, anti-inflammatory, and neuroprotective effects, among others (Wohlfeiler et al. 2021; Wu et al. 2020; Xie et al. 2020; Xu et al. 2014; Yu XiaoNa et al. 2020; Yu et al. 2013; Pandita et al. 2020). These properties contribute to overall health and well-being by combatting neurodegeneration and hepatic damage, highlighting beetroot's multifaceted benefits for human health.

India's diverse climate ensures the availability of all varieties of fresh fruits and vegetables. It ranks second in fruits and vegetable production in the world, after China. As per National Horticulture Database (2st Advance Estimates) published by National Horticulture Board 2023–2024, India produced 212.39 million metric tons of vegetables. The area under cultivation of fruits stood at 7.15 million hectares while vegetables were cultivated at 11.24 million hectares. India is the second largest producer of fruits and vegetables in the world next only to China.

Carrot, beetroot, and radish are the major root vegetable crops grown in India, during cool season. These root crops majorly have two categories: tropical types are annual, whereas the temperate types are biennial with distinguishable traits. These root crops have a wide range of variability for color from black and red to yellow, white, and orange with different shades. Indian formers are majorly growing theses crops during cool season (October to January) but breeders from Indian Agricultural Research Institute have developed the year-round and off-season cultivation of carrot and radish. Present production and area of cultivation in India are around 0.122 lakh ha and 2.374 MT (carrot) and 0.200 lakh ha and 3.098 MT (radish) (NHB-2023, 2nd Advance estimation).

All these crops have good export potential and processing importance. Carrots are consumed either raw or cooked and processed into value-added products, namely, canned carrots, chips, candy, kheer, halwa, powder, juice, beverages, preserve products, preparation of soups, beverages, wine, stews, curries, pies, and jams as blending agents (Raees-ul and Prasad 2015). Similarly, radish is also used to prepare various processed products, including infused pickles, salted sauerkraut, mixed pickles, crushed pickles and kanji a fermented probiotic drink that is excellent for gut health and offers many other benefits (Zappia et al. 2023).

The key impacts on vegetable production in India due to climate change, invasive pests and diseases, drought, and soil salinity include:

- Production Imbalance: Climate variability and pest invasions can lead to inconsistent yields across regions, creating imbalances in production.
- Dependency on Synthetic Chemical Fertilizers: Excessive reliance on synthetic fertilizers may degrade soil health and contribute to environmental pollution.
- Increase in Secondary and Micronutrient Deficiencies: Changes in soil composition and nutrient availability can result in deficiencies of secondary and micronutrients in crops, affecting their nutritional value.
- Pesticide Residue Accumulation: Higher pest pressure may necessitate increased pesticide use, leading to chemical residues in harvested produce, posing health risks.
- Unscientific Water Management: Inefficient water management practices can exacerbate water scarcity and affect crop growth and productivity.
- Reduced Productivity: Climate-related stresses and pest infestations can decrease crop productivity, impacting farmer livelihoods and food security.
- Decreased Produce Quality: Environmental stressors may compromise the quality of harvested vegetables, affecting market value and consumer satisfaction.
- Gene Pool Loss: Climate change and pest invasions may threaten the genetic diversity of vegetable crops, leading to the loss of valuable traits and adaptations.
- Environmental Pollution: Intensive agricultural practices, including pesticide and fertilizer use, can contribute to soil, water, and air pollution, harming ecosystems and human health.
- Social and Economic Disparities: Vulnerable communities, particularly smallholder farmers, may bear the brunt of climate-related impacts, exacerbating social and economic inequalities.

Breeding objectives for carrot, radish, and beetroot in India include:

- Enhanced Yield, Earliness, and Adaptability: Developing varieties with higher yields, early maturity, and adaptability to diverse agroclimatic conditions.
- Root Size, Shape, and Appearance: Optimizing root size, shape, and appearance to meet consumer preferences and market demands.
- Stress Tolerance: Breeding for tolerance to biotic and abiotic stresses such as pests, diseases, drought, and salinity.
- Reduced Bolting and Root Imperfections: Minimizing premature bolting and undesirable root traits like excessive cracking, forking, scars, and hairs.
- Preservation of Shelf Life: Improving postharvest characteristics to extend the shelf life of harvested vegetables and reduce food waste.
- Sustainability and Environmental Resilience: Promoting environmentally sustainable breeding practices that reduce dependency on synthetic inputs and enhance resilience to climate change.
- Socioeconomic Equity: Ensuring that breeding efforts benefit all stakeholders in the agricultural value chain, including smallholder farmers and consumers, to promote social and economic equity.
- Protocol standardization for year-round production practices.
- Delayed bolting, smooth and shiny roots, flat/convex crown, and easy to uproot.

- Free from forking, cracking, root scars, and shoulder greening.
- Protocol standardization for year-round production practices.
- Identification and utilization of self-incompatible (Sporophytic SI) and cytoplasmic male sterile (CMS) lines.
- Development of varieties and hybrids resistance and tolerance biotic and abiotic stresses with better adaptability.
- Standardization of protocol for haploid and double-haploid breeding to develop elite parental inbreds for rapid hybrid development (Table 14.3).

Table 14.3 Mapping of QTL for different horticultural traits by using diverse mapping populations

Trait	Mapping population	Technique	Molecular markers	QTLs/genes	Chromosome no.	References
Carrot (*Daucus carota L.*)						
Root shape	6 parent diallel population and F_2 mapping population	Genotype by sequencing	SNPs	One major QTL	Ch2	Turner et al. (2018)
	103 open-pollinated varieties	Genotype by sequencing	SNPs	One major QTL	Ch1	Podgorni et al. (2019)
	$F_{2:3}$	QTL mapping	SNPs	Two QTLs and genes associated with OFP-TRM and IQD pathway were identified within QTL support intervals	Ch 2 and Ch 6	Vega et al. (2024)
Root size and shape	749 accessions	GWAS	SNPs	Many small effect QTLs	–	Brainard et al. (2021)
Root morphological traits	A backcross inbred line (BIL) population derived from a cross between the wild species *Daucus carota* ssp. *carota* 'Songzi' and the orange cultivar 'Amsterdam forcing'	QTL mapping	SNPs	Root length: RL2-1 RL2-2 Root shoulder diameter: RSD2-1 RSD2-2 Root middle diameter: RMD3-1 RMD2-1 Root tip diameter: RTD2-1	Ch 1 Ch 2 Ch 2 Ch 2 Ch 3 Ch 2 Ch 2	Ou et al. (2022)
Root pigment:						
Orange	F_2	QTL analysis	AFLP and STS (Sequence Tagged Site)	*Y* and *Y2* loci	Ch 5 and Ch 7, respectively	Just et al. (2009)
	Doubled haploid, Nantes-type carrot (DH1)	Fine mapping	SNPs	*Y* candidate gene (*DCAR_032551*)	Ch 5	Iorizzo et al. (2016)
	F_2	GWAS and QTL analysis	SNPs	Total of 11 QTLs detected by both the association and QTL analysis	–	Shibaya et al. (2022)

(continued)

Table 14.2 (continued)

Trait	Mapping population	Technique	Molecular markers	QTLs/genes	Chromosome no.	References
Purple	F_2 population derived from a heterozygous P1p1 F_1 intercross between B1896 (true-breeding p1p1 yellow-rooted F_4M inbred derived from the cross PI173687 9 B493) and B7262 (P1P1 purple exterior with orange inner phloem and xylem inbred from the same cross)	Molecular mapping	SSRs and AFLP markers	P1 locus	Ch 3	Yildiz et al. (2013)
	F_3–F_4	SNP-based linkage map and QTL analysis	SNPs	RTPE (Reverse Through Phenotype Experiment) qtl	Ch 3	Cavagnaro et al. (2014)
	F_2	RNA-seq	SNPs	*DcMYB7, DcMYB8, DcMYB9, DcMYB10, DcMYB11*, and *DcMYB12*	Ch 3	Iorizzo et al. (2019)
	'Purple Haze', a carrot cultivar with purple root and nonpurple petiole	Molecular mapping	SNPs and SSRs	*DcMYB113*	Ch 3	Xu et al. (2020)
Plant evolution related:						
Organ development and root tissue patterning	F_2 population (*D. carota subsp. commutatus* × line 2874B)	Molecular mapping and QTL analysis	DArT markers	*DcAHLc1*	Ch 2	Podgorni et al. (2017)
Male sterility:						
Early flowering habit and fertility restorer	F_2 progeny, derived from a self-pollinated individual F1 plant from the intercross between the annual cultivar 'Criolla INTA' and a petaloid male sterile biennial carrot	Molecular mapping	SSR, RAPD, AFLP, and SCAR	*Vrn1* *Rf1*	Ch 2 Ch 9	Alessandro et al. (2013)
Petaloid trait	Segregated population obtained from a cross between a petaloid strain (YSN6MS) and a fertile individual	QTL-seq	RFLPs	A major QTL, *DcMADS2*, found at the center of the peak region is considered as the possible candidate gene	Ch 4	Fujii and Shiroto (2019)

(continued)

Table 14.2 (continued)

Trait	Mapping population	Technique	Molecular markers	QTLs/genes	Chromosome no.	References
Radish (*Raphanus sativus* L.)						
Root shape	F_2 population derived from cross between 'Huang-he hong-wan' and 'Utsugi-gensuke'	Linkage and QTL analysis	AFLPs, Brassica SSRs and SLG (S locus glycoprotein)-cleaved amplified polymorphic sequence (CAPS)	Three QTLs	Ch 3, Ch 8, and Ch 9	Tsuro et al. (2008)
	F_2 populations (1902, 1908, and 1909) derived from the crossing of five radish cultivars with highly divergent root shapes	QTL-seq	SNPs	Seven QTLs (*rs1.1, rs2.1, rs4.1, rs4.2, rs5.1, rs7.1*, and *rs7.2*)	Ch 1, Ch 2, Ch 4, Ch 5, and Ch 7	Wei et al. 2022
Root color	F_2	Linkage and QTL analysis	AFLPs, Brassica SSRs and SLG-CAPS	One QTL	Ch 11	Tsuro et al. (2008)
Red skin color	F_2	Ultra-high-density genetic map	SNP markers	*RsMYB90*	Ch 7	Luo et al. (2020)
	179 cultivated radish genotypes	GWAS	SNP markers	*RsPAP2*	Ch 2	Fan et al. (2020)
Green skin and green/red flesh	Advanced inbred lines	RNA-seq	–	*RsGLK2.1*	–	Ying et al. (2023)
Flowering traits:						
Bolting time	Two radish-inbred lines NH-JS1(late bolting) NH-JS2(early bolting)	RNA-seq	–	*RsFLC* *RsSOC1* (flowering time enhancer gene)	–	Jung et al. (2016)
	(early bolting genotype) × (late bolting genotype) F2 populations	QTL-seq and RNA-seq	SNP markers	3 QTLS: qBT2, qFB7.1, and qFB7.2 *RsFLC1* and *RsFLC2* were the candidate gene for qBT7.2 and qBT2 locus, respectively	Ch 2 and Ch 7	Jin et al. (2024)
Number of seeds per silique and silique length	F_2	Genotype by sequencing	SNP	qNSPSr3 and qNSPSr6 qSLr4	Ch 3 and Ch 6 Ch 4	Kim et al. (2022)
Lobed leaf shape	F_2 population	QTL-seq	InDel markers	*Rs390250*, orthologous to reduced complexity (RCO) gene of *A. thaliana*	Ch 7	Guo et al. (2021)

(continued)

Table 14.2 (continued)

Trait	Mapping population	Technique	Molecular markers	QTLs/genes	Chromosome no.	References
Fertility restorer	F_2	Bulk segregant analysis and RNA-seq	Intron length polymorphism (ILP) and cleaved amplified polymorphic sequence (CAPS) marker	*Rfd1* locus	Ch 3	Lee et al. (2014)
Bolting time and flowering time	$F_{2:3}$	QTL analysis	–	qRBT1 and qRFT1 Candidate genes: *Rs314940*, *Rs315000*, *Rs315310*, and *Rs315960*	Ch 6	Ma et al. (2021)

The core of plant breeding is the selection of better types among variants, in terms of yield and quality of edible parts; ease of cultivation, harvest, and processing; tolerance to environmental stresses; and resistance against pests. Each of these aspects of agronomic or food value are complex in nature. Understanding these traits using traditional breeding methods is more difficult (Breseghello and Coelho 2013; Hartl et al. 1997). Traditional breeding methods were based on the complementarity between parental characteristics. However, little or nothing was known about which part of the genome came from each parent. This situation changed with the advent and dissemination of molecular marker technologies, genomics and phenomics techniques, which made it possible to monitor the transmission of chromosome/traits segments in the progeny (Tang et al. 2010).

During the last three decades, genomics has emerged as an important tool in accelerating the output of new and improved varieties in many root crops (Bisbis et al. 2018; Bhat et al. 2016a, b). In the case of root crops, such methodologies have been recorded very minimum due to the partial genome sequencing, gene tagging, fine mapping and QTL mapping of important traits, lack of dense linkage maps availability, and markers-trait information (Ashkani et al. 2015). In spite of these problems, during the last decade, there have been a few reports on genomics in root crops related to useful agronomic traits. Since this breakthrough, researchers have focused on alternative approaches to improve QTL mapping and fine mapping in radish and carrot by using GWAS, bulked segregant analysis (BSA), and RNA-seq experiments, which have proven to be more effective than traditional mapping and selection methods (Simon 2019).

14.2 Root Architecture and Related Traits

Carrot

The commercial produce of carrot is the storage root, which is an enlarged, swollen base of the taproot, including the hypocotyls. It is composed of four distinct sections: peel (periderm), cortex or storage root (phloem, food-conducting tissue), cambium, and central core (xylem, water-conducting tissue). Cultivated carrots

(*Daucus carota subsp. sativus* Schubl. and Martens) are broadly classified into two groups based on the presence of root pigments, i.e., carotene group (*D. carotassp. sativus var. sativus*) and anthocyanin group (*D. carota* in Persian regions (Iran and Afghanistan)) (Kalia et al. 2023). According to Hort-Science Vegetable Cultivar Descriptions for North America, four carrot market classes are named, namely, 'Imperator', 'Nantes', 'Chantenay', and 'Danvers' (Mou 2022). Root shape traits including length, width, and curvature of the shoulders and tip play a crucial role in categorizing carrot cultivars into market classes. While traditionally assessed subjectively, these traits are now analyzed using digital imaging pipelines (Budahn et al. 2009; Campos et al. 2016; Turner et al. 2018; Brainard et al. 2021; Vega and Goldman 2023).

Over the course of domestication, selective breeding played a role in shaping the array of root shapes observed in the collection of carrot varieties (Wu et al. 2018; Ellison 2019; Geoffriau and Simon 2020). Macko-Podgorni and colleagues (2017) discovered a genetic variation on *ch-2* that exhibited indications of selection, effectively differentiating between cultivated and wild carrot varieties. Improved methods for assessing shoot and root characteristics in carrots would be beneficial for research aimed at pinpointing genetic loci responsible for these traits. So far, the focus of genetic studies in carrots has predominantly revolved around storage root pigmentation, particularly anthocyanin content and carotenoid accumulation, as evidenced by studies conducted by Yildiz et al. (2013). Two research studies have delved into the genetic regulation of carrot root size and shape: one conducted by Turner et al. in 2018 and the other by Macko-Podgórni et al. in 2020. However, only one of these studies performed genome-wide association studies (GWAS), and neither explored the predictive potential of genomic estimated breeding values (GEBVs) for traits related to root shape. The suggested gene, *DcAHLc1*, is categorized within the AT-hook nuclear motif found in plant regulatory genes, known for its role in governing root tissue patterning. Likewise, Turner et al. (2018), employing an image analysis pipeline to investigate root morphology, observed the colocalization of *quantitative trait loci (QTLs)* on chromosomes 1, 2, and 7 associated with interrelated carrot root traits. This implies the potential genetic linkage and quantitative inheritance control over these traits. Linking the OFP-TRM (Ovate Family Proteins (OFPs) and TONNEAU1 Recruiting Motif (TRM)) and International Quality Distribution (IQD) regulon to root shape control has been explored in radish (*Raphanus sativus*) (Wang et al. 2020). The plant-specific OFP-TRM and IQD regulatory pathway play a crucial role in shape patterning and have been extensively researched across various plant organs such as fruit, leaves, stems, and tubers (van der Knaap et al. 2014; Li et al. 2023). In their study, Vega et al. (2024) investigated how members of the OFP-TRM and IQD regulon overlap with root shape *quantitative trait loci (QTLs)*. Understanding the genetic architecture of root traits has significant implications for breeding decisions and can unveil new opportunities in crop improvement.

A novel set of QTL has been identified for four of the most relevant morphological components of root market class in carrot. A highly significant SNP on chromosome 2 associated with root fill was identified (Brainard et al. 2022). Regarding length, the gene *DCv3_Chr5.21023*, situated at the peak on chromosome 5, is identified as a piezo-type mechanosensitive ion channel according to predictions. This gene has been associated with the ability of roots to vertically penetrate the soil profile (Mousavi et al. 2023). A gene located at the peak on chromosome 9, known as *DCv3_Chr9.36166*, is identified as a homolog of protein terminal ear1. This protein has been linked to the regulation of root growth mediated by abscisic acid (Wang et al. 2018a, b).

This observation aligns with the sole other published GWAS on carrot root traits conducted by Macko-Podgórni et al. in 2020. Their study identified a quantitative trait locus (QTL) on chr-1, explaining approximately 10% of the phenotypic variation in maximum width. Concerning marker density, the plateau of the exponential association between single-nucleotide polymorphism (SNP) density and predictive accuracy was reached with a relatively modest number of markers. In genome-wide association studies (GWAS), employing a diverse panel encompassing extensive historical recombination demands a dense array of molecular markers distributed throughout the genome to optimize linkage disequilibrium (LD) between markers and *quantitative trait loci (QTLs)* (Brainard et al. 2022). In a study utilizing a high-density bin map, a total of 43 quantitative trait loci (QTLs) were identified across 16 morphological traits in carrots. Notably, five of these QTLs (designated as RL2-1, RSD2-2, DMC2-1, iSM2-1, and iSM2-2) exhibited stability over the course of 2 years (Ou et al. 2022). The four consistent QTLs associated with root length (RL) and root width (designated as RL2-1, RSD2-2, iSM2-1, and iSM2-2) were found to be situated on the chromosome previously documented in studies by Macko-Podgórni et al. in 2017 and Turner et al. in 2018. The inheritance patterns governing the shape and size of storage roots in carrots may be regulated by candidate genes located on chromosome 2 (Macko-Podgórni et al. 2017; Turner et al. 2018). Two clusters of QTLs were identified on chromosome 2. One was located in a 34.2–35.6 cm region compressing five QTLs related to RL, root shoulder diameter (RSD), dry matter content (DMC), iSM, and iMT. The other was located in a 44.4–45.0 cm region compressing nine QTLs related to Main/Lateral Length (MLL), root length (RL), (root shoulder diameter (RSD), root middle diameter (RMD), plant weight (PW), and dry matter content (DMC). Orizzo and his team utilized single-nucleotide polymorphisms (SNPs) to examine the genetic makeup and evolutionary relationships between wild and cultivated carrots in their research.

The study compared root tissue transcriptomes of Eastern (UHSBC-23-1) and Western (UHSBC-100) carrot cultivars to understand gene expression differences during storage root development and vascular cambium tissue patterning. It found 3544 differentially expressed genes (DEGs) at $p < 0.05$ significance. Functional annotation and gene ontology classification unveiled 97 functional categories, including 33 biological processes, 19 cellular components, 45 metabolic processes, and 26 KEGG (Kyoto Encyclopedia of Genes and Genomes pathways (Chaitra Kulkarni et al. 2023).

A 109-point linkage map was created for carrots, including various molecular markers such as RFLPs, RAPDs, AFLPs, and SAMPL, along with three phenotypic loci (P1, >2, and Rs). The map aimed to understand the linkage relationships between these markers and storage root traits (Vivek and Simon 1999). Through linkage mapping, phylogenetic analysis, and comparative transcriptome (RNA-seq) analysis on carrot roots displaying varying purple pigmentation, potential regulators affecting pigmentation presence/absence and anthocyanin concentration in the root phloem were pinpointed. Among these, a MYB transcription factor known as *DcMYB7* and two cytochrome *CYP450* genes with suggested flavone synthase activity emerged as promising candidates (Bannoud et al. 2019). The study explored phloem pigmentation, identifying two dominant genes controlling the trait. It also analyzed *QTL* for phloem anthocyanins in two regions of Chromosome 3 and investigated two dominant genes regulating purple root pigmentation in a similar genetic context (Cavagnaro et al. 2014). Transcriptome assembly from four genetic backgrounds produced 58,751 contigs and singletons, aiming to study new carrot anthocyanin genes. Identified 114 polymorphic SSRs and 20,058 SNPs (coverage depth ≥ 20×). Among verified markers, 84% SSRs and 82% SNPs were polymorphic. Moreover, 25% of genotyped SNPs showed polymorphism in two diverse mapping populations (Bannoud et al. 2019). A study found a genetic region linked to carrot root shape using GBS. About 81,748 SNPs were used for association mapping after filtering and imputation. An mixed linear model (MLM) identified a significant QTL on chromosome 1 for shoulder diameter. This QTL, within a 600 kb region, showed high LD, indicating domestication selection (Macko-Podgórni et al. 2019). Cytokinins play a vital role in the growth and development of carrot plants, especially in root development. The second stage of root growth exhibits the highest levels of cytokinins, while levels remain stable in petioles and leaves. Expression of *DcCYP735A2* is notably high during this stage in roots, indicating its significant contribution to the increased cytokinin levels (Wang et al. 2015). From carrot genotype transcriptome analysis, 160,227 transcripts formed 32,716 unigenes averaging 1453 bp in length. Among these, 4818 unigenes showed differential expression across four developmental stages. Notably, 87 hormone-related genes were identified, with their roles extensively examined (Wang et al. 2015).

During carrot root development, the genes responsible for l-galactose-1-phosphate phosphatase and l-galactono-1,4-lactone dehydrogenase exhibit a strong correlation with ascorbic acid (AsA) levels, indicating their significant role in the regulation of AsA biosynthesis and recycling (Wang et al. 2015). The study focused on identifying, expanding, and analyzing the evolution of homeobox genes in carrots, particularly their expression during root development. Transcriptome and real-time quantitative polymerase chain reaction (RT-qPCR) data suggested the involvement of WOX and KNOX subfamily members in this process (Que et al. 2018). In a study comparing root development between wild and cultivated carrots, differential expression was observed in 3285, 4637, and 570 genes. Upregulated clusters were enriched with genes related to cell cycle regulation, cell division, vascular tissue development, water transport, and sugar metabolism, as highlighted by Machaj et al. in 2018.

Radish

Radishes are significant root vegetable crops, contributing to approximately 2% of global vegetable production, as reported by Kopta and Pokluda in 2013. Genomic tools and plant resources are vital for radish research and breeding. Since 2014, radish genomes have been sequenced and improved. Initially, one wild radish and three cultivated radish genotypes were sequenced using Illumina and/or 454 platforms, as described by Y.-M. Jeong et al. in 2016. High-throughput sequence-based marker techniques, such as single-nucleotide polymorphism (SNP) markers, competitive allele-specific polymerase chain reaction (PCR), and restriction-site-associated sequencing, have been developed from genome-wide association studies (GWAS). These markers are utilized for genotyping, high-density genetic mapping, and gene identification across numerous accessions. Structural variations in the regulatory element and SNPs in the coding region of an MYB transcription factor gene, governing radish root skin color, were identified through GWAS of our core collection. InDel markers and Mass ARRAY-SNP typing markers, developed from accession variation data, effectively confirmed gene genotypes. In the study of domestication in Asian radish cultivars, 153 candidate domestication regions (CDRs) containing 512 genes were identified via SNP analysis. These genes are linked to root architecture, cell wall composition, sugar metabolism, and glucosinolate biosynthesis. Gene expression profiling during root development revealed advantages of domestication, such as a primary taproot with reduced lateral branches, less rigid cell walls, and enhanced taste (Kim et al. 2014). In the study on taproot thickening in radish, 2606 differentially expressed genes (DEGs) were identified between large acicular and medium obovate accessions. These DEGs are associated with pathways including "phenylpropanoid biosynthesis," "glucosinolate biosynthesis," and "starch and sucrose metabolism." Association analysis highlights *miR393a-bHLH77, miR167c-ARF8,* and *miR5658-APL* as potential key factors in taproot type variation. The findings propose a putative regulatory model for taproot thickening and development, as described by Xie et al. in 2020. Association mapping revealed *RsMyb1* as a candidate gene for radish root color determination, with higher expression in red or purple cultivars compared to white ones. *RsMyb1* primarily governs anthocyanin biosynthesis and is vital for root color determination, especially for marker-sssisted selection (MAS) applications (Yi et al. 2018). *Quantitative trait loci (QTLs)* for morphological traits in *Raphanus sativus* were identified. Four QTLs were found for root weight (RW), including 2 on linkage group 8 (LG8) (*qRW2 and qRW3*), 1 on LG2 (*qRW1*), and LG9 (*qRW4*). QTLs were also detected for root length (RL) and root diameter (RD), with *qRL1* located in the R block of LG1 and *qRL4* in the J block of LG8. For RD, *qRD1* in the F block and *qRD5* in the J block were detected in 2 out of the 3 years (Yu et al. 2016). In both *B. oleracea* and *B. rapa*, a phenomenon of QTL clustering affecting different traits within the same genetic interval is observed (Kennard et al. 1994; Song et al. 1995). This colocalization is attributed to either pleiotropy, where a single gene or closely linked loci influence multiple traits, or to a shared mechanistic basis in inheritance (Srinivas et al. 2016). QTLs qAD1, qAD2, qAD3, and qAD4, identified here, positively influenced the umbrella-shaped radish of the "835" allele, enhancing upper part size and light absorption capacity. These QTLs represent valuable genetic resources for selecting high-quality yield in radish (Yu et al. 2016). AIR encodes a protein linked to a family characterized by a proline-rich or glycine-rich N-terminus and a hydrophobic region (Neuteboom et al. 1999). It plays roles in auxin responses, root development, and root hair cell differentiation. Auxin response factors bind to auxin response promoter elements, orchestrating transcriptional responses to auxin and contributing to root development and primary root processes (Wilmoth et al. 2005). Root growth and development are vital for water and nutrient uptake, influenced by cell division rate, cell differentiation, and cell expansion and elongation (Scheres et al. 2002). In a study (GWAS) examining the genomes of 66 nonred-skinned (nR) and 34 red-skinned (R) radish accessions, three nonsynonymous single-nucleotide polymorphisms (SNPs) were pinpointed within the third exon of *RsMYB1.1* (Kim et al. 2024). CLE peptides, which are related to *CLAVATA3/EMBRYO SURROUNDING REGION*, constitute a category of small molecules crucial for plant growth and development. From the radish genome, a comprehensive set of 48 *RsCLE* genes has been identified (Dong et al. 2023).

Beetroot

In a wide-ranging GWAS involving 977 sugar beet accessions, 170,750 high-quality SNPs were identified using GBS. Within the 25 kb linkage disequilibrium range, several genes were associated with taproot traits like RGD, RS, CS, and FC, including *NRT1/PTR FAMILY 6.3, nudix hydrolase 15, TRANSPORT INHIBITOR RESPONSE 1, MYB77,* and *ERF014*. For aboveground traits such as HC, PT, PL, C, and FLT, genes like UDP-glycosyltransferase 79B6, NAC domain-containing protein 7, F-box protein *At1g10780, PHO1, CPR1, MYB77*, and alcohol acyltransferase 9 were implicated. Additionally, genes like *AP-2* complex subunit mu, trihelix transcription factor *ASIL2*, and late embryogenesis abundant protein 18 were linked to PQ variation. These genes play roles in hormone response, nitrogen and phosphorus transportation, secondary metabolism, fertilization, and embryo maturation (Liu et al. 2023).

Resequencing 306 sugar beet individuals aimed to analyze agronomic trait associations and construct genetic networks. The study assessed population structure, genetic diversity, and selectively scanned genomic regions, uncovering six new genes related to key agronomic traits. DFAX2 and P5CS were linked to skin roughness, while *FRO5, GL24,* and *PPR91* were associated with root and sugar yield. Additionally, the pleiotropic gene POLX affected growth vigor, plant height, crown size, flesh coarseness, and sugar yield (Duan et al. 2014; Gan et al. 2019, 2022; Jeong et al. 2016; Wang et al. 2023). To improve genomic prediction of rhizomania resistance in sugar beet, epistatic effects were incorporated. Initially, all SNPs were used for prediction, with each SNP assessed individually using machine learning. SNPs were then incrementally added to the model based on their importance to identify the optimal number for accuracy. Genomic prediction also considered SNP pairs, with feature selection based on the variable importance of individual SNPs (Lange et al. 2023).

14.3 Pigment Development Genetics and Genomics Studies

Carrot

An association mapping study targeting carotenoid biosynthesis genes in carrots identified significant associations between 109 SNPs in 17 candidate genes and carotenoid contents/color components. Notably, total carotenoids and β-carotene were linked to *ZEP, PDS*, and *CRTISO* genes, while α-carotene correlated with

CRTISO and *PTOX* genes. These findings highlight the involvement of *PDS/PTOX* and *ZEP* genes in carotenoid accumulation, offering new insights into the carotenoid pathway in nonphotosynthetic organs of carrots (Junaid et al. 2023a, 2023b; Kaneko et al. 2007; Kopta et al. 2013; Jourdan et al. 2015). The *Or* gene's discovery highlights its role in β-carotene and provitamin A carotenoid accumulation in carrots and brassicas. A QTL on chromosome 7, linked to a carotene hydroxylase gene akin to Arabidopsis' Lut5, influences carotenoid levels. Two alleles, *CHc* and *CHw*, were found in a segregating population. *Or* and *CH* genes likely played pivotal roles in carrot domestication, affecting carotenoid flux and β-carotene accumulation (Coe et al. 2023).

Fine mapping efforts reduced the genomic region of interest to 650 kb including 72 genes. A gene *Di19* involved in photomorphogenesis and homolog of the *bHLH36* transcription factor involved for carotenoid metabolism in major food crops. A single large effect quantitative trait locus (QTL) on the distal arm of chromosome 7, overlapped with the previously identified β-carotene accumulation QTL, Y_2. These findings characterize the gene underlying β-carotene accumulation in carrot (Ellison et al. 2017). QTL analysis for pigment content on a carotenoid biosynthesis function map in carrot was studied, Two major interacting loci, *Y* and Y_2 on LG2 and LG5 were identified (Just et al. 2009). Three dominant loci, *Y, Y1*, and *Y2*, regulate the distribution of α and β-carotene in xylem/phloem tissue. The Y2 mutant specifically governs the low carotene content in the xylem ("core") of storage roots, particularly in high carotene orange backgrounds (Kumar et al. 2013; Li et al. 2009; Ma et al. 2015; 2021a, 2021b; Macko -Podgorni et al. 2021; Simon et al. 1997). Purple carrots exhibit anthocyanin pigmentation across the entire root or in specific tissue patterns. This pigmentation can manifest within the phloem, either in the outer phloem (OP), also known as the cortex, and inner phloem (IP), or solely within the OP. Studies have identified *DcMYB7, DcMYB113*, and a *MADS-box* gene (*DCAR_010757*) as primary candidates responsible for anthocyanin pigmentation in carrots (Bannoud et al. 2021). Population genetics identifies genetic markers for carrot domestication and uncovers origins of high-carotenoid orange carrots. Selection for high α- and β-carotene orange phenotype relies on three recessive genes: *REC, Or*, and *Y2* loci. These genes regulate pigment accumulation by governing interactions among the carotenoid biosynthetic pathway, photosynthetic system, and chloroplast biogenesis (Coe et al. 2023). Both association and QTL analysis identified eleven QTLs. Among them, two potential candidate genes were proposed: an Orange (*Or*) gene associated with visual color evaluation, α- and β-carotene contents, and a chromoplast-specific lycopene β-cyclase (*CYC-B*) gene (Shibaya et al. 2022).

Radish

A high-density genetic map was built for Yunnan red radish using specific-locus amplified fragment (SLAF) sequencing, revealing loci controlling anthocyanin pigmentation. The F_2 population was utilized alongside SLAF-seq, resulting in a comprehensive genetic map spanning 1230.16 cM. Further analysis of gene structure and expression validated the findings. Specifically, *RsMYB1.3* within *QAC2* emerged as a probable candidate gene influencing the pigmented/nonpigmented trait in Yunnan red radish (Tao et al. 2022). Transcriptional analysis indicate higher expression of *RsDFR, RsANS, RsUFGT, RsF3H, RsCHS3*, and *RsF3′H* genes in red fleshy radishes compared to other varieties. Notably, *RsUFGT* is crucial for anthocyanin accumulation, supported by studies by Gao et al. (2019). Structural genes and the *AtPAP1* homolog control anthocyanin production with varied expression across red radish varieties, as observed in studies by Gao et al. (2019). InDel markers were created based on gene differences in *YAAS-WR1* and *YAAS-RR1*, as reported by Tao et al. (2022). A highly detailed genetic map has offered valuable insights into genome organization, recombination patterns, and the genetic basis of taproot skin color in radish (*Raphanus sativus* L.). Analysis through map-based cloning has identified the *R2R3-MYB* transcription factor *RsMYB90* as a key candidate gene responsible for determining taproot skin color in radish (Luo et al. 2020). Researchers employed QTL-seq and traditional linkage analysis to map genes related to purple skin in radish taproots. They identified a single dominant gene, *Rsps*, through segregation analysis. Whole-genome sequencing and linkage mapping narrowed the Rsps locus to a 238.51-kb region containing 18 genes. Interestingly, *RsMYB1.1* and *RsMYB1.4*, known for red skin, were located on different chromosomes. These findings illuminate the complex regulation of anthocyanin biosynthesis in radish, offering insights for molecular breeding to improve taproot appearance and anthocyanin content (Liu et al. 2019a, b). Gene structure analysis revealed that *RsF3'H* has no introns, while most anthocyanin biosynthesis genes (ABGs) display asymmetrical exon patterns, except *RsSAM*. Anthocyanin accumulation in red radish leaves and colored cultivar taproots increases gradually during development, with a significant rise at 30 days after sowing (DAS) and peak levels reached at maturity, as reported by Muleke et al. in 2017. A pioneering GWAS uncovered the vital role of the *RsPAP2* gene in red-skinned *Raphanus sativus* L. It identified a key SNP on chromosome 2 linked to the *R2R3-MYB* transcription factor responsible for anthocyanin pigment 2 (*PAP2*) production. This study showcases GWAS's efficacy in identifying radish candidate genes. It also suggests that *RsPAP2*, potentially working with *RsTT8*, can regulate gene transcripts associated with anthocyanin biosynthesis, enhancing its production (Fan et al. 2020). An SNP mutation in the *RsPP* gene alters petal color from purple to white in radish (Liu et al. 2021).

Beetroot

The study by Eagen and Goldman (1996) assessed RAPD marker frequencies across cycles of recurrent selection for pigment concentration and percent solids in red beet. The *R-Y-bl* genomic region plays a significant role in the genetic regulation of betalain biosynthesis in table beet, as highlighted by Goldman and Austin (2000).

14.4 Mapping of Genes for Flowering and Male Sterility: Carrot and Radish

Carrot

The mapping of the cola-locus linked to flower development and fertility was accomplished. Two MADS-box genes (*DcMADS3* and *DcMADS5*) crucial for flowering and reproduction, along with three other genes (*DcAOX2a, DcAOX2b,* and

DcCHS2) important for male reproduction, were identified at separate loci not cosegregating with the cola-locus (Budahn et al. 2014: Nothnagel et al. 2014). In the molecular mapping of carrot, two key traits were identified: early flowering (*Vrn1* gene) and restoration of petaloid cytoplasmic male sterility (*Rf1* gene). These traits are controlled by single dominant genes. *Vrn1* is located on chromosome 2 between markers at 0.70 and 0.46 cM, while *Rf1* is on chromosome 9 between markers at 4.38 and 1.12 cM. These genes are vital for carrot domestication and breeding (Alessandro et al. 2013). In 2017, Ou et al. explored the impact of temperature, light intensity, and photoperiod on carrot bolting. They identified 45 genes linked to flowering time, noting that *LATE ELONGATED HYPOCOTYL (LHY)* and *CONSTANS-LIKE 2 (COL2)* were more expressed in short days than long days.

The results are consistent with a model of two genes (*Vrn-A* and *Vrn-B*), with three alleles controlling the vernalization requirement (VR) in carrot (Budahn et al. 2014). In 2021, Wohlfeiler and colleagues investigated the vernalization requirement (VR) in carrot cultivars sourced from various origins. They induced vernalization at different durations of cold exposure (20, 30, 40, 60, and 90 days) and varied temperatures (around 5 and 15 degrees Celsius). Kalia et al. 2019, examined the morphometric traits and organizational characteristics of cytoplasmic ATP9 gene sequences in newly converted cytoplasmic male sterile (CMS) lines (BC4–7) of tropical carrot. The mitochondrial *atp6* gene exhibits variations in length, copies, and expression levels between male sterile and fertile lines of carrot (*Daucus carota* L.) (Nan and Gao et al. 2019; Öztürk Gökçe et al. 2022; Que et al. 2019; Tan et al. 2018). During floral development, carrot petaloid cytoplasmic male sterile (CMS) and maintainer lines exhibit significant gene expression differences. Notably, 16 *MADS-box* genes, categorized into A, B, C, and E classes, show altered expression. Reduced *DcPI* and *DcAG*-like gene expression at stage T2 in petaloid CMS implies energy deficiency-induced inhibition of B- and C-class *MADS-box* genes, leading to stamen-to-petal conversion. This petaloidy acts as intrinsic stress, upregulating *HSP* and *MADS-box* genes at stages T3 and T4 in the petaloid CMS line, potentially yielding fertile revertants (Liu et al. 2019a, b). SNP markers associated with the Rf (fertility restorer) gene locus were identified, leading to the development of a genetic map for this gene in carrots. Through GBS, three markers closely linked to the Rf locus were converted into simple probe markers: *DcSNP-ch9_3342931, DcSNP-ch9_4822988,* and *DcSNP-ch9_6233708* (Duran and Ipek 2022). The mitochondrial *atp9* genes exhibit differences between petaloid male-sterile and male-fertile carrots in terms of heteroplasmy status, involvement in recombination, post-transcriptional processing, and accumulation of RNA and protein products (Maeda et al. 2005; Malik et al. 1999; Meyer et al. 1999; Mousa et al. 2004; Muleke et al. 2017; Szklarczyk et al. 2014).

Radish

Molecular mechanisms that determine the differences in flowering times between two radish lines, *Arabidopsis* flowering-time (*Ft*) genes were identified in the radish. *Ft* enhancer genes, including *RsSOC1*, a key floral integrator, showed an appropriate expression to the late-bolting phenotype (Jung et al. 2016). Six genes (*Rs018140, Rs018950, Rs019220, Rs020080, Rs020590,* and *Rs021450*) for flower color and four genes (*Rs314940, Rs315000, Rs315310,* and *Rs315960*) for bolting and flowering times showed differing expression in parental lines, highlighting genetic complexity. Candidate genes within QTL regions offer potential for radish breeding and functional studies (Ma et al. 2021). De novo transcriptome analysis in radish identified critical genes, including *CONSTANS (CO), CIRCADIAN CLOCK ASSOCIATED 1 (CCA1), EARLY FLOWERING 3 (ELF3), LATE ELONGATED HYPOCOTYL (LHY)*, and *GIGANTEA* (GI). These genes are involved in rhythmic processes such as photoperiodic flowering, UV-B protection, cell elongation, bolting, and flowering (Nie et al. 2016a, b). A genome-wide study identified and characterized 254 flowering genes in radish, with an evolutionary analysis. Increased expression of *AGL19* and *AGL24* genes was observed, which can activate *LFY* and *AP1* expression, ultimately leading to flowering (Wang et al. 2017). An extensive examination of the *MADS-box* gene family, particularly focusing on key *RsMADS* genes, was performed throughout the radish genome. Li et al. (2016) investigated their involvement in flowering and floral organogenesis. Transcriptomic analysis identified DEGs associated with bolting and flowering in radish, including key genes *FT, CO, SOC1, FLC,* and *LFY*, profiled via RT-qPCR. Additionally, 24 miRNA-DEG pairs were implicated in radish bolting and flowering (Nie et al. 2016a, b). RSS0119, a marker linked to flowering time, was found to be highly correlated with flowering time in the F2 population derived from crossing Inbred line 13–66 (P1), a late-flowering Japanese cultivar 'Ninengo', with Inbred line 15–184 (P2), an early flowering Chinese cultivar 'Maer'(Wang et al. 2018a, b). Jin et al. (2024) conducted fine mapping and analyzed candidate genes for the *qBT2* and *qBT7.2* loci, which control bolting time in radish. Wang et al. (2018a, b) documented a naturally occurring insertion in the *RsFLC2* gene linked to the late-bolting trait in radish. Hu et al. (2018) analyzed the CONSTANS-like gene family genome-wide, found that *RsaCOL-02* and *RsaCOL-04* expression increased significantly during vernalization, whereas *RsaCOL-10* expression decreased notably. Lee et al. (2014) generated a high-resolution linkage map of *Rfd1*, a restorer-of-fertility locus for cytoplasmic male sterility in radish, using a combination of bulked segregant analysis and RNA-seq. They developed 21 molecular markers positioned within a 2.1 cM interval, including the *Rfd1* locus. In radish development, HRM markers were explored for distinguishing between two Ogura cytoplasmic male sterility restorer genes. A selection of 117 SNPs within a candidate region on chromosome 5 was made for the *Rft* gene, leading to the successful development of 14 HRM markers (Quiroz- Iturra et al. 2022; Satoh et al. 1992; Singh 2021; Smallwood et al. 1999, Wang et al. 2019; Ahn et al. 2023). In 2021, Kim et al. investigated how an insertion mutation within putative enhancer regions of the *MYB26*-like gene causes inhibition of anther dehiscence, leading to a novel genic male sterility.

14.5 Mapping of Genes for Biotic and Abiotic Traits: Carrot and Radish

Carrot

Le et al. (2019) identified *rQTLs* in carrots against *A. dauci*, noting new R gene clusters on chromosomes Chr_3 and Chr_7. These clusters comprised six genes (*DcRG_082* to *DcRG_087*) and four genes (*DcRG_233* to *DcRG_236*), respectively. Additionally, genes with rare or unknown domain combinations, like CK or CL, were implicated in plant resistance, such as *DcRG_169* and *DcRG_192* (CTNL class) acting as TMV (Tobacco Mosaic Virus) resistance protein N-like isoforms X1 and X2. Furthermore, *DcRG_079* and *DcRG_222* were predicted as putative EDR2s, sharing ~80% amino acid identity with the *A. thaliana EDR2* resistance gene (*AT4G19040*; Vorwerk et al. 2007). The distribution of predicted carrot R genes varied across nine chromosomes, ranging from 19 on Chr_6 to 52 on Chr_3. Notably, *RLP* and *KIN* gene clusters containing structurally similar R genes, like *DcRG_113* to *DcRG_119* (RLP) on Chr_3 and *DcRG_258* to *DcRG_262* (KIN) on Chr_8, suggested common evolutionary origins. As detailed for gray mold caused by *Botrytis cinerea*, Baranski (2008) demonstrated the efficacy of chitinase, specifically CHIT36, in combating *A. radicina* in transgenic plants. *CHIT36*, derived from *T. harzianum*, exhibited antifungal properties in vitro, resulting in a 50% reduction in *A. radicina* infection. Wally et al. (2009a) investigated the impact of the *NPR1* gene from Arabidopsis on systemic acquired resistance (SAR) against *A. radicina*. They observed a significant decrease in taproot lesion severity (by 50%) and foliar necrotic spots (by over 33%) in transgenic plants expressing^{NPR1}. Additionally, Wally and Punja (2010) explored resistance mechanisms in a transgenic carrot line (P23) that overexpressed the rice cationic peroxidase *OsPrx114*. They found that *OsPrx114* enhanced resistance to necrotrophic foliar pathogens by slightly increasing the expression of pathogenesis-related *(PR)* genes and promoting lignin formation in outer periderm tissues of taproots, particularly following *A. radicina* inoculation.

Radish

In the genome-wide association studies (GWAS) conducted on Fusarium wilt (FW) resistance using genotyping by sequencing, 44 single-nucleotide polymorphisms (SNPs) were identified as significantly associated with FW resistance. Moreover, the analysis revealed twenty potential candidate genes linked to this trait (Lee et al. 2021). There is evidence supporting the broadened host range of *Fusarium oxysporum f.sp. raphanin* (Garibaldi et al. 2006). In their study, Ma et al. (2024) pinpointed a clubroot-resistant germplasm from a radish core collection by employing rigorous criteria. Although they did not encounter any fully immune accessions, the research unveiled six highly resistant accessions (*R41, R245, R270, R280, R325,* and *R326*) demonstrating significant resistance against Clubroot disease. The investigation into *Fusarium oxysporum* disease resistance in *Raphanus sativus* involved quantitative trait locus (QTL) analysis and fine mapping using GRAS-Di technology. Within this study, two YR-QTLs were pinpointed, situated on chromosome R2 and R7, respectively. Notably, the QTL on chromosome R7 was termed *ForRs1* and identified as the primary QTL. Fine mapping efforts further delineated the *ForRs1* locus, narrowing it down to a 195 kb region (Ezeah et al. 2023).

A thorough genome-wide analysis revealed how genes like *MAPK, DREB, ERF, AP2, GST, Hsf,* and *Hsp* are involved in signal transduction, metabolism, and biosynthesis pathways, crucial for radish's response to heat stress. These pathways play vital roles in reducing stress-induced damage and enhancing heat tolerance (Wang et al. 2018a, b). A research investigation into the salt tolerance of Japanese wild radishes, 'Hamadaikon' (*Raphanus sativus f. raphanistroides* Makino), examined various traits including seed germination, plant height, root length, and fresh weight under saline conditions. Results indicated that superior levels of germination and growth in NaCl were evident at 25 °C when compared to 20 °C (Sugimoto 2009).

Beetroot

A genome-wide association study on drought tolerance traits in sugar beet germplasms at the seedling stage identified 11 significantly associated loci on chromosomes 2, 3, 5, 7, and 9 from 108,946 SNPs, using a mixed linear model (MLM). Additionally, combining genome-wide association analysis with RT-qPCR revealed 13 genes significantly differentially expressed in drought-tolerant extreme materials (Li et al. 2023).

14.6 Conclusion

Carrot (*Daucus carota ssp. Sativus* L.), Radish (*Raphanus sativus* L.), and Beetroot (*Beta vulgaris subsp. vulgaris* L.) are among the most globally cultivated root vegetables, valued for their processing, export, and economic significance to farmers. Despite their long-storage shelf life, these crops are highly susceptible to various pests, diseases, and environmental stresses, emphasizing the need for disease resistance and stress tolerance traits. With accessible and cost-effective genome sequencing platforms, breeders can address these challenges and provide solutions for future breeding programs. Techniques such as QTL mapping (including traditional, biparental, and advanced methods like association mapping, GWAS studies, fine/candidate gene mapping, GBS, RNA-seq experiments, positional cloning, and transformation to elite genotypes), alongside generating high-density SNP maps and marker-trait linkage studies, offer promising avenues for unraveling the genetic basis of complex traits in vegetable crops. While GWAS has effectively mapped candidate genes and underlying QTLs to facilitate breeding efforts, there remains a lack of substantial reports on QTLs for disease resistance, abiotic stress resistance, and secondary metabolites in root crops like carrot, radish, and beetroot. Utilizing transcriptome-wide and metabolome-wide association studies (TWAS and MWAS, respectively), alongside Meta-GWAS, could help bridge these knowledge gaps and provide insights into the genetic basis of various physiological, phenotypic, biochemical, and stress resistance traits in root crops. The application of these tools holds promise for advancing breeding efforts in the genomics era for carrot, radish, and beetroot.

References

Ahn HI, Han K, Yang HB, Lee ES, Lee YR, Kim J, Kim DS (2023) Development and investigation of HRM markers to discriminate two ogura cytoplasmic male sterility restorer genes in radish. Agronomy 14:43. https://doi.org/10.3390/agronomy14010043

Alessandro MS, Galmarini CR, Iorizzo M, Simon PW (2013) Molecular mapping of vernalization requirement and fertility restoration genes in carrot. Theor Appl Genet 126:415–423. https://doi.org/10.1007/s00122-012-1989-1

Ashkani S, Rafii MY, Shabanimofrad M, Miah G, Akhtar MS, Nasehi A (2015) Molecular breeding strategy and challenges towards improvement of blast disease resistance in rice crop. Front Plant Sci 6:156521. https://doi.org/10.3389/fpls.2015.00886

Bannoud F, Ellison S, Paolinelli M, Horejsi T, Senalik D, Fanzone M, Cavagnaro PF (2019) Dissecting the genetic control of root and leaf tissue-specific anthocyanin pigmentation in carrot (*Daucus carota* L.). Theor Appl Genet 132:2485–2507. https://doi.org/10.1007/s00122-019-03366-5

Bannoud F, Sofia C, Shelby E, Douglas S, Sebastian GT, Massimo I, Philipp WS, Pablo FC (2021) Genetic and transcription profile analysis of tissue-specific anthocyanin pigmentation in carrot root phloem. Gene 12(10):1464. https://doi.org/10.3390/genes12101464

Baranski R (2008) Genetic transformation of carrot (Daucus carota) and other Apiaceae species. Transgenic Plant J 2(1):18–38

Begum RA, Sohag K, Abdullah SMS, Jaafar M (2015) CO_2 emissions, energy consumption, economic and population growth in Malaysia. Renew Sust Energ Rev 41:594–601. https://doi.org/10.1016/j.rser.2014.07.205

Benard D, Punja ZK (1995) Role of Pythium spp. in the development of cavity spot on carrots in British Columbia. Can J Plant Pathol 17:31–45. https://doi.org/10.1080/07060669509500717

Bhat JA, Ali S, Salgotra RK, Mir ZA, Dutta S, Tyagi A, Mushtaq M (2016a) Genomic selection in the era of next generation sequencing for complex traits in plant breeding. Front Genet 7:224364. https://doi.org/10.3389/fgene.2016.00221

Bhat PN, Meegan CA, Von Kienlin A, Paciesas WS, Briggs MS, Burgess JM, Zhang B (2016b) The third fermi GBM gamma-ray burst catalog: the first six years. Astrophys J Suppl Ser 223:28. https://doi.org/10.3847/0067-0049/223/2/28

Bisbis MB, Gruda N, Blanke M (2018) Potential impacts of climate change on vegetable production and product quality—a review. J Clean Prod 170:1602–1620. https://doi.org/10.1016/j.jclepro.2017.09.224

Boiteux LS, Della Vecchia PT, Reifschneider FJB (1993) Heritability estimate for resistance to *Alternaria dauci* in carrot. Plant Breed 110:165–167. https://doi.org/10.1111/j.1439-0523.1993.tb01230.x

Bonnet A (1983) *Daucus carota* L. subsp. *dentatus* Bertol., géniteur de résistance à l'oïdium pour l'amélioration de la carotte cultivée. Agronomie 3:33–38

Brainard SH, Bustamante JA, Dawson JC, Spalding EP, Goldman IL (2021) A digital image-based phenotyping platform for analyzing root shape attributes in carrot. Front Plant Sci 12:690031. https://doi.org/10.3389/fpls.2021.690031

Brainard SH, Ellison SL, Simon PW, Dawson JC, Goldman IL (2022) Genetic characterization of carrot root shape and size using genome-wide association analysis and genomic-estimated breeding values. Theor Appl Genet:1–18. https://doi.org/10.1007/s00122-021-03988-8

Breseghello F, Coelho ASG (2013) Traditional and modern plant breeding methods with examples in rice (Oryza sativa L.). J Agric Food Chem 61:8277–8286. https://doi.org/10.1021/jf305531j

Budahn H, Peterka H, Mousa MA, Ding Y, Zhang S, Li J (2009) Molecular mapping in oil radish (*Raphanus sativus* L.) and QTL analysis of resistance against beet cyst nematode (*Heterodera schachtii*). Theor Appl Genet 118:775–782. https://doi.org/10.1007/s00122-008-0937-6

Budahn H, Baranski R, Grzebelu D, Kiełkowska A, Straka P, Metge K, Nothnagel T (2014) Mapping genes governing flower architecture and pollen development in a double mutant population of carrot. Front Plant Sci 5:504. https://doi.org/10.3389/fpls.2014.00504

Bush WS, Jason H (2012) Genome-wide association studies. PLoS Comput Biol 8(12):e1002822. https://doi.org/10.1371/journal.pcbi.1002822
Cai X, Xu C, Wang X, Wang S, Zhang Z, Fei WQ (2018) Construction of genetic linkage map using genotyping-by-sequencing and identification of QTLs associated with leaf color in spinach. Euphytica 214:1–11. https://doi.org/10.1007/s10681-018-2312-2
Campos MD, Nogales A, Cardoso HG, Kumar SR, Nobre T, Sathishkumar R, Arnholdt-Schmitt B (2016) Stress-induced accumulation of *DcAOX1* and *DcAOX2a* transcripts coincides with critical time point for structural biomass prediction in carrot primary cultures (*Daucus carota* L.). Front Genet 7:1. https://doi.org/10.3389/fgene.2016.00001
Cavagnaro PF, Iorizz M, Yildiz M, Senalik D, Parsons J, Ellison S, Simon PW (2014) A gene-derived SNP-based high resolution linkage map of carrot including the location of QTL conditioning root and leaf anthocyanin pigmentation. BMC Genomics 15:1–17. https://doi.org/10.1186/1471-2164-15-1118
Chopra RN (1933) Indigenous drugs of India
Christianson CE, Jones SS, du Toit LJ (2015) Screening carrot germplasm for resistance to *Xanthomonas hortorum* pv. *carotae*. HortScience 50:341–350. https://doi.org/10.21273/HORTSCI.50.3.341
Coe K, Bostan H, Rolling W, Turner-Hissong S, Macko-Podgórni A, Senalik D, Iorizzo M (2023) Population genomics identifies genetic signatures of carrot domestication and improvement and uncovers the origin of high-carotenoid orange carrots. Nature Plants 9:1643–1658. https://doi.org/10.1038/s41477-023-01526-6
Cooper C, Isaac S, Jones MG, Crowther T, Smith BM, Collin HA (2004) Morphological and biochemical response of carrots to Pythium violae, causative agent of cavity spot. Physiol Mol Plant Pathol 64:27–35. https://doi.org/10.1016/j.pmpp.2004.04.005
Devi S, Sharma PK, Behera TK, Jaiswal S, Boopalakrishnan G, Kumari K, Dey SS (2022) Identification of a major QTL, *Parth6. 1* associated with parthenocarpic fruit development in slicing cucumber genotype, Pusa Parthenocarpic Cucumber-6. Front Plant Sci 13:1064556
Dong J, Wang Y, Xu L, Li B, Wang K, Ying J, Liu L (2023) *RsCLE22a* regulates taproot growth through an auxin signaling-related pathway in radish (*Raphanus sativus* L.). J Exp Bot 74:233–250. https://doi.org/10.1093/jxb/erac406
Duan Y (2014) Genetic analysis and QTL fine mapping of the resistance to black rot in radish (*Raphanus sativus* L.) germplasm. Dissertation, Chinese Academy of Agricultural Sciences
Duangjit J, Causse M, Sauvage C (2016) Efficiency of genomic selection for tomato fruit quality. Mol Breed 36:1–16. https://doi.org/10.1007/s11032-016-0453-3
Duran ST, Ipek A (2022) Identification of SNP markers linked to *Rf* locus in carrot using GBS. Turk J Agric For 46:898–907. https://doi.org/10.55730/1300-011X.3051
Eagen KA, Goldman IL (1996) Assessment of RAPD marker frequencies over cycles of recurrent selection for pigment concentration and percent solids in red beet (*Beta vulgaris* L.). Mol Breed 2:107–115. https://doi.org/10.1007/BF00441426
Ellis PR (1999) The identification and exploitation of resistance in carrots and wild Umbelliferae to the carrot fly, *Psila rosae* (F.). Integr Pest Manag Rev 4:259–268. https://doi.org/10.1023/A:1009609414192
Ellis PR, Hardman JA, Crowther TC, Saw PL (1993) Exploitation of the resistance to carrot fly in the wild carrot species *Daucus capillifolius*. Ann Appl Biol 122(1):79–91. https://doi.org/10.1111/j.1744-7348.1993.tb04016.x
Ellison S (2019) Carrot domestication. In: Simon P, Iorizzo M, Grzebelus D, Baranski R (eds) The carrot genome. Springer, Cham, pp 77–91
Ellison S, Senalik D, Bostan H, Iorizzo M, Simon P (2017) Fine mapping, transcriptome analysis, and marker development for *Y2*, the gene that conditions β-carotene accumulation in carrot (*Daucus carota* L.). G3 Genes Genomes Genetics 7:2665–2675. https://doi.org/10.1534/g3.117.043067
Ezeah CSA, Shimazu J, Kawanabe T, Shimizu M, Kawashima S, Kaji M, Okazaki K (2023) Quantitative trait locus (QTL) analysis and fine-mapping for *Fusarium oxysporum* disease

resistance in Raphanus sativus using GRAS-Di technology. Breed Sci 73:421–434. https://doi.org/10.1270/jsbbs.23032

Falconer DS, Mackay TFC (1996) Introduction to quantitative genetics, 4/e. Addison-Wesley, Harlow Essex

Fan Y, Zhou G, Shabala S, Chen ZH, Cai S, Li C, Zhou M (2016) Genome-wide association study reveals a new QTL for salinity tolerance in barley (*Hordeum vulgare* L.). Front Plant Sci 7:204013

Fan L, Wang YXL, Tang M, Zhang X, Ying J, Liu L (2020) A genome-wide association study uncovers a critical role of the *RsPAP2* gene in red-skinned *Raphanus sativus* L. Hortic Res 7. https://doi.org/10.1038/s41438-020-00385-y

Fedoroff NV, Battisti DS, Beachy RN, Cooper PJ, Fischhoff DA, Hodges CN, Zhu JK (2010) Radically rethinking agriculture for the 21st century. Science 327(5967):833–834

Frese L, Weber WE (1984) Eine Methode zur Unterscheidung von Zuchtmaterial der Möhre mit partieller Resistenz gegen Meloidogyne hapla/A method for the differentiation of breeding material of carrot with partial resistance to *Meloidogyne hapla*. Zeitschrift für Pflanzenkrankheiten und Pflanzenschutz/J Plant Dis Protect 91:396–403

Fujii S, Shiroto Y (2019) Molecular identification of the causal locus for the petaloid phenotype in *Daucus carota*. Breed Sci 69(1):186–188. https://doi.org/10.1270/jsbbs.18120

Gabelman WH, Goldman IL, Breitbach DN (1994) Field evaluation and selection for resistance to aster yellows in carrot (*Daucus carota* L.). J Am Soc Hortic Sci 119:1293–1297

Gahlaut V (2019) Multi-locus genome wide association mapping for yield and its contributing traits in hexaploid wheat under different water regimes. Sci Rep 9(1). https://doi.org/10.1038/s41598-019-55520-0

Gan C, Deng X, Cui L, Yu X, Yuan W, Dai Z, Yao M, Pang W, Ma Y, Yu X, Choi SR (2019) Construction of a high-density genetic linkage map and identification of quantitative trait loci associated with clubroot resistance in radish (*Raphanus sativus* L.). Mol Breed 39:1–12. https://doi.org/10.1007/s11032-019-1020-5

Gan C, Yan C, Pang W, Cui L, Fu P, Yu X, Qiu Z, Zhu M, Piao Z, Deng X (2022) Identification of novel locus *RsCr6* related to clubroot resistance in radish (*Raphanus sativus* L.). Front Plant Sci 13:866211. https://doi.org/10.3389/fpls.2022.866211

Gao J, Li WB, Liu HF, Chen FB (2019) De novo transcriptome sequencing of radish (Raphanus sativus L.) fleshy roots: analysis of major genes involved in the anthocyanin synthesis pathway. BMC Mol Cell Biol 20:1–12. https://doi.org/10.1186/s12860-019-0228-x

Garibaldi A, Gilardi G, Gullino ML (2006) Evidence for an expanded host range of *Fusarium oxysporum f. sp. raphani*. Phytoparasitica 34:115–121. https://doi.org/10.1007/BF02981311

Geoffriau E, Simon PW (2020) Carrots and related Apiaceae crops, vol 33. CABI

Goldman IL, Austin D (2000) Linkage among the R, Y and Bl loci in table beet. Theor Appl Genet 100:337–343. https://doi.org/10.1007/s001220050044

Guo C, Song Y, Zhang X, Xiong Q, Liu K (2021) Fine mapping and identification of the lobed-leaf gene in radish (*Raphanus Sativus* L.). https://doi.org/10.21203/rs.3.rs-636204/v1

Hartl DL, Clark AG, Clark AG (1997) Principles of population genetics, vol 116. Sinauer associates, Sunderland, p 542

Hoban S, Kelley JL, Lotterhos KE, Antolin MF, Bradburd G, Lowry DB, Whitlock MC (2016) Finding the genomic basis of local adaptation: pitfalls, practical solutions, and future directions. Am Nat 188:379–397

Howard RJ, Williams PH (1976) Methods for detecting resistance to *Pythium* and *Rhizoctonia* root diseases in seedling carrots. Plant Dis Rep 60:151–156

Hu T, Wei Q, Wang W, Hu H, Mao W, Zhu Q, Bao C (2018) Genome-wide identification and characterization of CONSTANS-like gene family in radish. PLoS One 13:e0204137. https://doi.org/10.1371/journal.pone.0204137

Huang Y, Li MY, Wang F, Xu ZS, Huang W, Wang GL, Ma J, Xiong AS (2015) Heat shock factors in carrot: genome-wide identification classification and expression profiles response to abiotic stress. Mol Biol Rep 42:893–905. https://doi.org/10.1007/s11033-014-3826-x

Iorizzo M, Ellison S, Senalik D, Zeng P, Satapoomin P, Huang J, Bowman M, Iovene M, Sanseverino W, Cavagnaro P, Yildiz M (2016) A high-quality carrot genome assembly provides new insights into carotenoid accumulation and asterid genome evolution. Nat Genet 48(6):657–666. https://doi.org/10.1038/ng.3565

Iorizzo M, Cavagnaro PF, Bostan H, Zhao Y, Zhang J, Simon PW (2019) A cluster of MYB transcription factors regulates anthocyanin biosynthesis in carrot (*Daucus carota* L.) root and petiole. Front Plant Sci 9:1927. https://doi.org/10.3389/fpls.2018.01927

Jeong YM, Kim N, Ahn BO, Oh M, Chung WH, Chung H, Mun JH (2016) Elucidating the triplicated ancestral genome structure of radish based on chromosome-level comparison with the Brassica genomes. Theor Appl Genet 129:1357–1372. https://doi.org/10.1007/s00122-016-2708-0

Jin Y, Luo X, Li Y, Peng X, Wu L, Yang G, Zhang W (2024) Fine mapping and analysis of candidate genes for qBT2 and qBT7. 2 locus controlling bolting time in radish (*Raphanus sativus* L.). Theor Appl Genet 137:4. https://doi.org/10.1007/s00122-023-04503-x

Jourdan M, Gagné S, Dubois-Laurent C, Maghraoui M, Huet S, Suel A, Geoffriau E (2015) Carotenoid content and root color of cultivated carrot: a candidate-gene association study using an original broad unstructured population. PLoS One 10:e0116674. https://doi.org/10.1371/journal.pone.0116674

Junaid MD, Öztürk Gökçe ZN, Gökçe AF (2023a) Investigation of drought induced biochemical and gene expression changes in carrot cultivars. Mol Biol Rep 50:349–359. https://doi.org/10.1007/s11033-022-08050-4

Junaid MD, Öztürk ZN, Gökçe AF (2023b) Drought stress effects on morphophysiological and quality characteristics of commercial carrot cultivars. Turk J Bot 47:111–126. https://doi.org/10.55730/1300-008X.2750

Jung WY, Park HJ, Lee A, Lee SS, Kim YS, Cho HS (2016) Identification of flowering-related genes responsible for differences in bolting time between two radish inbred lines. Front Plant Sci 7:1844. https://doi.org/10.3389/fpls.2016.01844

Just BJ, Santos CA, Yandell BS, Simon PW (2009) Major QTL for carrot color are positionally associated with carotenoid biosynthetic genes and interact epistatically in a domesticated× wild carrot cross. Theor Appl Genet 119:1155–1169. https://doi.org/10.1007/s00122-009-1117-z

Kalia P, Mangal M, Singh S, Chugh C, Mishra S, Chaudhary S (2019) Morphological and molecular changes on cytoplasmic male sterility (CMS) introgression in Asiatic carrot (*Daucus carota* L.). Planta 250:507–518. https://doi.org/10.1007/s00425-019-03185-4

Kalia P, Singh BK, Bhuvaneswari S, Patel V, Selvakumar R (2023) Carrot: breeding and genomics. Vegetab Sci:221–230. https://doi.org/10.61180/vegsci.2023.v50.spl.08

Kamei A, Tsuro M, Kubo N, Hayashi T, Wang N, Fujimura T, Hirai M (2010) QTL mapping of clubroot resistance in radish (*Raphanus sativus* L.). Theor Appl Genet 120:1021–1027. https://doi.org/10.1007/s11032-019-1020-5

Kaneko Y, Kimizuka-Takagi C, Bang SW, Matsuzawa Y (2007) Radish. In: Kole C (ed) Genome mapping and molecular breeding in plants. Springer, Heidelberg, pp 141–160. https://doi.org/10.1007/978-3-540-34536-7_3

Kennard WC, Slocum SS, Figdore TC, Osborn MC (1994) Genetic analysis of morpholocigical variation in *Brassica oleracea* using molecular markers. Theor Appl Genet 87:721–732. https://doi.org/10.1007/BF00222898

Khan AS, Yu S, Liu H (2012) Deformation induced anisotropic responses of Ti–6Al–4V alloy Part II: a strain rate and temperature dependent anisotropic yield criterion. Int J Plast 38:14–26

Kim DH, Kim SJ, Jeong SI, Cheon CJ, Kim SY (2014) Antiadipogenic effects of red radish (*Raphanus sativus* L.) sprout extract in 3T3-L1 preadipocytes. J Life Sci 24:1224–1230. https://doi.org/10.5352/JLS.2014.24.11.1224

Kim J, Park S, Lee J, Han K, Jang Y, Chae W (2022) A preliminary study for identifying quantitative trait loci associated with seed production in radish using genotyping-by-sequencing. Horticulturae 8(3):268. https://doi.org/10.3390/horticulturae8030268

Kim J, Jang H, Mun JH (2024) Effect of structural variation in the promoter region of *RsMYB1. 1* on the skin color of radish taproot. Front Plant Sci 14:1327009. https://doi.org/10.3389/fpls.2023.1327009

Kirtikar KR, Basu BD (1935) Indian medicinal plants. Vander Vossen and Sambas, p 1993

Kopta T, Pokluda R (2013) Yields, quality and nutritional parameters of radish (*Raphanus sativus*) cultivars when grown organically in the Czech Republic. Hortic Sci 40:16

Korte A, Farlow A (2013) The advantages and limitations of trait analysis with GWAS: a review. Plant Methods 9:1–9. https://doi.org/10.1186/1746-4811-9-29

Kraus C (1992) Untersuchungen zur Vererbung von Resistenz und Toleranz gegen Meloidogyne hapla bei Möhren, unter besonderer Berücksichtigung von Daucus carota ssp. azoricus Franco. Doctoral dissertation, Universitaetsbibliothek Hannover

Kulkarni CC, Cholin SS, Bajpai AK, Ondrasek G, Mesta RK, Rathod S, Patil HB (2023) Comparative root transcriptome profiling and gene regulatory network analysis between Eastern and Western carrot (*Daucus carota* L.) cultivars reveals candidate genes for vascular tissue patterning. Plants 12:3449. https://doi.org/10.3390/plants12193449

Kumar SR, Anandhan S, Dhivya S, Zakwan A, Sathishkumar R (2013) Isolation and characterization of cold inducible genes in carrot by suppression subtractive hybridization. Biol Plant 57:97–104. https://doi.org/10.1007/s10535-012-0250-8

Lamb KP (1953) Observations on yield and varietal susceptibility of some carrot varieties to insect attack in the field. N Z J Sci Technol Sect A 34:351

Lange TM, Heinrich F, Kopisch-Obuch F, Keunecke H, Gültas M, Schmitt AO (2023) Improving genomic prediction of rhizomania resistance in sugar beet (*Beta vulgaris* L.) by implementing epistatic effects and feature selection. F1000Research 12:280

Le Clerc V, Aubert C, Cottet V, Alternaria dauci Yovanopoulos C, Piquet M, Suel A, Briard M (2019) Breeding for carrot resistance to without compromising taste. Mol Breed 39:1–15. https://doi.org/10.1007/s11032-019-0966-7

Lee YP, Cho Y, Kim S (2014) A high-resolution linkage map of the Rfd1, a restorer-of-fertility locus for cytoplasmic male sterility in radish (*Raphanus sativus* L.) produced by a combination of bulked segregant analysis and RNA-Seq. Theor Appl Genet 127:2243–2252. https://doi.org/10.1007/s00122-014-2376-x

Lee ON, Koo H, Yu JW, Park HY (2021) Genotyping-by-sequencing-based genome-wide association studies of fusarium wilt resistance in radishes (*Raphanus sativus* l.). Gene 12:858. https://doi.org/10.3390/genes12060858

Leiss KA, Cristorfori G, van Steenis R, Verpoorte R, Klinkhamer PGL (2013) An eco-metabolomic study of host plant resistance to western flower thrips in cultivated, biofortified and wild carrots. Phytochemistry 93:63–70. https://doi.org/10.1016/j.phytochem.2013.03.011

Leunov VI, Sokolova LM, Beloshapkina OO, Khovrin AN (2021) Resistance of carrots to *Alternaria* sp., *Fusarium* sp. and factors influencing it. In: IOP conference series: earth and environmental science, vol 624, no 1. IOP Publishing, p 012010

Li HS (2009) Genetic dissection of resistance to turnip mosaic virus and black rot in radish (Raphanus Sativus L.). Dissertation, Chinese Academy of Agricultural Sciences

Li C, Wang Y, Xu L, Nie S, Chen Y, Liang D, Liu L (2016) Genome-wide characterization of the MADS-box gene family in radish (*Raphanus sativus* L.) and assessment of its roles in flowering and floral organogenesis. Front Plant Sci 7:211403. https://doi.org/10.3389/fpls.2016.01390

Li J, Liu X, Yao Q, Xu L, Li W, Tan W, Liu D (2023) Tolerance and adaptation characteristics of sugar beet (*Beta vulgaris* L.) to low nitrogen supply. Plant Signal Behav 18:2159155. https://doi.org/10.1080/15592324.2022.2159155

Liu S, Jia M, Chen J, Wan H, Dong R, Nie S, Yu Q (2019a) Removal of bound polyphenols and its effect on antioxidant and prebiotics properties of carrot dietary fiber. Food Hydrocoll 93:284–292. https://doi.org/10.1016/j.foodhyd.2019.02.047

Liu T, Zhang Y, Zhang X, Sun Y, Wang H, Song J, Li X (2019b) Transcriptome analyses reveal key genes involved in skin color changes of 'Xinlimei'radish taproot. Plant Physiol Biochem 139:528–539

Liu D, Wei X, Sun D, Yang S, Su H, Wang Z, Yuan Y (2021) An SNP mutation of gene RsPP converts petal color from purple to white in radish (*Raphanus sativus* L.). Front Plant Sci 12:643579

Liu P, Yuan W, Fu J, Jiang Z, Hayashi H, Neubig G (2023) Pre-train, prompt, and predict: a systematic survey of prompting methods in natural language processing. ACM Comput Surv 55:1–35. https://doi.org/10.1145/3560815

Luo X, Xu L, Wang Y, Dong J, Chen Y, Tang M, Fan L, Zhu Y, Liu L (2020) An ultra-high-density genetic map provides insights into genome synteny, recombination landscape and taproot skin colour in radish (*Raphanus sativus* L.). Plant Biotechnol J 18(1):274–286. https://doi.org/10.1111/pbi.13195

Lynch JP (1998) The role of nutrient-efficient crops in modern agriculture. J Crop Prod 1:241–264

Ma J, Xu Z-S, Wang F, Xiong A-S (2015) Isolation, purification and characterization of two laccases from carrot (*Daucus carota* L.) and their response to abiotic and metal ions stresses. Protein J 34:444–452. https://doi.org/10.1007/s10930-015-9639-5

Ma Y, Chhapekar SS, Rameneni JJ, Kim S, Gan TH, Choi SR, Lim YP (2021) Identification of qtls and candidate genes related to flower traits and bolting time in radish (*Raphanus sativus* L.). Agronomy 11(8):1623. https://doi.org/10.3390/agronomy11081623

Ma Y, Chhapekar SS, Lu L, Oh S, Singh S, Kim CS, Kim S, Choi GJ, Lim YP, Choi SR (2021a) Genome-wide identification and characterization of NBS-encoding genes in *Raphanus sativus* L. and their roles related to *Fusarium oxysporum* resistance. BMC Plant Biol 21:1–17. https://doi.org/10.1186/s12870-020-02803-8

Ma Y, Chhapekar SS, Lu L, Yu X, Kim S, Lee SM, Gan TH, Choi GJ, Lim YP, Choi SR (2021b) QTL mapping for Fusarium wilt resistance based on the whole-genome resequencing and their association with functional genes in *Raphanus sativus*. Theor Appl Genet 134(12):3925–3940. https://doi.org/10.1007/s00122-021-03937-5

Ma Y, Wang H, Song J, Yang W, Jia H, Agerbirk N, Zhang X (2024) Identification of clubroot-resistant germplasm in a radish (*Raphanus sativus* L.) core collection. Agronomy 14:157. https://doi.org/10.3390/agronomy14010157

Machaj G, Bostan H, Macko-Podgórni A, Iorizzo M, Grzebelus D (2018) Comparative transcriptomics of root development in wild and cultivated carrots. Gene 9:431. https://doi.org/10.3390/genes9090431

Mackay TF, Stone EA, Ayroles JF (2009) The genetics of quantitative traits: challenges and prospects. Nat Rev Genet 10:565–577

Macko-Podgórni A, Machaj G, Stelmach K, Senalik D, Grzebelus E, Iorizzo M, Simon PW, Grzebelus D (2017) Characterization of a genomic region under selection in cultivated carrot (*Daucus carota subsp. sativus*) reveals a candidate domestication gene. Front Plant Sci 8:238757. https://doi.org/10.3389/fpls.2017.00012

Macko-Podgórni A, Stelmach K, Kwolek K, Senalik D, Simon PW, Grzebelus D (2019) A genome-wide association study reveals a genomic region associated with the carrot root shape. In: II international symposium on carrot and other apiaceae 1264, pp 129–134. https://doi.org/10.17660/ActaHortic.2019.1264.15

Macko-Podgórni A, Stelmach K, Kwolek K, Machaj G, Ellison S, Senalik DA, Simon PW, Grzebelus D (2020) Mining for candidate genes controlling secondary growth of the carrot storage root. Int J Mol Sci 21(12):4263

Macko-Podgórni A, Machaj G, Grzebelus D (2021) A global landscape of miniature inverted-repeat transposable elements in the carrot genome. Gene 12:859. https://doi.org/10.3390/genes12060859

Maeda K, Kimura S, Demura T, Takeda J, Ozeki Y (2005) *DcMYB1* acts as a transcriptional activator of the carrot phenylalanine ammonia-lyase gene (*DcPAL1*) in response to elicitor treatment, UV-B irradiation and the dilution effect. Plant Mol Biol 59:739–752. https://doi.org/10.1007/s11103-005-0910-6

Malik MK, Slovin JP, Hwang CH, Zimmerman JL (1999) Modified expression of a carrot small heat shock protein gene, Hsp17. 7, results in increased or decreased thermotolerance. Plant J 20(1):89–99. https://doi.org/10.1046/j.1365-313X.1999.00581.x

McDonald MR (1994) Cavity spot of carrot (*Pythium* spp.): etiology, epidemiology and control. PhD dissertation, University of Guelph, Ontario, 314 pp

Meyer K, Keil M, Naldrett MJ (1999) A leucine-rich repeat protein of carrot that exhibits antifreeze activity. FEBS Lett 447:171–178. https://doi.org/10.1016/S0014-5793(99)00280-X

Mora F, Quitral YA, Matus I, Russell J, Waugh R, Del Pozo A (2016) SNP-based QTL mapping of 15 complex traits in barley under rain-fed and well-watered conditions by a mixed modeling approach. Front Plant Sci 7:187078. https://doi.org/10.3389/fpls.2016.00909

Mou B (2022) Vegetable cultivar descriptions for North America List 28 2022. HortScience 57(8):949–1040. https://doi.org/10.21273/HORTSCI.57.8.949

Mousa M, Kandeel N, Mohamed M, Nasr MA, Budahn H, Peterka H (2004) Mapping QTL (s) for resistance to the beet cyst nematode (Heterodera schachtii Schm.) in radish (Raphanus sativus L.). Ass Univ Bull Environ Res 7:1–18. https://doi.org/10.21608/auber.2004.150707

Mousavi SN, Parchami M, Ramamoorthy SK, Soufiani AM, Hakkarainen M, Zamani A (2023) Bioconversion of carrot pomace to value-added products: Rhizopus delemar fungal biomass and cellulose. Fermentation 9:374. https://doi.org/10.3390/fermentation9040374

Muleke E, Fan L, Xu L, Zhu X, Zhang W, Liu L (2017) Coordinated regulation of anthocyanin biosynthesis genes confers varied phenotypic and spatial-temporal anthocyanin accumulation in radish (*Raphanus sativus* L.). Front Plant Sci 8:268743

Nan H, Gao LZ (2019) Genome-wide analysis of *WRKY* genes and their response to hormone and mechanic stresses in carrot. Front Genet 10:363. https://doi.org/10.3389/fgene.2019.00363

National Horticulture Board 2023–24. https://www.nhb.gov.in/

Neuteboom LW, Ng JM, Kuyper M, Clijdesdale OR, Hooykaas PJ, Van Der Zaal BJ (1999) Isolation and characterization of cDNA clones corresponding with mRNAs that accumulate during auxin-induced lateral root formation. Plant Mol Biol 39:273–287. https://doi.org/10.1023/A:1006104205959

Nie S, Li C, Wang Y, Xu L, Muleke EM, Tang M, Liu L (2016a) Transcriptomic analysis identifies differentially expressed genes (DEGs) associated with bolting and flowering in radish (*Raphanus sativus* L.). Front Plant Sci 7:682. https://doi.org/10.3389/fpls.2016.00682

Nie S, Li C, Xu L, Wang Y, Huang D, Muleke EM, Liu L (2016b) De novo transcriptome analysis in radish (*Raphanus sativus* L.) and identification of critical genes involved in bolting and flowering. BMC Genomics 17:1–16. https://doi.org/10.1186/s12864-016-2633-2

Nimmakayala P, Tomason YR, Abburi VL, Alvarado A, Saminathan T, Vajja VG, Reddy UK (2016) Genome-wide differentiation of various melon horticultural groups for use in GWAS for fruit firmness and construction of a high resolution genetic map. Front Plant Sci 7:225562. https://doi.org/10.3389/fpls.2016.01437

Nothnagel T, Budahn H, Krämer I, Schliephake E, Schreyer L, Krämer R (2014) Resistance to Asparagus virus 1 in the Wild Relative *Asparagus amarus*. J Phytopathol 162:180–189

Ou C, Sun T, Liu X, Li C, Li M, Wang X, Zhuang F (2022) Detection of chromosomal segments introgressed from wild species of carrot into cultivars: quantitative trait loci mapping for morphological features in backcross inbred lines. Plants 11:391. https://doi.org/10.3390/plants11030391

Öztürk Gökçe ZN, Gökçe AF, Junaid MD, Chaudhry UK (2022) Comparative transcriptomics of drought stress response of taproot meristem region of contrasting purple carrot breeding lines supported by physio-biochemical parameters. Funct Integr Genomics 22:697–710. https://doi.org/10.1007/s10142-022-00868-2

Pandita D, Pandita A, Pamuru RR, Nayik GA (2020) Beetroot. In: Antioxidants in vegetables and nuts-properties and health benefits, pp 45–74. https://doi.org/10.1007/978-981-15-7470-2_3

Parsons J, Matthews W, Iorizzo M, Roberts P, Simon P (2015) *Meloidogyne incognita* nematode resistance QTL in carrot. Mol Breed 35:1–11. https://doi.org/10.1007/s11032-015-0309-2

Pavan S, Marcotrigiano AR, Ciani E, Mazzeo R, Zonno V, Ruggieri V, Lotti C, Ricciardi L (2017) Genotyping-by-sequencing of a melon (*Cucumis melo* L.) germplasm collection from a secondary center of diversity highlights patterns of genetic variation and genomic features of different gene pools. BMC Genomics 18:59. https://doi.org/10.1186/s12864-016-3429-0

Pink DAC (1993) Beetroot- *Beta vulgaris subsp. vulgaris*. In: Genetic improvement of vegetable crops. Pergamon, pp 473–477. https://doi.org/10.1016/B978-0-08-040826-2.50036-9

Pobożniak M, Gaborska M, Wójtowicz T (2021) Resistance and tolerance of ten carrot cultivars to the hawthorn-carrot aphid, *Dysaphis crataegi* Kalt., in Poland. PLoS One 16(3). https://doi.org/10.1371/journal.pone.0247978

Pryor B, Davis RM, Gilbertson RL (2000) A toothpick inoculation method for evaluating carrot cultivars for resistance to *Alternaria radicina*. HortScience 35:1099–1102

Que Q, Wang G-L, Feng K, Xu Z-S, Wang F, Xiong A-S (2018) Hypoxia enhances lignification and affects the anatomical structure in hydroponic cultivation of carrot taproot. Plant Cell Rep 37:1021–1032. https://doi.org/10.1007/s00299-018-2288-3

Que F, Hou XL, Wang GL, Xu ZS, Tan GF, Li T, Xiong AS (2019) Advances in research on the carrot, an important root vegetable in the Apiaceae family. Hortic Res 6:69. https://doi.org/10.1038/s41438-019-0150-6

Quiroz-Iturra LF, Simpson K, Arias D, Silva C, González-Calquin C, Amaza L, Handford M, Stange C (2022) Carrot *DcALFIN4* and *DcALFIN7* transcription factors boost carotenoid levels and participate differentially in salt stress tolerance when expressed in *Arabidopsis thaliana* and *Actinidia deliciosa*. Int J Mol Sci 23:12157. https://doi.org/10.3390/ijms232012157

Raees-ul H, Prasad K (2015) Nutritional and processing aspects of carrot (*Daucus carota* L)-a review. South Asian Journal of Food Technology and Environment 1:1–14. https://doi.org/10.46370/sajfte.2015.v01i01.01

Rikkerink EH, Oraguzie NC, Gardiner SE (2007) Prospects of association mapping in perennial horticultural crops. In: Association mapping in plants. Springer, New York, pp 249–269

Rockman MV (2012) The QTN program and the alleles that matter for evolution: all that's gold does not glitter. Evolution 66:1–17

Santos CA, Simon PW (2002) Some AFLP amplicons are highly conserved DNA sequences mapping to the same linkage groups in two F_2 populations of carrot. Genet Mol Biol 25:195–201. https://doi.org/10.1590/S1415-47572002000200013

Santure AW, Poissant J, De Cauwer I, Van Oers K, Robinson MR, Quinn JL, Slate J (2015) Replicated analysis of the genetic architecture of quantitative traits in two wild great tit populations. Mol Ecol 24:6148–6162

Satoh S, Sturm A, Fujii T, Chrispeels MJ (1992) cDNA cloning of an extracellular dermal glycoprotein of carrot and its expression in response to wounding. Planta 188:432–438. https://doi.org/10.1007/BF00192811

Scheres B, Benfey P, Dolan L (2002) Root development. In: The Arabidopsis book/American Society of Plant Biologists, p 1. https://doi.org/10.1199/tab.0101

Shibaya T, Kuroda C, Tsuruoka H, Minami C, Obara A, Nakayama S, Isobe S (2022) Identification of QTLs for root color and carotenoid contents in Japanese orange carrot F_2 populations. Sci Rep 12:8063. https://doi.org/10.1038/s41598-022-11544-7

Siddiqui ZA, Nesha R, Varshney A (2011) Response of carrot cultivars to *Meloidogyne incognita* and *Pectobacterium carotovorum* subsp. *carotovorum*. J Plant Pathol 93:503–506. https://www.jstor.org/stable/41999025

Sidorova T, Miroshnichenko D (2013) Transgenic carrot expressing thaumatin II gene has enhanced resistance against *Fusarium avenaceum*. Acta Agric Scand 974:123–130. https://doi.org/10.17660/ActaHortic.2013.974.14

Simlat M, Stobiecki M, Szklarczyk M (2013) Accumulation of selected phenolics and expression of *PAL* genes in carrots differing in their susceptibility to carrot fly (*Psila rosae* F.). Euphytica 190:253–266. https://doi.org/10.1007/s10681-012-0836-4

Simon HA (2019) The Sciences of the Artificial, reissue of the third edition with a new introduction by John Laird. MIT Press

Simon PW, Rubatzky VE, Bassett MJ, Strandberg JO, White JM (1997) B7262, purple carrot inbred. HortScience 32:146–147

Singh BK (2021) Radish (*Raphanus sativus* L.): breeding for higher yield, better quality and wider adaptability. In: Advances in plant breeding strategies: vegetable crops: volume 8: bulbs, roots and tubers, pp 275–304. https://doi.org/10.1007/978-3-030-66965-2_7

Slate J, Santure AW, Feulner PG, Brown EA, Ball AD, Johnston SE, Gratten J (2010) Genome mapping in intensively studied wild vertebrate populations. Trends Genet 26:275–284

Smallwood M, Worrall D, Byass L, Elias L, Ashford D, Doucet CJ, Holt C, Telford J, Lillford P, Bowles DJ (1999) Isolation and characterization of a novel antifreeze protein from carrot (*Daucus carota*). Biochem J 340:385–391. https://doi.org/10.1042/bj3400385

Song KK, Slocum TC, Osborn (1995) Molecular marker analysis of genes controlling morphological variation in *Brassica rapa* (syn. *campestris*). Theor Appl Genet 90:1–10. https://doi.org/10.1007/BF00220989

Srinivas R, Deviprasad AG, Manonmani HK (2016) Elimination of inhibitory effects of chlorpyrifos and quinolphos on radish and green gram seed germination by bioremediation of contaminated soil: a comparative study. Int J Curr Microbiol App Sci 5:26–42. https://doi.org/10.20546/ijcmas.2016.502.003

Stapley J, Reger J, Feulner PG, Smadja C, Galindo J, Ekblom R, Bennison C, Ball AD, Beckerman AP, Slate J (2010) Adaptation genomics: the next generation. Trends Ecol Evol 25(12):705–712

Stelfox D, Henry AW (1978) Occurrence of rubbery brown rot of stored carrots in Alberta. Can Plant Dis Surv 58:87–91

Sugimoto K (2009) Evaluation of salt tolerance in Japanese wild radishes (Raphanus sativus f. raphanistroides Makino). Bull Minamikyushu Univ A 39:79–88

Sukumaran S, Li X, Li X, Zhu C, Bai G, Perumal R, Yu J (2016) QTL mapping for grain yield, flowering time, and stay-green traits in sorghum with genotyping-by-sequencing markers. Crop Sci 56:1429–1442

Sweet JB, Beale RE, Wright IR (1989) Cavity spot disease in six carrot cultivars treated with a metalaxyl and thiram fungicide. Tests of agrochemicals and cultivars 10. Ann Appl Biol 114:38–39

Szklarczyk M, Szymański M, Wójcik-Jagła M, Simon PW, Weihe A, Börner T (2014) Mitochondrial a *tp9* genes from petaloid male-sterile and male-fertile carrots differ in their status of heteroplasmy, recombination involvement, post-transcriptional processing as well as accumulation of RNA and protein product. Theor Appl Genet 127:1689–1701. https://doi.org/10.1007/s00122-014-2331-x

Takaichi M, Oeda K (2000) Transgenic carrots with enhanced resistance against two major pathogens, *Erysiphe heraclei* and *Alternaria dauci*. Plant Sci 153:135–144. https://doi.org/10.1016/S0168-9452(99)00254-X

Tan GF, Wang F, Zhang XY, Xiong AS (2018) Different lengths, copies and expression levels of the mitochondrial atp6 gene in male sterile and fertile lines of carrot (*Daucus carota* L.). Mitochondrial DNA Part A 29:446–454. https://doi.org/10.1080/24701394.2017.1303492

Tang D, Kang R, Cheh CW, Livesey KM, Liang X, Schapiro NE, Lotze MT (2010) HMGB1 release and redox regulates autophagy and apoptosis in cancer cells. Oncogene 29:5299–5310. https://doi.org/10.1016/j.stem.2010.03.015

Tao J, Li S, Wang Q, Yuan Y, Ma J, Xu M, Sun Y (2022) Construction of a high-density genetic map based on specific-locus amplified fragment sequencing and identification of loci controlling anthocyanin pigmentation in Yunnan red radish. Hortic Res 9:uhab031. https://doi.org/10.1093/hr/uhab031

Tibbitts TW, Setiamihardja R, Palzkill DP, Olszyk DM (1983) Calcium-related leaf necrosis of seedling carrots. J Am Soc Hortic Sci 108(6):1017–1019. https://doi.org/10.21273/JASHS.108.6.1017

Tomlinson JA (1965) Rep Natn Veg Res Stn for 1964:70

Tsuro M, Suwabe K, Kubo N, Matsumoto S, Hirai MO (2008) Mapping of QTLs controlling root shape and red pigmentation in radish, *Raphanus sativus* L. Breed Sci 58(1):55–61. https://doi.org/10.1270/jsbbs.58.55

Turner SD, Ellison SL, Senalik DA, Simon PW, Spalding EP, Miller ND (2018) An automated image analysis pipeline enables genetic studies of shoot and root morphology in carrot (*Daucus carota* L.). Front Plant Sci 9:1703. https://doi.org/10.3389/fpls.2018.01703

Van der Knaap E, Chakrabarti M, Chu YH, Clevenger JP, Illa-Berenguer E, Huang Z, Keyhaninejad N et al (2014) What lies beyond the eye: the molecular mechanisms regulating tomato fruit weight and shape. Front Plant Sci 5:89256. https://doi.org/10.3389/fpls.2014.00227

Vega A, Goldman IL (2023) Planting density does not affect root shape traits associated with market class in carrot. HortScience 58:996–1004. https://doi.org/10.21273/HORTSCI17232-23

Vega A, Brainard SH, Goldman IL (2024) Linkage mapping of root shape traits in two carrot populations. G3 Genes Genomes Genetics 14(4):jkae041. https://doi.org/10.1093/g3journal/jkae041

Vivek BS, Simon PW (1999) Linkage relationships among molecular markers and storage root traits of carrot (Daucus carota L. ssp. sativus). Theor Appl Genet 99:58–64. https://doi.org/10.1007/s001220051208

Vorwerk S, Schiff C, Santamaria M, Koh S, Nishimura M, Vogel J, Somerville S (2007) EDR2 negatively regulates salicylic acid-based defenses and cell death during powdery mildew infections of *Arabidopsis thaliana*. BMC Plant Biol 7:1–14. https://doi.org/10.1186/1471-2229-7-35

Wally O, Punja ZK (2010) Enhanced disease resistance in transgenic carrot (Daucus carota L.) plants over-expressing a rice cationic peroxidase. Planta 232(5):1229–1239

Wang Y, Shen W, Chan Z, Wu Y (2015) Endogenous cytokinin overproduction modulates ROS homeostasis and decreases salt stress resistance in Arabidopsis thaliana. Front Plant Sci 6:166381. https://doi.org/10.3389/fpls.2015.01004

Wang Q, Zhang Y, Zhang L (2018a) A naturally occurring insertion in the *RsFLC2* gene associated with late-bolting trait in radish (*Raphanus sativus* L.). Mol Breed 38:1–12. https://doi.org/10.1007/s11032-018-0897-8

Wang R, Mei Y, Xu L, Zhu X, Wang Y, Guo J, Liu L (2018b) Differential proteomic analysis reveals sequential heat stress-responsive regulatory network in radish (*Raphanus sativus* L.) taproot. Planta 247:1109–1122. https://doi.org/10.1007/s00425-018-2846-5

Wang Q, Zheng P, Zhang L (2019) Identification and classification of S haplotypes in radish (*Raphanus sativus*). Plant Breed 138:121–130. https://doi.org/10.1111/pbr.12664

Wang Y, Jiali Y, Yang Z, Liang X, Wanting Z, Meng N, Yuelin Z, Liwang L (2020) Genome-wide identification and functional characterization of the cation proton antiporter (CPA) family related to salt stress response in radish (Raphanus sativus L.). Int J Mol Sci 21:8262. https://doi.org/10.3390/ijms21218262

Wang H, Fang XM, Sutar PP, Meng JS, Wang J, Yu XL, Xiao HW (2021) Effects of vacuum-steam pulsed blanching on drying kinetics, colour, phytochemical contents, antioxidant capacity of carrot and the mechanism of carrot quality changes revealed by texture, microstructure and ultrastructure. Food Chem 338:127799. https://doi.org/10.1016/j.foodchem.2020.127799

Wang L, Zhang Z, Han P, Liang Y, Zhang H, Fu Z, Bai C (2023) Association analysis of agronomic traits and construction of genetic networks by resequencing of 306 sugar beet (*Beta vulgaris* L.) lines. Sci Rep 13:15422. https://doi.org/10.1038/s41598-023-42182-2

Watson MT, Falk BW (1994) Ecological and epidemiological factors affecting carrot motley dwarf development in carrots grown in the Salinas Valley of California. Plant Dis 78:477–481

Wei Q, Wang J, Wang W, Hu H, Yan Y, Bao C, Hu T (2022) Identification of QTLs controlling radish root shape using multiple populations. Horticulturae 8(10):931. https://doi.org/10.3390/horticulturae8100931

Wellenreuther M, Hansson B (2016) Detecting polygenic evolution: problems, pitfalls, and promises. Trends Genet 32:155–164

Wilmoth JC, Wang S, Tiwari SB, Joshi AD, Hagen G, Guilfoyle TJ, Reed JW (2005) *NPH4/ARF7* and *ARF19* promote leaf expansion and auxin-induced lateral root formation. Plant J 43:118–130. https://doi.org/10.1111/j.1365-313X.2005.02432.x

Wohlfeiler J, Alessandro MS, Cavagnaro PF, Galmarini CR (2021) Gradient of vernalization requirement in carrot cultivars from diverse geographical origins. Crop Sci 61:3373–3381. https://doi.org/10.1002/csc2.20526

Wood W, Carden L (2014) Elusiveness of menstrual cycle effects on mate preferences: comment on Gildersleeve. Haselton, and Fales

Wu S, Zhang B, Keyhaninejad N, Rodríguez GR, Kim HJ, Chakrabarti M, Illa-Berenguer E, Taitano NK, Gonzalo MJ, Díaz A (2018) A common genetic mechanism underlies morphological diversity in fruits and other plant organs. Nat Commun 9:4734. https://doi.org/10.1038/s41467-018-07216-8

Wu C, Qiu Y, Duan Y, Guo Y, Wang H, Zhang X, Song J, Li X (2020) Construction of high-density linkage map and identification of QTLs associated with resistance to black rot in radish (*Raphanus sativus*) by RAD sequencing. Plant Breed 139(3):660–671. https://doi.org/10.1111/pbr.12794

Xie Y, Ying J, Xu L, Wang Y, Dong J, Chen Y, Liu L (2020) Genome-wide sRNA and mRNA transcriptomic profiling insights into dynamic regulation of taproot thickening in radish (*Raphanus sativus* L.). BMC Plant Biol 20:1–14. https://doi.org/10.1186/s12870-020-02585-z

Xu L, Jiang QW, Wu J, Wang Y, Gong YQ, Wang XL, Limera C, Liu LW (2014) Identification and molecular mapping of the RsDmR locus conferring resistance to downy mildew at seedling stage in radish (Raphanus sativus L.). J Integr Agric 13:2362–2369

Xu ZS, Yang QQ, Feng K, Yu X, Xiong AS (2020) DcMYB113, a root-specific R2R3-MYB, conditions anthocyanin biosynthesis and modification in carrot. Plant Biotechnol J 18(7):1585–1597. https://doi.org/10.1111/pbi.13325

Yamamoto E, Shirasawa K, Kimura T, Monden Y, Tanaka M, Isobe S (2020) Genetic mapping in autohexaploid sweet potato with low-coverage NGS-based genotyping data. G3 Genes Genomes Genetics 10:2661–2670. https://doi.org/10.1534/g3.120.401433

Yano K, Yamamoto E, Aya K, Takeuchi H, Lo PC, Hu L, Matsuoka M (2016) Genome-wide association study using whole-genome sequencing rapidly identifies new genes influencing agronomic traits in rice. Nat Genet 48:927–934. https://doi.org/10.1038/ng.3596

Yi G, Kim JS, Park JE, Shin H, Yu SH, Park S, Huh JH (2018) MYB1 transcription factor is a candidate responsible for red root skin in radish (*Raphanus sativus* L.). PLoS One 13:e0204241. https://doi.org/10.1371/journal.pone.0204241

Yildiz M, Willis DK, Cavagnaro PF, Iorizzo M, Abak K, Simon PW (2013) Expression and mapping of anthocyanin biosynthesis genes in carrot. Theor Appl Genet 126:1689–1702. https://doi.org/10.1007/s00122-013-2084-y

Ying J, Wang Y, Xu L, Yao S, Wang K, Dong J, Ma Y, Wang L, Xie Y, Yan K, Li J (2023) RsGLK2. 1-RsNF-YA9a module positively regulates the chlorophyll biosynthesis by activating RsHEMA2 in green taproot of radish. Plant Sci 334:111768. https://doi.org/10.1016/j.plantsci.2023.111768

Yu XiaoNa YX, Lu Lu LL, Ma YinBo MY, Chhapekar SS, Yi SoYoung YS, Lim YongPyo LY, Choi SuRyun CS (2020) Fine-mapping of a major QTL (Fwr1) for fusarium wilt resistance in radish. Theor Appl Genet 133:329–340. https://doi.org/10.1007/s00122-019-03461-7

Yu X, Choi SR, Ramchiary N, Miao X, Lee SH, Sun HJ, Kim S, Ahn CH, Lim YP (2013) Comparative mapping of Raphanus sativus genome using Brassica markers and quantitative trait loci analysis for the Fusarium wilt resistance trait. Theor Appl Genet 126:2553–2562. https://doi.org/10.1007/s00122-013-2154-1

Yu X, Choi SR, Dhandapani V, Rameneni JJ, Li X, Pang W, Lim YP (2016) Quantitative trait loci for morphological traits and their association with functional genes in *Raphanus sativus*. Front Plant Sci 7:171104. https://doi.org/10.3389/fpls.2016.00255

Zappia A, Spanti A, Princi R, Imeneo V, Piscopo A (2023) evaluation of the efficacy of antioxidant extract from lemon by-products on preservation of quality attributes of minimally processed radish (*Raphanus sativus* L.). Antioxidants 12:235. https://doi.org/10.3390/antiox12020235

Zhu C, Gore M, Buckler ES, Yu J (2008) Status and prospects of association mapping in plants. Plant Genome 1

15 Genome-wide Association Study (GWAS) and QTL-Based Improvement of Lettuce (*Lactuca* spp.)

Khaled F. M. Salem, Tilak Chandra, Essam Fathy El-Hashash, and Amira A. Ibrahim

Abstract

Lettuce, a widely cultivated annual crop belonging to the Asteraceae family, is extensively utilized in salads, sandwiches, and various other culinary applications. It exhibits a diverse array of types and forms, characterized by a range of colors, textures, margins, and leaf surfaces. Although over a hundred species of Lactuca have been documented, only a limited number can be successfully hybridized through traditional breeding methods. Conventional breeding practices integrate genotypes from various genetic sources to increase genetic variation and select for desirable traits, including high yield, early maturity, improved quality, pest resistance, and environmental stress tolerance. While innovative biotechnological approaches are gaining traction, traditional agricultural practices remain vital for enhancing lettuce genotypes with preferred characteristics.

K. F. M. Salem (✉)
Department of Biological Sciences, Faculty of Science and Humanities, Shaqra University, Ad-Dawadimi, Saudi Arabia

Department of Plant Biotechnology, Genetic Engineering and Biotechnology Research Institute (GEBRI), University of Sadat City, Sadat City, Egypt
e-mail: khaledfathi@su.edu.sa; khaled.salem@gebri.usc.edu.eg

T. Chandra
Division of Agricultural Bioinformatics, ICAR-Indian Agricultural Statistics Research Institute, New Delhi, India

E. F. El-Hashash
Agronomy Department, Faculty of Agriculture, Al-Azhar University, Cairo, Egypt
e-mail: dressamelhashash5@gmail.com; dressamelhashash@azhar.edu.eg

A. A. Ibrahim
Botany and Microbiology Department, Faculty of Science, Arish University, El-Arish, Egypt
e-mail: amiranasreldeen@yahoo.com; amiranasreldeen@sci.aru.edu.eg

S. S. Dey et al. (eds.), *GWAS and QTL Mapping in Horticultural Crops*,
https://doi.org/10.1007/978-981-95-2306-1_15

Therefore, underlying footprint cohorts and their integration to assisted breeding are crucial to understand to realize breeding vision. Current technologies offer solutions to these challenges to several extent; however, further advancements are necessary to satisfy the increasing demand for superior product appearance, quality, biosafety, and reduced economic and environmental costs. This chapter explores the diversity of lettuce germplasm and its applications in conventional breeding by leveraging genome-wide association studies and quantitative trait locus (QTL) mapping to facilitate lettuce improvement through advanced tools and technologies-assisted breeding.

Keywords

Lettuce · Genetic diversity · Germplasm genotyping · Genome-wide association studies · Quantitative trait loci · Conventional and modern breeding

15.1 Introduction

Lettuce, a widely cultivated annual crop in the Asteraceae family, is widely utilized in salads, sandwiches, and multiple industrial applications, such as cooking oil and as a sedative (Peng and Simko 2023). It exhibits a variety of types and forms, characterized by a range of colors, textures, margins, and surfaces. As a self-fertilizing leafy vegetable with a chromosome number of $2n = 18$, lettuce features crisp leaves and is a staple in various culinary applications. It can thrive in diverse climates and soil conditions but often requires an optimum climate to reach maximum production efficiency (Lebeda et al. 2007). Although more than one hundred species of Lactuca exist, only a limited number can be hybridized through traditional methods. It is highly adaptable, with a deep, horizontal root system that efficiently absorbs water and minerals; its short stems support leaves that cluster in a dense rosette formation (Lebeda et al. 2007; Damerum et al. 2020). Its leaves may be downy, savoy, or curled, presenting a spectrum of colors from yellow to dark green (Still 2007). The elongation of the stem marks the transition from the late vegetative phase to the onset of the reproductive stage, culminating in the formation of an inflorescence comprised of numerous flower heads. Each flower produces ribbed seeds, covered in pappus hair, that are available in various colors. Freshly harvested seeds rapidly enter dormancy, with most strains exhibiting some degree of thermal dormancy (Still 2007; Lebeda et al. 2007). It is typically grown from seed under both irrigated and rain-fed conditions. Conventional breeding integrates genotypes from diverse genetic sources to enhance genetic variation and select for desirable agronomically useful traits such as high yield, early maturity, improved quality, pest resistance, and tolerance to environmental stresses. While innovative biotechnological methods are gaining traction, traditional agricultural practices remain crucial for the enhancement of lettuce genotypes with preferred characteristics. Due to its perishable and delicate nature, it is particularly vulnerable to postharvest losses caused by various biotic and abiotic factors. Although current technologies can help address these challenges, further advancements are required to meet the increasing demand for superior product appearance, quality, biosafety, and reduced economic and environmental costs. Consequently, this chapter explores the diversity of lettuce germplasm and its applications in conventional breeding by employing genome-wide association studies and quantitative trait locus (QTL) mapping to facilitate improvements in lettuce through advanced tools and technologies.

15.2 Requirement of Improvement: Optimal Breeding Strategies

Retailers and consumers increasingly demand high-quality, blemish-free fresh vegetable produce, yet there is growing concern regarding pesticide use and residues in leafy vegetables (Hand et al. 2003). While contemporary breeding techniques that leverage advancements in DNA technology have garnered the interest of lettuce breeders, traditional breeding methods remain the most significant and predominant approach for developing new lettuce varieties with desirable traits. The primary objective of lettuce breeding is to establish genotypes capable of thriving in diverse environments. As global challenges evolve, considerations must include crop adaptability, abiotic and biotic stress resilience, quality enhancement, and yield expansion (Hartman et al. 2014; McCabe et al. 2001). To address these challenges, breeding strategies can be employed to select lines with the necessary genes and quantitative trait loci that govern critical traits (Hartman et al. 2014). For instance, traditional lettuce production methods utilizing impatiens necrotic spot virus-resistant germplasm provide valuable resources for developing lettuce cultivars resistant to this virus (Richardson et al. 2024). Postharvest challenges in lettuce, including enzymatic discoloration and rapid deterioration, pose significant issues. However, integrating advanced breeding technologies such as genome-wide association studies (GWAS) and QTL mapping can mitigate these losses and accelerate cultivar development (Peng and Simko 2023). These breeding methods are often less time-consuming and technically not complex, yet they are most effective when combined with molecular analyses to assess hybridity and screen progeny plants (Patella et al. 2019). Damerum et al. (2020), Hassan et al. (2021), and Peng and Simko (2023) comprehensively highlight various modern genetic and genomic tools and techniques for producing new lettuce cultivars, including conventional breeding, mutagenesis, genetic engineering tools such as gene transformation and genome editing, germplasm diversity and conservation, genetic resource preservation, biotechnological advancements, crop cultivation methods, and molecular biology applications in lettuce improvement. The integration of phenotyping and genotyping data through traditional and molecular marker-assisted breeding methods is essential for the development of new horticultural crop varieties (Patella et al. 2019). Moreover, a comprehensive database has been developed devoted to lettuce that encompasses reference genomes, resequencing variants, a complete miRNAome, extensive metabolite profiles, and epigenetic data detailing the genome-wide chromatin accessibility landscape (Guo et al. 2023). It additionally features omics-driven hierarchical accessible functions for analysis, integrated tools for data sharing, and a summary of current research advancements related to various breeding and omics aspects of lettuce.

15.3 Lettuce Genetic Resources: Footprint for Association Analysis

The preservation and utilization of plant genetic resources in breeding programs are crucial, as they significantly contribute to long-term solutions for food and nutrition security (Zohary 1991). However, these resources are often underutilized due to widespread challenges and insufficient quantification and classification. It is essential to maintain, evaluate, and characterize commercially valuable traits to support effective crop breeding programs. Genetic diversity in horticultural crops is vital for selecting optimal genotypes that enhance production (Van Treuren et al. 2012). Understanding the relationships among different traits forms the foundation for selecting products and components aimed at improvement.

Significant differences can be observed among diverse genotypes of lettuce (Fig. 15.1), indicating their potential for enhancing yield and quality traits (Hassan et al. 2021). These are often considered and maintained in gene banks as extensive collections of germplasm in the form of wild species, landraces, and exotic lines,

Fig. 15.1 Morphology of some common Lactuca species and their relatives. (Source: https://www.reddit.com/r/coolguides/comments/j61rd7/types_of_lettuce)

providing a diverse array of gene pools based on the geographical distribution of desirable agronomic traits. Hybridizing commercial cultivars with genetically modified national genotypes introduces new alleles, thereby increasing genetic diversity and incorporating desirable traits. Breeders aim to enhance lettuce for characteristics such as stress tolerance and high yield by employing both modern genomic tools and traditional breeding methods, which can succeed only by their integration with genetic resources (Lebeda et al. 2019).

15.4 GWAS: Roadmap to Breeder's Trait Discovery

Lettuce is well-known for its high level of genetic variability, attributed to its complex domestication history and polyphyletic origin (Wani et al. 2020). In lettuce breeding, genetic diversity is an invaluable resource for enhancing cultivars with desired traits. It is essential for effective breeding and the development of novel cultivars of this commercially significant crop. Pedigree analysis and multivariate analysis of a wide range of plant traits can be employed to assess genetic diversity (Van de Wiel et al. 2010). The genetic diversity within the lettuce gene pool provides novel features and advantageous alleles, facilitating the development of superior varieties (Van Treuren et al. 2012). Through domestication and targeted breeding programs, numerous lettuce cultivars have been produced with desirable characteristics, such as larger leaves and heads, improved taste and texture, and reduced latex content (Hartman et al. 2014). Nevertheless, a significant marker trait association for such a trait is the need of the hour. Its accessions can be classified into five primary horticultural categories: butterhead, crisphead (iceberg), leaf, romaine (cos), and stem, primarily based on morphological traits. These varieties exhibit significant differences in characteristics such as size, shape, texture, growth rate, and leaf size (Park et al. 2021). To facilitate the identification of various lettuce genotypes as genetic donors in different breeding programs and to prevent confusion regarding cultivar identity associated with morphological traits and multiple cultivar registrations, the genetic diversity of lettuce has been thoroughly investigated (Van de Wiel et al. 2010).

Although genetic diversity may decrease as a result of selective breeding, our understanding of this phenomenon in lettuce remains limited. Genotyping-by-sequencing has proven effective for analyzing plant species with large genomes, such as lettuce, and has emerged as a high-throughput, cost-effective method for genome-wide single-nucleotide polymorphisms (SNP) identification (He et al. 2014; Sthapit Kandel et al. 2020). Therefore, it is a cost-effective method for discovering genetic variants across the genome, yielding a compressed genomic representation (Park et al. 2021). Genome-wide association studies (GWAS) utilizing mixed linear model analyses have identified significant quantitative trait loci on four lettuce chromosomes, with chromosome 8 being the most prominent, explaining 6.7% of the total phenotypic variation for resistance to bacterial leaf spot (Sthapit Kandel et al. 2022). Genetic factors influencing lettuce resistance to Sclerotinia minor, identified through genome-wide association mapping, often correlate with anthocyanin content. Notably, mapped QTL for resistance co-locate with the red leaf color trait, showing lower disease incidence in red and dark red accessions (Simko et al. 2023). GWAS have also identified 5311 expression quantitative trait loci (eQTL) regulating 4105 genes, including nine associated with flavonoid biosynthesis and six linked to leaf color variation, particularly affecting anthocyanins (Zhang et al. 2017).

Research on the genotype of red lettuce leaf (RLL) genes has revealed that the expression levels of RLL1 (bHLH) and RLL2 (MYB) regulate high anthocyanin accumulation in red leaf lettuce cultivated under artificial light conditions (Wada et al. 2022). Additionally, a major locus controlling the polymorphism between dark- and pale-green leaves in lettuce was identified through GWAS, which was fine-mapped to an interval of 5375 bp on chromosome 4, containing a gene homologous to GLK in Arabidopsis. A recombinant inbred line mapping population study identified small-effect QTL that accounted for 12.3% and 25.2% of total tip burn variation under glasshouse and field conditions, respectively. This research emphasizes the importance of phenotyping in production-system-specific environments to study environmentally sensitive traits, highlighting genetic markers and germplasm resources for tip burn-resistant lettuce production (Beacham et al. 2023). Single-nucleotide polymorphisms (SNPs), the most common sequence variations in a genome, have been extensively utilized to assess the genetic diversity and population structure of germplasm collections (Sthapit Kandel et al. 2020; Park et al. 2021). Furthermore, Park et al. (2021) reveal genomic regions which have been found to be associated with delayed bolting, suggesting that delayed bolting may be positively selected to reduce phenotypic variance. Moreover, significant associations for salt tolerance (Das et al. 2024) and for various other useful agronomic traits have been dissected (Shim et al. 2019; Kim et al. 2021).

15.5 QTL Exploration: Mapping Loci for Agronomical Traits

Mapping quantitative trait loci (QTL) in lettuce is an invaluable tool for breeders, as it identifies specific genetic regions associated with important traits such as yield, quality, and disease resistance (Simko et al. 2013). QTL in lettuce have been linked to multiple threats such as biotic stresses (Christopoulou et al. 2015; Simko et al. 2013), abiotic stresses (Jenni et al. 2013; Hartman et al. 2014), and qualitative-quantitative traits (Zhang et al. 2007; Atkinson et al. 2013), highlighting their potential for crop improvement. Since the advent of DNA markers, in vitro molecular replication techniques have been employed to dissect various agronomical traits in lettuce (Rafalski and Tingey 1993). Various DNA markers, including AFLP, RAPD, RFLP, SSRs, SCAR, and SNPs, have been developed to construct genetic maps for lettuce enhancement. Regions of QTL that harbor genes for specific quantitative traits, such as plant height, can be detected using DNA tags and genetic mapping techniques (Collard et al. 2005). Once QTL and their associated DNA markers for important agricultural traits are confirmed, these markers can be utilized in marker-assisted selection (MAS) within breeding programs (Collard and Mackill 2008).

For instance, recent research has identified positions of quantitative parameters associated with lettuce seedling root growth (Roberts et al. 2020). Landry et al. (1987) created a lettuce linkage map using 41 RFLP loci and three phenotypic markers. Kesseli et al. (1994) developed a genetic association map incorporating RFLP and RAPD markers in *L. sativa*, and Jeuken et al. (2001) produced an interspecific AFLP map for lettuce that included F2 populations of *L. sativa* × *L. saligna*.

Improved lettuce cultivars can be developed by leveraging the genetic diversity present in lettuce and its wild relatives, particularly concerning plant morphology, quality, and resilience to abiotic stressors. The inheritance of numerous traits beneficial to the fresh-cut industry has been elucidated through quantitative trait loci (QTL) mapping. This knowledge facilitates the development of molecular breeding techniques aimed at accelerating cultivar advancement (Simko 2013). Lettuce breeding programs utilize marker-assisted selection (MAS) to enhance precision and reduce the time required for identifying genotypes with favorable gene combinations. Therefore, these two genomic insights allow for the use of marker-assisted selection, where specific markers linked to beneficial traits are used to select parent plants, ultimately speeding up the breeding cycle and improving the chances of success in creating superior lettuce varieties. Conclusively, the identified QTL linked to biotic stress-related traits confer resilience against pathogens or herbivores, enabling the development of lettuce varieties with enhanced resistance to diseases, reducing reliance on chemical pesticides, and promoting sustainable agricultural practices (Table 15.1).

Table 15.1 Dissected QTL/genes/markers from the studied population influencing diverse agronomical traits in lettuce

Population type	Identified QTL/genes	Influenced trait	References
RIL	–	Disease and pest resistance	Hand et al. (2003)
RIL/90	*qDM2.1*, *qDM5.1*, and *qDM9.1*	Downey mildew resistance	Simko et al. (2013)
RIL/161	*qFLT7.2*, *qBFr2.1*, and *LsFT*	Bolting and flowering earliness	Rosental et al. (2021)
F_2/96	*qFOL7.1* and *qFOL8.1*	Fusarium wilt resistance	Seki et al. (2020)
BIL/29	*Rbq1*, *Rbq2*, *rbq3*, *rbq4*, *rbq5*, *rbq6* and *rbq7*	Non-host resistance	Jeuken et al. (2008)
RIL/114	Multiple	Architecture and domestication traits	Hartman et al. (2013)
F_2	Multiple	Downey mildew resistance	Parra et al. (2021a, b)
RIL/90	*pm-1.1*, *pm-2.1*, *pm-2.2*, and *pm-7.1*	Powdery mildew resistance	Simko et al. (2014)
RIL/382	Multiple	Tipburn resistance	Macias-Gonzalez et al. (2019)
F_5/254	Multiple	Postharvest discoloration	Atkinson et al. (2013)

(continued)

Table 15.1 (continued)

Population type	Identified QTL/genes	Influenced trait	References
RIL/234	*qSs5, qSs8, qSs9, qBc7, qBc9,* etc.	Quantitative resistance	Pink et al. (2022)
RIL/248	*qDMR2.1, qDMR4.1,* and *qDMR5.1*	Downey mildew resistance	Parra et al. (2021a, b)
F_2/163	BLS (Xcvr)and locus y	Leaf spot resistance	Wang et al. (2016)
495 accessions	Multiple	Leaf spot resistance	Sthapit Kandel et al. (2022)
RIL/60	*MYB* transcription factors	Antioxidant properties	Damerum et al. (2015)
RIL/209	*Aquaporin, LEA, ABRE,* and *GST*	WUE and Yields	Damerum et al. (2021)
RIL/127	*dtb7.1, dtf7.1, dts7.1, ln7.1,* etc.	Stem growth and flowering time	Lee et al. (2021)
RIL/113	Multiple	Shelf life	Zhang et al. (2007)
$F_{2:3}$/100	Multiple	Root architecture	Johnson et al. (2000)
RIL/102	Multiple	Taste and flavor traits	Chadwick et al. (2024)
496 accessions	*qLDR7.3, qLRD4.2,* etc.	Lettuce drop fungal infection	Simko et al. (2023)
RIL/162	*qLDR1.1* and *qLDR5.1*	Resistance to lettuce drop	
240 accessions	locus 1–5	Flavonoid biosynthesis	Zhang et al. (2017)
500 accessions	*qRLC2.1, RLL3, RLL2, qRLC5.2, qRLC7.2, RLL4, ANS,* etc.	Necrotic leaf spot resistance	Simko et al. (2023)
F_2/218	RLL1 and RLL2	Leaf color	Su et al. (2020)
133 accessions	GST	Red coloration	Wada et al. (2022)
134 accessions	GLK	Leaf development	
RIL/95	*qDM2.2, qDM4.1, qDM4.2, qDM7.1,* and *qDM9.2*	Downy mildew resistance	Simko et al. (2015)
RIL/152	*qTPB5.2,* etc.	Multiple agronomical traits	Jenni et al. (2013)
F_2/125	*qSTP5*	Leaf miner resistance	Kandel et al. (2021)
RIL/356	*qRC9.1, qRS2.1,* and *qLS7.2* etc.	Salt tolerance	Wei et al. (2014)
RIL	*qTPB5.2*	Tip burn resistance	Macias-Gonzalez et al. (2021)
RIL/236	*qFLT4.1, qFLT9.4,* etc.	Photoperiod sensitivity	Han et al. (2021)
F_2/126	RBQ1, RBQ2, and RBQ	Downey mildew resistance	Jeuken and Lindhout (2002)
F_2/126	Multiple	Quantitative resistance	Den Boer et al. (2013)
RIL	Multiple	Abiotic stress-responsive	Hartman et al. (2014)

(continued)

Table 15.1 (continued)

Population type	Identified QTL/genes	Influenced trait	References
F_8/125	PRL-01, LRND-01, etc.	Root growth	Roberts et al. (2020)
BIL/28	*Rbq1, Rbq2, rbq3–7, Rbq8–15,* etc.	Non-host resistance	Zhang et al. (2009)
RIL/89	Slg4.1	Seed longevity	Schwember and Bradford (2010)
$F_{2:3}$/200	*qVR3-2.1, qVR3-4.1,* etc.	Agronomic traits	Simko et al. (2022a, b)
RIL/131	Multiple	Seed germination	Hayashi et al. (2008)
RIL	Multiple	Tip burn resistance	Beacham et al. (2023)
RIL/103	Htg6.1 etc.	Seed and seedling traits	Argyris et al. (2005)
RIL/163	*qXCR2.1* etc.	Bacterial leaf spot resistance	Sandoya et al. (2019)
RIL/169	Chla M 1.1, Lpa1.1, etc.	GxE interaction	Hayashi et al. (2012)

15.6 QTL Monitoring Biotic Stresses

The first comprehensive mapping of disease resistance loci in lettuce using GWAS, providing the most thorough analysis of QTL for field resistance to downy mildew (Simko et al. 2022a, b). A QTL for powdery mildew resistance was identified on linkage group 2, explaining 15% of the variation, while QTL for susceptibility were located on linkage groups 1, 4, and 8, indicating potential areas for further research for resistance breeding (Rauscher et al. 2009). Two other QTL for lettuce resistance to downy mildew, identified from a cross between two legacy cultivars, were revealed using a mixed-model approach, providing broad resistance under field conditions (Simko et al. 2013). *Fusarium* wilt, a significant lettuce disease in tropical and subtropical regions caused by *Fusarium oxysporum f. sp. lactucae,* was analyzed in F2 individuals from a cross between highly resistant and partially resistant varieties, revealing two significant QTL on linkage groups 7 and 8, which provide elevated race-specific resistance (Seki et al. 2020). Quantitative resistance to downy mildew has been achieved through the use of backcross inbred lines of cultivated lettuce, signifying the introgression of genomic segments from wild lettuce (Jeuken et al. 2008). Downy mildew, a major lettuce disease, is managed using resistant cultivars containing dominant resistance genes. F2 segregating populations derived from near-isogenic lines (NIL) were genetically mapped to one or two loci in the lettuce reference genome (Parra et al. 2021a, b). Quantitative trait loci for powdery mildew resistance in a cross between two genotypes explained 35–42% of the phenotypic variation. Notably, these QTL are not located in known hotspots of lettuce resistance genes, and disease resistance alleles originated from both parents, indicating that even highly susceptible accessions can contribute alleles for powdery mildew resistance (Simko et al. 2014). Genetic variation in quantitative resistance against fungal pathogens in lettuce and its wild relatives has revealed mapping loci conferring resistance and predicting molecular mechanisms, identifying five genetic loci that influence quantitative resistance, and explaining 5–10% of the variation (Pink et al. 2022). A genetic linkage map was developed to map bacterial leaf spot (BLS) resistance and seed color traits, with the dominant resistance gene for BLS (Xcvr) mapped on linkage group 2 and the gene locus for seed color identified on linkage group 5 (Wang et al. 2016). Lettuce drop, caused by *Sclerotinia minor* and *Sclerotinia sclerotiorum*, is a significant disease in lettuce. Two major QTL for resistance to lettuce drop were consistently detected across all experiments, explaining 11–25% of the phenotypic variation (Mamo et al. 2019). Downy mildew in lettuce, caused by *Bremia lactucae*, significantly impacts productivity. Five quantitative trait loci were located in the Dm50 chromosomal region, with the most substantial effect attributed to qDM7.1, which conferred resistance across multiple field experiments and accounted for up to 32.9% of the total phenotypic variance (Simko et al. 2015). The leaf miner *Liriomyza trifolii* causes considerable damage to leafy vegetables like lettuce through stippling and tunneling. A quantitative trait loci analysis identified a novel locus, stippling, on linkage group 5, flanked by two SNPs, which accounted for over 13% of the total phenotypic variance (Kandel 2015; Kandel et al. 2021). Resistance genes (Dm genes) in wild Lactuca species provide race-specific resistance, but their effectiveness is compromised due to pathogen adaptation. Multiple QTL have been mapped that mediate resistance against diverse races, offering race-specific complete resistance (Jeuken and Lindhout 2002). Downy mildew is one of the most significant diseases impacting lettuce production globally, with substantial economic consequences. A genome-wide association study identified 63 quantitative trait loci associated with downy mildew resistance, with genomic regions commonly linked to resistance located on chromosomes 4 and 5, although QTL were identified on all chromosomes (Simko et al. 2022a, b). Three regions exhibiting quantitative resistance to downy mildew in *Lactuca saligna* have been delineated into seventeen sub-QTL, with plant-stage-dependent effects that either reduce or promote infection (Den Boer et al. 2013). Non-host resistance in lettuce to downy mildew, using a pathosystem model, focused on the contribution of QTL to non-host resistance at various stages of the lettuce life cycle. Sixteen QTL were identified, with only two being effective across all developmental stages, while the others were effective only at specific stages (Zhang et al. 2009). *Verticillium* wilt, a severe soil-borne disease in lettuce, is best managed through host resistance. Three QTL for resistance to this pathogen were identified on linkage groups LG 2 and LG 4, explaining approximately 10% of the total trait variation, along with a significant effect on red leaf color (Simko et al. 2022a, b). Bacterial leaf spot, a disease affecting lettuce globally, has been studied through quantitative trait loci analysis in two recombinant inbred line populations (Sandoya et al. 2019). By identifying QTL associated with traits that enhance tolerance to these environmental challenges, breeders can focus on developing varieties that thrive under adverse conditions.

15.7 Cohorts Regulating Quality Traits

Metabolic spectroscopy offers a rapid alternative for identifying stable QTL associated with sugars and bitter sesquiterpene lactones, which can aid in developing breeding markers for flavor and nutritional traits as well as candidate genes for key metabolites (Chadwick et al. 2024). An intraspecific linkage map was also developed from a high-density genetic map constructed from seven mapping populations, using markers from two or more populations across nine chromosomal linkage groups, focusing on the genetic basis of postharvest discoloration traits (Truco et al. 2007; Atkinson et al. 2013). Wild relatives of crops may exhibit allelic variation in genes that influence beneficial phytonutrient concentrations. A set of quantitative trait loci has been identified for traits such as antioxidant potential, carotenoid content, total chlorophyll content, and specific phenolic compounds, which regulate phytonutrient biosynthetic pathways (Damerum et al. 2015). Intraspecific recombinant inbred mapping populations derived from a cross between green and red leaf lettuce identified and mapped QTL related to traits such as days from sowing to bolting, flowering time, seed-setting, total number of leaves, plant height, and branches of the main inflorescence over two consecutive years (Lee et al. 2021). In total, 107 QTL were detected across all nine linkage groups, with only five QTL being common to both environments. Several regions on the genome exhibited co-localization of multiple QTL (hotspots), with relatively little overlap between developmental hotspots and those associated with shelf life (Zhang et al. 2007). Two qualitative loci, RLL1 and RLL2, were identified through allele frequency analysis in two contrasting pools, along with four polymorphic genes that regulate red leaf color in lettuce, which have undergone disruptive selection since domestication (Su et al. 2020). Among the predicted genes in the identified chromosomal regions, transcription factors such as quinone oxidation-reduction (redox) enzymes were recognized as the most promising candidate genes involved in regulating postharvest discoloration (Peng et al. 2024). Currently, there are few molecular markers associated with postharvest characteristics that extend shelf life. Evidence suggests that pre-harvest leaf features are likely correlated with postharvest phenotypes (Chase et al. 2024). QTL related to the biophysical properties of the leaf cell wall, such as break strength, plasticity, and elasticity, demonstrated a strong correlation with shelf life through analyses of shelf-life QTL, leaf biophysical properties, and leaf developmental traits (Zhang et al. 2007). These attributes were associated with the epidermal cell area, supporting the hypothesis that the small, robust cell-walled lettuce ideotype and pre-harvest leaf characteristics are linked to the postharvest shelf life of leafy crops. RNA interference-based knockdown of xyloglucan endotransglucosylase/hydrolase gene expression resulted in smaller leaves and leaf cells due to cell wall thinning (Chase et al. 2024). The genetic basis for antioxidant content in lettuce was investigated by measuring oxygen radical absorbance capacity and chlorophyll content in recombinant inbred lines at two production sites. Main effect QTL were identified, although no main effects were observed; instead, epistatic interactions were noted in one pair of marker intervals for each trait (Hayashi et al. 2012). In a nutshell, by identifying specific genetic regions associated with desirable quality characteristics, breeders can more effectively select for improved varieties.

15.8 QTL Involved in Growth and Development

Bolting, a critical stage in the plant life cycle, can lead to quality and economic losses in lettuce. A major QTL for bolting and flowering time was located on chromosome 7, with several small-effect loci on distinct chromosomes associated with bolting and flowering (Rosental et al. 2021). The genetic architecture of a recombinant inbred line population derived from a cross between Lactuca sativa and *Lactuca serriola* was examined, identifying a few genomic regions with major and intermediate QTL that control the transition from wild to cultivated crisphead lettuce (Hartman et al. 2013). Genetic mapping of lettuce populations under long- and short-day conditions revealed two distinct sets of QTL that control day length-independent and photoperiod-sensitive flowering time (Han et al. 2021). Assessments of above-ground biomass at the end of the rosette stage revealed a mosaic of abiotic stress QTL across the genome, with two clusters reflecting general growth responses that co-locate with previous studies on leaf shape and flowering time (Hartman et al. 2014). Additionally, a QTL affecting stalk length in stalk lettuce was identified, contributing to phenotypic variation (Zixue et al. 2022). Lettuce seeds exhibit poor shelf life and thermal inhibition above 25 °C. A major QTL on chromosome 4 (Slg4.1) accounts for 23% of the phenotypic variation in seed viability, with the Salinas allele improving longevity compared to conventionally stored control seeds (Schwember and Bradford 2010). The germination of seeds under varying environmental conditions was investigated using a recombinant inbred line population, revealing 38 QTL from two seed maturation environments, with variation attributed to individual final germination rates (Hayashi et al. 2008). A pleiotropic locus (Htg6.1) was found to regulate traits related to germination and seedling growth in cultivated lettuce, accounting for 25–23% of the phenotypic variation in high-temperature germination across two different growth populations (Argyris et al. 2005). Therefore, mapping quantitative trait loci in lettuce for quality traits is beneficial for improving growth and development for useful traits such as plant size, leaf production, and maturation time.

15.9 QTL Governing Abiotic Stresses

Two major QTL were detected for tip burn on linkage groups 1 and 5, contributing to scalable resistance and demonstrating pleiotropic effects between leaf crinkliness and tip burn (Macias-Gonzalez et al. 2019). The major QTL for tip burn resistance, designated qTPB5.2, can account for up to 70% of the phenotypic variance in tip burn incidence, a significant physiological disorder associated with calcium deficiency in lettuce, leading to leaf necrosis and unmarketable crops (Macias-Gonzalez et al. 2021). Genomic regions controlling traits such as carbon isotope discrimination, transpiration, stomatal conductance, leaf temperature, and yield under both well-watered and drought conditions were found within a QTL hotspot on chromosome 8. This hotspot is associated with promising candidate genes, including aquaporins, late embryogenesis abundant proteins, abscisic acid-responsive element binding proteins, and glutathione S-transferases (Damerum et al. 2021). An interspecific cross that evaluated root architectural traits and water acquisition differences between wild and cultivated lettuce identified approximately thirteen quantitative trait loci, accounting for 28–83% of the phenotypic variation for these traits (Johnson et al. 2000). Three major clusters of QTL revealed pleiotropic effects or tight linkages between tip burn incidence and severity, head type, stem length, head firmness, and ribbiness. Notably, *qTPB5.2* was significant, explaining 38–70% of the variation in tip burn incidence, making it a potential candidate gene for breeding tip burn resistance through marker-assisted selection (Jenni et al. 2013). Cultivated lettuce is more sensitive to salinity stress than its wild relatives, potentially due to differences in root architecture and sodium uptake. QTL related to root development, growth under salt stress conditions, and ion accumulation were identified, including three major QTL with several linked markers (Wei et al. 2014). Variation in transplanted lettuce due to soil and environmental conditions significantly limits production and yield optimization; quantitative trait loci associated with enhanced root growth traits accounted for 60.2% of the genetic variation for this trait (Roberts et al. 2020). By pinpointing genetic regions associated with tolerance to several environmental factors, breeders can select for traits that improve lettuce survival and productivity under challenging conditions.

15.10 Conclusion and Prospects

Lettuce is a common and affordable vegetable with roots in ancient Egyptian civilization. It is rich in essential nutrients, including vitamin A, potassium, calcium, and folic acid. Given its commercial importance, considerable research has focused on domesticating various lettuce varieties. Genome-wide association studies have identified quantitative trait loci associated with both biotic and abiotic stress, leading to enhanced tolerance in lettuce plants. Future research should aim to find significant associations between genotypes and desirable phenotypes to improve both the qualitative and quantitative attributes of lettuce, such as increasing vitamin content and regulating mineral absorption through osmotic balance in extracellular and intracellular environments. Additionally, further studies could investigate the potential benefits of consuming lettuce, particularly its role in supplementing essential nutrients like fiber, sugars, proteins, vitamins, and minerals on a global scale.

References

Argyris J, Truco MJ, Ochoa O, Knapp SJ, Still DW, Lenssen GM, Schut JW, Michelmore RW, Bradford KJ (2005) Quantitative trait loci associated with seed and seedling traits in Lactuca. Theor Appl Genet 111:1365–1376

Atkinson LD, McHale LK, Truco MJ, Hilton HW, Lynn J, Schut JW, Michelmore RW, Hand P, Pink DA (2013) An intra-specific linkage map of lettuce (Lactuca sativa) and genetic analysis of postharvest discolouration traits. Theor Appl Genet 126:2737–2752

Beacham AM, Hand P, Teakle GR, Barker GC, Pink DA, Monaghan JM (2023) Tipburn resilience in lettuce (Lactuca spp.)–the importance of germplasm resources and production system-specific assays. J Sci Food Agric 103(9):4481–4488

Chadwick M, Swann JR, Gawthrop F, Michelmore R, Scaglione D, Jose-Truco M, Wagstaff C (2024) Mapping taste and flavour traits to genetic markers in lettuce Lactuca sativa. Food Chem Mol Sci 9:100215

Chase K, Belisle C, Ahlawat Y, Yu F, Sargent S, Sandoya G, Begcy K, Liu T (2024) Examining preharvest genetic and morphological factors contributing to lettuce (Lactuca sativa L.) shelf-life. Sci Rep 14(1):6618

Christopoulou M, McHale LK, Kozik A, Reyes-Chin Wo S, Wroblewski T, Michelmore RW (2015) Dissection of two complex clusters of resistance genes in lettuce (Lactuca sativa). Mol Plant-Microbe Interact 28(7):751–765

Collard BC, Mackill DJ (2008) Marker-assisted selection: an approach for precision plant breeding in the twenty-first century. Philos Trans R Soc B Biol Sci 363(1491):557–572

Collard BC, Jahufer MZZ, Brouwer JB, Pang ECK (2005) An introduction to markers, quantitative trait loci (QTL) mapping and marker-assisted selection for crop improvement: the basic concepts. Euphytica 142:169–196

Damerum A, Selmes SL, Biggi GF, Clarkson GJ, Rothwell SD, Truco MJ, Michelmore RW, Hancock RD, Shellcock C, Chapman MA, Taylor G (2015) Elucidating the genetic basis of antioxidant status in lettuce (Lactuca sativa). Hortic Res 2

Damerum A, Chapman MA, Taylor G (2020) Innovative breeding technologies in lettuce for improved post-harvest quality. Postharvest Biol Technol 168:111266

Damerum A, Smith HK, Clarkson GJJ, Truco MJ, Michelmore RW, Taylor G (2021) The genetic basis of water-use efficiency and yield in lettuce. BMC Plant Biol 21(1):237

Das MK, Park S, Adhikari ND, Mou B (2024) Genome-wide association study of salt tolerance at the seed germination stage in lettuce. PLoS One 19(10):e0308818

Den Boer E, Zhang NW, Pelgrom K, Visser RG, Niks RE, Jeuken MJ (2013) Fine mapping quantitative resistances to downy mildew in lettuce revealed multiple sub-QTLs with plant stage dependent effects reducing or even promoting the infection. Theor Appl Genet 126:2995–3007

Guo Z, Li B, Du J, Shen F, Zhao Y, Deng Y, Kuang Z, Tao Y, Wan M, Lu X, Wang D (2023) LettuceGDB: the community database for lettuce genetics and omics. Plant Commun 4(1)

Han R, Lavelle D, Truco MJ, Michelmore R (2021) Quantitative trait loci and candidate genes associated with photoperiod sensitivity in lettuce (Lactuca spp.). Theor Appl Genet 134:3473–3487

Hand P, Kift N, McClement S, Lynn JR, Grube R, Schut JW, Arend AVD, Pink DAC (2003) Progress towards mapping QTLs for pest and disease resistance in lettuce

Hartman Y, Hooftman DA, Eric Schranz M, van Tienderen PH (2013) QTL analysis reveals the genetic architecture of domestication traits in Crisphead lettuce. Genet Resour Crop Evol 60:1487–1500

Hartman Y, Hooftman DA, Uwimana B, Schranz ME, van de Wiel CC, Smulders MJ, Visser RG, Michelmore RW, van Tienderen PH (2014) Abiotic stress QTL in lettuce crop–wild hybrids: comparing greenhouse and field experiments. Ecol Evol 4(12):2395–2409

Hassan MN, Mekkawy SA, Mahdy M, Salem KF, Tawfik E (2021) Recent molecular and breeding strategies in lettuce (Lactuca spp.). Genet Resour Crop Evol 68:3055–3079

Hayashi E, Aoyama N, Still DW (2008) Quantitative trait loci associated with lettuce seed germination under different temperature and light environments. Genome 51(11):928–947

Hayashi E, You Y, Lewis R, Calderon MC, Wan G, Still DW (2012) Mapping QTL, epistasis and genotype× environment interaction of antioxidant activity, chlorophyll content and head formation in domesticated lettuce (Lactuca sativa). Theor Appl Genet 124:1487–1502
He J, Zhao X, Laroche A, Lu ZX, Liu H, Li Z (2014) Genotyping-by-sequencing (GBS), an ultimate marker-assisted selection (MAS) tool to accelerate plant breeding. Front Plant Sci 5:484
Jenni S, Truco MJ, Michelmore RW (2013) Quantitative trait loci associated with tipburn, heat stress-induced physiological disorders, and maturity traits in crisphead lettuce. Theor Appl Genet 126:3065–3079
Jeuken M, Lindhout P (2002) Lactuca saligna, a non-host for lettuce downy mildew (Bremia lactucae), harbors a new race-specific Dm gene and three QTLs for resistance. Theor Appl Genet 105:384–391
Jeuken M, Van Wijk R, Peleman J, Lindhout P (2001) An integrated interspecific AFLP map of lettuce (Lactuca) based on two L. sativa× L. saligna F2 populations. Theor Appl Genet 103:638–647
Jeuken MJW, Pelgrom K, Stam P, Lindhout P (2008) Efficient QTL detection for nonhost resistance in wild lettuce: backcross inbred lines versus F 2 population. Theor Appl Genet 116:845–857
Johnson WC, Jackson LE, Ochoa O, Van Wijk R, Peleman J, St. Clair DA, Michelmore RW (2000) Lettuce, a shallow-rooted crop, and Lactuca serriola, its wild progenitor, differ at QTL determining root architecture and deep soil water exploitation. Theor Appl Genet 101:1066–1073
Kandel R (2015) Identification of QTLs for Liriomyza trifolii resistance and two agronomic traits in lettuce. (Doctoral dissertation,. University of Florida)
Kandel R, Lu H, Sandoya GV (2021) Identification and mapping of quantitative trait loci for resistance to Liriomyza trifolii in romaine lettuce cultivar 'Valmaine'. Sci Rep 11(1):998
Kesseli RV, Paran I, Michelmore RW (1994) Analysis of a detailed genetic linkage map of Lactuca sativa (lettuce) constructed from RFLP and RAPD markers. Genetics 136(4):1435–1446
Kim SH, Kim DG, Kim JB, Lee HY, Jang YA, Jang SW, Kim TS (2021) Genome-wide association study for the identification of useful SNPs related to the productivity in the KNOU lettuce core collection, pp 49–50
Labeda A, Ryder EJ, Grube R, Dolezalova I, Kristkova E (2007) Lettuce (Asteraceae; Lactuca spp.) In: Singh RJ (ed) Genetic resources, chromosome engineering, and crop improvement. Vegetable crops
Landry BS, Kesseli RV, Farrara B, Michelmore RW (1987) A genetic map of lettuce (Lactuca sativa L.) with restriction fragment length polymorphism, isozyme, disease resistance and morphological markers. Genetics 116(2):331–337
Lebeda A, Křístková E, Kitner M, Majeský Ľ, Doležalová I, Khoury CK, Widrlechner MP, Hu J, Carver D, Achicanoy HA, Sosa CC (2019) Research gaps and challenges in the conservation and use of North American wild lettuce germplasm. Crop Sci 59(6):2337–2356
Lee ON, Fukushima K, Park HY, Kawabata S (2021) QTL analysis of stem elongation and flowering time in lettuce using genotyping-by-sequencing. Gene 12(6):947
Macias-González M, Truco MJ, Bertier LD, Jenni S, Simko I, Hayes RJ, Michelmore RW (2019) Genetic architecture of tipburn resistance in lettuce. Theor Appl Genet 132:2209–2222
Macias-González M, Truco MJ, Han R, Jenni S, Michelmore RW (2021) High-resolution genetic dissection of the major QTL for tipburn resistance in lettuce, Lactuca sativa. G3 11(7):jkab097
Mamo BE, Hayes RJ, Truco MJ, Puri KD, Michelmore RW, Subbarao KV, Simko I (2019) The genetics of resistance to lettuce drop (Sclerotinia spp.) in lettuce in a recombinant inbred line population from Reine des Glaces× Eruption. Theor Appl Genet 132:2439–2460
McCabe MS, Garratt LC, Schepers F, Jordi WJ, Stoopen GM, Davelaar E, van Rhijn JHA, Power JB, Davey MR (2001) Effects of PSAG12-IPT gene expression on development and senescence in transgenic lettuce. Plant Physiol 127(2):505–516
Park S, Kumar P, Shi A, Mou B (2021) Population genetics and genome-wide association studies provide insights into the influence of selective breeding on genetic variation in lettuce. Plant Genome 14(2):e20086
Parra L, Nortman K, Sah A, Truco MJ, Ochoa O, Michelmore R (2021a) Identification and mapping of new genes for resistance to downy mildew in lettuce. Theor Appl Genet 134:519–528

Parra L, Simko I, Michelmore RW (2021b) Identification of major quantitative trait loci controlling field resistance to downy mildew in cultivated lettuce (Lactuca sativa). Phytopathology 111(3):541–547

Patella A, Palumbo F, Galla G, Barcaccia G (2019) The molecular determination of hybridity and homozygosity estimates in breeding populations of lettuce (Lactuca sativa L.). Gene 10(11):916

Peng H, Simko I (2023) Extending lettuce shelf life through integrated technologies. Curr Opin Biotechnol 81:102951

Peng H, Luo Y, Teng Z, Zhou B, Pearlstein D, Wang D, Turner E, Nou X, Wang TT, Tao Y, Fonseca JM (2024) Genome-wide association mapping reveals loci for enzymatic discoloration on cut lettuce. Postharvest Biol Technol 207:112577

Pink H, Talbot A, Graceson A, Graham J, Higgins G, Taylor A, Jackson AC, Truco M, Michelmore R, Yao C, Gawthrop F (2022) Identification of genetic loci in lettuce mediating quantitative resistance to fungal pathogens. Theor Appl Genet 135(7):2481–2500

Rafalski JA, Tingey SV (1993) Genetic diagnostics in plant breeding: RAPDs, microsatellites and machines. Trends Genet 9(8):275–280

Rauscher G, Hayes R, Simko I (2009) QTL mapping of resistance to powdery mildew in lettuce. Phytopathology 99:S107

Richardson KL, Nayak S, Hasegawa DK, Eriksen RL (2024) Evaluation of lettuce germplasm for resistance to impatiens necrotic spot virus. Euphytica 220(3):33

Roberts J, Broadley MR, Pink D, Hand P, Lynn J, Monaghan JM (2020) Quantitative trait loci (QTLs) linked with root growth in lettuce (Lactuca sativa) seedlings. Mol Breed 40:1–9

Rosental L, Still DW, You Y, Hayes RJ, Simko I (2021) Mapping and identification of genetic loci affecting earliness of bolting and flowering in lettuce. Theor Appl Genet 134(10):3319–3337

Sandoya GV, Maisonneuve B, Truco MJ, Bull CT, Simko I, Trent M, Hayes RJ, Michelmore RW (2019) Genetic analysis of resistance to bacterial leaf spot in the heirloom lettuce cultivar Reine des Glaces. Mol Breed 39:1–11

Schwember AR, Bradford KJ (2010) Quantitative trait loci associated with longevity of lettuce seeds under conventional and controlled deterioration storage conditions. J Exp Bot 61(15):4423–4436

Seki K, Komatsu K, Hiraga M, Tanaka K, Uno Y, Matsumura H (2020) Identification of two QTLs for resistance to Fusarium wilt race 1 in lettuce (Lactuca sativa L.). Euphytica 216(11):174

Shim E, Kim D, Kim H, Lee S, Kim H, An M, Ko S, Yi S, Jung J (2019) Genome wide association studies on agronomic traits in lettuce (Latuca sativa L.) based on genotype-by-sequencing, p 206

Simko I (2013) Marker-assisted selection for disease resistance in lettuce. Transl Genomics Crop Breed Biotic Stress 1:267–289

Simko I, Atallah AJ, Ochoa OE, Antonise R, Galeano CH, Truco MJ, Michelmore RW (2013) Identification of QTLs conferring resistance to downy mildew in legacy cultivars of lettuce. Sci Rep 3(1):2875

Simko I, Rauscher G, Sideman RG, McCreight JD, Hayes RJ (2014) Evaluation and QTL mapping of resistance to powdery mildew in lettuce. Plant Pathol 63(2):344–353

Simko I, Ochoa OE, Pel MA, Tsuchida C, Fonti Forcada C, Hayes RJ, Truco MJ, Antonise R, Galeano CH, Michelmore RW (2015) Resistance to downy mildew in lettuce 'La Brillante'is conferred by Dm50 gene and multiple QTL. Phytopathology 105(9):1220–1228

Simko I, Peng H, Sthapit Kandel J, Zhao R (2022a) Genome-wide association mapping reveals genomic regions frequently associated with lettuce field resistance to downy mildew. Theor Appl Genet 135(6):2009–2024

Simko I, Puri KD, Dhar N, Peng H, Subbarao KV (2022b) Mapping quantitative trait loci for lettuce resistance to Verticillium dahliae race 3, plant development, and leaf color using an ultra-high-density bin map constructed from F2 progeny. PhytoFrontiers™ 2(3):257–267

Simko I, Sthapit Kandel J, Peng H, Zhao R, Subbarao KV (2023) Genetic determinants of lettuce resistance to drop caused by Sclerotinia minor identified through genome-wide association mapping frequently co-locate with loci regulating anthocyanin content. Theor Appl Genet 136(9):180

Sthapit Kandel J, Peng H, Hayes RJ, Mou B, Simko I (2020) Genome-wide association mapping reveals loci for shelf life and developmental rate of lettuce. Theor Appl Genet 133:1947–1966

Sthapit Kandel J, Sandoya GV, Zhou W, Read QD, Mou B, Simko I (2022) Identification of quantitative trait loci associated with bacterial leaf spot resistance in baby leaf lettuce. Plant Dis 106(10):2583–2590

Still DW (2007) Lettuce. In: Vegetables. Springer, Berlin/Heidelberg, pp 127–140

Su W, Tao R, Liu W, Yu C, Yue Z, He S, Lavelle D, Zhang W, Zhang L, An G, Zhang Y (2020) Characterization of four polymorphic genes controlling red leaf colour in lettuce that have undergone disruptive selection since domestication. Plant Biotechnol J 18(2):479–490

Truco MJ, Antonise R, Lavelle D, Ochoa O, Kozik A, Witsenboer H, Fort SB, Jeuken MJW, Kesseli RV, Lindhout P, Michelmore RW (2007) A high-density, integrated genetic linkage map of lettuce (Lactuca spp.). Theor Appl Genet 115:735–746

Van de Wiel CCM, Rajičić TS, Van Treuren R, Dehmer KJ, Van der Linden CG, Van Hintum TJ (2010) Distribution of genetic diversity in wild European populations of prickly lettuce (Lactuca serriola): implications for plant genetic resources management. Plant Genetic Resour 8(2):171–181

Van Treuren R, Coquin P, Lohwasser U (2012) Genetic resources collections of leafy vegetables (lettuce, spinach, chicory, artichoke, asparagus, lamb's lettuce, rhubarb and rocket salad): composition and gaps. Genet Resour Crop Evol 59:981–997

Wada KC, Inagaki N, Sakai H, Yamashita H, Nakai Y, Fujimoto Z, Yonemaru JI, Itoh H (2022) Genetic effects of Red Lettuce Leaf genes on red coloration in leaf lettuce under artificial lighting conditions. Plant-Environ Interact 3(4):179–192

Wang Y, Lu H, Hu J (2016) Molecular mapping of high resistance to bacterial leaf spot in lettuce PI 358001-1. Phytopathology 106(11):1319–1325

Wani MS, Tantray YR, Jan I, Singhal VK, Gupta RC (2020) Lactuca L.: world distribution and importance

Wei Z, Julkowska MM, Laloë JO, Hartman Y, de Boer GJ, Michelmore RW, van Tienderen PH, Testerink C, Schranz ME (2014) A mixed-model QTL analysis for salt tolerance in seedlings of crop-wild hybrids of lettuce. Mol Breed 34:1389–1400

Zhang FZ, Wagstaff C, Rae AM, Sihota AK, Keevil CW, Rothwell SD, Clarkson GJ, Michelmore RW, Truco MJ, Dixon MS, Taylor G (2007) QTLs for shelf life in lettuce co-locate with those for leaf biophysical properties but not with those for leaf developmental traits. J Exp Bot 58(6):1433–1449

Zhang NW, Lindhout P, Niks RE, Jeuken MJW (2009) Genetic dissection of Lactuca saligna nonhost resistance to downy mildew at various lettuce developmental stages. Plant Pathol 58(5):923–932

Zhang L, Su W, Tao R, Zhang W, Chen J, Wu P, Yan C, Jia Y, Larkin RM, Lavelle D, Truco MJ (2017) RNA sequencing provides insights into the evolution of lettuce and the regulation of flavonoid biosynthesis. Nat Commun 8(1):2264

Zixue Z, Ruining L, Qingyu Z, Lei L, Yuli Z (2022) Genome-wide identification of differential simple sequence repeats in stalk lettuce and its use for Qtl mapping

Zohary D (1991) The wild genetic resources of cultivated lettuce (Lactuca sativa L.). Euphytica 53:31–35

Genome-wide Association Study (GWAS) and QTL-Assisted Improvement of Squash (*Cucurbita pepo* L.)

16

Khaled F. M. Salem, Tilak Chandra, Essam Fathy El-Hashash, and Amira A. Ibrahim

Abstract

Pumpkin and squash are widely utilized across various industries, including culinary, ornamental, snacking, and seed oil production. Despite their substantial genetic diversity, productivity and genetic improvement in these crops are constrained by several factors, including auxiliary traits, notably the emergence of new insect pests, pathogens, and abiotic stresses driven by climate change, as well as the growing demand for nutritious food. These challenges have necessitated a shift in breeding objectives and methodologies. Conventional breeding alone is insufficient to meet the increasing global demand, exacerbated by population growth, resource depletion, and plateauing yield trends. Consequently, the future of pumpkin and squash breeding lies in integrating traditional breeding

K. F. M. Salem (✉)
Department of Biological Sciences, Faculty of Science and Humanities, Shaqra University, Ad-Dawadimi, Saudi Arabia

Department of Plant Biotechnology, Genetic Engineering and Biotechnology Research Institute (GEBRI), University of Sadat City, Sadat City, Egypt
e-mail: khaledfathi@su.edu.sa; khaled.salem@gebri.usc.edu.eg

T. Chandra
Division of Agricultural Bioinformatics, ICAR-Indian Agricultural Statistics Research Institute, New Delhi, India

E. F. El-Hashash
Agronomy Department, Faculty of Agriculture, Al-Azhar University, Cairo, Egypt
e-mail: dressamelhashash@azhar.edu.eg

A. A. Ibrahim
Botany and Microbiology Department, Faculty of Science, Arish University, El-Arish, Egypt
e-mail: amiranasreldeen@sci.aru.edu.eg

S. S. Dey et al. (eds.), *GWAS and QTL Mapping in Horticultural Crops*,
https://doi.org/10.1007/978-981-95-2306-1_16

techniques with advancements in bioinformatics and molecular biology, which can enhance the efficiency of breeding programs by predicting and estimating genetic variations linked to desirable phenotypic traits. Two promising approaches that can significantly advance modern breeding efforts are genome-wide association studies (GWAS) and the mapping of quantitative trait loci (QTL) for traits of interest, both of which are crucial for ensuring food and nutrition security. In this chapter, we critically examine how these techniques have been applied to squash breeding, highlighting their challenges and prospects.

Keywords

Squash · Genetic diversity · Recombinant inbred line · Genome-wide association studies · Quantitative trait loci · Conventional breeding

16.1 Introduction

Pumpkins and squashes, members of the Cucurbitaceae family, are cultivated for their fruits and seeds, exhibiting wide adaptation across both temperate and tropical regions (Hosan et al. 2021). Their considerable diversity in fruit color, size, shape, leaf morphology, plant stature, shelf life, and response to changing food patterns has driven breeders to develop and improve cultivars based on evolving market demands. As self-pollinating species, pumpkins and squashes are used globally for food and medicinal purposes, owing to their natural bioaccumulation of essential vitamins and antioxidants (Ibrahim et al. 2023). The edible parts of the plant—flesh, seeds, and peel—are rich in primary and secondary metabolites, including proteins, carbohydrates, monounsaturated fatty acids, carotenoids, tocopherols, tryptophan, delta-7-sterols, and other bioactive phytochemicals. Additionally, the seeds, a byproduct of fruit processing, are cost-effective and can be incorporated into various food products to enhance their nutritional profile (Villamil et al. 2023). These seeds are particularly enriched with a range of tetracyclic and triterpenoid compounds, which possess immunosuppressive, antibacterial, antioxidant, and anti-ulcer activities, as well as properties that promote bladder and sphincter muscle function (Dhakad et al. 2023). Thus, pumpkins and squashes are regarded as a valuable yet often underappreciated edible crop, contributing significantly to global food and nutritional security. Moreover, pumpkins and squashes have the potential to be revolutionary crops, offering balanced nutrition and greater adaptability to poor soils and adverse climatic conditions compared to many other staple crops. Despite their versatility and high nutrient content, breeding for these crops remains in the early stages of development. This is partly due to the challenges of cross-pollination between different varieties, which can lead to unintended hybridization. Additionally, their long growing seasons and diverse environmental requirements complicate the timing of pollination and seed development. The tendency to produce male and female flowers at different rates, coupled with low seed set, further limits the number of viable offspring available for breeding programs (Mazourek et al. 2021; Ibrahim et al. 2023).

Fortunately, a diverse range of landraces and variants of *C. pepo* exists, exhibiting distinct genomes and chromosomal counts (Alavilli et al. 2022; Ibrahim et al. 2023). To enhance genetic diversity and promote the expression of desirable traits, cross-breeding common genotypes with locally adapted varieties can be an effective strategy (Lee et al. 2022). Squash breeders and geneticists are focused on improving key traits, crop characteristics, and tolerance to biotic and abiotic stresses through the application of advanced genomic techniques. Among these, genome-wide association studies (GWAS) and quantitative trait locus (QTL) mapping hold particular promise in overcoming breeding challenges by identifying specific genetic regions linked to traits of interest, such as flowering time, fertility, and resistance to cross-pollination (Alavilli et al. 2022; Lee et al. 2022; Ibrahim et al. 2023). By pinpointing these loci, breeders can more efficiently select individuals with desirable traits, minimizing the influence of environmental variability and unwanted genetic backgrounds. For instance, GWAS can facilitate the identification of genes involved in carotenoid synthesis, which can help optimize the nutritional content of squash (Reddy et al. 2023). Similarly, QTL mapping can reveal loci responsible for traits like sterility or seed set, allowing for the selection of plants with higher fertility rates (Wang et al. 2020; Lee et al. 2023). Given the extensive phenotypic diversity in squash, which includes a wide array of agronomic traits, the genetic basis of this diversity has been the focus of considerable research. Genomic sequencing projects have played a key role in identifying and localizing QTL associated with important traits such as yield, fruit quality, morphology, vegetative growth, and disease resistance. Together, these genomic tools can accelerate breeding by providing molecular markers for traits that are challenging to assess phenotypically, thereby enhancing the efficiency and precision of squash breeding programs. This chapter reviews the mapping approaches used for the genetic dissection of complex traits in the four major cucurbit species, summarizing the QTL identified for each species.

16.2 GWAS: Variants Associated with Economic Traits

Genetic resources for squash species have historically been limited; however, the availability of draft genomes, annotations, web-based tools, and other genomic data is rapidly increasing, significantly accelerating squash breeding initiatives. Advances in molecular marker-assisted technologies, such as genotyping-by-sequencing (GBS) and genome-wide association studies (GWAS), are playing a pivotal role in driving innovation in squash breeding. In addition to single-nucleotide polymorphism (SNP) markers, simple sequence repeats (SSRs) have highlighted *C. pepo* as having the highest genetic diversity within the Cucurbitaceae family (Gong et al. 2008). GWAS is a powerful method for assessing hundreds of thousands of genetic variants across multiple genomes to identify those linked to specific traits (Eltaher et al. 2021, 2022; Alavilli et al. 2022). These studies have yielded robust associations for a variety of traits, including disease resistance, quality, and quantitative characteristics, with the number of identified associations expected to grow as sample sizes increase. The discovery of genome-wide SNPs and core marker sets for assessing genetic variation in cultivated pumpkin has provided valuable molecular tools for squash breeding (Nguyen et al. 2020; Alavilli et al. 2023). These efforts have utilized both historical and contemporary genotype data from USDA germplasm collections of *Cucurbita pepo*, *C. moschata*, and *C. maxima*, revealing genetic signals for several important traits, including the bush gene in *C. pepo* (Hernandez et al. 2023). Similarly, four major fruit-type cultivars have been identified within *C. moschata*, with significant differentiation patterns observed in specific chromosomal regions. Additionally, 15 loci associated with pumpkin seed traits have been mapped, with 32 genes previously linked to seed size regulation in Arabidopsis and Oryza sativa located within the identified linkage disequilibrium intervals (Lee et al. 2023).

GWAS analysis has revealed significant carotenoid variation in squash, identifying several SNPs associated with key processes such as plastid biogenesis, protein ubiquitination, transcriptional regulation, fruit ripening, and carotenoid transport regulation (Reddy et al. 2023). Furthermore, GWAS has proven valuable for screening germplasm accessions, particularly those from East Asia, to explore genetic diversity and trait associations. A study conducted on four traits, fruit shape, fruit rind color, leaf lobe, and leaf silver patch identified significant association signals for these traits, explaining 4–23% of the total phenotypic variation (Lee et al. 2022). Chilling stress is another major abiotic constraint affecting the global winter cultivation of cucurbit crops. A bulked-segregant analysis-assisted next-generation sequencing study identified single-nucleotide polymorphisms (SNPs) in three genes, revealing the existence of chilling-tolerant and chilling-susceptible lines (Alavilli et al. 2023). Additionally, several new candidate genes, including *AP2/ERF*, were identified as being involved in resistance to powdery mildew (Alavilli et al. 2023). Given the resource-intensive nature of squash breeding due to the plant's biology, smaller breeding populations, and limited greenhouse generation use genomic and trait-based selection approaches hold considerable promise. Genomic selection, in particular, is a promising technology for improving breeding efficiency. It has demonstrated predictive ability in whole-genome models for fruit quality traits, although with low to moderate predictive accuracy. For example, three rounds of genomic recurrent selection were conducted on a butternut squash population, using a fruit quality index derived from a multi-trait genomic selection model. This approach led to significant improvements in the fruit quality index and changes in correlated traits (Hernandez et al. 2020). In a nutshell, GWAS provides valuable insights into the genetic basis of traits, making it a powerful tool for accelerating the improvement of squash through assisted breeding. By leveraging significant associations, breeders can achieve faster, more precise, and more sustainable breeding outcomes.

16.3 Mapping QTL for Agronomical Traits

Quantitative trait locus (QTL) mapping is a powerful tool for understanding the genetic basis of complex traits in plants like squash (Zhong et al. 2017). The *Cucurbita* genus has developed its first SNP-based genetic map, a critical tool for advancing biological research. Markers located in gene coding regions are instrumental in understanding physiological processes, thereby accelerating the breeding of new squash varieties and supporting future mapping and diversity studies (Esteras et al. 2012). Marker-assisted breeding has become an essential technique for improving economic traits in modern squash breeding. Markers such as RFLP, RAPD, AFLP, SSR, and ISSR are widely used in various aspects of squash breeding, including germplasm management, genetic map development, hybrid breeding, purity testing, and QTL mapping (Xingzhi et al. 2010). The first SNP-saturated map for squash species, anchored to a physical map, has been reported, and several QTL associated with traits such as early flowering, fruit shape, fruit length, rind color, and flesh color have been identified, along with corresponding candidate genes (Montero-Pau et al. 2017). Distinct mapping populations have been used to identify loci linked to various traits, including dwarfism (Xiang et al. 2018), resistance to Tomato leaf curl New Delhi virus (Saez et al. 2016, 2020), fruit and agronomic characteristics (Del Valle Echevarria et al. 2020), both quantitative and qualitative traits (Esteras et al. 2012; Montero-Pau et al. 2017; Zhong et al. 2017), and seed-related traits (Wang et al. 2020) (Table 16.1).

16.4 QTL Governing Biotic and Abiotic Threats

Squash is susceptible to a variety of diseases caused by fungal, oomycete, and viral pathogens, making host resistance a critical component of integrated disease management and a major objective for plant breeders. This section discusses key *Cucurbita* diseases, including powdery mildew, downy mildew, Phytophthora rot, and viruses transmitted by aphids and whiteflies, with a focus on the exploitation of germplasm for the development of resistant cultivars (Sabharwal et al. 2024). A molecular and morphological marker map of *Cucurbita* genetics was constructed using random amplified polymorphic DNA (RAPD) markers, comprising 148 RAPD markers across 28 linkage groups. This map includes markers for five morphological traits, such as the B gene for fruit yellowing before anthesis, the M gene for silver mottling of leaves, and a locus that controls the intensity of rind color in mature fruit (Brown and Myers 2002). Phytophthora capsici Leonian, the pathogen responsible for squash blight, root rot, fruit rot, and crown rot, poses a significant global threat. Whole-genome resequencing of resistant and susceptible bulk segregants has identified single-nucleotide polymorphism (SNP)-based QTL significantly associated with resistance to this devastating disease (Ramos et al. 2020). One such QTL, QtlPC-C13, located on linkage group 13, explained 17.92–21.47% of the phenotypic variation in the population. Within the confidence interval of

Table 16.1 QTL mapping results for squash traits highlight how the identification of specific loci can accelerate breeding by pinpointing genes that control key quality traits

Influenced traits	Population type and numbers	Identified QTL/gene and numbers	Associated marker(s)	References
Qualitative/ quantitative traits	F_2/146	59	SNP	Esteras et al. (2012)
Fruit-related traits	F_2/200	29	SNP	Zhong et al. (2017)
Fruit-related traits	RIL/120	48	SNP	Montero-Pau et al. (2017)
Ovary color and carotenoid content	RIL/92	*qαcr2.1, qlut4.1,* etc.	SSR	Kazminska et al. (2018)
Cold tolerance	F_2/166	CI-1-1, qCI-4-1, etc.	SSR	Xu et al. (2017)
	F_2/300	*qCII-1, qCII-2*	SNP	Liao et al. (2024)
Fruit-associated traits	RIL	26	SSR, SNP, and DArT	Kazminska et al. (2020)
	F_2/200	fl2.1, fth1.1, etc.	SNP	Han et al. (2022)
Dwarf phenotype	F_2	3	SSR and EST-SSR	Xiang et al. (2018)
	F_2	*CpDw*	SNP	Ding et al. (2021)
Tomato leaf curl New Delhi virus resistance	F_2	Major	SNP	Saez et al. (2020)
Fruit and agronomic traits	BC_1	–	SNP	Del Valle et al. (2020)
Seed traits	F_2	10	SLAF	Wang et al. (2020)
Phytophthora crown rot resistance	F_2/168	*QtlPC-C04, QtlPC-C11, QtlPC-C14*	SNP	Ramos et al. (2020)
	F_2/83	*QtlPC-C13*, etc.	SNP	Michael et al. (2021)
	F_3/176	Multiple	SNP	Vogel et al. (2021)
	F_2	*QtlPC-C13*	SNP	Michael et al. (2021)
Yellow mosaic virus resistance	F_2/174	QtlZYMV-C04, QtlZYMV-C08, etc.	SNP	Shrestha et al. (2021)
Dwarf vine traits	F_2/186	*qCmB2* etc.	SNP	Zhang et al. (2015)
Trichome density	F_2/886	CpTD3.1, CpTD15.1, etc.	InDel, SNP	Wang et al. (2023)
Agronomic traits	F_2/88	12	SNP	Hashimoto et al. (2020)

(continued)

Table 16.1 (continued)

Influenced traits	Population type and numbers	Identified QTL/gene and numbers	Associated marker(s)	References
Powdery mildew resistance	F_2/177	–	SNP	Holdsworth et al. (2016)
Early flowering	F_2/200	qfffn5-a, qfmfn1-a, etc.	SNP	Abbas et al. (2022a, b)
	F_2	CpDFF4.1, CpDFF11.1, CpDFF20.1, etc.	InDel, SNP	Qu et al. (2023)
Silver leaf disorder	F_2/93	–	RAPD, SSR	Kabelka and Young (2010)
Seed width	F_2/193	–	AFLP, SSR	Tan et al. (2013)
Fruit length	F_2	CmoFL1	SNP	Zhou et al. (2024)
Hull-less seed traits	F_2/143	Qtlhull-less-C12, etc.	SNP	Meru et al. (2022)
	F_2/160	Cphl-1	SNP	Kaur et al. (2022)
Environmental sex expression	F_2/162	qfffn6.1, qnff6.1, etc.	InDel, SNP	Abbas et al. (2022a, b)
Peel color	F_2/177	*KNAT6*	SNP	Wang et al. (2024a, b)
Begomovirus Resistance	F_2/229	qMI-Sq/To7.1, etc.	SNP	Verma et al. (2023)
Large fruit traits	F_2/102	CmaCh17G011340, CmaCh04G029660	SNP	Pan et al. (2022)

This approach enables a deeper understanding of the impact of these genes on phenotypic traits, facilitates the localization of genes/QTL, and examines the relationships between them. By considering the genetic diversity within germplasm collections, this strategy also allows for the consolidation of target genes into a single, optimized genotype for breeding

QtlPC-C13, approximately nine candidate disease resistance gene homologs were identified (Michael 2021; Michael et al. 2021).

Quantitative trait locus (QTL) analysis using recombinant inbred lines (RILs) derived from a cross between two inbred lines with different fruit morphotypes identified loci for eight traits, located in ten regions across eight chromosomes, which explained up to 41.4% of the observed phenotypic variation (Kazminska et al. 2020). QTL-seq analysis identified four significant QTL associated with resistance to zucchini yellow mosaic virus (ZYMV) on chromosome 2, with seven markers embedded within the QTL intervals (Shrestha et al. 2021). Whole-genome resequencing revealed five genomic regions with significant allele frequency differences between resistant and susceptible bulks. Disease severity was best explained by a four-QTL model, which included loci on chromosomes 4, 5, 8, and 19 identified via bulked-segregant analysis sequencing (BSA-Seq), along with an additional locus on chromosome 19, which had a small-to-moderate effect, accounting for 28–35% of the phenotypic variance (Vogel et al. 2021). A linkage map developed using simple sequence repeat (SSR) and restriction site-associated DNA (RAD) markers successfully identified 12 QTL related to flowering and fruit traits, along with one phenotypic trait locus, explaining moderate to high levels of phenotypic variation (Hashimoto et al. 2020). Introgression mapping based on cultivars has identified the wild species-derived Pm-0 locus as the major resistance gene for powdery mildew in squash (Holdsworth et al. 2016). A molecular study on chilling tolerance in pumpkin rootstocks identified three QTL that explained approximately 0.3–20% of the phenotypic variation in cold tolerance (Xu et al. 2017).

Additionally, two major QTL for cold tolerance in pumpkin, qCII-1 and qCII-2, were identified, with *CmoERF017* emerging as a key candidate gene for positively regulating the pumpkin response to low-temperature stress (Liao et al. 2024). Squash silverleaf, a significant physiological disorder, affects productivity by reducing yield potential, and molecular markers associated with resistance have been delineated (Kabelka and Young 2010). Pumpkin production is also facing substantial losses due to begomoviruses, such as squash leaf curl China virus and tomato leaf curl New Delhi virus. A QTL-seq analysis identified a qMI-Sq/To7.1 region on chromosome 7, highlighting the global concern of viral resistance (Verma et al. 2023). Moreover, two QTL were identified on chromosome 7, accounting for 10–24% of the phenotypic variation. These QTL encode a protein of the TPX2 family, which may serve as a candidate gene for mottled leaf in zucchini, as determined by integrated annotation and alignment (Wang et al. 2024a, b).

16.5 QTL Governing Growth, Development, and Quality Traits

A high-density genetic map was utilized to identify genomic regions associated with dwarf vine, a key agronomic trait for squash. Three quantitative trait loci (QTL) were identified, one of which explained 21.4% of the phenotypic variation. This QTL was linked to a gene encoding *gibberellin 20-oxidase*, which was found to have a deletion in its promoter region in bush-type lines, suggesting its crucial role in improving productivity (Zhang et al. 2015). Fine mapping of this QTL revealed that the *gibberellin 2-oxidase* gene, originating from *CpDw*, is responsible for the dwarf phenotype, with the first identified SNP potentially affecting enzyme activity (Ding et al. 2021). Genetic mapping efforts have also been used to map the ovary color locus and identify QTL for high carotenoid content and flesh color using a recombinant inbred line (RIL) mapping population (Kazminska et al. 2018). In another study, an F_2 population of inbred lines was used to construct a genetic map and identify QTL underlying fruit-related traits. The map included 2413 bins, with an average genetic distance of 0.94 cM, covering seven traits: fruit length, diameter, shape index, weight, flesh thickness, seed cavity size, and total soluble solids content (Han et al. 2022). Trichomes, although a valuable model for studying cell differentiation and proliferation, are also problematic for zucchini field workers, causing skin irritation and complicating pollination and fruit harvesting. QTL CpTD3.1 and CpTD15.1 were identified through QTL-seq and genetic map-based analyses, with CpTD15.1 being the major-effect QTL, found to be stable across different environmental conditions (Wang et al. 2023). Early flowering is a critical trait for pumpkin breeding, and the identification of QTL and genes controlling this trait has been a research priority. Two QTL related to the first female flowering node and four related to the first male flowering node were detected using a high-density genetic map. These QTL were linked to two candidate genes, including a flowering locus T-like protein homologous to the FLOWERING LOCUS T gene, which regulates flowering time in various crops (Abbas et al. 2022a, b). Additionally, a major locus located in a 117-kb candidate region on chromosome 20, identified through composite interval mapping, revealed the Cp4.1LG20g08050 gene encoding a RING finger protein. This gene may serve as a candidate for regulating early flowering in zucchini (Qu et al. 2023).

A genetic linkage map was constructed using polymorphic loci to identify QTL for seed width traits. Composite interval mapping revealed four QTL located across three linkage groups, explaining between 2.87% and 29.68% of the phenotypic variation for this important trait (Tan et al. 2013). Fruit length is a key economic trait that impacts both yield and appearance. A candidate gene, CmoFL1, encoding an E3 ubiquitin ligase, was identified as potentially associated with cell division and negatively regulating fruit length, underscoring its significance in determining fruit yield and appearance (Zhou et al. 2024). Pumpkin seeds, a valuable source of vegetable oil, protein, healthy fatty acids, and minerals, are in high demand, particularly for their hull-less form, which is preferred by the food industry. A QTL-seq analysis identified a single QTL on chromosome 12 (Qtlhull-less-C12) significantly associated with the hull-less trait. Several candidate genes, including a NAM-domain-containing protein and Fiber Protein 11, were found within the QTL interval (Meru et al. 2022). The hull-less pumpkin seed, commonly used in the oil, nut, and baking industries, eliminates the need for de-hulling. A new genomic region, named Cphl-1, was identified on chromosome 12, containing several SNPs, as revealed by polymorphic KASP assays, which contribute to the hull-less trait (Kaur et al. 2022). High-density linkage mapping using SLAF-seq technology was employed to identify QTL for female flowering node position and the number of female flowers. Major QTL were detected with phenotypic variation explained (PVE) values of 30.5% and 22.9%, respectively. These QTL were linked to protein-coding genes involved in sex expression, photoperiodic flowering, and hormone responses (Abbas et al. 2022a, b).

Fruit peel color is an important agronomic trait, and genetic mapping plays a crucial role in molecular breeding and understanding the regulatory mechanisms behind it. A key locus, CmaMg, was mapped on chromosome 11, with CmaKNAT6 as a candidate gene. This gene encodes a homeobox protein, and its deletion or premature transcriptional termination influences the peel color of mature fruit (Wang et al. 2024a, b). The Atlantic Giant pumpkin, known for producing the world's largest fruit, was studied through genome-wide resequencing of an F2 population. Thirteen genes were selected as candidates for fruit size, including two phytohormone-related genes, CmaCh17G011340 and CmaCh04G029660. These genes provide valuable insights into the genetic mechanisms underlying large fruit formation in *Cucurbita*, and the developed genetic map, along with KASP markers, offers tools for marker-assisted breeding (Pan et al. 2022). Additionally, a genome-wide association study (GWAS) identified three major allelic effects associated with fruit size in the mammoth group of *Cucurbita*. A missense variant in the homeobox-leucine zipper protein ATHB-20-like, located on chromosome 8, was significantly associated with fruit weight, while another missense variant in the translocation protein SEC62 was linked to fruit diameter. These findings highlight the complex genetic basis of fruit size, which is controlled by several molecular mechanisms, including endoreduplication, hormonal regulation, the CLV-WUS signaling pathway, MADS-box family genes, and the ubiquitin-proteasome pathway (Reddy et al. 2022). Altogether, QTL mapping is a valuable tool for improving squash breeding by providing insights into the genetic underpinnings of important agronomical traits. Furthermore, QTL mapping enables more precise, targeted breeding, helping squash breeding programs to develop varieties that meet both agronomic and consumer needs more efficiently and sustainably.

16.6 Conclusion and Prospects

Squash is a distinctive fall vegetable, renowned for its unique fruit shape and highly valued for its culinary versatility. Genome-wide association studies (GWAS) and quantitative trait locus (QTL) mapping are powerful tools that address key challenges in squash breeding, including issues related to genetic diversity, flower synchronization, and fertility. These approaches enable researchers to identify specific loci associated with traits of interest, such as the timing of male and female flower development, seed set, and sterility. GWAS, in particular, facilitates the identification of genetic variants linked to complex traits, allowing breeders to select for genotypes with optimized flowering patterns, reduced sterility, or improved cross-pollination compatibility. QTL mapping complements this by pinpointing loci responsible for quantitative traits, enabling the development of molecular markers for selecting favorable alleles within breeding populations. By incorporating these methods, breeders can reduce reliance on traditional phenotypic selection and mitigate the influence of environmental variability. This enhances the precision and efficiency of breeding programs by facilitating early-stage selection and minimizing the impact of undesirable genetic interactions.

References

Abbas HMK, Huang H, Wu T, Wang R, Du H, Lu S, Xue S, Yao C, Jin Q, Zhong Y (2022a) High-density genetic mapping identified a major locus for environmental sex expression in pumpkin (Cucurbita moschata Duch.). Hortic Plant J 8(5):593–601

Abbas HMK, Zhou YY, Huang HX, Qamar MTU, Wang MM, Lu S, Xu C, Liu Z, Wu TQ, Zhong YJ (2022b) QTL mapping, whole genome resequencing, and marker-assisted selection provide basics of early flowering in pumpkin. Plant Breed 141(2):266–276

Alavilli H, Lee JJ, You CR, Poli Y, Kim HJ, Jain A, Song K (2022) GWAS reveals a novel candidate gene CmoAP2/ERF in pumpkin (Cucurbita moschata) involved in resistance to powdery mildew. Int J Mol Sci 23(12):6524

Alavilli H, Lee JJ, You CR, Lee SC, Song K (2023) Identification of a novel candidate gene for chilling tolerance in pumpkin (Cucurbita moschata) using whole-genome resequencing. J Plant Biol 66(4):317–330

Brown RN, Myers JR (2002) A genetic map of squash (Cucurbita sp.) with randomly amplified polymorphic DNA markers and morphological markers. J Am Soc Hortic Sci 127(4):568–575

Del Valle Echevarria AR, Campbell A, Radovich TJ, Silvasy T, Moore S, Kantar MB (2020) Quantitative trait loci (QTL) analysis of fruit and agronomic traits of tropical pumpkin (Cucurbita moschata) in an organic production system. Horticulturae 6(1):14

Dhakad GG, Nannaware MS, Wagh GV, Shrirao AV, Kochar NI, Chandewar AV (2023) Review on medicinal use of pumpkin seed on human health. Res J Pharmacol Pharmacodyn 15(2):59–65

Ding W, Wang Y, Qi C, Luo Y, Wang C, Xu W, Qu S (2021) Fine mapping identified the gibberellin 2-oxidase gene CpDw leading to a dwarf phenotype in squash (Cucurbita pepo L.). Plant Sci 306:110857

Eltaher S, Baenziger PS, Belamkar V et al (2021) GWAS revealed effect of genotype × environment interactions for grain yield of Nebraska winter wheat. BMC Genomics 22:2. https://doi.org/10.1186/s12864-020-07308-0

Eltaher S, Sallam A, Emara HA et al (2022) Genome-wide association mapping revealed SNP alleles associated with spike traits in wheat. Agronomy 12:1469. https://doi.org/10.3390/agronomy12061469

Esteras C, Gómez P, Monforte AJ, Blanca J, Vicente-Dólera N, Roig C, Nuez F, Picó B (2012) High-throughput SNP genotyping in Cucurbita pepo for map construction and quantitative trait loci mapping. BMC Genomics 13:1–21

Gong L, Pachner M, Kalai K, Lelley T (2008) SSR-based genetic linkage map of Cucurbita moschata and its synteny with Cucurbita pepo. Genome 51(11):878–887

Han X, Min Z, Wei M, Li Y, Wang D, Zhang Z, Hu X, Kong Q (2022) QTL mapping for pumpkin fruit traits using a GBS-based high-density genetic map. Euphytica 218(8):106

Hashimoto T, Kubo N, Nishimura K, Nagano AJ, Sasaki A, Nakamura Y, Mimura Y (2020) Quantitative Trait locus analysis in squash (Cucurbita moschata) based on simple sequence repeat markers and restriction site-associated DNA sequencing analysis. Horticulturae 6(4):71

Hernandez CO, Wyatt LE, Mazourek MR (2020) Genomic prediction and selection for fruit traits in winter squash. G3 Genes Genomes Genetics 10(10):3601–3610

Hernandez CO, Labate J, Reitsma K, Fabrizio J, Bao K, Fei Z, Grumet R, Mazourek M (2023) Characterization of the USDA Cucurbita pepo, C. moschata, and C. maxima germplasm collections. Front Plant Sci 14:1130814

Holdsworth WL, LaPlant KE, Bell DC, Jahn MM, Mazourek M (2016) Cultivar-based introgression mapping reveals wild species-derived Pm-0, the major powdery mildew resistance locus in squash. PLoS One 11(12):e0167715

Hosen M, Rafii MY, Mazlan N, Jusoh M, Oladosu Y, Chowdhury MFN, Muhammad I, Khan MMH (2021) Pumpkin (Cucurbita spp.): a crop to mitigate food and nutritional challenges. Horticulturae 7(10):352

Ibrahim AA, Salem KF, Abdein MA, Ramadan SM (2023) Advances in summer squash (Cucurbita pepo L.) molecular breeding strategies. In: Smart plant breeding for vegetable crops in post-genomics era. Springer Nature, Singapore, pp 163–215

Kabelka EA, Young K (2010) Identification of molecular markers associated with resistance to squash silverleaf disorder in summer squash (Cucurbita pepo). Euphytica 173:49–54

Kaur B, Garcha KS, Bhatia D, Khosa JS, Sharma M, Mittal A, Verma N, Dhatt AS (2022) Identification of single major QTL and candidate gene (s) governing hull-less seed trait in pumpkin. Front Plant Sci 13:948106

Kazminska K, Hallmann E, Rusaczonek A, Korzeniewska A, Sobczak M, Filipczak J, Kuczerski KS, Steciuk J, Sitarek-Andrzejczyk M, Gajewski M, Niemirowicz-Szczytt K (2018) Genetic mapping of ovary colour and quantitative trait loci for carotenoid content in the fruit of Cucurbita maxima Duchesne. Mol Breed 38:1–14

Kazminska K, Hallmann E, Korzeniewska A, Niemirowicz-Szczytt K, Bartoszewski G (2020) Identification of fruit-associated QTLs in Winter Squash (Cucurbita maxima Duchesne) using recombinant inbred lines. Gene 11(4):419

Lee ES, Lee YR, Lee O, Yang HB, Lee HE, Han K, Kim DS (2022) Genetic diversity and genome-wide association study of pumpkins (Cucurbita Moschata) originating from East Asia. Available at SSRN 4160303

Lee CY, Wang JF, Chang CH, Tung CW (2023) Analyzing genomic variation in cultivated pumpkins and identification of candidate genes controlling seed traits. Plant Genome 16(4):e20393

Liao Y, Liu X, Xu N, Chen G, Qiao X, Gu Q, Wang Y, Sun J (2024) Fine mapping and identification of ERF transcription factor ERF017 as a candidate gene for cold tolerance in pumpkin. Theor Appl Genet 137(10):1–17

Mazourek M, Hernandez C, Fabrizio J (2021) Reconsidering approaches to selection in winter squash improvement: improved quality and breeding efficiency. Plant Breed Rev 44:247–272

Meru G, Fu Y, Shrestha S, Michael VNE, Dorval M, Mainviel R (2022) Genomic position and markers associated with the hull-less seed trait in pumpkin. Plants 11(9):1238

Michael VNE (2021) Genetic mapping and characterization of resistance to phytophthora Capsiciun squash. Doctoral dissertation, University of Florida

Michael VNE, Fu Y, Shrestha S, Meru G (2021) A novel QTL for resistance to phytophthora crown rot in squash. Plants 10(10):2115

Montero-Pau J, Blanca J, Esteras C, Martínez-Pérez EM, Gómez P, Monforte AJ, Cañizares J, Picó B (2017) An SNP-based saturated genetic map and QTL analysis of fruit-related traits in Zucchini using Genotyping-by-sequencing. BMC Genomics 18:1–21

Nguyen NN, Kim M, Jung JK, Shim EJ, Chung SM, Park Y, Lee GP, Sim SC (2020) Genome-wide SNP discovery and core marker sets for assessment of genetic variations in cultivated pumpkin (Cucurbita spp.). Hortic Res 7

Pan L, Wang M, Yang Y, Chen C, Dai H, Zhang Z, Hua B, Miao M (2022) Whole-genome resequencing identified QTLs, candidate genes and Kompetitive Allele-Specific PCR markers associated with the large fruit of Atlantic Giant (Cucurbita maxima). Front Plant Sci 13:942004

Qu SP, Dan YANG, Yu HY, Chen FY, Wang KX, Ding WQ, Xu WL, Wang YL (2023) QTL analysis of early flowering of female flowers in zucchini (Cucurbita pepo L.). J Integr Agric 22(11):3321–3330

Ramos A, Fu Y, Michael V, Meru G (2020) QTL-seq for identification of loci associated with resistance to Phytophthora crown rot in squash. Sci Rep 10(1):5326

Reddy UK, Natarajan P, Abburi VL, Tomason Y, Levi A, Nimmakayala P (2022) What makes a giant fruit? Assembling a genomic toolkit underlying various fruit traits of the mammoth group of Cucurbita maxima. Front Genet 13:1005158

Reddy UK, Lopez-Ortiz C, Talavera-Caro AG, Natarajan P, Tomason Y, Alaparthi S, Levi A, Nimmakayala P (2023) GWAS resolves molecular mechanisms underlying natural variation for carotenoids in Cucurbita maxima Duchesne. Sci Hortic 312:111881

Sabharwal P, Thakur S, Shrestha S, Fu Y, Meru G (2024) Breeding and genetics of resistance to major diseases in Cucurbita—a review. Crop Sci

Saez C, Martínez C, Ferriol M, Manzano S, Velasco L, Jamilena M, López C, Picó B (2016) Resistance to Tomato leaf curl New Delhi virus in Cucurbita spp. Ann Appl Biol 169(1):91–105

Saez C, Martínez C, Montero-Pau J, Esteras C, Sifres A, Blanca J, Ferriol M, López C, Picó B (2020) A major QTL located in chromosome 8 of Cucurbita moschata is responsible for resistance to Tomato leaf curl New Delhi virus. Front Plant Sci 11:207

Shrestha S, Michael VNE, Fu Y, Meru G (2021) Genetic loci associated with resistance to zucchini yellow mosaic virus in squash. Plants 10(9):1935

Tan XingZhi TX, Ge Yu GY, Xu WenLong XW, Cui ChongShi CC, Qu ShuPing QS (2013) Construction of genetic linkage map and QTL analysis for seed width in pumpkin (Cucurbita maxima)

Verma N, Garcha KS, Sharma A, Sharma M, Bhatia D, Khosa JS, Kaur B, Chuuneja P, Dhatt AS (2023) Identification of a major-effect quantitative trait loci associated with begomovirus resistance in Cucurbita moschata. Phytopathology 113(5):824–835

Villamil RA, Escobar N, Romero LN, Huesa R, Plazas AV, Gutiérrez C, Robelto GE (2023) Perspectives of pumpkin pulp and pumpkin shell and seeds uses as ingredients in food formulation. Nutr Food Sci 53(2):459–473

Vogel G, LaPlant KE, Mazourek M, Gore MA, Smart CD (2021) A combined BSA-Seq and linkage mapping approach identifies genomic regions associated with Phytophthora root and crown rot resistance in squash. Theor Appl Genet 134:1015–1031

Wang Y, Wang C, Han H, Luo Y, Wang Z, Yan C, Xu W, Qu S (2020) Construction of a high-density genetic map and analysis of seed-related traits using specific length amplified fragment sequencing for Cucurbita maxima. Front Plant Sci 10:1782

Wang Y, Wang G, Lin D, Luo Q, Xu W, Qu S (2023) QTL mapping and stability analysis of trichome density in zucchini (Cucurbita pepo L.). Front Plant Sci 14:1232154

Wang C, Ding W, Chen F, Zhang K, Hou Y, Wang G, Xu W, Wang Y, Qu S (2024a) Mapping and transcriptomic profiling reveal that the KNAT6 gene is involved in the dark green peel colour of mature pumpkin fruit (Cucurbita maxima L.). Theor Appl Genet 137(10):1–18

Wang K, Wang X, Zhang L, Chi Y, Luo Y, Xu W, Wang Y, Qu S (2024b) Morphological analyses and QTL mapping of mottled leaf in zucchini (Cucurbita pepo L.). Int J Mol Sci 25(5):2491

Xiang C, Duan Y, Li H, Ma W, Huang S, Sui X, Zhang Z, Wang C (2018) A high-density EST-SSR-based genetic map and QTL analysis of dwarf trait in Cucurbita pepo L. Int J Mol Sci 19(10):3140

Xingzhi T, Chongshi C, Shuping Q (2010) Advances in marker-assisted breeding of cucurbita squash. In: China cucurbits and vegetables

Xu Y, Guo SR, Shu S, Ren Y, Sun J (2017) Construction of a genetic linkage map of rootstock-used pumpkin using SSR markers and QTL analysis for cold tolerance. Sci Hortic 220:107–113

Zhang G, Ren Y, Sun H, Guo S, Zhang F, Zhang J, Zhang H, Jia Z, Fei Z, Xu Y, Li H (2015) A high-density genetic map for anchoring genome sequences and identifying QTLs associated with dwarf vine in pumpkin (Cucurbita maxima Duch.). BMC Genomics 16:1–3

Zhong YJ, Zhou YY, Li JX, Yu T, Wu TQ, Luo JN, Luo SB, Huang HX (2017) A high-density linkage map and QTL mapping of fruit-related traits in pumpkin (Cucurbita moschata Duch.). Sci Rep 7(1):12785

Zhou Y, Zhao M, Shen Q, Zhang M, Wang C, Zhang Y, Yang Q, Bo Y, Hu Z, Yang J, Zhang M (2024) Genetic mapping reveals a candidate gene CmoFL1 controlling fruit length in pumpkin. Front Plant Sci 15:1408602

A Compendium of Advances on GWAS and QTL Atlas for Accelerating Trait Discovery in Eggplant

17

Tilak Chandra, Sarika, Shyam Sundar Dey, Mir Asif Iquebal, and Dinesh Kumar

Abstract

Vegetable crops serve as major nutrient sources for humanity and have been extensively cultivated over thousands of years of domestication. Among the primary vegetables belonging to the nightshade family, eggplants occupy a significant position in terms of both cultivation and global consumption. They are widely consumed fresh by diverse populations and are utilized as raw materials in various culinary applications and processing industries. The increasing global population and climate crisis underscore the urgency to elevate the demand for this vegetable. However, many desirable traits in eggplants are polygenic, often exhibiting complex inheritance patterns. Consequently, vegetable breeders encounter multiple challenges in improving eggplant traits. A comprehensive understanding of the expression frequencies of the genes and alleles that govern specific quantitative traits and phenotypes is essential. In this chapter, we discuss two prominent genetic tools and their applications in eggplant breeding. These tools provide insights into the genetic foundations of agronomically desirable traits, such as yield, nutritional value, disease resistance, and adaptability. This strategic approach represents a significant advancement in genetic analysis and has proven invaluable for identifying candidate genes. Additionally, quantitative trait locus (QTL) mapping in eggplant aids breeders by pinpointing specific genomic regions associated with traits of interest, including yield, disease resistance, and growth rate. The identification of these loci is instrumental for trait refinement through targeted selection, marker-assisted selection, and the

T. Chandra · Sarika · M. Asif Iquebal · D. Kumar
Division of Agricultural Bioinformatics, ICAR-Indian Agricultural Statistics Research Institute, New Delhi, India

S. S. Dey (✉)
Division of Vegetable Science, ICAR-Indian Agricultural Research Institute, New Delhi, India

S. S. Dey et al. (eds.), *GWAS and QTL Mapping in Horticultural Crops*,
https://doi.org/10.1007/978-981-95-2306-1_17

enhancement of trait combinations by predicting breeding outcomes. The integration of these two technologies enhances the precision and effectiveness of breeding strategies, ultimately leading to improved genotypes through trait discovery-assisted breeding.

Keywords

Eggplant · Germplasm accessions · Linkage group · Genome-wide association studies · Quantitative trait loci · Molecular markers · Trait discovery

17.1 Introduction

Eggplant (*Solanum melongena* L.) is a berry-producing vegetable crop and ranks as the third most significant species among Solanaceous crops, with extensive applications in both industrial and culinary sectors worldwide. This economically important fruit was domesticated in India and Southeast Asia (Wei et al. 2020a, b). The eggplant genome spans approximately 1.17 Gb across 12 chromosomes, containing 36,582 protein-coding genes alongside a variety of repetitive elements, with long terminal repeat (LTR) retrotransposons being the most prevalent (Wei et al. 2020a, b). Eggplant exhibits substantial diversity among landraces and its close relatives (Tiwari et al. 2023). Due to its extensive phenotypic variability, comparative analyses have identified three major types of genetic variations: single-nucleotide polymorphisms (SNPs), insertions/deletions (indels), and structural variants (SVs) (Bhanushree et al. 2018; Li et al. 2019; Wei et al. 2020a, b). Therefore, identifying agronomically relevant trait-regulating markers is crucial for future research endeavors. Additionally, asymmetric structural variations are prevalent in the eggplant genome and are often found in regulatory regions of potential protein-coding genes, presenting new opportunities for dissecting novel trait-specific elements. Genome-wide association studies (GWAS) and quantitative trait locus (QTL) identification are promising tools for examining these perturbations throughout the genome and their regulatory elements (Ro et al. 2022; Tiwari et al. 2023). Although the construction of high-resolution genetic maps and map-based QTL/gene mining in eggplant has lagged behind other crops in the Solanaceae family, such as tomato and potato, this chapter discusses recent advances and prospects of two advanced molecular breeding platforms: GWAS and QTL mapping. These technologies aim to revitalize breeding applications for eggplant trait improvement. Moreover, QTL studies highlight the syntenic importance of genome sequences and the development of ultra-dense physical maps, enhancing our understanding of QTL orthology in related crops (Rinaldi et al. 2016). Consequently, these molecular footprints play critical roles in eggplant improvement by identifying genetic variants linked to desirable biotic, abiotic, and quality-related traits. By leveraging genetic diversity across eggplant populations, researchers can associate specific genomic regions with phenotypic traits, while QTL mapping focuses on the relationship between trait variation and genetic markers. Collectively, these approaches can facilitate the identification of candidate genes and markers suitable for marker-assisted selection, thereby accelerating eggplant breeding initiatives and opening promising avenues for future research.

17.2 Eggplant Breeding Initiative vis- à-vis Prevalent Constraints

Eggplant, a major vegetable crop in tropical and subtropical regions, suffers significant losses due to biotic and abiotic stresses. Breeding efforts could lead to substantial advancements in several key traits (Tiwari et al. 2023). These include (i) enhancing disease and pest resistance to combat common threats such as bacterial wilt and fruit borers, (ii) improving fruit quality by focusing on attributes such as size, shape, color, and texture, which can increase consumer appeal, (iii) increasing yield, along with enhancing tolerance to abiotic stresses like drought and salinity, (iv) considering the nutritional content of the fruit, with the aim of raising levels of antioxidants and vitamins, (v) optimizing plant architecture for improved growth habits and enhancing fruit shelf life to boost marketability, (vi) selecting for adaptability to various environmental conditions to ensure that new varieties can thrive across different regions, and (vii) addressing the presence of prickles on eggplant, including those on the leaves, stems, and fruit calyxes, which complicate cultivation, harvesting, and transportation. These prickles not only create challenges during harvesting but also lead to labor inefficiencies and can cause scratches on the skin of the fruits during transport, collectively representing an undesirable agronomic trait.

17.3 Genetic Resources and Their Importance in Eggplant Trait Discovery

As a significant vegetable crop grown globally, with India as its origin, eggplant is renowned for its diverse fruit and plant morphology. Consequently, a large number of wild relatives are being utilized in breeding programs to address various constraints (Gramazio et al. 2023). The application of genomic techniques and a wide array of molecular markers facilitates efficient marker-assisted selection (MAS), enabling the identification of genes or quantitative trait loci (QTL) of interest while eliminating undesirable traits. This approach conserves time and resources compared to conventional breeding methods. Breeding eggplant materials that perform well under low nitrogen fertilization is crucial for mitigating environmental degradation, reducing human health risks, and lowering production costs. *Solanum elaeagnifolium*, a wild relative of eggplant, offers valuable genetic variation for introgression breeding. The diverse genetic resources available are invaluable for genome-wide association studies (GWAS) and QTL mapping due to their rich diversity in phenotypic traits, including fruit shape, color, size, and disease resistance (Portis et al. 2014) (Fig. 17.1).

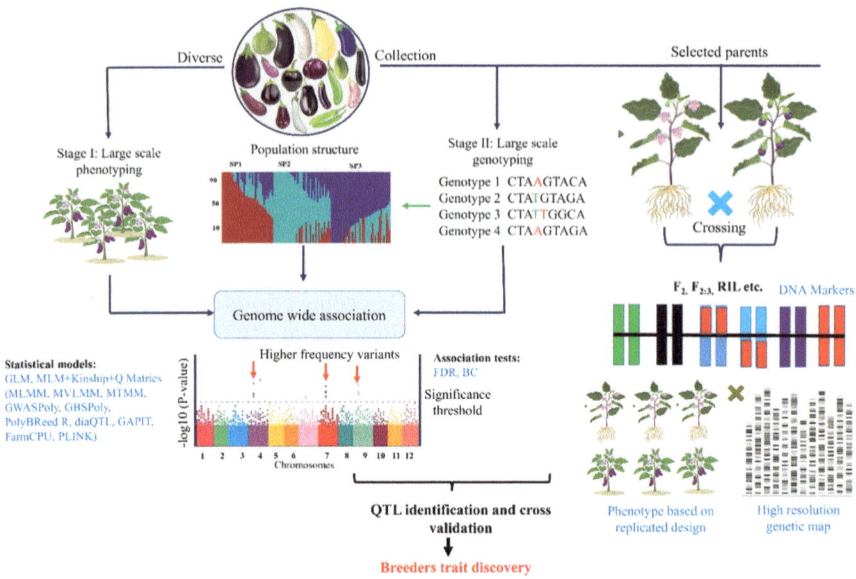

Fig. 17.1 Roadmap for eggplant trait discovery utilizing two modern breeding technologies. A diverse array of genetic resources, including landraces, breeding cultivars, and exotic lines, is currently employed to explore functional omics footprints. Genome-wide association studies (GWAS) identify these footprints and their significant associations with desired traits of interest. Alternatively, two diverse parental lines are selected and crossed to create an experimental population. This population is then aligned with a high-resolution genetic map, integrated with high-throughput phenotyping platforms in a replicated design, facilitating functional breeding and trait discovery

Access to various eggplant accessions and landraces enhances the genetic variation necessary for identifying associations between specific genetic markers and traits of interest. This diversity also facilitates the development of molecular markers, such as single-nucleotide polymorphisms (SNPs) and simple sequence repeats (SSRs), which are critical for tracking traits during mapping studies. Advances in genomic tools, such as high-throughput sequencing, further enhance the accuracy and efficiency of these analyses, helping to elucidate the genetic architecture of complex traits. The insights gained from these studies can directly inform breeding programs by identifying favorable alleles linked to important traits, thereby improving the development of resilient eggplant cultivars. Ultimately, research on these genetic resources not only contributes to crop improvement but also supports the conservation of genetic diversity essential for future agricultural sustainability.

17.4 Genome-Wide Association Analysis: Mapping Variations for Breeders' Traits

Eggplant is a crucial vegetable crop globally, and accessions have been selected based on single-nucleotide polymorphism (SNP) markers significantly associated with various traits, including days to maturity, flower size, fruit width, harvest fruit color, leaf prickles, and stem prickles (Ro et al. 2022). A detailed list of significant studies on eggplant has been provided in Table 17.1. Similarly, genome-wide association studies (GWAS) identified 20 SNPs significantly linked to total phenolic content, with five SNPs associated with the gene encoding *IRX12 laccase-4* on chromosome 10 (Ro et al. 2023). A mixed linear model has been widely applied, revealing 56 SNP-phenotypic associations distributed across nine chromosomes, which segregated into two major clusters based on geographical provenance and fruit type (Cericola et al. 2014). GWA analysis validated many known controlling loci and detected numerous new marker-trait associations for agronomic traits by comparing the map positions of these SNPs with loci derived from conventional linkage mapping (Portis et al. 2015). A study involving approximately 141 accessions identified simple sequence repeat (SSR) markers associated with nine fruit traits. The analysis revealed two subgroups with extensive long-range linkage disequilibrium (LD) associations for eight phenotypic traits, although no association was found with fruit glossiness (Ge et al. 2013). Comprehensive GWAS indicates that fruit shape selection influences the genetic structure and diversity of eggplants, revealing a narrow genetic background in round- and oval-fruited varieties (Liu et al. 2019). Multi-parent advanced generation inter-cross (MAGIC) populations serve as valuable breeding materials for the genetic dissection of complex plant traits. In the MAGIC population, two variants in the *APRR2* gene, predicted to significantly impact fruit chlorophyll pigmentation, were identified, with a large deletion observed in two reference genomes of accessions lacking chlorophyll in the fruit peel (Arrones et al. 2022). Furthermore, a genome-wide association study found strong associations with two *MYB* and *COP1* genes through the analysis of a set of filtered SNPs exhibiting low residual heterozygosity (Mangino et al. 2022).

Overall, GWAS is instrumental in trait discovery for eggplant by identifying specific genetic markers, such as SNPs, associated with desirable traits like fruit color, size, and disease resistance. By elucidating the genetic architecture of these traits, GWAS facilitates marker-assisted selection, enabling breeders to efficiently develop varieties with targeted characteristics. Additionally, GWAS enhances our understanding of the link between genotype and phenotype, highlights underutilized genetic diversity, and aids in improving nutritional content. It also identifies genetic factors for disease resistance, helping to breed varieties capable of withstanding climate change stresses.

Table 17.1 Genome-wide association studies in eggplant with functional relevance to agronomical traits

No of accessions	Kind of markers	Number of markers	Linked candidate gene	Function(s)	Associated traits	References
288	SNPs	114,981	*Acetylserotonin O-methyl transferase*	Melatonin biosynthesis	Agro-morphological traits	Ro et al. (2022)
224	SNPs	29,183	*IRX12 laccase-4*	Lignin biosynthesis	Phenolics and antioxidant contents	Ro et al. (2023)
191	SNPs	>10,000	–	–	Fruit color	Cericola et al. (2014)
191	SNPs	–	*Pectinesterase*, Glycine-rich protein, etc.	Multiple pathways	Morphological traits	Portis et al. (2015)
141	SSRs	105	–	–	Fruit traits	Ge et al. (2013)
377	SNPs	93,582	*SUN, OVATE* homolog	Auxin distribution	Shape selection	Liu et al. (2019)
420	SNPs		*APRR2*	Transcription factor	Color palette	Arrones et al. (2022)
420	SNPs		*MYB113/COP1*	Anthocyanin biosynthesis	Anthocyanin Pigmentation	Mangino et al. (2022)

17.5 Mining QTL: A Regulatory Cohort

17.5.1 Agronomical Advantageous and Domestication-Related Traits

A molecular genetic linkage map for eggplant was developed using tomato molecular markers, encompassing approximately 233 markers and revealing the conservation of extensive collinear regions (Doganlar et al. 2002a, b). A detailed list of significant mapped QTL in eggplant has been tabulated in Table 17.2. Single marker mapping identified three quantitative trait loci (QTL): *qsfwpu1, qsfwpu2*, and *qsfwpu3*, each exhibiting a phenotypic variance of over 10% and located on linkage groups LG1 and LG4, underscoring their significance in eggplant breeding programs (Bhanushree et al. 2020). Analysis of an interspecific F_2 population suggested that domestication-related traits in the Solanaceae family have been influenced by mutations in a limited number of target loci with substantial phenotypic effects (Doganlar et al. 2002a, b). An interspecific population derived from a cross between cultivated eggplant and its wild relative revealed conserved loci affecting traits such as leaf lobing, flowering time, and trichome density, indicating potential orthologs in Solanaceous plants (Frary et al. 2003). An F_2 eggplant population identified multiple location-specific QTLs explaining 4–93% of phenotypic variance across different environments (Portis et al. 2014; Mangino et al. 2022). Introgression lines (ILs) represent valuable resources for the breeding and genetic analysis of important traits. A single marker-defined chromosomal segment from a wild eggplant relative detected approximately ten stable QTLs, with five associated with higher mean values for plant- and flower-related traits and one related to flower and four to fruit traits (Mangino et al. 2020). In a study involving 2200 F_2 individuals from resistant and susceptible materials, ten QTL sites were identified using DEEP-BSA analysis, with *SMEL4_05g015980.1* and *SMEL4_05g016110.1* emerging as key candidate genes. These genes contained a non-synonymous mutation in the exonic region and a variant in the promoter region identified through SNP annotation (Xiao et al. 2024). Parthenocarpy, the ability to set fruit without pollination, is essential for fruit development under unfavorable conditions. Linkage maps based on polymorphic markers revealed two QTLs on chromosomes 3 and 8, which control parthenocarpy and were mapped using selective markers, emf21H22 and emh11J10, on either side of Cop8.1 in populations derived from various parental combinations (Miyatake et al. 2012). Additionally, a set of molecular markers linked to hotspots of QTLs influencing various morphological traits across different experimental locations was identified. These hotspots, affecting traits such as plant hairiness, prickliness, and pigmentation, were located on chromosomes 3, 6, and 10, possibly reflecting pleiotropic effects (Frary et al. 2014). A high-resolution genetic map was created using length amplified fragment sequencing to discover single-nucleotide polymorphisms (SNPs) and linkage groups. A cohort of QTLs associated with stem height and fruit and leaf morphology traits was detected, explaining significant phenotypic variance across nine linkage groups (Wei et al. 2020a, b).

Table 17.2 Mapped significant QTLs and their relative location with linked markers regulating desirable agronomical advantageous traits in eggplant to assist future breeding for sustainable production amid climatic perturbations

Population/ numbers	Associated marker (s)	Mapped QTL	Chromosomal location	Linked markers/transcripts	Associated traits	References
F_2/168	SSR	*qsfwpu1, sfwpu2 and sfwpu3*	1, 4	–	Fruit weight	Bhanushree et al. (2020)
F_2	–	*Fw2.1, fl2.1,* etc.	Multiple	–	Domestication traits	Doganlar et al. (2002a, b)
F_2	RFLP	*llob6.1, dtf1.1, dtf2.1*	–	–	Morphological traits	Frary et al. (2003)
F_2		fcl1/2E04.ML, pedlE01.ML	Multiple	Disease-resistant protein, cellulose synthase	Growth and development	Portis et al. (2014)
Introgression lines			Multiple		Fruit traits	Mangino et al. (2020)
F_2/156	SNP, SSR, etc.	FomE02.1, FomE11.1, Vd20E08.1, Vd40E05.1, Vd40E09.1	Multiple	MBS/LRR resistant protein	Fungal wilt resistance	Barchi et al. (2018)
F_2/156	SNP, SSR, etc.	FrucolE05, NasE05, etc.	Multiple	UDP-glucosyltransferase and other glycosyltransferases	Sugar contents	Toppino et al. (2016a, b)
F_2/2200	SNPs	*EBWR9*	5, 8, 9, and 11	Transmembrane receptor-like kinases	Wilt resistance	Xiao et al. (2024)
F_2/161	SSR	*emk03O04, emf11A03*	6	–	Phomopsis blight resistance	Narayanswami et al. (2023)
F_2/156	SSR, SNP, etc.	*Rfo-Sa1*	–	–	Vascular wilt resistance	Toppino et al. (2016a, b)
RIL/178	AFLP/SSR	*Ers1*	Multiple	–	Bacterial wilt resistance	Lebeau et al. (2013)
F_2	SSR/SNP	*Cop3.1, Cop8.1*	3, 8	emf21H22 and emh11J10	Parthenocarpy regulation	Miyatake et al. (2012)

$F_{2:3}$	SSR/SNP	*FM1*	2, 4	–	Fusarium wilt resistance	Miyatake et al. (2016)
F_2/156	SNP	–	Multiple	–	Anthocyanin accumulation	Barchi et al. (2012a, b)
F_2 and $F_{2:3}$	SNP	–	6	*SmARF10B*	Leaf prickle development	Ke et al. (2024)
F_2/58	AFLP, RFLP, and COSII	Multiple	3, 6, 10	Transcription factors	Morphological traits	Frary et al. (2014)
F_2/121	SNP	Msh, fl, fd, fs, etc.	1, 3, 5, 6, 7, 8, 10, 11, and 12	Multiple	Morphological traits	Wei et al. (2020a, b)
F_2/308	SNP	*qPC.12*	12	WUSCHEL-related homeobox-like protein	Prickles development	Qian et al. (2021)
F_2/35	SNP	*Pl*	6	Nudix hydrolase, GATA transcription factor, ARF10B	Prickles lack trait	Miyatake et al. (2020)
F_2	SNP/InDels	*fl3.1*	3	IQD1 protein	Fruit length	Pang et al. (2024)
RIL/164	SNP	P-ACH3.1 etc.	Multiple	–	Nutritional traits	Sulli et al. (2021)
F_2/200	SNP	–	10	WD repeat-containing protein	Pigmentation-related traits	Guan et al. (2022)
IL/16	SNP	wmh3, pfb1, dfb4 etc.	Multiple	*SUN, OVATE, and YABBY*	Fruit shape-related traits	Mangino et al. (2021)
F_2	SNP/SSR	–	–	–	Agronomical key traits	Barchi et al. (2012a, b)
IL/10	–	ph2.HN, ey2.HN etc.	Multiple	Fructose-bisphosphate aldolase 3, PAL, etc.	Nitrogen fertilizer responsiveness	Rosa-Martinez et al. (2023)
RIL/168	SNP	FomE02, FomE11.1	2, 11	Multiple disease resistance protein	Fusarium resistance	Tassone et al. (2022)

(continued)

Table 17.2 (continued)

Population/ numbers	Associated marker (s)	Mapped QTL	Chromosomal location	Linked markers/transcripts	Associated traits	References
F_2	SNP/InDels		8, 10	*ANS* and *MYB1*	Anthocyanin biosynthesis	You et al. (2023)
F_2/2794	SNP	–	8	*APRR2*	Rind color	Fang et al. (2023)
F_2/308	SNP/InDels	*TGW9.1*	9	Putative ubiquitin ligase	Thousand-grain weight	Qian et al. (2022)
F_2	SNP	*EBWR10*	10	*nsLTPs*	Bacterial wilt resistance	Gong et al. (2024)
DH/132	SNP	*Bwr-6*	6	–	Polygenic resistance	Salgon et al. (2018)
RIL/180	AFLP, SSR, etc.		2, 5, 9	*ERs1*	Resistance to the Ralstonia spp.	Salgon et al. (2018)
F_2	SNP	dr1.1, dr2.1, and dr6.1	1, 2, 6	*Smechr0201082*	Seed dormancy	Ai et al. (2024)

A set of 164 F_6 recombinant inbred lines derived from two eggplant lines identified several metabolomic quantitative trait loci (mQTLs) associated with the accumulation of metabolites, with one or more major mQTLs explaining ≥10% of the phenotypic variation (Sulli et al. 2021). A high-density recombination bin-based genetic map of eggplant was developed using 200 F_2 plants from an interspecific cross. This map, along with 172 $F_{2:3}$ lines, identified a major QTL with pleiotropic effects on leaf vein color (LVC) and fruit pericarp color (FPC), explaining 65% and 55% of the phenotypic variation in LVC and FPC, respectively (Guan et al. 2022). Additionally, a subset of 384 SNPs was used to genotype an F_2 intraspecific mapping population, resulting in a genetic linkage map with 415 markers. This map facilitated the identification of QTLs for fruit weight, maximum fruit diameter, fruit shape, and adaxial leaf lamina anthocyanin across two locations (Barchi et al. 2012a, b). Introgression lines were selected to assess the effects of nitrogen fertilization on various traits, leading to the identification of 36 QTLs associated with most traits—12 specific to high nitrogen treatment, 17 specific to low nitrogen treatment, and 7 stable across treatments, conclusively differentiated based on nitrogen regimes (Rosa-Martinez et al. 2023). The recombinant inbred line (RIL) population was analyzed for *Fusarium oxysporum f. sp. melongenae* (Fom) resistance, identifying two major QTLs on chromosomes CH02 and CH11 linked to full and partial resistance traits. Differentially enriched regions were identified between resistant and susceptible progeny in the genomic regions underlying both QTLs (Tassone et al. 2022). Three QTLs controlling anthocyanin biosynthesis in eggplant peel were localized to chromosomes 8 and 10, with genes *ANS* and *MYB1* as potential candidates for regulating anthocyanin biosynthesis (You et al. 2023). Rind color in eggplant is a significant economic trait affecting consumer preferences. A dominant gene governs green peel color, with a candidate gene (*EGP19168.1*) on chromosome 8; a deletion SNP in white-skinned eggplant led to a premature termination codon, emphasizing the genetic basis of rind color variation (Fang et al. 2023). Thousand-grain weight (TGW) is another crucial trait affecting breeding costs and variety promotion. A major QTL, TGW9.1, is strongly associated with TGW, explaining 20.51% of the phenotypic variance. Among the various annotated genes in the TGW9.1 region, one candidate gene, *SMEL_009g329850*, encodes a putative ubiquitin ligase tightly controlling this trait (Qian et al. 2022). In a separate study, F2 populations from paternally inbred eggplant seeds were used to identify regulatory sites for seed dormancy. Three QTLs (*dr1.1, dr2.1,* and *dr6.1*) were found on eggplant chromosomes, confirming the expression of nine anchored candidate genes involved in seed dormancy regulation (Ai et al. 2024).

17.5.2 Biotic and Abiotic Constraints

Mining quantitative trait loci (QTL) for biotic and abiotic stresses in eggplant is essential for developing resilient varieties. This process aids in identifying genetic resistance to pests and diseases, enhancing tolerance to environmental stresses such as drought and salinity, and improving overall yield. *Fusarium* and *Verticillium* are soil pathogens that cause vascular wilts in eggplant, leading to significant yield losses. Major and minor QTLs influencing responses to both pathogens, along with putative resistance orthologous genes, have been primarily located within the Rfo-Sa1 locus (Barchi et al. 2018). An F_2 population was generated for genetic analysis and mapping of QTLs through bulk segregant analysis, revealing two markers (emk03O04 and emf11A03) as potential links to resistance genes. These markers serve as differentiating markers for resistant and susceptible plants, co-segregating with the resistance gene (Narayanswami et al. 2023). A major QTL (PVE $\geq$ 10%) for *Fusarium* resistance was identified at the Rfo-Sa1 locus by assessing the population's resistance to *Fusarium* and *Verticillium*, with markers linked to these QTLs potentially useful for marker-assisted breeding (Toppino et al. 2016a, b). Conversely, *Fusarium* wilt, a soil-borne disease, poses a significant threat to eggplant production. Three eggplant germplasms—LS1934, LS174, and LS2436—exhibit high resistance to the pathogen; however, their resistance loci have yet to be mapped. Quantitative trait locus analyses in $F_{2:3}$ populations identified a resistance locus, FM1, located at the end of chromosome 2, which is proximal to previously reported resistance loci from the eggplant ally Rfo-Sa1, as well as another locus from LS2436 located in the middle of chromosome 4 (Miyatake et al. 2016). Furthermore, a high degree of resistance against *Ralstonia* strains was exhibited by an F_6 population of recombinant inbred lines, developed with 119 markers across 18 linkage groups. The dominant gene, *ERs1*, was responsible for controlling resistance; however, a significant QTL involved in disease progression delay was detected on another linkage group (Lebeau et al. 2013). Notably, the rhizosphere bacterial community in eggplant is influenced by the QTL EBWR10, which controls bacterial wilt resistance by recruiting *Bacillus* species and altering activities of catalase (CAT), polyphenol oxidase (PPO), and phenylalanine ammonia-lyase (PAL), resulting in increased concentrations of NO, H_2O_2, and O_2^- (Gong et al. 2024). Eggplant cultivation is challenged by various diseases, including bacterial wilt caused by the Ralstonia solanacearum species complex (RSSC), which involves both broad-spectrum and strain-specific QTLs (Salgon et al. 2018). Two strains, PSS4 and R3598, bypass the previously reported EBWR9 bacterial wilt resistance QTL. Ten and three QTLs of resistance were detected and mapped, with their influence significantly affected by environmental conditions. The most stable QTLs were identified on chromosomes 3 and 6, which are putatively syntenic with BW-resistance QTLs in tomato (Salgon et al. 2018). By utilizing molecular markers linked to beneficial traits, breeders can establish more precise and efficient breeding programs. This approach not only promotes sustainable cultivation practices but also ensures that eggplant production can meet the growing demands for food security.

17.5.3 Berry Quality and Quantity

Mining quantitative trait loci (QTL) for berry quality and quantity in eggplant is crucial for improving both the marketability and nutritional value of the crop. By identifying the genetic factors that influence traits such as size, color, flavor, and firmness, breeders can develop eggplant varieties that align with consumer preferences and enhance culinary applications. QTL for fruit metabolic content in breeding lines has been characterized to include basic qualitative data, health-related compounds, and peel anthocyanins. A major QTL (PVE $\geq$ 10%) was identified for most traits, with putative orthologs in other Solanaceae crops providing valuable insights into metabolic content (Toppino et al. 2016a, b). Restriction-site-associated DNA sequencing and high-throughput sequencing were employed to identify functional single-nucleotide polymorphism (SNP) markers that differentiate between two parental lines of an eggplant mapping population. A total of 347 de novo SNPs and 84 anchoring markers were assembled into 12 major and one minor linkage group, spanning 1390 cM, which can be used to elucidate the genetic basis of seven traits related to anthocyanin content (Barchi et al. 2012a, b). Quantitative trait loci sequencing (QTL-seq) was utilized to map the fruit length-related locus *fl3.1* on eggplant chromosome 3. This region was further fine-mapped to identify *SmeFL* as the most likely candidate gene responsible for the observed differences in fruit length (Pang et al. 2024). Additionally, understanding the QTL associated with fruit yield facilitates the optimization of production, enabling farmers to achieve higher outputs. Overall, this research supports the development of superior eggplant cultivars that are both attractive to consumers and economically viable for growers.

17.6 Prickle Formation: A Coherent Evil

Mining quantitative trait loci (QTL) for prickle formation in eggplant is beneficial for enhancing both plant health and consumer appeal. On one hand, prickles can deter pests and herbivores, thereby reducing the need for chemical pest control and promoting more sustainable farming practices. Conversely, their presence on leaves, stems, and fruit calyxes poses challenges during cultivation, harvesting, and transportation, making them an undesirable agronomic trait. The genetic mechanisms underlying prickle formation in eggplant have been investigated through a cross between wild and cultivated varieties, revealing a key genetic locus on chromosome 6. The candidate gene *ARF10B* has been predicted to play a role in leaf prickle development; downregulation via RNA interference (RNAi) resulted in reduced leaf prickle size and density, confirming its involvement in prickle development (Miyatake et al. 2020; Ke et al. 2024). Similarly, *ARF18* has been identified as a potential gene within the PRICKLE LOCUS, which is crucial for regulating the presence or absence of prickles in eggplant (Li et al. 2024). A genetic linkage bin map created from a cross between a wild landrace and a cultivated variety unveiled a major QTL controlling prickle development, designated as *qPC.12*, located on chromosome 12. This QTL encodes a WUSCHEL-related protein, identified as the functional gene for prickle formation (Qian et al. 2021). Additionally, the prickle (*Pl*) gene locus on chromosome 6 governs the absence of prickles, with a primer pair amplifying an insertion/deletion variant that can be utilized for marker-assisted selection of the no-prickly phenotype (Miyatake et al. 2020; Li et al. 2024). Understanding the genetic basis of prickle formation allows breeders to selectively cultivate varieties with an optimal level of prickles that provide protection without negatively impacting harvestability or marketability.

17.7 Conclusion and Challenges

Eggplant is vital for food security due to its superior nutritional value, providing essential vitamins and minerals while being low in calories. Promoting the cultivation and consumption of eggplant can enhance food security, support local economies, and encourage sustainable agricultural practices. Genome-wide association studies (GWAS) are invaluable for eggplant breeding, as they identify genetic markers linked to desirable traits such as disease resistance, yield, and fruit quality, while also aiding in the elimination of undesirable traits. Additionally, quantitative trait loci (QTL) mapping is crucial for eggplant breeding, as it identifies specific regions of the genome associated with important traits. The integration of GWAS and QTL mapping allows for more precise identification of genetic variants linked to desirable traits, facilitating improved marker-assisted selection. Overall, this synergy accelerates the eggplant breeding process and enhances the development of resilient, high-yielding eggplant varieties with special attributes.

References

Ai J, Wang W, Hu T, Hu H, Wang J, Yan Y, Pang H, Wang Y, Bao C, Wei Q (2024) Identification of quantitative trait loci and candidate genes controlling seed dormancy in eggplant (Solanum melongena L.). Gene 15(4):415

Arrones A, Mangino G, Alonso D, Plazas M, Prohens J, Portis E, Barchi L, Giuliano G, Vilanova S, Gramazio P (2022) Mutations in the SmAPRR2 transcription factor suppressing chlorophyll pigmentation in the eggplant fruit peel are key drivers of a diversified colour palette. Front Plant Sci 13:1025951

Barchi L, Lanteri S, Portis E, Valè G, Volante A, Pulcini L, Ciriaci T, Acciarri N, Barbierato V, Toppino L, Rotino GL (2012a) A RAD tag derived marker based eggplant linkage map and the location of QTLs determining anthocyanin pigmentation

Barchi L, Rotino GL, Toppino L, Valè G, Acciarri N, Ciriaci T, Portis E, Lanteri S (2012b) SNP mapping and identification of QTL for horticultural key breeding traits in eggplant (Solanum melongena L.). In: International symposium on biotechnology and other omics in vegetable science, vol 1145, pp 9–16

Barchi L, Toppino L, Valentino D, Bassolino L, Portis E, Lanteri S, Rotino GL (2018) QTL analysis reveals new eggplant loci involved in resistance to fungal wilts. Euphytica 214:1–15

Bhanushree N, Saha P, Tomar BS, Lyngdoh YA, Gurung B, Meena RD, Munshi AD, Saha ND, Ghoshal C (2018) Genetic analysis and identification of molecular marker linked to the gene for fruit skin colour in eggplant (Solanum melongena L.). Vegetab Sci 45(02):149–153

Bhanushree N, Saha P, Tomar BS, Gopala Krishnan S, Gurung B, Ghoshal C, Das Munshi A (2020) Single marker analysis and mapping of QTLs governing fruit weight in eggplant (Solanum melongena L.). Proc Natl Acad Sci India Sect B Biol Sci 90:769–775

Cericola F, Portis E, Lanteri S, Toppino L, Barchi L, Acciarri N, Pulcini L, Sala T, Rotino GL (2014) Linkage disequilibrium and genome-wide association analysis for anthocyanin pigmentation and fruit color in eggplant. BMC Genomics 15:1–15

Doganlar S, Frary A, Daunay MC, Lester RN, Tanksley SD (2002a) A comparative genetic linkage map of eggplant (Solanum melongena) and its implications for genome evolution in the Solanaceae. Genetics 161(4):1697–1711

Doganlar S, Frary A, Daunay MC, Lester RN, Tanksley SD (2002b) Conservation of gene function in the Solanaceae as revealed by comparative mapping of domestication traits in eggplant. Genetics 161(4):1713–1726
Fang H, Wang P, Wang W, Peng J, Zheng J, Zhu G, Zhong C, Yu W (2023) Fine mapping and identification of SmAPRR2 regulating rind color in eggplant (Solanum melongena L.). Int J Mol Sci 24(4):3059
Frary A, Doganlar S, Daunay MC, Tanksley SD (2003) QTL analysis of morphological traits in eggplant and implications for conservation of gene function during evolution of solanaceous species. Theor Appl Genet 107:359–370
Frary A, Frary A, Daunay MC, Huvenaars K, Mank R, Doğanlar S (2014) QTL hotspots in eggplant (Solanum melongena) detected with a high resolution map and CIM analysis. Euphytica 197:211–228
Ge HY, Liu Y, Zhang J, Han HQ, Li HZ, Shao WT, Chen HY (2013) Simple sequence repeat-based association analysis of fruit traits in eggplant (Solanum melongena). Genet Mol Res 12(4):5651–5663
Gong C, Wang Z, Li Z, Sun B, Luo W, Luo S, Chen S, Mai P, Li Z, Li Y, Wang Y (2024) A QTL of eggplant shapes the rhizosphere bacterial community, co-responsible for resistance to bacterial wilt. Hortic Res 11(2):uhad272
Gramazio P, Alonso D, Arrones A, Villanueva G, Plazas M, Toppino L, Barchi L, Portis E, Ferrante P, Lanteri S, Rotino GL (2023) Conventional and new genetic resources for an eggplant breeding revolution. J Exp Bot 74(20):6285–6305
Guan W, Ke C, Tang W, Jiang J, Xia J, Xie X, Yang M, Duan C, Wu W, Zheng Y (2022) Construction of a high-density recombination bin-based genetic map facilitates high-resolution mapping of a major QTL underlying anthocyanin pigmentation in eggplant. Int J Mol Sci 23(18):10258
Ke C, Guan W, Jiang J, Huang L, Li H, Li W, Lin Y, Lin L, Xie X, Wu W, Gao W (2024) Map-based cloning of LPD, a major gene positively regulates leaf prickle development in eggplant. Theor Appl Genet 137(10):216
Lebeau A, Gouy M, Daunay MC, Wicker E, Chiroleu F, Prior P, Frary A, Dintinger J (2013) Genetic mapping of a major dominant gene for resistance to Ralstonia solanacearum in eggplant. Theor Appl Genet 126:143–158
Li D, Qian J, Li W, Jiang Y, Gan G, Li W, Chen R, Yu N, Li Y, Wu Y, Kang D (2019) Genome sequence and analysis of the eggplant (Solanum melongena L.). BioRxiv:824540
Li S, He Y, Li D, Shi S, Wang Y, Tang X, Ge H, Liu Y, Chen H (2024) Fine mapping an AUXIN RESPONSE FACTOR, SmARF18, as a candidate gene of the PRICKLE LOCUS that controls prickle absence/presence on various organs in eggplant (Solanum melongena L.). Sci Hortic 327:112874
Liu W, Qian Z, Zhang J, Yang J, Wu M, Barchi L, Zhao H, Sun H, Cui Y, Wen C (2019) Impact of fruit shape selection on genetic structure and diversity uncovered from genome-wide perfect SNPs genotyping in eggplant. Mol Breed 39:1–13
Mangino G, Plazas M, Vilanova S, Prohens J, Gramazio P (2020) Performance of a set of eggplant (Solanum melongena) lines with introgressions from its wild relative S. incanum under open field and screenhouse conditions and detection of QTLs. Agronomy 10(4):467
Mangino G, Vilanova S, Plazas M, Prohens J, Gramazio P (2021) Fruit shape morphometric analysis and QTL detection in a set of eggplant introgression lines. Sci Hortic 282:110006
Mangino G, Arrones A, Plazas M, Pook T, Prohens J, Gramazio P, Vilanova S (2022) Newly developed MAGIC population allows identification of strong associations and candidate genes for anthocyanin pigmentation in eggplant. Front Plant Sci 13:847789
Miyatake K, Saito T, Negoro S, Yamaguchi H, Nunome T, Ohyama A, Fukuoka H (2012) Development of selective markers linked to a major QTL for parthenocarpy in eggplant (Solanum melongena L.). Theor Appl Genet 124(8):1403–1413
Miyatake K, Saito T, Negoro S, Yamaguchi H, Nunome T, Ohyama A, Fukuoka H (2016) Detailed mapping of a resistance locus against Fusarium wilt in cultivated eggplant (Solanum melongena). Theor Appl Genet 129:357–367
Miyatake K, Saito T, Nunome T, Yamaguchi H, Negoro S, Ohyama A, Wu J, Katayose Y, Fukuoka H (2020) Fine mapping of a major locus representing the lack of prickles in eggplant

revealed the availability of a 0.5-kb insertion/deletion for marker-assisted selection. Breed Sci 70(4):438–448

Narayanswami B, Tomar BS, Akhtar J, Behera TK, Ellur RK, Sevanthi RMV, Sarika J, Saha P (2023) Genetic analysis and identification of SSR marker linked to Phomopsis blight resistance in eggplant (Solanum melongena L.). Plant Breed 142(5):700–710

Pang H, Ai J, Wang W, Hu T, Hu H, Wang J, Yan Y, Wu X, Bao C, Wei Q (2024) Fine mapping of QTL-fl3. 1 reveal SmeFL as the candidate gene regulating fruit length in eggplant (Solanum melongena L.). Vegetab Res 4(1)

Portis E, Barchi L, Toppino L, Lanteri S, Acciarri N, Felicioni N, Fusari F, Barbierato V, Cericola F, Valè G, Rotino GL (2014) QTL mapping in eggplant reveals clusters of yield-related loci and orthology with the tomato genome. PLoS One 9(2):e89499

Portis E, Cericola F, Barchi L, Toppino L, Acciarri N, Pulcini L, Sala T, Lanteri S, Rotino GL (2015) Association mapping for fruit, plant and leaf morphology traits in eggplant. PLoS One 10(8):e0135200

Qian Z, Zhang B, Chen H, Lu L, Duan M, Zhou J, Cui Y, Li D (2021) Identification of quantitative trait loci controlling the development of prickles in eggplant by genome re-sequencing analysis. Front Plant Sci 12:731079

Qian Z, Ji Y, Li R, Lanteri S, Chen H, Li L, Jia Z, Cui Y (2022) Identifying quantitative trait loci for thousand grain weight in eggplant by genome re-sequencing analysis. Front Genet 13:841198

Rinaldi R, Van Deynze A, Portis E, Rotino GL, Toppino L, Hill T, Ashrafi H, Barchi L, Lanteri S (2016) New insights on eggplant/tomato/pepper synteny and identification of eggplant and pepper orthologous QTL. Front Plant Sci 7:1031

Ro N, Haile M, Kim B, Cho GT, Lee J, Lee YJ, Hyun DY (2022) Genome-wide association study for agro-morphological traits in eggplant core collection. Plants 11(19):2627

Ro N, Haile M, Ko HC, Cho GT, Lee J, Kim B, Lee S, Kim SH (2023) Genome-wide association study of phenolic content and antioxidant properties in eggplant germplasm. Gene 14(7):1315

Rosa-Martínez E, Villanueva G, Şahin A, Gramazio P, García-Martínez MD, Raigón MD, Vilanova S, Prohens J, Plazas M (2023) Characterization and QTL identification in eggplant introgression lines under two N fertilization levels. Hortic Plant J 9(5):971–985

Salgon S, Raynal M, Lebon S, Baptiste JM, Daunay MC, Dintinger J, Jourda C (2018) Genotyping by sequencing highlights a polygenic resistance to Ralstonia pseudosolanacearum in eggplant (Solanum melongena L.). Int J Mol Sci 19(2):357

Sulli M, Barchi L, Toppino L, Diretto G, Sala T, Lanteri S, Rotino GL, Giuliano G (2021) An eggplant recombinant inbred population allows the discovery of metabolic QTLs controlling fruit nutritional quality. Front Plant Sci 12:638195

Tassone MR, Bagnaresi P, Desiderio F, Bassolino L, Barchi L, Florio FE, Sunseri F, Sirangelo TM, Rotino GL, Toppino L (2022) A genomic BSAseq approach for the characterization of QTLs underlying resistance to Fusarium oxysporum in eggplant. Cells 11(16):2548

Tiwari JK, Karkute SG, Tiwari SK, Singh M (2023) Brinjal: breeding and genomics. Vegetab Sci:166–176

Toppino L, Barchi L, Lo Scalzo R, Palazzolo E, Francese G, Fibiani M, D'Alessandro A, Papa V, Laudicina VA, Sabatino L, Pulcini L (2016a) Mapping quantitative trait loci affecting biochemical and morphological fruit properties in eggplant (Solanum melongena L.). Front Plant Sci 7:256

Toppino L, Barchi L, Valentino D, Portis E, Lanteri S, Rotino GL (2016b) QTLs mapping for Fusarium oxysporum and Verticillium dahliae resistance in eggplant (Solanum melongena L.)

Wei Q, Wang W, Hu T, Hu H, Wang J, Bao C (2020a) Construction of a SNP-based genetic map using SLAF-Seq and QTL analysis of morphological traits in eggplant. Front Genet 11:178

Wei Q, Wang J, Wang W, Hu T, Hu H, Bao C (2020b) A high-quality chromosome-level genome assembly reveals genetics for important traits in eggplant. Hortic Res 7

Xiao XO, Lin W, Nie H, Duan Z, Liu K (2024) QTL mapping for bacterial wilt resistance in eggplant via bulked segregant analysis using genotyping by sequencing. Agronomy 14(6):1159

You Q, Li H, Wu J, Li T, Wang Y, Sun G, Li Z, Sun B (2023) Mapping and validation of the epistatic D and P genes controlling anthocyanin biosynthesis in the peel of eggplant (Solanum melongena L.) fruit. Hortic Res 10(2):uha

GWAS and QTL Mapping in Broccoli

18

Shrawan Singh, Raman Selvakumar, Monica Sharma, and Anamika Chandel

Abstract

Broccoli (*Brassica oleracea*) is a highly nutritious cruciferous vegetable, renowned for its abundant content of essential vitamins, minerals, and bioactive compounds that promote optimal health and nutrition security. Strikingly, it is an excellent source of vitamin C, which plays a critical role in immune function and skin integrity, as well as vitamin K, which is indispensable for bone metabolism and blood coagulation. Moreover, broccoli is a rich source of folate, potassium, and dietary fiber, all of which are integral to cellular processes, cardiovascular health, and gastrointestinal function. The breeding of broccoli, however, is complicated by its complex genetic makeup, biennial growth cycle, and sensitivity to environmental variations. Genomic tools such as genome-wide association studies (GWAS) and quantitative trait locus (QTL) mapping are pivotal for enhancing agronomic traits in broccoli. These methodologies facilitate the identification of genetic loci associated with key agronomical traits, including disease resistance, yield, quality, and abiotic and biotic stress tolerance. Both GWAS and QTL mapping contribute to a more comprehensive understanding of the genetic architecture underlying complex traits, thereby enabling the development of targeted and efficient breeding strategies. Ultimately, the application of these technologies is critical for developing broccoli varieties with enhanced performance across diverse environmental conditions, thereby improving productivity, qual-

S. Singh (✉) · R. Selvakumar · A. Chandel
Division of Vegetable Science, ICAR-Indian Agricultural Research Institute, New Delhi, India
e-mail: shrawansingh@iari.res.in

M. Sharma
Department of Plant Pathology, College of Horticulture & Forestry, Dr YS Parmar University of Horticulture & Forestry (Solan), Hamirpur, Himachal Pradesh, India

S. S. Dey et al. (eds.), *GWAS and QTL Mapping in Horticultural Crops*,
https://doi.org/10.1007/978-981-95-2306-1_18

ity, and resilience—factors that are essential in addressing the challenges posed by global climate change and increasing food demands.

Keywords

Broccoli · Genotyping · Molecular markers · Genome-wide association studies · Quantitative trait loci · Glucosinolates

18.1 Introduction

Broccoli (*Brassica oleracea* L. var. *italica*), a member of the Brassicaceae family, is a widely cultivated and nutritionally significant vegetable, valued for its hypertrophic reproductive structures, including the floral head and stalk. These edible parts are abundant in essential nutrients such as vitamins A and K, calcium, and magnesium. Additionally, they contain bioactive compounds, including glucosinolates, sulforaphane, selenium, and flavonoids, which are linked to anticancer effects and other health benefits (Kumar and Srivastava 2016; Luo et al. 2020; Vallejo et al. 2003; Shapiro et al. 2001; Massie et al. 1996). The *italica* group of broccoli originates from the domestication and cultivation of *Brassica oleracea* (CC genome; $2n = 18$) in the Mediterranean region. Historical records trace the cultivation of *B. oleracea* mustard plants back to Hellenic culture in the sixth century BCE (Buck 1956). The term "broccoli" is derived from a colloquial Latin term that refers to the projecting shoots of cabbage-family plants (Buck 1956). Over the past two millennia, selective breeding from *B. oleracea* has resulted in the development of various broccoli landraces, primarily in Italy (Buck 1956; Ciancaleoni et al. 2013; Maggioni et al. 2017; Branca et al. 2017; Stansell and Björkman 2020).

The variety *Vrocculi o Sparaceddi* is considered one of the earliest domesticated forms of wild brassica, serving as the progenitor of modern broccoli (Stansell and Björkman 2020). Significant advances in broccoli cultivation, particularly in the development of heading varieties, were made over the past 300 years by Danish and English horticulturists (Buck 1956). Initially, broccoli was consumed and cultivated mainly in Italy, primarily as sprouting broccoli (Stansell and Björkman 2020). The development of calabrese broccoli, characterized by large, compact heads resembling cauliflower, contributed to its global popularity (Stansell and Björkman 2020; Li et al. 2019). Broccoli was introduced to the United Kingdom in the 1700s and to the United States in the 1800s, where its consumption surged post-World War II (Li et al. 2019). In the 1980s, it was introduced to China, where it has since become widely cultivated and consumed. China now leads the world in broccoli production, with a cultivation area exceeding 80,000 hectares (Branham and Farnham 2019). As scientific evidence continues to highlight its health benefits, the demand for broccoli is steadily increasing, especially in major production regions like China, the United States, and India (Branham and Farnham 2019; Kumar and Kandpal 2017; Vallejo et al. 2004).

The genetic enhancement of broccoli has been significantly advanced by the application of modern molecular biology tools, particularly genome-wide association studies (GWAS) and quantitative trait locus (QTL) mapping. These high-throughput genomic techniques allow for a deeper understanding of the genetic architecture of important agronomic traits in broccoli, enabling the identification of key genetic markers associated with traits such as yield, nutritional content, disease resistance, and tolerance to abiotic stress (Li et al. 2019; Branham and Farnham 2019).

GWAS, by analyzing variations across the entire genome, help pinpoint specific loci linked to traits of interest, such as glucosinolate content, a bioactive compound known for its anticancer properties (Luo et al. 2020). For example, studies have identified loci associated with higher levels of these beneficial metabolites, allowing for the development of varieties with enhanced health benefits. Similarly, GWAS have been used to identify genetic variations that contribute to disease resistance, such as resistance to Fusarium wilt and black rot, two common pathogens that impact broccoli production (Luo et al. 2020). Quantitative trait locus (QTL) mapping, on the other hand, provides more detailed insights into the genetic determinants of complex traits. By analyzing segregating populations, researchers can identify specific genomic regions that control traits such as head size, texture, and stress tolerance. These traits are crucial for broccoli's commercial success, particularly as growers face challenges from climate change and increased pathogen pressures. A notable example includes the identification of QTLs related to drought resistance, which has become increasingly relevant as water scarcity threatens agricultural productivity (Wu et al. 2024).

Integrating GWAS and QTL mapping provides a powerful approach to pinpoint genetic regions that influence multiple traits simultaneously, thus accelerating breeding efforts. For instance, a study combining both methods revealed loci associated with both yield and stress resistance, highlighting the possibility of simultaneously improving multiple agronomic traits (Stansell and Björkman 2020). Moreover, advances in high-throughput sequencing and bioinformatics tools have enhanced the precision of these approaches, enabling breeders to identify potential genetic markers with greater accuracy and reliability.

The integration of these molecular tools into broccoli breeding programs offers several advantages. It allows for marker-assisted selection (MAS), where breeders can select individuals with desirable traits based on their genetic profiles, thus accelerating the breeding process. Furthermore, with the help of molecular breeding, it is possible to improve traits that are difficult to select for through traditional breeding methods, such as disease resistance and abiotic stress tolerance. As a result, these molecular approaches contribute to the development of broccoli varieties that are not only more resilient but also nutritionally enriched, better meeting the evolving demands of consumers and the challenges posed by environmental factors (Branham and Farnham 2019; Li et al. 2019). Through the application of GWAS, QTL mapping, and other genomic tools, the genetic improvement of broccoli is entering a new era, one in which precise genetic information can be harnessed to meet global food security and nutritional challenges. These molecular approaches offer the potential for improving broccoli not just for its yield but also for its health-promoting properties, making it a vital crop in future agricultural systems (Kumar and Srivastava 2016; Vallejo et al. 2004).

18.2 GWAS in Broccoli: A Tool for Trait Identification

Genome-wide association studies (GWAS) have revolutionized plant breeding by enabling the identification of genetic variants associated with specific phenotypic traits. In broccoli, GWAS has been particularly instrumental in identifying genetic loci linked to key agronomic traits such as head size, disease resistance, and tolerance to environmental stressors. One of the major areas of focus in broccoli GWAS is the identification of loci associated with bioactive compounds, which are central to broccoli's health benefits (Luo et al. 2020).

Kumar and Srivastava (2016) applied GWAS to identify loci linked to glucosinolate content in broccoli, a group of sulfur-containing compounds that have been shown to possess anticancer properties. Similarly, Luo et al. (2020) integrated GWAS with transcriptomic data to uncover significant loci associated with the accumulation of bioactive compounds like flavonoids and glucosinolates. This integration of genomics and transcriptomics enabled the identification of candidate genes involved in the biosynthesis of these important metabolites. Such findings provide breeders with molecular markers that can be used to select for broccoli varieties with enhanced nutritional and medicinal value.

In a study by Branham and Farnham (2019), GWAS was used to identify loci related to heat tolerance, a critical trait for improving broccoli productivity in warmer climates. These findings are especially important as global warming continues to pose a threat to crop yields. By identifying loci that confer heat tolerance, breeders can develop heat-resistant broccoli varieties, ensuring continued production in regions experiencing rising temperatures. Moreover, GWAS has facilitated the identification of loci linked to disease resistance, a major concern in broccoli cultivation. Fusarium wilt, black rot, and downy mildew are just a few examples of diseases that can severely impact broccoli yield. Studies utilizing GWAS have identified genetic loci associated with resistance to these diseases, providing valuable targets for the development of disease-resistant cultivars (Luo et al. 2020).

18.2.1 QTL Mapping in Broccoli: Locating Genes for Key Traits

Quantitative trait locus (QTL) mapping has been a complementary approach to GWAS in identifying genetic loci associated with complex traits in broccoli. While GWAS typically focuses on identifying associations between genetic variants and traits in natural populations, QTL mapping is often used to study segregating populations derived from crosses between distinct genotypes. QTL mapping has provided valuable insights into the genetic basis of traits such as head development, disease resistance, and stress tolerance. Gardner et al. (2016) used QTL mapping to identify regions of the genome associated with total phenolic content and antioxidant capacity in broccoli. These traits are important for the health benefits of broccoli, as phenolic compounds and antioxidants play a key role in preventing oxidative stress and reducing the risk of chronic diseases. The identification of QTLs for these traits is crucial for breeding broccoli varieties with enhanced nutritional qualities.

In addition to phenolic compounds, aliphatic glucosinolates are another group of important bioactive compounds in broccoli. QTL mapping has been successfully employed to identify loci associated with the accumulation of these compounds, which have been shown to possess anticancer and anti-inflammatory properties (Brown et al. 2015). For example, Brown et al. (2015) identified a QTL on chromosome 5 that governs the accumulation of 2-propenyl and 3-methylsulfinylpropyl glucosinolates. These findings provide breeders with valuable genetic markers for selecting broccoli varieties with higher glucosinolate content, thereby enhancing the crop's health-promoting properties.

QTL mapping has also been used to identify loci associated with abiotic stress tolerance in broccoli. Drought and salinity are two major stress factors that can significantly reduce broccoli yield. In a study by Wu et al. (2024), QTL mapping was used to identify loci associated with drought tolerance in broccoli, highlighting the importance of this trait for regions facing water scarcity. Similarly, QTLs for salt tolerance have been mapped, providing targets for breeding salt-tolerant broccoli varieties that can thrive in saline soils.

18.2.2 Integrated Approaches: Combining GWAS and QTL Mapping

The integration of GWAS and QTL mapping has proven to be a powerful strategy for identifying genetic loci associated with complex traits in broccoli. By combining these two approaches, researchers can identify genetic variants that are both statistically significant and functionally relevant, accelerating the development of improved broccoli cultivars. One example of such an integrated approach is the work of Stansell and Björkman (2020), who used both GWAS and QTL mapping to identify loci associated with head development and disease resistance in broccoli. This integration allowed for the identification of multiple loci for key traits, providing breeders with a more comprehensive understanding of the genetic basis of these important characteristics. Moreover, the combination of these methods has enabled the identification of epistatic interactions between loci, further enhancing our understanding of how multiple genes interact to control complex traits. In a more recent study, Li et al. (2019) combined QTL mapping and GWAS to identify loci related to drought tolerance and other abiotic stress responses in broccoli. This integrated analysis revealed multiple candidate genes involved in regulating stress tolerance pathways, which could be targeted in breeding programs focused on improving the resilience of broccoli to environmental stressors.

18.3 GWAS and QTL Mapping in Broccoli Improvement

18.3.1 Plant Architecture

Plant architecture is a complex trait associated with the stem and leaf morphologies, which include plant height, leaf size, leaf shape, leaf angle, petiole length, and lateral shoot growth. This significantly influences planting density, yield, and the overall quality of broccoli (Hale et al. 2007). Numerous quantitative trait loci (QTLs) associated with stem and leaf characteristics pertinent to plant architecture have been reported (Lin et al. 2013, 2019; Branham and Farnham 2017, Lan and Paterson, 2000; Huang et al. 2021a, b). The QTLs linked to leaf lamina width were reported on linkage groups C01 and C07 (Lan and Paterson 2000), stem width on LG5 (Lin et al. 2013), leaf apex on C06 and C07, leaf shape on C03, leaf length on C07-09, wing petiole length on C07, and various traits such as lobe number, wing number, leaf shape, and lamina petiole length on linkage group C03 (Walley et al. 2012).

Stansell et al. (2019) mapped several QTLs for leaf morphology and lateral shoot development such as four QTLs for leaf apex and two QTLs for leaf margin. They identified leaf-associated hotspot genomic regions namely *Lea3* on C03 0.7–1.7 Mb and *Lea7* on C07 37.0–39.5 Mb. Bo7g093130, a homolog of the *GRF1-INTERACTING FACTOR 1* (*GIF1*) gene in a major QTL *LA_C07@36.6*, may contribute to the narrow leaf shape, while an ortholog of a *LATE MERISTEM IDENTITY1* (*BoLMI1, Bo3g002560*) near the major *LM_C03@0.7* accounts for over 40% of leaf margin phenotype (Stansell et al. 2019). They also identified three QTLs, i.e., *LT_C03@5.9*, *LT_C04@15.0*, and *LT_C09@9.0*, on C03, C04, and C09 for lateral shoot growth. Huang et al. (2021a, b) reported major QTLs for plant height, outer petiole length, and leaf width as *phc1* on C01, *plc6–2* on C06, and *lwc3–1* on CO3, respectively. Treccarichi et al. (2021) examined SSR loci including *BoABI1, BoAP1, BoPLD1, BoTHL1*, and *PBCGSSRBo39* and revealed significant correlations between the heterozygosity index (H) of the *BoPLD1* gene and inflorescence characteristics of broccoli and cauliflower. Specifically, a higher level of homozygosity at the *BoPLD1* locus negatively impacted curd traits and overall yield in both broccoli and cauliflower. Thus, suggesting the application of GWAS and QTL mapping in broccoli, targeting genomic regions that govern essential traits, thereby enhancing breeding efficiency.

18.3.2 Stem Development

Hollow stem in broccoli is an undesirable physiological disorder characterized by the formation of a hollow cavity in the center of the fleshy stem and fractures within the internal stem tissues. As the soft, fleshy stem constitutes a part of the edible broccoli head, the presence of a hollow stem reduces marketable yield by decreasing the proportion of edible stalks and increasing susceptibility to secondary pathogen infections and decay in both stems and florets (Boersma et al. 2013; Bohuon et al. 1996).

Environmental factors such as excessive nitrogen application and warm temperatures significantly contribute to the incidence of hollow stem, though genetic factors also play a pivotal role in its manifestation. Yu et al. (2019a, b, c, d) identified nine QTLs associated with hollow stem formation on chromosomes C02, C03, C05, C06, and C09. Among these, QHS.C09-2 was identified as a major QTL, accounting for 14.1% of the phenotypic variance, highlighting its potential importance in breeding efforts to reduce the occurrence of hollow stem (Yu et al. 2019a, b, c, d).

18.3.3 Heading

Broccoli develops edible reproductive structures characterized by the proliferation and arrested development of floral buds (Björkman and Pearson 1998). Breeding programs place a primary focus on optimizing floral head quality, which is considered the most critical agronomic trait. While some genes and loci implicated in head formation have been identified through forward and reverse genetic approaches, the complete genetic mechanisms underlying this trait remain largely elusive (Duclos and Bjoerkman 2008). Efforts to investigate the genetic basis of head formation in broccoli have included the identification of homologs of *Arabidopsis thaliana* floral meristem identity genes such as *LEAFY* (LFY), *APETALA1* (AP1), and *CAULIFLOWER* (CAL). These studies suggest that *BoCAL* and *BoAP1* play roles in curd development in cauliflower—a closely related subspecies to broccoli, differing primarily in the developmental stage of the reproductive meristem at harvest (Fujime and Okuda 1996; Carr and Irish 1997; Smith and King 2000). However, subsequent research highlights the complexity of head formation in both broccoli and cauliflower, suggesting that this process is not solely regulated by floral meristem identity genes (Fujime and Okuda 1996). Beginning in the 1990s, researchers initiated the construction of genetic maps by crossing broccoli cultivars and inbred lines with diverse genetic materials, including broccoli landraces, cabbage, cauliflower, kale, and Chinese kale. These efforts aimed to identify loci associated with essential traits such as disease resistance, head morphology, nutritional quality, and flowering or maturation time (Camargo et al. 1995; Bohuon et al. 1996; Kianian and Quiros 1992; Slocum et al. 1990; Brown et al. 2007).

Several quantitative trait loci (QTLs) linked to head traits, including head weight, height, width, and floret dimensions, have been identified. However, early genetic maps were often inconsistent due to variations in germplasm, marker systems, linkage group nomenclature, and the lack of a *Brassica oleracea* reference genome prior to 2014 (Lin et al. 2013; Brown et al. 2007; Walley et al. 2012). For instance, Stansell et al. identified key QTLs influencing head quality, such as BU_C04@51.5, BR_C09@49.5, HC_C09@48.8, HU_C09@48.8, HE_C09@47.7, and OQ_C09@49.5, using a double-haploid (DH) mapping population derived from 'Early Big' (a broccoli DH line) and TO1000DH3 (a non-heading Chinese kale). These loci were localized to genomic regions of approximately 49 Mb on chromosome C09, which include *FLOWERING LOCUS C* (*FLC*) homologs (*Bo9g173400* and *Bo9g173370*). These *FLC*-containing regions were identified as significant hotspots, collectively accounting for over 40% of the observed variation in heading traits (Lin et al. 2018; Stansell et al. 2019). Further studies have identified QTLs associated with subtropical adaptation, such as qCQ-2, qCQ-3, and qCQ-6. Specific haplotypes of candidate genes, including *BoFLC3* within the qCQ-3 interval and *PERIANTHIA* (PAN), a bZIP transcription factor required for *AGAMOUS* activation within qCQ-6, have been proposed to confer adaptation to high temperatures and short day lengths (Lin et al. 2018). In addition to QTLs linked to head traits, mapping efforts have also focused on bud morphology, providing critical insights into the genetic basis of head-related characteristics (Lin et al. 2018; Stansell et al. 2019). These findings offer valuable genetic resources and breeding materials to improve broccoli cultivars, particularly in the context of environmental adaptability and agronomic performance. The integration of genetic mapping, QTL identification, and candidate gene analysis is advancing our understanding of the genetic factors controlling broccoli head quality and adaptation to diverse environments. Future breeding programs can leverage these insights to develop varieties with improved floral head traits and enhanced resilience to climatic challenges.

18.3.4 Head Color

Broccoli is a rich source of chlorophyll (green) and anthocyanin (purple or violet), which contribute to its health value through their antioxidant properties. These pigments are mainly present in sepals, which constitute a major part of unopened flower buds of broccoli. The anthocyanin accumulation in purple colored Cole crops was due to transcription factors such as R2R3 MYB and basic helix-loop-helix (bHLH) as well as WD40 proteins which represent the three major families of anthocyanin regulatory proteins (Yuan et al. 2009; Chiu et al. 2010). Liu et al. (2022) reported two QTLs as tightly correlated with anthocyanin biosynthesis ($p < 0.05$) on C07 (*BoPur7.1*) and C09 (*BoPur9.1*). The *BoPur9.1* region contained 14 genes including *Bo9g174880*, *Bo9g174890,* and *Bo9g174900* which showed high homology with *AT5G07990* (gene encoding flavonoid 3′ hydroxylase). This was identified as a candidate gene for *BoPur9.1*. The purple mutant was designated with the symbol *Pur* for the Purple allele. They also reported multiple biomarkers represented potential targets of marker-assisted selection for use in anthocyanin-rich varieties in broccoli. The purple sepal trait was mapped using doubled haploid (DH) population and SLAF sequencing and identified three QTLs with a major locus, *qPH.C01–2*, located on LG1, and two loci, *qPH.C01–4* and *qPH.C01–5*, located near *qPH.C01–2* (Reshma et al. 2018; Yu et al. 2019a, b, c, d). Liu et al. (2020a, b) reported that the key structural genes associated with anthocyanin biosynthesis, such as *PAL, C4H, 4CL, CHS, CHI, F3H,* and *DFR,* were downregulated in shaded conditions in purple broccoli. *BoMYB6*, *BoMYBL2*, and three *BobHLH* genes (*BoTT8_5-1, BoTT8_5-2*, and *BoEGL5-3*) are key in regulating anthocyanin levels under light conditions. *CRY3* and *HY5* were triggering anthocyanin biosynthesis through the regulation of MYB transcription factors.

For anthocyanin biosynthesis in broccoli, two quantitative trait loci (QTLs), *BoPur7.1* and *BoPur9.1*, were identified on chromosomes 7 and 9, respectively. This identification was achieved through QTL-seq bulk segregant analysis of an F_2 population derived from a cross between BT 126, a variety rich in anthocyanins, and SN 60, which lacks this compound (Yu et al. 2023). Subsequent high-resolution mapping of *BoPur9.1* narrowed the QTL region to 73 kb, encompassing 14 candidate genes, of which *Bo9g174880, Bo9g174890,* and *Bo9g174900*, homologous to flavonoid 3-hydroxylase (*AT5G07990)*, were identified as significant candidates involved in anthocyanin biosynthesis (Liu et al. 2022). Gene expression analysis revealed that the expression level of *BoF3'H* was significantly elevated in BT 126 compared to SN 60, suggesting its crucial role in anthocyanin accumulation (Liu et al. 2022). Moreover, the anthocyanin content in BT 126 was found to be influenced by environmental factors such as temperature, indicating the complex genetic and environmental interactions affecting anthocyanin production (Liu et al. 2022).

18.3.5 Flowering

Treccarichi and coworkers identified three specific allele variants, namely P1_155, P2_165, and P4_288 from SSR markers *BoAP1, BoTHL1*, and *BoPLD1*, respectively, which are strongly associated with increased inflorescence size. These markers allow for faster, cost-effective selection of genotypes with hypertrophic inflorescences, aiding in the development of new cultivars with improved biotic and abiotic stress resistance, organoleptic qualities, and nutritional value (Treccarichi et al. 2023).

Flowering is a critical agronomic trait in broccoli, influencing crop maturity, head quality, hybrid seed production, and adaptation to specific geographic regions. The timing of flowering is regarded as a polygenic trait controlled by multiple quantitative trait loci (QTLs). To identify the QTLs and genes underlying this trait, segregating populations were developed through crosses between broccoli and various germplasms, including broccoli × cabbage, broccoli × Chinese kale, and intra-broccoli crosses. Studies employing such populations have consistently identified multiple QTLs associated with flowering, particularly in crosses like broccoli × cabbage and broccoli × Chinese kale (Lin et al. 2018; Branham and Farnham 2017; Kennard et al. 1994; Camargo and Osborn 1996; Bohuon et al. 1998; Rae et al. 1999).

While numerous QTLs contribute to flowering time, most findings emphasize the influence of one or a few major QTLs (Lin et al. 2018; Branham and Farnham 2017; Kennard et al. 1994; Camargo and Osborn 1996; Bohuon et al. 1998; Rae et al. 1999).

Research from the late 1990s identified several QTLs governing flowering time using populations derived from broccoli (non-vernalization type) × cabbage (vernalization type) and broccoli (late-flowering type) × Chinese kale (early-flowering type) (Kennard et al. 1994; Camargo and Osborn 1996; Bohuon et al. 1998; Rae et al. 1999). However, results have sometimes varied even with similar populations, as seen in two studies involving broccoli × cabbage crosses. These discrepancies are likely attributable to differences in germplasm sources and environmental conditions (Okazaki et al. 2006; Shu et al. 2018).

For instance, Okazaki et al. identified six QTLs influencing flowering time in Japan (February–July 2001, Niigata), with a major QTL located in the *BRMS215-F2-R4b* interval, explaining 36.8% of phenotypic variance. The candidate gene *BoFLC2* within this region was suggested to regulate flowering (Okazaki et al. 2006). Similarly, Shu et al. employed QTL-seq and traditional linkage mapping in a broccoli × cabbage population cultivated in Beijing (spring 2013–winter 2014). They identified a significant QTL, *Ef2.1*, on chromosome C02 (2.65–2.68 Mb), which explained 51.5% of the variance for early flowering. The *GROWTH-REGULATING FACTOR 6* homolog (*BolGRF6*) was proposed as a likely candidate gene (Shu et al. 2018).

More recently, Stansell et al. (2019) pinpointed QTLs associated with days to maturity (DM_C03@6.4, DM_C09@50.0) and days to flowering (DF_C03@6.4, DF_C09@50.0). Among these, the major QTLs DM_C09@50.0 and DF_C09@50.0, located near 50 Mb on chromosome C09, collectively accounted for approximately 50% of the observed phenotypic variation (Stansell et al. 2019). Broccoli's sensitivity to high temperatures, which inhibit vernalization and disrupt floral meristem development, poses a significant challenge for its cultivation in warm climates. Lin et al. (2018) explored non-vernalization-responsive QTLs associated with subtropical adaptation using tropical broccoli accessions from Taiwan. They identified a key QTL, *qDCI-3*, which is linked to tolerance of high ambient temperatures and short photoperiods. The putative gene *BoFLC3*, located within this QTL, was proposed as an alternative mechanism for regulating flowering in tropical and temperate conditions (Branham and Farnham, 2017).

18.3.6 Male Sterility Systems in Broccoli

Broccoli demonstrates pronounced heterosis, with most commercial varieties being F_1 hybrids. Historically, the production of F_1 hybrids relied on self-incompatibility systems; however, male sterility-based breeding systems now dominate this process (Stansell and Bjorkman 2020; Li et al. 2019; Reshma et al. 2018). Male sterility mechanisms in broccoli are categorized into cytoplasmic male sterility (CMS) and genic male sterility (GMS) (Han et al. 2019). Among these, Ogura CMS has become the most widely studied and applied system due to its advantages, including complete male gamete abortion, maternal inheritance, and ease of transferability (Han et al. 2018, 2019).

Ogura CMS originated as a natural mutation in radish (*Raphanus sativus*) populations (Ogura 1968). The sterility is caused by the mitochondrial *orf138* gene, which can be fully restored by the nuclear gene *RFO (PPR-B)* (Uyttewaal et al. 2009; Yamagishi et al. 2021). Efforts to introduce this sterility system into *Brassica oleracea* through distant hybridization and protoplast fusion initially faced challenges. Early CMS variants (CMSR1 and CMSR2) retained excessive radish cytoplasmic content, resulting in undesirable traits such as yellowing at low temperatures, deformed floral morphology, and poor seed set (Bannerot et al. 1974; Sigareva and Earle 1997; Shu et al. 2016a, b).

18.3.6.1 Advancements in Ogura CMS Utilization

In the late 1990s, the American company Asgrow used asymmetric protoplast fusion to reduce radish mitochondrial content, creating CMSR3. This variant resolved the issues associated with earlier versions and became widely adopted for seed production in *B. oleracea* crops (Shu et al. 2016a, b; Wang et al. 2011). CMSR3 has since been incorporated into many elite parent lines and dominates hybrid seed production (Shu et al. 2016a, b; Wang et al. 2011). Specific *orf138* PCR markers have been developed to assist in marker-assisted selection (MAS) for the CMS system (Yao et al. 2009).

Shu et al. (2016a, b) advanced CMS research by developing mitochondrial markers capable of differentiating CMS types. Using six *orf138*-related and two simple sequence repeat (SSR) markers, they categorized 39 CMS broccoli accessions into five groups. Notably, CMSR3 accounted for 79.49% of CMS accessions in China (Shu et al. 2016a, b).

18.3.6.2 Genic Male Sterility (GMS) as an Alternative

The development of genic male sterility (GMS) systems provides a promising alternative to CMS. A dominant GMS (DGMS) resource, 79–399–3, identified in cabbage (*B. oleracea*), has been extensively utilized for hybrid seed production (Han et al. 2018; Sampson 1966; Ruffio-Chable et al. 1993; Fang et al. 1997; Ji et al. 2017). DGMS-based systems offer higher seed quality and yield compared to CMS-based systems (Han et al. 2019). However, their application in broccoli remains limited due to the reliance on tissue culture to maintain and propagate homozygous DGMS lines, which is inefficient for large-scale hybrid seed production (Han et al. 2019; Shu et al. 2016a, b).

Despite these limitations, DGMS systems are preserved as an alternative in broccoli breeding. In recent decades, numerous DGMS lines have been developed, supported by molecular markers for efficient MAS (Han et al. 2019; Shu et al. 2016a, b). Shu et al. introduced SSR markers linked to the male-sterile gene, with scaffold10312a demonstrating an accuracy of ≥96.43% (Shu et al. 2016a, b). These markers enable the early selection of sterile plants in breeding lines, such as *DGMs8554* and *DGMs94174*, using polyacrylamide gel-based differentiation.

Han and his colleagues further advanced DGMS breeding by developing a high-throughput kompetitive allele-specific PCR (KASP) marker known as K6. This marker, applicable across all *B. oleracea* crops, allows the identification of homozygous DGMS plants within selfing progenies of heterozygous lines, bypassing traditional test crossing (Han et al. 2019). Its efficiency lies in allele-specific fluorescence detection using systems like the Applied Biosystems Viia 7 real-time PCR, significantly reducing labor and time requirements.

18.3.6.3 Reutilization of CMS Resources in Broccoli Breeding

Efforts to enhance CMS systems in *B. oleracea* have focused on transferring the Ogura CMS restorer gene (*RFO*) from radish to rapeseed and subsequently to broccoli (Yu et al. 2016; Liu et al. 2020a, b). Liu et al. employed interspecific hybridization and backcrossing to introduce *RFO* into broccoli, utilizing foreground markers (*BnRFO-AS2F/BnRFO-NEW-R*) for identifying *RFO*-positive hybrids and background SSR markers to assess genetic backgrounds (Liu et al. 2020a, b). By analyzing polymorphic loci in BC_2 progenies, individuals closer to broccoli's genetic background were identified, demonstrating progress in achieving functional *RFO*-positive lines suitable for hybrid seed production (Liu et al. 2020a, b).

The development of novel male-sterile lines through genetic engineering offers a rapid and efficient pathway to diversify resources for hybrid seed production. Chen et al. successfully engineered male-sterile transgenic broccoli by employing RNA interference (RNAi) targeting the jasmonic acid pathway gene *BoiDAD1F* (Chen et al. 2010). These genetically modified plants exhibited male sterility under standard growth conditions but could regain fertility upon treatment with exogenous jasmonic acid (JA), making them particularly suitable for a two-line hybrid seed production system (Chen et al. 2010).

Similarly, Li et al. demonstrated that overexpression of the microRNA *bol-miR171b* in broccoli led to almost complete male sterility. Notably, this genetic modification also resulted in an increase in chlorophyll content, potentially offering dual benefits for hybrid seed production and agronomic performance (Li et al. 2018).

18.3.7 Glucosinolates

Broccoli contains several essential secondary metabolites, including glucosinolates (like sulforaphane), carotenoids, phenolic acids, and flavonoids, which play significant roles in its nutritional and health benefits. Genetic studies have identified multiple loci and genes that regulate the accumulation of these compounds in broccoli, and research models based on Arabidopsis have proven valuable for understanding these processes in broccoli as well. Notable genes involved in the biosynthesis of these metabolites include cytochrome *P450 79F1 (CYP79F1), cytochrome P450 83A1 (CYP83A1), UDP-glucosyltransferase 74B1 (UGT74B1)*, and *sulfotransferase 18 (ST5b)*, among others.

Several genetic mapping efforts have helped pinpoint loci associated with the variation in these metabolites. For example, Sotelo et al. (2014) identified 82 significant QTLs for glucosinolate accumulation in different broccoli tissues, such as leaves, seeds, and flower buds, with QTL9.2 being a key contributor to glucosinolate variation. Brown et al. (2014) constructed a genetic linkage map and identified 14 QTLs associated with the accumulation of various glucosinolates in florets, particularly noting that locus GSL12 on chromosome C09 accounts for around 40% of progoitrin variation. Similarly, Li et al. (2018) identified 18 QTLs for sulforaphane metabolism in broccoli florets, with 6 of these being consistently observed across different environments.

Further research on carotenoid and phenolic compound variation has also been undertaken. Using a SNP array, Brown et al. (2015) discovered three QTLs for carotenoid variation in broccoli florets, while Gardner et al. (2016) identified 23 loci involved in total phenolic concentration and its individual components in broccoli. Additionally, in the BolTBDH mapping population, 33 QTLs were identified that regulate phenolic content in leaves, flower buds, and seeds. Transcriptome analyses have further revealed differentially expressed genes involved in glucosinolate metabolism, contributing valuable insights into how these compounds are synthesized in various broccoli tissues, including seeds, sprouts, and byproducts. These studies provide essential genetic resources for improving broccoli's secondary metabolite content through breeding, helping to enhance its nutritional value and health benefits (Sotelo et al. 2014; Brown et al. 2014; Li et al. 2018; Gardner et al. 2016).

Recent advancements in genetic engineering have focused on enhancing the levels of anticancer metabolites, particularly glucosinolates and their bioactive derivative sulforaphane, in broccoli by targeting biosynthetic and regulatory genes (Zuluaga et al. 2019; Li et al. 2019; Kim et al. 2020; Cao et al. 2021; Baskar and Park 2015). Zuluaga et al. demonstrated that overexpressing *BoMYB29* in the doubled haploid (DH) line AG1012 significantly upregulated the aliphatic glucosinolate biosynthetic pathway, leading to elevated levels of methylsulphinylalkyl glucosinolates, including glucoraphanin, a precursor of sulforaphane (Zuluaga et al. 2019).

Li et al. identified and isolated two *tryptophan synthase beta* subunit (*TSB*) genes, *BoTSB1* and *BoTSB2*, from broccoli and created overexpression lines in *Arabidopsis*. These lines exhibited increased accumulation of tryptophan, indole-3-acetic acid (IAA), and indole glucosinolates, providing a promising target for glucosinolate enhancement. However, no transgenic broccoli plants were produced in this study (Li et al. 2019).

Research into *BroMYB28* revealed its potential role in glucoraphanin biosynthesis. While its function was uncertain until 2019, Agrobacterium-mediated transient overexpression of *BroMYB28* in broccoli successfully demonstrated its role, leading to glucoraphanin accumulation (Kim et al. 2020; Baskar and Park 2015).

Cao et al. advanced this field by generating transgenic broccoli lines overexpressing *MAM1*, *FMOGS-OX2*, and *Myrosinase*, both independently and in combination. Individually, these transgenes increased sulforaphane levels by 1.7- to 3.4-fold (*MAM1*), 1.6- to 2.7-fold (*FMOGS-OX2*), and 3.7-fold (*Myrosinase*). Transgenic plants with the combined overexpression of all three genes demonstrated a synergistic effect, with sulforaphane content enhanced by 1.86- to 5.5-fold compared to wild-type plants (Cao et al. 2021).

18.3.8 Revolutionizing Broccoli Genomics

Advancements in methodologies and the decreasing costs of high-throughput next-generation (NGS) and third-generation sequencing (TGS) technologies have revolutionized genomic research and crop improvement, including broccoli. In 2014, the draft genome of *Brassica oleracea*, based on NGS short reads, was made publicly available (Parkin et al. 2014; Liu et al. 2014). Subsequently, in 2018, the first high-quality broccoli (HDEM) reference genome was published, leveraging TGS nanopore long reads and optical mapping technologies (Belser et al. 2018). These foundational studies elucidated critical insights into genome duplication and gene divergence while enabling the prediction of genes associated with phytochemical biosynthesis and morphological traits. Moreover, they provided a valuable framework for the development of high-density molecular markers (Peng et al. 2018; Brown et al. 2014, 2015).

Techniques such as bulked-segregant analysis combined with whole-genome resequencing (BSA-seq) have facilitated rapid mapping of genes and quantitative trait loci (QTLs) alongside candidate gene identification (Branham and Farnham 2017; Branham and Farnham 2019). Omics-based approaches have further contributed by identifying differentially expressed genes and miRNAs linked to key traits (Chen et al. 2015; Lin et al. 2015, 2019; Tian et al. 2014; Torres-Contreras et al. 2018; Gao et al. 2014; Lee et al. 2017; Li et al. 2019). High-throughput strategies have also enabled the application of kompetitive allele-specific PCR (KASP) markers for fingerprinting essential broccoli germplasm (Shen et al. 2021), as well as the exploration of genetic diversity and population structure in broccoli cultivars (Stansell and Bjorkman 2020; Huang et al. 2021a, b). Additionally, these advancements have shed light on the genomic and morphological aspects of domestication syndromes in calabrese broccoli landraces, hybrids, and sprouting broccoli (Stansell and Bjorkman 2020).

18.3.8.1 Molecular Markers Identified in Broccoli

Molecular marker technologies, such as RFLP, RAPD, AFLP, SSR, SCAR, SRAP, and SNP, are fundamental tools in Brassica research, playing a critical role in identifying key traits, constructing genetic maps, mapping genes, and analyzing germplasm diversity. These technologies have proven invaluable for mapping important agricultural traits, including self-incompatibility, male sterility, floret morphology, and overall crop quality (Lv et al. 2014; Dhaka et al. 2017; Gan et al. 2019).

For instance, Lin et al. (2013) created a genetic linkage map using 202 AFLP markers, spanning nine linkage groups over 1034.5 cM, and identified 73 QTLs linked to seven broccoli traits. This map provides essential insights into trait variation and will support more efficient selection in breeding programs. Brown et al. (2014) employed 547 nonredundant SNP markers to map QTLs for carotenoid accumulation in broccoli flowers. Their map, spanning 984.1 cM, uncovered three key QTLs, contributing to the genetic understanding of carotenoid biosynthesis and nutritional enhancement in broccoli. Similarly, Walley et al. (2012) used SSR and AFLP markers to construct a genetic map of 946.7 cM, identifying several QTLs associated with agronomic traits, including ball weight, stem diameter, and weight loss during storage. These findings highlight the utility of molecular markers in improving both the yield and quality of Brassica crops.

More recently, Yu et al. (2019a, b, c, d) leveraged the TO1000 reference genome to identify loci associated with hollow stem traits in broccoli, an important physiological disorder that affects marketable yield and quality. This work, combined with advancements in high-density genotyping and transcriptomics (Yan et al. 2023; Li et al. 2023), has significantly enhanced the precision of QTL mapping and gene identification in Brassica crops. Collectively, these studies offer critical insights into the genetic architecture of key traits in Brassica species, providing valuable resources for enhancing breeding programs aimed at improving disease resistance, nutritional content, and overall crop performance. Furthermore, the integration of these molecular tools with modern genomic techniques has the potential to revolutionize Brassica crop breeding, facilitating more efficient selection for traits of economic importance. Marker sequence analysis of the HDEM reference genome revealed that the target *Pp523* region for single dominant locus for downy mildew resistance is 49.29–50.68 Mb on chromosome 8 (Belser et al. 2018).

18.3.8.2 Used Molecular Markers

Molecular markers have been instrumental in speeding up the breeding process with more accuracy. In broccoli, Giovannelli et al. (2002) identified eight RAPD (random amplification of polymorphic DNA) markers linked to downy mildew resistance at cotyledon and true leaf stage. They converted two markers ($UBC3596_{620}$ and $OPM16_{750}$) to SCAR markers and mapped at 6.7 and 3.3 Cm to the locus. Farnham et al. (2002) mapped the downy mildew resistance locus *Pp523* in broccoli and developed flanking RAPD markers OPK17_980 and AFLP marker AT.CTA_133/134, with genetic distances of 3.1 cM and 3.6 cM, respectively. The *Ppa523* locus was reported to be syntenic to the top arm end of chromosome 1 (Farhino et al. 2004).

Resistant to clubroot is lacking in *B. olearcea*, and mining of resistance loci/genes from its close relatives, such as turnip, radish, and rapeseed, is used. For clubroot, a major resistance locus in broccoli has been mapped using RFLP markers (Figdore et al. 1993). In kale, two QTLs for clubroot-resistance were mapped using RAPD markers (Grandclément and Thomas 1996), while Rocherieux et al. (2004) mapped nine genomic regions anchoring clubroot-resistance genes with several types of markers. Two clubroot-resistance QTLs, pb-3 and pb-4, were mapped in cabbage (Brassica oleracea) using RFLP and AFLP markers (Voorrips 1995). Xie et al. (2022) cloned four genes including *Crr1*, *CRa*, *CRd*, and *CRb*^{kato}. Of these, *Crr1* and *CRa* belong to the Toll/interleukin-1 receptor-nucleotide-binding-leucine-rich repeat (TIR-NB-LRR) gene family and are found mainly in turnip and Chinese cabbage. They introgressed *CRa* gene from *B. rapa* to *B. olearcea* by distant hybridization and MAS using *CRa*-specific markers SC2930-Q-F. Rocherieux et al. (2004) generated $F_{2:3}$ population for clubroot using resistant kale and susceptible broccoli. They constructed a genetic map using RFLP markers. Nagaoka et al. (2010) also developed mapping population by crossing Anju (resistant DH line) with GC (susceptible DH line) and identified five CR-QTLs including *pb-Bo(Anju)1*,

PbBo(Anju)2, *PbBo(Anju)3*, and *PbBo(Anju)4* from resistant parent and *pb-Bo(GC)1* from GC. They identified linked primers for *CRa*(SC2930-T-FW/SC2930-RV and SC2930-Q-FW/SC2930-RV), *CRb*(KBrH129J18R), *CRc*(B50-C9-FW/-RV and B50-6R-FW/-RV), *CRk*(HC688-4-FW/-7-RV), *Crr3*(OPC11-2S), *Crr1*(BRMS-088), and *Crr2* (BRMS-096) in 41 cruciferous genotypes (Xie et al. 2022).

Resistance to black rot in *B. oleracea* has been reported as quantitative in nature, with varying numbers of genes involved (Vicente et al. 2002). Doullah et al. (2011) constructed linkage map of 320 cM using cleaved amplified polymorphic sequences (CAPS) and sequence-related amplified polymorphism (SRAP) analysis with disease rating of F_3 families obtained from a susceptible broccoli and resistant cabbage (Green commet P09 × Reiho P01) and identified two genomic regions on LG 2 and LG 9 significantly associated with resistance to black rot. The genetic map was improved by Tonu et al. (2013), who identified three QTLs namely *XccBo(Reiho)1*, *XccBo(Reiho)2*, and *XccBo(Reiho)1* for resistance to black rot and the major QTL, i.e., *XccBo(Reiho)2* was from parent Reiho.

The inheritance of downy mildew is a dominant character controlled by a single locus. Seven resistance genes namely *CRa*, *CRb*, *CRc*, *CRk*, *Crr1*, *Crr2*, and *Crr3* impart resistance to clubroot of cruciferous genotypes (Xie et al. 2022). Single locus "*f*" governs resistance against black rot in cabbage (Williams et al. 1975), which was mapped by Camargo et al. (1995) as *wg8a9b_wg4d79* on LG9 in cabbage line Badger Inbred-16 using F_3 families from Badger Inbred-16 (resistant) × Broccoli 'OSU Cr-7' (susceptible) and RFLP markers.

18.3.8.3 Germplasm Used in Developing the Population

The mode of inheritance of field resistance to downy mildew (*Peronospora parasitica* (Pers. ex Fr.) Fr.) was studied by crossing resistant plants of broccoli 'OL 87125' (HRI, Wellesbourne, UK) with susceptible plants of broccoli 'OL 87098' (HRI, Wellesbourne, UK) (Coelho and Monteiro 2003). Although downy mildew resistance is available in broccoli, it can also be explored from cauliflower. Singh et al. (2022) identified resistant sources in cauliflower such as 'BR-2', '3-5-1-1', 'CCm', '395aa', 'GPMT', 'DC-300', 'DC-360', 'PC-1', 'BR-1', 'AL-15', '18–19', and 'KT-2'. They developed F_2 and $F_{2:3}$ populations and investigated single dominant gene for downy mildew resistance. Al-Lami et al. (2023) screened 150 diverse Brassicaceae lines including three broccoli namely Red Arrow (27.1% disease incidence), Purple Sprouting (32.3%), and Di Ciccio (53.7%) against *A. brassicae*. Zhang et al. (2024) screened 173 broccoli lines/hybrids against *Alternaria alternata* and identified 58 lines as resistant including two B230 and B238 as highly resistant. However, *Brassica oleracea* is highly susceptible for *Alternaria brassicicola and B. brassicae*, but only a moderate level of resistance sources has been reported so far. Singh et al. (2023) screened 38 wild *Brassica* species and reported strong resistance in *Biscutella*. *Capsellabursa pastoris* (L.) Medik. *Camelina sativa* (L.) Crantz, *Diplotaxis erucoides* (L.) DC, and *Diplotaxis gomez-campoi* Mart.-Laborde. Al-lami et al. (2023) also reported *Camelina sativa* and *D. erucoides* as the most resistant against *A. brassicae*. These wild species can be explored through pre-breeding procedure for introgression of Alternaria leaf spot resistance in broccoli.

Black leg (*Leptosphaeria maculans*) is a serious problem in brassica crops, and no strong resistance has been reported so far. This pathogen develops typical gene-for-gene interaction with the host. The *Rlm* genes which impart resistance to *Brassica oleracea* are originally from *B. rapa*, *B. juncea*, or *B. nigra*. Dilmaghani et al. (2012) screened 3392 accessions and could identify only six accessions with moderate resistance. Robin et al. (2017) identified two resistant lines in Korean cabbage germplasms. Degrave et al. (2021) identified a new single major gene *Rlm14* for blackleg resistance in *Brassica oleracea*.

In two separate studies, Jiang et al. developed transgenic broccoli plants overexpressing the cytosolic ascorbate peroxidase gene (*BoAPX*) and the WRKY transcription factor gene (*BoWRKY6*), both of which conferred enhanced resistance to downy mildew caused by *Hyaloperonospora parasitica* (Jiang et al. 2016a, b).

BoAPX-overexpressing plants demonstrated significantly higher resistance to *H. parasitica* infection compared to wild-type plants. This increased resistance was accompanied by reduced electrical conductivity, indicative of lower cellular damage, and elevated ascorbate peroxidase (APX) enzyme activity. Notably, these plants also exhibited improved tolerance to heat stress, highlighting the dual protective role of *BoAPX* against biotic and abiotic stresses (Jiang et al. 2016a, b).

Similarly, *BoWRKY6*-overexpressing broccoli plants showed enhanced resistance to downy mildew, although the degree of resistance varied among lines. Two lines, BWK14 and BWK31, exhibited particularly high levels of resistance, suggesting potential for targeted breeding or further genetic studies to optimize this trait (Jiang et al. 2016a, b).

18.3.8.4 Candidate Gene Identification

Advancements in broccoli genomics and breeding have significantly expanded the understanding of key traits and their underlying genetic mechanisms. Gardner et al. (2016) identified two stable QTLs (no. 7 and no. 52) linked to total phenolic content and antioxidant capacity, revealing candidate genes in the phenylpropanoid biosynthesis pathway and MYB transcription factors. Brown et al. (2015) mapped a QTL on chromosome 5 associated with aliphatic glucosinolates (2-propenyl and 3-methylsulfinylpropyl), aligning with glucosinolate biosynthetic genes such as GSL-ALK, GSL-PRO, and GSL-ELONG using SNP markers. Recent studies by Wu et al. (2024) further enhanced these findings through a chromosome-level genome assembly of broccoli using PacBio HiFi reads and Hi-C technology, producing a 613.79-Mb genome. Their research demonstrated that overexpressing BoMAM1 increased the accumulation of C4-glucosinolates, highlighting its role as a critical enzyme in glucosinolate biosynthesis.

In stress tolerance, He et al. (2021) revealed bolTLP1's interaction with key stress response proteins, such as RD2 and VOZ2, elucidating its role in abiotic and biotic resilience. Lin et al. (2018) linked specific haplotypes of BoFLC3 and PAN to variations in curd induction and adaptation to high temperatures and short day lengths. Research by Rahim et al. (2019) on anthocyanin biosynthesis identified structural and regulatory genes, including BoPAL, BoDFR, and BoTT8, which were highly expressed in purple broccoli hypocotyls. They developed a 14-bp InDel marker in the BoTT8 promoter to differentiate between green and purple lines, facilitating targeted breeding. Liu et al. (2022) advanced this work by identifying distinct molecular markers for green and anthocyanin-rich purple varieties.

Furthermore, Lin et al. (2015) discovered pleiotropic QTLs on linkage groups (LGs) 1, 3, and 5 influencing head traits such as height, weight, and floret width. The QTL linked to AFLP d-64 (S) on LG 3 was particularly notable for enhancing head development under high-temperature conditions. These integrated genomic, molecular, and phenotypic studies provide a robust foundation for improving broccoli through precision breeding and stress-resilient cultivar development.

18.3.9 Abiotic Stress

18.3.9.1 Heat Stress

Broccoli production faces significant challenges as demand for larger cultivation areas rises and the need for consistent yields becomes more pressing in the context of climate change and extreme weather conditions (Branham and Farnham 2019; Branham et al. 2017). As a cool-season crop, broccoli flourishes within a temperature range of 15–23 °C during the early stages of floral development (Farnham and Bjorkman 2011). However, when temperatures surpass 25 °C, several negative effects on broccoli quality arise, such as:

1. Many broccoli varieties require vernalization at temperatures below 23 °C, with higher temperatures impeding proper head formation.
2. For non-vernalization-dependent germplasms, temperatures above 30 °C lead to unfavorable characteristics like bracing, uneven head formation, irregular bud sizes, discoloration, and the development of brown beads, all of which diminish marketability.
3. High temperatures during head development can significantly reduce overall yield (Branham and Farnham 2019; Björkman and Pearson 1998; Heather et al. 1992).

To address these concerns, significant work has been made in creating heat-tolerant broccoli cultivars and determining the genetic components that influence heat tolerance. In the United States, projects such as the National Institute of Food and Agriculture (NIFA) Project (No. 2010-51181-21062) and the USDA Vegetable Brassica Research Project (CRIS No. 6080-21000-019-00D) (Farnham and Bjorkman 2011; Atallah et al. 2014; Ward et al. 2015) have helped to promote sustainable broccoli production in heat-prone areas, particularly along the East Coast. Researchers in Taiwan, China, and Indonesia are working to adapt broccoli farming to subtropical and tropical temperatures, which requires the creation of heat-tolerant types (Lin et al. 2018; Astarini et al. 2020).

18.3.9.2 Genetic Basis of Heat Tolerance

Heat tolerance in broccoli is a polygenic trait influenced by several loci with beneficial effects (Branham and Farnham 2019; Branham et al. 2017). Recent studies have identified key quantitative trait loci (QTLs) linked to heat tolerance. For example, Lin et al. (2013) discovered 31 QTLs associated with head size and weight in broccoli cultivated under high-temperature conditions, with daytime temperatures averaging 36.4 °C and nighttime temperatures around 25.9 °C. Branham et al. (2017) used genotyping-by-sequencing on a doubled haploid (DH) population to create a high-density genetic map and identified five QTLs, including a significant epistatic interaction between QHT_C03 and QHT_C05, which accounted for 62.1% of the phenotypic variation. Furthermore, Branham and Farnham (2019) utilized whole-genome resequencing on bulked segregants from a DH population to find two additional QTLs for heat tolerance, one of which, QHT_C09.2, was shown to influence the relationship between heat tolerance and maturity.

Recent advances in functional genomics have provided valuable insights into the molecular mechanisms governing heat tolerance. For instance, a catalase gene from broccoli, associated with heat stress, was cloned and expressed in *Arabidopsis* to improve heat tolerance, though its specific role in preserving high-quality broccoli heads under elevated temperatures is still uncertain (Lin et al. 2010; Chiang et al. 2013). Furthermore, with the advent of next-generation sequencing technologies and the availability of reference genomes, studies have identified differentially expressed genes, microRNAs, and potential pathways involved in heat stress responses, further enhancing our understanding of the genetic basis of heat tolerance in broccoli (Chen et al. 2015; Lin et al. 2019).

18.3.9.3 Salinity Stress Tolerance

Wang et al. (2022) discovered nine PROPEP genes within the broccoli genome, with *BoPROPEP4* showing a strong response to salinity stress. The application of the synthesized peptide BoPep4, which is C-terminally encoded, significantly enhanced salt tolerance in broccoli sprouts. This was achieved by reducing the Na^+/K^+ ratio and promoting the biosynthesis of wax and cutin in the leaves. RNA-seq analysis further revealed that BoPep4 regulates stress response pathways, particularly those related to auxin and abscisic acid (ABA) signaling, as well as the biosynthesis of cutin and wax. These findings underline *BoPep4*'s essential role in the plant's defense against salinity stress.

In a related study, the ectopic overexpression of *bolTLP1* in transgenic *Arabidopsis* and broccoli lines demonstrated enhanced tolerance to both high salinity and drought conditions (He et al. 2021). The bolTLP1 gene likely governs stress tolerance through the regulation of phytohormone-mediated signaling pathways, such as ABA, ethylene, and auxin, and by modulating activities of hydrolases, oxidoreductases, and sulfur compound synthesis. Additionally, *bolTLP1* influences the differential expression of histone variants. Protein interaction studies revealed that *bolTLP1* physically interacts with key stress-response proteins, including RESPONSE TO DESICCATION 2 (RD2), RESPONSIVE TO DEHYDRATION 22 (RD22), VASCULAR PLANT ONE-ZINC FINGER2 (VOZ2), SM-LIKE1B

(LSM1B), and MALATE DEHYDROGENASE (MDH), suggesting its direct involvement in stress tolerance mechanisms.

Li et al. developed RNAi transgenic broccoli lines targeting the cellulose synthase gene *BoiCesA*, revealing that *BoiCesA* knockdown resulted in reduced cellulose content and significantly enhanced salt tolerance. This tolerance was associated with altered expression of stress-responsive genes (*BoiProH*, *BoiPIP2;2*, and *BoiPIP2;3*). However, these transgenic plants also exhibited notable phenotypic defects, including dwarfism and reduced leaf size (Li et al. 2017).

In three separate studies, Jiang et al. investigated the role of transcription factors in conferring salt stress tolerance. Overexpression of the C3H-type zinc finger genes *BoC3H* and *BoC3H4*, as well as the ethylene response transcription factor gene *BoERF1*, was shown to enhance salinity tolerance (Jiang et al. 2017, 2018, 2019). *BoC3H*-overexpressing lines demonstrated improved germination rates, increased dry weight and chlorophyll content under salt stress, and reduced cell death in leaves. These improvements were linked to decreased levels of hydrogen peroxide (H_2O_2), relative electrical conductivity (REC), and malondialdehyde (MDA), as well as enhanced free proline content and elevated activities of antioxidant enzymes such as catalase, peroxidase, and superoxide dismutase (Jiang et al. 2017). *BoC3H4*-overexpressing lines exhibited increased salinity tolerance, characterized by higher proline levels, reduced chlorophyll loss, and lower MDA and REC compared to wild-type plants. However, these lines were more susceptible to *Sclerotinia sclerotiorum*, possibly due to suppressed expression of the defense-related gene *BoPDF1.2* (Jiang et al. 2019). *BoERF1*-overexpressing lines showed enhanced germination rates and reduced chlorophyll degradation under salt stress, with less cell death in leaves, similar to *BoC3H*-overexpressing lines. Additionally, these transgenic plants displayed improved resistance to *Sclerotinia* stem rot, highlighting their dual role in abiotic and biotic stress tolerance (Jiang et al. 2018).

18.3.9.4 Drought Tolerance

A comprehensive genome-wide study on aquaporins in broccoli (*Brassica oleracea* var. *italica*) identified 65 aquaporin genes, emphasizing their essential role in maintaining water balance and overall plant homeostasis. These aquaporins were classified into subfamilies, including PIPs, TIPs, and NIPs, with a particular focus on gene duplication events, especially within the NIP subfamily, which could contribute to enhanced resilience against abiotic stresses. Notable aquaporin isoforms, such as BoiPIP1-2a, BoiPIP1-2b, and *BoiTIP1-2b*, showed high expression levels, suggesting their critical involvement in the transport of water and nitrogenous compounds. Tandem duplications in NIPs, particularly *BoiNIP2-1a, BoiNIP2-1b, BoiNIP3-1b, BoiNIP4-1d,* and *BoiNIP4-1f,* indicated functional diversification related to stress responses (Espinosa and Carvajal 2022).

Additionally, Chevilly et al. (2021) investigated the drought response of different broccoli cultivars, aiming to identify physiological and molecular markers of drought tolerance. Their comparative analysis revealed that drought-resistant cultivars exhibited higher levels of methionine and abscisic acid, while showing lower levels of urea, quinic acid, and gluconic acid lactone under drought conditions.

Furthermore, drought treatment led to an increase in essential amino acid levels in both the leaves and florets of broccoli, indicating the plant's adaptive mechanisms to water scarcity.

18.3.10 Genome Editing Techniques

Genome editing has emerged as a transformative tool for achieving precise and efficient modifications in the genomes of living organisms. Four principal classes of engineered nucleases underpin these advancements: meganucleases, zinc finger nucleases (ZFNs), transcription activator-like effector nucleases (TALENs), and CRISPR-associated protein (Cas9) systems. Among these, the CRISPR/Cas9 platform has gained prominence due to its cost-effectiveness, versatility, and precision, making it a highly adaptable approach for targeted genome modifications. Over the past decade, CRISPR/Cas9 has been extensively applied to modify genes in various plant species, demonstrating considerable success in enhancing desired traits (Liu et al. 2017).

While genome editing has shown promise in certain *Brassica oleracea* crops (Ma et al. 2019; Sun et al. 2020), its application to broccoli-specific genetic backgrounds has yet to achieve complete efficiency. To date, only one study has utilized the CRISPR/Cas9 system on broccoli-related plant material. This research focused on DH1012, a doubled haploid genotype derived from a cross between *B. oleracea var. alboglabra* (A12DHd) and *B. oleracea var. italica* (Green Duke GDDH33). The study targeted the *BolC.GA4.a* (*Bol038154*) gene, successfully inducing a dwarf phenotype (Lawrenson et al. 2015; Sparrow and Irwin 2014).

18.3.11 Challenges and Future Directions

Despite the significant progress made in the genetic improvement of broccoli through GWAS and QTL mapping, several challenges remain. One of the primary challenges is the complexity of the broccoli genome. As with many crops, broccoli exhibits a high level of genetic diversity, which can make it difficult to identify and validate genetic markers associated with traits of interest. Additionally, the polyploid nature of the *Brassica* genome adds another layer of complexity, requiring careful consideration in both GWAS and QTL mapping studies (Li et al. 2019).

Another challenge is the environmental influence on trait expression. Many traits, especially those related to yield and stress tolerance, are influenced by environmental factors such as temperature, soil quality, and water availability. This means that genetic studies must account for environmental variation, which can complicate the identification of stable QTLs and genetic markers. Future research will need to incorporate multi-environment trials to validate the genetic loci identified in GWAS and QTL mapping studies and ensure their applicability in diverse growing conditions (Kumar and Srivastava 2016). To overcome these challenges, researchers are increasingly turning to advanced genomic tools, such as high-throughput sequencing, genome assembly, and CRISPR-Cas9 gene editing. These technologies offer new opportunities for improving the accuracy and efficiency of genetic studies in broccoli, facilitating the development of improved cultivars with enhanced nutritional content and resistance to diseases and environmental stresses.

18.4 Future Perspectives

Recent advances in the molecular breeding of broccoli have focused on enhancing agronomic traits, secondary metabolite production, male sterility, tolerance to abiotic stresses, resistance to diseases, and insect deterrence. Marker-assisted selection (MAS) has facilitated the development of heat-stress and clubroot-resistant cultivars. However, significant challenges remain, including limited foundational research and an incomplete understanding of the genetic basis underlying many desirable traits. Future studies in broccoli molecular breeding should prioritize the following areas.

18.4.1 Identification of Functional Genes and Loci

Progress has been made in documenting markers and mapping genes/quantitative trait loci (QTLs) associated with favorable traits, with advances in sequencing technologies identifying candidate genes/QTLs. Integrative *omics* approaches, such as transcriptomics, proteomics, and metabolomics, have been employed to unravel the mechanisms regulating key traits. Despite these efforts, only a fraction of genes in broccoli have been cloned and functionally validated. Future efforts should focus on:

1. *Establishing Advanced Mapping Populations:* Developing secondary populations like near-isogenic lines (NILs), introgression lines (ILs), and chromosome segment substitution lines (CSSLs) for precise mapping and isolation of QTLs governing complex traits
2. *Leveraging Reverse Genetics:* Utilizing existing gene databases and advanced sequencing methods to identify additional functional genes for breeding purposes
3. *Validating Target Genes:* Using transient expression systems or genetic transformation to confirm the roles of genes identified through fine mapping or homology cloning

18.4.2 Enhancing Genetic Diversity with Landraces and *B. oleracea* Subspecies

Modern broccoli exhibits limited genetic diversity, leading to potential compromises in quality, yield, and resilience. Italian broccoli landraces and other *Brassica oleracea* subspecies serve as valuable reservoirs of genetic variability, offering traits such as diverse heading structures, varied flowering and maturity periods, elevated glucosinolate content, and enhanced disease resistance. Incorporating beneficial alleles from these genetic resources using MAS can enrich broccoli's genetic base and improve its commercial traits.

For example, the compact head structure of Calabrese broccoli—critical for transport and storage—has been significantly improved in recent decades, likely through genomic introgression from cauliflower (though no formal studies confirm this). Expanding the use of landraces and subspecies could similarly enhance other essential traits.

18.4.3 Incorporating Disease Resistance Genes from Related Species

Broccoli remains vulnerable to devastating diseases such as clubroot and black rot. To combat these challenges, resistance genes from allied species can be introduced via distant hybridization and MAS, employing gene pyramiding to achieve durable resistance.

- *Clubroot Resistance:* The clubroot pathogen *Plasmodiophora brassicae* manifests multiple physiological races, necessitating diverse resistance mechanisms. Currently, the CRa locus, effective against race 4, has been introgressed from *B. rapa*, but this is insufficient to address other races. Incorporating additional loci from species like turnip, radish, and rapeseed, and pyramiding these loci in broccoli would bolster resistance.
- *Black Rot Resistance:* High levels of resistance to black rot have been identified in the A and B genomes of Brassica species, whereas moderate resistance to clubroot exists in the C genome of cabbage. Transferring these resistance loci from *B. carinata* and cabbage into broccoli could enhance its resilience to these diseases.

18.4.4 Advancing Broccoli Breeding Using CRISPR/Cas9 Genome Editing

The CRISPR/Cas9 system has emerged as a transformative genome editing tool in crop improvement, successfully applied in rice, maize, soybean, and tomato for traits such as increased yield, disease resistance, herbicide tolerance, and optimized plant architecture. However, its application in broccoli remains unexplored. Future research should focus on:

- Developing CRISPR/Cas9 protocols tailored to broccoli's unique genetic and physiological characteristics
- Targeting specific genes associated with yield, stress resistance, and metabolite production for precise edits
- Combining CRISPR-based editing with conventional breeding to accelerate the development of superior broccoli cultivars

References

Al-Lami HF, You MP, Banga SS, Barbetti MJ (2023) Novel resistances provide new avenues to manage alternaria leaf spot (Alternaria brassicae) in canola (Brassica napus), mustard (B. juncea), and other brassicaceae crops. Plant Dis 107(2):372–381

Astarini IA, Defiani MR, Suriani NL, Griffiths P, Stefanova K, Siddique K (2020) Adaptation of broccoli (Brassica oleracea var. italica L.) to high and low altitudes in Bali, Indonesia. Biodiversitas J Biol Divers 21:5263–5269

Atallah SS, Gómez MI, Björkman T (2014) Localization effects for a fresh vegetable product supply chain: Broccoli in the eastern United States. Food Policy 49:151–159

Bannerot H, Boulidard L, Cauderon Y, Tempe J (1974) Transfer of cytoplasmic male sterility from Raphanus sativus to Brassica oleracea. Proc Eucarpia Meet Crucif 25:52–54

Baskar V, Park SW (2015) Molecular characterization of BrMYB28 and BrMYB29 paralogous transcription factors involved in the regulation of aliphatic glucosinolate profiles in Brassica rapa ssp. pekinensis. Comptes Rendus Biol 338:434–442

Belser C, Istace B, Denis E, Dubarry M, Baurens F-C, Falentin C, Genete M, Berrabah W, Chèvre A-M, Delourme R et al (2018) Chromosome-scale assemblies of plant genomes using nanopore long reads and optical maps. Nat Plants 4:879–887

Björkman T, Pearson KJ (1998) High temperature arrest of inflorescence development in broccoli (Brassica oleracea var. italica L.). J Exp Bot 49:101–106

Boersma M, Gracie AJ, Brown PH (2013) Evidence of mechanical tissue strain in the development of hollow stem in broccoli. Sci Hortic 164:353–358. https://doi.org/10.1016/j.scienta.2013.09.020

Bohuon EJR, Keith DJ, Parkin IAP, Sharpe AG, Lydiate DJ (1996) Alignment of the conserved C genomes of Brassica oleracea and Brassica napus. Theor Appl Genet 93:833–839. https://doi.org/10.1007/BF00224083

Bohuon EJR, Ramsay LD, Craft JA, Arthur AE, Marshall D, Lydiate DJ, Kearsey MJ (1998) The association of flowering time quantitative trait loci with duplicated regions and candidate loci in Brassica oleracea. Genetics 150:393–401

Branca F, Chiarenza GL, Cavallaro C, Gu H, Zhao Z, Tribulato A (2017) Diversity of Sicilian broccoli (Brassica oleracea var. italica) and cauliflower (Brassica oleracea var. botrytis) landraces and their distinctive bio-morphological, antioxidant, and genetic traits. Genet Resour Crop Evol 65:485–502

Branham SE, Farnham MW (2017) Genotyping-by-sequencing of waxy and glossy near-isogenic broccoli lines. Euphytica 213(4):84. https://doi.org/10.1007/s10681-017-1873-9

Branham SE, Farnham MW (2019) Identification of heat tolerance loci in broccoli through bulked segregant analysis using whole genome resequencing. Euphytica 215(34):1–12. https://doi.org/10.1007/s10681-019-2413-0

Branham SE, Stansell ZJ, Couillard DM, Farnham MW (2017) Quantitative trait loci mapping of heat tolerance in broccoli (Brassica oleracea var. italica) using genotyping-by-sequencing. Theor Appl Genet 130:529–538

Brown AF, Jeffery E, Juvik JA (2007) A polymerase chain reaction-based linkage map of broccoli and identification of quantitative trait loci associated with harvest date and head weight. J Am Soc Hortic Sci 132:507–513

Brown AF, Yousef GG, Chebrolu KK, Byrd RW, Everhart KW, Thomas A, Reid R, Parkin IAP, Sharpe AG, Oliver R, Kopsell DA (2014) High-density single nucleotide polymorphism (SNP) array mapping in Brassica oleracea: identification of QTL associated with carotenoid variation in broccoli florets. Theor Appl Genet 127:2051–2064. https://doi.org/10.1007/s00122-014-2360-5

Brown AF, Yousef GG, Reid RW et al (2015) Genetic analysis of glucosinolate variability in broccoli florets using genomeanchored single nucleotide polymorphisms. Theor Appl Genet 128:1431–1447. https://doi.org/10.1007/s00122-015-2517-x

Buck PA (1956) Origin and taxonomy of broccoli. Econ Bot 10:250–253

Camargo L, Osborn T (1996) Mapping loci controlling flowering time in Brassica oleracea. Theor Appl Genet 92:610–616

Camargo LEA, Williams PH, Osborn TC (1995) Mapping of quantitative trait loci controlling resistance of Brassica oleracea to Xanthomonas campestris pv. campestris in the field and greenhouse. Phytopathology 85(10):1296–1300

Cao H, Liu R, Zhang J, Liu Z, Fan S, Yang G, Jin Z, Pei Y (2021) Improving sulforaphane content in transgenic broccoli plants by overexpressing MAM1, FMOGS-OX2, and Myrosinase. Plant Cell Tissue Organ Cult (PCTOC) 146:461–471

Carr SM, Irish VF (1997) Floral homeotic gene expression defines developmental arrest stages in Brassica oleracea L. vars. botrytis and italica. Planta 201:179–188

Chen GJ, Cao BH, Xu F, Lei JJ (2010) Development of adjustable male sterile plant in broccoli by antisense DAD1 fragment transformation. Afr J Biotechnol 9:4534–4541

Chen C-C, Fu S-F, Norikazu M, Yang Y-W, Liu Y-J, Ikeo K, Gojobori T, Huang H-J (2015) Comparative miRNAs analysis of two contrasting broccoli inbred lines with divergent head-forming capacity under temperature stress. BMC Genomics 16:1026

Chevilly S, Costagliola G, Coque M, Miras-Moreno B, Arnao MB, Melgar JC (2021) Identification of distinctive physiological and molecular responses to salt stress among tolerant and sensitive cultivars of broccoli (Brassica oleracea var. italica). BMC Plant Biol 21:481. https://doi.org/10.1186/s12870-021-03263-4

Chiang C-M, Chen S-P, Chen L-FO, Chiang M-C, Chien H-L, Lin K-H (2013) Expression of the broccoli catalase gene (BoCAT) enhances heat tolerance in transgenic Arabidopsis. J Plant Biochem Biotechnol 23:266–277

Chiu LW, Zhou X, Burke S, Wu X, Prior RL, Li L (2010) The purple cauliflower arises from activation of a MYB transcription factor. Plant Physiol 154:1470–1480. https://doi.org/10.1104/pp.110.164160

Ciancaleoni S, Chiarenza GL, Raggi L, Branca F, Negri V (2013) Diversity characterisation of broccoli (Brassica oleracea L. var. italica Plenck) landraces for their on-farm (in situ) safeguard and use in breeding programs. Genet Resour Crop Evol 61:451–464

Coelho PS, Monteiro AA (2003) Inheritance of downy mildew resistance in mature broccoli plants. Euphytica 131:65–69

Degrave A, Wagner M, George P, Coudard L, Pinochet X, Ermel M, Gay EJ, Fudal I, Moreno-Rico O, Rouxel T, Balesdent MH (2021) A new avirulence gene of Leptosphaeria maculans, AvrLm14, identifies a resistance source in American broccoli (Brassica oleracea) genotypes. Mol Plant Pathol 22(12):1599–1612

Dhaka N, Rout K, Yadava SK et al (2017) Genetic dissection of seed weight by QTL analysis and detection of allelic variation in Indian and east European gene pool lines of Brassica juncea. Theor Appl Genet 130:293–307. https://doi.org/10.1007/s00122-016-2811-2

Dilmaghani A, Gladieux P, Gout L, Giraud T, Brunner P, Stachowiak A et al (2012) Migration patterns and changes in population biology associated with the worldwide spread of the oilseed rape pathogen *Leptosphaeria maculans*. Mol Ecol 21:2519–2533

Doullah M, Mohsin G, Ishikawa K, Hori H, Okazaki K (2011) Construction of a linkage map and QTL analysis for Black Rot resistance in Brassica oleracea L. Int J Nat Sci 1:1–6

Duclos DV, Bjoerkman T (2008) Meristem identity gene expression during curd proliferation and flower initiation in Brassica oleracea. J Exp Bot 59:421–433

Espinosa, Carvajal (2022) Genome-wide identification and biological relevance of broccoli aquaporins. Plant Genome 15(4):e20262. https://doi.org/10.1002/tpg2.20262

Fang Z, Sun P, Liu Y, Yang L, Wang X, Hou A, Bian C (1997) A male sterile line with dominant gene (Ms) in cabbage (Brassica oleracea var. capitata) and its utilization for hybrid seed production. Euphytica 97:265–268

Farinhó M, Coelho P, Carlier J, Svetleva D, Monteiro A, Leitão JM (2004) Mapping of a locus for adult plant resistance to downy mildew in broccoli (Brassica oleracea convar. italica). Theor Appl Genet 109:1392–1398

Farnham MW, Bjorkman T (2011) Breeding vegetables adapted to high temperatures: a case study with broccoli. HortScience 46:1093–1097

Farnham M, Wang M, Thomas C (2002) A single dominant gene for downy mildew resistance in broccoli. Euphytica 128:405–407

Figdore SS, Ferreira ME, Slocum MK, Williams PH (1993) Association of RFLP markers with trait loci affecting clubroot resistance and morphological characters in *Brassica oleracea* L. Euphytica 69(1–2):33–44. https://doi.org/10.1007/BF00021723

Fujime Y, Okuda N (1996) The physiology of flowering in brassicas, especially about cauliflower and broccoli. Acta Hortic:247–254

Gan C, Deng X, Cui L, Yu X, Yuan W, Dai Z, Yao M, Pang W, Ma Y, Yu X et al (2019) Construction of a high-density genetic linkage map and identification of quantitative trait loci associated with clubroot resistance in radish (Raphanus sativus L.). Mol Breed 39:1–12

Gao J, Yu X, Ma F, Li J (2014) RNA-seq analysis of transcriptome and glucosinolate metabolism in seeds and sprouts of broccoli (Brassica oleracea var. italic). PLoS ONE 9:e88804

Gardner AM, Brown AF, Juvik JA (2016) QTL analysis for the identification of candidate genes controlling phenolic compound accumulation in broccoli (Brassica oleracea L. var. italica). Mol Breed 36:81

Giovannelli JL, Farnham MW, Wang M, Strand AE (2002) Development of sequence characterized amplified region markers linked to downy mildew resistance in broccoli. J Am Soc Hortic Sci 127:597–601

Grandclément C, Thomas G (1996) Detection and analysis of QTLs based on RAPD markers for polygenic resistance to Plasmodiophora brassicae Woron in Brassica oleracea L. Theor Appl Genet 93:86–90. https://doi.org/10.1007/BF00225731

Hale AL, Farnham MW, Nzaramba MN, Kimbeng CA (2007) Heterosis for horticultural traits in broccoli. Theor Appl Genet 115:351–360. https://doi.org/10.1007/s00122-007-0569-2

Han F, Yuan K, Kong C, Zhang X, Yang L, Zhuang M, Zhang Y, Li Z, Wang Y, Fang Z et al (2018) Fine mapping and candidate gene identification of the genic male-sterile gene ms3 in cabbage 51S. Theor Appl Genet 131:2651–2661

Han F, Zhang X, Yuan K, Fang Z, Yang L, Zhuang M, Zhang Y, Wang Y, Liu Y, Li Z et al (2019) A user-friendly KASP molecular marker developed for the DGMS-based breeding system in Brassica oleracea species. Mol Breed 39:90

He L, Li L, Zhu Y, Pan Y, Zhang X, Han X, Li M, Chen C, Li H, Wang C (2021) BolTLP1, a thaumatin-like protein gene, confers tolerance to salt and drought stresses in broccoli (*Brassica oleracea* L. var. *Italica*). Int J Mol Sci 22:11132. https://doi.org/10.3390/ijms222011132

Heather D, Sieczka J, Dickson M, Wolfe D (1992) Heat tolerance and holding ability in broccoli. J Am Soc Hortic Sci 117:887–892

Huang J, Sun J, Liu E, Yuan S, Liu Y, Han F et al (2021a) Mapping of QTLs detected in a broccoli double diploid population for planting density traits. Sci Hortic 277:109835. https://doi.org/10.1016/j.scienta.2020.109835

Huang J, Liu Y, Han F, Fang Z, Yang L, Zhuang M, Zhang Y, Lv H, Wang Y, Jialei J et al (2021b) Genetic diversity and population structure analysis of 161 broccoli cultivars based on SNP markers. Hortic Plant J

Ji J-L, Yang L-M, Fang Z-Y, Zhuang M, Zhang Y-Y, Lv H-H, Liu Y-M, Li Z-S (2017) Recessive male sterility in cabbage (Brassica oleracea var. capitata) caused by loss of function of BoCYP704B1 due to the insertion of a LTR-retrotransposon. Theor Appl Genet 130:1441–1451

Jiang M, Jiang J, He C, Guan M (2016a) Broccoli plants over-expressing a cytosolic ascorbate peroxidase gene increase resistance to downy mildew and heat stress. J Plant Pathol 98:413–420

Jiang M, Liu Q-E, Liu Z-N, Li J-Z, He C-M (2016b) Over-expression of a WRKY transcription factor gene BoWRKY6 enhances resistance to downy mildew in transgenic broccoli plants. Australas Plant Pathol 45:327–334

Jiang M, Jiang J-J, Miao L-X, He C-M (2017) Over-expression of a C3H-type zinc finger gene contributes to salt stress tolerance in transgenic broccoli plants. Plant Cell Tissue Organ Cult (PCTOC) 130:239–254

Jiang M, Ye Z-H, Zhang H-J, Miao L-X (2018) Broccoli plants over-expressing an ERF transcription factor gene BoERF1 facilitates both salt stress and sclerotinia stem rot resistance. J Plant Growth Regul 38:1–13

Jiang M, Miao L, Zhang H, Zhu X (2019) Over-expression of a transcription factor gene BoC3H4 enhances salt stress tolerance but reduces sclerotinia stem rot disease resistance in broccoli. J Plant Growth Regul 39:1162–1176

Kennard WC, Slocum MK, Figdore SS, Osborn TC (1994) Genetic analysis of morphological variation in Brassica oleracea using molecular markers. Theor Appl Genet 87:721–732

Kianian SF, Quiros CF (1992) Generation of a Brassica oleracea composite RFLP map: linkage arrangements among various populations and evolutionary implications. Theor Appl Genet 84:544–554

Kim Y-C, Cha A, Hussain M, Lee K, Lee S (2020) Impact of Agrobacterium-infiltration and transient overexpression of BroMYB28 on glucoraphanin biosynthesis in broccoli leaves. Plant Biotechnol Rep 14:373–380

Kumar R, Kandpal K (2017) Influence of foliar fertilization of boron on broccoli (Brassica oleracea var. italica) in boron deficient soil of Doon Valley, India. Progress Hortic 49:65

Kumar P, Srivastava DK (2016) Biotechnological advancement in genetic improvement of broccoli (Brassica oleracea L. var. italica), an important vegetable crop. Biotechnol Lett 38(7):1049–1063. https://doi.org/10.1007/s10529-016-2155-x

Lan TH, Paterson AH (2000) Comparative mapping of quantitative trait loci sculpting the curd of Brassica oleracea. Genetics 155(4):1927–1954. https://doi.org/10.1093/genetics/155.4.1927

Lawrenson T, Shorinola O, Stacey N, Li C, Østergaard L, Patron N, Uauy C, Harwood W (2015) Induction of targeted, heritable mutations in barley and Brassica oleracea using RNA-guided Cas9 nuclease. Genome Biol 16:258

Lee Y-S, Ku K-M, Becker TM, Juvik JA (2017) Chemopreventive glucosinolate accumulation in various broccoli and collard tissues: microfluidic-based targeted transcriptomics for by-product valorization. PLoS ONE 12:e0185112

Li S, Zhang L, Wang Y, Xu F, Liu M, Lin P, Ren S, Ma R, Guo Y-D (2017) Knockdown of a cellulose synthase gene BoiCesA affects the leaf anatomy, cellulose content and salt tolerance in broccoli. Sci Rep 7:41397

Li H, Zhang QL, Li LH, Yuan JY, Wang Y, Wu M, Han ZP, Liu M, Chen CB, Song WQ et al (2018) Ectopic overexpression of bol-mir171b increases chlorophyll content and results in sterility in broccoli (Brassica oleracea L. var. italica). J Agric Food Chem 66:9588–9597

Li Z, Mei Y, Liu Y, Fang Z, Yang L, Zhuang M, Zhang Y, Lv H (2019) The evolution of genetic diversity of broccoli cultivars in China since 1980. Sci Hortic 250:69–80. https://doi.org/10.1016/j.scienta.2019.02.035

Li L, Tian Z, Chen J et al (2023) Characterization of novel loci controlling seed oil content in Brassica napus by marker metabolite-based multi-omics analysis. Genome Biol 24:141. https://doi.org/10.1186/s13059-023-02984-z

Lin K-H, Huang H-C, Lin C-Y (2010) Cloning, expression and physiological analysis of broccoli catalase gene and Chinese cabbage ascorbate peroxidase gene under heat stress. Plant Cell Rep 29:575–593

Lin KH, Chang LC, Lai CD, Lo HF (2013) AFLP mapping of quantitative trait loci influencing seven head-related traits in broccoli (*Brassica oleracea* var. *italica*). J Hortic Sci Biotechnol 88(3):257–268. https://doi.org/10.1080/14620316.2013.11512964

Lin H-H, Lin K-H, Chen S-C, Shen Y-H, Lo H-F (2015) Proteomic analysis of broccoli (Brassica oleracea) under high temperature and waterlogging stresses. Bot Stud 56:1–11

Lin Y-R, Lee J-Y, Tseng M-C, Lee C-Y, Shen C-H, Wang C-S, Liou C-C, Shuang L-S, Paterson AH, Hwu K-K (2018) Subtropical adaptation of a temperate plant (Brassica oleracea var. italica) utilizes non-vernalization-responsive QTLs. Sci Rep 8:13609

Lin CW, Fu SF, Liu YJ, Chen CC, Chang CH, Yang YW, Huang HJ (2019) Analysis of ambient temperature-responsive transcriptome in shoot apical meristem of heat-tolerant and heat-sensitive broccoli inbred lines during floral head formation. BMC Plant Biol 19:1–16. https://doi.org/10.1186/s12870-018-1613-x

Liu S, Liu Y, Yang X, Tong C, Edwards D, Parkin IAP, Zhao M, Perumal S, Yu J, Huang S et al (2014) The Brassica oleracea genome reveals the asymmetrical evolution of polyploid genomes. Nat Commun 5:3930

Liu X, Wu S, Xu J, Sui C, Wei J (2017) Application of CRISPR/Cas9 in plant biology. Acta Pharm Sin B 7:292–302

Liu C-Q, Li G-Q, Yao X-Q, Huang L, Wu X-Y, Xie Z-J (2020a) Characterization of Ogura CMS fertility-restored interspecific hybrids and backcross progenies from crosses between broccoli and rapeseed. Euphytica 216:1–12

Liu C, Yao X, Li G, Huang L, Xie Z (2020b) Transcriptomic profiling of purple broccoli reveal slight-induced anthocyanin biosynthetic signaling and structural genes. PeerJ 8:e8870. https://doi.org/10.7717/peerj.8870

Liu H, Zou M, Zhang B et al (2022) Genome-wide association study identifies candidate genes and favorable haplotypes for seed yield in Brassica napus. Mol Breeding 42:61. https://doi.org/10.1007/s11032-022-01332-6

Luo F, Fang H-X, Wei B-D, Cheng S-C, Zhou Q, Zhou X, Zhang X, Zhao Y-B, Ji S-J (2020) Advance in yellowing mechanism and the regulation technology of post-harvested broccoli. Food Qual Saf 4(2):107–113. https://doi.org/10.1093/fqsafe/fyz017

Lv H, Fang Z, Yang L et al (2014) Mapping and analysis of a novel candidate Fusarium wilt resistance gene FOC1 in Brassica oleracea. BMC Genomics 15:1094

Ma C, Zhu C, Zheng M, Liu M, Zhang D, Liu B, Li Q, Si J, Ren X, Song H (2019) CRISPR/Cas9-mediated multiple gene editing in Brassica oleracea var. capitata using the endogenous tRNA-processing system. Hortic Res 6:20

Maggioni L, Von Bothmer R, Poulsen G, Lipman E (2017) Domestication, diversity and use of Brassica oleracea L., based on ancient Greek and Latin texts. Genet Resour Crop Evol 65:137–159

Massie IH, Astley D, King GJ (1996) Patterns of genetic diversity and relationships between regional groups and populations of Italian landrace cauliflower and broccoli (Brassica oleracea L. var. botrytis L. and var. italica Plenck). Acta Hortic 407:45–54

Nagaoka T, Doullah MAU, Matsumoto S, Kawasaki S, Ishikawa T, Hori H, Okazaki K (2010) Identification of QTLs that control clubroot resistance in Brassica oleracea and comparative analysis of clubroot resistance genes between B. rapa and B. oleracea. Theor Appl Genet 120:1335–1346

Ogura H (1968) Studies on the new male sterility in Japanese radish, with special references on the utilization of this sterility towards the practical raising of hybrid seeds. Mem Fac AgricKagoshima Univ 6:40–75

Okazaki K, Sakamoto K, Kikuchi R, Saito A, Togashi E, Kuginuki Y, Matsumoto S, Hirai M (2006) Mapping and characterization of FLC homologs and QTL analysis of flowering time in Brassica oleracea. Theor Appl Genet 114:595–608

Parkin IAP, Koh C, Tang H, Robinson SJ, Kagale S, Clarke WE, Town CD, Nixon J, Krishnakumar V, Bidwell SL et al (2014) Transcriptome and methylome profiling reveals relics of genome dominance in the mesopolyploid Brassica oleracea. Genome Biol 15:R77

Peng L, Zhou L, Li Q, Wei D, Ren X, Song H, Mei J, Si J, Qian W (2018) Identification of quantitative trait loci for clubroot resistance in Brassica oleracea with the use of brassica SNP microarray. Front Plant Sci 9:822

Rae AM, Howell EC, Kearsey MJ (1999) More QTL for flowering time revealed by substitution lines in Brassica oleracea. Heredity 83:586–596

Rahim A, Afrin KS, Jung H-J, Kim H-T, Park J-I, Hur Y, Nou IS (2019) Molecular analysis of anthocyanin biosynthesis-related genes reveal BoTT8 associated with purple hypocotyl of broccoli (Brassica oleracea var. italica L.). Genome 62:253–266

Reshma V, Adarsh M, Dhanush K (2018) Development of hybrids in cole crops: a review. Plant Arch 18:1–11

Robin AHK, Larkan NJ, Laila R, Park J-I, Ahmed NU, Borhan H et al (2017) Korean *Brassica oleracea* germplasm offers a novel source of qualitative resistance to blackleg disease. Eur J Plant Pathol 149:611–623

Rocherieux J, Glory P, Giboulot A, Boury S, Barbeyron G, Thomas G, Manzanares-Dauleux MJ (2004) Isolate-specific and broad-spectrum QTLs are involved in the control of clubroot in Brassica oleracea. Theor Appl Genet 108:1555–1563

Ruffio-Chable V, Bellis H, Herve Y (1993) A dominant gene for male sterility in cauliflower (Brassica oleracea var. botrytis): phenotype expression, inheritance, and use in F1 hybrid production. Euphytica 67:9–17

Sampson DR (1966) Linkage of genetic male sterility with a seedling marker and its use in producing f1 hybrid seed of Brassica oleracea (cabbage, broccoli, kale, etc.). Can J Plant Sci 46:703

Shapiro TA, Fahey JW, Wade KL, Stephenson KK, Talalay P (2001) Chemoprotective glucosinolates and isothiocyanates of broccoli sprouts. Cancer Epidemiol Prev Biomark 10:501–508

Shen Y, Wang J, Shaw RK, Yu H, Sheng X, Zhao Z, Li S, Gu H (2021) Development of GBTS and KASP panels for genetic diversity, population structure, and fingerprinting of a large collection of broccoli (Brassica oleracea L. var. italica) in China. Front Plant Sci 12:655254

Shu J, Liu Y, Li Z, Zhang L, Fang Z, Yang L, Zhuang M, Zhang Y, Lv H (2016a) Detection of the diversity of cytoplasmic male sterility sources in broccoli (Brassica oleracea var. Italica) using mitochondrial markers. Front Plant Sci 7:927

Shu J, Liu Y, Li Z, Zhang L, Fang Z, Yang L, Zhuang M, Lv H (2016b) A generic SSR marker closely linked to a dominant genic male sterility gene (DGMs79-399-3) in broccoli (Brassica oleracea var. italica). Mol Breed 36:86

Shu JS, Liu YM, Zhang LL, Li ZS, Fang ZY, Yang LM, Zhuang M, Zhang YY, Lv HH (2018) QTL-seq for rapid identification of candidate genes for flowering time in broccoli x cabbage. Theor Appl Genet 131:917–928

Sigareva MA, Earle ED (1997) Direct transfer of a cold-tolerant Ogura male-sterile cytoplasm into cabbage (Brassica oleracea ssp. capitata) via protoplast fusion. Theor Appl Genet 94:213–220

Singh S, Kalia P, Meena RK, Sharma BB, Parihar BR (2022) Agro-morphological and molecular diversity analysis of new cytoplasmic male sterile lines in Indian cauliflower for their use in hybrid breeding. Sci Hortic 301:111107. https://doi.org/10.1016/j.scienta.2022.111107

Singh S, Sagar S, Rao M, Saroha S, Garg P, Prasad L (2023) Screening wild Brassica species against Alternaria brassicicola (Schw.) Wiltsh for breeding Alternaria leaf spot resistance in Brassica vegetables. Genet Resour Crop Evol:1–17

Slocum MK, Figdore SS, Kennard WC, Suzuki JY, Osborn TC (1990) Linkage arrangement of restriction fragment length polymorphism loci in Brassica oleracea. Theor Appl Genet 80:57–64

Smith LB, King GJ (2000) The distribution of BoCAL-a alleles in Brassica oleracea is consistent with a genetic model for curd development and domestication of the cauliflower. Mol Breed 6:603–613

Sotelo T, Soengas P, Velasco P, Rodríguez VM, Cartea ME (2014) Identification of metabolic QTLs and candidate genes for glucosinolate synthesis in Brassica oleracea leaves, seeds and flower buds. PLoS ONE 9:e91428

Sparrow PACHN, Irwin JA (2014) Brassica oleracea and B. napus. In: Methods in molecular biology, vol 1223. Springer, Berlin/Heidelberg, pp 287–297

Stansell Z, Björkman T (2020) From landrace to modern hybrid broccoli: the genomic and morphological domestication syndrome within a diverse B. oleracea collection. Hortic Res 7:1–17. https://doi.org/10.1038/s41438-020-0292-z

Stansell Z, Farnham M, Björkman T (2019) Complex horticultural quality traits in broccoli are illuminated by evaluation of the immortal BolTBDH mapping population. Front Plant Sci 10:1104. https://doi.org/10.3389/fpls.2019.01104

Sun B, Jiang M, Zheng H, Jian Y, Huang W-L, Yuan Q, Zheng A-H, Chen Q, Zhang Y-T, Lin Y-X et al (2020) Color-related chlorophyll and carotenoid concentrations of Chinese kale can be altered through CRISPR/Cas9 targeted editing of the carotenoid isomerase gene BoaCRTISO. Hortic Res 7:161

Tian Y, Tian Y, Luo X, Zhou T, Huang Z, Liu Y, Qiu Y, Hou B, Sun D, Deng H et al (2014) Identification and characterization of microRNAs related to salt stress in broccoli, using high-throughput sequencing and bioinformatics analysis. BMC Plant Biol 14:1–13

Tonu NN, Doullah A-U, Shimizu M, Karim M, Kawanabe T, Fujimoto R, Okazaki K (2013) Comparison of positions of QTLs conferring resistance to Xanthomonas campestris pv. campestris in Brassica oleracea. Am J Plant Sci 4:11–20

Torres-Contreras AM, Senés-Guerrero C, Pacheco A, González-Agüero M, Ramos-Parra PA, Cisneros-Zevallos L, Jacobo-Velázquez DA (2018) Genes differentially expressed in broccoli as an early and late response to wounding stress. Postharvest Biol Technol 145:172–182
Treccarichi S, Di Gaetano C, Di Stefano F, Gasparini M, Branca F (2021) Using simple sequence repeats in 9 brassica complex species to assess hypertrophic curd induction. Agriculture 11:622. https://doi.org/10.3390/agriculture11070622
Treccarichi S, Ben Ammar H, Amari M, Cali R, Tribulato A, Branca F (2023) Molecular markers for detecting inflorescence size of Brassica oleracea L. crops and B. oleracea complex species (n = 9) useful for breeding of broccoli (B. oleracea var. italica) and cauliflower (B. oleracea var. botrytis). Plants 12:407. https://doi.org/10.3390/plants12020407
Uyttewaal M, Arnal N, Quadrado M, Martin-Canadell A, Vrielynck N, Hiard S, Gherbi H, Bendahmane A, Budar F, Mireau H (2009) Characterization of Raphanus sativus pentatrico-peptide repeat proteins encoded by the fertility restorer locus for ogura cytoplasmic male sterility. Plant Cell 20:3331–3345
Vallejo F, Garcia-Viguera C, Tomas-Barberan F (2003) Changes in broccoli (Brassica oleracea L. var. italica) health-promoting compounds with inflorescence development. J Agric Food Chem 51:3776–3782
Vallejo F, Tomas-Barberan F, Ferreres F (2004) Characterisation of flavonols in broccoli (Brassica oleracea L. var. italica) by liquid chromatography–UV diode-array detection–electrospray ionisation mass spectrometry. J Chromatogr A 1054(1–2):181–193. https://doi.org/10.1016/j.chroma.2004.02.034
Vicente JG, Taylor JD, Sharp AG, Parkin IAP, Lydiate DJ, King GJ (2002) Inheritance of race-specific resistance to Xanthomonas campestris pv. campestris in Brassica genomes. Phytopathology 92:1134–1141
Voorrips RE (1995) Plasmodiophora brassicae: aspects of pathogenesis and resistance in Brassica oleracea. Euphytica 83:139–146
Walley PG, Carder J, Skipper E, Mathas E, Lynn J, Pink D, Buchanan-Wollaston V (2012) A new broccoli× broccoli immortal mapping population and framework genetic map: tools for breeders and complex trait analysis. Theor Appl Genet 124:467–484. https://doi.org/10.1007/s00122-011-1721-6
Wang Q, Zhang Y, Fang Z, Liu Y, Yang L, Zhuang M (2011) Chloroplast and mitochondrial SSR help to distinguish allo-cytoplasmic male sterile types in cabbage (Brassica oleracea L. var. capitata). Mol Breed 30:709–716
Wang A, Guo J, Wang S, Zhang Y, Lu F, Duan J, Liu Z, Ji W (2022) BoPEP4, a C-terminally encoded plant elicitor peptide from broccoli, plays a role in salinity stress tolerance. Int J Mol Sci 23:3090. https://doi.org/10.3390/ijms23063090
Ward B, Smith P, James S, Stansell Z, Farnham M (2015) Increasing plant density in eastern united states broccoli production systems to maximize marketable head yields. HortTechnology 25:330–334
Wu S, Li Q, Chen L, Ma X (2024) Overexpression of BoMAM1 (BolI0108790) in broccoli leads to high accumulation and high ratio of C4-GSLs. Sci Rep 14(1):1245. https://doi.org/10.1038/s41598-024-21550-3
Xie Q, Zhang S, Liu Y et al (2022) Germplasm enhancement and identification of loci for clubroot resistance in Brassicaceae crops. Genes 13:1600. https://doi.org/10.3390/genes13091600
Yamagishi H, Jikuya M, Okushiro K, Hashimoto A, Fukunaga A, Takenaka M, Terachi T (2021) A single nucleotide substitution in the coding region of Ogura male sterile gene, orf138, determines effectiveness of a fertility restorer gene, Rfo, in radish. Mol Genet Genomics 296:705–717
Yan L, Khan A, Rauf A et al (2023) Research progress on the utilization technology of broccoli. Front Plant Sci 14:1138700. https://doi.org/10.3389/fpls.2023.1138700
Yao X, Li Y, Xie Z, Liu L (2009) Identification of specific SRAP marker associated with cytoplasmic male sterility gene of broccoli. Mol Plant Breed 7:941–947

Yu H-L, Fang Z-Y, Liu Y-M, Yang L-M, Zhuang M, Lv H, Li Z-S, Han F-Q, Liu X-P, Zhang Y-Y (2016) Development of a novel allele-specific Rfo marker and creation of Ogura CMS fertility-restored interspecific hybrids in Brassica oleracea. Theor Appl Genet 129:1625–1637
Yu H, Wang J, Sheng X, Zhao Z, Shen Y, Branca F, Gu H (2019a) Construction of a high-density genetic map and identification of loci controlling purple sepal trait of flower head in Brassica oleracea L. italica. BMC Plant Biol 19:1–8. https://doi.org/10.1186/s12870-019-1831-x
Yu H, Wang J, Zhao Z, Sheng X, Shen Y, Branca F, Gu H (2019b) Construction of a high-density genetic map and identification of loci related to hollow stem trait in broccoli (*Brassica oleracea* L. *italica*). Front Plant Sci 10:45. https://doi.org/10.3389/fpls.2019.00045
Yu H, Wang J, Sheng X, Zhao Z, Shen Y, Branca F, Gu H (2019c) Construction of a high-density genetic map and identification of loci controlling purple sepal trait of flower head in Brassica oleracea L. italica. BMC Plant Biol 19:228
Yu H, Wang J, Zhao Z, Sheng X, Shen Y, Branca F, Gu H (2019d) Construction of a high-density genetic map and identification of loci related to hollow stem trait in broccoli (Brassica oleracea L. italica). Front. Plant Sci 10:10
Yuan G, Sun B, Yuan J, Wang Q (2009) Effects of different cooking methods on health-promoting compounds of broccoli (Brassica oleracea var. italica). J Zhejiang Univ Sci B 10:580–588. https://doi.org/10.1631/jzus.B0920051
Zhang Q, Branca F, Li N, Liu N, Ding Y (2024) Identification of black spot resistance in broccoli (Brassica oleracea L. var. italica) germplasm resources. Appl Sci 14(7):2883
Zuluaga DL, Graham NS, Klinder A, Kloeke AEEVO, Marcotrigiano AR, Wagstaff C, Verkerk R, Sonnante G, Aarts MGM (2019) Overexpression of the MYB29 transcription factor affects aliphatic glucosinolate synthesis in Brassica oleracea. Plant Mol Biol 101:65–79

MIX
Papier aus verantwortungsvollen Quellen
Paper from responsible sources
FSC® C105338
FSC
www.fsc.org

If you have any concerns about our products,
you can contact us on
ProductSafety@springernature.com

In case Publisher is established outside the EU,
the EU authorized representative is:
Springer Nature Customer Service Center GmbH
Europaplatz 3, 69115 Heidelberg, Germany

Printed by Libri Plureos GmbH
in Hamburg, Germany